Advances in Cartography and Geographic Information Engineering

Jiayao Wang · Fang Wu
Editors

Advances in Cartography and Geographic Information Engineering

测绘出版社
Surveying and Mapping Press

Springer

Editors
Jiayao Wang
Institute of Spatio-Temporal Big Data
Henan University
Kaifeng, China

Fang Wu
School of Surveying and Mapping
Information Engineering University
Zhengzhou, China

ISBN 978-981-16-0613-7 ISBN 978-981-16-0614-4 (eBook)
https://doi.org/10.1007/978-981-16-0614-4

Jointly published with Surveying and Mapping Press
The print edition is not for sale in China (Mainland). Customers from China (Mainland) please order the print book from: Surveying and Mapping Press.

This Springer imprint is published by the registered company Springer Nature Singapore Pte Ltd.
The registered company address is: 152 Beach Road, #21-01/04 Gateway East, Singapore 189721, Singapore

Foreword I

This book contains a collection of research papers that records the past 60 years in the development course of cartography in new China. The chapters are divided according to the subdisciplines within modern cartography. The progress of academic theory and technology is described with a focus on development. This book comprises a succinct summary of the research areas and provides detailed descriptions of the key technological elements. This is the first scholarly work of this stature too be published in the areas of cartographic literature since the founding of new China.

The 60-year cartographic history is important because it may be a major historical turning point and milestone in the history of cartography. With the advancement of science and technology, especially information technology (IT), cognitive science and life sciences, map workers opened the doors to exploring the following areas: understanding the mysteries of the map by regarding humans with objective existence as an interacting and interdependent whole; seeking answers to why maps have had such a tenacious vitality for a long time by researching special functions of maps in terms of spatial cognition, spatial thinking and spatial memory; and exploring how to use a map and what kind of map to use to describe network space or cyberspace since these elements have opened a new space for human survival, just as it was once our mission to describe geographical space. These events have resulted in promising prospects for map workers, and they occurred during this historical stage.

The birth of digital maps, such as electronic maps, represents another historic turning point during the mapping process. It was developed on the basis of information technology and is a synonym for geospatial data. Furthermore, it can be used to replace the position of the topographic map as the first long-term carrier of spatial information since it has more information sources, richer content, is more alive, easy to supplement, and contains updated data compared with conventional maps. Based on digital maps, electronic maps generated by using visualization technology are especially welcome by users because of their dynamic, multidimensional, multi-scale and diverse forms of expressions. The discovery of electronic maps and their attention-catching impact on traditional disciplines, market segmentation, management systems, copyrights and so on should not be ignored. In the wake of technical breakthroughs in electronic paper and display screens, along with the generation of cloud computing environments that reduce the degree of dependence on hardware and software, it is estimated that electronic maps are creating a new age.

There is a close relation between geographic information systems (GISs) and cartography. Although they have different goals, the fact that they depend on each other, support each other and promote each other is an outstanding and brilliant concept. It is with the participation of electronic maps that geographic information systems can interact with users. Likewise, only with a geographic information system platform can cartography perform in active service. These subjects live and develop in one subject and one society, and they have made important contributions to China's economic and national defense construction. They comprise a beautiful landscape in the 60-year historical evolution of cartography.

There were few major turning points over the thousands of years in the history of cartography. Pei Xiu's Six Principles of Geographic Description and Map Making (one of which is Drawing Square Grid with Chinese Unit of Length "li") in the Jin Dynasty is a significant contribution and symbol of ancient Chinese cartography. The Atlas of China in the Qing Dynasty represents the birth of the survey map in the late Qing Dynasty and the start of China's modern map. However, beyond the above turning points, cartography in China has stagnated for thousands of years. In recent years, a new mission of digital map development and construction was put forward almost immediately when we finished nationwide basic topographic surveying and mapping. This turning point means that we must upgrade and reconstruct the recently established geospatial framework alone. The growth surprised us, and it all occurred in the past 60 years. Thus, these 60 years represent a milestone stage in the history of cartography.

Professor Jiayao Wang had long been the chairman of professional committee of cartography and geographic information systems at the Chinese Society of Geodesy, Photogrammetry and Cartography (CSGPC). He has achieved great progress in talent training, scientific research and social service in the fields of cartography and geographic information science.

The collection of academic theses compiled by him gather together the results of 60 years of research in the field of cartography after the founding of new China. Experience has been gained, prompting a new direction of investigation. This research comprises an important contribution to the development of cartography in our country. This is a book with both academic value and historical literature significance, which is desirable and expected from experts and colleagues in the field of cartography. The experts and the members from the professional committee who participated in writing this book contributed much hard work, which is worthy of heartfelt thanks by colleagues and students alike.

June 2010

Jun Gao
Academician, China Academy of Sciences
Honorary Membership, Chinese Society of
Geodesy, Photogrammetry and Cartography

Foreword II

Abstract We are most pleased to provide our observations on this book: *Advances in Cartography and Geographic Information Engineering in China*, edited by Prof. Jiayao Wang, from Zhengzhou Information Engineering University, China. The book brings-together experts from Cartography and Geographic Information Engineering in China, who have reviewed and summarised six decades of developments and achievements in China.

It was with great interest that we read the contributed chapters from the book, which provided essential information related to advances in Chinese disciplines of Cartography and Geographic Information Engineering. To become better informed about the advances in the theories and praxis related to China's Geographic Information Engineering has enabled us to appreciate the foundations of current thinking and developments and to appreciate the continual progresses made by members of the academic and professional Cartography and Geographic Information Engineering research and development community in China.

In this contribution, we provide our observations about the book, and its collective chapters. As well, we provide remarks about the contributions that the outcomes of this 60 years of progress has afforded.

Observations

The following sections of this paper provide our observations of each chapter. Here, our intention is to summarise the developments that have taken place over the last 60 years and to provide some general comments. Following our observations, the final part of this paper provides some, general, overarching statements.

Cartography Theory

This chapter covers the development of cartographic theory and begins by looking at Eduard Imhof's concept relating to this theory in the 1950s. It follows cartographic theory through the 1960s (Theory, System Theory and Cybernetics), modern cartographic theory in the 1970s and 80s (digital cartography), the 1990s (computer applications—remote sensing, networks, GIS and virtual reality). The author of this chapter states that the development of cartographic theory in China over the last 60 years was in-step with the work on theoretical cartographies being conducted internationally. It is considered that the 'old theories' of cartography—map projection, cartographic generalization and cartographic semiotics theory—were addressed up until the 1970s, after which research focussed on digital cartographic theories (until the 1990s): the 'new theories'. After this period research was conducted on cartography in the information age. The most recent stage of theoretical research combines the 'old three theories' with 'new theory' and focuses on geographic information provision. The chapter illustrates the research activities related to theoretical cartography undertaken in China during these periods and provides numerous examples of advancing cartographic theory. The chapter concludes with sections that expand on future research issues and provides information about publications that illustrate the outcomes of cartographic theoretical research in China. The chapter provides the reader with comprehensive information about research into theoretical issues in cartography and the outcomes of that research. It allows for a greater appreciation of the theoretical advances in this research area and the contributions to international cartographic and GI Science knowledge.

Map Projections

This chapter focuses on map projection related achievements made by Chinese scholars between the 1950s and 2010s. After briefly listing worldwide map projections and related developments and ancient Chinese scholars' contributions, the chapter identified five major achievements in the research and application of map projections: (1) the translation and publication of many monographs on mathematical cartography and map projections; (2) the publication of reference books on map projections and mathematical cartography authored by Chinese scholars; (3) the publication of academic journals; (4) significant progresses in the 1950s–1980s related to theoretical, methodological, technical and practical aspects of map projections; and (5) further study on map projection selection and design, transformation, spatial dynamic projection, and the application of map projections to GIS. These achievements have enriched and improved the general theories of map projections and spatial referencing and developed new map projections. The chapter concluded with an overview of 10 representative publications by Chinese scholars on map projections.

Map Design

Research undertakings into map design by Chinese cartographers has led to the development of standards, specifications and schema for the design of effective topographic maps, thematic maps and atlases. The results from these research activities have been applied in the era of digital mapping, multimedia mapping and distributed mapping (via communication networks). The chapter describes the evolution of map and atlas design research, from manual map production ('traditional cartography'), to computer-aided mapping and related atlas production, to interactive mapping and electronic and atlases, linked to remotely-sensed geographic information, to, currently, computer-integrated map and atlas design techniques. The chapter provides information relating to the main research achievements from the 60 years of research and the research issues that still need to be addressed: strengthening contemporary map design and improving the 'intelligent' level of map design through the exploration of theoretical and technical issues. An overview of key publications—research and atlas focussed—is provided.

Cartographic Generalization

The chapter reviewed the evolution of cartographic generalization, and argued that cartographical generalization has evolved (1) from a subjective process to an objective process; (2) from qualitative to quantitative description; (3) from map model-based to model, algorithm and knowledge-based; (4) from full automation illustrations to realistic human-computer collaboration; (5) from the research of isolated and scattered model, algorithm and automatic generalization to a holistic approach. The chapter reviewed the main research achievements in cartographic generalization made by Chinese scholars between the 1950s and 2010s: (1) theory and system method of cartographic generalization in a digital environment; (2) models and algorithms of automatic cartographic generalization for basic and more complex features; (3) applications of fractal theory, wavelet theory, mathematical morphology, artificial intelligence methods, and computational geometry in cartographic generalization; (4) knowledge acquisition, representation and processing for cartographic generalization; (5) automatic generalization; (6) quality evaluation and control model for automatic generalization; and (7) the development of automatic cartographic generalization systems. The chapter concludes with an overview of 12 representative publications by Chinese scholars on cartographic generalization.

Thematic Cartography

Thematic cartography in China is seen to consist of four stages: initial stage (1950s–1960s), development stage (1970s–1980s), rapid development & deepening (1990s) and information development (beginning of C21 to now). The initial stage of thematic mapping saw the compilation of thematic maps of natural phenomena—geology, soils, vegetation, hydrology, fauna—and human geography—census mapping. The development stage focussed on the advancement of theories related to technical methods, the automation of cartographic processes and thematic map and atlas production, which included national atlases, including atlases based on aerial photography remote sensing information. The rapid development & deepening stage employed remote sensing, GIS (Geographic Information Systems), digital mapping and advanced surveying technology to automate thematic map compilation and production. As well as terrestrial maps, where production methods were increasingly standardised, celestial maps were also produced. The information development stage harnessed the technologies being made available through advances in remote sensing, GPS, GIS (for example *SuperMap*) and electronic communications. The chapter covers the wide spectrum of advances in thematic mapping theory and practice in China and makes available a comprehensive outline summary of the four stages of thematic cartography development. And, as per the previous chapters covered in this contribution, of key publications information is provided.

Geo-Information Tupu

The geo-information Tupu is described as a "group of digital maps, charts, graphs, curves, or images arranged in accordance with some regularities of index progressive changes or classification, which reflect the temporal-spatial information laws of Earth science". It is formed of three parts: graphs; descriptive parameters and mathematical models; and reorganisation and the visualization of object features as an outcome of modelling. It is argued by the author that the Tupu promotes the discipline area, it provides a new way of thinking about geoscience and, by providing a new theoretical pathway, it can enrich the theoretical, practical and technology of cartography. The chapter provides examples about how the Tupu can be applied. A synopsis of representative publications is provided. The concept of the Tupu is indeed most interesting, and it provides an excellent example of forward thinking from the cartographic community.

Geospatial Databases

This chapter focuses on geospatial databases, including 'the geographic spatial database management system' and 'the computer expression of geographic information, involving data model, data acquisition, database design, data engineering, data standards, data quality and data application'. The development of a map database system began in China in the 1980s. Since then, research has been undertaken into the computer data representation method of geographical spatial entities, spatial data models, data indexes, data manipulation and query language, data engines, geospatial database architecture, geospatial database security control, on database sharing, integration and fusion, and on the construction of major geospatial databases. As well, the chapter identified some issues for further studies and provided information about representative publications by Chinese scholars on geospatial databases.

Digital Cartography

This chapter focuses on digital cartography, the application of digital mapping technologies in cartography, which has dramatically changed the ways of map making and map use. In China, research on digital mapping technology began in the 1970s. The initial focus was on the exploration of basic theory, basic algorithms and the development and improvement of drawing equipment. The following decade saw the focus on perfecting digital cartographic systems through the application of electronic technology, and, in particular, using computer-aided cartographic systems. From the 1990s on, digital mapping technology was further developed and applied to updating large scale databases, implementing complete digital map production systems and producing multimedia electronic atlases. The chapter concluded by identifying the digital cartography issues in China that need to be addressed: interconnectivity and interoperability, map editing, map output and quality; the need for a general purpose digital mapping system; multi-source data integration and assimilation in map production; and software functions for automatic map design and generalization, compilation, and automation of data processing workflows.

Map Printing and Publishing

This chapter covers the progress in map printing and publishing—from photomechanical processes to digital imaging and publishing. It provides detail of the transition from colour separation techniques related to manual map production to digital publishing technologies that were implemented since the 1990s. The intervening years between these two types of printing and publishing methodologies saw advances, due to the implementation of digital technologies, of typesetting, screening, computer-to-platemaking, digital proofing and digital printing. More recently, electronic and on-line publishing, including digital map flow, quality control and production management, has been implemented. The chapter also provides information about various map publishing standards. A comprehensive overview of map printing and publishing developments and advances over the last 60 years has been provided.

Geographic Information Systems

This chapter focuses on Geographic Information Systems (GIS), which appeared in China between the late 1970s and early 1980s. GIS development in China can be classified into three stages: early preparation 1970s and 1980s; accelerated development and industrialization 1990s, and popularization and standardization 2000+. The chapter identified the current major problems related to GIS in China. These include the emphasis on technology and application, rather than on theory, there being no government agency to manage and guide the GIS industry, and the need for training to be attuned to market demand. The chapter also provided an overview of 47 representative publications by Chinese scholars on the topic area.

Virtual Geographic Environment

This chapter covers virtual geographical environments (VGE)—the application of virtual reality technology to three-dimensional, dynamic and interactive geographical space. The initial exploration of VGE in China focused on the visualization and simulation of geographical environments. This was followed by integration of VGE with geographical models. Currently, from 2005, new theories and techniques have been studied. Many institutions in China are engaged in VGE research. This is being done in the areas of spatial cognition, VGE construction, a virtual Earth system, geospatial multi-source data acquisition and integration, construction and management of VGE models, display and interaction, distributed VGE, artificial intelligence in VGE and the application of VGE in various fields.

Geographic Ontology

This chapter considers the concept of ontology as applied to the semantic understanding of geographic information. Through a greater knowledge about the contextual meaning of geographic information it is anticipated that wider application of spatial and temporal data can be supported.

The chapter begins with an overview of the development of the study of ontology and its application to geographic information. In the development of geographic databases there is a need for a clear understanding of definitions relating to classifications of entities and their attributes. In addition to the textual relationships between these entities, there is an additional requirement for spatial relationships the topological configurations such as adjacency and connectivity, to be defined. These are important when considering geographic information retrieval and querying. The chapter discusses how spatial ontologies has enabled what was previously a limited textual database formalised method of querying and data retrieval, easier methods for end users to interact with geographic databases without specialised geospatial knowledge. Queries can now be made using a natural language, rather than users needing specialised querying languages. The chapter concludes by discussing the future need to extend research based on the semantic web applied to real-world geographic applications.

Spatial Data Mining and Knowledge Discovery

Extracting meaningful information from large geographic databases using spatial data mining and knowledge discovery forms the basis for Chap. 16. Through the development of algorithms to extract specific information from these large and often complex geographic databases, meaningful and useful information can be acquired. As technological developments advance, "big data" about many day-to-day activities carried out by individuals can be collected at the location where the transactions took place. This is increasingly important today, as data about varied geographic objects and processes are acquired at increasing intervals. Through the adoption of probability based, and statistical analyses, and utilising innovative geovisualisation approaches enables information specialists to extract information in both time and space, regarding the behaviour of the individuals generating this data. This information can then be used for marketing, or public/private service provisioning.

Spatial Data Uncertainty and Quality Control

The acquisition of geographic data, including location, shape, size and contextual topological information such as proximity and adjacency to other geographic objects, carries an element of uncertainty. Geographic objects are represented as points, lines and polygons in vector formatted databases, whilst they are arranged as a series of contiguous cells of the same shape and size in a raster format. In the raster format, each cell is assigned a value based on the geographic phenomenon occurring at that location. With the raster format, the size of the cell gives some indication of the level of uncertainty in the data. However, in the vector format this is not the case and no such uncertainty can be easily ascertained. This chapter considers the concept of uncertainty and how it can be managed in a geographic database. Whilst "error" is simply an estimate of how far an object is from what is true, uncertainty is a broader concept where there is a lack of knowledge, or vagueness about the data under consideration. The chapter discusses the achievements made by the GIS community towards greater understanding and representation of uncertainty.

Standardization of Cartography and Geographic Information

This chapter discusses the important topic of standardization and quality control of geographic information and cartographic output. This is an increasingly important issue in a global world, where much data is shared across nations, particularly when considering global scale issues threatening our planet, including that of sea-level change, global warming and pollution control.

In order for China to participate in this exchange, as it modernises from what was essentially an agrarian society towards an information society, there is a corresponding need for China's geographic data to meet global data standards. The chapter sees standardization comprising three components: technical standards; management standards; and, working standards forming the basis for data quality. These standards have been established at several administrative levels including international standards, regional standards, national standards, industry standards, local standards, and enterprise standards.

The development of topographic mapping and symbology, geographic names, marine charting, thematic mapping and cartographic terms are discussed with respect to China's development in the geographic sciences. The chapter further reviews the spatial reference systems that have been adopted within China.

History of Cartography

In the introduction to this paper, the author states that "Cartographic research is about the knowledge of real world (i.e., what to draw), the knowledge of seeker or observer, or the knowledge of art maker (i.e. map author), the knowledge of the map itself as a natural object, and user knowledge, etc. The history of cartography should also be concerned about the process that maps created and used as graphic symbols'" This is indeed a wonderful introduction to a chapter that covers the History of Cartography. The author then provides an overview of the history of Chinese cartography. This provides a unique insight into Chinese cartographic developments and innovations over many Dynasties. As well, the chapter covers various discoveries of ancient maps in China, and describes these mapping products, and their importance. Then the chapter summarizes the main research achievements by scholars studying Chinese map history and associated cartographers, and the publications and atlases that have been produced to report on the research findings. For someone unfamiliar with the history of Chinese mapping and cartographers, this chapter offers the opportunity to glean valuable information.

Summary and Conclusion

This book provides a timely reminder about the advances in theoretical and practical cartography and GI Science in general and the contributions to these advances made by Chinese scholars and practitioners. Indeed, the chapters in the book provide a key information resource that brings-together information relating to the continuous research and developments carried out in China over the last 60 years. The chapters document the progression of research in each of the decades and the achievements made as part of these research endeavours. And, as illustrated in the various chapters, the research undertaken has been applied to the wide range of cartographic and GIScience issues that have been identified as key issues needing to be addressed.

The outcomes of research illustrated in this book underscore the need for a contemporary discipline in cartography and GIScience to be underpinned by continual advances to theoretical and practical knowledge. The foundations provided by this knowledge were established in China some 60 years ago and the on-going activities of colleagues from Chinese research institutes and production establishments supports the Chinese and international cartographic and GIScience communities.

We commend Prof. Jiayao Wang and his contributing authors for consolidating information about advances in cartography and GIScience research and development over the last six decades and making this knowledge more widely available to international colleagues.

William Cartwright
School of Science RMIT University
Melbourne, Australia

Colin Arrowsmith
Flinders University and University of South Australia
Adelaide, Australia

Gang-Jun Liu
School of Science RMIT University
Melbourne, Australia

William Cartwright is Professor of Cartography in the School of Science at RMIT University, Australia. He holds a Doctor of Philosophy from the University of Melbourne and a Doctor of Education from RMIT University. He has six other university qualifications—in the fields of cartography, applied science, education, media studies, information and communication technology and graphic design. He is the author of over 350 academic papers. His major research interest is the application of integrated media to cartography and the exploration of different metaphorical approaches for depicting geographical information. He was President of the International Cartographic Association (2007–2011) and Chair of the Joint Board of Geospatial Information Societies (now UN-GGIM: Geospatial Societies) (2011–2014). He is Co-editor of the International Journal of Cartography. In 2013 he was made a Member of the Order of Australia (AM) for service to cartography and geospatial science as an academic, researcher and educator. In 2017 he was made an Honorary Fellow of the International Cartographic Association.

Colin Arrowsmith is Associate Professor at Flinders University and at the University of South Australia. He holds a Doctor of Philosophy from RMIT as well as two masters' degrees and a bachelor's degree from the University of Melbourne, and a Graduate Diploma of Education from Hawthorn Institute of Education. Colin has authored more than 80 refereed publications and eight book chapters in the fields of GIS, tourism analysis and film studies. Colin's research interests include geospatial science education, the application of spatial information systems, including geographic information systems (GIS), to investigating the impact of tourism on nature-based tourist destinations, tourist behaviour, as well as investigating the issue of managing micro-historical data within GIS utilising cinema data. Colin is a member of the Surveying and Spatial Sciences Institute of Australia and a Fellow of the Royal Geographical Society.

Gang-Jun Liu is Senior Lecturer of Geomatics in the School of Science at RMIT University, where he teaches GIS-based Spatial Analysis, Spatial Statistical Analysis and Human Geography and supervises research projects for both Masters' and Ph.D. candidates. He is an Adjunct Professor at Henan University, an Adjunct Professor at Nanjing Normal University, and an Adjunct Research Professor at Henan Institute of Geographical Research. Gang-Jun holds a Doctor of Philosophy from Monash University, a Postgraduate Certificate in Spatial Analysis from Peking University and a bachelor's degree from Henan University. He is developing and applying GIS-based techniques and approaches to modelling, analysis and visualization of geographical phenomena, patterns and processes, at fine resolution, to support urban and regional sustainability. He has nearly 40 years' work experience in both government and tertiary institutes and has authored more than 40 refereed publications and four book chapters in the application of GIS based spatial analytical workflows.

Foreword III

This book was written to commemorate the 60th anniversary of the founding of the People's Republic of China. Over the past 60 years, China's map science has moved forward and grown stronger along with the advancement of new China. This book records the historical developments from traditional cartography to digital cartography, and further, the grand goal of advancing toward information cartography has been recorded.

However, the six decades represent a blip in the history of human civilization. Extraordinarily large changes have taken place in the field of cartography, which is a traditional discipline with a history nearly as long as the world's earliest culture.

Traditional analogs, such as the long history of manual mapping, have been replaced by digital and computer-aided mapping. Qualitative mapping methods based on long-time experience have been replaced by mathematical models, algorithms, knowledge and knowledge-based reasoning methods. Whether or not cartography is a science has long been debated; however, recently, cartography has been acknowledged by educational circles and industry as a science that describes the Earth's data field and flow of information. Cartography has formed into a map science and geographic information engineering (geographic information system) discipline with an advanced theoretical system, technology and products system and service system, and it is now listed as among the national key disciplines in China.

Progress and achievements in cartography have been immense in the past six decades following the founding of the People's Republic of China. This progress is reflected in the following ways: the means of map production has fully changed from the by-hand method to the computer-digitized method. Digital and integrated cartography and publishing has become a basic mapping production technology in China. The Chinese series scale topographic map covering the whole country has been completed. Chinese series scale map spatial databases and small-scale maps of countries surrounding China and world map databases have been built. A foundational geospatial data framework has been constructed for Chinese digital development. A comprehensive study of automatic cartographical generalization of spatial data has obtained substantial breakthroughs, and the theories and methods for automatic generalization of spatial data were built. This progress has laid a solid foundation for the final realization of intelligent mapping. Editing and publishing of various national thematic atlases and provincial (autonomous regions, direct-controlled municipalities) atlases have reached a new climax after the compilation and publishing of the national atlas (five volumes) of the People's Republic of China. These atlases can be seen as the landmark results of interdisciplinary fusion based on cartography. As an extension and deepening of spatial analysis and geoscience analysis, research on spatial data mining and knowledge discovery has made important progress, and practical applications have made a very good start. Research on uncertainty and spatial data quality evaluation and control has continued to deepen and gradually merged into the spatial data production process. As a functional expansion and extension of cartography, geographic information system (GIS) technology developed very quickly, its software system with independent intellectual property rights has successively been developed, and service-oriented and information sharing properties are becoming increasingly prominent. As a new growth point in cartography, virtual geographic environment (VGE) technology based on 3-D visualization technology is more practical and generalizable and has developed toward the integration of VGE and GIS. Theoretical exploration has deepened and become more systematic, a large number of high-level monographs and teaching materials have been published, and a theoretical system that takes geospatial cognition as the core of the cartography and geographic information engineering disciplines has been formed. Mapping discipline construction has made remarkable achievements, and cartography and geographic information engineering have become subjects and national key disciplines. These developments and achievements have drawn worldwide attention, which makes the community proud. China is now included among the countries that have advanced in the development of the cartography and geographic information engineering disciplines.

The Professional Committee of Cartography and Geographic Information Systems at CSGPC has developed and prepared this book to reflect the sixty years of development and achievement in this discipline to commemorate the 60th anniversary of the founding of the People's Republic of China. This span of time saw the rapid development of subjects, and few have had this experience; thus, it has been very difficult for a group with only a few people to complete this task. Based on this consideration, we have chosen more than 20 domestic experts to write thematic manuscripts in accordance with the unified writing outline. Deputy editor and the editor-in-chief are responsible for examining and modifying this book. Since the depth of investigation and the results obtained are not completely collaborative, the length of each subject differs, which reflects the development situations of each individual theme. This way of organization ensures short-term writing efforts.

The book consists of the following seventeen themes: (1) overview; (2) cartography theory; (3) map projection; (4) map design; (5) cartographic generalization; (6) thematic cartography; (7) geo-information Tupu (summary mapping); (8) geographical spatial database; (9) digital mapping; (10) map printing and publishing; (11) geographic information systems; (12) virtual geographical environment; (13) geographic ontology; (14) spatial data mining and knowledge discovery; (15) spatial data uncertainty and quality control; (16) mapping and geographic information standardization; and (17) history of cartography. Each chapter is in accordance with the unified directory structure, which includes an introduction, development process, major study achievements, representative works (including map work), problems and prospects, and references at the end of the chapters. Therefore, it is easy to read and make further queries.

The editor-in-chief is Prof. Jiayao Wang, and the deputy editor is Prof. Fang Wu. The thematic manuscripts were written by Jiayao Wang, Yufen Chen, Xiaohua Lv, Guangxia Wang, Fang Wu, Nan Jiang, Qingwen Qi, Tiejun Cui, Qun Sun, Ruizhi Shi, Yixin Hua, Gang Wan, Hongwei Li, RongQin Lan, Changqing Zhu, Jingtong Jiang, and Min An. All authors made great contributions. Gratefully acknowledged to Dr. Rongqin Lan, who spent nearly two years in translating all the original Chinese texts, typing and typesetting. Heartfelt thanks Xiaoyu Wang, doctoral student from the School of Earth and Environmental Sciences, University of Queensland, for her technical revising and editing of the translated manuscripts, as well as her valuable suggestions in the success of this effort. Sincerest thanks to Information Engineering University and Henan University for funding the publication of this book. Large amounts of literature were used in the writing process. Therefore, we owe them our sincerest thanks.

The book is of important learning value for scientists and technicians who are engaged in teaching, researching and engineering. In particular, it can help graduate students realize 60 years of Chinese advances in the subject of cartography and geographic information engineering. This book can serve as an excellent reference for graduate students carrying out topic research and dissertation writing.

Realistically, it is very difficult to write such a book. This book represents only a preliminary summary of the progress and achievements in this subject over the past 60 years and is still not entirely in-depth, systematic or comprehensive.

Being subject to the limits of our abilities and due to objective conditions, inadequacy is unavoidable in this book, and we look forward to your comments.

November 2020

Compilation Committee of Advances in Cartography and Geographic Information Engineering in China

About This Book

This book reviews and summarizes the developments and achievements in cartography and geographic information engineering subjects in China over the past 60 years after the founding of the People's Republic of China. It consists of 17 thematic chapters contributed by subject matter experts to present the progress and results in cartographic theory, map projection, map design, cartographic generalization, thematic cartography, geo-information Tupu (summary mapping), geospatial database, digital cartography, map printing and publishing, geographic information system, virtual geographical environments, geographic ontology, geospatial data mining and knowledge discovery, geospatial data uncertainty and quality control, mapping and geographic information system standardization, and the history of cartography. The chapters in this book comprehensively reflect cartography as a traditional discipline that has almost the same long history as the world's first culture, which has experienced extraordinary changes. Each chapter is in accordance with a unified directory structure and includes an introduction, development process, major study achievements, problems and prospects, representative works (including map work), and references. The structure is tight, and the layering is distinct; thus, it is easy to read.

The book can be used as a reference, which has important learning value for scientists and technicians who are engaged in teaching, research and engineering, to learn about the scientific properties, technical properties and engineering properties of cartography and geographic information engineering subjects. This book can help the younger generation of science and technology personnel (especially graduate students) to learn the 60-year history of Chinese advancements in the subjects of cartography and geographic information engineering, as well as understand the latest tendencies and the development direction for the future.

Contents

Chapter 1
Overview

Jiayao Wang

1.1 Introduction

China is one the first countries in the world to develop map science. Map science is a magnificent part of the history of ancient China. Early Chinese scientists, such as PEI Xiu in the Weijin period (224–273 AD), JIA Dan in the Tang Dynasty (750–805 AD), SHEN Kuo in the Song Dynasty (1031–1095 AD), ZHU Siben in the Yuan Dynasty (1273–1333 AD) and LUO Hongxian in the Ming Dynasty (1504–1564 AD) have made great historical contributions to map science. During the reigns of Emperors Kangxi and Qianlong, China introduced the western scientific mapping method to implement nationwide, large-scale geographic longitude and latitude surveying and mapping of the country.

Through the compilations created by *Huang Yu Quan Lan Tu* (*Map of China in the Emperor Kangxi Reign of Qing Dynasty*) and *Da Qing Hui Dian Yu Tu* (*Map of China in the Emperor Guangxu Reign of Qing Dynasty*), the map science of China was raised to a new level. *Hai Guo Tu Zhi* (*An Illustrated Gazetteer of Maritime Countries*) was compiled by WEI Yuan (1794–1859) in the late Qing Dynasty, and this compilation adapted the geographic latitude and longitude and map projection methods instead of the traditional Chinese method of Ji Li Hua Fang (Drawing Square Grids with Chinese Units of Length "li"). These were creative efforts toward creation of a world atlas in China's map science history.

For decades, from the Qing Dynasty to the period of the Republic of China, topographic map surveying and mapping were conducted intermittently and separately, and only a quarter of the national area of China was covered. Some middle- or small-scale maps and a 1:10 00000 map based on international map subdivisions and other general-purpose maps, as well as general atlases were compiled. Of these compiled

J. Wang (✉)
Henan University, Kaifeng, China
e-mail: wrongpu@163.com

J. Wang and F. Wu (eds.), *Advances in Cartography and Geographic Information Engineering*, https://doi.org/10.1007/978-981-16-0614-4_1

maps, only the ShenBao Maps published in 1934 had better quality, higher precision, and a more widespread influence. However, in general, China lagged behind developed countries in terms of map science.

The founding of the People's Republic of China in 1949 marked the rebirth of map science in China. At the 60th anniversary of the founding of the People's Republic of China, map science has experienced six decades of magnificent history.

1.2 Development Process

Following the founding of the People's Republic of China, map science development has generally experienced three stages: traditional cartography, digital cartography and cartography informatization.

1.2.1 Formation of Traditional Cartography

In the late 1950s and early 1960s, traditional cartography was formed after more than 10 years of construction and development since new China was founded. The achievements are demonstrated in the following aspects.

1. **Setting up surveying and mapping organizations, agencies and colleges to cultivate talent**

The Military Bureaus of Surveying and Mapping, which are directly under the Ministry of Defense and the State Bureau of Surveying and Mapping, were established for military missions and state affairs in 1950 and 1956, respectively. The People's Liberation Army Academy of Surveying and Mapping was created in 1946 in the liberated areas of northeastern China. In addition, the Wuhan Surveying and Mapping Institute (Wuhan Institute of Surveying and Mapping) was founded in 1956. Cartography majors have been offered in the geography department of Nanjing University since 1957. These institutes have become the training bases for senior technical personnel in the country and military.

2. **The plan for surveying and mapping in China was back on track**

A plan to complete the basic national map was developed based on a uniform geodetic coordinate system (Beijing geodetic coordinate system 1954), a uniform map scale system (the scales of the national map are 1:25,000, 1:50,000, 1:100,000, 1:200,000, 1:500,000 and 1:1,000,000; the scales of the local engineering map are 1:500, 1:1000, 1:5,000 and 1:10,000), a uniform map projection (the Gauss–Krüger projection of 6^0 zones and the Krasovsky ellipsoid were used for map scales from 1:25,000 to 1:500,000; and the Gauss–Krüger projection of three-degree zones for map scales from 1:500 to 1:10,000), a uniform sheet numbering system, and uniform cartographic symbols (including map symbols and map decoration specification). The new topographic maps were uniformly created. The accuracy and artistic quality of the maps obviously improved, and the maps were rich in content. Therefore, these maps could satisfy the requirements of national economic construction, defense construction and scientific research. Furthermore, the technical force grew stronger through these efforts.

3. **Placing more attention on scientific research in cartography and advancement of the mapping industry**

Since many of the issues we face in mapping practice require science and technological solutions, the cartography department at the Zhengzhou Institute of Surveying and Mapping and the Wuhan Institute of Surveying and Mapping attached great importance to scientific research while undertaking the teaching task. Moreover, a cartography research team was set up at the Institute of Geography of the Chinese Academy of Sciences (CAS) in 1954 and expanded to a research office in 1959; in Wuhan city, a surveying and mapping research office was established in 1957 and renamed the Institute of Surveying and Mapping in 1959; and the Institute of Foreign Geographic Name Translation was set up in 1959. There have been an increasing number of scientific researchers committed to cartography. These researchers have obtained a series of theoretical results using a variety of methods and techniques in the discipline of cartography. For example, projection methods were used for the world map, Asia map, China map, China provinces map and China regional map; cartographic generalization, including the map content generalization principle and its methods, map decoration (symbols and indications), and printing of maps (chromium gel image method) has been used, which promoted the development of China's mapping industry. Since then, China began to prepare the 1:1,000,000 scale map, published a 1:4,000,000 Southeast Asian situation map, planned compilation of the National General Atlas of the People's Republic of China, and began to prepare the provinces (autonomous regions) atlas.

4. **Maturation of the internal and external conditions promoted the formation of traditional cartography**

Considering the external conditions, some subjects related to cartography, such as geography, surveying and printing, have become relatively complete theories and technologies, laying a foundation for the formation of traditional cartography. Simultaneously, considering the internal conditions, cartography itself accumulated a rich history in the process of map production over a long period of time. These conditions were summarized and concluded by cartographers from various countries during different phases, which formed the basis for the systemic and complete techniques, methods and theories in map production. Map projection, map compilation, map decoration and map printing, which are the branch disciplines of cartography, have stabilized. Traditional cartography is said to be the accumulation of achievements and scientific summaries of cartography from the late 1950s to the early 1960s.

During this period, cartography was known as traditional cartography, which was defined as the study of the technique, technology and theory of mapping. In terms of mapping theory, the research core included map projection, map generalization, map content representation and symbol systems. These research cores are now known as the three old theories. Map-making technology mainly comprised the compilation manuscript, original printing and map plate-making printing technologies used in the process of map production. In terms of the craft of map-making, map production techniques, especially map printing techniques, were the main methods that were studied. Obviously, the goals of traditional cartography included map-making and product output. In this case, it is appropriate that traditional cartography is defined as "the art, science and technology of map-making".

1.2.2 From Traditional Cartography to Digital Cartography

Traditional cartography is the basis of map production, and traditional cartography successfully guided map production for more than 10 years following the formation of new China. However, traditional cartography has three obvious flaws. First, traditional cartography gives priority to experience but ignores the construction and research of the basic theory. Second, traditional cartography gives priority to contact with disciplines directly related to the subject itself, ignoring the connection with other higher-level subjects. Third, traditional cartography gives priority to map production but ignores the study of mapping applications, especially research on cognitive activities specific to map makers themselves and laws governing the user's cognitive activities.

Traditional cartography is said to be a relatively closed system. Thus, it is very difficult or even impossible to obtain substantial progress in cartography. This lack of progress forced cartographers to consider moving out of the closed system of traditional cartography to seek sources of further developing cartography from external systems and a greater hierarchy. All this progress occurred recently in the 1950s after appearance of three scientific systems theories (i.e., information theory, systematicness and cybernetics), and electronic computers were born. This progress not only has decisive significance to the development of modern engineering technology, but once again, this progress also completely changed the scientific landscape of the world and the way contemporary scientists consider the theory of relativity and quantum mechanics. This progress also confirms the direction of cartographic development.

When cartographic scientists realized the defects in traditional cartography and moved out of the traditional closed system of cartography, the next steps were technological revolution and theoretical innovation. These next steps represent the advent of the digital cartography era. The main aspects of this new era are outlined as follows.

1. **Computer technology and electronic publishing technology for maps offer a new means for cartography development**

Abroad, research on computer-aided cartography began in the 1950s, and the research underwent the initial phases of equipment development and software design, as well as advanced stages from experimental trials to more extensive applications in the 1970s. Simultaneously, the cartography research office at the Institute of Geography of the CAS was the first to begin studying computer-aided thematic mapping and computer-aided map plotting based on importing technology, digestion and absorption. The Zhengzhou Institute of Surveying and Mapping appointed selected teachers to Wuhan University and Nanjing University to study electronic computers and computer-aided mapping technology twice in 1972 and 1974, which strengthened the foundation of teaching staff. The institute began to set up a computer mapping laboratory in 1976, they completed China's first computer-aided relief map in 1978, they set up China's first majors of computer mapping and finally began to enroll undergraduates in 1979. Soon afterward, Wuhan Science and Technology University of Surveying and Mapping created the computer-aided mapping major.

Following the 1980s, electronic computers were continually replaced by new computers, some new high-speed and high-precision mapping equipment were applied, and increasing focus was placed on research of computer-aided mapping software. Therefore, the map database was built to meet the needs of large-scale computer-aided mapping. Using a single map database or departmental information system, a multifunctional, multipurpose and comprehensive cartographic information system (CIS) was developed. The electronic publishing system for maps was born in the 1990s. Digital cartographic software based on MicroStation was developed to import hardware at the Zhengzhou Institute of Surveying and Mapping. This software was combined with an electronic publishing system to form an integrated digital cartography and publishing system. This system had become the dominant mode of map production for both the state and military. Integrated digital cartography and publishing systems technology based on computers has replaced traditional manual mapping and publishing technology. Digital cartography extends business through diversification into products, such as digital line graphs (DLGs), digital raster maps (DRGs), digital orthographic maps (DOMs) and digital elevation models (DEMs). All these products became mainstream products, which is a great leap forward in terms of development and an important milestone in cartographic history.

2. **The advancement of cartographic technology requires a new theory**

With cartographic technological progress, especially following the systems theory, cybernetics and theory of transmission, a major problem has been how to combine the systems theory, control theory and the theory of transmission and research cartography as a function of the overall system. In 1969, Kolacny, a cartographer in the Czech Republic, proposed a model for map information transmission systems. As a complete process, this model illustrated the linkage between the map maker and map user. The model caused concern and drew the attention of international academia, including cartographers. Various map information transmission models were proposed. Further research focused on the three transformations (geographic information known by cartographers, maps and geographical environment images converted from map readers) in the process of map information transmission, map information loss and transmission efficiency, which were part of the process of information transmission, controllability of map information transmission and its control process of the control model, and so on. This process reflects all the characteristics of cartography in the information transmission system (Wang and Chen 2000).

What role would a human's cognition play in the process of information transmission? This question is a spatial cognitive theory problem that should be involved in map information transmission and digital information transmission. Cognitive science is a new science that was formally established in 1979. Cognitive science was introduced into cartography, and then, the concept of spatial cognition was put forward by China's scholars at the end of the 1980s and early 1990s. Spatial cognition is thought to be an important area of research in cognitive science. This area of science studies how people understand their existing environment, including related position, spatial distribution, dependencies and the change rules of many factors and phenomena. Studies go further, delving into the structure of the human cognitive system, the spatial cognitive ability of human beings, spatial cognition and mental mapping, and the basic process of mapping spatial cognition, including perception, representation, memory and thinking. The theoretical and practical significance of mapping spatial cognition in cartographic expert systems and geographic information systems (GISs) was revealed (Gao 1991; Wang and Chen 2000).

How can the information transmission of map symbols and their spatial cognitive efficiency be improved? This concept is a matter of applied theory. In 1967, French cartographer Bertin published the book *Graphic Semiotics* and proposed the visual variables of graphic symbols and the theory of visual perception effects for the first time. The six visual variables (shape, size, direction, brightness, density and color) and various visual perception effects (associative perception, ordered perception, quantitative perception, qualitative perception, dynamic perception and stereo perception) have received attention from Chinese cartographers when used in the composition. An experimental method of map perception and map design was studied by Gao (1984). Research and experimentation on map visual perception, especially electronic map visual perception, was also carried out (Chen and Chen 1999). Therefore, theoretical research on map visual perception using digital mapping technology was promoted.

For spatial information transmission in digital cartography, it is necessary to study theories regarding the uncertainties in spatial data and data quality control. Since the mid-1990s, many scholars have focused on spatial data uncertainty factors and forms, the processing method of spatial data uncertainty, the transmission mechanism of spatial data uncertainty in the process of digital cartography, and the method for spatial data quality control. Upon entering the 21st century, a series of theoretical achievements have been made in this research field. For example, the book, *Principles of Spatial Data and Spatial Analyses Uncertainty* (Shi 2005), published by the Science Press in China, comprehensively analyzed the sources of uncertainties in spatial data and spatial analysis. This book deeply addresses the model of spatial data uncertainty, the relation model of uncertainty and the model of spatial analysis uncertainty, visualization of modeling uncertainty and uncertainty of metadata. A doctoral dissertation, 'Research on establishment and application of the DEM precision model' (Wang 2005), brought forward the system contents of DEM error analysis and precision modeling, as well as deeply studying the DEM accuracy evaluation model based on terrain features, the error propagation model of DEM linear modeling, level of detail (LOD) error model and precision evaluations in the cases of 5 terrain types, including loess, hills, middle mountains, mountains and glaciers, and the precision evaluation of a DEM fusion model with features (such as roads and so on). Finally, a DEM accuracy evaluation system was designed, and a prototype was implemented.

3. **The integrated technology of digital cartography and map publishing effectively promotes surveying, mapping production and spatial data infrastructure construction**

The advancement of mapping technology and the deepening of cartographic theory are bound to promote map production development. Since the 1990s, under the guidance of cartographic theory, the integrated technology of digital cartography and map publishing has been used to fulfill the production of national series scale topographic maps and various scales of military collaboration maps, joint operations maps, aeronautical charts and nautical charts. This integrated technology was also used to design, compile and publish a 1:3,000,000 Geography Map of the People's Republic of China, a 1:5,000,000 World Geography Map, the National Atlas of the People's Republic of China, the National Economic Atlas, the Population Atlas of China, the Atlas of Chinese War History, the Atlas of the Jiangsu Province, the Atlas of the Zhejiang Province, the Atlas of the Jiangxi Province, the Atlas of the Xinjiang Uygur Autonomous Region, the Atlas of Chongqing city and atlases for other provinces and cities. The richness of these products in terms of contents, novel expressions, fine graphic lines and beautiful printing is like nothing that has been previously seen. These maps reflect the unity of scientific character, artistic quality and practicality. Simultaneously, 1:50,000 1:250,000, 1:50, 1:1,000,000 scales of the cartographic database of China, 1:3,000,000 cartographic databases of the Chinese mainland and its surrounding areas, 1:5,000,000 and 1:14,000,000 global map databases, marine surveying and mapping databases, and digital DOM databases were also established. Thus, a preliminary framework of Chinese and global spatial data infrastructure was constructed.

With the development of computer-aided cartographic and spatial database technology and the establishment of various databases, professional electronic maps (atlases), multimedia maps, online maps, electronic navigation maps, and other mobile navigation maps, the visualization system emerged at a historic moment. Other high-level applications are also popular, such as navigation electronic maps (Beijing Four Dimensional Graphical Navigation Information Technology Co., LTD.,

China Map Publishing House) and China Electronic Maps (Beijing Lingtu Software Technology Co., LTD., People's Traffic Video Electronic Publishing House).

1.2.3 The Concept of Cartography Informatization Is Proposed

Technically speaking, digital cartography hit a large milestone. However, the concept of traditional cartography remained limited.

There are also three obvious flaws in digital cartography. First, it has been a closed system in a digital environment, as the final goals are digital map production and output, and the production process of digital cartography is an isolated system. Second, digital cartography attaches significant importance to digital mapping technology research and ignores the independent innovation of digital cartographic theory research. Many theories also remain regarding the terms and concepts of digital cartography, as there is a lack of in-depth discussions, experimental demonstrations and actual applications. Third, prioritizing the products of digital maps still ignores application services, especially research on comprehensive application services.

Therefore, we have three basic understandings of digital cartography. First, digital cartography was developed in the late 1970s and early 1980s, so the history of digital cartography is just over 30 years. Changing the technique from manual mapping to digital mapping, which laid a foundation to further develop cartography in the information age, is a significant milestone. Second, traditional cartography has formed an adapted basic theory after a long period of information precipitation and accumulation. However, due to the relative youth of digital cartography, it has not yet formed its own basic theoretical systems that are both inherited and innovative. Third, digital development is only the instrument, and digital, information and knowledge services are the real objectives of development. It is well known that informatization and digitization are not entirely separate: if there is no digitization, there can be no informatization. Cartography informatization will extend and deepen digital cartography.

The only way to move out of digital cartography seems to be through cartography informatization. That is, the path forward must move out the closed system of digital cartography and seek continued development of the deep structure of the outside system. This path is considered to be the 'data-information-knowledge' process. Continually extending, expanding and deepening informatization and knowledge will aid in the arrival of the cartography information era. The main features are as follows.

1. **The focus of cartography has been transferred from information acquisition to intensive information processing**

Chen (1991), a member from the CAS, pointed out that the primary problem in developing cartography is the profound issue of map information. Map workers have made great efforts toward solving this issue. Map workers have even applied virtual contours, interpolation isotherms and derived boundary methods to supplement the information.

The satellite remote sensing technology that was developed in the 1960s has become the major technical means for collecting Earth observation data. After entering the 1990s, Earth observation data collection has advanced in the direction of multisensor, high-resolution and multitemporal remote sensing. Remote sensing information has become the main information source of cartography. Additionally, satellite navigation and positioning systems that can provide precise, all-weather, real-time, worldwide information completely meet the requirements of mapping and enable updating of large-, medium- and small-scale maps. Currently, driven by the integration of Earth observation information collection, processing and service, it is not only possible to transfer the focus of cartography from information collection to intensive information processing but it also necessary. This process has three main aspects.

(1) **The improvement of the cartography modeling method-Intelligent spatial data processing toward cartographic generalization**

Cartographic generalization is the most challenging and creative research field in cartography. Since the 1950s, as the focus of cartography has gradually shifted, and the field has received increasing attention. Cartography has evolved from the process of qualitative description to quantitative expression, from manual methods to automatic cartographic generalization based on modeling, algorithms and knowledge, and from the goal of fully automated cartographic generalization to human-machine cooperation. Automatic cartographic generalization has progressed from experiments with single features and separate models or algorithms to a whole process (all elements, the entire process and controllable) with process control and quality evaluation. In other words, cartography has undergone an evolution from simple to complicated, from part to whole and from digital to intelligent. Therefore, cartography already has the basic conditions for engineering and industrialization in the process and quality control system of automatic cartographic generalization (Wang 2008a, b).

Corresponding books have been published in different periods, such as the Chinese translation of *Mathematical Statistical Method in Map-Making* (Bocalov, an original author from the former Soviet Union), *Cartographic Generalization* (original author Topfer Germany, 1982), *Modeling Method of Cartographic Data Processing* (Wang and Zou 1992), *Mathematical Methods of General Cartography* (Zhu and Xu 1990), *Principles and Methods for Automatic Generalization of Digital Maps* (Wang and Wu 1998), *Multiscale Spatial Data Representation and Automatic Generalization* (Wu 2003), *Basic Theory and Technology of Map Generalization* (Wu 2004), *Spatial Information Intelligent Processing Methods for Map Generalization* (Wu et al. 2008a), and so on.

(2) **Deepening of the cartographic analysis function-Spatial data mining and knowledge discovery (SDMKD)**

In the traditional cartography era, acquiring knowledge from a map involves visual readings, measurements, calculations, analyses and deductions. With the advent of computers for digital mapping and GISs, measurements, calculations and analyses can be realized by computers. Spatial analysis has become the main function of GISs. However, the level of accessible knowledge is only surficial or shallow in the two cases above. Spatial data mining and knowledge discovery (SDMKD) is a highly attractive and challenging research area for humans facing the problem of processing large amounts of data. SDMKD is the key to transforming data into information and furthering knowledge. SDMKD is a nontrivial process that distinguishes effective, innovative and potentially useful and ultimately understandable patterns from the amounts of data. The purpose of SDMKD is to support spatial decision-making. The complexity and challenge of solving the problem has drawn increasing attention from Chinese scholars. Among the research, '*Theory and Application of Spatial Data Mining*' (Li et al. 2006) is the most representative academic work and basically reflects the research of this field in China. The project *Spatial Data Mining Technique and Application Engineering* (Liu et al.) won the first prize in science and technology progress issued by the Chinese Association of Surveying and Mapping in 2008. In addition, several research projects have been conducted. For example, 'spatial clustering analysis with obstacle constraints based on swarm intelligence' (Zhang 2007), 'spatial data mining of space discrete points' (Wang 2008a, b), 'spatial data mining based on fuzzy association rules' (Xiong et al. 2005), and so on. Some studies have made significant progress. These include 'SDMKD' from the Innovation Project of the CAS and 'Research on spatial data online analysis and spatial data mining' from the Open Fund Project of Visual and Auditory Information Processing State Key Laboratory. SDMKD will lead to the emergence of a wide variety of knowledge maps. These maps would greatly enrich the varieties and content of thematic maps and make map services seem more targeted and more human.

(3) **Map services become even more prominent**

The essence of cartography informatization is service. In the guiding ideology, geospatial information service is the goal. On the technical side of the network service, geographic spatial information services apply web service/grid service, popularization and universal service based on open standards and protocols. Product applications have the characteristics of various service modes, such as paper map services, digital map services, electronic map services, location-based services, embedded web services and network map services. These product applications have made good achievements over the years.

2. **Expansion and extension of the cartographic function—geographic information systems**

Before computers existed, people mainly used paper maps, series maps and atlases to obtain geographic knowledge, and many types of measurements and analyses have made on these maps. The previous maps, series maps, and atlases together can be regarded as non-computerized (contents are solidified) GISs. Since the wide use of computers began, especially the development of computer-aided cartographic technology and map database technology, people have attempted to employ special tools to collect, store, manage, analyze and use geospatial information. These attempts led to the introduction, development and application of GISs in the late 1970s and early 1980s. Although China lagged behind more developed countries by more than 10 years, GISs rapidly developed (Wang 2008a, b). More than 30 years have passed, and GIS software has made significant progress in China through the introduction of foreign advanced software (ArcInfo, MapInfo, and so on) to the independent development of software (such as MapGIS, SuperMap, GeoStar and military geographic information systems (MGISs)). GISs have been used in a wide range of fields, including resource management, urban planning and management, intelligent transport, environmental monitoring and management. GISs have become the foundational platforms to support the construction of digital cities, digital provinces and autonomous regions, digital China, digital rivers, digital oceans, and digital Earth. The software architecture of a GIS extends from the stand-alone version (fragmented information silo) to the network version. In the network environment, great changes have also taken place in the architecture of GISs, developed from host-based GIS, desktop GIS, web GIS, distributed GIS (DGIS) and open GIS (OpenGIS). The development mode of GIS software went from feature pack to integrations, modules, and components. GIS functions advanced from management to analysis and further to decision-making support. Research on digital terrain analysis (Zhou and Liu 2006) and spatial analysis modeling (Zhu and Shi 2006) deepened and became more practical. In particular, the publishing of a series of teaching books on GISs from Nanjing Normal University marked the overall great progress in GIS theories, methods and technologies.

Currently, Chinese educational and industrial communities focus on the key techniques of geospatial information sharing and spatial data interoperability based on web services and resource sharing and cooperation based on grid services.

Under the financial aid of the National Science-Technology Support Plan Projects and the National '863' Project, practical results have been obtained in realizing geospatial information sharing and spatial data interoperability based on web services. Web services can support different protocols and data formats, such as the simple object access protocol (SOAP), web services description language (WSDL) and unified description and integration (UDDI) protocol. The key information resource sharing technologies and functional corporations based on grid services have made some progress.

GIS functions and data are distributed and heterogeneous in a grid environment. Data access and integration, as well as service migration/composition based on user requests on a one-stop service system will become a new model of geospatial information service. Thus, the user can enjoy the service without the need to acquire data. The internally tightly coupled GIS software does not need to be installed, and the user can enjoy the GIS function service.

3. **The new growth point of geography—spatial information visualization and the virtual geographic environment**

Originally, the map was a visual product. So-called visualization refers to applying computer image processing technology to graphically display the complex scientific phenomena and the natural landscape, as well as very abstract concepts that were studied to understand the phenomenon, discover laws and spread knowledge.

The application of visualization theory and technology in cartography began worldwide in early 1993. This application mainly explores how to effectively apply theory and technology to spatial data visualization in the field of computer graphics. For GISs, visualization technology has been far beyond the scope and level of traditional symbolic representation and visual variable notation. We have entered a stage to explore the visual effect and the function of the visual tools under the conditions of a dynamic environment, temporal-spatial transformations, and multidimensional, interactive maps. More importantly, visualization is a kind of spatial cognition behavior. While providing insight into the complex process of spatial data analysis and displaying the multidimensional and multitemporal data and process, visualization can effectively improve and enhance the capacity of geographical environment information transmission and aid in understanding and finding the relationship between natural phenomena and cultivating the ability of a human's imagery thinking.

Virtual reality (VR) technology is, in fact, the most effective application of visualization technology; the purpose is to construct a virtual geographical environment.

The research field in China began in the mid-1990s, starting with a study of 3D terrain visualization (Xu 1990; Gao et al. 1999). Later, research more often involves using 3DS animation to observe the Earth's surface along the preset path, interactively using OpenGL software on a computer or workstation to observe a three-dimensional virtual terrain simulation in real time, using the Performer and MultiGen (three-dimensional modeling software) in the SGL workstation to accomplish Earth surface modeling and real-time display, or applying VRML (virtual reality modeling language) to spread virtual landscape information on the Internet. Upon entering the 21st century, VR technology has developed toward being more universal and practical. In recent years, spatial information visualization software has emerged endlessly but has been increasingly integrated with GIS applications. These technologies can be divided into two kinds, i.e., digital map browsing software and commercial game software. The former focuses on scheduling and rendering of large-scale geographic data. Data are usually in distributed storage on a different server and should be transmitted to the client through the network by using a special caching mechanism to increase the speed of browsing for the client. The latter focuses on the rendering of small-scale geospatial data to achieve very realistic effects by making full use of the new functions of graphics hardware. Therefore, spatial information visualization and VR technology have increased interests in geographical environment information based on graphs and pictures. Globally integrated and multiresolution 3D visualization human-machine interfaces have become the default mode for people to understand the geographical environment.

4. **Deepening cartographic theory—cartography informatization requires a scientific theoretical system**

Cartography informatization involves inheritance and development based on traditional cartography and digital cartography. The scientific theoretical system of cartography informatization should certainly include the deepening and upgrading of traditional cartography and digital cartography theory.

Cartography informatization is a new definition that has been widely discussed in recent years. Why on Earth should we put forward cartography informatization? What is cartography informatization? What is the relationship between cartography informatization and traditional cartography, as well as digital cartography? How can a cartography informatization discipline system be built, including a theoretical system? Such problems have been put forward but have not received in-depth study, although they draw higher attention from academic circles.

In China, the State Bureau of Surveying and Mapping and the Institute of Surveying and Mapping have organized a conference on informatization surveying and mapping systems. *'Proceedings of Informatization Surveying and Mapping'* was published by the Surveying and Mapping Press in 2008 with funding from the Book Publishing Foundation of Surveying and Mapping Science and Technology. Among these publications, 'On informatization characteristics of cartography and the idea of the theory and technology system' (Wang 2008a, b) explores the formation, development conditions, contents and features of cartography informatization, as well as proposes a theory system and technical system and discusses the key research issues. This publication was based on the individual analysis of traditional cartography and digital cartography according to their formation conditions, research object and features, defects and the way toward further development. 'The characteristics of basic geographic information service and related strategy in the new period' (Zhou and Liu 2008) suggests that along with the advancement of basic geographic informatization, the basic geographic information service constantly reflects the new characteristics of the information age, including diversification of basic geographic information products, network services, personalized information services, public services, and collaborative services.

Cartography informatization requires construction of a scientific system of its own, but this task is a long-term undertaking, and the current study is only the beginning of this undertaking.

1.3 Main Research Achievements

1.3.1 Cartography and Map Publishing Technology Fully Realizes Digitalization and Integration

The development of computer technology has brought a large revolution to cartography technology. Equipment development and software design for computer-aided mapping began in the 1970s in China. In the late 1980s, a computer-aided mapping software system was developed. In 1991, China's first set of practical color map publishing systems were created. Computer-aided cartography completely changed the status of the difficult and complicated manual mapping that had lasted for one thousand years. Map printing and publishing progressed from wet photomechanical reproduction and dry plate photography technology to digital photography. Color separation technology was extended from manual techniques to color separation photography and color separation software. Graphic layout and large version imposition technology advanced from manual methods to computer processing. Screening technology developed from amplitude modulation screening and frequency modulation screening to hybrid screening technology. Plate-making technology developed from the photographic transfer process and film printing to computer (digital)-to-plate. Proofing technology developed from artificial overprint proofing, film proofing and chemical proofing to digital proofing. Printing techniques developed from lithographic printing and offset printing to digital printing (computer to print). Printing process management developed from independent process management to digital whole process management. Since the 1990s, map production has fully implemented the transformation from manual simulation methods to computerization and digitalization methods. Digital and integrated mapping and publishing technology has become a basic technology for map production. Especially when combined with GISs, databases and remote sensing technology, map production efficiency is significantly improved, map content is enriched, map varieties are increased, and map currency is enhanced. Digital and integrated mapping and publishing technology is universally used in the production of the National Series Scale Topographic Map, Great National Atlas, Atlas of Provinces (Autonomous Regions, Municipalities), and all kinds of thematic maps (atlas). This technology is a milestone in map production in China (Wang and Wu 1998).

1.3.2 Implementation of National Basic-Scale Topographic Map Coverage and Preliminary Completion of the Basic Geographical Spatial Database for China and Worldwide

Surveying for China's 1:50,000 topographic maps began in 1951, and nationwide large-scale surveying began in 1956. Surveying and updating have been carried out simultaneously since 1964. By 2003, the 1:50,000 topographic maps covered an area of 760 square kilometers. In 2006, we started a project called the '1:50,000 National Topographic Mapping for the Blank Area of Western China', covering an area of 260 square kilometers. By 2010, 1:50,000 topographic maps had fully covered China's land area. Topographic map surveying began at a scale of 1:1,000,000 in 1978. This map is one of the major basic-scale topographic maps for provinces (or autonomous regions and municipalities) and basically covers the main grain and cash crop production regions in the plains and hills. The 1:5,000, 1:2,000 and larger-scale topographic maps basically cover China's urban areas (Tang 2003). The 1:1,000,000, 1:500,000, 1:250,000 and 1:100,000 topographic maps covered all of China's land area. Computer-aided mapping and database technology are used to build all kinds of geospatial databases, such as 1:1,000,000-, 1:500,000-, 1:250,000- and 1:50,000-scale basic geospatial databases, multiscale marine surveying and mapping databases and aviation chart databases; 1:3,000,000 scale databases of China and neighboring countries; 1:5,000,000-scale world map databases; large-scale DOM databases; 1:10,000-scale provincial (or autonomous region and municipalities level) databases; and finally, large-scale urban geospatial information databases that are under construction (Cui 2007). The success of the large digital engineering construction project has laid a solid foundation for China's geospatial data framework, including digital Earth, digital China, digital provinces and autonomous regions, digital cities, digital rivers, digital oceans, and so on. This database can provide reliable and timely spatial data for national and regional economic planning, disaster prevention and mitigation, irrigation construction, post-disaster reconstruction, design and construction of major projects, digital battlefield construction, development of information weapons and equipment, command automation systems, and so on.

1.3.3 Studies on Automatic Generalization of Geospatial Data Have Achieved Substantial Breakthroughs and Developments

As the most challenging and creative research area of cartography and geographic information engineering, cartographic generalization studies have always received close attention from scholars both at home and abroad.

In the 1950s, China began investigating the basic theory, method and specific application of cartographic generalization. A method to determine the cartographic generalization index with mathematical statistics was explored in the 1970s (Fan 1978). The emphasis was placed on the model and algorithm of cartographic generalization, human-machine cooperation and expert systems, and the relationship between them in the 1990s (Peng 1998). Since entering the 21st century, researchers have focused on all elements and whole process methods of automatic cartographic generalization based on algorithms, models and knowledge, especially those dedicated to intelligent cartographic generalization and automatic generalization process control and quality evaluation based on cartographic generalization chains. Substantial breakthroughs have been made. After decades of study, cartographic generalization has developed from a subjective process to an objective scientific method, from a qualitative description to a quantitative description, from a cartographic model to an automatic generalization based on an algorithm, model and knowledge, and from the pursuit of fully automated generalization to the realization of human-computer interaction and collaboration. The change from the automatic generalization test of a single feature to the automatic generalization of the whole (i.e., all features, entire process and controllable) with process control and quality design has been profound. In addition, this progress has basically established the system of theory, method and technology for spatial data automatic generalization. The research results reached an international advanced level. It is very helpful to simulate the human being's mode of thinking in the process of cartographic generalization with computers. Additionally, it objectively and correctly reflects the characteristics of map information processing by human beings or with computers and will realize optimal human-machine collaboration (Wang 2008a, b). The research achievements of automatic cartography generalization laid a solid foundation of theories, methods and techniques for using large-scale digital map data to produce smaller-scale maps and for automatic derivation of multiscale spatial databases based on large-scale databases and all-in-one updates, as well as multiscale spatial data representation in GIS.

1.3.4 The Compilation and Publishing of the Great National Atlas, Regional Atlases and All Kinds of Thematic Atlases Has Become a Monument in Cartographic History

The compilation and publishing of the atlas is one of the landmark projects of the cartography and geographic information engineering disciplines.

The Great National Atlas of China is a national key scientific research project approved by China's state council in the early 1980s. Five volumes, including general, natural, agricultural, economic and historical atlases, have been published. Among them, the National Agricultural Atlas was first published in 1990, the National Economic Atlas was published in Chinese and English versions at the same time in 1993 and in an electronic (CD-ROM) edition in 1995, the National General Atlas was published in 1995 and an electronic version (CD-ROM) was released in 1997, the National Natural Atlas was published in Chinese and English versions at the same time in 1998 and an electronic version (CD-ROM) was released in 2006, and the National Historical Atlas was divided into three copies and was scheduled for publishing beginning in 2009. Five volumes of the Great National Atlas of China as a whole are the complete assembly of the Chinese natural, economic, social and historical factors. These five volumes are an overview and summary of the research achievements of China in the field of geosciences, biological systems, environmental science, economics and history. These atlases not only provide an important scientific basis for the overall planning and macroscopic decision-making of China's economic construction and social development but also provide accurate maps and abundant materials for scientific research and teaching. These maps drove the publishing of all kinds of regional atlases and thematic atlases in China. The release of the provincial (autonomous region, municipalities) atlas and the corresponding electronic (CD-ROM) version have been pushed to a new pinnacle. The map data of all thematic atlases are richer. Among the provincial (autonomous region, municipalities) atlases, the Atlas of the Jiangsu Province, the Atlas of the Zhejiang Province, the Atlas of the Jiangxi Province, the Atlas of the Jilin Province, the Atlas of the Xinjiang Uyghur Autonomous Region, the Atlas of the Chongqing Municipality, etc., belong to typical works because of their advanced design concepts, selected contents and expression methods, and high levels of printing and binding quality. The Atlas of Shenzhen was awarded the only 'Cartography Award for Outstanding Achievement' at the 19th International Cartography Association conference. Many thematic atlases have outstanding features, such as the Atlas of Chinese Warfare History, the Atlas for the Natural Protection in China, the Atlas of China's Population, the Administrative Division Atlas of the People's Republic of China, the Standard Atlas of Geographic Names of Administrative Zones of the People's Republic of China, the Atlas of Remote Sensing Monitoring of Land Resources and Ecological Environment Around Beijing, the Atlas for the New Beijing Great Olympics, the Atlas for the Wenchuan Earthquake Disaster, and all kinds of image atlases of cities. All kinds of electronic maps, multimedia electronic maps, network electronic maps and mobile navigation electronic maps emerged at the right moment for specialized applications. The user scope is more popular. In particular, to celebrate the 60th anniversary of the founding of the People's Republic of China, a series of large atlases for China's different provinces have been edited and published for the first time. The total is 34 copies, one for each province (autonomous regions, municipalities and special administrative regions). Compiling and publishing the Great National Atlas, the Regional Atlases of China and the thematic maps and atlases is a milestone in the development history of cartography in China and has become one of the symbols of our age.

1.3.5 Evolutionary Process of Geographic Information System Software from Imports to Domestic Self-development

GIS is the development and extension of cartographic function. GIS is the result of computer-aided mapping and mapping databases but is beyond them. GISs in China started in the 1980s and originally imported foreign software (such as Arc Info) to construct application systems. Then, we worked on domestic software step-by-step and published the first teaching materials on GISs (Huang and Tang 1990). The software rapidly developed in the 1990s, and domestic GIS software was put into use. Since entering the 21st century, GIS software products have greatly increased. This evolution started with an integrated GIS basic software platform and grew to present a series of GIS software products, including basic platform software, special tool software and application software. The typical software products are MapGIS series software, SuperMap series software, GeoStar series software, and military and other special series of GIS software. This software has been widely used in many fields, including resources, environment, traffic, communication, energy, agriculture, forestry, water conservancy, defense, public security and aerospace, digital provinces, digital cities, digital rivers and oceans, and other fields. At the same time, a large amount of teaching materials on GISs have been published (Wang 2001). With the rapid development of computer network technology, GIS has changed from the traditional stand-alone version to the network version. The application of GIS expanded to all areas and broad geographic areas; there were many different types of distributed and heterogeneous GISs. These GISs were created and maintained using different software platforms or database systems according to the application requirements from different business organizations, governments, enterprises and individuals. According to this pattern, every user of a GIS needs to build a large database and install the complete and tightly coupled functions of GIS software and have his/her own spatial data storage and processing equipment. This requirement will result in redundant construction and considerable waste of information, storage and calculation resources, leading to spatial data inconsistencies and influencing the effect of spatial analysis and decision-making support. In addition, the unshared information and isolated system will make it difficult to solve important and complex problems involving many domains and multidisciplinary departments (Wang 2009). The emergence of web services and grid service technology provides beneficial conditions for solutions to the above problems. These technologies laid a solid foundation to realize geospatial information sharing and spatial data interoperability based on web services and further information resource sharing and problem solving in a dynamic and collaborative environment based on grid services. The research of GridGIS has recently received more attention.

GridGIS has been developed under the Chinese High-tech R&D (863) Program and National Science and Technology Support Plan during the state 10th five-year plan and the state 11th five-year plan, mainly focused on solving two problems: first, how to realize the connectivity and interoperability among legacy (existing) systems in grid environments, and second, how to build a new system according to the requirements of deployment models, development models, running modes and service modes in the grid environments (Wang 2009). In recent years, research in this field has made rewarding progress in China.

1.3.6 Generalization and Practical Application of Spatial Information Visualization and Virtual Geographic Environment Technology

Spatial information visualization and virtual geographic environment technology is a new and growing part of cartography. China's earliest spatial information visualization research was used in 3D terrain visualization (Xu 1990). Later, many people used the existing graphics software and animation software to produce geospatial information visualization products, for example, using the 3DMAX animation to obtain geospatial observations along a specific path, using OpenGL software to realize real-time interaction and stereoscopic observation of three-dimensional terrain simulation in computers or workstations, using Performer and MultiGen (three-dimensional modeling software) to complete geospatial modeling and real-time display on the SGI workstation (Liao 1996), and so on. Developed on the basis of spatial information visualization technology, the virtual geographical environment is a kind of advanced human-computer interaction system generated by computers. This environment is a cognitive environment based on visual perception, including hearing, touch and smell. This environment enables a person to engage with a world based on computer graphics, and this environment is created by a head-mounted 3D solid displayer, data glove and stereo earphone, giving a person an immersive feeling by implementing observation, touching, and detection tests (Gao et al. 1999). Of course, such equipment is expensive. Only a few famous universities in our country have this technology at present. Therefore, the application range of this technology is limited. At present, research on spatial information visualization and virtual geographic environment technology mainly focuses on three aspects: digital Earth browser software, distributed virtual geographic environment platforms, and integration with GIS and its practical application. Digital Earth browser software such as foreign Google Earth, World Wind, Skyline, Earth Viewer 3D, Arc Globe, etc., and domestic Geo Globe focus on large-scale geospatial data (terrain, texture, border, node, and 3D building model) scheduling and rendering. Data are usually in distributed storage on a different server and must be transferred to the client through the network using a special cache mechanism to increase the speed of browsing on the client's end.

Virtual geographic environment platforms use high-level architecture (HLA), for example, to build a distributed virtual geographic environment (DVGE) system framework (Xu et al. 2005). The integration of GIS and virtual geographic environment technology is a trend. Therefore, we can fully utilize both the multidimensional dynamic expression ability of virtual geographic environment technology and the advantages of GIS data processing and spatial analysis. Currently, there are primary products from virtual geographic information systems (VGISs). The theoretical study of virtual geographic environments has made many accomplishments (Su and Sheng 1997).

1.3.7 Spatial Data Mining and Knowledge Discovery Are Already Becoming Practical from Theoretical Research

SDMKD is a nontrivial process that distinguishes or extracts effective, innovative, potentially useful, and ultimately understandable patterns (knowledge) from massive spatial datasets. Traditional map application and analysis acquires knowledge and rules of the geographical environment through visual reading and simple measurement. SDMKD is largely limited by a person's knowledge and experience and the measurement tool and method. Spatial analysis of GIS mainly involves performing graphics operations. The functions of extracting and finding useful information and knowledge hidden in massive amounts of geospatial data are still relatively weak. Therefore, SDMKD is the development of a traditional map application and analysis in the digital map environment and the extension and deepening of the GIS spatial analysis function. SDMKD adapts to the developing trends and needs of information-age cartography that emphasize information processing rather than information acquisition. SDMKD studies have only been conducted in our country in the last 10 years. However, the studies on processing models and spatial analysis of mapping data in the 1980s–1990s, i.e., the basic modeling schema of 'Data Preprocessing-Design and Building of Digital Model-Data Processing-Design and Building of Map Model-Interpretation and Application of Map Model', can be seen as the primary research results of SDMKD (Wang et al. 1992). In the late 1990s, especially since the turn of the century, SDMKD research showed the following characteristics (Wang et al. 2006; Bi et al. 2008). First, a series of theoretical successes have been obtained. One of the most representative works is the *Spatial Data Ming and Application* by Di (2000). In detail, the book discussed the space data cleaning, available theory and method of spatial data mining, spatial statistical analysis theory of image texture, geo-rough space, cloud model, data field, spatial data mining based on the concept lattice, landslide detection and data mining, spatial data mining based on the inductive learning and rough set, spatial clustering and knowledge mining, image mining based on spatial statistics, and so on. Meanwhile, the development of a spatial data mining system was introduced. Second, spatial analysis combined with spatial data mining and knowledge discovery broadened the research perspective of SDMKD (Chen et al. 2003). Both methods have the advantage of GIS spatial data management and processing and enrich the function of GIS. Third, SDMKD research is becoming increasingly practical. Some application results were obtained; one of the most representative results is using spatial data mining technology for data integration, analysis, evaluation and optimization, and decision-making support in the field of sustainable land use and management.

1.3.8 Spatial Data Uncertainty and Quality Control Research Have Made Significant Progress

The concept of spatial data uncertainty is more generalized than the measurement error. Measurement error processing already has a set of mature theories and methods. With the wide application of GISs developed on the basis of computer-aided mapping and map databases, the error in the spatial data is not only from the process of surveying and mapping but also exists in the whole process of GIS information acquisition, collection, storage, processing, analysis and application. Hence, studying the cause of GIS spatial data uncertainty and its performance, handling method and transmission mechanism has received much attention. The research in China began in the mid-1990s and has reached its highest level since entering the new century (Guo et al. 2003). For a period of time, the research mainly focused on three basic theoretical questions. First, the confidence region of the basic geometric elements of the spatial data was analyzed. That study mainly considered the confidence region of point, line and multilateral (surface), which are the most basic geometric elements of GIS spatial data. Among these elements, the confidence region of the point is the key. Starting from the perspective of the error ellipse, an oval confidence region is given for the point, and then, confidence regions consist of ellipse and line segments for the established line segment and polygon (Guo et al. 2006). Second, the position uncertainty was considered by taking spatial data elements as a whole. That is, the 'ε-band' width was determined according to the average information entropy of the marginal distribution of all linear elements and then extended to the error entropy donut of the surface elements. Third, the statistic measurement model of the disfigurement rate of attribute data accuracy in GIS was researched. The stratified (data layer) sampling method was used to investigate the disfigurement rate model of the attribute data, and then, the accuracy of attribute data was measured through the disfigurement rate model. There are several representative research results in this field. The *Principles of Spatial Data and Spatial Analyses Uncertainty* (Shi 2005) comprehensively analyzed the sources of uncertainty in spatial data and spatial analysis. That study deeply addressed the model of spatial data uncertainty, the relation model of uncertainty and the model of spatial analysis uncertainty, the visualization of modeling uncertainty and the uncertainty of metadata, the theory of uncertainty and its applications in spatial data mining and cadastral data quality control. In addition, we propose a GIS data quality information service system based on web services. The thesis 'Research on establishment and application of DEM precision model' (Wang and Zhu 2005) put forward the content system of DEM error analysis and the precision model with the aim to address the existing problems; this thesis deeply studied the DEM accuracy evaluation model based on terrain features, the error propagation model of DEM linear modeling, the LOD error model and precision evaluation in the case of 5 types of terrain, including loess, hills, middle mountains, mountains and glaciers, and the precision evaluation of a fusion model of a DEM with features (such as roads, etc.). Finally, a DEM accuracy evaluation system was designed, and a prototype was implemented. In particular, the application of the space data uncertainty theory to quality control and evaluation in spatial data production has made many accomplishments in recent years and played an important role in ensuring the reliability of spatial data applications (Wu et al. 2008b).

1.3.9 Theoretical Research of Cartography and Geographic Information Engineering Changes from Traditional to Modernized

Cartography theory has always been widely considered by China's cartographic academia and industry. In the era of traditional cartography, cartography theory studies have focused on three aspects (Wang 2001), i.e., map projection, cartographic generalization and map symbol (representation). Among them, map projection is the mathematical foundation of a map. Zhou proposed a new map projection theory with a distortion ellipse by successfully applying a mathematics method (Zhou 1957). FANG explored new map projection methods according to Zhou's theory and designed many new map projections (Fang 1983). Cartographic generalization is a perpetual subject of cartography. Cartographic generalization mainly studies the basic principles and methods of cartographic generalization and how to determine the generalization index by using mathematical statistics, as well as its applications (Gao and Zhang 1965). The map symbol is the most basic expression form of a map. The map symbol mainly includes studies of the rule of symbol composition and other design issues. Map projection, cartographic generalization and map symbols are commonly known as the 'old three theory' of traditional cartography. In the era of digital cartography, using computers, it is easy to solve new projection design and calculation problems that could not be solved with a slide rule or hand-operated calculator in the past. Hence, new map projections emerge endlessly. The mathematical foundation of the map can be established through computer drawing. Map projection transformation among various map projections can also be implemented by a computer. Principles of *Mathematical Cartography and Map Projection Transformation Principles and Methods* (Yang et al. 2000) was the masterpiece in terms of research results at that time. That study was generally acknowledged as leading at the international level and became the pride of China's cartographic community. From cartographic generalization research came the emergence of the quantifiable generalization index and mathematical generalization methods. Mathematical methods, such as mathematical statistics, graph theory, fuzzy theory, neural networks, fractal theory and mathematical morphology, are widely used in the single feature of computer cartographic generalization. Theoretical research has been conducted and summarized (Wang 2001). The map symbol function emphatically studies graphic visual variables and the visual perception effect and its applications in map design (Chen 1996; Liao 2007). On this basis, theoretical research on the geographic information transmission mode, cartographic model and map spatial cognition was carried out. After entering the 21st century, the concept of cartography informatization was put forward.

The theories and methods of map projection transformation have matured and are widely used in map production (atlas). Cartographic generalization theory focuses on intelligent process control and quality evaluation with characteristics of all its features, the entire process and controllability. As a map language, map symbol systems study sentence, semantic and pragmatic rules. The 'old three theory' is reborn. Based on multimode spatial-temporal integrated cognition theory developed for maps, GISs and virtual geographic environments and by taking the spatial information transmission process (including information acquisition, processing and services) as an integrated system, a theory system including map models, visual perception, geographic ontology and spatial reasoning methods has been formed. This formation marked the progress of theoretical research and the great advancements of engineering technology in the cartography and geographic information engineering disciplines.

1.3.10 The Discipline and Professional Education Have Formed a Complete System and Entered the Key National Disciplines

The cartographic education career of new China can be traced to the northeast School of Surveying and Mapping in 1946, which existed before the national liberation. In 1952, this school was moved to Beijing and its name was changed to the Institute of Surveying and Mapping. The Wuhan Surveying and Mapping Institute (Wuhan Institute of Surveying and Mapping) was founded in 1956 by the local government. Since then, the cartographic discipline and major housed by the above two schools have been officially listed in the national higher education system. Thus, the preliminary development stage of the cartographic discipline and major occurred in the 1950s and 1960s, the adjustment and development stage took place in the 1970s and 1980s, and the rapid development stage occurred in the 1990s. At present, the discipline and major education system of cartography and geographic information engineering (engineering) or cartography and GISs (science) in China is complete. These educational disciplines entered the national key disciplines and formed a complete education training system that includes secondary school students, undergraduates and master's and doctoral students who are distributed in surveying and mapping engineering, geology and mineral resources, as well as normal colleges and universities; these schools have trained a large number of students and personnel with high and intermediate-levels of technical expertise for the country and the army, and this education has satisfied the needs of all kinds of talent for the national economy and national defense construction. Among them, there many people in governments and senior army cadres, as well as memberships in the CAS, the Chinese Academy of Engineering and the International Academy for Europe and Asia (IAEA); they are young academic leaders and the backbone of science and technology. These colleges and universities are also the scientific research base of cartography and geographic information engineering. A number of key national and provincial laboratories and engineering research centers have been built. A large number of research projects have been undertaken and completed, and the results have been applied in the construction of the national economy and national defense, especially projects under the financial aid of the National Natural Science Foundation of China, the National High-tech Research and Development Projects (863), and the National Science-Technology Support Plan Projects. A series of independent innovation successes have been achieved, and students have won the National Awards of Natural Sciences and National or Provincial S&T Progress Award. These projects have played a leading role in the development of these subjects in China.

1.4 Problems and Prospects

Since new China was founded 60 years ago, the field of cartography has developed from traditional cartography to digital cartography and is moving toward cartography informationization at the core of geospatial information services. Currently, cartography (map science) subjects have been developed into cartography (map science) and geographic information engineering (GISs) and have become national key disciplines, which is the most glorious stage in the history of cartography in China. However, with the rapid development of spatial information technology, communication and network technology, and spatial data processing technology, the development of cartography and geographic information engineering is facing many opportunities and challenges.

First, objectively speaking, the research of cartography and geographic information engineering science and technology in China was committed to the tracking, introduction and absorption of foreign advanced technology and less independent innovation. Undoubtedly, these methods have helped us to achieve progress in the science and technology of cartography and geographic information engineering in our country and catch up with the advanced level of developed countries. However, if we want to reach or even exceed the advanced level of developed countries, we must be committed to independent innovation, especially original innovation. Otherwise, it will be impossible to build the theory, technology and application service system for cartography informationization and realize a second leap in the development of cartography. Hence, it is necessary to greatly enhance the capacity of independent innovation in cartography and geographic information engineering. Independent innovation is the driving force for the whole evolution of cartography and geographic information engineering science and technology.

Second, information science is the foundation of today's social activities. Information acquisition, transmission, storage, processing and utilization technology are important parts of human activity. As the two technical backbones of information science, electronic computer and communication network technology has rapidly developed. The development goal of computers in the 21st century includes faster computing speeds, a higher degree of system integration, greater storage capacities, smaller sizes and multi-functions. The rapid development of communication network technology will change the way people work, live and interpersonally communicate. The traditional Internet realized the interconnection of computer hardware. The World Wide Web (Web) achieved communication in the form of a Web page. Currently, grid technology uses the high-speed Internet to link different geographically distributed resources, including computers, databases, storage and software, for full sharing of all information resources. The core idea is that the Internet is one computer. Considering such a situation, China's cartography and geographic information engineering disciplines may face many new problems. How do we deal with terabytes of space data with multiple users and quick and efficient information access? How can spatial data be transformed into spatial information and further into knowledge? How can the problems of information resource sharing and collaborative work in the grid environment be solved? How can we build a new generation of GISs, i.e., grid geographic information systems (Grid GIS), and provide diversified map services? And so on.

Third, the problem to solve using cartography and geographic information engineering science and technology is an information source. In the long developmental history of cartography, expeditions, field surveying, mapping and aerial photography were once used as means of map information acquisition, and these methods have promoted the development of cartography in different historical stages. Although remote sensing has become the main technical means of information acquisition from Earth observations since the 1960s, today, GIS spatial data mainly come from the digitalization of maps. The first reason is that image recognition technology lags behind; useful mass satellite image data have not yet been transferred into usable information. The second reason is that we cannot automatically extract vector data from images, as there is a lack of interaction between image data and vector data (Li 2004). The third reason is that the fusion of vector data and image data in GIS is far from perfect for the unreasonable administrative system of satellite images. Therefore, cartography and geographic information engineering disciplines should strive to research and solve the three aspects of this problem.

Additionally, the construction of digital society is the foundation of national modernization. Digital society is a broad concept, including digital Earth, digital countries, digital provinces (autonomous regions and municipalities) digital cities, digital rivers, digital oceans, and so on. All these digital concepts are currently in the process of construction. Among them, digital cities are the most representative or typical. A digital city brings together spatial information technologies (including GIS, remote sensing (RS) and global navigation satellite system (GNSS)) or computers, modern communication and network technology and information security technology to provide geographic information, information technology and information systems services for its enterprises, governments and citizens. Digital cities can visually reproduce the resource distribution situation of real cities by means of powerful system software and mathematical models on digitally integrated platforms consisting of urban, natural, social and economic factors. Digital cities can be used to analyze, simulate and research all kinds of schemes of planning, construction and management for real cities and to promote information sharing, communication and integration for users from different departments and at different levels. Digital city construction demands the full use of cartography and geographic information engineering science and technology, especially GIS technology. However, it should be said that the current gap is very large. The problems facing the cartography and geographic information engineering discipline during the new period are as follows. How can digital city information infrastructure be provided with spatial data and technical support? How can digital city application systems be provided with integrated GIS, MIS and OA software support? How can digital cities be constructed with geographic information sharing platforms? How can efficient geographic information services be provided for governments, enterprises and the public? And so on.

It is necessary to summarize the advances, problems and lack of advances in cartography in China since new China was founded 60 years ago and to analyze the development of spatial information science and technology and the discipline. With the arrival of the information age, the informatization of surveying and mapping has the characteristics of real-time global geospatial information acquisition, automated or intelligent spatial information processing, and popular spatial information services in a networked environment. Confronted with a new choice and challenge, the cartography and geographic information engineering discipline has much space for development (Wang 2008a, b).

1.4.1 The Innovative Subject System Will Be Improved and Perfected Based on Spatial Cognition with the Spatial Information Transmission Process as a System

The proposition that our discipline must be based on spatial cognition with the spatial information transmission process as a system reflects that service will be taken as the core idea in the information era. Spatial cognition is an important research field in cognitive science. Many studies on map spatial cognition have been conducted both at home and abroad. Spatial information transmission is a special research field that consists of information transmission. Many explorations on map information transmission modes have been performed both at home and abroad. Spatial information transmission models under digital environments have also been discussed.

By comparing the spatial cognitive system and geographical information system, we can discover a similar principle in the spatial information processing (handling) system. Spatial cognition is an epistemology; space information transmission is a methodology. The former runs through the entire course of the latter. Others, such as visual perception, geographic ontology, the map model and the spatial data scale theory, all belong to the applied theory used to correctly understand the geographic environment in the process of spatial information transmission. Many studies have been performed to date. Therefore, an innovative theoretical system of the discipline is likely to be constructed through further improvement and development by China's theoretical research workers in the cartography and geographic information engineering disciplines. The key is to move in the correct direction of research based on the overall discipline theory system, to plan research content in a scientific way and to improve the research methodology. Through independent innovation, we must build an integrated software system for digital mapping and publication with independent intellectual property rights based on the import, digestion and absorption of both hardware and software in digital cartography and publishing.

Integrating digital cartography with a publishing system (film output of color separation screening and digital plate-making) has been completed in China based on import, digestion and absorption of advanced foreign hardware and software. Although secondary software development and research works have been conducted according to the practical demand of mapping and publication, the whole system is subject to the limitations of MicroStation and its independent intellectual property rights. In recent years, we used this system to accomplish a great deal of mapping and production tasks. It is possible to further develop a set of microcomputer-based mapping software systems, including map design (general design, color design, symbol design and representation method design), map data processing (digital cartographic generalization, data processing of thematic mapping) and prepress editing and publishing (film output of color separation screening, digital plate-making). In fact, such small cartographic systems exist at present. It is important to make an overall design with reference to foreign hardware, to propose a solution of home-manufactured software and to develop a unified specification and interface. It is necessary to create a map design, map compilation (data processing), and prepress editing and publishing that are componentized and modularized products. Then, the general cartographic system and thematic cartographic system will be built.

1.4.2 The Intelligent Process Control and Quality Evaluation of Spatial Data Automatic Generalization Will Enter Its Application Stage

Intelligent generalization of spatial data is the most challenging and creative research area in the cartography and geographic information engineering disciplines. A large number of research results with great theoretical and technical value have been obtained by scholars in academic circles both abroad and domestically. However, due to the complexity of spatial data automatic generalization, the present research results do not apply to actual cartographic production and application, and many problems need to be solved. The purpose of process control and quality evaluation of spatial data automatic generalization is to ensure the quality of spatial data generalization. The focus of future research will be as follows (Wang 2008a, b). Innovation of ideological concepts will contribute to the innovative research and production applications of spatial data automatic generalization, and the research history in this area is proof that this is an effective method. The process of realizing spatial data automatic generalization has gone from simple to complex, from local to global and from digital to intelligent, and this process is far from being complete. Models and algorithms of spatial data automatic generalization must be continually improved and optimized with the advances in people's cognition and the development of science and technology. One major aspect of future research is to improve the intelligence level of automatic generalization. Knowledge engineering of cartographic generalization should be considered an important task. Objectively speaking, there are necessary bases for the engineering and industrialization of automatic generalization processes and quality control systems. Therefore, we should work with map production units to build engineering and industrialization bases. Once we have done these things, we will be able to develop an intelligent spatial data automatic generalization system (as a part of the whole digital cartographic system) with independent intellectual property rights. By that time, the following four problems can be solved. One problem is the use of large-scale digital map data for the production of smaller-scale maps. The second is the automatic derivation of a smaller-scale database from a large-scale basic spatial database and the all-in-one update of multiscale spatial databases. The third problem is to adapt to the needs of visualization and presentation of multiscale spatial data in GIS. The fourth is to extract user-required spatial data from distributed heterogeneous databases to build a spatial data warehouse.

1.4.3 The Virtual Geographic Environment Will Be Combined or Integrated with a Geographic Information System

GIS involves the expansion and extension of cartographic functions, while VGE is a new point in cartography. Each of them developed quickly, and the combination or integration of the two would be a solid foundation. In view of the interactive relationship between GIS and VGE, GIS has powerful capabilities for storage, management, processing and analysis of mass spatial data, while VGE provides the multidimensional and dynamic visualization of spatial information and the function of real-time interactive manipulation. Thus, the combination and integration of the two methods is an inevitable trend that helps to integrate the advantages of each other. As early approximately 10 years ago, the problem of integrating GIS with VGE was put forward (Su and Sheng 1997). The DVGE system framework constructed by using the HLA can be applied to the integrated management of multidimensional geographic information and multimedia integration, innovative scientific research, distributed collaborative planning, and design and decision-making (Xu et al. 2005). However, to realize the combination and integration of GIS and VGE, we must further develop a framework and a platform for the integration of GIS and VGE; research the web service registration, discovery and composition technologies under a network/grid environment; exploit an adaptive symbol system for the visual presentations of multiscale spatial data; and explore technical problems, such as the spatial-temporal modeling and simulation oriented evolution model. Only in this way will the function of the integration of GIS and VGE be more powerful, and the comprehensive utilization efficiency will be higher.

1.4.4 The Research Focus of Spatial Data Mining and Knowledge Discovery Will Convert from Theory to Application and Lead to the Emergence of a New Knowledge Map-a New Type of Spatial Data Processing Product

SDMKD expand and deepen spatial analysis and belong to one of the spatial data processing processes. To date, there has already been much research and many achievements. There have been some preliminary spatial data mining function modules specializing in processing spatial data both at home and abroad, such as GeoMiner, a spatial data mining prototype system based on the MapInfo platform being developed by Simon Fraser University in Canada, and the integrated human-computer interaction spatial-temporal data mining system developed by Wuhan University of China.

Overall, however, the research achievements in this field at home and abroad involve more scientific theory than practical application. Many scales of spatial databases, digital orthographic image databases, digital elevation model databases, and various professional (thematic) databases have been built in China. There will be many good opportunities for development and application of SDMKD. The existing data services may not be enough to satisfy the demands of users, especially policy-makers. There is a pressing need to obtain the knowledge hidden in data. SDMKD is actually an inevitable trend in spatial information technology development and social demand. The aim of data mining is to identify interesting and understandable knowledge and to further generate a visual knowledge map. As a kind of spatial data processing product, it is crucial to solve the imminent bottleneck of excessive data while knowledge is scarce. Next, we need to focus on the following aspects. Further study the SDMKD model and algorithm and develop many specific tool boxes for advanced knowledge discovery; explore spatial data and their complex relationships, such as the strong spatial-temporal correlation among spatial data, the information correlation relationship across different spatial and temporal scales, different organization hierarchies and different technology domains; obtain association knowledge, distribution pattern and spatial correlation; and research the visualization methods of knowledge and the automatic generation of all kinds of knowledge maps. These aspects will help to promote the practical application of SDMKD.

1.4.5 The Research Focus of Spatial Data Uncertainty and Spatial Analysis Will Shift to the Spatial Data Quality Evaluation and Control System

Because of the complexity and fuzziness of the real world and the limitations of human cognition and expression ability, spatial data and spatial analysis will inevitably present some uncertainties. Research on uncertainty theory in the processes of spatial data collection, processing, analysis and application is directly relevant to quality control in the processes of spatial data production, analysis and application. The spatial data uncertainty problem is a problem of spatial data quality. It is directly related to the reliability of digital map products and applications and the reliability of GIS spatial analysis. At present, systematic theoretical results have been obtained in the field, but they do not form any spatial data quality evaluation and control system and cannot be effectively used in the whole process of spatial data production and application. The uncertainty of the spatial data and spatial analysis results from the production and application of spatial data, and uncertainty theory is bound to return to the spatial data production and application process. To effectively guarantee the reliability of the production quality and application of spatial data, future research must focus on the spatial data quality content (elements) and evaluation index system in the whole process of spatial data production and application, the spatial data quality evaluation and control model, the data quality standards for the whole process of spatial data production and application, and construction of a spatial data quality evaluation and control system.

1.4.6 Web Services and Grid Services Will Become the Mainstream Mode of Current and Future Geographic Information Services

Along with the development of computer and network communication technology, GIS architecture has gone through several stages during the development process, i.e., host-based GIS, desktop GIS, WebGIS, distributed GIS and open GIS, while the development mode of GIS software has undergone a series of changes from the GIS function package, integrated GIS, and modular GIS to component GIS. There are some problems in the current WebGIS, such as the relatively fixed configuration of data and functions, relatively simple processing functions, and lack of interoperability among systems. In reality, cross-platform interoperability, resource sharing and collaborative work have not yet be achieved. On the one hand, driven by the ideas and technologies of open GIS, DGIS and component GIS (ComGIS), geospatial information sharing and spatial data interoperability, which are based on the web service standard of the SOAP, WSDL and unified description and integration (UDDI) and other standards, will become the mainstream mode of geographic spatial information services in both the present and future. On the other hand, the emergence and application of grid technology creates conditions in the broad sense of realization of resource sharing and collaborative problem-solving (work). There will be a super network consisting of a sensor network, working platform network and grid by adding a data service, functions service, calculation service, storage and resources services and a sensor service to the registry. Then, there will be a true sense of real-time dynamic GIS. Under the environment of GridService, the functions and data of GIS will be distributed; a one-stop service platform with data access and integration (DAI) services responding to user requests and service migration and service composition will become a mainstream service mode. The user will be able to enjoy the data service without the need for any data. Users can enjoy all GIS functions without any need to possess or install tightly coupled GIS software and services. Additionally, GIS software development models based on components will become mainstream. The core of both WebService and GridService is service. There are both similarities and differences between these services types. WebService is a stateless service, while GridService is a stateful service. Now, both services tend to merge with each other. They are the current and future main technology to realize networked/grid-enabled, popular and universal applications of spatial information services.

We review the progress in order to summarize achievements, find problems and consider the future to solve problems and promote development. Therefore, we can attempt to summarize the past and shape the future with scientific attitudes.

To put the vision of this subject into practice, we must firmly grasp the important weapon of ideological liberation, invigorate academic ideas, create a favorable academic atmosphere, and dare to be different when facing these academic issues. Additionally, we must seize the fundamental principle of independent creativity to innovate the theoretical system, technology system and service system of this discipline. Moreover, we must attach great importance to talent cultivation and innovation platform construction, especially leading technological talent and high-level open laboratories, freeing them to exploit their talent. As long as follow and implement the "scientific development" prompted by President Hu Jintao, over another 10 to 15 years, we will be able to solve the existing problems based on today's achievements, and our vision of the future can become a reality. By then, cartography and geographic information engineering disciplines may realize the second leap forward in development.

1.5 Representative Publications

(1) ***Cartographic Reference Manual*** (Lu Quan, Yu Cang. Surveying and Mapping Press, July 1988)

The book consists of eleven chapters. Chapter 1 includes several new arguments on modern cartographic theory, such as cartographic theory structure, views about maps as spatial information carriers, and academic viewpoints related to basic cartographic theory, including map transmission theory, symbolism and information-pattern-epistemology. Chapter 2 introduces a newly developed classification method based on the traditional map classification method, including introduction of the map classification system in the *Thesaurus of Surveying and Mapping Science and Technology* according to the needs of automatic computer retrieval. Chapters 3 and 4 briefly discuss the history of cartography in China, provides readers with an understanding of the country's historical legacy of mapping, and discusses adaptations to the international new trend of studying ancient maps and the history of maps. In particular, the fourth chapter introduces the mapping situation during the period from the late Qing Dynasty to the Republic of China and the thematic mapping in China in recent years. Chapters 5 and 6 systematically summarize and introduce cartography-related international organizations and their activities and relevant publications both at home and abroad. These chapters conveniently enable readers to obtain clues to further explore the international advanced level. Chapter 7 mainly introduces the use of electronic calculators in map projection calculations. Chapter 8 covers significant ground with discussions of thematic maps, including navigation and aeronautical charts, and shows the further development trend from map compilation to thematic mapping. Additionally, Chapter 8 also reflects the tendency that since topographic map compilation involves precision measurements and error and the application of mathematical statistics methods, the theory of topographic map compilation is transitioning from qualitative to quantitative descriptions. Chapters 9, 10 and 11 systematically generalize and summarize all kinds of technologies and management methods in classical mapping, and these chapters introduce all aspects of new types of maps and mapping technologies.

(2) ***Introduction to Maps*** (Yin Gongbai, Wang Jiayao, Tian Desen, Huang Caizhi. Surveying and Mapping Press, December 1990)

The book systematically and comprehensively expounds all kinds of knowledge related to maps and introduces the development trend of maps, new concepts of modern cartography and new theories and technologies. This book is divided into 10 chapters. This chapter covers maps and cartography, introducing the definition of a map and its basic characteristics, the content, classification and function of the map, and the concept of cartography and its connection with other subjects is expounded. Chapter 2 is the mathematical basis of the map, and this chapter introduces the basic concept of map projections from the shape and size of the Earth, coordinate systems and elevations and its control points, describes the projection method and applications of Gauss-Krüge projection and conformal conic projection, and discusses the problems of map orientation, map scale, map subdivision and sheet numbering system. Chapter 3 covers map language and includes three main sections, i.e., map symbol, map color and map annotation. Chapter 4 discusses the general map and introduces the types and contents of general maps and describes the representation methods of natural geographical features, social and economic features and external margin elements. Chapter 5 presents the thematic map and introduces the characteristics, types and contents of the thematic map, as well as the representation of thematic map elements, such as point, line and shape, and comparisons are made among the various representations. Chapter 6 comprises the atlas and series map and information about atlases and series maps, respectively, are introduced, including definitions, classification, characteristics, captions and the geographical names index. Chapter 7 covers the map compilation and printing methods and mainly discusses the four aspects of the map compilation method, commuter-aided mapping, RS mapping and map printing. Chapter 8 discusses the main map works, with introductions of the main map products of China and foreign countries. Chapter 9 covers map analyses and applications, where map analyses are reduced to four methods, including visual analysis, map diagram analysis, mathematical statistics analysis, and mathematical model analysis; additionally, the applications of maps in the aspects of scientific research, national economy and national defense construction are emphatically introduced. Chapter 10 discusses the history of cartography, and based on the map origin aspect, the development course of cartography in ancient, modern and contemporary China is explored; further, the development track of ancient and modern cartography abroad is introduced, and the main tasks and goals of cartography in China over the next period are put forward.

(3) ***Map Application Study*** (Huang Wanhua, Guo Zhengxiao, Zhao Yongjiang, Li Linghui. Xi-an Map Press, November 1999)

The book is a total of 12 chapters. This chapter comprises the introduction, which mainly presents the architecture, research objects and research contents of map application studies and analyzes the existing problems and main tasks in the field of map application studies. Chapter 2 covers the formation and development of a map application study, and in this chapter, the three development stages of map application studies are mainly introduced, providing a brief history of map application studies abroad and the research work and contributions by both domestic and foreign scholars. Chapter 3 discusses the map function theory, including the basic characteristics, functions, roles and main usages of the map and the means and measures to improve the efficient use of the map. Chapter 4 covers visual map recognition and reading, including an introduction to the expression methods and representation methods of map language and map content; then, the recognition of map projections, cartographic generalization, map recognition and reading, the principle, routine and content of visual map recognition and reading, etc., are discussed. Chapter 5 discusses the map quality evaluation and mainly includes the factors and contents, standard and routine methods of map quality evaluations and the preparation of evaluation reports. Chapter 6 comprises map measurement and calculation and mainly discusses the determination of the ground point's position and elevation according to the topographic map, as well as the methods of distance measurement, area measuring and calculation and the Earth's surface calculation, ground slope measurement and volume calculation. Chapter 7 covers the map content analysis method, which mainly includes the diagram method, diagram analytic method, mathematical analysis, mathematical statistics analysis and information theory analysis. Chapter 8 presents digital mapping technology and application, including the digital map overlap analysis, map information description and presentation. Chapter 9 discusses geographical information systems - expansion and extension of map function, which generally includes introduction of the concept, function and application of GISs. Chapter 10 covers cartographic methodology, which includes the content and function of cartographic methodology, such as the method of using a single map, the method of using a variety of maps, the combined use of the map method and RS method, and this chapter presents some practical applications of cartographic methodology. Chapter 11 comprises the accuracy and reliability of map use and research, including an analysis of the factors that affect accuracy, the calculation of map accuracy and its influencing factors, the calculation of technical accuracy and its influencing factors, cartographic generalization and its impact on the accuracy and reliability of map study, and so on. Chapter 12 includes an investigation of map users and utilization efficiency, in which the content, manner and means of surveying, as well as the organization, implementation, and writing of the report are introduced, and the present situation of foreign map users and relevant surveys are described.

(4) ***Cartography*** (Zhu Guorui. Wuhan University press, January 2004)

The book is composed of 17 chapters that are divided into six parts: map and cartography, map projection, map data and map symbol, design of map graphics, color and annotation, map design and map compilation, map publishing and printing, as well as its analysis and application. Chapter 1 and 2 introduce basic knowledge of the map and mapping process and the basic concept and development of the cartography trend, respectively. Chapters 3, 4 and 5 introduce the basic theory of map projection, several commonly used map projections, such as conic projection, azimuthal projection, cylindrical projection, pseudoconical projection, pseudocylindrical projection, pseudoazimuthal and polyconic projection, and the application and transformation of map projections, respectively. Chapters 6, 7 and 8 introduce map data, map symbols and the representation method of map content, respectively. Chapters 9, 10 and 11 introduce map graphics design, map color design and map annotation design, respectively. Chapters 12, 13, 14 and 15 present cartographic generalization and cartographic mathematical models, including regression models, square root models, hierarchical models and classification models, map editing and compilation and atlas compilation. Chapters 16 and 17 introduce the electronic map and electronic publishing of the map and map analyses and applications, respectively.

(5) ***The Principle of Cartography*** (included in Chinese college and university teaching material for the 21st century) (Ma Yaofeng, Hu Wenliang, Zhang Anding, Chen Fengzhen. Science Press of China, June 2004)

The book completely and systematically introduces the essence of the map, including the theory, technology and method of map-making and map use. This book contains nine chapters. The first chapter is an introduction, which mainly presents the basic concept of maps and cartography and synoptically introduces the methods of map-making, a brief history and the cartographic progress. The second chapter introduces the mathematical foundation of the map, with brief introductions of the Earth ellipsoid, geodetic control, map scale, map projection and its basic principle, several kinds of commonly used map projections, map projection discrimination and choice and the automatic generation and transformation of map projections. The third chapter covers map language, which mainly discusses the characteristics, classification, quantified expression and visual variables of map symbols, as well as the effects of visual perception, map symbol design, map color and design, and map annotation. The fourth chapter covers map generalization, with brief introductions of the map generalization essence, the influential factors of map generalization and the principles and fundamentals of map generalization; simultaneously, automatic map generalization is also roughly introduced. The fifth chapter presents a general map, with introductions of the basic concept of a general map, the representation methods of natural geographical features and socioeconomic features, map orientation and topographic map query, and includes a description of the present situation of the national fundamental geographic information database. The sixth chapter discusses the thematic map, which mainly includes introductions of the thematic map basic concept, the characteristics of thematic elements and its presentation methods, and a briefly introduction to the characteristics and types of atlases and electronic atlases. The seventh chapter covers map design and production, including the general process of map compilation, the design of general maps and thematic maps, and new technologies involved in map plate-making, printing and map production. The eighth chapter discusses modern mapping technology, which mainly includes introductions to the hardware and software of computer-aided mapping and databases, map editing and production, RS mapping, GISs mapping, electronic map systems, and so on. The ninth chapter is composed of map analyses and applications and mainly introduces the basic concept, technologies and methods of map analysis, as well as several kinds of geographical element analysis methods, topographic map reading and field applications.

(6) ***Principles and Methods of Cartography*** (national finely designed course materials of China) (Wang Jiayao, Sun Qun, Wang Guangxia, Jiang Nan, lv Xiaohua. Science Press of China, the first edition was printed in March 2006, and the second printing occurred in July 2007)

The book systematically and completely introduces the principles and methods of cartography. There are a total of 18 chapters divided into six sections: introduction, mathematical foundation of map, representations of map contents and elements, cartographic generalization of map contents, technologies and methods of modern cartography and map analyses and applications. This chapter and Chap. 2 introduce the map and cartography, respectively. Chapters 3, 4 and 5 discuss the basic principles of map projection, commonly used map projections, map mathematical foundation design and map projection transformation. Chapters 6, 7 and 8, 9, 10 through 11 introduce information sources and their processing, as wel as map symbol designs, designs of the map overall effect, content representation of the general map, content representation of the thematic map, and the characteristics of special maps and the representation method. Chapters 12 and 13 discuss the basic theory of cartographic generalization and cartographic generalization methods for all kinds of map content elements, respectively. Chapters 14, 15, through 16 introduce digital mapping and map databases, digital mapping technology, and other new mapping technologies, such as multimedia electronic maps and Internet maps. Chapters 17 and 18 discuss map analyses and map applications, respectively.

(7) ***A New Cartography Course*** (a Nationally Planned Textbook of Regular Higher Education for "the 11th Five-year National Plan") (Mao Zanyou, Zhu Liang, Zhou Zhanao, Han Xuepei. Higher Education Press, April 2008)

The book is the basic teaching material for the geography field. The first edition was published in 2000; this book is the second edition. The book is divided into 10 chapters and covers the basic problems of cartography. Among these chapters, the first chapter is an introduction to the definition and characteristics of a map, the function of the map and its classification, the history of the map and the current development, the mapping method, and definitions of cartography and related disciplines. The second chapter covers geoids and map projections, with a focus on geoids, geodetic systems, map projections and map scale. The third chapter discusses map data, which involves terrestrial observation data, multisource RS data and global positioning system (GPS) data, as well as map data processing and map databases. The fourth chapter covers cartographic generalization, which outlines the nature, influencing factors, content and methods of cartographic generalization and its current development, and compares manual and automatic generalization. The fifth chapter presents map symbolization, which includes the map symbol (map language), map symbol classification and quantified expression, visual variables and colors of map symbols, the mental perception of symbols and graphics, and map annotations. The sixth chapter is map representation, which discusses the representations of geographical data that appear in point, linear or polygon distributions, the expressions of 3D spatial information and contour lines, and the dynamic representation of geographic information. The seventh chapter presents map editing, where first, general maps and thematic maps are introduced, and then, thematic map editing and design, atlas editing and RS photomaps are introduced. The eighth chapter covers digital mapping, which mainly includes introductions of the theory and technique of digital cartography, the digital mapping method and GIS-based digital mapping. The ninth chapter comprises map reproduction, which mainly includes the traditional lithography process, electronic publishing and prepress system, the development of modern map reproduction methods and the publishing management of map production. The tenth chapter includes map analysis, which separately introduces the mathematical methods of map analysis, reading analysis, graphical analysis, composite analysis and map information about Tupu (a geoinformatics graphic analysis).

References

Bi, S., Geng, H., & Lu, G. (2008). Domestic research actualities and technical system discussion about spatial data mining. *Geomatics World, 6*(1), 21–27.

Chen, J., Fu, Z., & Bian, F. (2003). Mining spatial association rules with spatial analysis. *Computer Engineering, 29*(11), 29–31.

Chen, S. (1991). *The flow of information and cartography*. Chinese Yearbook of Cartography (pp. 1–11). China Map Publishing House.

Chen Y. (1996). A preliminary study of map language. *Military Surveying and Mapping*, (2), 40–42.

Chen, Y., & Chen, Y. (1999). Application of visual perception theory on electronic map design. *Journal of Institute of Surveying and Mapping, 16*(3), 218–221.

Cui, T. (2007). *Principle of geographic spatial database*. Science Press.

Di, K. (2000). *On spatial data mining and knowledge discovery*. Wuhan University Press.

Fan, X. (1978). *Study on the density and the selection index of Chinese residents*. The Institute of Surveying and Mapping.

Fang, B. (1983). *Different points weft polyconic projection and its application: Map projection* (pp. 255–275). Mapping Press.

Gao, J. (1991). *Map spatial cognition and cognitive cartography*. Chinese Cartography Yearbook (pp. 12–18). China Map Publishing House.

Gao, J. (1984). Experimental method of map feeling and map design. In *The progress of cartography: China surveying and mapping institute of new technology of cartography teaching class lectures in the summer of 1984*.

Gao, J., Xia, Y., & You, X. (1999). *Application of virtual reality in terrain environment simulation*. PLA press.

Gao, J., & Zhang, T. (1965). *Topographic map compilation*. Institute of Surveying and Mapping.

Guo, T., Wang, J., & Wei, H. (2003). A study of confidence regions of the fundamental factors in GIS. *Acta Geodaetica et Cartographica Sinica, 32*(2), 164–167.

Guo, T., Wang, J., & Wang, G. (2006). Error model of curves in GIS and its digitizing test. *Geomatics and Information Science of Wuhan University, 31*(1), 78–81.

Huang, X., & Tang, Q. (1990). *Introduction to geographic information system*. Higher Education Press.

Li, D. (2004). On integrated earth observation systems of geomatics. *Science of Surveying and Mapping, 29*(3), 1–2.

Li, D., Wang, S., & Li, D. (2006). *Theory and application of spatial data mining*. Science Press.

Liao, D. (1996). *The research and practice of real time dynamic virtual goblin generation technology.* Master's Degree Thesis of the Institute of Surveying and Mapping.
Liao, K. (2007). *Introduction to earth information science.* Science Press.
Peng, R. 1998. *Automatic generation of double lane highway sign.* Digital mapping technology and digital map production (pp. 244–247). Xi'an Map Press.
Shi, W. (2005). *Spatial data and spatial analysis uncertainty principle.* Science Press.
Su Y., & Sheng, Y. (1997). Integration of advanced visual and geographic information systems. *Earth Information Science, 2*, 33–38.
Tang, H. (2003). Introduction to the situation of national rapid survey and mapping. *Geographic Information World, 1*(3), 39–43.
Wang, G. (2005) *Research on the establishment and application of DEM precision model.* Doctoral Dissertation of the Information Engineering University.
Wang, G., & Zhu, C. (2005). LOD modeling and accuracy analysis of terrain information. *Acta Geodaetica et Cartographica Sinica, 34*(3), 228–232.
Wang, J. (2001). *Principle of spatial information system.* Science Press.
Wang, J. (2008a). The progress and trend of automatic generalization of spatial data. *Journal of Geomatics Science and Technology, 25*(1), 1–7,12.
Wang, J. (2008b). *A visualization of information characteristics and cartography theory and technology system* (pp. 130–138). Information mapping. Surveying and Mapping Press.
Wang, J. (2009). Geographic information system in the development and development of geographic information system. *Chinese Engineering Science, 11*(2), 10–16.
Wang, J., & Chen, Y. (2000). *The theory of cartography.* PLA press.
Wang, J., Fan, Y., & Han, T. (1992). *General map generalization principle.* Surveying and Mapping Press.
Wang, J., & Wu, F. (1998). *The integrated principle and method of digital map generalization.* PLA press.
Wang, J., Zhu, Y., & Wu, M. (2006). Grid and geography information system based on grid. *Journal of Zhengzhou Institute of Surveying and Mapping, 23*(1), 1–7.
Wang, J., & Zou, J. (1992). *Model method for data processing of map drawing.* PLA press.
Wu, F. (2003). *Multi scale representation of spatial data and automated generalization.* PLA press.
Wu, F., Qian, H., & Deng, H. (2008a). *Intelligent processing of spatial information for automatic map generalization.* Science Press.
Wu, F., Liu, P., & Jin, C. (2008b). Analysis on the quality elements of vector map data. *Geomatic Science and Engineering, 28*(2), 1–5.
Wu, H. (2004). *Research on the basic theory and technology of map generalization.* Surveying and Mapping Press.
Xiong, P., Zhu, T., & Huang, T. (2005). Algorithm of mining fuzzy associate rules in anomaly detection. *Geomatics and Information Science of Wuhan University, 30*(9), 841–845.
Xu, B., Huang, J. & Lin, H. (2005). Architecture of distributed virtual geographic environment based on HLA. *Editorial Board of Geomatics and Information Science of Wuhan University, 30*(12), 1096–1099.
Xu, Q. (1990). *Research and practice of 3D terrain visualization technology.* Master's Degree Thesis of the Institute of Surveying and Mapping.
Yang, Q., Snyder, J. P., & Tobler, W. R. (2000). *Map projection transformation principles and application.* Taylor and Francis.
Zhang, X. (2007). *A Study of Spatial Clustering with Constraints Based Swarm Intelligence.* Doctoral Dissertation of the Information Engineering University.
Zhou, C. (1957). According to the problem of the new projection of the deformed ellipse. *Journal of Surveying and Mapping, 1*(1), 3–16.

Zhou, Q., & Liu, X. (2006). *Digital terrain analysis*. Science Press.
Zhou, X., & Liu, R. (2008). *Characteristics and countermeasures of basic geographic information service in the new period* (pp. 145–152). Information mapping. Surveying and Mapping Press.
Zhu, C., & Shi, W. (2006). *Spatial analysis modeling and principle*. Science Press.
Zhu, G., & Xu, Z. (1990). *Mathematical methods in common map drawing*. Surveying and Mapping Press.

Chapter 2
Cartographic Theory

Yufen Chen

2.1 Introduction

Maps have the same long history as mankind. However, cartography has only been regarded as a science for a short period of several decades. For a long time, cartography was without its own theoretical system, and at best, it was regarded as a technology or engineering science by scientific academia and even by some cartographers. The subject status of cartography has not yet been established. Cartography did not truly become a science until the emergence of the theoretical cartographic concept, and the research of modern cartography theory had provided a theoretical basis for cartography in the middle of the 1950s.

Theory can play a largely leading role in the development of a discipline. In the late 1950s and early 1960s, cartography was formed as an independent science, often called traditional cartography. During this period, a systemic and integrated technique, method, craft and theory about map making had already been formed in the field of cartography. This information included the previous summaries of theories and technologies about map making and map applications. Traditional cartography usually adopts manual compilation that has more processes and longer cycles. Technical innovation is mostly focused on map editing and compilation, solution design and testing of map printing, and techniques involved in the process of map compilation and production. Theoretical research on cartography covered the technical summary. There have been systematic studies on the so-called three old theories, including map projection, cartographic generalization and cartographic semiotics theory. Overall, the research on cartographic theory, especially basic theory, is far from sufficient.

The concept of the theory of cartography was put forward by Swiss cartographer Imhof in 1956. Cartography is divided into two research fields: theoretical cartography and practical cartography. The theoretical research on cartography has new

Y. Chen (✉)
Information Engineering University, Zhengzhou, China

J. Wang and F. Wu (eds.), *Advances in Cartography and Geographic Information Engineering*, https://doi.org/10.1007/978-981-16-0614-4_2

developments in terms of scope. Since the 1960s, under the impact of information theory, system theory and cybernetics, as well as with the development of new cartographic technology and the mutual understanding between disciplines, some emerging disciplines and many branches have appeared in cartography. In particular, the theoretical cartography proposal has resulted in a step forward in the research of cartographic theory. In this period, some of the modern cartographic theories, such as cartographic information theory, cartographic transmission theory, cartographic modeling theory, cartographic perception theory and cartographic semiotics theory, were proposed. The research on the modern cartographic theory in international cartographic circles reached its peak in the late 1970s and early 1980s, and thus, a cartographic discipline system was established. This research led to a new concept called modern cartography (digital cartography), which presented some new characteristics that are different from traditional cartography. Modern cartographic theory is a breakthrough from the closed system of traditional cartography that gives priority to experience and takes product output as the main goal, forming an open interdisciplinary system that integrates cross disciplines as a whole, such as systems theory, information theory and cybernetics based on earth system science.

After the 1990s, in cartography, the application of information technology, such as computer technology, remote sensing technology, network and mobile communication technology, geographic information system technology and virtual reality technology helped traditional cartography develop towards the cartography of the information age. The research on cartographic theory transformed from an experience paradigm to a map model and computing paradigm.

The development trajectory of cartographic theory in the 60 years since the founding of new China is basically consistent with the international cartographic communities; however, there are some lags in time.

The development course of cartographic theory in China over the past 60 years can be roughly divided into three stages: traditional cartographic theory research, digital cartographic theory research and theoretical research of information age cartography.

The research of traditional cartographic theory generally started at the beginning of the founding of new China and continued until the 1970s. In this period, map projection, cartographic generalization and cartographic semiotics theory were the three subjects of traditional cartographic theory, which are collectively called the three old theories. The research on digital cartographic theory was generally implemented from the 1970s to the 1990s. Since the Reform and Opening-up Policy was carried out in China in 1978, a group of cartographic scholars went abroad to attend international conferences and participate in international academic exchanges. Chinese cartography had never experienced such a good period. The cartographic field reached an overall development and has made remarkable achievements, which completed the transformation from traditional cartography to digital cartography. Additionally, theoretical cartography research began turning to digital cartographic theory from traditional cartographic theory. Research on cartographic theory has attracted the attention of cartographic scholars in our country, and many research results have been obtained. This research played an important role in the development of cartography in China.

The research of cartographic theory of the information age generally began after the 1990s. The application of information science and technology in cartography prompted Chinese cartographic experts to think about 'cartography the in 21st century' or 'cartography of the information age'. The discussion on 'the system of transforming information cartography' contributed to the development from digital cartographic theory to information age cartographic theory.

2.2 Development Process

2.2.1 The Research of Traditional Cartographic Theory (from the Early Stages of the People's Republic of China to the Late 1970s)

For nearly 30 years from the founding of new China to the late 1970s, cartographic theory has been moving forward, and its theoretical status has been confirmed by international cartographic academic circles. Cartographic theory research has entered the stage of modern cartographic theory research. However, our country rarely had contact with foreign countries at that time. Therefore, the research on cartographic theory in China was basically still in the stage of traditional cartographic theory during this period.

1. **Map projection theory**

In the early 1950s, the original theory of discovering new map projections with distortion ellipses was put forward by Zhou (1957). Later, as research continued, more new map projections were developed by several cartographers. For instance, conformal or equal-spaced conical projection with total-equal area and graphic interpretation methods in the search for azimuthal map projections were proposed by Hu and Chow (1180), Hu (1962); graphic interpretation methods in seeking cylindrical map projections with predefined distortion were explored by Gong (1964); double azimuthal projection was put forward by Li (1963); pseudoazimuthal projections and their applications for a whole map of China were researched by Liu (1963); the use of numerical methods for establishing optional conical projections was put forward by Yang (1965); the design and analytical calculation method of polyconic projection in unequal graticules was put forward by Zhong (1965), and so on. Many new map projections appeared during this time, and they are rational applications of the inverse solution of modern map projections (Wu and Liao 1981). As new map projections increase, the need to classify map projections according to the inherent law of map projections and research on map projection families was proposed. For example, a map projection classification method according to the inherent law of map projection was studied by Li (1979), and the Gauss-Krüger projection family was proposed by Yang (1980). Notably, this research is the unity of analysis and synthesis and is a synthesis that is based on analysis.

2. **Cartographic generalization theory**

In the process of general map compilation, Chinese cartographers always attach importance to developing cartographic generalization. In cartographic generalization, we must correctly handle the relationship between geographical authenticity and geometric accuracy, as well as the relationship between map load and map readability, and the relationship between typical law representation and regional characteristics. How can these relationships be addressed? More attention was paid to the qualitative description in the past. That is, words were used to explain how much should be selected and the size of the graphic generalization, and some generalization indicatrix and representative patterns were completed for demonstration. This progress played a role in the advancement of cartography at that time. However, it is often not easy to master in practice due to a lack of quantitative analysis of cartographic generalization. In the late 1950s and early 1960s, some researchers began to use mathematical statistics, probability theory and graphic calculation methods for the index of selection. For instance, Zhang (1958) discussed the criteria for selecting the contents of settlements on an atlas according to an analysis of the early China atlas. Lu and Ju (1964) summarized the index that reflects river and pond density and explored how to use the index to measure river and pond density. Chen (1964) used a graphic calculation method to compute the selection criteria of the cartographic generalization of relief. Since the 1960s, there have been some studies on cartographic symbols for general maps according to the characteristics of the natural landscape in China. A series of topographic schemes were proposed. The cartographic generalization theory of the six elements of the topographic map was systematically explored. A quantitative analysis of the cartographic generalization index and the load of the map was carried out. A theoretical method for the application of mathematical statistics in cartography was proposed. For example, Jiang and Yu (1966) discussed the significance of loess landform representation in a large-scale topographic map, and according to the characteristics and graphics of various types of loess landforms, they proposed the principle for cartographic representation and generalization and prepared some kinds of sample atlases for the generalization of relief forms as a reference.

3. **Map symbol system theory and map content representation**

During the period of traditional cartographic theory study, the theoretical research on map symbol design, landscape representation method, map color, thematic map and atlas compilation was conducted, and related research achievements were obtained. For example, Min (1957), through research on map legends and special map symbols, discussed the design principle of the map schema. One of the research focuses of map design in that period was geomorphology representation, which mainly studied the representation method of small-scale maps and thematic maps, especially the application of relief hill shading, the basic problem of shading combinations with different light sources, and the experimental study on the selection of terrain map height. The corresponding principles and methods were proposed, and a color gradient scale for the landform representation of China's small-scale geomorphologic map was initially established (Shi 1964; Liu 1964; Zhang 1963; Wu 1965). Colors on the map not only play a role as the legend but are also an important factor of the symbols and annotations. The natural map not only has a wider variety of colors than general maps, but the colors are very important means to reflect the geographical landscape. Ding (1966) studied the colors of the natural map. The study of the compilation of thematic maps and atlases mainly focused on exploring a variety of thematic map representation methods (Zhang 1963). Aiming at the problems of China's cartography having a low technological level and narrow basis at that time, the overall design of the regional atlas and some technical problems were discussed by Liu (1963).

Overall, in the 30 years from the beginning of the founding of the PRC to the end of the 1970s, the theoretical research on cartography in China was mainly conducted by summarizing practical experience. Some research results have been obtained in theoretical studies on map projection theory, map generalization theory and map symbol systems (representations), and some results have been put into applications. However, a complete system of cartographic theory has not yet been formed.

2.2.2 The Research of Digital Cartographic Theory (Late 1970s to Late 1990s)

Along with the development of computer technology and its application in cartography, cartography has achieved advanced developments from traditional cartography to digital cartography. The three old theories of cartography, which include map projection theory, cartographic generalization theory and cartographic symbol system, gained new life in the digital environment and conditions.

At the same time, due to the introduction of information science and interdisciplinary research and integration, a series of new cartography theories appeared.

1. **"Old three theory" research in the digital cartographic environment**

Study on map projection theory. With the emergence of increasingly newer map projections, we need to solve the problem of the transformation between different map projections in cartographic practice. The map projection computation was realized by the "calculating ruler" or "log table" before the electronic computer was used for cartography. However, because the transformation between map projections is a complicated workload calculation problem, it is difficult to solve the problem with the traditional "calculating rule" or "logarithmic table". After entering the digital era of cartography, computer applications in cartography became increasingly popular; the complex map projection transformation problem can be solved by using electronic computers. In this context, the study of map projection transformation obtained ample theoretical results (Wu 1979; Wu and Liao 1981; Yang 1982, 1984; Hu 1982; Liu 1985). During this period, a number of new map projections were studied, and the map projection classification based on the new situation was also explored.

Study on cartographic generalization theory. Since the need for automated cartographic generalization in the environment of digital mapping and the application of mathematics, knowledge engineering science and technology development have provided favorable conditions, the research on automated cartographic generalization models, algorithms, knowledge, and so on, made remarkable progress in this period. The application experiment was carried out in the automatic generalization of the individual elements. The research on the algorithm of the mutual relationship of the elements also received attention. The expert system was introduced into cartographic research, and cartographic expert systems became a hot topic of research for a time (Zhang and Zhang 1991; Fei and Guo 1993; Sun 1991; Hua 1991; Wang 1993).

Study on cartographic symbol system theory. Since computer-aided mapping has become a basic mapping method, the simplification of map symbols, the construction and establishment of the graphical element and symbol (automatic drawing) library, automatic call of symbols in the process of digital map visualization, the principle of automatic generation of symbols, and so on, became the focus of the theoretical research on map symbol systems, and these issues were addressed successfully (Jiang 1986).

2. **Study on the new theory of cartography**

At the end of the 1970s to the late 1990s, Chinese scholars began modern cartographic theory research.

Liao and Lu were the earliest to introduce cartographic information theory, cartographic communication theory, cartographic perception theory and cartographic model theory to Chinese cartographers (Liao 1980, 1983; Lu 1980). These researchers offered a good starting point for carrying out the theoretical research of modern cartography in China. Later, Gao published a series of papers about the theory of cartography, which more comprehensively and thoroughly promoted research (Gao 1982, 1984a, b). During this period, the focus of cartography theory research in China mainly probed the modern cartography theory, such as cartographic information theory, cartographic information communication theory, cartographic model theory, cartographic semiotics and cartographic visual perception theory.

(1) **Cartographic information theory and cartographic communication theory**

Influenced by psychology theories and methods, British cartographer Keates first defined the concept of cartographic communication in 1964 and suggested linking cartographic communication and cartographic information theory through the map. In the same year, the French Moles also published a paper in French entitled "information theory and cartographic information", which for the first time combined information theory with cartography and put forward the theory of cartographic information. The main research contents include the technical and theoretical issues on map graphics display, conversion, transmission, storage, processing and utilization of spatial information; the concept of cartographic information, cartographic information capacity and the method for measuring the information content of a map; and how to study the design of map symbols from the perspective of information. It provides a quantitative index for the quantitative analysis and evaluation of cartographic generalization and a map, as well as laying a theoretical basis for computer-aided mapping. Since the 1980s, Chinese scholars have conducted research on cartographic information theory. First, aiming at the special nature of cartographic information, they delved into the concept of cartographic information, and the characteristics of cartographic information were explored based on the concept of information theory (Mao 1987; Tong 1990; Liu 1992a). The measurement of map information content is the focus of the research on cartographic information theory. With in-depth study, research on the measurement method of map information content was carried out. The experiment of measuring the map information content was carried out, which provided a reference for the evaluation of the map and the improved design of the map (Zhu and Huang 1985; He 1987; Zou 1991; He et al. 1996; Liu 1992a, b).

The proposed map information theory has resulted in a request for the study of the cartographic information transmission process, thus forming cartographic communication theory. Since the first cartographic information communication mode was put forward by the Czech cartographer Kolacny in 1969, the problem has been discussed in the cartographical field. Cartographic communication, as the basic theory of modern cartography, has been recognized by the majority of people, various cartographic communication models have been proposed, and a special committee on cartographic communication has also been established in ICA. Chinese scholars have published many papers on international cartographic communication theory and proposed corresponding research proposals (Liao 1980; Lu 1980). In view of the control process of cartographic information transmission, a control model for the process of cartographic information communication was proposed, and the basic conditions for realizing cartographic information communication were discussed (Wang 1989; Lu 1989; Zhong 1989).

The study of cartographic information communication theory has changed the thinking modes of cartographic scholars. That is, researchers began to consider the relationship between map production and map use from the perspective of cartographic information communication. This change is an important indicator of the advancement from traditional cartography to modern cartography.

(2) **Map visual perception theory**

In 1967, the French scholar Bertin published the famous "graphic symbol" and studied information processing and communication methods based on visual language. The book introduced semiotics into cartography by first presenting the concept of visual variables and the visual perception effect of graphic symbols, and the book cited the map as a special case.

Since then, the international cartographic community has begun to study map visual perception theory. Map visual perception theory is a newly formed modern cartographic theory that applies theoretical and experimental methods from physiology, psychology and psychophysics to the study of cartography. The main research content includes the visual perception process, visual variables and visual perception effects and the physiological and psychological factors of map visual perception. Starting from the study of human vision, map visual perception theory involves the experimental research of ways to design the map to be accepted by readers through physical, psychological and physiological methods. The map visual perception experiment is mainly based on the experimental method of mind and matter science to find the relationship between the stimulus and the visual responses to various map graphics.

Gao (1984a, b) was the first scholar in China to introduce and engage in the study of map visual perception theory. The research results of 'the experimental method of map perception and map design" discussed map visual variables and map visual perception effects and the experimental method of map design. This research area serves as an important reference for domestic cartography textbooks and related research papers. Mao (1989) and his 'Gestalt principles in map visual perception theory'listed the impacts of the Gestalt principles on the integrity, constancy, graphics and background, contour and subjective contour, sense of order and quantitative perception, stable pattern and figural after-effects in cartography; the researcher also summarized the applications of the principle of Gestalt in map design. In 'the study of the visual illusion in cartography', He (1991) analyzed the common phenomenon of visual illusions in mapping and discussed how to apply the laws and principles of optical illusion phenomenon scientifically and correctly to cartographic generalization, map design and map drawing. The theory of cartographic perception provides a scientific basis for map design and plays an important role in actual map production.

(3) **Cartographic semiotics theory**

Cartographic semiotics theory is an important part of the theory of modern cartography that uses the fundamental concepts and principles of semiotics to study the characteristics, laws and nature of map symbols. The theory is composed of three parts: the syntax of the map, the semantics of the map and the pragmatics of the map. The syntax of the map addresses the structure and arrangement of map symbols and the principle of their composition and transformation. The semantics of the map touch upon the relationship between the map symbols and the elements they represent. The pragmatics of the map relate to map use, map perception and map language training in both general and special cases.

Chinese scholars mainly focus on the essence of map symbols, visual variables and the composition law of the symbols, map language, and so on, and several papers have been published. Zhu (1987), for example, discussed the inherent basic rules of the design of map symbols, proposed stipulative principles and the equivalence principle of map symbols, studied the connotation and extension of map symbols and their relationship with each other, and proposed a new map symbol classification principle. Yu (1995) discussed the essence of map symbols from the perspective of philosophy, compared the similarities and differences between map symbols and other symbols, and thus explored the multiple functions of map symbols in the map-maker's ideological expression, cognition, thinking, culture and aesthetics.

Considering the different understandings of the visual variables of map symbols by foreign scholars, Chinese scholars have written their own views. For example, Wang (1990) proposed taking the color, shape, size, direction, brightness and structure as the symbol variable. Chen (1995a, b, c, d) studied the adaptability of Bertin's visual variable to map symbols, presented the meaning of visual variables used to construct map symbols, discussed the visual variables of the spatiotemporal maps in view of their characteristics, and considered that the visual variables of spatial-temporal maps should be composed of 6 visual variables and 3 dynamic variables. Chen (1989; 1995a, b, c, d) studied the composition of map symbols, deemed that the composition of the map symbol is a combination of the composition factors (visual variable) according to certain principles and rules, and summarized the composition law of map symbols. In addition, map language is also the research focus of this period. A study of map grammar using the linguistic method was carried out by Zhu (1996); it is considered that the concept of map language is a logical concept model, and syntax is the spatial structure of this model. Finally, the organization rules and the language layer of the map grammar structure were presented. Chen (1996) explored the grammatical structure and basic features of map language from the perspective of linguistics and considered that the map symbol systems and their usage rules can be called map languages; similar to natural language, map language also has its own language rules; the basis of natural language is the letter, and the basis of the map language is the visual variable of the map symbol.

(4) **Map spatial cognition theory**

Spatial cognition is the process of acquiring new spatial knowledge by procuring, processing, storing, transmitting and interpreting spatial information. It is an important research field of cognitive science and has been widely regarded as an important research subject in the fields of psychology, cartography, geography, computer science and artificial intelligence. Map spatial cognition is the use of a recognizable spatial environment, that is, through the map representation of the graphics space to understand the geographical space. Gao (1991) introduced the research methods of cognitive science to cartography, with the definition of spatial cognition being from the angle of cartography, and Jun proposed two important concepts: mental map and cognitive mapping; Jun also discussed the cultural significance of the cognitive function of map space, the cognitive scientific basis of the map design expert system, and the relationship between the spatial cognition of humans and the geographic information system. Starting from the relationship between a human and the map, the cognitive theory of map space takes the map reader and the map as a whole in the information exchange, studies on the characteristics of the map as a spatial cognitive tool in the condition of digital mapping, and the process and law that humans use maps for spatial cognition. Research content involves three main aspects, including the basic concept of map space cognition theory, the cognitive experiment of the map, and the application of the cognitive theory of map space.

The basic concepts of the cognitive theory of map space include mental maps and cognitive mapping.

The mental map refers to the 'abstract alternative' to the cognitive environment, which is formed in the minds of people who acquire spatial information through various means. The formation process of the mental map is the process of environmental information processing, which is cognitive mapping. Based on the analysis of the basic concepts of the mental map and cognitive mapping, the characteristics of the mental map were discussed, and the important role of the mental map in the map design and establishment method of the mental map were presented by Gao (1991), Chen (1995a, b, c, d), Wang and Chen (2000), and Chen (2000a). According to the comparison of human spatial cognition and the function of the map, the advantages and significance of the map as a spatial tool were explored by Gao (1991), Chen and Ye (1996), Chen (2000a). By comparison with the paper map, the unique function of the electronic map as a spatial cognitive tool was investigated by Chen (1996, 2000b, 2001). The map-based cognitive process of geographical space was discussed, which includes the perception process, the imagery process, the memory process and the thinking process (Wang and Chen 2000).

Cognitive experiments are an important research method in the cognitive theory of map space. Chen (2000a) explored the experimental method of map space cognition in her doctoral dissertation. A set of software for the visual cognition experiment of the electronic map was compiled with Visual Basic 5.0. Visual cognitive experiments were carried out on the electronic map. The method and procedure of the visual cognition experiment of electronic maps are introduced in her paper.

In this period, studies on the theoretical and practical significance of map spatial cognition theory for cartography and geographic information engineering mainly focused on the applications of map spatial cognition in cartographic expert systems (Wang and Chen 2000). The cartographic expert system is a knowledge acquisition, representation and utilization system based on cartographic expert knowledge and experience. Among them, knowledge representation is the core, and its wide application involves the theory and method of spatial cognition. The applications of spatial cognition theory in generative rules, predicate logic, semantic networks, and frame representation were analyzed.

Compared with international cartographic circles, modern Chinese cartography theory research in this period was relatively late, although the research content involved several main aspects of modern cartography theory research, but the research results were not rich enough, and the combination for the actual map production remained inadequate. Nevertheless, through the study on the theory of modern cartography, China's emphasis on cartographic theory research was unprecedented, and it played a very important role in the development of modern cartography.

2.2.3 *Study on the Theory of Cartography in the Information Era (Late 1990s to Present)*

With the development of information science and technology and its application in cartography and geographic information engineering, cartography has entered the information era, its function has been extended, and there is a new growth point in the discipline. Hence, the characteristics of cartography theory research in this period are the combination of 'three old theories' and a 'new theory' and obviously have the characteristics of 'information'.

1. **Multimode space-time comprehensive cognitive theory**

Cartography in the digital era regards the digital mapping process as a system, with the output of the digital map products being the ultimate goal. Spatial cognition research is mainly based on maps, especially paper maps. After entering the information age, the geographic information system and the virtual geographic environmental information system have been widely used, which has resulted in the need for spatial cognition research based on GIS and virtual geographic information systems. In view of this situation, some scholars put forward the concept of 'multi-mode space-time comprehensive cognition', which shows that the research on spatial cognition theory in the information age in China is gradually deepening.

In the case that GIS has been widely used in various professional fields, departments and regions, it has many advantages in recognizing the geographical environment through a geographic information system. In addition to a spatial cognitive function based on the map (electronic map), through the convenient, fast measurement and spatial analysis function of GIS, we can obtain a deep level of knowledge that is not easy or even possible to obtain with only the graphics experience. In fact, a person's spatial cognition system and a geographic information system are both information processing systems. Their processes are characterized by information input, encoding, memory, decision-making and results output. From this perspective, geographic information systems are a more advantageous tool than maps (paper) (Wang and Chen 2000). This is why many researchers have carried out spatial cognitive research with electronic maps (the basic visual form of geographic information systems) (Chen 2001). However, it is regrettable that even though everyone is using geographic information systems, relevant reports on the systematic study of spatial cognition based on geographic information systems have not been seen.

The virtual geographic environment is a 'virtual space' that can be entered. It can also be used as a tool for human spatial cognition. Because this virtual environment allows users to be personally on the scene and able to support interactive operations, there are many advantages to using human cognition in the geographical environment. For this reason, Gao and Wan (2002) believed that virtual geographic space was not only a new type of map that allows users to be on the scene but can also realize planning and project ahead of schedule (namely, the virtual future), which allows users to experience, evaluate and understand the environment. Wan et al. (2005) discussed spatial cognition in virtual terrain environments from several aspects, such as the spatial cognition of maps, multiple sensory modalities, human-computer interaction technology, incongruity perception, performance and evaluation. In the doctoral dissertation of Wan (2006), the advantages and disadvantages of two-dimensional and three-dimensional visualization were verified by cognitive experiments. It was considered that the 2-dimensional, 3-dimensional and 4-dimensional (dynamic) visualizations have their own functions, as well as different effects and different inputs. The availability of the virtual geographic environment has also been evaluated. Wan et al. (2008) carried out a series of experimental studies on spatial cognition based on paper maps, electronic maps, remote sensing images and virtual geographic environments. Unfortunately, there is no systematic study on spatial cognition based on the virtual geographic environment.

In summary, since maps, geographic information systems and virtual geographic environments are spatial cognition tools for humans, it is natural to propose a multi-mode (map-based, geographic information system and virtual geographic environment) space-time comprehensive cognitive system. We emphasized here that this system is a multi-mode problem, and each system does not replace the other; a dynamic cognition in time and space that is not static; and a comprehensive cognition (that is, the complementary advantages and environmental complex) rather than a single tool or individual elements in a complex environment. Obviously, the systematic study of multimode space-time comprehensive cognitive problems will expand, extend and deepen the research on the spatial cognitive theory of maps.

2. **Research on the perception theory of electronic maps in the information age**

Research on map visual perception based on the paper map was carried out. However, since the electronic map display and visual perception environment are so different from the visual perception environment of paper maps, the map design principle of the visual perception of paper maps is not fully applicable to the design principle of electronic maps. Therefore, map visual perception theory must be further developed under the condition of digital mapping to meet the needs of the visual perception of electronic maps.

Chinese studies on the perception theory of electronic maps in the information age have mainly focused on the experimental method of the audio visual perception of electronic maps, the design of tourism map symbols, and the color design of network maps.

Chen and Chen (1999) compared the interpreted blueprints and visual perception environment of electronic maps with those of paper maps, summarized the characteristics of the visual perception of electronic maps, and discussed the contents and methods of the visual perception of electronic maps. A perception experiment of the grading of brightness changes with different hues on the electronic map was designed by using the experimental method of response time to test a subject's visual perception experience with different colors and different brightness levels on the computer screen (Chen et al. 2000). An experimental study on the adaptation effects of different colors as the background color and the background color of the electronic map was carried out (Chen 2000b).

A series of master's degree theses are related to the visual perception experiment of the electronic map. For example, Ling et al. (2005) studied the cognitive factors that influence the user interface design of map visualization systems, developed cognitive experimental software for user interfaces, and provided a theoretical reference and tool for the user interface design of electronic map visualization systems (Ling 2005; Ling et al. 2005). Xie (2006) conducted a visual perception experiment with a single symbol and the whole effect of the symbolization based on the existing symbol library of military electronic topographic maps and set up a military electronic topographic map symbol table based on the experimental results. Using the online cognitive test method, Wu (2007) experimentally studied the following three aspects: the best choice of symbols on the tourism network map, the visual perception experience of individual symbols and the visual perception effect of the whole map, finally establishing a tourism network map symbol system.

Liu (2008) conducted an experiment on the visual perception of the user interface of a multimedia electronic atlas and its related factors. Sun (2008) used EVC to develop visual perception experiment software for a PDA mobile map based on a simulator and carried out visual perception experiments on the symbol and color of the mobile map. Li (2008) carried out a visual perception experiment on the symbol and color of urban tourism electronic maps. Wang (2009) analyzed the key issues in the design of electronic maps, investigated the basic procedure and the statistical method of experimental data for electronic maps, developed a general experimental platform for electronic maps, and carried out experiments on the visual perception of the annotations of electronic topographic maps.

Han et al. (2006) used the questionnaire method, focus group method, confusion discrimination investigation experiment and subjective assessment survey to investigate and analyze the influence of visual and psychological factors on the design of tourist map symbols and to explore the method of designing tourist map symbols from the perspective of visual perception and psychological cognition. Xie et al. (2008) applied the theory of map visual perception to study the color matching problem of the network map and proposed the color matching principle of the network map.

3. **Research on map symbol language**

In the information age, research on cartographic semiotics has fundamentally changed. These studies mainly focus on map linguistics in terms of electronic map symbols and digital mapping, including the linguistics mechanism of map symbols, the visual variable and composition rule of map symbols, standardization of map language, and so on. For example, the linguistic features of spatial information and the mechanism of automatic understanding are studied comprehensively and deeply by Du (2001). In view of the application of multimedia and visualization in maps and its effect on map language, 10 forms of modern map language and the essential features of map language are discussed. From the perspective of linguistics, the linguistic mechanism of map symbols is analyzed, and the semantic concept model of map symbols is proposed. The linguistic features of topographic map symbols are expounded by using the methods and theories of linguistics to map language (Wang 2000b; Hu an Yan 2008; Su and Chen 2007, and others). By studying the visual variable of the 3D map symbol, Xu et al. (2006) presented the visual parameters of the 3D map symbol, which is composed of three aspects: state, dynamic change and operation. By means of the cognitive experiment method, they verified the application and effect of these symbol parameters in information expressions, such as quality, quantity, relation and dynamic features. Zhou et al. (2008) studied the internal structure and composition rules of map symbols, discussed the speech and semantic structure of topographic map symbols, and looked at the main forms of modern map language, namely, visual symbols, auditory and tactile symbols, as well as the essence of map language, namely, visual variables and acoustic and tactile variables.

According to the characteristics of the electronic map display and reading, Xie et al. (2008) investigated the symbol system of electronic topographic maps using the theory of map visual perception and the theory of cartographic semiotics and decomposed all symbols into minimal elements according to the needs of building a map symbol library and presented the composition rule of point, line and surface symbols. In view of the existing problem of map expression language, Ding et al. (1999) discussed the principle of standardizing and normalizing map language and its management and storage technology in theory. Based on the traditional paper map standard and the related international standard, Wang et al. (2003) proposed feature classification and its combinatorial characteristics for the complex geographical environment in terms of the cognitive principle of geographical environment elements and graphic symbol processing method. It conforms to the visual rules of map symbols and at the same time, adapts to the draft standard of the electronic map symbol system and symbol library at the level of digital mapping technology. Xie and Chen (2006) analyzed the difference between the electronic map and paper map according to the paper map standard and proposed the basic principles, contents and system structure of electronic topographic map symbol system construction through a user cognition experiment. Jiang et al. (2007) discussed the design principle of the electronic map symbol system based on multiple display modes and studied the symbol design of electronic maps in different modes.

4. **Theoretical research on deep processing of geospatial data**

The focus of cartography is transferring from information acquisition to information intensive processing. This is a major feature of cartography in the information era. There are three aspects of problems that have been proposed or are being studied. That is, spatial information generalization, spatial relations reasoning, spatial data mining and spatial data uncertainty.

In the era of information cartography, spatial information generalization is mainly manifested as the automatic generalization of spatial data and integrated intelligent quality control and evaluation. Previous research on cartographic generalization models and algorithms mainly delves into the intelligent methods of automatic cartographic generalization and process control based on knowledge and generalization chains (Qian 2006); intelligent processing of spatial information for automatic map generalization (Wu et al. 2008); multiscale representation and automatic generalization of spatial data (Wu 2003); multiscale modeling of spatial data and its visualization (Li and Wu 2005); quality assessment models for automatic cartographic generalization (Wu et al. 2009); map cluster (group) goal description and automatic generalization (Yan and Wang 2009); and so on. These studies satisfy the need for the automatic establishment of small-scale databases based on large-scale map databases and the integrated update of multiscale databases, the production of small-scale maps, multiscale visualization of spatial data in GIS, and spatial data mining and knowledge discovery.

Spatial relation reasoning is an important aspect of spatial information processing. It is also required for solving the problem of the deep-going spatial information service. Spatial relationships have a certain ambiguity. To study spatial relation reasoning, we should describe the spatial relationship first. A further study on the fuzzy description of spatial relations and the combination of reasoning was conducted by Du et al. (2007). The depth and detailed description of spatial direction relations and the combination of reasoning were deeply studied by Wang (2009) in his doctoral thesis. Guo et al. (2007) studied the combination of spatial reasoning and cartographic generalization and proposed the theory of spatial reasoning and progressive cartographic generalization. In addition, Yan and Guo (2001, 2002a, b, 2003) and many other scholars have conducted in-depth studies on the description and reasoning of spatial relations.

Spatial data mining and knowledge discovery is another important research field that uses the deep processing of spatial data to explore the translation relations of data, information, and knowledge and to support decision-making, and this field has been highly studied by many scholars. Di (2000) carried out the study of spatial data mining and knowledge discovery earlier in China. Li et al. (2006) systematically studied and summarized the theory and application of spatial data mining. Zhou (2004) and Zhang (2007) studied spatial clustering analysis based on swarm intelligence with constraints. Bi (2008) studied spatial data mining and its application in vector graphics features.

Spatial data uncertainty exists throughout the process of spatial information acquisition, data production, processing and deep processing. On the one hand, the uncertainty model of spatial data is studied; on the other hand, the quality control of spatial data is also explored to a certain extent. According to data types, the main research scope includes the uncertainty of positional uncertainties of spatial data and its visual expressions, the spatial data attribute uncertainty model and variables, the uncertainty of the digital elevation model (DEM) and the quality control of map data.

2.3 Main Research Achievements

The past six decades, particularly the past three decades since the reform and opening up, have witnessed remarkable achievements at different development stages, including traditional cartography, digital cartography and information cartography. The most prominent achievements are as follows:

2.3.1 There Is a Deep Understanding of the Characteristics of Modern Cartography

For a long time, Chinese academic circles have been paying close attention to the study of the scientific features of cartography. The research has spanned three stages of development from traditional cartography and digital cartography to information cartography. Chen (1964) was the earliest in China to study the modern characteristics and growth point problem of cartography. Based on past historical experience, he believed that the development of cartography is closely related to social needs and the level of productivity; the development of cartography also reflects that it essentially belongs to both technical and regional science; from a general survey of progress and achievements in the various stages, on the one hand, it reflects the level of mapping technology at that time, and on the other hand, it reflect the extent of the mathematical foundation and geographical knowledge. By analyzing the development and changes in cartography after entering the twentieth century and especially in the nearly 20 years since the establishment of new China, Shupeng put forward 5 characteristics of cartography at the time, that is, map design standardization and normalization, high levels of generalization based on in-depth analysis, an extension to reflect the two poles of macro and micro phenomena, advances in airborne photogrammetry, the introduction of new technologies of plastics and electronics, and so on. Accordingly, he proposed 9 basic problems as the main growth point for Chinese cartography. After a lapse of 16 years, Liao (1980) followed Chen Shupeng and summarized the characteristics of modern cartography into 3 aspects by considering the international scope, i.e., the research of cartography theory was greatly enhanced, thematic map and comprehensive cartography was extensively advanced and in-depth, remote sensing and remote surveying became an important data source of maps, and the automatic compilation of maps developed rapidly. The studies have touched on some of the core issues in digital cartography. Gao (1986) studied the main feature of modern cartography starting from the characteristics of cartography. He stated that the map is the model of the objective world; a map must be adapted to the graphic feeling ability of humans; and a map is the carrier of spatial information. Based on the above information, he divided the main development characteristics of modern cartography into four aspects. First, cartography is an interdisciplinary science and has been applied and studied in many scientific departments. Second, crossing sciences provides a powerful tool for the theorization of cartography. Third, cartography has become a quantitative science in production, research and application. Fourth, computer technology broadens the applications of cartography. At the same time, he gave the following definition of cartography: it is a comprehensive science centered on map information transmission and aims to explore the theoretical essence of a map and its production technology and application method; he pointed out that the term cartography can be replaced by map science in a certain sense. That is, cartography has the main characteristics of comprehensive science. Undoubtedly, these new assertions have had a positive effect on the further development of cartography. Ding and Zhang (1993) pointed out that cartography has become a comprehensive science involving many domains and discussed the characteristics of modern cartography from 5 aspects. Among them, there are some helpful suggestions, such as the crossing, penetration and complement between cartography and adjacent disciplines, which is increasingly frequent, and the differentiation and integration of the internal discipline of cartography. Based on reviewing the research and progress on China's cartographic technology in the twentieth century, Chen (2001) proposed some ideas about the future of map science. He considered that our map workers should actively promote the informationization and structural readjustment of the cartographical industry, innovate map designs and map making from scientific conception, develop the electronic atlas market as cartographic science is coming into the knowledge and internet era, welcome the opportunity for economic globalization, the rise of information technology, and the social needs of distance education. He also proposed that we should develop exploration research on Geo-Information TUPU as a starting point for cartographical innovation. This is a forward-looking way of thinking about the new era of map science and influences the development of cartography in the new period.

2.3.2 The Preliminary Establishment of the Scientific System of Cartography

Research on the problems of the structural system of cartography abroad began in the 1970s and began in the 1980s in China. In China, Liao (1983) discussed the first modern cartographic system, given the opinion that the modern cartography system consists of three parts: theoretical cartography, cartography and application cartography. The research contents of each part are described. He argued that theoretical cartography is the theoretical basis of cartography, including the concept of cartography, cartographic information theory, cartographic model theory, cartographic transmission theory, mathematics principle of cartography, map symbol, cartographic perception theory, theory of cartographic generalization and integrated cartographic theory. Cartography is map compilation method and technology, including general cartography, thematic cartography, remote sensing cartography, computer-aided mapping, map making and map printing science. Application cartography is also known as the map application principle and method, including the basic function of a map, the evaluation standard and method of a map, the methodology of map analysis, maps using steps, map analysis and use method, automatic analysis and processing system of map information and the actual application of a map. Gao (1986) deemed that the theoretical structure of cartography includes three basic levels, that is, basic theory, applied foundation and technical method. As a basic theory of cartography, cartographic communication theory is the first level. Cartographic information theory, cartographic semiotics, cartographic model theory and cartographic perception theory belong to the second level, which is the level of application. The third level involves the technical methods that directly serve map production and is the most practical part of cartography, including map design, map projection, mapping technology, map reproduction, and so on. Wu (1990) also has a similar view and discourse. Wang and Chen (2000) discussed the modern scientific framework of cartography and divided cartography into three parts: theoretical cartography, cartography and applied cartography. Theoretical cartography is the first part of the scientific system of modern cartography, including the basic theory and application theory. The history of cartography also belongs to the category of theoretical cartography. Among them, basic theory is the theory related to overall cartography and has general guiding functions for the development of cartography; the core of this theory is cartographic spatial cognitive theory and geographic information communication theory. Applied theory is a bridge to connect basic theory and cartography. This theory directly guides the process of mapping. The main content of this theory includes cartographic information theory, cartographic model theory, cartographic perception theory and map semiotics. Cartography, which belongs to the technology and engineering science and technology categories, involves the whole process of mapping, and its main content includes the digital design of maps, digital cartographic generalization and digital direct printing and plate making. Applied cartography is the application part of the scientific system of cartography; the main content includes the analysis and application of the simulation map, digital map, analysis and application of electronic map, and so on. At the same time, Jiayao put forward that there is a close relationship between the scientific system of cartography and a higher level of science. This system has two outer layers: (1) cognitive science, earth and environmental science, mathematics, linguistics, psychology, information science and system science, and (2) natural science, technology and engineering sciences and social and human sciences.

2.3.3 The Extension and Deepening of the Research on Spatial Cognition Theory

Spatial cognition is the research field that studies cognitive science. The formal establishment of cognitive science occurred in 1979. The purpose was to explore how people carry out information processing in the implementation of cognitive activities.

Gao (1991) attempted to make use of the scientific research methods of cognition in cartography studies, given a definition of spatial cognition considering cartography, and considered that it is the study of how people know their living environment, including their relevant positions and dependency relationships with all kinds of things and phenomenon, as well as their changing rules; this study also expounded two main concepts in cartographic spatial cognition, namely, mental mapping and cognitive mapping. Wang and Chen (2000) deemed that the cognitive process of cartographic spatial cognition is the same as human cognition, which includes the sensation process, imagery process, memory process and thinking process of cartographic spatial cognition.

In addition to paper maps, there are electronic maps, especially multimedia electronic maps. They have many characteristics that are quite different from the paper map. Wang and Chen (2000) studied the relationship between cartographic spatial cognition and geographic information systems; in terms of the analysis and comparison between the human spatial cognition system and GIS, the working principles of both are the same and both belong to an information processing system that can input information, encode, memorize, make decisions, and output results. For this reason, geographic information systems can simulate the processing of geographical environment information flow in the human brain. In fact, an intelligent geographic information system is a simulation of the human spatial cognitive ability and cognitive process; this study area is a spatial cognitive approach that is based on GIS and considers imagery thinking (pattern recognition), cognitive operation (problem solving) and decision-making support. Chen (2001) considered that spatial cognition based on paper maps lacked the function of acquisition, storage and representation of spatial information through human vision, feeling and hearing in the past; he also discussed the characteristics and main contents of spatial cognition based on electronic maps and proposed that the core research content of electronic map spatial cognition is the process of establishing mental maps based on different electronic maps and the thinking process and cognitive strategies of users with different spatial cognitive abilities when using electronic maps. Furthermore, Gao and Wan (2002) used the virtual geographic environment (geographical virtual space) as a new window of spatial cognition and considered that as the spatial cognitive tool, this space was the natural and reasonable extension and expansion of maps in the digital age; it was a new spatial cognitive tool that was supported by the digital map. This environment would extend the method of cartographic spatial cognition and accord well with the characteristics of human spatial cognition.

2.3.4 *Cartographic Visual Perception Theory Has Been Successfully Applied in the Practice of Cartography and Geographic Information Engineering*

In the applied research of theoretical cartography, map visual perception theory should be one of the research fields that most directly guides the practice of cartography and geographic information engineering, which is because more than 85% of the geographical space information is obtained by human visual perception. Gao (1984a) deemed that the visual perception of maps is a very complicated process that is influenced by many factors and involves many aspects of the problem. Regardless of what type of map is used, in the process of cognition, it cannot be separated from the following steps: detection, identification, recognition and interpretation. Many results have been obtained in studies on the psychological factors of visual perception, the visual variables and the perception effect of map symbols (Gao 1984a; Mao 1989; He 1991; Chen 1995a, b, 2000b). The guiding role of map visual perception theory in the practice of cartography and geographic information engineering is feasible. Through a case study of the *Officer Atlas* design and based on summarizing the research and experimental results at home and abroad, Wang (1991, 1993) proposed that map design should be taken as system engineering; he also expounded the representation method of the map symbol system based on the map visual perception theory; Jiayao analyzed the scaling method of cartographic data, types of maps, the relationship between symbol types and visual variables, studied the system theory method of symbol design and summarized the factors that are directly involved in the design of map symbols into 8 aspects, namely, map content, geographic data (spatial data and feature), the usage of maps, the visual perception of mapping information, the visual variables, the perception law of psychophysics, the traditional association and the standard, production technology and results. The most compelling case of the direct guidance role of map visual perception theory to cartography and geographic information engineering is the publication of a large number of high level map works, such as the national atlas, the provincial (autonomous regions, municipalities directly under the central government) atlas and the large wall map. The theory and method of map visual perception have been widely used. This progress shows that the level of map design has been greatly improved and the overall level has reached the internationally advanced ranks. In addition to the five volumes of the *Great National Atlas*, the *Atlas of Chongqing City*, *Atlas of the Jiangsu Province*, *Atlas of the Zhejiang Province*, *Atlas of the Jiangxi Province*, *Atlas of the Jilin Province*, *Atlas of Shenzhen City* and many thematic atlases have been published, and these maps have reached a very high level. Under the condition of modern mapping and publishing technology, the level of a map is largely determined by the level of design. The rapidly increased levels of the Chinese atlases marks the improvements in map design. Undoubtedly, the theory and method of map visual perception play an important role in map design.

2.3.5 *The Concept of Cartography in the Information Era Was Put Forward to Theoretically Prepare for the Further Development of Cartography in the New Era*

Studies on the concept of cartography in the information age began in the early 1990s and gradually deepened. In an article entitled 'Information Flow and Cartography', Chen (1991) pointed out that cartography as a mature discipline must respond in a timely manner to the opportunity and challenge of the information age, not only in engineering technology but also in the fields of application and theoretical research. The sense of time lost or self appreciation would be not consistent with the historical law of the development of cartography or the mainstream of modern academic thought. It is believed that cartography in the 21st century will be more active in the field of earth science and information science. In an article entitled 'Cartography into the 21st Century', Gao (1996) presented and described three pairs of new terms regarding maps, namely, real maps, virtual maps, static maps, dynamic maps, plane maps, stereo maps and accessible maps, and explained their connotations and interrelations; Jun explored the present situation on the frontiers of cartography, including geography information systems, spatial information visualization, virtual reality and cartographic information systems; he hoped to form a framework, namely, the new face of cartography in the 21st century and again stressed the need to use the new theory and technology to strengthen the system and focus on interdisciplinary research to find the growth point of the subject and new frontier discipline. Wang (2000a) provided a brief introduction to the meaning and task of cartography as science and technology in the information age and then analyzed the characteristics of cartography in the information age from four aspects, i.e., first, the shift in the focus of cartography to the user-oriented deep processing of geographic information and its practical end product; second, the extension and expansion of cartographic functions to the geographic information system; third, the new growth point of cartography - spatial data visualization and virtual reality technology; and fourth, the automation of map production - historical changes in the map production mode.

With the deepening of investigation, Gao (2004) reckoned that traditional cartography can be summarized as three kinds of relationships: field and map, map and reader, reader and field, and the framework of the cartography discipline can be called the "Cartography Triangle"; the emergence of the digital map expanded the field of vision and the scope of the service of traditional cartography, and since obvious changes have taken place in the field of cartography, the "Cartography Triangle" has been replaced by "Cartographic Tetrahedron". Modern cartography should explore six kinds of relations. Apart from the above three relationships, three other kinds of relationships should be considered: digital maps and maps, digital maps and fields, and maps and readers. Therefore, Jun concluded that the shift from "Cartography Triangle" to "Cartographic Tetrahedron" reflects the development and change in cartography in the information age. In recent years, "information mapping" has been widely discussed in Chinese academic circles and industry. In his article entitled

'Ideas on the Characteristics and the Theory and Technology System of Information Cartography', Wang (2008) summarized the main defects of traditional cartography and digital cartography and the way out for further development. Then, He focused on the formation of information cartography and its development condition and connotation; he also analyzed the characteristics of information cartography from the four aspects of concept, technology, products and services modes, discussed the theory system of information cartography and believed that we should attach more importance to the relationship between informatization and digitalization and the relationships among data, information and knowledge, including spatial cognition theory, graphic symbol variables and visual perception theory, model theory, spatial data uncertainty theory, spatial data mining and knowledge discovery theory, spatial data assimilation and geo ontology theory. Correspondingly, Jiayao established the technology system of information cartography, including integrated technology for data acquisition, intelligent technology for data processing and popular and universal technology for information services on network grids. Meanwhile, the author put forward the main research problems of information cartography, that is, the construction and perfection of the theoretical system of information cartography, which is focused on spatial cognition; the automatic generalization of spatial data and the quality control and evaluation of cartographic generalization; the automation of spatial data mining and knowledge discovery; and the popularization and pervasiveness of spatial information services based on Web/Grid.

Of course, research on information cartography is still in the preliminary stage and will continue to deepen in the future.

2.4 Problems and Prospects

Cartography theory is a research field of great scope, while construction of the theoretical system of cartography is very complex. There are many problems in the present study.

2.4.1 The Problems We Now Face

(1) **There are different opinions on the theoretical system of map science (cartography)**

Zhang and Zhu (1990) considered that along with the production and use of maps adapting to various needs, the theory of cartography has gradually formed in mapping practice. At the same time, the development of emerging disciplines, such as information science, pattern theory, system theory and perception theory, as well as their penetration into the cartography subject, has promoted the development of the theory of cartography. The researchers argued that these developments also caused some cognitive disputes related to cartographic theory. Hence, some problems worth further discussion were put forward. For example, what is the unique basic theory of the cartography discipline? What is the theory to apply to other new theories to study problems in cartography? And so on. In view of the discussions on the theoretical issues of map cartography at home and abroad, these researchers put forward that map projection, map generalization and the map symbol system comprise the basic theory of cartography, while the applications of information theory and other new emerging disciplines belong to the applied theory of cartography. These ideas are quite different from the views of most scholars from the current academic circles of cartography. In contrast, more scholars today regard the map spatial cognition theory, map information communication theory and so on as the basic theory of cartography (because there is global importance throughout the whole process of cartography), map model theory, map visual perception theory and so on; the applied theory of cartography (because they directly guide map design and production) incorporates the so-called three old theories of map projection, cartographic generalization and the map symbol system. Of course, these can be further studied in depth and discussed in the future.

(2) **Theoretical research on cartography is not sufficiently systematic**

After many years of study and experiments, considerable research results have been achieved in the field of cartographic theory, such as map spatial cognition, map information communication, cartographic models, map visual perception, spatial data mining and knowledge discovery, and spatial data uncertainty. However, the research content is not sufficiently systematic. For example, as the name suggests, map spatial cognition could be implemented on the basis of a map. There is much research in this area. However, spatial cognition research based on geographic information systems (the extension and expansion of cartography functions) and virtual geographic environments (a new growth point of cartography) is rare. No one has engaged in the research of space-time comprehensive cognition with multiple modes (based on maps, geographic information systems, and virtual geographic environment systems). Even though spatial cognition research based on maps remains surficial, an in-depth study of the process and mechanism of spatial cognition is lacking, as is experimental studies on the various processes of spatial cognition. There are no persuasive achievements. A similar situation also occurred in the field of geographic information communication. At present, there are still not enough studies on the core of various cartography theories and their positions in the global and whole process of cartography, logical relations between each other, etc. In this case, it is very difficult to build the theoretical system of cartography with scientific completeness and logic.

There are two reasons for the lack of systematic research on cartographic theory. First, few people have engaged in cartography theory research until now. Researchers should have profound knowledge and understanding of cartography as well as a keen eye for cartography and the related new and emerging subjects and technologies (such as cognitive science, system science and information science). At the same time, researchers should have a systematic and profound understanding of the development of maps and cartography and certain practical experience, which is a relatively high standard. Therefore, the number of such experts is very small, and it is difficult to form a stable research team. Second, the basic theoretical research has not received the attention it deserve; applying for research projects is difficult; and project funds are few and cannot easily support continued use. This led to the research of cartography theory not being sustained for a long time, and the depth of research is generally not adequate. Therefore, systematic research results have not been obtained in this field.

(3) **The guiding role of research results on practice is still not obvious**

Compared with theoretical research in the era of traditional cartography, research in the information era has drawn more attention from Chinese scholars. The materials obtained on the Internet make the data acquisition efficient, and the experimental methods based on computers make the analysis of experimental data more convenient and accurate. These results encourage Chinese scholars to conduct more experimental explorations. However, in view of the current research situation, the experimental research means are still relatively weak; the experimental method and statistical method of experimental data are relatively simple; the analysis of the experimental results is not deep enough; and the consideration of experimental variables is not sufficiently comprehensive. More is needed to explain the phenomenon than simply practical applications. This is especially true in basic theory research, such as map spatial cognition and map information communication. All of these factors led to cartographic theory research failing to play its role in the practice of cartography. This problem must be considered in future research.

2.4.2 Prospects for the Future

Although cartographic theory is also facing many problems, it has made considerable progress and achieved a series of research results. This is very important for the development of the future. In looking at the future prospects in the development of cartographic theory research, we believe the following information to be accurate:

(1) **Centering on multimode space-time comprehensive cognition, the theoretical system of cartography in the information era will gradually become established and perfected**

The deepening of information cartographic theory and the construction and perfection of the theory system stresses that we should take multimode spatial cognition as the core with the following considerations.

On the one hand, as mentioned above, in the information age, humans know the geographical environment and use geographical conditions based on the map (of course, this is basic), geographic information system (the expansion and extension of cartographic functions) and virtual geographic environmental information system (a new growth point in cartography). Therefore, not only must we study the process, mechanism and characteristics of spatial cognition based on maps, GIS and virtual geographic environment but we must also study multimode space-time comprehensive cognition. Spatial cognition runs throughout the whole process of cartography and geographic information engineering and holds the key to the problems. Spatial cognition reflects all the characteristics of cartography that exist as a system and reveals the essence of the process from the builders to the users in cartography and geographic information engineering, as well as involving the global and whole process of cartography. It is reasonable to deepen the theoretical research of information cartography based on the core of multimode space-time comprehensive cognition.

On the other hand, when constructing the theoretical system of information cartography, we emphasize the core of multimode space-time comprehensive cognition because by its very nature: it is both epistemological and methodological and is the leading theory of the theoretical system of information cartography. Other theories simply involve the application theory of multimode space-time comprehensive cognitive theory. For example, vision is the main organ for the perception of geographic information, and visual perception is a very important method of spatial cognition. Vision is highly sensitive to graphic symbol variables, so the effect of visual perception is very significant. For that reason, visual perception theory is an important part of the information system of cartographic theory; the model is an abstraction and expression of the objective things in the real world, essentially belonging to the category of spatial cognition methodology and is also the basis and spatial analysis tool for the automation of mapping (including cartographic generalization). Spatial data mining and knowledge discovery, in essence, is the deepening of spatial cognition theory and has the characteristics of epistemology and methodology, where the problem to be solved is to transform data into information and further into knowledge, and its fundamental purpose is to serve as decision support. Geographic ontology is a formal and explicit description of the shared conceptualization of the geographic information field, with a strong function for heterogeneous semantic conversion, important significance for the research of many frontier fields in GIS, such as geographic information sharing, semantic interoperability, distributed cooperative work, spatial data mining, knowledge discovery, and so on, and is the direct representation of the epistemology and methodology of spatial cognition. The uncertainty theory of spatial data is determined by the complexity and ambiguity of the real world and the ability of human beings to express the real world, which is directly related to quality control of the production and application of spatial data, the reliability of GIS spatial analysis and geographic information products, and the similarities between the perceptive geographical world and the objective geographical world. Obviously, the construction and perfection of the theoretical system of information cartography on the core of multimode space-time comprehensive cognition will be the research focus in the future and can also achieve the expected results.

(2) Studies on the various branches of the theoretical system of information cartography based on the core of multimode space-time comprehensive cognition will be deeper and more scientific

As has been pointed out, at present, there are still many problems in the research of the theoretical system of cartography, such as the research of each of the components is not thorough enough, most of the research results are general concepts, and the scientific conclusions after experimental verification, with less and practice testing. Entering the information age, we should pay special attention to the research of basic theory and the application theory of information cartography, strengthen the scientific research means and environmental construction, and make full use of the existing infrastructure construction results (such as a variety of scale map databases and multi-resolution remote sensing image databases); additionally, geographic information system and virtual geographic environments should be used for further research on the process, mechanism, difference, relation and characteristics of spatial cognition. The research on the role of spatial cognition, graphic variables and the features of visual perception in each procedure of geographic information transmission are based on these three platforms. The research on the adaptability of the map model, especially the cartographic generalization model, algorithm and knowledge for automatic mapping with different scales (scale dependent model), different regions and different practical uses as well as automatic construction of generalization chain and generalization process control and quality evaluation should become a focus. The research of the principles and methods of spatial data mining and the knowledge discovery model, algorithm and knowledge map based on geographic information database, geographic information system and the virtual geographic environment platform, as well as the practical application and decision support function of spatial data mining and knowledge discovery will become important focuses. Research on the spatial data uncertainty model and process data quality control in the process of data acquisition, processing, analysis and application, research on the theoretical model of spatial data assimilation in the situation of inconsistent scales, semantics and times, and research on the formal description of the semantic ontology and semantic transformation rules will need to become focuses in the future. Additionally, research on the continuous spatialization model of non-spatial (location) information (such as social, economy and humanity, etc.), and others will become important. There is currently basic experience in the research of these problems. In the new period, it is possible to obtain valuable results in these research areas.

2.5 Representative Publications

(1) ***The theory of modern cartography*** (Tian Desen. Surveying and Mapping Press, Beijing, 1991)

This book systematically introduces new concepts, theories, and research contents, such as the cartographic model theory, cartographic information communication theory and cartographic perception theory, which were put forward in the development of modern cartography, and this book discusses the practical significance of these theories. These theories represent the summary of map production activities and the deepening of the understanding of cartography. They reveal the development law of cartography. Research on modern cartography theory will have a positive impact on the development of Chinese cartography. The whole book is divided into 7 chapters. The main contents include the development of modern cartography, the research and development of basic cartographic theory, the cartographic model, information theory and its application to cartography and cartographic information representation and measure, cartographic information communication and cartographic perception.

(2) ***Theoretical cartography*** (Wang Jiayao, Chen Yufen. The People's Liberation Army Press, Beijing, 2000)

This book systematically summarizes the historical development path of cartography, describes cartography and its scientific system, introduces new theories of modern cartography, such as map spatial cognitive theory, map model theory, map information theory, map information transmission theory, cartographic perception theory and cartographic semiotics, and discusses their practical significance in cartography. These new theories constitute a theoretical system of modern cartography. They are the theoretical summary of cartography through long-term developments and production practices and the deepening and sublimation of people's cognition of cartography in the information era and will have a positive effect and influence on the development of modern cartography. The book has 8 chapters. Its main contents include the development history path of cartography, cartography and its scientific system, mapping spatial cognitive theory, map model theory, map information theory, map information communication theory, cartographic perception theory and map semiotics.

(3) ***Research on Cartography and Geographic Information Engineering*** (Wang Jiayao. Science Press of China, Beijing, July 2005)

The book is a collection of papers from the first author and his MS and PhD graduate students and reflects the author's achievements in the research of modern cartography and geographic information engineering science and technology over many decades. This book contains papers published since the 1980s, the national report he wrote and submitted to the international cartographic association (ICA) on behalf of the Chinese Academy of Surveying and Mapping the *Development Report of Cartography Discipline* written for *Blue Book of Surveying and the Mapping Science in China* on behalf of the Cartography and Geographic Information System (GIS) Professional Committee, China Academy of Surveying and Mapping, and so on.

The book consists of five parts: the first part generally includes nine papers, reflects the progress of modern cartography and the geographic information engineering science and technology in China, as well as its position in the international community; the second part includes 19 papers and introduces the research on the basic theory of modern cartography and geographic information engineering science and technology, as well as presenting the author's research results in the field of map design theory, digital map and image processing and recognition theory and the geographic information engineering theory; the third part covers the research on the automatic generalization of the digital map, which includes 15 papers, and presents the research results of the author in terms of the basic theory and methods of automatic generalization for digital maps, and the digital mapping system; the fourth part covers the geographic information system technology and application, including 24 papers, which gives the research results of the author on the technology of geographical information system, spatial analysis and digital earth, and the integration of 3S technology; finally, the fifth part is the abstract of academic writings and doctoral dissertations and introduces 7 works of the author and 19 doctoral theses that he supervised.

References

Bi, S. (2008). *Spatial data mining and its application of vector graphics feature*. Science Press.

Chen, Z. (1964). Cut density coefficient formula and its application of landforms in the preparation of general chart. *Acta Geodaetica et Cartographica Sinica, 7*(2), 139–154.

Chen, Y. (1995a). Visual variable for map symbol. *Journal of the Institute of Surveying and Mapping, 12*(2), 145–148.

Chen, Y. (1995b). Spatio-temporal maps and visual variables. *Journal of the Institute of Surveying and Mapping, 12*(3), 214–217.

Chen, Y. (1995c). Mental map and it's role in map design. *Journal of the Institute of Surveying and Mapping, 12*(4), 145–148.

Chen, Y. (1995d). Composition regulation of the topographic symbol. *Military Surveying and Mapping,* (4), 52–55.

Chen, Y. (1996). Map language. *Military Surveying and Mapping,* (2), 40–42.

Chen, Y. (2000a). *Research on cognitive theory of map space*. Doctoral Dissertation of the Information Engineering University.

Chen, Y. (2000b). A color matching experiment on electronic maps. *Journal of the Institute of Surveying and Mapping, 17*(1), 53–56.

Chen, S. (2001). Prospective considerations for cartography science. *Developments in Surveying and Mapping, 26*(1), 1–6.

Chen, Y. (2001). Spatial cognition research on electronic maps. *Progress in Geography, 20*(supplement), 63–68.

Chen, Y., & Chen, Y. (1999). Application of visual perception theory on electronic map design. *Journal of the Institute of Surveying and Mapping, 16*(3), 218–221.

Chen, S. (1964). Some modern characteristics and growth point of cartography. In *Selected papers of the conference of 1962 drawing (The First Volume)* (pp. 1–23). Chinese Industry Press.

Chen, S. (1991). *The flow of information and cartography. Chinese Yearbook of Cartography* (pp. 1–11). China Map Publishing House.

Chen, Y., & Ye, X. (1996). Map: Important tool for spatial cognition. *Military Surveying and Mapping,* (1), 50–52.

Chen, Y. (1989). *On construction rules of atlas symbolization.* Master's Degree Thesis of the Institute of Surveying and Mapping.

Chen, Y., Chen, G., & Zhu, F. (2000). A value grading perception experiment on electronic maps. *Journal of Surveying and Mapping Institute, 17*(3), 207–209.

Di, K. (2000). *Spatial data mining and knowledge discovery.* Wuhan University Press.

Ding, C. (1966). Color of natural map. *Bulletin of Surveying and Mapping,* (3), 37–41.

Ding, J., & Zhang, R. (1993). The development of contemporary cartography and its discipline system in china. *Acta Geodaetica et Cartographica Sinica, 22*(4), 299–303.

Ding, S., Li, M., & Li, Y. (1999). The discussion about the map language standardization and normalization. *Jiangsu Surveying and Mapping, 22*(1), 46–48.

Du, S., Wang, Q., & Qin, Q. (2007). *Fuzzy description of spatial relations and combination reasoning.* Science Press.

Fei, L., & Guo, Q. (1993). Design of an intelligent comprehensive system of topographic map. *Journal of Wuhan Technical University of Surveying and Mapping,* (S1), 1–5.

Gao, J. (1982). Study on the theory of European cartography. *Military Surveying and Mapping,* (2), 52–55 & (3), 51–53.

Gao, J. (1984a). Experimental method of map feeling and map design. In *The progress of cartography.* Chinese Institute of Surveying and mapping new technology in the summer of 1984 lecture class teaching materials of cartography.

Gao, J. (1984b). Structure and professional education of map science. *Journal of the Institute of Surveying and Mapping, 00,* 97–108.

Gao, J. (1986). Map: Cartography theory, characteristics and structure of science. *Map,* (1), 4–10; & (2), 1–7, 20.

Gao, J. (1991).*Spatial cognition and cognitive map cartography—Cartography new exploration in the field of science and technology and culture. Chinese Cartography Yearbook* (pp. 12–18). China Map Publishing House.

Gao, J. (1996). The 21st century cartography. In *Zhengzhou surveying and mapping institute fifty anniversary symposium* (pp. 1–7).

Gao, J. (2004). Cartographic tetrahedron: Explanation of cartography in the digital era. *Acta Geodaetica et Cartographica Sinica, 33*(1), 6–11.

Gao, J., & Wan, G. (2002). New window of spatial cognition: Review of ten years of geographic virtual space. In *Geographic spatial information technology and its application—Mapping the 2002 academic annual meeting proceedings China Association* (pp. 22–23). Chengdu Map Press.

Guo, Q., Huang Y., & Zheng, C. et al. (2007). *Spatial reasoning and incremental map generalization.* Wuhan University Press.

Gong, J. (1964). A graphical analysis method for specifying the deformation distribution of cylindrical projection. *Bulletin of Surveying and Mapping,* (1), 1–5.

Han, Y., Huang, C., & Yang, M. (2006). Research and practice on the application of experimental psychology to the travel map's symbol design: A case study of the confusing differentiation experiment. *Science of Surveying and Mapping, 31*(6), 149–150, 23.

He, Z. (1987). A study of the method for measuring the information content of map. *Journal of Wuhan Technical University of Surveying and Mapping, 12*(1), 70–80.

He, Z. (1991). A study of optical illusions in mapmaking. *Journal of Wuhan Technical University of Surveying and Mapping, 16*(3), 22–30.

He, Z., Bat, T., & Teng, Y. (1996). Determining varied information content of map with information method. *Journal of Wuhan Technical University of Surveying and Mapping, 21*(1), 64–68.

Hu, P. (1982). Semi-numerical method of map projection transformation. *Acta Geodaetica et Cartographica Sinica,* (3), 191–202.

Hu, Y. (1962). A graphical analysis method for the exploration of azimuthal projection. *Acta Geodaetica et Cartographica Sinica, 5*(2), 141–153.
Hu, Y., & Chow, Y. (1958). Conformal or equal-spaced conical projection with total-equal area. *Journal of Surveying and Mapping, 2*(3), 195–202.
Hu, Z., & Yan, H. (2008). Analysis on linguistics mechanism for cartographic symbols and its application. *Geography and Geo-Information Science, 24*(1), 17–20, 34.
Hua, Y. (1991). *Thematic map design expert system PC-MAPPER*. Doctoral Dissertation of the Institute of Surveying and Mapping.
Jiang, X. (1986). The positional symbol system of computer-assisted thematic mapping and automatic drawing of derived symbols. *Acta Geodaetica et Cartographica Sinica, 15*(3), 196–207.
Jiang, M., & Yu, J. (1966). Mapping representation of all kinds of loess landforms on large scale topographic maps. *Acta Geodaetica et Cartographica Sinica, 9*(1), 44–62.
Jiang, N., Dai, Y., Zheng, Z., & Lihua, X. (2007). Design on multi-display patterns of the electronic map's symbol system. *Science of Surveying and Mapping, 32*(3), 20–22.
Li, G. (1963). Double azimuth projection. *Acta Geodaetica et Cartographica Sinica, 6*(4), 279–302.
Li, C. (1979). On the classification of map projections. *Acta Geographica Sinica, 34*(2), 139–155.
Li, D. (2008). *Interactive design of urban tourism electronic map*. Master Degree Thesis of the Information Engineering University.
Li, L., & Wu, F. (2005). *Multi-scale representation model of spatial data and its visualization*. Science Press.
Li, D., Wang, S., & Li, D. (2006). *Theory and application of spatial data mining*. Science Press.
Liao, K. (1980). The characteristics and trends of modern cartography and cartography on China's future development views. In *The third national academic conference papers* (I) (pp. 1–16). Surveying and Mapping Press.
Liao, K. (1983). On the modern cartographic system. *Acta Geographica Sinica, 38*(1), 80–89.
Ling, Y. (2005). *Research on adaptive user interface of map visualization system*. Master Degree Thesis of the Information Engineering University.
Ling, Y., Chen, Y., & Wang, Y. (2005). Cognitive basis of adaptive user interface for cartographic visualization system. *Acta Geographica Sinica, 34*(3), 277–282.
Liu, D. (1963). The overall design of atlases and some technical problems. *Bulletin of Surveying and Mapping*, (2 supplement), 1–7.
Liu, H. (1985). Numerical method of isometric projection transformation. *Acta Geodaetica et Cartographica Sinica*, (1), 61–72.
Liu, Z. (1964). The use of shading in the political map. *Bulletin of Surveying and Mapping*, (3), 16–18.
Liu, H. (1992a). Exploring the scientific implication of map information. *Journal of the Institute of Surveying and Mapping, 9*(2), 51–55.
Liu, H. (1992b). The study of the method of measuring map information. *Journal of the Institute of Surveying and Mapping, 9*(3), 49–55.
Liu, J. (2008). *Research on user interface design of multimedia electronic atlas*. Master's Degree Thesis of the Information Engineering University.
Liu, J., & Li, G. (1963). Application of pseudo azimuthal projection of China graph. *Acta Geodaetica et Cartographica Sinica, 6*(2), 104–119.
Lu, Q. (1980). The theory of modern cartography theory development and the research of our task mapping. In *The third national academic conference papers (I)* (pp. 17–34). Surveying and Mapping Press.
Lu, L. (1989). Basic conditions for the realization of cartographic information communication. *Map*, (4), 7–8, 58.
Lu, S., Ju, X. (1964). The method of measuring the density of rivers and lakes on topographic maps. *Acta Geodaetica et Cartographica Sinica, 7*(4), 292–300.
Mao, J. (1987). The concept and content of map information. *Areal Research and Development, 6*(3), 65.

Mao, Z. (1989). The principle of gestalt psychology in the theory of cartographic perception. *Map,* (4), 9–13.

Min, B. (1957). Map schema (illustrations or special symbols) design principles. *Journal of Surveying and Mapping, 1*(2), 131–145.

Qian, H. (2006). *Intelligent research on automatic drawing synthesis and process control.* Doctoral Dissertation of the Information Engineering University.

Shi, Z. (1964). The production of "Chinese shading map" of the new taste. *Journal of Surveying and Mapping, 7*(1), 58–72.

Su, L., Chen, Y. (2007). Linguistic characteristics of symbols of topographic maps. *Science of Surveying and Mapping, 32*(5), 34–35, 38.

Sun, Q. (1991). *The establishment of the expert system for military thematic maps and the realization of the automatic generation subsystem of mathematical foundation.* Doctoral Dissertation of the Information Engineering University.

Sun, Y. (2008). *The design and implementation of PDA tourist map.* Master's Degree Thesis of the Information Engineering University.

Tian, D. (1991). *The theory of modern cartography.* Surveying and Mapping Press.

Tong, Y. (1990). Map information characteristics from the information society. *Areal Research and Development, 9*(1), 47–48.

Wan, G. (2006). *Study on the construction of the theory and technology of humanoid virtual geographical environment.* Doctoral Dissertation of the Information Engineering University.

Wan, G., Gao, J., & You, X. (2005). Several spatial cognition problems in virtual terrain environment simulation. *Science of Surveying and Mapping, 30*(2), 48–50.

Wan, G., Gao, J., & Liu, Y. (2008). Research on cognitive map formation based on reading experiments. *Journal of Remote Sensing, 12*(2), 339–346.

Wang, J. (1989). On the controllability of map information transmission. *Military Mapping Album,* 12–14.

Wang, M. D. (1990). *Map language structure* (pp. 57–60). Chinese cartography yearbook. China Map Publishing House.

Wang, J. (1991). "Atlas" of the design characteristics of officers. *Journal of the Institute of Surveying and Mapping, 8*(3), 33–42.

Wang, J. (1993). Introduction to modern military map drawing. *Journal of the Institute of Surveying and Mapping,* (1 supplement), 36–66.

Wang, J. (2000a). Cartography in information era. *Engineering of Surveying and Mapping, 9*(2), 1-5.

Wang, J. (2000b). New research on map linguistics. *Bulletin of Surveying and Mapping,* (5), 21–22, 28.

Wang, J. (2008). A visualization of information characteristics and cartography theory and technology system. In *Informatization of surveying and mapping* (pp. 130–138). Surveying and Mapping Press.

Wang, L. (2009). *Experimental study and practice of electronic map.* Master's Degree Thesis of the Information Engineering University.

Wang, J., & Yufen, C. (2000). *Theoretical cartography.* PLA press.

Wang, J., Wang, H., & Chen, X. (2003). Study on standardization of map symbols in digital mapping. *Earth Information Science, 5*(2), 16–19.

Wu, H. (1965). Principles and methods of selecting the height of the terrain map. *Acta Geodaetica et Cartographica Sinica, 8*(2), 126–144.

Wu, Z. (1990). Map information and its calculation method. *Areal Research and Development, 9*(4), 53–55.

Wu, F. (2003). *Multi scale representation and automated generalization of spatial data.* PLA press.

Wu, Z. (2007). *Design and research of tourism network map.* Master's Degree Thesis of the Information Engineering University.

Wu, Z., & Liao, K. (1981). The development of cartography over the past 30 years in the people's republic of china. *Acta Geodaetica et Cartographica Sinica, 10*(1), 1–8.

Wu, F., Qian, H., & Deng, H. (2008). *Intelligent processing of spatial information for automatic map generalization.* Science Press.
Wu, F., Deng, H., & Qian Hai, Z. (2009). *Quality assessment models for automated map generalization.* Science Press.
Wu, Z. (1979). How to transform the coordinates of one kind of map projection point to another kind of map projection point coordinate problem. *Acta Geographica Sinica,* (1), 55–68.
Xie, C. (2006). *Study on the symbol system of electronic topographic map.* Master's Degree Thesis of the Information Engineering University.
Xie, C., Chen, Y. (2006). Research on the electronic topographic map symbol system. *Journal of Zhengzhou Institute of Surveying and Mapping, 23*(3), 215–218, 221.
Xie, C., Chen, Y., & Ke, X. (2008). Research on the composition regulation of the electronic topographic symbol. *Beijing Surveying and Mapping,* (1), 35–37, 19.
Xu, Z., Ai, T., & Wei, Y. (2006). On visual variables of 3D map symbol. *Geomatics and Information Science of Wuhan University, 31*(6), 557–560.
Yan, H., & Guo, R. (2001). On classification of spatial direction relations. *Engineering of Surveying and Mapping, 10*(4), 13–15.
Yan, H., & Guo, R. (2002a). Research on formal description of directional relation based on voronoi diagram. *Science of Surveying and Mapping, 27*(1), 24–27 & 27(3), 4–7, 12.
Yan, H., & Guo, R. (2002b). On fundamental problems of directional relationships. *Acta Geodaetica et Cartographica Sinica, 31*(4), 357–360.
Yan, H., & Guo, R. (2003). Research on formal description model of directional relationships. *Acta Geodaetica et Cartographica Sinica, 32*(1), 42–46.
Yan, H., & Wang, J. (2009). *Map group objective description and automated generalization.* Science Press.
Yang, Q. (1965). Using the numerical method to establish the conical projection of arbitrary character. *Acta Geodaetica et Cartographica Sinica, 8*(4), 295–318.
Yang, Q. (1980). Study on Gauss-kruger projection family. In *Proceedings of map projection* (pp. 158–172). Surveying and Mapping Press.
Yang, Q. (1982). Research on numerical transformation of isometric projection. *Acta Geodaetica et Cartographica Sinica,* (4), 268–282.
Yang, Q. (1984). Research on the third coordinate transformation of map projection. *Journal of Institute of Surveying and Mapping,* (1), 109–121.
Yu, L. (1995). The philosophical levels of map symbol and the exploration of its information function. *Acta Geodaetica et Cartographica Sinica, 24*(4), 259–266.
Zhang, K. (1958). On the choice of the residents' points on the map in our country. *Bulletin of Surveying and Mapping,* (12), 546–550.
Zhang, K. (1963). Preliminary study on various representations of special map. *Acta Geodaetica et Cartographica Sinica, 6*(3), 195–205.
Zhang, X. (2007). *Research on clustering analysis based on the clustering analysis of the constrained conditions.* Doctoral Dissertation of the Information Engineering University.
Zhang, W., & Zhang, X. (1991). Cartography and expert system. *Maps,* (1), 46–48.
Zhang, K., & Zhu, G. (1990). On the theoretical system of cartography. *Journal of Wuhan Technical University of Surveying and Mapping, 15*(2), 28–33.
Zhong, Y. (1965). Not equal geographic design and analytic calculation methods of conical projection. *Acta Geodaetica et Cartographica Sinica, 8*(3), 218–236.
Zhong, J. (1989). The role of mental map in cartographic communication. *Journal of the Institute of Surveying and Mapping,* (1), 74–78.
Zhou, C. (1957). A new projection based on the deformation ellipse. *Journal of the Measurement and Mapping,* (1), 1–14.
Zhou, H. (2004). *Research on spatial data mining.* Doctoral Dissertation of the Information Engineering University.
Zhou, Z., Liu, H., & Li, S. (2008). Modern cartographic language. *Engineering of Surveying and Mapping, 17*(2), 6–8.

Zhu, D. (1987). A research on the nature of map symbols. *Journal of Wuhan Technical University of Surveying and Mapping, 12*(2), 55–63.
Zhu, Y. (1996). Map grammar study. *Bulletin of Surveying and Mapping,* (3), 22–27.
Zhu, G., & Huang, C. (1985). On the measurement of map information of settlements. *Journal of Wuhan Surveying and Mapping Science and Technology University, 22*(4), 89–96.
Zou, J. (1991). On the measuring of cartographic information amount. *Journal of the Institute of Surveying and Mapping,* (4), 72–77.

Chapter 3
Map Projection

Xiaohua Lv

3.1 Introduction

Map projection is a foundational theory and an important part of cartography. It occupies an important and leading place in the development of cartography in China and is the most successful branch of cartography for its use of modern mathematical methods.

The surface of the Earth is an undevelopable curved surface, but a map is usually a continuous two-dimensional plane. For this reason, when using a map to represent a part or the whole of the Earth's surface, it is inevitable to yield a kind of insurmountable contradiction between a sphere and flat surface. To solve the problem of transformation from the non-developable Earth's surface to a continuous map plane, the map projection method is used.

Map projection is the main content of the cartographic mathematics foundation, the base and skeleton of a map, the actual embodiment of the scientific nature and accuracy of a map and is the chief problem encountered before the map is compiled. It is the map projection that makes a map rigorous, scientific, accurate and measurable. The projection is one of the important features of the map, which is different from photo and landscape paintings and others. In a broad sense, the map projection is an important part of the spatial datum of geographic information, the basis for realizing spatial information location and visualization, and the basic framework of geospatial data. Geometrical perspective was initially applied as a map projection method to solve the contradiction between the sphere and plane. This method is based on the principle of perspective. The term map projection is derived from using a geometric perspective method to project the graticule of meridians and parallels on the Earth's surface to a plane in its early development stage. The geometric perspective method can only realize some simple transformations from sphere to plane. This method has many limitations and cannot easily meet the needs of compiling various types

X. Lv (✉)
Information Engineering University, Zhengzhou, China

J. Wang and F. Wu (eds.), *Advances in Cartography and Geographic Information Engineering*, https://doi.org/10.1007/978-981-16-0614-4_3

of maps. Since the emergence of the branch subject of mathematics analysis, it has been widely used in map projection. The map projection problem is usually studied by a mathematical analysis method based on relating the graticule of meridians and parallels on the Earth's surface to the corresponding graticule of meridians and parallels on the plane.

In fact, there are few map projections that completely use the so-called projection of the principle of geometry. Instead, the vast majority use mathematical analysis methods to solve the transformation from the Earth's surface to the plane. Therefore, map projection is also called mathematical cartography. Obviously, the word projection cannot fully reflect the research content of map projection. If interpreted literally, projection contains only the geometric perspective. However, the term has been used for a long time and has spread widely. Using the name projection does not hinder its development, while the development of the discipline also gives new and more extensive connotations to the term projection. Although map projection is used to solve the contradiction between the sphere and plane, it is impossible to completely and accurately represent all parts of the Earth on a plane, that is, a distortion exists. In general, a length deformation, area deformation and angle deformation exist.

Therefore, the field of map projection mainly studies the theory, method and application of projecting the Earth ellipsoid (or spherical) onto the map plane and the deformation law of map projection. This research area also studies the conversion between different map projections and measurements on maps. The main tasks of map projection are outlined as follows: study the theory and method for establishing the graticule of meridians and parallels with different map uses, which includes the calculation and plotting of the projection coordinates and their deformation, as well as some practical issues related to map arrangement and map subdivision; projections of specific curves of the Earth ellipsoid (sphere) onto a plane and its depiction method, as well as problems, such as the measurement of the elements on map; and the projection method suitable for solving the geometric problems of a plane or line in space must meet the application needs of multiple disciplines.

3.2 Development Process

Map projection was born between the sixth century BC and the fifth century BC. At that time, map projection was mainly used to draw the map of the heavens. For example, the ancient Greek astronomer Thales (BC 640–546) was the first to use gnomonic projection for celestial maps. The great geographical discovery expanded the geographical concept of the Earth in the sixteenth century, and cartography was also flourishing during that time. The outstanding Dutch cartographer Mercator (AD 1512–1594) put forward a conformal cylindrical projection and used it to compile a map of the world. In the nineteenth century, the German mathematician Gauss (AD 1777–1855) proposed the equiangular transverse cylindrical projection-Gauss projection and applied it to large-scale topographic maps. The general principle of the deformation of map projection was described in detail, and at the same time, an approximate calculation method for the conformal projection was pointed out by French mathematician Tissot. Chybishev (1821–1894) pointed out that the most appropriate projection for representing a part of the Earth's surface on a flat map is having a projection border line with the same length proportion. This discussion is very important for the exploration of new projections. From the beginning of the twentieth century to the 1950s, Kavraisky developed equidistant conic projections and equivalent conic projections for the Soviet Union and designed a pseudocylindrical projection with arbitrary properties. Urmaev is the author of the *Principles of Mathematical Cartography* and *Methods to Explore New Projection* and other books, in which he discussed the basic theory of map projection and proposed a means to explore new map projections based on known deformation distributions. Ginzburg prepared a series of projections, such as azimuthal projections, pseudocylindrical projections, polyconic projections, and pseudoazimuthal projections with isallolines, such as an oval or ellipse, and presented two valuable generalized formulas for azimuthal projections. Volkov is the author of *Principle and Method of Measurement on Maps*, systematically expounding the problem of map projections.

Cartography has a long history in China. In the period of the Jin Dynasty, PEI Xiu (AD 224–273) created a scientific method of map compilation, the so-called Six Principles in Mapmaking, that is, Fen Lv (scale), Zhun Wang (direction), Dao Li (distance), Gao Xia (relative height or elevation), Fang Xie (rise and fall of a gradient) and Yu Zhi (conversion between the distance in reality and that on the map). It was not until the late Ming Dynasty that the graticule of meridians and parallels appeared on the Chinese map. Nationwide latitude and longitude measurements were carried out in the Qing Dynasty, and map projection was used in the compilation of *Huang Yu Quan Lan Tu* (*Map of China in the Emperor Kangxi Reign of Qing Dynasty*) and *Da Qing Yi Tong Yu Tu* (*Atlas of China in Qing Dynasty*). In the early years of the Republic of China during the period of the Northern Warlords, map projection was not applied to the 1:50,000 scale basic terrain map, but polyhedral projection was used in the 1:100,000, 1:500,000 and 1 1,000,000 scale maps. During the Guomindang period, a 1:50,000 scale map used the Lambert conformal conic projection, dividing the country into eleven zones by latitude, each with a separate projection; a 1:1,000,000 scale map subdivision applied the modified polyconic projection for one of a million international map subdivisions. A small number of books on projection theory have been published.

Since new China was founded, significant progress has been made in the research of the theory and application of map projection under the joint efforts of scientific and technological workers in the field of surveying and mapping, and a relatively complete system of theory, method and engineering application of map projection and personnel training system with Chinese characteristics gradually formed.

From the early days of the founding of the PRC to the middle of the 1950s, map projection research mainly learned from the former Soviet Union. During this period, a number of map projection textbooks, monographs and reference books of the Soviet Union were translated into Chinese, such as *Mathematical Cartography*, *Practice Materials of Mathematical Cartography*, *Methods of Map Projection*, *Examples of*

Mathematical Cartography, Collection of Nomographs of Mathematical Cartography, Similar Elements Transformation Theory and Its Application in Mathematical Cartography and Map Compiling, and the reference book *Cartographic Table.* During this period, a Chinese professor, Jun FANG, published his book *Map Projection* as a reference book for Chinese cartographers. For National Fundamental Scale Maps, the mathematical foundation gradually established a unified system and had a unified geodetic coordinate system, unified scale system, unified projections, and a unified sheet division and numbering system.

Beginning in the late 1950s, the academic journal *Acta Geodaetica et Cartographica Sinica* was founded by the Chinese Society of Geodesy, Photogrammetry and Cartography (CSGPC) and published a succession of papers on map projection, and other journals, such as *Bulletin of Surveying and Mapping* and *Collected Translations of Surveying and Mapping,* often published articles about map projection. Several textbooks or works, such as *Mathematical Cartography* and *Map Projection,* have been edited and published in China's surveying and mapping universities and secondary schools. They systematically expounded the general theory of map projection, showing new research results in the field of map projection at home and abroad, as well as gaining practical experience in the design of mathematic map bases in China. These books provided a chance for Chinese cartographers to further study and exchange information.

From the late 1950s to the end of the 1980s, the map projection field developed rapidly, from research on the basic theory of map projection to the exploration and application of new map projections; remarkable achievements have been obtained. A number of important research results have been obtained in the aspects of the research of map projection general theory, exploration of new map projections, building of the mathematical foundation for regional maps and atlases, selection and application of map projections, map projection transformations, and so on. During this period, a number of theoretical and practical papers were published, such as The Initial Analysis of Contradictory Movement in Map Projection, On the Classification of Map Projection, Pseudoazimuthal Projection and Its Application to the Whole Map of China, Gauss-Krüger Projection Family, Polyconic Projection Family, and so on; *Mathematical Cartography* and *The Principle of Map Projection* and other books have been edited and published, and some reference books have appeared, such as the Tables for Regional Map Projection, *Projections for Small-Scale Map, "Table for the Calculating Map Projection, Coordinates Table of Krueger-Gauss,* and so on. Some results were also obtained in the selection and design of map projections, including *The Selection and Design of Map Projections for the Atlas of the People's Republic of China, On the Projections of the Provincial Atlas of People's Republic of China* and *On the Projections of the Provincial (autonomous regions) Atlas,* and so on. These studies basically constructed the theory, method and application system of map projection with Chinese characteristics.

Since the 1990s, with the development of spatial remote sensing technology, computer technology, and GIS theory and technology, digital maps have appeared, and map production technology, publishing technology and application technology have undergone revolutionary changes, gradually forming the subjects of cartography and geographic information engineering. The historical change in the map production mode, the unprecedented abundance of map compilation information, the increasing use of maps and the increasing diversity of map types and products all drive the further study of map projection selection and design, as well as map projection transformation. Multisource data fusion, geographic spatial database construction, GIS spatial analysis and representation all promote the depth of research and the breadth of application in terms of spatial datum integration, spatial data processing and so on. Map projection, as the basic theory of cartography, has made considerable progress. New breakthroughs have been made in map projection design, map projection transformation, spatial dynamic projection, and the application of map projection in GIS. During this period, a number of commercial GIS tool software platforms and a variety of special types of GIS have been launched at home and abroad, and these GISs contain a function module of map projection or a map projection module. Map projection is an important part of GIS; a number of papers, teaching materials and monographs, such as *Map Projection, Principles and Methods of Map Projection Transformation*, have been published. In particular, Professor YANG Qihe devoted all of his efforts to completing a great work, *Map Projection Transformation—Principles and Applications*. Based on the development of computer science, spatial information science and spatial remote sensing technology, the book breaks through the research content and method of traditional mathematical cartography, focusing on the latest research results in the field of map projection transformation in both China and abroad. As an international authority in the research field in the twentieth century, this researcher remains ahead in the whole map projection transformation field worldwide. The book was published by Taylor & Francis in 2000.

3.3 Main Research Achievements

3.3.1 General Theories of Map Projection Are Constantly Enriched and Improved

During the 60 years since the establishment of new China, Chinese cartographers have unceasingly explored how to perfect and enrich the general theory of map projection and systematically discussed many basic theoretical issues, such as map projection equations, map projection distortion and its calculation, map projection deformation ellipses and their application, map projection conditions and map projection classification. Many research results have been obtained, and an important theoretical foundation was laid for the application and development of map projection.

The book, *The Principles of Mathematical Cartography* (Wu et al. 1989), discusses the coordinate line, line element, and surface element on the Earth ellipsoid and their expression forms based on arbitrary spatial surfaces, as well as in depth discussions of the basic problems, such as the Earth curvature, the geodesic line and its differential equations and isometric coordinates.

In the aspects of the map projection equation, various forms of differential equations of map projection were derived. By introducing isometric latitude q, a second-order elliptic type partial differential equation for conformal map projection were derived, that is, $\frac{\partial^2 \ln v}{\partial q^2} + \frac{\partial^2 \ln v}{\partial \lambda^2} = 0$, and this equation provides an important theoretical basis for the exploration of new projections and conformal map projection transformation. The paper 'Comments on the map projection functions' (Dang 1960) put forward that all conformal projections can be expressed with the linear combination of homogeneous harmonic polynomials; at the same time, the numerical integration method (difference method, finite element method, and so on) can be directly applied to the map projection calculation. Based on this basic principle, the Bernoulli type differential equation was derived according to the condition of conical projection, that is, $\frac{d\rho}{d\phi} + P(\phi) \cdot \rho = Q(\phi) \cdot \rho^n$.

From the aspects of map projection deformations, the book, *The Principles of Mathematical Cartography* (Wu et al. 1989), systematically discussed the definition and measurement of the length deformation, area deformation and angle deformation, explored the equations and their properties in depth in terms of the maximum deformation of azimuth, the maximum deformation line of azimuth, the maximum deformation line of length, and the isalloline of length, and it was concluded that the maximum deformation line of length is the envelope along the direction of the extreme value length proportion or the principal direction; the book also discussed in detail the properties and expression methods of deformed ellipse on the projection plane (the conjugate diameter is the oblique axis with a θ angle intersected by latitude and longitude lines, and with the principal bearing as the major semi-axis and semi-minor axis); and further, the book presented the measurement method of the length deformation along all directions of a point on a projection plane, and the measuring methods of area deformation, shape deformation and the deformation of a whole region. Liu (1965) discussed the deformation transformation law of commonly used projections and considered that the conversion between area deformation and angle deformation is actually the conversion between the area proportion and the shape deformation. For a normal cylindrical projection, area proportion and shape deformation are converted along the equilateral hyperbola in the coordinate system, while the deformation conversions of azimuthal projection and conic projection are along the two quadric conical surfaces in the *K*-*P*-μ2 coordinate system and *K*-*P*-*n* coordinates, respectively. The deformation conversion law between different projections is further revealed.

In the classification of a map projection, the traditional classification method based on the properties of map projection deformation and the shapes of latitude and longitude lines in the normal axis projection has been surpassed, and an intrinsic feature classification method according to the pole position of the spherical coordinate system, the parameter attributes in the projection equation and the isometric direction of the deformed elliptical has been given. At the end of the 1970s, Li (1979) analyzed the inherent law of projection deformation and proposed a new method of map projection classification on the basis of the function expression of $a = b^K$ and $a = Kb$, that is, starting from projective invariants and according to whether there is an equal-length direction and whether the equal-length direction is the only one at any point in the projection region; the map projection was divided into the elliptic, parabolic and hyperbolic types, and subdivided into three kinds of common projections of conformal projection, equidistant projection, and equal area projection for each type. This classification method not only expresses the hierarchical relationship but also makes clear the adjacent relationship, and it can reveal the ways and methods to explore the new projection.

In addition, Li (1963) projects the surface of the Earth in a certain way on the auxiliary sphere and then onto the plane with some methods. One of the most significant projections is the dual azimuthal projection with equidistant-perspective projection, that is, it projects the Earth's surface onto the auxiliary sphere with a radius of KR using the equidistant projection method, and then, the auxiliary sphere is projected onto the plane by using the perspective projection method; the author also suggested a scheme of double azimuthal projection with equidistant-perspective projection for the whole Chinese territory. Hua (1983b) pointed out that when projecting from an ellipsoid's surface onto a sphere, the projection constants are the same except for the integral constant regardless of whether the conditions are conformal, equidistant, or equal-area deformations; no matter what deformation condition there is, the projection length proportions on the latitude lines can be regarded as the same within a certain width of the latitude zone. The paper gives the viewpoint that a geodetic distance can be regarded as a spherical distance in a certain width of the latitude zone, and the geodetic azimuth angle can be regarded as a spherical azimuth. Yang (1984b) derived the general formula for the projection of ellipsoids on a sphere $\phi = f(B)\lambda = \alpha \cdot l$, analyzed the extreme value property of the length proportion of latitude lines n, and concluded that when $n_0 < 1$, there is an extreme value for conformal and equal-area projection; when $n_0 = 1$, there is no extreme value for the conformal, equidistant, or equal-area projection (except $B_0 = 0°$ and $B_0 = 90°$). This finding provides a theoretical basis for the practical application of the projection of ellipsoidal surfaces on spheres. Li (1987b) described the origin of the projection $m = n^k$ and its reasoning process; Guozao also stated that under the condition of $m = n^k$, the area proportion formula of orthogonal projection is $P = mn = n^{1+k}$, and the angular maximum deformation formula is $\sin\frac{\omega}{2} = \frac{n^{1-k}-1}{n^{1-k}+1}$; the projection series of $m = n^k$ was graphically illustrated. The paper filled the gap in map projection by creating more abundant, systematic and complete map projection types and further improved the general theory of map projection.

In addition to the study of the basic theory, research on the development trend of map projection and the modern significance of natural dialectics is the important driving force of map projection development. Wu (1963) discussed and pointed out the new direction of map projection on the basis of summarizing the research results of map projection; Wu (1965) applied the principles of dialectical materialism to analyze the contradictory movements within map projections to enhance the epistemology and methodology of map projection research. The paper, 'The direction and task of mathematical cartography' proposed that with the development of automated mapping, traditional mapping methods must be combined, that is, the graticule of meridians and parallels, as well as all kinds of information on the Earth's surface must be processed and finally represented on the map plane. In recent years, several papers, such as 'The evolution of the subject character of map projection' and 'Map projection and information superhighway', have been published; further research on the theory system of map projection has been facilitated.

3.3.2 Established a Unified System for the Mathematical Foundation of National Fundamental Scale Maps

A topographic map is a national fundamental scale map and is important basic data for economic construction and national defense construction; the mathematical foundation should be scientific and accurate; otherwise, it will directly affect the scientific nature and usability of the map.

In the 1950s, the Krueger-Gauss projection began to be used as the mathematical basis of measurement and the production of topographic maps in China, and the Krassowski ellipsoid was used as the basis of the calculations. The scales of less than 1:25,000 and more than 1:500,000 in topographic maps use 6° zones, with a maximum length of deformation of up to 1.38‰; the scales of more than 1:25,000 use 3° zones, that is, using the same Krueger-Gauss projection with 3° zones to ensure a higher accuracy of the map. The 1:1,000,000 maps use modified polyconic projection in terms of international regulations, the length deformation is not more than ±0.07%, and the angular deformation is less than 5′. A unified system of topographic map scale series has been established. There are six scales of nationwide maps, including 1:25,000, 1:50,000, 1:100,000, 1:200,000, 1:500,000 and 1:1,000,000; there are other scales of local maps for engineering use, such as 1:100, 1:500, 1:1,000, 1:5,000 and 1:10,000. The unified system of topographic map scale series, their sheet range and the number of adjacent scale maps are shown in Table 3.1.

According to the national standard, *Map Sheet Division and Numbering of National Basic Scale Topographic Maps*, formulated in 1991, there has been a large change in the numbering method. The 1:1,000,000 scale topographic maps turned the column–row type into the row-column type. The sheet numbering of the six kinds of scale topographic maps of 1:10,000–1:500,000 are based on the sheet numbering of the 1:1,000,000 map. The numbering is made up of a 10-digit code, of which the top three digits are the row number (1 digit) and column number (2 digits) of the 1:1,000,000 map. The fourth digit is the character code of scale. The first 3 digits of the remaining six are the row number of the sheet (numerical code), and the last 3 digits are the column number of the sheet (numerical code). Zero should precede a row or column code if there are fewer than three digits.

To establish the mathematical foundation of the topographic map, various calculation tables adapted to the needs of China were translated and compiled. For example, the *Gauss-Krüger Coordinates Table* (China Industry Press, 1963) records once

Table 3.1 .

Scales		1:1,000,000	1:500,000	1:250,000	1:100,000	1:50,000	1:25,000	1:10,000
Sheet range	Difference in long	6°	3°	1°30′	30′	15′	7′30″	3′45″
	Difference in lat	4°	2°	1°	20′	10′	5′	2′30″
The number of adjacent scale maps		1	4 1	16 4 1	144 36 9 1	576 144 36 4 1	2304 576 144 16 4 1	9216 2304 576 64 16 1

every 5′ for latitudes between 16° and 56° and records once every 7.5′ for longitudes between 0° and 3.5°, including the X and Y coordinates table, the table of meridian convergence angle, the table of border size and sheet area of topographic map, and auxiliary tables. These tables have been used for border decoration of a 1:10,000–1:200,000 scale map. Regarding the Gauss-Krüger projection zone coordinates transformation, Yangshan JIN, a Chinese engineer, derived a new zone transformation formula and compiled the *Gauss-Krüger Projection Zone Coordinates Transformation Table Between the 6° Zones* and the 3° Zones. To improve the accuracy of the coordinates of points in the grids of adjacent zones, raise work efficiency and change the traditional manual operation method, professor YANG Qihe designed, compiled and published the *Tables of Coordinate Transformation for Gridlines of Adjacent Zones of the Gauss Projection* (Surveying and Mapping Press, Beijing, 1980). The table provides the coordinate transformation values of points on the vertical kilometer lines and horizontal kilometer lines of the gridlines of adjacent zones for 1:25,000, 1:50,000, and 1:100,000 scale maps. According to the table, we can directly use a coordinate plotter to plot the marginal points and gridlines both on its own zone and on the adjacent zones, improving the work efficiency. To accommodate the application of computers in the compilation of topographic maps, professor YANG Qihe proposed an analytical method and numerical method of Gauss-Krueger projection transformation as well as the transformation method among the Gauss-Krüger projection, Mercator projection and conformal conic projection of 1:1,000,000 topographic maps; he also calculated and compiled the *Constant Coefficient Table for Coordinate Transformation of Adjacent Zones of the Gauss-Krueger Projection between the 6° Zones* and the 3° Zones, the *Constant Coefficient Table for Coordinate Transformation among Gauss-Krueger Projection, Mercator Projection and Conformal Conic Projection*, wrote the BASIC programs for the calculation of zone transformation, realized the automatic calculation of the sheet elements of topographic maps, and published a book, the *Principles and BASIC Program for Conformal Projection Transformation* (Yang, Surveying and Mapping Press, Beijing, 1987), in which the constant coefficient formula and constant coefficient table of the conformal projection, which were adapted to the computer, were derived.

In 1978, China decided to use conformal conic projection as the mathematical foundation of 1:1,000,000 topographic maps. The projection constants derived according to the condition that the absolute values of the scale error at the edge parallels and on the central parallels should be equal. In the range of 1:1,000,000 topographic maps, the projection length distortion at the edge and on the central parallels is approximately ±0.030%, and the area deformation is approximately two times the length of deformation. Two standard latitude lines lie on the edge latitude lines inside approximately 35′.

Because of the large length deformation of the Gauss-Krueger projection in the low latitude region, to improve the accuracy of the mathematical foundation of the topographic map, Chinese cartographers proposed the Cylindrical Transverse Conformal Projection with Two Standard Meridians (Li 1983b), in which the rectangular coordinate formula, meridian convergence angle formula and length proportion formula of the projection were deduced and compared with the commonly used Gauss-Krueger projection and universal transverse Mercator projection. The "Cylindrical Transverse Conformal Projection with Two Standard Meridians" makes elliptic cylindrical surfaces intersect with the two meridians symmetrical with the central meridian on the Earth's surface, with elliptic cylindrical surfaces passing through the Earth's north and south poles and the Earth's surface between the two secant meridians outside the elliptic cylindrical surface. Therefore, the distortion of the projection zone is at a minimum in the low latitude region, and an improvement is noted in high latitude areas. This projection is a good choice for large-scale topographic maps. YANG Qihe studied the Gauss-Krueger projection family (Yang 1983b), which considered the combination of conformal projections with zone division along the meridians as the Gauss-Krueger projection family, derived the general formula of the Gauss-Krueger projection family, analyzed three different forms of conformal projections along the central meridian with the proportion of length $m_0 = f(B)$, explored the intrinsic relation between the Gauss-Krueger projection and its derivative projection, as well as the relation between the selected parameters and the projection arrangement in the conformal projections system with $m_0 = 1 - \mathrm{q} \sin^2 KB$.

3.3.3 Innovation and Development of New Map Projection Exploration and Application

In the 60 years since the foundation of new China, to better adapt to the needs of cartography in teaching, scientific research and production, national economic construction and national defense construction, a series of innovative achievements have been made in the exploration of new types of map projections, and they have been successfully applied in practice.

Based on the distortion properties of projection, the shape of parallels and meridians after projection, the distortion distribution of projection and the other conditions, theories and methods for exploring new types of map projection have been put forward by Chinese cartographers. With a focus on exploring the general method of conformal projection, the variation method and finite difference method of the Poisson and Laplace equations, the method of exploring the orthogonal projection according to $q_x^2 + q_y^2 = \frac{1}{\nu^2}$, the method of establishing new projections by using numerical analysis, and the exploration of new projections meet the special purpose requirements.

The paper 'Problems with the discovery of the distortion ellipse on hand' (Zhou 1957) proposed a generalized projection function condition $\mathrm{a} = \varphi(b)$, and the projection conditions $\mathrm{a} = \mathrm{b}^K$ and $\mathrm{a} = K\mathrm{b}$ were used to find the new projections in the azimuthal projection, cylindrical projection and conic projection. By extending the traditional projection condition to a function condition, the common method that first predefines the shape of the parallels and meridians and then introduces the projection condition and other supplementary conditions to obtain the corresponding map projection is reformed. This theory is innovative. The paper 'Conformal or equal-spaced conical projection with total-equal area' (Hu and Zou 1958) presented a new method to determine the projection constant under the condition of maintaining the total-equal area of the specified mapping region. That is, assuming that the distortion of the total area is $P = \sum(\mathrm{p} - 1)\,\mathrm{dF}$, the basic conditions for the conical conformal projection with a total-equal area and the equal-spaced conical projection with a total-equal area are $\sum_{i=1}^{n} (\frac{\alpha^2 K^2}{r_i^2 U_i^{2\alpha}} - 1)\Delta F_i = 0$ and $\sum_{i=1}^{n} \left(\frac{\alpha(C-S_i)}{r_i} - 1\right)\Delta F_i = 0$, respectively, and then, their projection constants can be derived according to specific conditions. The paper 'On graphic interpretation methods in seeking azimuthal projections' (Hu 1962) presented a graphical analysis method for the inverse solution of map projection expressions by designing the radius curve or the changing curve of the arbitrary distortion (length, area or angle). In essence, the method involves predefining the distortion size of several positions of polar distance in a given mapping region to determine the formulas of projection radius ρ with the polynomial of polar distance Z. The paper 'Pseudoazimuthal projections and their applications for a whole map of China' (Liu and Li 1963) discusses how to let distortion isograms approach the geometrical contour of the mapping area to the greatest extent, and the significance of each parameter in the projection and the method of determining the parameters are analyzed and presented. In the paper 'The use of numerical methods for establishing optional conical projections' (Yang 1965), based on the general formula of conic projection, the author applies a numerical method to calculate the coordinates and distortion values of each point of a certain distortion distribution in a given area according to the unique properties of the projection. The basic theoretical problem of establishing an optional conical projection is solved. It is a typical application of the inverse solution of the map projection. The paper 'Cylindrical transverse conformal projection with two standard meridians' (Li 1983a) and the paper 'A modified transverse cylindrical conformal projection for 1:1,000,000 scale topographic maps' (Li 1983a) discuss two kinds of conformal projection schemes for zone division along meridians.

The variety of generalized formulas of map projection is a prominent feature of exploring new projections. The paper 'Double azimuthal projection' (Li 1963) provides a comprehensive generalization of nearly all commonly used azimuthal projections and reveals their internal relations. The paper 'Research on azimuthal projections' (Yang 1983a) reveals some new features of the gnomonic projection applied to navigation. The paper 'Research on the polyconic projection family' (Yang 1983b) explores the combination of approximate conformal polyconic projection, and the inherent relation and regularity of this kind of projection and its geometrical characteristics are revealed. The paper, 'A discussion of three modified projections-modified equidistant azimuthal, conic and cylindrical projections' (Yang 1983c), presents three modified equidistant projections whose distortion distribution differs from three regular equidistant projections, and they are quite applicable to the whole map of China.

In seeking world map projection methods, many papers have been published, including 'The design and analytical calculating method of polyconic projection in unequal graticules' (Zhong 1965), 'A polyconic projection with unequally spaced parallel lines and its applications' (Fang 1983), 'A polyconic projection with two spherical arcs as border meridians–research on the design method of polyconic projection with predefined distortion distribution' (Zhong 1983), and so on. Some of studies have been widely used in world map design. For example, the Polyconic Projection with Meridional Intervals on Decrease Away From Central Meridian by Tangent was used in the new version of a 1:14,000,000 map of the world published in 1976 by the Sino Maps Press.

In the paper, 'On conformal projection maintaining the length of any desired curve on the earth ellipsoid true to scale' (Cheng 1990), the existence of the conformal projection maintaining the length of any desired curve on the Earth ellipsoid as true to scale was proven, and the theory and method of establishing the projection was studied. By introducing the definition of the envelope of a single parameter surface family, it can be proven that the envelope of a single parameter surface family is a curve surface, conical or cylindrical surface formed by a tangent line of space curve; if a projection that maintains the length of any desired curve on the Earth ellipsoid as true to scale and the length proportion on curve L is minimal, then the curvature of the curve on the projection plane is the geodetic curvature of the curve; and other important theorems can be proven in this way. The conformal projection and oblique elliptical cylinder conformal projection maintaining the length of one constant bearing line on the Earth ellipsoid as true to scale were established for the first time. It is fundamental to develop conformal space projections. The paper, 'On conformal space projection' (Cheng 1991), proposes the concept of establishing a perfect conformal space projection that maintains a certain curve length true to scale based on a numerical method and discusses the Chebyshev projection with an arbitrary border. This is currently one of the most effective ways to solve the difficult problem of the Chebyshev projection.

For the purpose of compiling urban tourism maps, the paper, 'Cartographic projection system for variable scale maps' (Hu 1987), enlarges and highlights the key part of the city through the flexible use of the close-up shot from movies, while properly tailoring the suburbs. As a result, each part of the map has a different scale. It is beyond the scope of conventional projection and improves the function of this kind of map. To satisfy the special requirement of the mathematics foundation in map design and overcome the disadvantages of a certain projection, it is necessary to join together other projections and form a combination of projections. One combination method joins two kinds of projections closely together by taking certain parallels or meridians as borders. Another combination method depicts a certain region of the Earth's surface somewhere with projection A and then depicts the other region of the Earth's surface in another location with projection B. Finally, parallels and meridians are drawn in the joints between the two projection areas using a numerical method. In the paper, 'On combinatorial azimuthal projections' (Yang 1988), an azimuthal projection method with different segmented functions is proposed, the joint processing and boundary conditions are discussed, and a geometric interpretation of the combined azimuthal projection is given. To meet the requirements of using the change in scale to present the change in quantity features of the presented object when making a thematic map, a multi-focus projection with any number of projection centers is designed by using the continuous change in scale in radial directions from the projection center to the surrounding areas. In the paper, 'CAC-based general multifocal projection' (Wang et al. 1992), based on directly designing the variation law of scale, a kind of multifocus map projection method with flexibility, generality and controllability in the environment of computer-aided mapping (CAC) is proposed. Some practical forms of $F(R)$ for different requirements are studied, such as the power function form $FR = \alpha(1 + R)^{\beta}$ and segmented function form. It provides a convenient and effective way to meet the needs of users and to obtain a better effect of the variable scale map. In the paper, 'A kind of adjustable map projection with the "magnifying glass" effect' (Wang et al. 1993), considering the limitation of the variable scale map projection that lacks effective control over scale change, a kind of adjustable map projection with a 'magnifying glass' effect is proposed, and several shapes of the 'magnifying glass', such as round, non-circular and multi-focus are designed. This projection can effectively improve the expression of the map.

In addition, to meet the special requirements of the mathematical foundation for the expression of some special elements, oblate equal-area projection, retroazimuthal projection, double azimuthal projection, double equidistant projection and multiperspective azimuthal projection are designed and applied. Oblate equal-area projection is a kind of equal area projection in which distortion isograms are approximately elliptical or rectangular, or a modified equal area projection is used; for the retroazimuthal projection, the azimuth angle of any point on the projection plane to the projection center always stays in the right direction; the double azimuthal projection keeps the azimuth angle from two fixed points to any point in the right direction; the double equidistant projection maintains the correct distance between two fixed points to any point. These projections are important for navigation, direction finding, positioning, ranging and locating.

In addition to the latitude and longitude lines, there are some other special curves, such as the rhumb line (conformal course), great circle route, small circular route, isogonal line (curve of equal bearing) and hyperbolic line on the ellipsoid of the Earth. These curves are often used to determine the position of the moving target, which has important applications in navigation and aviation and is usually called the position line. Since the 1970s, the equations and characteristics of the position line, projection calculation, graphics depiction and its applications have been studied deeply. The paper, 'A discussion of the circle arc grid on the Mercator chart' (Hua 1983a), studies and discusses in detail the basic theory of constructing the circle arc grid on chart, the distortion of the circle arc grid on the map with the Mercator projection and depicting and calculating the circle arc grid on a chart. The paper, 'Computation method for positioning by long-range hyperbolic navigation system' (Hua 1990), proposed that the received time difference or phase difference after skywave correction should be converted into the geodetic distance difference, which can be used as a spherical distance difference to calculate a ship's approximate position in the form of geodetic latitude and longitude. As a correction, the difference must be added to the previous time difference or phase difference on the Earth's surface to be an initial value of the next iteration until the accuracy requirement is met. A software and long-range hyperbolic navigation map based on the research results can be directly applied to the positioning of ocean voyages and marine surveying. 'The theory of the position line and positioning navigation software system' thoroughly studied the theory, method and application of the position line.

To adapt to the development of spatial remote sensing technology and the processing of satellite remote sensing image data, the map projection has opened up a new research direction, known as space projection. This research direction aims to study the projection method for recording the geographic information acquired by space-borne sensors on the image plane. Space projection should consider various factors, such as the shape of the Earth, the satellite motion, orbit precession, and rotation of the Earth. Time *t* is a projection parameter, which is different from the traditional map projection. It is a four-dimensional space dynamic projection adapted to dynamic sensor imaging. Alden P. Colvocoresses (1974) designed a dynamic model called the Space Oblique Mercator Projection (SOM), a new map projection designed for use in Landsat mapping by the U.S. Geological Survey. In later work, REN Liucheng conducted a series of explorations on the theory of space projection, space azimuthal projection, space cylindrical projection and space conic projection and published 'Problems of the continuous cartography of satellite image by using the method of strip division of space projection' (Ren et al. 2005), 'The research on the space oblique azimuth projection' (Ren et al. 2006), and the book, *Theory of Space Projection and Its Applications in Remote Sensing Technology* (Ren 2003, Beijing, China: Science Press).

With the continuous advancement of the Chinese lunar exploration project, to provide a horizontal reference for the processing, expression and application of lunar space information and thematic information, preliminary research on the mathematical foundation of lunar mapping has been conducted in China. In the paper, 'On the Theory and Method of Lunar Projection' (Chen et al. 2006), aiming at China's moon program, by taking the surface of the ellipsoid as the mathematical surface of the moon, the properties and distortion patterns of the traverse tangent cylindrical conformal projection, the normal secant conical conformal projection and the circumcenter perspective projection of lunar mapping are discussed. The paper, 'Design and selection of projection for lunar mapping' (Lv et al. 2008) proposed the basic principle for the design of the lunar map projection system, and the basic framework of multilevel and multitype lunar map projections for all or certain regions of the moon, and the large and medium scales are primarily formed.

3.3.4 The Design of Map Projection Schemes for Regional Maps and Atlases is More Scientific

The selection and design of projection schemes for small- and medium-scale regional maps and atlases has always been the focus of map projection research. It is an important manifestation of the theoretical research results and applications of map projection. It reflects the current development level of mathematical cartography in China to a certain extent.

The design of projections for small-scale general maps is combined with the regional characteristics of China and the mapping production requirements. To display China properly on a map of the world and decrease the size of the distortion of the geometrical contour of the whole world, several kinds of world map projections were analyzed and designed. In the early 1950s, the compilation of the world political communications map used the 1950 Polyconic Projection designed by the former Soviet's Central Institute of Surveying and Mapping Science, with the east longitude 150° as the central meridian, the distortion of most of the land area is less than 60% and the angular deformation is not more than 60°; several major continents, such as Asia, Europe, North and South America, and Africa, maintain a good proportion in an area comparison, and the distortion shape is not overly large. According to the map's purpose and use, the 1:14,000,000 map of the world published in 1976 by the Sino Maps Press used the Polyconic Projection with Meridional Intervals on Decrease Away From the Central Meridian by Tangent. This projection completely represents the Pacific and Atlantic Oceans, but the graticule repeats at a rate of 60° of longitude difference; the central meridian used the east LONG 120°, so China is located in the middle of the sheet, the shape is correct, and the areal deformation is appropriate; a complete North and South American continent is located in the East, there is no significant visual distortion in the shape of the world's major continents, and the maximal areal deformation is no greater than 2–3 times; the whole projection distortion distributed uniformly, and the areal deformations at the same latitude are approximately equal.

The compilation of the Asian topographic map and political map used the azimuthal equal-area oblique projection. The projection center is located at north latitude 40° and east longitude 90°. The angular deformation throughout Asia does not exceed 10 degrees; in most parts of China, the angular deformation does not exceed 2°; the length deformation in most parts of China is not more than 1.5% and not more than 10% on the border of Asia. In addition, the compilation used the equidistance oblique azimuthal projection. The projection center is located at north latitude 40° and east longitude 90°, and for the Bonne projection, the standard parallel is at north latitude 40 degrees and the central meridian is at east longitude 80°. European maps were designed with oblique equal area azimuthal projections, conformal conic projections or equidistance conic projections, while North America, South America, Africa and other map designs mainly applied the oblique equal area azimuthal projection.

There are many projection schemes designed for the map of the whole Chinese territory, including the political map of China's most southern part of the territory. This map uses the oblique azimuthal conformal projection or oblique equidistant azimuthal projection; the projection center is located at north latitude 30° and east longitude 105°; or the oblique pseudoazimuthal projection is used, where the projection center is located at north latitude 35° and east longitude 105°. For the oblique azimuthal conformal projection, the length deformation of the center point is 2%, and the length deformation of the vast regions is within 2%. Some marginal areas could be confined to approximately 5%. The areal deformation is twice as large as the length deformation. For the oblique equidistant azimuthal projection, no deformation occurred at the center point. It maintains a correct distance from the projection center to an arbitrary point. The length deformation in vast areas is less than 2%, the local marginal area is approximately 4%, and the angular deformation in most areas is less than 2°. For oblique pseudoazimuthal projection, the length deformation equals the areal deformation, the shapes of the length distortion isograms are roughly similar to those for the borderline of China, the length deformation in vast areas is less than 1%, the local marginal area is approximately 3%, and the angular deformation in most areas is less than 1°. If the South China Sea Islands are displayed as illustrations, the conformal secant conic projection or equidistant secant conic projection is used for the map of the whole Chinese territory. In the situation of the conical conformal projection with the standard parallel between north latitude 27° and north latitude 47°, the length deformation in the vast region between north latitude 19° and north latitude 51° is less than ±2%, the maximum deformation of the North reaches 3.5%, and the areal deformation is twice as large as the length.

China's provincial (autonomous regions) map uses the normal secant conical conformal projection. The standard parallel can be selected individually for the compilation of a single map of a province (autonomous region) or several provinces (regions). For example, the conformal conic projection with a standard parallel between north latitude 41° and north latitude 51° is used for the maps of the Liaoning province, Heilongjiang province and Jilin province. In this situation, the length deformation in the vast region between north latitude 39° and north latitude 53° is less than ±0.4%, and the maximum deformation in the north reaches 0.5%. If compiling an atlas, a unified standard parallel is used for a large administrative region. The provinces alone no longer choose the projection. Therefore, the mathematical foundation of each administrative region's maps is uniform and comparable.

To facilitate the selection and design of the map projection and the establishment of the mathematical foundation of a map, as well as to meet the needs of compiling various types of maps at home and abroad, China's cartographers have successively calculated and published several collections of calculation tables for map projections. The reference book, *Tables for Calculating Map Projection,* by Fang (China Surveying and Mapping Press, Beijing, 1977) provides various constant values of the ellipsoid of the Earth and examples of commonly used map projections, including the values of the curvature radius M, N, R of the ellipsoid of the Earth, the value of the parallel radius r, the arc length of meridian S_m, the arc length of parallel S_p, the trapezoidal area of the ellipsoidal surface F and include computational examples of the normal conical projection, azimuthal projection, cylindrical projection, and Gauss-Krueger projection. Another reference book, *Map and Projections for Small-Scale Maps* (Cartographic Department of Wuhan Institute of Surveying and Mapping, 1978), offers tables of coordinate calculation results for approximately 24 kinds of map projections for the map of the whole Chinese territory and maps of every region in China. The process and method of establishing the mathematical foundation of the map and the schematic diagram of the graticule of meridians and parallels of various kinds of projections are presented for convenience in practical use. Nomographs of commonly used projections were made to improve the measurement accuracy. These factors promote the standardization of establishing a map mathematical foundation. The book, *Tables of Regional Map Projections,* aims to offer rapid projection coordinate results to meet the needs of compiling all types of internal or external regional maps and to provide reference data for making a reasonable plan for regional map projection. In the book, many good map projection schemes have been collected from the regional wall charts and atlases inside and outside military publishing units since the founding of the P. R. of China. The book contains many new research results of the regional map projection, including map projection schemes and projection coordinate results from the map of the whole Chinese territory, provincial (autonomous regions) maps, maps of several provinces and maps of the sea area of China, as well as maps of world countries, continent maps, and some special maps. This book can provide projection coordinate results for both special regions and arbitrary mapping areas extending along meridians or parallels.

The atlas is among the important works concentrated on reflecting the development level of cartographic science. The selection and design of a map projection is an important link in the system engineering of atlas design. In the past 60 years, the projection selection and design of the Chinese Atlas has become increasingly scientific and reasonable, and many new projection schemes have been used, laying the foundation for the scientific and practical applications of atlases. Several papers were published, such as 'The selection and design of projections for the great atlas of the People's Republic of China', 'On the map projection of provincial atlas of the People's Republic of China', 'On the map projection of provincial (autonomous regions) atlas' and 'Several problems on the mathematical foundation of general atlas'. These papers provide important guidance for the design and establishment of the mathematical foundation for the national atlas and provincial (autonomous regions) atlas. The paper, 'The design of the mathematical foundation of the geographical map of China's provinces (autonomous regions)' (Gong 1981), suggests that the conic projection with two standard parallels is suitable for China's provinces (except the South China Sea Islands) according to their geographical features and range of uses. The approximate calculation formula for the two standard parallels is $\phi_1 = \phi_S + 0.16(\phi_N - \phi_S)$, $\phi_2 = \phi_N - 0.12(\phi_N - \phi_S)$. This formula provides a projection scheme for provincial (autonomous regions) geographic maps that are designed especially for wall charts, and their scales are 1:250,000, 1:500,000, 1:750,000 and 1:1,000,000. According to the projection scheme, the length deformation in each provincial (autonomous) region is less than $\pm 0.5\%$. This scheme can fully satisfy the requirements of the map for use in high precision measurements, as well as in scientific and technical publications.

The *Officer Atlas* is a medium-sized comprehensive military reference atlas published in the early 1990s. By audaciously innovating the selection and design of projections, some projection schemes with distinguishing features are adopted for world maps and thematic maps. Equivalent difference latitude parallel polyconic projection, normal cylindrical conformal projection, equal-area pseudocylindrical interrupted projection and composite equidistant azimuthal projection were used for the world map series. Oblique equal-area azimuthal projection, Urmayev pseudo-cylindrical projection, and pseudoazimuthal projection with oval-shaped distortion isograms were used for the continental and oceanic map series, respectively. The pseudoazimuthal projection with three petal-shaped distortion isograms, the oblique conformal projection, and the normal conformal conical projection were used for the map of the whole Chinese territory and provincial map series, respectively. There are many types of projections and new design ideas to give the map a clear and scientific expression and to improve the scientific value of the map works.

In the past 30 years, the access and application of computer technology in cartography have created favorable conditions for the projection selection of the atlas and the establishment of a mathematical foundation. The design and calculation of complex projections, data conversion between a newly compiled map and the original map with different map projections, and so on, are no longer considered or are unavoidable factors in the compilation of the atlas. The automatic establishment of the mathematical foundation of the map is basically realized. The paper, 'The methods of engraving and drawing the map geographical meshes on the electronic computers' (Li 1987a) introduces the general method of automatically drawing the map geographical meshes in computer-aided mapping, discusses the processing methods of the unbounded projection function, multi-point correspondence, boundary line, cutting and so on, as well as designing the common procedure of geographical mesh cutting, rectangular cutting and circular cutting, and is successfully applied to production. A number of map projection tool software programs have been developed by relevant specialties in colleges or universities at home. In terms of the systematic and practical principles, the Map Projection and Transformation Software Package (LV Xiaohua et al., Institute of Surveying and Mapping, Zhengzhou Information Engineering University) can be used to implement the calculation of commonly used projections and has a set of OCX controls for further development by users. This software package provides an important tool for the calculation of complex projections, the automatic establishment of the mathematical foundation of the map, the conversion of map data, and the unification of the spatial information datum. Therefore, the design and application of the new and scientific map projection scheme can be realized smoothly. The project won the Science and Technology Progress Award in 2005.

In compliance with international practice, Chinese charts usually adopt the Mercator projection, while aviation maps use conformal conic projections with double standard latitudes. For the research progress on the projection of charts and aviation maps and the establishment of their mathematical foundation, see other chapters.

3.3.5 The Research Achievements of Map Projection Transformation Are Remarkable

Map projection transformation is a new research field that has developed along with computer technology in the last 30 years, and it is an important branch of mathematical cartography.

In traditional analog mapping, demands for cartographic material are relatively simple and only include the paper map of the corresponding scale, control survey results and other documents. To transfer the basic cartographic material to the geographical meshes of newly compiled maps, we usually apply photographic collage, grid transfer or rectifier transfer methods to realize the transformation of the map projection. However, this transformation is difficult to achieve when the difference between the two kinds of projections is relatively large, and the accuracy cannot be guaranteed.

When the map production changes into digital mapping, along with the development of modern measurement technology, the method of cartographic material support, demand and usage changes greatly. That is, cartographic information is diverse and complex, analog data are basically replaced by digital data, and map projection transformation is becoming increasingly important and urgent. In addition, the development and application of GIS, the construction of geographic spatial databases, the processing of satellite remote sensing images and so on also presented new requirements for map projection transformation. Map projection transformation is an unavoidable theoretical and practical problem in the transformation of traditional, manual mapping to digital mapping. Transformation is a basic problem in spatial data processing that is related to geo-information science, surveying and mapping science and technology, as well as other fields.

Map projection transformation explores the theory and method of transforming one kind of map projection point's coordinates to another kind of map projection point's coordinates. In the broad sense, transformation can be understood as the research of the theory, method and application of spatial data processing and the transformation between spatial points and planar points. In a narrow sense, transformation can be seen as a method to establish the one-to-one correspondence relationship between points on two plane fields.

The most remarkable achievements in the research of map projection transformation at home and abroad belong to the Chinese. As representatives of the older generation of cartographers, prof. YANG Qihe and prof. WU Zhongxing systematically carried out research on the theory and method of map projection transformation.

Many of the research results that have advanced the field of cartography have been achieved in the aspects of the general theory of map projection, the mathematical method of map projection transformation, the analytic transformation of map projection, numerical transformation of map projection, conformal projection transformation, and so on. In the paper, 'How to transform coordinates of points from one kind of projection to another' (Wu 1979), for the first time, the basic equation of the coordinate transformation from one kind of map projection point to another was presented as $X = F_1(x, y), Y = F_2(x, y)$, and the basic principle of the analytic method and numerical method of map projection transformation is discussed. In the paper 'Research on the numerical transformation between conformal projections' (Yang 1982b), based on conformal projection theory and complex function theory by introducing the equivalent latitude q, the author put forward an orthomorphic polynomial that is suitable for the conformal equiangular projection, i.e., $X = X_0 + \sum_{k=1}^{n} (a_k P_k - b_k Q_k)$ and $Y = Y_0 + \sum_{k=1}^{n} (b_k P_k + a_k Q_k)$; the formula of the expansion points chosen on the central meridian can be derived as a conclusion if the formula of the coordinates relation of two conformal projections on the central meridian is given; then, the formula of the transformation relation between two conformal projections will be uniquely determined; the basic methods of direct transformation and inverse transformation for the orthomorphic polynomial according to the direct solution and the least squares method were investigated; some factors affecting the accuracy of coordinate transformation, such as the order of orthomorphic polynomial, the size and shape of transformation region, and the arrangement of common points, were analyzed. In the paper 'Research on the transformation of map projections in computer-aided cartography' (Wu et al. 1983), the inverse method, the direct method, the synthesis method, the numerical method and the numerical-analytical method for map projection transformation were discussed more systematically; the analytical transformation between the Gauss-Krueger projection, Mercator projection and conformal conic projection were deeply discussed, and the practical application of the numerical transformation method was also discussed. The paper, 'Research on the third kind of coordinate transformation of map projections' (Yang 1984a), proposed a third kind of coordinate transformation of map projection, that is, $\{B_i, y_i\} \Leftrightarrow \{l_i, x_i\}$ and $\{B_i, x_i\} \Leftrightarrow \{l_i, y_i\}$, which is essentially the problem of coordinate transformation on border lines, and the analytical formula of the third kind of coordinate transformation of the Gauss-Krueger projection and conic conformal projection were derived. The third kind of coordinate transformation further enriched the research content of map projection transformation. In the paper, 'Complex function and conformal projection' (Cheng 1985), a series of analytical function expressions for conformal projection were derived according to the necessary and sufficient conditions of the conformal projection-Cauchy Riemann equation. For example, the expression for the Mercator projection is $x + iy = c(q + i\lambda)$, for the normal conformal azimuthal projection, the expression is $\omega = -e^{-(q+i\lambda)}$, and for the Gauss-Krueger projection, the expression is $\sin(\frac{x-\Delta x}{R} + i\frac{y}{R}) = th(\alpha q - \ln k + i\alpha\lambda)$; through these expressions, the coordinate transformation among various projections has the advantage of clear and simple calculation. The paper, 'Research on the transformation of the finite element between conformal projections' (Li 1985), derived the differential equation of the conformal projection transformation and proved that the transformation from one conformal projection to another conformal projection can be reduced to solve the Dirichlet problem of the Laplasse equation; then, as a basic idea, the finite element method for solving the Dirichlet problem was introduced to solve the problem of the extreme value of the equivalent function instead of solving the partial differential equation, and the continuous solution to the problem is discretized into the calculation of the approximate values on the grid node; finally, the transformation accuracy of the method is evaluated using several examples.

In addition, there are other theses, including 'On the numerical method for transforming the zones of the Gauss projection' (Yang 1982a), 'The semi-numerical method for the map projection transformation' (Hu 1982), 'The numerical method of the transformation of the conformal projection' (Liu 1985), 'A comprehensive appraisal of the numerical transformation method for map projection' (Lv et al. 2002), and two books, *Map Projection Transformation Principles and Methods* (Yang, The People's Liberation Army press, 1990) and *Map Projection Transformation–Principles and Applications*' (Yang et al., Taylor & Francis Group, 2000). The two books systematically discuss the theoretical methods and applications of map projection transformation. These books are abundant in content, have reliable theories and comprehensive methods and will be widely applicable since they combine theory with practice. The publication of these two books marks that China has reached the internationally advanced level in the research of map projection transformation.

3.4 Problems and Prospects

After entering the twenty-first century, the information tide with information processing technology and network communication technology as the core is raging like a storm. Human society has entered the information age. The high integration of a series of modern information technologies, such as satellite remote sensing, global positioning systems, geographic information systems, and digital transmission networks, and the rise of geospatial information science formed on the basis of information science and earth system science has certainly had a profound impact on the development of map science and technology in the new century.

With the development of the discipline of cartography and geographic information engineering, the traditional theory and method of static, two-dimensional and vector map projection has made it difficult to describe its own development, and it is difficult to adapt to the demand for geospatial information in the new century.

Map projection studies include the depiction of spatial information on a certain mapping surface (a flat or curved surface). It is the theory and method of spatial data processing. The main task is to establish a unified grid (grid or grid flat surface) for multisource spatial data. In essence, map projection is the location model and basic framework of spatial information. Therefore, we must continue to study the theory and application system of map projection in the information age, including the research object, main task, role and position of map projection, and the new map projection method to further expand the application range and improve the application level.

Maps not only result from the human understanding of the objective environment but also from the tools people use to further understand the environment and things in the environment. The rapid development of science and technology and the sustainable development of human society will face many global issues, such as crustal deformation monitoring, earth dynamics research, global plate movement, earthquake monitoring, and global disaster climate change. Therefore, we must create innovative map projections for large regions, such as the oceans, the continents and the world, to better meet the precise and vivid expression of certain thematic elements.

In the internet age, map products have a variety of forms, such as paper maps, electronic maps, and network maps. The map will become more widespread and popular. At present, there are more than 400 kinds of map projections. The large number of projection types provides great space for the choice of map projections. However, it also brings many difficulties to the application of map projections. For this reason, we should search for simple, practical and novel map projections while providing standard map projections to meet the needs of different purposes and different user groups.

With the continuous development of GIS applications, the integration and analysis of multisource spatial information will become a main application mode. This is the inevitable result of GIS for practical applications. Although numerous GIS software programs at home and abroad have partial functions of map projection, the involved parameters and application conditions of the projection are different. Obviously, the nonstandard usage of functions and all kinds of mistakes directly affect the quality of data and visual expression. For this reason, we must further standardize the map projection in GIS and define the function and application condition of the map projection according to the development and application of GIS.

Digital construction, such as digital earth, digital city, digital river, etc., and the application of large-scale GIS involves space for local small area expansion to the national or even global range. The volume of GIS data is growing; data sources are becoming more diverse and complex. It is difficult to realize the seamless walk-throughs of spatial data with the continuous visualization of large areas; there is no uniform, tight, continuous spatial information positioning foundation, and the efficiency and precision of a spatial measurement is difficult to improve (Hu et al. 2001). It is necessary to further carry out research on spatial mathematical foundations for large-scale GIS and digital earth.

The spatial datum of the geographic information includes many factors, such as the reference ellipsoid, the geodetic coordinate system, the land elevation system, the depth system of the ocean, the map projection and the projection zone. The spatial datum of geographical information in every country and region of the world is numerous, complex, and of different types. Even in the same country, the spatial reference is not exactly the same in different historical periods or in different regions of the same period. Research on the theory and method of the fusion of geographic information spatial datum is an important way to realize the integration and transformation of different spatial datum geographic information for global, national, regional or city scales and realize the scientific organization and resource sharing of geographic information. This is basic work for the establishment of a global geospatial information assurance system and the most effective way to make good use of information resources.

The development of spatial remote sensing technology provides up-to-date images for map revisions. Dynamic space projection is expected to be a mathematical basis for remote sensing image processing. We must continue to deepen the research on space projection theories, methods and applications, further explore the principle of space projection aimed at the imaging mechanism of different sensors, extend the application of space projection in remote sensing image geometric precision rectification and precise positioning of target objects, and provide technical support for remote sensing image analysis, map content updating and high precision mapping.

To realize the multilevel scientific expression of various kinds of information acquired by lunar exploration satellites from the surface of the moon and provide a mathematical basis for the processing and mapping of the lunar space information and thematic information, we must actively carry out studies on the selection and design of lunar projection systems. It is of great theoretical significance and practical value as well as a realistic urgent need to choose landing sites, route design for the lunar rover and construction of the experimental base and launch base of the deep space detector of the moon.

Undeniably, as long as the analog product of the geographical space environment, the map, exists, map projection as the spatial mathematical foundation of map is indispensable; as the theoretical basis of cartography, map projection remains the essential tool for the processing, transmission and interpretation of spatial information in the past, present and future.

3.5 Representative Publications

(1) ***Science of Map Projections*** (FANG Bingyan. Map Press, Beijing, 1979)

The whole book is divided into 3 parts, with 16 chapters altogether. The first part includes the chapters one through four, which mainly expound on the basic theory of map projection and the basic knowledge for the study of this course. The second part consists of chapters five through thirteen, which mainly describe the principle and method of establishing various map projections. Other chapters belong to the third part and mainly introduce the applications of map projection in map compilation. Among them, the fourteenth chapter introduces the selection and design of map projection, the fifteenth chapter introduces the discrimination of map projection, and the sixteenth chapter introduces the coordinate transformation of map projection. In the book, in addition to the commonly used map projection in China, the projection scheme of foreign countries and the projection of the past in China are also introduced.

(2) ***Map Projections*** (WU Zhongxing, Surveying and Mapping Press, Beijing, 1980)

The book discusses in detail the basic theory of map projection, the map projections used in topographic maps, regional maps and other maps and their practical applications. The book has 6 chapters. The first chapter covers the basic theory of map projection. The research objects of map projection, map projection distortion and its calculation, map projection conditions, map projection classification and so on are discussed in detail. The second chapter covers the map projection of the topographic map, discusses the projection requirements of the topographic map, the coordinate formula, length ratio formula and meridian convergence angle formula of equal-angle transversal elliptic cylindrical projection (Gauss projection), and the application of the Gauss projection in the topographic map. The third chapter covers the commonly used map projection for regional maps, discusses the general formula of azimuthal projection, cylindrical projection, conic projection and a variety of specific projections, and gives examples of applications. The fourth chapter covers the map projection for other kinds of maps, focusing on polyconic projection, pseudocylindrical projection, pseudoconical projection and composite projection. The fifth chapter describes the design of the mathematical foundation for regional maps, the general principles of map projection selection, and the method and procedure of designing the mathematical foundation for regional maps. The sixth chapter covers map measurement and introduces the measurement method of point location, line length, curve length, area and azimuth on the map.

(3) ***Mathematical Foundation of the Chart*** (HUA Tang, Marine Assurance Department of the PLA Navy Command, TianJin, 1985)

The book has four parts and 17 chapters.

The first part, Mercator projection and its derivative projections, includes the Mercator projection, several issues in the calculation of Mercator projection, planimetric map, Miller projection, the measurement of Mercator chart, etc., and falls into 5 chapters. The second part includes a total of 6 chapters, and the main contents include several kinds of position lines on the Mercator projection plane, such as the conformal projection of the ellipsoid's surface on the spherical surface–the application of the Gauss second law; the representation of the rhumb line, isogonal line and great circle on the Mercator projection plane and their relationships; the position line (fringe loops) in the Mercator chart, the circular arc grids in the Mercator chart, the medium- and short-range hyperbolic navigation map, and the long-distance hyperbolic navigation map. The third part is the transformation between the Gauss-Krueger projection and Mercator projection, which includes the Gauss-Krueger projection, the relations of the Gauss-Krueger projection and Mercator projection, and the transformation between the Gauss-Krueger projection and Mercator projection and falls into 3 chapters. The fourth part is gnomonic projection, which is divided into 3 chapters, including the properties and formula of gnomonic projection, some problems in the calculation of gnomonic projection and the measurement on the map of gnomonic projection.

(4) ***Principles and BASIC Program for Conformal Projection Transformation*** (YANG Qihe, Surveying and Mapping Press, Beijing, 1987)

Conformal projection is a kind of map projection commonly used in surveying and mapping. In particular, it is widely used as a mathematical basis for topographic maps, charts, and aerial maps. Based on the research of the theory and method of map projection transformation, the book systematically introduces the principle of conformal projection transformation and its BASIC programs to meet the needs of PC-1500 pocket computer applications. The book is composed of 5 chapters. The first chapter systematically discusses the theory and method of the conformal projection transform, which is the theoretical basis of the conformal projection transform. The second chapter provides 4 kinds of commonly used conformal projection formulas and their subroutines and examples of inverse conversion and discusses the direct conversion and neighboring zone coordinate conversion of two kinds of conformal projections. The third chapter discusses the constant coefficients conversion method and program of conformal projection, provides the constant coefficient table for the conversion between the commonly used conformal projections and the constant coefficient table of coordinate transformation with the 3° zone or 6° zone. The fourth chapter provides the BASIC programs and some examples of the numerical transformation between any two conformal projections, the direct and inverse numerical transformations between conformal projections, and the numerical transformations between the local coordinate system of the Gauss-Krüger projection and the national coordinate system.

(5) ***Principles of Mathematical Cartography*** (WU Zhongxing, YANG Qihe, Surveying and Mapping Press, Beijing, 1989)

The book aims to explore the principles and methods of map projection. First, the theory of surface and various curve coordinates are introduced, and then, the basic theory of map projection is discussed. Thus, the different ways and methods to explore map projection are described. At the same time, the Gauss-Krueger projection and its derivative projections in particular are discussed. Finally, some issues, including the automatic drawing of the graticules of parallels and meridians of map projection and the automatic establishment of the mathematical foundation of maps, are introduced. The book has 14 chapters, which mainly include the surface coordinate system and curvilinear coordinates, the general theory of map projection, the projection of the ellipsoid's surface on the sphere, methods for exploring map projection according to the deformation property of a projection, methods for exploring map projection according to the shape of the parallel and meridian after projected, methods for exploring map projection according to the distortion distribution of projection, methods for exploring map projection according to other conditions, the Gauss-Krueger projection and its derivative projections, projection of some special curves on the Earth's surface (or spherical), lunar map projections, space map projection, theory and method of map projection transformation, automatic setting of the mathematic basis, measurement of the map mathematical element, and so on.

(6) ***Map Projection Transformation Principle and Method*** (YANG Qihe, The People's Liberation Army press, Beijing, 1990)

The book consists of a total of 17 chapters and systematically discusses the basic theory and method of map projection transformation. Chapters one through three cover the basis and preparatory knowledge of the book, with brief introductions to the general theory of map projection, commonly used map projection equations, mathematical methods of map projection transformation, and so on. Chapters four and five discuss the three kinds of commonly used projections, three types of pseudo projections, polyconic projections and the analytical conversion of double projections. Chapters six and seven systematically discuss the analytical transformations of the conformal projection, the oblique cylindrical conformal projection and the oblique conic conformal projection analytic transformation. Chapter eight discusses in detail the theory and method of the numerical transformation of map projection. Chapter nine systematically discusses the theory and method of the numerical transformation of the conformal projection. Chapter ten introduces the third type of coordinate transformation of the map projection. In the eleventh chapter, the theory and method of the neighbor strip's coordinate conversion of the Gauss-Krueger projection is discussed in detail. In the twelfth chapter, the theory and method of the mathematical basis conversion of topographic maps are discussed in detail. Chapter thirteen is about the coordinate transformation of the position line and the target point on the projection plane; it is an example of combining the position line theory and projection transformation theory. Chapters fourteen and fifteen introduce the coordinate transformation of variable scale map projection and single space photograph projection and coordinate transformation of space dynamic projection, respectively; these chapters are part of the important content in the development of modern mathematical cartography. Chapters sixteen and seventeen systematically discuss the computer-aided map projection transformation and the principle and method of establishing the mathematical foundation of maps based on computers; these chapters are part of the applications of map projection transformation.

(7) ***Map Projection*** (LI Guozao, YANG Qihe, HU Dingquan, The People's Liberation Army press, Beijing, 1993)

The whole book is made up of 17 chapters that introduce the basic formulas of the Earth's ellipsoid, spherical coordinates and several curve equations on the sphere, basic theory of map projection, depicting the Earth's ellipsoid on the sphere, azimuthal projection, cylindrical projection, conic projection, pseudoazimuthal projection, pseudocylindrical projection, Hammer-Aitoff projection, pseudoconic projection, polyconic projection, Gauss-Krueger projection and its derivative projection, the application of Gauss-Krueger projection in topographic maps, the selection of map projections and the design of mathematical foundations for regional maps, map projection transformation, computer-aided setting of map mathematics foundations, and map projections for special purposes.

(8) ***Map Projection Transformation—Principles and Applications*** (YANG Qihe, John P. Snyder, Waldo R. Tobler. Taylor & Francis Group, 2000).

The book, which is based on the original book *Map Projection Transformation Principle and Method* by Prof. YANG Qihe (1990) was rewritten and revised by the first author, and John P. Snyder, a cartographer of the U.S. Geological Survey, and Waldo R. Tobler, a professor of University of California. As an international authoritative work on the research direction of this subject in the twentieth century, the book was published by Taylor & Francis Group in 2000.

There are 13 chapters in this book. Chapter 1 introduces fifty years of advancement in map projection studies in China. Chapter 2 outlines various map projection equations. Chapter 3 addresses the general theory of map projections, such as the Tissot theory. Chapter 4 introduces the analytical transformation for azimuthal, cylindrical and conic projections, etc. Chapter 5 discusses the analytical transformation for the conformal projection. Chapter 6 discusses numerical transformation. Chapter 7 introduces the numerical transformation for the conformal projection. Chapter 8 discusses the so-called 'third type of coordinate transformations', including linear interpolation. Chapter 9 concentrates on zonal transformations for the transverse Mercator (Gauss-Krüger) projection. Chapter 10 presents several new map projections, and Yang's new projections are the subject. Variable-scale and composite projections are the subject of Chap. 11. Chapter 12 is about position lines, namely, the small circle and the great circle of the sphere on the projection plane, similarly of ellipses and hyperbolas on the sphere, spherical resection and intersection. The book closes with the up-to-date topic of spatial information positioning systems in Chap. 13. A related topic is treated in Appendix 6.

There are also six appendixes in this book that give the inverse transformation of algebraic series, constant coefficient tables of zone transformation for the Gauss-Krüger projection, examples of numerical transformation, transformation for the mathematical elements of topographic maps, the position line and position navigation software system, and the bibliography of Chinese literature on projections and other references.

(9) ***Theory of Space Projection and Its Applications in Remote Sensing Technology*** (REN Liucheng, China: Science Press, Beijing, 2003)

The book discusses and sets up a relatively complete system theory for the imperfect spatial projection theory, explores the theory and method of space projection suitable for satellite remote sensing images, and introduces some useful studies on the application of spatial projection in remote sensing image processing and remote sensing mapping. As a result, the research foundation of spatial projection theory and application is initially established.

In theory, the author focuses on the basic theory of space projection, spatial azimuth projection, spatial cylindrical projection and spatial oblique conic projection; in application, the focus is on the geometric precision correction for remote sensing images with missing control points by using the spatial projection method, accurate positioning of target objects, and the application of spatial projection in mapping and other issues. The whole book is divided into 8 chapters. Chapter 1 is the introduction; Chap. 2 is the basic theory of spatial projection; Chap. 3 is the spatial azimuth projection; Chap. 4 is the spatial cylindrical projection; Chap. 5 is the space oblique Mercator projection; Chap. 6 is the space oblique conic projection; Chap. 7 is the application of spatial projection in remote sensing image processing; and Chap. 8 is the application of spatial projection in remote sensing image mapping.

(10) ***Atlas of Map Projections*** (HU Yuju, GONG Jianwen, Surveying and Mapping Press, Beijing, 2006).

The atlas is a collection of the graphics and data on the map projection geographic mesh at home and abroad. The third edition was revised based on the 1985 and 1991 editions.

The atlas introduces in detail a variety of map projections and the deformation of projections, graphics of various commonly used projections, the projection of some special shapes, and so on. Dominated by figures, combined with a brief text explanation, this atlas enables readers to obtain perceptual understanding and general knowledge of map projection. The content mainly includes general knowledge of map projection, conic projection, cylindrical projection, azimuthal projection, pseudoconic projection, pseudocylindrical projection, pseudoazimuthal projection, polyconic projection, other projections, projection transformation, and ancient Chinese map projection. The appendix contains commonly used projection formulas and map projections in China.

The atlas is rich in content; the data were entirely collected; the atlas is illustrated with pictures and includes text; and the character is concise and straightforward. Compared with the previous two editions, this atlas adds some new map projections that were designed for thematic maps abroad and at home, such as the azimuthal projection with the 'magnifying glass' effect, star-shaped projection and Captain Vere grid projection. At the same time, some ancient Chinese maps are expressed in a grid, such as the Map of Tracks of Yu (Stone Maps of the Chinese Water Systems in the Song Dynasty), the Enlarged Map of China (Ming Dynasty), the Great Universal Geographic Map (Ming Dynasty) and the Map of China in the Qianlong Reign (Qing Dynasty).

Acknowledgements Modification proposals were proposed by Prof. LIU Peilan of Resources and Environmental Science Institute, Wuhan University.

References

Chen, Q., Zheng, Y., & Su, M. (2006). On the theory and method of lunar projection. *Bulletin of Surveying and Mapping, (4)*, 26–30.

Cheng, Y. (1985). Complex function and conformal projection. *Acta Geodaetica et Cartographica Sinica, 14*(1), 51–60.

Cheng, Y. (1990). On conformal projection maintaining the length of any desired curve on the earth ellipsoid true to scale. *Acta Geodaetica et Cartographica Sinica, 19*(2), 110–119.

Cheng, Y. (1991). On conformal space projection. *Acta Geodaetica et Cartographica Sinica, 20*(1), 36–45.

Dang, S. (1960). A note on the map projection function. *Acta Geodaetica et Cartographica Sinica, 4*(1), 44–48.

Fang, B. (1983). The unequal difference weft polyconic projection and its applications. In *Proceedings of Map Projection* (pp. 255–275). Surveying and Mapping Press.

Gong, J. (1981). The design of the mathematics foundation of the province (area) of our country. *Bulletin of Surveying and Mapping,* (6), 28–33.
Hu, P. (1982). The semi-numerical method for the map projection transformation. *Acta Geodaetica et Cartographica Sinica, 11*(3), 191–202.
Hu, P., Wu, Y., & Yang, C. (2001). Research on system space mathematical base of large-scale GIS and the digital earth. *Geomatics and Information Science of Wuhan University, 26*(4), 296–302.
Hu, Y. (1962). On graphic interpretation methods in seeking azimuthal projections. *Acta Geodaetica et Cartographica Sinica, 5*(2), 141–153.
Hu, Y. (1987). Cartographic projection system for variable scale maps. *Journal of Wuhan Technical University of Surveying and Mapping, 12*(2), 47–54.
Hu, Y., & Zou, Y. (1958). Conformal or equal-spaced conical projection with total-equal area. *Acta Geodaetica et Cartographica Sinica, 2*(3), 195–202.
Hua, T. (1983a). On the research Mercator chart of circular grid. In *Proceedings of Map Projection* (pp. 346–362). Surveying and Mapping Press.
Hua, T. (1983b). Discussion and application of the ellipsoidal surface in the earth plane. *Acta Geodaetica et Cartographica Sinica, 12*(2), 134–151.
Hua, T. (1990). Computation method for positioning by long-range hyperbolic navigation system. *Acta Geodaetica et Cartographica Sinica, 19*(3), 179–186.
Li, C. (1979). On the classification of map projections. *Acta Geographica Sinica, 34*(2), 139–155.
Li, G. (1963). Double azimuth projection. *Acta Geodaetica et Cartographica Sinica, 6*(4), 279–302.
Li, G. (1983a). Cylindrical transverse conformal projection with two standard meridians. In *Proceedings of Map Projection* (pp. 137–150). Surveying and Mapping Press.
Li, G. (1983b). An improved normal transverse cylindrical conformal map projection and its application in 1:1 000k map making. In *Proceedings of Map Projection* (pp. 151–157). Surveying and Mapping Press.
Li, G. (1987a). The methods of engraving and drawing the map geographical meshes on the electronic computers. *Journal of the Institute of Surveying and Mapping,* (1), 33–46.
Li, G. (1987b). On $m= n^k$ orthogonal projection. *Acta Geodaetica et Cartographica Sinica, 16*(2), 149–157.
Li, J. (1985). A research on transformation of finite element between conformal projections. *Acta Geodaetica et Cartographica Sinica, 14*(3), 214–221.
Liu, J. (1965). On the deformation transformation of circular cylinder, bearing and conical projection. *Acta Geodaetica et Cartographica Sinica, 8*(3), 210–217.
Liu, J., & Li, G. (1963). Pseudo azimuthal and its applications to the map of China. *Acta Geodaetica et Cartographica Sinica, 6*(2), 104–119.
Liu, H. (1985). The numerical method of transformation of conformal projection. *Acta Geodaetica et Cartographica Sinica, 14*(1), 61–72.
Lv, X., Deng, S., & Niu, X. (2008). Design and selection of projection on lunar mapping. *Geomatics & Spatial Information Technology, 31*(6), 155–157.
Lv, X., & Liu, H. (2002). A comprehensive appraisal of numerical transformation method for map projection. *Journal of the Institute of Surveying and Mapping, 19*(2), 150–153.
Ren, L., Lv, X., & Zhang, X. (2005). Problems of continuum cartography of satellite image by using the method of strip division of space projection. *Journal of Institute of Surveying and Mapping, 22*(2), 139–141.
Ren, L., Lv, S., & Lv, X. (2006). The research on the space oblique azimuth projection. *Acta Geodaetica et Cartographica Sinica, 35*(1), 35–39.
Wang, Q., & Hu, Y. (1992). A CAC-based general multi-focal projection. *Journal of Wuhan Technical University of Surveying and Mapping, 17*(4), 18–25.
Wang, Q., & Hu, Y. (1993). A kind of adjustable map projection with" magnifying glass" effect. *Acta Geodaetica et Cartographica Sinica, 22*(4), 270–278.
Wu, Z. (1963). The status quo and development trends of mathematical cartography. *Acta Geodaetica et Cartographica Sinica, 6*(1), 39–43.

Wu, Z. (1965). A preliminary analysis of the marxist materialist dialectics of contradiction in Map Projection. *Acta Geodaetica et Cartographica Sinica, 8*(2), 115–123.
Wu, Z. (1979). How to transform coordinates of points from one kind of map projection to another. *Acta Geographica Sinica, 34*(1), 55–68.
Wu, Z., & Yang, Q. (1983). A research on the transformation of map projections in computer-aided cartography. In *Proceedings of Map Projection* (pp. 287–322). Surveying and Mapping Press.
Wu, Z., & Yang, Q. (1989). *The principles of mathematical cartography*. Surveying and Mapping Press.
Yang, Q. (1965). Using the numerical method to establish the conical projection of arbitrary character. *Acta Geodaetica et Cartographica Sinica, 8*(4), 295–318.
Yang, Q. (1982a). On the numerical method for transforming the zones of Gauss projection. *Acta Geodaetica et Cartographica Sinica, 11*(1), 19–31.
Yang, Q. (1982b). A research on numerical transformation between conformal projections. *Acta Geodaetica et Cartographica Sinica, 11*(4), 269–282.
Yang, Q. (1983a). Gauss-Kruger study of projection. In *Proceedings of Map Projection* (pp. 158–172). Surveying and Mapping Press.
Yang, Q. (1983b). The study of azimuthal map projections. In *Proceedings of Map Projection* (pp. 75–95). Surveying and Mapping Press.
Yang, Q. (1983c). The study of polyconic map projection family. In *Proceedings of Map Projection* (pp. 215–229). Surveying and Mapping Press.
Yang, Q. (1983d). Discussion on the three kinds of improved projections: Equidistant azimuthal projection, equidistant polyconic projection and equidistant cylindrical projection. In *Proceedings of Map Projection* (pp. 186–199). Surveying and Mapping Press.
Yang, Q. (1984a). A research on the third kind of coordinate transformation of map projections. *Journal of the Institute of Surveying and Mapping,* (1), 109–121.
Yang, Q. (1984b). On the general formulas for the projection from ellipsoidal surface onto spherical surfaceand the characteristics of extremity. *Acta Geodaetica et Cartographica Sinica, 13*(3), 225–236.
Yang, Q. (1988). About combined position projection. *Bulletin of Surveying and Mapping, 2,* 32–34.
Zhong, Y. (1965). Design and analytical calculation method for multi cone projection of non-equal parts. *Acta Geodaetica et Cartographica Sinica, 8*(3), 218–236.
Zhong, Y. (1983). A polyconic map projection keeping the connection of the great circle meridian: Discussion on the designing method of polyconic projection with predetermined deformation distribution. In *Proceedings of Map Projection* (pp. 230–254). Surveying and Mapping Press.
Zhou, C. (1957). According to the problem of the new projection of the deformed ellipse. *Acta Geodaetica et Cartographica Sinica, 1*(1), 1–14.

Chapter 4
Map Design

Guangxia Wang

4.1 Introduction

Map design refers to creating a comprehensive plan for the technical specification of the map, the overall structure, the mathematical foundation, the map content and means of expression, the map symbol and color, the production craft and so on. Map design must be based on the visual perception theory and map design principles, aiming at the purposes of the map and the requirements of the user. In general, some principle provisions would be provided in the forms of a map design document, map compilation specifications, map legends or symbols, and so on.

Map design is essentially the process of creating a map. That is, the representation of the colorful and real geographic world on a map. Map design has always been the focus of scientific research on surveying and mapping and is the core problem in cartography. The transference process from the ground to the map consists of two stages: the first is the perception of the objective world, and the second is the formal design for the expression of the perception.

Map design mainly uses modeling methods to understand the drafted objects of the objective world. That is, generalizations are made regarding the drawing objects according to the needs of different users, and the scientific contents of the map are established. The use and regional characteristics of the map and map scale limits must be made clear based on a reduced map model. A person's ability to understand, analyze and generalize are important to semantically select, process, and classify the objective factors and phenomena, and the content fitting for the expression and level of detail must be confirmed. That is, it must be decided what to express and to what extent. Therefore, many factors, such as people's cognitive ability and application needs, as well as the social environment and technical characteristics may play a key role in map design.

G. Wang (✉)
Information Engineering University, Zhengzhou, China

J. Wang and F. Wu (eds.), *Advances in Cartography and Geographic Information Engineering*, https://doi.org/10.1007/978-981-16-0614-4_4

A symbolic method is primarily used in formal design to express people's understanding, to process the graphics of drawing objects and to outline the whole picture of the map. The problems that need to be resolved at this stage are outlined as follows: How do we express various objects and phenomena with map symbols? How can we design symbols and drawings using graphic semiotics and color theory? How do we use graphics technology and crafts to ensure the effectiveness of the map product? At this point, the cartographer's quality, such as the design ability of symbols, the cartographic generalization knowledge of map content, the cultivation of the map aesthetics, the drawing experience and skills, the mapping technology level and ability, etc., may play key roles.

It can be seen from the process and content of map design that not only the objective factors, which include the use of maps, map scale, and the regional characteristics of mapping, but also the subjective factors, such as the map representation methods, the human cognitive ability, the human visual perception and feelings, and the mapping and printing technology will affect map design. Therefore, map design has been the research focus and a difficult problem in the field of map science. It is also one of the bottlenecks restricting automated map production. At present, the way of human thinking and creative inspiration is not clear, nor is the simulation of these operations with a computer. Thus, the map design is also full of challenges and creativity. For 60 years, a large number of theoretical and technical methods and innovations have been carried out in this field by Chinese cartographers, and a series of standards, such as general map compilation specification and schema have been formed. These specifications have been used to guide the compilation and update the national series of scale topographic maps. A high level of general maps, thematic maps and comprehensive atlases were designed and compiled. The corresponding computer-aided map design system and digital mapping system were researched and built, and many multimedia electronic maps and network maps for the country and certain regions were produced and published. These maps provide solid theoretical and methodological guidance for future map design.

4.2 Development Process

For 60 years, the types of cartographic products have increasingly diversified from the static paper map to the dynamic electronic map, multimedia map, and network map; furthermore, an accessible virtual geographic environment has been created. Map contents cover wide ranges from the land to the ocean, from the underground to the air and further to outdoor and indoor spaces. The map production process evolved from manual to computer-aided mapping and then to full digital mapping based on the map database. The changes in map contents, types and production technologies will inevitably lead to the changes in map design theory, map representation method, map design process technology and other aspects. The development of map design, according to the map design method and the main object of map design, can be divided into the following 5 stages.

4.2.1 *The Manual Map Design Phase with a General Map as the Mainstay*

This stage refers to the period from the founding of new China in 1949 to the beginning of the 1960s. Maps, especially general maps, were mainly designed and produced by hand in this stage. Paper maps were the only form of map, emphasizing the design skills and artistic quality of map works.

This period was just 10 years after the founding of new China (1949–1959). It was also in this 10 years that traditional cartography (before the end of the 1950s and early 1960s) developed and was perfected. Cartography in this period formed as a complete system of map production theory, method, technology and craft.

Map design in traditional cartography aimed at the use of technology to solve the problem of map content abstraction and generalization from the ground surface to the map, map projection, map representation method, map-making art and method and so on. The emphasis was on the technological level of map-making and production and the effect of the map product, i.e., design skills and the art of map work, while focusing on whether the design results met cartographic requirements.

Under the guidance of this kind of thought, the requirements of experience, technology and art level of the design personnel are higher. Designers are required to be able to design and draw a map in compliance with regulations according to the uses of the map, to design map symbols according to the characteristics of color, to choose a reasonable representation method according to the characteristics of features, to make a plan of drawing operation and the print process according to the technical requirements of manual editing, drawing, photographs, reprints, coating, plate making, printing, and so on. To meet those requirements, by the guidance of experts, designers complete many map readings, analyses, designs and practices for a long time and explore the "skills" and "rules" in the process of map design. Designers need to summarize and refine the technologies and methods that can be sensed but not expressed in words to determine what is the regularity of practical significance. Therefore, map design in this period is mainly based on experience and lacks a complete set of design theory and method guidance. The process and results of the design are inevitably subjective, and the design quality depends on the experience, art and skills of the designer. However, this period covers the early days of China's founding, and national construction, especially economic construction, requires many investigations and map productions. To meet the needs of different industries, a large number of general maps need to be designed and produced. Cartographers conducted a large number of investigations and summarized the modeling process from the ground to the map, including what geographical features should be represented in the map? How are the contents of expression sorted? How is each type of object classified? And so on. That is, to study the simplified or recognized abstract model and the assortment and classification model of objects (or phenomena), a number of national wide water systems, a mountain range assortment and classification charts, as well as population density classification maps were compiled. In the aspect of representation method, after many experiments and statistical analyses, a set of topographic map symbols, including symbol shape, size and color, was designed based on the characteristics of the map elements, color, symbol, reading habits, and so on. It is consistent with the map content (Min 1957). After research, the contents representation principles and methods of the general map in China were basically formed. In addition, a series of common map works were designed, and among them, the design and compilation of China's topographic map and geographic map is representative.

(1) **Design of the topographic map**

There are two characteristics of topographic maps in this period. First, the map symbol is not only very detailed and clear but also has vivid content. The map symbol can fully reflect the real situation of the ground and its roles in the national economy; second, compared with the previous map, the topographic map has improved the timeliness, completeness and artistry of content and representation methods (Wu 1959).

The timeliness is reflected in the aspect that most of the topographic maps at that time were based on aerial photographs. The content that was not shown in the photo had been acquired through field annotation or measurement, such as the population of a residential area, the depth of a river and its flow rate, road traffic conditions, the variety and diameters of trees, and so on. At the same time, a currency check system was initially established to maintain the timeliness of the map. The completeness is reflected in the full consideration of use requirements and convenience. For example, the design of drawings is very complete: a slope ruler was attached outside the border to facilitate the determination of the slope of the terrain; a kilometer network was plotted for the convenience of measuring the area of the ground for the direction and distance between points, as well as the precise coordinates of a point; the three-north direction line was added to distinguish map orientation; and the serial number of the adjoining sheet was labeled on the four edges of the map for the purposes of map matching. The artistry was shown in the two aspects of color design and map surface decoration. First, natural colors are used for the ground features in the design of the map to represent each element of the map, so there is no need to use text to identify the water, vegetation, landscape and other natural elements. Conversely, bright, unnatural colors are used to express the very important geographical elements and to reflect the hierarchical structure of the map elements, and as a result, all kinds of topographic maps were published in multiple colors. This not only can increase the artistry but can also improve map legibility and increase map information loading. Second, the design expectations for map surface decoration are very rigorous and fine; there are strict requirements for the sizes of the map symbols, line number thickness, arrangement of fonts, map layout, and so on. In particular, the use of the scribing method in the 1960s makes the map symbols more sophisticated and beautiful.

(2) **Design of the small-scale general map**

Based on the compilation of topographic maps, with a large number of geographical survey and research results, the design of small-scale general maps began to occur. In this period, a variety of small-scale general maps were compiled. The representative results are outlined as follows: 1:1,500,000 National Situation Map, 1:4,000,000 China Relief Map, 1:1,200,000 general map and 1:1,400,000 Southeast Asia Situation Map. Among them, the 1:1,500,000 National Situation Map has beautiful colors and better print quality, which laid a foundation for the subsequent small-scale general map. The 1:4,000,000 China Relief Map uses the two-layer planar method to represent a network of rivers, including primary and secondary tributaries; the contour description is a vivid reflection of the landform, and the special landform symbols are more expressive.

In this period, the production technique for the topographic map and small-scale general map was usually the traditional method, that is, the original publication was obtained through the map compilation and fair drawing procedures.

For a large-scale map with simple content, the cartographic generalization involves a little work, so the method "compilation together with drafting, mapping at one time" was realized in the late 1950s and early 1960s. At the same time, topographic map compilation for parts of provinces also started, and the large atlas and thematic atlas were also prepared. These maps provided important guidance and practical basis for the follow-up compilation of general maps, thematic maps and atlases.

In the practice of designing and compiling general maps, cartographers began to study the quantity and quality standards of mapping for different scale maps, how to fully reflect the geographical features on the map, the cartographic generalization of map content, and other issues. Cartographers stressed designing a reasonable map representation method in accordance with the law of geographic research and improving the objectivity and expression accuracy of ground objects and geographical phenomena. In addition, corresponding research results were obtained.

4.2.2 The Manual Map Design Phase with the General Atlas as the Mainstay

This stage roughly refers to the 1960s to 1970s. In China, to meet the needs of national economic construction, science and culture in mapping, a large number of geological survey and mapping results, especially the published series scale topographic map and general map, provided abundant cartographic data for the design and compilation of the atlas; thus, the compilation of a large national atlas began. The provinces and autonomous regions in which conditions permitted mapping also successively compiled regional atlases.

China's large national atlas is a state key scientific research project co-implemented by the Chinese Academy of Sciences and the State General Bureau of Surveying and Mapping, which began in 1958, and the atlas consists of four volumes: general, nature, economy, and history. In 1981, the State Council once again listed the national atlas as a state key scientific research project. *The People's Republic of China Agricultural Atlas* was added as the fifth volume in addition to the original four volumes. These volumes were successively published in the 1990s.

The design content and process of the large national atlas including *the People's Republic of China General Atlas* are very complex, involving the overall structural design, the expression method of each element, the symbolic system, the cartographic generalization index, colors, the processing scheme, and so on. Therefore, the design of this atlas was of great scientific significance and application value but also had a significant international impact. The design idea, theory, method and compilation technology were strongly advanced due to the research on the theory and method of map design, and this promoted the compilation of the China's provincial atlas (Liu 1963) and provided theoretical and technical guidance for the preparation of regional Atlases.

In addition, design and editing of the *the People's Republic of China Atlas* began at the end of 1972; after 8 years, bookbinding and a stylebook were completed in the first half of 1980; and finally, the atlas was printed in Octavo by the Chinese Map Publishing House in 1981 (generally referred to as the Octavo, China atlas). This atlas is the largest folio and most detailed Chinese atlas created since the founding of new China.

Overall, through the design and compilation of the general atlas and the research on the comprehensive index of map content, map expression methods, map layout design, color design, and so on, the general rules of map design were discovered and summarized, which laid a solid foundation for the further research on the theory and method of map design.

4.2.3 Computer-Aided Map Design Phase with the Regional Atlas as the Mainstay

This stage generally refers to the 1980s, which is considered to be a period of great technological breakthroughs. In the field of cartography, the technology of computer-aided mapping began to replace manual mapping technology, which provided a new technical means of map compilation. China's reform and opening up had just begun, and national economic construction and national defense construction were undergoing rapid development and various types of maps and atlases were needed to meet the needs of all aspects of economic and transportation construction. Therefore, the objective of map design during this period was creation of regional atlases, that is, provincial (autonomous regions and municipalities directly under the central government) atlases. This period was the second pinnacle in atlas compilation at home, and a series of provincial atlases, including the Shanxi Province Atlas and Shanghai City Atlas were designed and completed.

In addition, with the development of the national economy, the preparation of all kinds of city maps and atlases was scheduled. Most of the existing large-scale urban maps were compiled before the 1970s. After entering the 1980s, urban construction developed rapidly and changed significantly. It was urgent to develop a new batch of urban maps to meet the needs of urban planning, construction and management. At the same time, to implement urban planning in a organized and comprehensive way, a map must reflect the status of the city in detail. Therefore, the compilation of the current situation maps in some cities began in this period. The problem of using color infrared aerial photograph for compiling the current situation map of city was discussed (Pan 1989).

The design and preparation of a large number of provincial atlases met the needs of economic construction, especially urban construction, and promoted the theory and method of map design. Prof. GAO Jun (an academic of the Chinese Academy of Sciences) recommended the theory of cartography from Europe and the United States in 1982 and first introduced the cartographic visual perception theory in China in 1984; he proposed that cartographic visual perception theory should be used to guide map design to improve the communication efficiency of the map (Gao 1984). This proposal marked the deepening of research on the related theories and methods of map design in China, such as cartographic perception theory, cartographic communication theory and cartographic semiotics theory. After that, map design gradually transformed from purely empirical and results-oriented methods to the map design theory, as well as the map user's needs and the visual perception effect.

4.2.4 The Human-Computer Interaction Map Design Phase with the Electronic Map and Comprehensive Atlas as the Mainstays

The human-computer interaction map design stage occurred just after the 1990s. In this period, full digital mapping technology and multimedia technology became mainstream technology. At the same time, multi-resolution, multi-temporal satellite image data covering the whole country, as well as the globe, became important information sources for general and thematic maps. The types of cartographic products expanded.

In the 1990s, the electronic map editing and publishing system came out and broke the traditional division boundaries of mapping and publishing. This system marked a new turning point, where the traditional manual production mode was replaced with the emergence of entirely digital mapping and publication. An integrated digital mapping and publishing system has the functions of map design, map compilation, color separation netting and film output and has been applied to produce a number of maps and atlases. The new technology of direct digital map plate making also led to fundamental changes in mapping technology. The compilation and publication of maps and atlases realized the fundamental transformation from the traditional process to full digitalization and automation.

(1) **Research on the theory of map design**

In the field of map design theory research, the applied study of cartographic visual perception theory, cartographic spatial cognition theory, cartographic transmission theory and cartographic semiotics theory began in the 1990s. The research on cartographic expert system technology has also progressed. Expert systems with a certain use value emerged in succession. These systems provided a new technical means for the design of thematic maps, the selection of map projects, the rapid creation of the design scheme and the rapid generation of statistical maps. In addition, the scientific nature, accuracy and design efficiency of the map design was improved, and map designs changed from manual mode to the intelligent design of human-computer interactions (Zhang 1990; Gao 1993).

In terms of map representation, symbol design, legend design, and color design, some scholars expounded the basic structure and characteristics of the thematic map legend system from the viewpoint of system theory (Liu et al. 1992); the concept of symbolic constitutional elements was put forward based on map transmission theory, and its design pattern was summarized as "Selection, Matching, Superposition, Strengthening, Weakening, Stretching and Compression". In addition, an example was provided to illustrate the design pattern and process of symbolic constitution elements (Ma 1995). Some experts believed that a deeper understanding of map symbols and better development of its information function from only the perspective of practical technology was not enough. The essence, characteristics, functions and application rules of the map symbol should be analyzed at the height of ideology, culture and philosophy. It was considered that the essence of the map symbol was the artificial referent of the objective entity and the expression medium of the cartographer's ideology. Compared with other symbols, there are similarities and dissimilarities. For that reason, the multiple functions of map symbols in human expression, cognition, thinking, culture and aesthetics were discussed, and the aesthetic study of the map was included for the beauty of the map and to explore the issues of creative thinking and methodology. The role of artistic thinking in map creation, the relationship between map science creation and the creation of beauty were also discussed. At the same time, the research content of map aesthetics and art education for map workers were described (Yu 1990, 1995).

(2) **Research on electronic map design**

As a new form of expression, the electronic map has its own characteristics and rules in map design and expression. In the design of the electronic map, the visual variable and its perceptual effect is one of the key research fields of theoretical cartography. Through much research and practice, relevant results have been achieved and used to guide the design of the electronic map.

The emergence of the electronic map has changed the way of reading the map. Considering the dynamic, interactive and multimedia characteristics of the electronic map, the display mechanism, expression scope, representation method, visual variable and perceptual effect of the map have changed greatly. The roaming function of the map, which goes beyond the limits of reading on a sheet map, expands the scope of mapping. The dynamic nature of the map extends its expressive pattern, taking map develop from static browsing to dynamic interaction and so on. The research on the modern map has been beyond the scope of the Earth and extends to the study of the relationship between other planets and stars, as well as various aircraft tracks in the sky. Therefore, how to scientifically design the electronic map of the screen, including the dynamic symbol composition rule of the electronic map, visual variables, dynamic mapping technology, and so on, as well as a series of theoretical and methodological issues, require further studied. In addition to the visual variables and the perception effect, the visual perception research on the electronic map also includes the visual perception effect of legend design (online legend, legend arrangement, animation map legend), the perception effect of the series screen electronic map, the animated map and the 3D map (Chen et al. 1999). Based on the Bertin symbolic parameter system, four dynamic parameters were extended by Ai (1998), that is, the interval length, change rate, change order, and rhythm. These results indicate that the method of map representation has been changed from a simple static method to hybrid representation methods that include a combination of static and movement, a combination of pictures and texts and a combination of multimedia. The performance of the map and the effect of the information transmission have been improved. Based on analyzing the development status and multimedia electronic map trends at home and abroad and aiming at the special geographical region of the Arctic, Pang et al. (1999) used the MapInfo platform to explore a new means of making a multimedia electronic map based on a color map electronic publishing system. The result is of greater use value to electronic map production. Xu et al. (2007) thought that the design principle of the network map was embodied in interface design, layer display design, symbol design, annotation design and color design. Network map design involves many aspects, such as the network service platform, network map data source, network map expression form, network map design technology process, and so on. Wang et al. (2008) put forward the technical requirement of the design and expression of the network electronic map from the viewpoint of map cognition.

(3) **Study on the urban atlas design**

Due to the rapid development of cities, urban construction, urban development and planning, urban tourism, urban external contacts, and others, need to prepare a series of maps and atlases, various electronic versions of the atlas and a multimedia electronic atlas.

Compared with the previous atlas, these urban atlases had breakthroughs in terms of design features and representation methods. The design of the "*Beijing Tourism Atlas*" broke through the monotonous representation of only a flat map. This tourist attraction map used an aerial photo map, perspective view map, topographic shading map, bird's eye view and sketch map to represent three-dimensional phenomena on a two-dimensional map according to the different natural and humanistic landscape characteristics of tourist attractions, causing the elements expressed on the flat map to produce the stereoscopic effects and have a strong sense of reality (Qian 1992). At the same time, modern advanced technologies and methods, such as a remote sensing analysis system and geographic information system, were applied to the design and compilation of the atlas, especially the thematic analysis atlas. This has revolutionized the printing technique of the atlas and promoted the development of modern atlas compiling technology. Yin (1992) also introduced the method and skill of making an urban planning atlas using an integrated map compiling, designing and printing system.

The production of the urban atlas during this period was mainly based on the map database, which used computer-aided mapping technology. For example, Zou et al. (1998) discussed in detail the method of digital mapping production based on MicroStation and pointed out that digital mapping is the only way to realize the foundational transformation of map production toward digital, rapid, modern, and automated mapping. He et al. (1999, 2003) described the creation method for the electronic atlas and introduced the computer-aided mapping and plate making techniques and atlas processing. Hong et al. (2006) introduced the method of using new technology to compile a geographic base chart, a group of thematic maps, and a group of general maps.

Furthermore, the content, structure and development trend of the urban atlas were also discussed, it was pointed out that the content of the urban atlas should show the "present" and "future", and the development direction of a city; the structure of an atlas should have individuality, highlighting the characteristics of the city; it needs to achieve the unification of content and form for a specific map; the color design should be scientific to reflect the regularity; it requires the development of a series of large-scale city maps to be conducive to facilitating the planning and construction of a city (Ding 1992). The research of the content, function, representation and symbol system of an urban map should be strengthened for modern urban mapping to make it develop toward comprehensive mapping (Zhan 1993).

(4) **Study on electronic atlas design**

The electronic atlas emerged as a new cartographic work at the end of the 1990s. On the one hand, the multimedia electronic atlas became a new means for the transmission of geographic information. On the other hand, it put forth the concept of "geography" and "space" for traditional electronic reading materials, which makes the expression and transmission of geographical information more direct and effective. In addition to the characteristics of the traditional atlas, due to the application of computer technology, multimedia technology and visualization technology, the electronic atlas is easier to modify and update, the production cycle is shorter, and the cost is lower. The flexibility, selectivity, reality and dynamic nature of the electronic map, with automatic map subdivision, automatic establishment of mathematical foundation, automatic generalization of map data, map database storage and management, map symbol library, and other functions, frees the user from the limitations of the map sheet and fixed scale. In addition, spatial data can be extracted from elements and layers according to the practical requirements. More importantly, image generation technology can be used to visually represent some scientific phenomena, natural landscape and even abstract concept in a vivid, intuitive, and dynamic approach. It is convenient for people to understand, judge and analyze from different viewpoints and perspectives. At the same time, the analysis function of the atlas can be used for the deep processing of information to obtain conclusive thematic maps or thematic data sets; so the functions and values of the atlas should be fully exhibited. These kinds of works (electronic version) are outlined as follows: '*National Nature Atlas of the People's Republic of China*', '*National Economy Atlas of China*', '*National General Atlas of China*' and '*Population Atlas of China*', '*Hong Kong Electronic Atlas*', '*Shenzhen Electronic Atlas*', and so on. These atlases have various functions, such as a zoom window, dynamic representation, query and search, statistical analysis, superposition and comparison, cartometry, and so on. They are gradually developing towards an integrated geographic information system.

Compared with the previous paper atlas, the electronic atlas has undergone great changes in various aspects of the concept, characteristics, classification, composition, production technique and so on. From the aspect of the map display mechanism, the electronic atlas database storage structure and management technology, automatic map subdivision technique, automatic generalization of map content and multi-scale display technology need to be studied. From the aspect of data expression, in addition to consideration of the characteristics of a single electronic map, the completeness, logic, and consistency of the map symbol need to be studied. Zhang (1996) introduced the application of the electronic map (atlas) at home and abroad, and discussed the development trend of the electronic map (atlas); based on the practical design and production of several urban multimedia electronic atlases, Xu et al. (2003) summarized the basic ideas for the designing and realization of an urban multimedia electronic atlas; Wang et al. (2005) discussed the automatic map subdivision of the electronic atlas, database storage structure and management technique, automatic generalization and multi-scale display technology; Shi et al. (2008) proposed a super media network information organization structure based on 'graph group-thematic-map sheet' to reflect the basic geographical features of a region, and they described the content framework of an integrated electronic atlas for an urban area that considers the distribution law of all kinds of natural and social economic factors and their relationships to be the main content.

(5) **Study on the image atlas design**

Satellite image data has become an important data source for the general map and thematic mapping and promotes the design and preparation of an image map (atlas), and so on. In the process of the design and compilation of an image atlas, how to retain the integrity of image information and overlay more cultural information is a problem that requires an atlas designer to address. According to the practical needs, designers have put forward some design ideas, such as "highlight the map and supplement with the image" (*Shenzhen Photo Atlas* as a typical case), "highlight the image and supplement with the map" (*Image Atlas of Shanghai City* as a typical case) and the "map and image fusion with each other" (*Image Atlas of Suzhou City* as a typical case). The vector representation principle of the image map (atlas) features and the design principle of the symbol system and image map color were summarized, and many image atlases have been designed and compiled, such as the *Shenzhen City Photo Atlas* (China Map Publishing House, 1999), *Lanzhou City Image Atlas* (Gansu People's Publishing House, 2003), *Nanjing City Image Atlas* (Chengdu Map Publishing House, 2004), *Hong Kong Street* (compiled and published by the Survey and Mapping Office of the Lands Department, government of the Hong Kong Special Administrative Region, 2005), and so on. At the same time, through the design of *Image Atlas of Shanghai City* (Huang 1992) and *Image Atlas of Suzhou City* (Wang et al. 2007), the theory, method and technology of image atlas production based on a high resolution aerial (remote sensing) image were studied and tested, the principles and methods of the overall design, color design, symbol design and others for the image map (atlas) were put forward to guide the making of an image map (atlas).

(6) **Study of the national atlas design**

In this period, the compilations of the *Great National Atlas of China* were completed progressively: *The National Agricultural Atlas of People's Republic of China* was completed in 1990; *The National Economic Atlas of People's Republic of China* was completed in 1993; and *The National General Atlas of the People's Republic of China* was completed in 1995. The national atlas reached the international level of cartography at the time in terms of content selection, overall design, expression method, technique and technology, as well as in other aspects. This atlas promoted the research of the theory and method of map design and provided guidance for the compilation of the provincial atlas (the expression contents and methods of each atlas have been described in detail in the map works).

4.2.5 Human-Computer Cooperative Map Design Phase with the Coexistence of Many Varieties of Maps

Since entering the 21 century, with the further improvement and extensive application of GIS technology, as well as the application of virtual reality technology, visualization technology, and multimedia technology in cartography, great changes have occurred in the map production process; meanwhile, the variety of maps has been greatly expanded, so that the digital map, electronic map, dynamic map, three-dimensional map, accessible map and the like have emerged in succession (Gao 1997). Map design must be carried out on the visual human-machine collaboration platform, which is based on the integration of various technologies. At the same time, in the production of an electronic atlas, it can quickly produce a paper map (atlas) and establish a map information system. Therefore, map design in this period can be called the human-computer cooperative map design phase with the coexistence of many varieties of maps. The main objective is the design of an electronic map, network map and large comprehensive atlas (system). The aim is to improve the human and computer collaborative map design, and collectively accomplished the complex geographic information fusion, process and display, while emphasizes the scientific nature, accuracy, systematicity and authenticity of the design results.

After decades of development, China's cartography has caught up with the advanced countries in the world. Especially in the past 10 years, cartography has entered a period of rapid development. Mapping technology has realized the historical transformation from traditional manual mapping to digital and integrated mapping. In addition, the product has realized the transformation from single paper map to digital map, electronic map, network map, geographic information system, virtual geographic environment and so on. The theory, technology and method of map design have changed greatly. These changes force us to study the cartographic experts' way of thinking in the production and design of a map, to pay attention to whether the results of design meet the requirements of users, and to consider people and map as a whole; that is to say, we are not only concerned about the process from spot to map but also concentrate on the relationship between reader and map. As a result, we need to explore the theory and method of map content design, expression mode design, representation design, layout and its effect design to adapt to the new changes in map demand. This change in the design concept lets us pay more attention to the guiding role of a human's cognitive ability and map perception effect on map design and gradually places map design under the guidance of theory instead of experienced supervisors.

For map design, the goals are to design a map to let the users of the map obtain more information and knowledge, make people have the feeling of pleasure, make the map reflect well-known information and support the discovery of unknown information and knowledge. In a word, how can readers quickly and effectively obtain the necessary information and knowledge from the map? This requires study from the user's perspective on the problems, such as the visual perception characteristics of user groups, the map usage, the personalized and professional map features, the human's cognitive ability and so on. These problems involve the purpose of the map, the level of knowledge, thinking and judgment, aesthetic quality of map designers and map users, and other factors. Therefore, map design is a process of creative thinking and abstraction; not only are scientific theories (map spatial cognition theory, map information transmission theory, visual perception theory, semiotics, map aesthetics, psychology, physiology, artificial intelligence and other fields of knowledge) required for guidance, but advanced technologies, such as computer-aided mapping technology, map database technology, virtual reality technology, graphics and image processing technology, visualization technology and so on, are also required. Only with rich experience, broad geographical knowledge, considerable cultural and artistic accomplishment, proficiency in the basic theory and process of cartography, and a high spatial cognition ability can a map designer be qualified for map design, map production, and map innovation. Undoubtedly, these requirements create higher standards for map designers.

4.3 Main Research Achievements

4.3.1 The Study of the Topographic Map System

Since the founding of the People's Republic of China, a large number of general maps have been compiled. Among them, the national series of scale topographic maps formed the standard and systematic results. The national maps maintain scientific unity and connection in map projection, sheet numbering, coordinate system, and elevation system; additionally, there is a symbol system with relatively fixed content and basically similar symbols. A large-scale topographic map was compiled and published in 1952 and revised in 1953 and 1958; then, the compilation specifications and schema for large-scale topographic maps was formed. In addition, China's topographic map system, including 1:50,000, 1:100,000, 1:200,000, 1:500,000, and 1:1,000,000 topographic maps, was put forward.

With the increase in map applications and the development of mapping technology, on the basis of the full study of the requirements of various kinds of maps, the original 1:200,000 topographic map was withdrawn in 1980, while a new series of scale topographic maps were formed, including 1:10,000, 1:25,000, 1:50,000, 1:100,000, 1:250,000, 1:500,000, and 1:1,000,000 topographic maps. These maps are divided into three levels of large, medium and small according to the scale, i.e., 1:10,000, 1:25,000, 1:50,000, and 1:100,000 are large scale topographic maps, 1:250,000 and 1:500,000 are medium scale maps, and 1:200,000, 1:500,000, and 1:1,000,000 are small scale topographic maps. Currently the most commonly used maps are 1:10,000, 1:50,000 and 1:250,000.

In the past 60 years, studies and practices on the classification and grading of topographic map content, the expression method of all elements and the symbols and colors have been carried out by cartographers. The compilation of different scale of topographic maps nationwide was completed. Among them, the representative work is the millionth scale topographic map (also known as the new 1:1,000,000 map), which was compiled and published in 1959, a total of 64 sheets. The characteristics of the millionth scale topographic map were analyzed and studied by Liu et al. (1958) and Gao (1963). In addition, the expression method and its comprehensive characteristics of the water system on the topographic map was discussed in detail by Wu (1959). Based on the practical and theoretical summary of map compilation, many scholars put forward many amendments to the map schema and symbols at that time. For example, Qian (1985) conducted a preliminary test on the military value of China's series scale topographic map symbols and pointed out that these maps were appropriate for both military and civilian use; the maps also changed frequently and had more supplementary provisions, so that the mapping products based on these maps were not uniform enough; finally, improvements were suggested. Huang (1964) discussed the symbol representation of topographic maps in the desert region and proposed amendments. These study results provided technical and methodological guidance for the production of topographic maps and enabled the topographic map system and schema symbols to be gradually standardized and made more scientific.

Since the 1990s, to meet the needs of digital production and information construction, China has established the corresponding basic map database for 1:10,000, 1:50,000 and 1:250,000 scale maps, which were regularly updated; China also formed a geographic information encoding standard for the digital environment and map symbol storage regulations. In addition, a series of studies on the standardization, normalization and serialization of topographic map symbols and design have been carried out. Yang (1993) proposed a three bit digital code that was arranged in layers, classes and categories for large scale topographic map (1:500 to 1:2,000) scheme symbols, which not only covers all the symbols but also includes marginal decoration, map surface lettering, and even symbols of cadastral surveys; Dingguo also suggested the improvement of some topographic symbols to adapt to the needs of computer graphics. These factors are conducive to the standardization and normalization of digital terrain surveying and digital cadastral surveying. Yin (1997) discussed the factors affecting the quality of the national basic topographic map and suggested improvements to the design of the national basic topographic map in China based on the investigation and analysis of China's national basic topographic map and its corresponding compilation specification and schema; these suggestions were in reference to the design experience of foreign topographic maps. The mathematical definition of scale symbol, non-scale symbol and semi-scale symbol in the map was improved by Huang et al. (2006), and the conditions for the transformation of these three kinds of map symbols were discussed. These results have important reference values to the further standardization, normalization and serialization of topographic map design, to improve the fine degree of topographic map symbols and notes, to improve map color design and map decoration design and to increase the layering and expression of a map.

4.3.2 The Research on the Design of Small Scale General Map

On the basis of topographic map compilation, since the 1960s, a variety of small scale general maps have been compiled and published, such as 1:1,500,000 *National Situation Map*, 1:4,000,000 *China Relief Map*, 1:2,000,000 *General Map*, 1:4,000,000 *Southeastern Asia Situation Map*, 1:2,500,000 *Whole Chinese Territory Map*, 1:3,000,000 *The People's Republic of China Map* and 1:5,000,000 *World Geographic Map*. In addition, the design principles of the small scale general map, the expressive content and the expression method were deeply studied, and some research results have been obtained. Among them, the China Relief Map is worth mentioning as an important tool for scientific research, especially for geographical research, and has been highly valued by cartographic scholars for the theoretical and practical results in the geomorphology representation method. As a representative figure of the relief shading method, Shi (1964) performed an experiment and summarized the expression method of the 1:4,000,000 *China Hypsometric Shading Map*, and he discussed the shading representation and color decoration of the geomorphic features on the small scale relief map. Wu (1965) studied the design features of the terrain height meter and put forward some principles and methods for the selection of a terrain height table. These research results provide the basis and technical method for the following 3D terrain representations. With the improvement in map printing technology and the widespread use of the halftone original decoration method, Liu (1965) put forward the principle and method for improving the production and replication technology of the shading original. Zhang (1966) summarized five methods of making a hypsometric map, where the proposed "iterative duplication method" not only can solve the problem of producing flat color edition but also solve the absolute transparency of each hypsometric negative; thus, the quality of plate making was significantly improved. At the same time, he proposed that the strip masking method should be used for making a hypsometric edition whenever possible to save time during retouching and allow layer and layer and layer and water be more closely matched. This study showed a new prospect for the improvement of the making hypsometric map method. These results provide a wealth of practical references for the scientific expression of small scale maps, especially terrain maps, and provide guidance for the design of a series of small scale maps (atlases). According to the theoretical research and experience, the following conclusions are obtained. Shading can enhance and improve the expressiveness of the landform morphology and regional landforms; it is feasible to represent a landform with hill shading as a primary method, and the contouring and hypsometric methods are secondary. Finally, the scheme of height grading and color table for the landform expression of small scale terrain maps in China was preliminarily designed and established.

With the application of computer technology, new achievements in the design, compilation and representation methods of small scale general maps have been made. Wu (1988) provided the formula for calculating the gray level of the terrain surface in the projection plane and the shading algorithm for the natural landscape. Xu et al. (2002) introduced the background of the establishment of the 1:4,000,000 *Administrative Map of the People's Republic of China* and emphatically explained the use of compiling the data and characteristics of the map content, technical route design and map products. Li et al. (2002) researched the relief representation method for the 1:3,000,000 *Geographic Map of the People's Republic of China*; he described the selection of the relief representation method and obtained results with the help of a computer; finally, he summarized the technology and method of computer-based relief shading for a small scale geographical wall map.

These studies have shown that using the digital mapping technical route and process with computer mapping as the main body, supplemented by local manual transition plotting, is feasible. In the whole process of map design and compilation, the procedures, including map content design, computer production, binding design, step-by-step transition transfer and so on, were implemented synchronously and alternately according to the program flowchart and were finally unified and combined. This is the main technical method for the design of small scale general map during this period.

4.3.3 The Study on the Overall Map Design

(1) Map symbol study

Since there is a map, there is a map symbol. Along with the progress of society and the development of map production, the application scope of the map expands day by day, and the contents need to be expressed more and more. People are constantly exploring the methods of content representation and the design of map symbols.

Map symbol design is an important part of map design. Research results in this area are outlined as follows: the system theory and method of map symbol design considering 8 factors (Wang 1993); the relationship between map symbols and characters; the significance of text symbols in the map; the evolution of nature and function of the character symbol after transplanting to map; the three types of symbols and texts on the map, such as original text symbols, text annotation symbols and text graphic symbol, and their different functions and characteristics (Yu 1996); the study on the constitution and rule of map symbols (Ling 1997a, b); the characteristics of map annotation, including the subordination of position, the dispersion of arrangement, and the hierarchy and classification of contents, as well as the map notation font, font size, text arrangement, and text color (Ling et al. 2007); and so on.

With the application of computer technology in mapping, the construction of the map symbol library and the composition rule of the electronic map symbol become an important issue in the study of map symbols, and relevant results have been achieved. Based on the theory of map transmission, Ma (1995) put forward the concept and the design pattern of symbol composition elements. This concept provides a new vision for the design of the graphic output module based on a map database, which can be used as the basis and operation mode of symbol design. Ai (1998) believed that the design principle of the traditional map symbol is based on Bertin‘s visual parameter system and to meet the requirement of map symbol design under the new technology conditions, dynamic visual variables of symbols need to be extended based on the original visual variable system to adapt to the function and application of a dynamic map in the fields of course response, real-time tracking and process monitoring. According to the definition of topology and graph theory and the visual threshold of symbol configuration, Zhong et al. (2001) proposed the concept of a lower limit (limit circle) for the clear depiction and representation of symbols. By comparing the surface area of an object on the projected plane with a limit circle, the mathematical definitions of scale symbols, non-scale symbols and semi-scale symbols were derived. Thus, the basic concepts of the three kinds of map symbols can be quantitatively described with precise mathematical expressions. Zhu (2006) put the time dimension into the dynamic map so that the electronic map experienced a new development in the form of visual expression. At the same time, this change also presented a new challenge to the problem of dynamic map visualization. The advent of the information age means that the traditional two dimensional information cannot meet the new requirements. However, the development of the computer makes it possible to display and describe the three dimensional geometric features and attributes of objects. Fu (2007) discussed the application characteristics of the traditional two-dimensional (electronic) map symbols in a three-dimensional visualization environment and the dialectical relationship between the two, and he explored some principles of the geographic information expression and map symbol design in the three-dimensional visualization environment.

With the development of geographic information technology and digital technology, making an electronic map has become increasingly simple and convenient. The applications are growing in popularity, and digital network communication is increasingly widespread. It is imperative to promote the guidelines or standards of the electronic topographic map symbol to guarantee the uniformity and accuracy of cognized geographic information. Wang et al. (2003) suggested that the electronic map symbol system and the symbol entity should adapt to the change in digital technology and the visual environment; through his studies, a draft standard symbol library of the basic map for universal use in geographic information system software was created, and much effective testing, supplementation, modification and perfecting was done for relevant maps. The result has been published on the website of the China spatial information network.

Through these studies, the composition rules of map symbols are explained in theory. These rules provide mathematical tools for the scientific and rational design of symbols and the quantitative calculation of information loads. Thus, the experience and behavior of map symbol design is more scientific, and the theory and technology of symbol standardization and normalization has been promoted.

(2) Study on map color design

As a kind of information carrier to describe and study the living environment of humans, maps combine science and art as an organic whole. With the extensive use of maps and the continuous improvement of peoples' aesthetic abilities in mapping, an increasing amount of attention has been paid to map aesthetics. Map color design is an important part of map aesthetics and plays an important role in the beautification of the map, guiding readers to appreciate and study the map. The task of map color design is to use color to express the quantity and quality characteristics of geographical phenomena, that is, to establish a one-to-one mapping relationship between color and geographic feature data. Since a map is a scientific expression of geographical phenomena, the establishment of this kind of mapping is different from the creation of pure artistic works in which the artist has more freedom; mapping is subject to a series of mapping principles and color habits. However, map color design has never had a complete system of standards, and the practice of using color is also different from charts, which created great difficulties in map design work.

For a long time, people have carried on with research and practice for the map color characteristics, the constitutive law of the color-coded standard, the color design features of map symbols, and the characteristics and rules of the whole color design of map layout. At the same time, the characteristics and rules of color design for the electronic map, map design expert system and so on, were discussed, and a series of research results were obtained. Zhong (1989) studied the constitutive law of the map color standard and analyzed the basic elements of color; he showed that the total number of colors in a map color standard depends on three factors and pointed out that certain regularity of coloration exists in various color standards and ink chromatography; thus, we can use these rules to create the color design according to a plant's own equipment or specific mapping task. Jiang (1991) preliminarily explored the map design expert system and clarified that a map design expert system should be composed of five parts, namely, the knowledge base, database, reasoning machine, explained part, and knowledge acquisition. This study result provides a reference for research on the map color design expert system. Taking the map design, especially the color design, as the research object, You et al. (1996) considered that color design as a specific research process is very suitable for the implementation of visualization methods; at the same time, he analyzed the advantages and strategies of the visualization method to solve this problem; finally, through examples, he discussed three problems, including the mapping of color space, the interaction between user and color space, the mapping between data and color space, and so on. Ling et al. (2000) studied color using the map boundary and pointed out that the bounding lines on a map comprise a smaller portion compared with the color block, so its role is often neglected; however, color use greatly influences the entire color effects of a map and should be considered as one of the most important aspects of map color design. Wang (2003) discussed the powerful images of color combination and considered that as a visual symbol, color is not a kind of isolated ornament once it is assembled on the map; different color combinations will result in different psychological effects, as the audiences' feeling of map work chiefly depends on mastery and use of tones. Hou (2004) described the different forms of map aesthetics, such as the appearance, drawing materials, composition of a picture, symbols, annotations and colors, and he discussed the principles and methods for the color design of a map in terms of point, line and surface symbols. Zhou (2006) suggested that like all works of art, a map requires giving prominence to the subject and having a clear hierarchy, as well as obtaining natural harmony and being rich and vivid; to achieve this effect, it is important to deal with the problems regarding color selection and color matching.

Due to the restrictions of the display device, in the process of the symbol design of an electronic map, the use of three visual variables, such as shape, size and pattern, is limited. Thus, the use of a color scheme is more important. Chen (2000) designed a color perception experiment for the electronic map - A Color Matching Experiment on Electronic Maps, and a method and idea was provided for the color matching experiment of the electronic map. Guo et al. (2004) automatically configured the colors in the national administrative region map according to the contrast harmonic theory of color design and realized the automatic implementation of color design in an administrative map. The results show that in the visual environment, the color effect of a map designed by automatic color implementation is better than that by a human with no or only little cartographic knowledge. They also explored the color design of the color hill-shading map based on visual imagery and designed a set of feasible color schemes for the color hill-shading map according to the characteristics of China's terrain, as well as laying the foundation for the establishment of hill-shading map knowledge engineering in the digital environment.

(3) Study on map aesthetics and art design

The aim of map aesthetics design is to beautify the map in the form of aesthetics that are suitable for map application to improve the information content of the map and increase the reader's interest; this enhances the transmission efficiency of information and gives the map the beauty of the present time. It is necessary to study map aesthetics and the art of map style design to develop the map design, especially large map works, thematic maps and atlas designs, to a higher level.

The artistic beauty of the map is mainly reflected in the appearance, content and color. Research results in this area are mainly listed as follows: Yu (1989, 1990) emphasized the aesthetic problems in cartographic research and discussed the necessity of research and education in map technology aesthetics from three aspects of the period request for map aesthetics, the aesthetic value and practical value, and the expression form. His papers summarized the map aesthetics rules that can guide map design work, and he discussed the relationship between the activity of mapping science and technology and the aesthetic concept and creation of beauty. He also studied the relationship between the characteristics of map aesthetics and the map visual perception effect, as well as the way to achieve complete reunification of formal beauty and functional beauty, including the characteristics and methods of external map decoration. The aesthetics of map design and map production were discussed from the perspective of technology aesthetics by You (1990). He pointed out that designers should be concerned about the fashion trends of color and product style, and it is feasible to use the technology aesthetics as the core theory to realize the art of the map, because it can guide the map design and production and meet the aesthetic requirement of the current society. Guo (1991) believed that the aesthetic characteristics of the map are the adaptability of the map to readers. Readers create the feeling of map beauty. It is only when readers adapt to the various representations of a map that they can find a sense of beauty in the map. Map contents must be determined according to the subject to be expressed by a map and to convey the correct information to readers at the same time. The adaptability of map style mainly refers to the appearance, the basic material, the composition, the symbol and so on. Wang (2004) believes that, in addition to a specific theme and rich contents, a successful map should also have beauty in form; many map designers do everything possible in the pursuit of exquisite beauty in form and a remarkable creative idea, and the purpose is to enable readers to easily obtain all kinds of information in the form of beauty, as well as produce visual aesthetic pleasure in the reading process. Yang (2007) studied the specific principles and methods of map symbol and area color design; he considered that although the design of graphics, text and numbers is important, the color design is the most effective way to reflect the inner scale and law of the artistic beauty of a map.

The results of these studies indicate that designers should grasp and apply the principle of formal beauty in long term practice, having owned the perfect knowledge structure and keeping their vision widened and their knowledge will broaden. They must also care for the development of modern science and technology, culture, art and other related fields; only then can they understand and use the formal beauty at a higher level. Lu (1988a) stressed that the thinking mode of a map designer should have acute, random and jumping characteristics, as well as the ability to think horizontally, comprehensively and philosophically. In addition to professionals, the development of contemporary cartography needs more well-rounded designers. Map designers having a wide range of hobbies and understanding other area, such as literature, painting, music, calligraphy, photography and so on, may plays an important role in map design.

(4) Study on the applications of the map visual variable and overall design

The information transfer of the visual map is the embodiment of a map author's design idea in the visual perception of map readers. Therefore, in the process of information transfer for a whole map, the visual perception effect of the reader is the key to the success or failure of information transmission, and it should be an important basis for evaluating the quality of map design and production.

For example, in the applications of multilayer planar map design, map color design, layered landscape design, map symbol design, map title design, map frame design, and so on, reasonable use of visual illusion can not only reduce the noise in the process of map information transmission and improve the map information transmission effect but can also improve the expressiveness of the map and enhance the loading function of map information. He et al. (1991) introduced the rules and principles of common visual illusion in mapping and preliminarily discussed their applications in mapping. Chen (1995) discussed the method of establishing the mental map and its characteristics from the viewpoint of cartography. Song (2002) studied the human's visual perception and main factors affecting the perception effect, and to a certain extent, he analyzed the determination of map content selection and generalization index and the relationships among map symbol design, color design, layout, decoration design and visual perception. Some of problems that require attention in map design were proposed as a result.

4.3.4 The Study on Atlas Design

Here, we mainly introduce the studies on the overall design of the comprehensive atlas, urban atlas and electronic atlas.

(1) Study on the overall design of the comprehensive atlas

Since the 1960s, China began the compilation of a national atlas, and five large national comprehensive atlases were successively published until the 1990s. Through practice, a series of studies were carried out on the overall structure of the comprehensive atlas, including the representation methods of various elements, symbol system, color design, process scheme design, and so on. The regularity of summary results regarding the design idea, map contents, map representation method, layout design, color design and so on, effectively promoted the study of the design theory and method of the map, as well as the compilation of provincial atlases (Liu 1963). These studies provided theoretical and technical guidance for the preparation of a regional atlas.

The upsurge of making an integrative provincial atlas appeared in the promotion of the national atlas compilation. These atlases are called the first generation of provincial atlases. After the reform and opening up, China had set off the second wave of provincial atlas compilation. Most of these atlases were general and thematic atlases, and integrative atlases were compiled less often and are called the second generation of provincial atlases. Since the end of twentieth century, the content, form and preparation methods of provincial atlases have constantly improved; the comprehensive atlas of provinces, municipalities and autonomous regions, which have been published or are being compiled, are known as the third generation of provincial atlases (hereinafter referred to as the third generation of provincial atlases).

Through the compilation practice of the comprehensive atlas, a series of research results, such as the basic principle, overall design, expression method and preparation process in the design of the comprehensive atlas and regional atlas, were summarized. For example, Zhou (1965) discussed the characteristics of compiling a regional atlas and the principles of the compilation process design; at the same time, several commonly-used regional atlas compilation process schemes were analyzed and evaluated according to the clipping, content standard compilation quality, printing effect and drawing time, and the best scheme for the production of the general geographic map in a regional atlas was given. Mao et al. (1982) discussed the structure design, use of materials, content representation, relationship between elements and map decoration of the *Atlas of the People's Republic of China*, and the characteristic of combining science and art provided a reference for the preparation of other atlases. Li (1987) discussed the principles and methods for the preparation of a river basin atlas. He stressed that the compilation of a river basin atlas can be used to summarize the scientific results of the investigation and study of Chinese rivers from various aspects, and they have important significance in the study of rivers, resource exploration, development, utilization, cities along rivers, and construction of ports, shipping, tourism and even natural ecological protection. Lu (1988b) summarized the characteristics of the atlas and put forward constructive suggestions for some problems in the compilation process of the China Atlas. She thought that for an atlas with short timeliness and little range, the use of computer mapping technology and loose-leaf binding should be considered; for a highly specialized or small drawing area atlas, that is, a small circulation atlas or an atlas that is not publicly issued, dual color printing and loose leaf binding should be adopted. Meanwhile, she pointed out that cartographers must fully understand readers and make full use of the characteristics of the atlas. These are the key factors to improving the quality of China's atlases.

Fu (1991) discussed the characteristics of atlas design, development and compilation in the information age according to the understanding of concept, process technology, application efficiency, and so on. He proposed that when using the remote sensing analysis system, geographic information systems, and other advanced technologies and methods to compile an atlas, especially the thematic analysis atlas, we must organically combine remote sensing, geographical information system, computer assisted cartography and map color printing technology into an automatic assembly line to reform atlas printing technology and develop modern atlas compilation technology. Combined with the design of *Officer Atlas*, Wang (1991) expanded upon the design features of the atlas from six aspects of the overall design idea, architecture pattern, selection of contents, expression method, cooperation system and printing process. The study results provide an important theoretical basis and technical methods for the compilation of a comprehensive atlas. Lu (1997) discussed the editorial and design characteristics of the national general atlas, i.e., the presentation content, the explicit guiding ideology of introductory map design, the scientific definition solution and the orientation problems of an introductory map. Looking at the national atlases of the world, there is no fixed pattern in the editing and design of introductory maps, and there is a large difference in the number of map sheets and contents. Also, some focus on one aspect of the content. Combined with the compilation of the *Naval Officer Atlas*, Jia et al. (2002) summarized the atlas's basic design ideas including the basic content and principle of the overall design; he also presented the structure model of the atlas, as well as the projection selection, scale design, map subdivision method, and so on; further, he proposed that the architectural organization and design of the atlas should follow the principle of structural integrity, macro to micro principles and the principle of internal logicality. The structural integrity of the atlas mainly refers to the integrity of the main content and auxiliary content. The auxiliary content of a comprehensive atlas shall include a table of contents, index, legend, explanation, and appendix; and the main contents must include a natural map, social economic map, human map, and so on. Pang (2007) believed that an atlas is different from ordinary books because it is a visual map to show the reader, with more use of symbols, colors and lines to show the content of things and the relationships among various things. Thus, for the expression technique to be recognized by readers, it should be reasonable and scientific and directly affect people's understanding of the contents of the map, as well as have a direct impact on the use value of the atlas. He also studied the overall design content, the basic principle and the evaluation methods of the atlas. Bai (2007) and Wang (2008) summarized the design ideas and methods of the provincial and regional atlas according to the compilation of the Uyghur version of the *Atlas of Xinjiang Uyghur Autonomous Region* and the *Atlas of the Status of Fujian Province*; They also studied the nature of the third generation of provincial atlases, the compiling principle, selection of contents, data compilation, contents and expressions, compiling characteristics, and so on and put forward the corresponding design ideas. These findings offered guidance for the theory and a method of preparation for a general regional atlas. Wang (2008) put forward three issues regarding the overall design of a regional comprehensive atlas, that is, the content system design based on the concept of regional scientific development, the map information representation design based on cartographic theory, and the compilation process design based on databases and GIS. Additionally, the relevant design thoughts were discussed, which provided a reference for the preparation of the general regional atlas.

Under the guidance of atlas design theories, beginning from 1990, the compilations of the *Great National Atlas of China* and many high level comprehensive regional atlases were completed and published one after another. Representative works are listed as follows: *The National Agricultural Atlas of the People's Republic of China, The National Economic Atlas of the People's Republic of China, Atlas of the Jiangsu Province, Atlas of the Zhejiang Province, Atlas of the Jiangxi Province, Atlas of Chongqing City, Atlas of the Xinjiang Uyghur Autonomous Region, Atlas of the Jilin Province, Atlas of the Status of Fujian Province*, and so on. Compared with the original atlases, these atlases have been greatly improved in terms of content, representation method, color use, and symbol design. The general maps have fine symbol design, harmonious use of color, and a strong sense of three-dimensional terrain; the representation method of the thematic map is novel, the thematic information is prominent, the symbol design is vivid and intuitive, the color is highlighted, the level is distinct, and the status and development characteristics of the thematic phenomena in different regions are well expressed. These changes demonstrated the characteristics of the atlas as a product of scientific research, economic planning, and scientific tools for external propaganda.

(2) Study on the overall design of the city atlas.

The city is the center of politics, economy, culture and transportation, and cities have an important status in social development and human activities. With the rise of the information revolution, the combination of map science, spatial aerial remote sensing, computer technology, and so on has led to the renewal of the traditional concept of urban mapping. Some city atlases and city image atlases have since appeared.

The main study results on city atlases in China are outlined as follows. According to the compilation of the *Atlas of Shanghai City*, Zou (1990) studied the function of a comprehensive city atlas and the basic principles of map compilation; he emphasized that the topic and content of an atlas are based on the urban geographical science system, which aims to reflect the nature of a city, and the focus was on practical application. Qi et al. (1991) analyzed the compilation and printing characteristics of the *Atlas of Xi'an City* in terms of content selection, internal structure, pattern design, decoration, printing, and so on. Yang et al. (1992) discussed the overall design idea and design characteristics of *Atlas of Xi'an City*, as well as the mathematical foundation, format, expression form, content structure, arrangement principle, geographical base map and compilation process. Ma et al. (1994) discussed the design principles and characteristics of the city atlas, illustrating the important guidance and promotional role of the city atlas in city modernization, scientific management and planning. He et al. (1999) described in detail the characteristics of *Atlas of Shenzhen City*, such as overall design, content arrangement, basic geographic base map compilation, color design, layout design, general map series compilation, thematic map series compilation, production technology of electronic versions, and so on.

The main study results on city image atlases in China are outlined as follows. Huang (2002) described the design idea and compilation of *Image Atlas of Shanghai City*, and the process of making an image map using a computer; he showed the advantages of computer aided mapping technology in improving the quality of map works and production efficiency, as well as reducing production costs, etc.; Yuming described the difference between an image map and traditional line map in the expression of spatial information. Based on the design and compilation of the *Image Atlas of Suzhou city*, Wang et al. (2007) focused on several key technical problems, such as the overall design of the atlas, the expressed contents and representation method, data selection and processing, digital proofing, new process of computer direct plate making, and so on. The research has important reference value for the theory, method and technology of image atlas production based on high resolution aerial (remote sensing) images.

On the basis of research and experiments, a large number of urban atlases have been prepared and published, and they are important reference tools for urban construction and planning. At the same time, a variety of electronic atlases and multimedia electronic atlases were also established, which reflect the characteristics of the times in terms of map representation methods. Among them, the *Atlas of Shenzhen City* achieved the Outstanding Cartography Achievement Award awarded by ICA (International Cartographic Association) in 1999. This is the first time that one of China's map works won this award.

(3) Study on the overall design of the electronic atlas

Based on the publishing of a paper version of the atlas, the overall design of the electronic atlas was studied and some achievements were obtained. By comparing the electronic atlas with the traditional atlas, Liu (2003) considered that the design of an automatic map subdivision was more dynamic and interactive and mainly affected by map scale, geographical features of the mapping region and map projection. The core of automatic map subdivision is the calculation of the subdivision; the purpose is to retain the elements within the scope of a separated sheet. Xu et al. (2003) discussed the function of the multimedia electronic atlas of a city and pointed out that the multimedia electronic atlas of a city is no longer just a map product; it can be viewed as an information system-urban information system. Wang et al. (2005) discussed the automatic map subdivision techniques of electronic atlases, the database storage structure, management techniques and the automatic generalization and multi-scale display technologies. Based on the development of an electronic atlas of Wuwei city, Shi et al. (2008) developed and studied the design and integration of urban electronic atlas to serve the government and the national economic construction department, putting forward a hypermedia network information organization structure based on Series-Thematic map-Map sheet to reflect the basic geographical features of the region; they also designed the content framework of the integrated electronic atlas of the city area with the distribution rules of all kinds of natural and social economic factors and their relationships as the main content; finally, they designed the functions of attribute query, dynamic map display, multimedia automatic link, 3D flight and so on, as well as user identification and information interaction interfaces.

4.4 Problems and Prospects

4.4.1 Research on the Theory and Application of Modern Map Design Should Be Strengthened

In traditional cartography, technically, map design mainly solved expression and technical realization problems, such as map content abstract and generalization in the processes of field to map, map projection, map representation, map making process and so on. However, in modern cartography, a significant feature of map design is not only the focus on the process from field to mapping but focusing more on the relationship between reader and map. In other words, it stressed that the user is the judge and consumer of the map, so human and map must be studied as a whole. The theoretical and practical problems in map content design, expression model design, representation method design, map layout configuration and effect design are explored to adapt to changes in new map requirements. These design concept changes in map design place more emphasis on the guiding role of people's cognitive ability and people's map perception effect on map design; additionally, map design will gradually develop from passive design guided by experience to active design guided by theory.

4.4.2 The Intelligence Level of Map Design Should Be Improved

At present, since the spatial cognitive mechanism of the brain remains a "black box", a map designer can design a beautiful map, but the knowledge and thinking process used in this process are not fully understood. Therefore, there are many theoretical and technical problems to be studied and solved in the field of map design. In theory, as the map design involves human vision and visual perception, it has close contact with frontier science, such as some theories and methods in the field of cognitive science and cognitive psychology; also, it must explore how to promote the image thinking ability of the brain by the 'graphic language' of the map, enhancing the spatial cognitive ability, graphics memory, understanding ability and so on. Therefore, the research on map use, map perception and spatial cognition will be the important theoretical issues in the future. It involves not only continuing to study the theoretical problem of the traditional paper map but also the study of the theoretical issues of new types of maps, such as digital maps, electronic maps, multimedia maps and atlases. In terms of techniques and methods, with the application of computer technology, artificial intelligence technology, data mining, knowledge discovery technology and so on in the discipline of cartography and map design mainly studies how to use data mining and knowledge discovery technology to construct a new map expression content system and form a 'knowledge map'. At present, map works and their service objects are still very limited, and the content of map works contains little information. We must study how to deeply process a large number of resources and environmental information with the viewpoints and methods of system theory to provide users with an intelligent knowledge base, deep processing and practical final products, as well as update the values of map design, with a focus on practicality and science. We must research the data organization and hierarchical display mechanism, interface and map expression for map products, especially electronic maps, multimedia maps and electronic atlases. We must research how to use virtual reality technology and visualization technology to build a real time and realistic virtual geographic environment with the support of a spatial database. Also, we must research how to use life science and technology, as well as artificial intelligence technology to establish an intelligent map design system, and so on.

4.5 Representative Publications

(1) ***The Compilation and Design of the Map*** (GAO Jun. Zhengzhou Institute of Surveying and Mapping, Zhengzhou, Henan, China, Sept., 1977)

The whole book consists of 12 chapters. Chapter one, an overview, describes the role of map compilation and design in map production, and the general contents of map compilation and design. In the second and third chapters, the general map is summarized; the key is the type and use of the topographic map. The fourth chapter, the overall design of the map, describes the process of overall design and the specific content of each process. From chapters five through seven, the research contents and methods of map compilation data, regional research and geographic name translation are described, respectively. The eighth and ninth chapters describe the representation method of the various content elements in general maps, focusing on the terrain representation. In the tenth chapter, the principles and methods of map symbol design and font selection are introduced. In the eleventh chapter, the principle and method of programming for the map production process are introduced, and the typical map making process is given. Chapter 12 introduces the final results of the map design, that is, the types of map compilations and design documents, as well as their content and writing methods.

(2) ***Principles of Map Design*** (CHEN Yufen, JIANG Nan. The People's Liberation Army Press, Beijing, China, Aug., 2001)

The book comprehensively and systematically introduces the basic theories and technical methods involved in map design, focusing on the new theories and core technologies in map design, while considering the traditional map design theories and methods.

The whole book consists of nine chapters. Chapter one covers the map and map design. Chapter two covers the basic theories of map design. Chapter three discusses the overall design of the map. Chapter four discusses determining the map content. Chapter five is the design of the content representation method for the general map. Chapter six covers the design of the content representation method for the thematic map. Chapter seven covers map annotation and translations of geographic names. Chapter eight discusses the design of the map production process. Chapter nine is the design of electronic maps.

(3) ***Map design and compilation*** (ZHU Guorui. Wuhan University Press, Wuhan, China, Oct., 2001)

The book is divided into 4 parts and has a total of 15 chapters. The first part consists of four chapters. Chapter one covers the map and cartography; chapter two discusses the map symbol and map representation; chapter three is the mathematical foundation of the map; and chapter four describes the map making process. The second part of the book consists of two chapters, where chapter five covers the geographical variables and graphical representation, and chapter six discusses map generalization. The third part of the book consists of five chapters, where chapter seven covers the map design documents; chapter eight is the map layout design; chapter nine is the cartographic region and mapping data; chapter ten discusses map compilation manuscripts and publishing preparation; and chapter eleven covers the design and production of an atlas. The fourth part of the book consists of four chapters; chapter twelve is the general map; chapter thirteen is the natural map; chapter fourteen is the social and economy maps; and chapter fifteen is the special map.

(4) ***Map design and compilation*** (WANG Guangxia et al. Surveying and Mapping Press, Beijing, China, Jun., 2011)

This book comprehensively introduces the theories, technologies, methods and applications of map design and compilation. The whole book consists of 11 chapters and an appendix. The first chapter is the introduction, which introduces the concepts of the map, map design and compilation, as well as the research content and development; the second chapter introduces the theoretical basis of map compilation design; the third chapter describes the content of map editing and design and the preparation work; the fourth through seventh chapters describe the theories and methods of the overall design, map symbol design, color design and map representation design; the eighth through tenth chapters expound upon the basic concepts and characteristics of map graph generalization and geographic information generalization, as well as the methods of cartographic generalization and the concepts and methods of electronic map multi-scale representation; chapter eleven analyses the characteristics of typical map product design and compilation; and the appendix lists the contents and requirements in the map design and compilation practices.

(5) ***Atlas of the Yellow River Basin*** (Chief Editor: ZHANG Zhengming, LI Hongjie; preparation unit: Yangtze River Water Conservancy Committee of Water Resources Department; publishing units: China Map Publishing House; time of publication: December 1989)

This atlas is a large comprehensive atlas for the river basins in China. The objective form of the atlas is to reflect the history of the Yellow River, the natural geographical environment of river basins, the characteristics of water and sediment resources of the Yellow River, as well as harnessing these resources and the development achievements and scientific research results; this atlas provides a scientific basis for understanding and studying of the Yellow River, as well as controlling and developing the Yellow River and developing the basin economy.

As a standard 8 folio atlas, the content includes six groups, such as general maps, historical, social and economic maps, natural conditions and resources maps, control and exploitation maps, and tributaries of the river; there is a total of 92 pictures, 260 thousand commentary texts, and more than 100 color photos. The atlas gives priority to the map, which is combined with text, color photos and statistical charts; additionally, this atlas is published in two kinds of forms: bound book and loose leaf.

The atlas uses the data from fixed stations (sites), the latest research results and a great deal of maps to reflect the distribution of water resources in the Yellow River from different aspects (different series, measured and natural runoff, surface water, groundwater, and so on). As a representation method, according to the characteristics of the water and sediment in the Yellow River, in addition to displaying the water quantity increment with the width, the atlas also vividly displays the variance in sediment concentration using different colors. The DC figure is an important part of the atlas, and the selection principle is according to the size of the catchment area with consideration for the amount of incoming water and sediment; there is a total of 14 pictures of 25 tributaries. To reflect the evolution law of the deposited silt at the estuary mouth, in addition to the estuary regional map, modern change charts in the Yellow River estuary and sediment diffusion charts in the coastal area were compiled. The geographical base map of the atlas takes the national topographic map as the basic data and takes the Standard Map of Province-level Administrative Boundary, as well as the relevant provinces (autonomous regions) map data, and so on as supplementary reference materials. The data of the Yellow River estuary and its nearby coastline were updated according to the satellite image at that time.

(6) ***Atlas of Qinghai-Tibet Plateau*** (Chief Editor: LIAO Ke; deputy editor: LV Renwei, LIN Kangtai; preparation unit: Institute of Geography, Chinese Academy of Sciences; publishing units: Science Press; publishing time: December 1990)

The atlas is a summary of the scientific expedition and research results in the Qinghai-Tibet Plateau and has high scientific value and application value; the design concept and scientific content have reached an advanced international level. As an excellent piece of scientific literature, from the viewpoint of art, this atlas shows the dynamic evolution of a unique geographical region, which vividly depicts the historical and geographical relations between humans and nature. The atlas scientifically utilizes the various representation methods of a thematic map and is creatively combined with examples. By making full use of the map, aerial photos, satellite images, topographic surveying, mapping information, picture examples, this atlas completely and visually reflects the complicated natural conditions and rich natural resources, as well as the development and utilization prospects of the Qinghai-Tibet Plateau. The atlas not only shows the characteristics of the natural environment on the Qinghai-Tibet Plateau but also reveals the formation and evolution of the plateau and its impact on the environment and human activities. The atlas includes the introduction, geology, topography, climate, hydrology, soil, vegetation, animals, resource utilization, and partition general map for a total of 10 map groups and 233 color maps. There are also some charts, satellite remote sensing images and ground photographs, and 100,000 words of explanatory text, as well as a gazetteer with 7,000 geographical names. It is a standard 8 folio atlas, and all maps, charts, explanatory text, and place names in this atlas were published for the first time. In addition, the 1/4 maps are the research achievements from the process of atlas compilation.

(7) ***The officer Atlas*** (Chief Editor: GAO Jun; deputy editor: WANG Jiayao; preparation unit: Headquarters of the General Staff; publishing units: The People's Liberation Army Press; publishing time: December 1992)

This atlas is a comprehensive atlas for use by military cadres, and it mainly provide the political, economic, military, and other comprehensive geographic environmental information of the world, continents, oceans, and China, as well as neighboring countries, large districts, key regions and choke points at strategic and campaign levels.

The atlas is composed of seven parts: an introductory map, world map, continental and oceanic map, China map, map of neighboring countries, map of the history of the Chinese people's Revolutionary War and an appendix. The introductory map plays a guiding role for the atlas, including the universe and outer space, the human understanding and exploration of the universe, the earth, the human understanding and mapping of the earth and its military function and significance. The world map comprehensively presents global political, military, and economic situations and the natural environment. The continental and oceanic map comprehensively shows the politics, nature, resources and economic situations of the continents and oceans, highlighting key areas and strategic routes, and selected example battles are also shown. The China map is the focus of the atlas; it includes the whole China map, which shows China's politics, nature, resources, economy, transportation, humanities and other conditions from an overall standpoint; the sea area chart shows China's vast sea area; the group of maps with large districts show the units reflecting the regional geographical environmental characteristics, terrain, natural conditions, economic strength, human resources and transportation maps and provincial maps. The map of China's neighboring countries, including China's land and sea neighbors, reflects the basic geographical environment. The map of the history of the Chinese people's Revolutionary War shows typical battles that reflect the relationship between combat operations and the regional geographical environment.

The content of the atlas is abundant with an innovative design, bright colors, vivid symbols and pictures, and the accompanying words are also excellent; so this atlas has strong practicability and high artistic quality. In the layout of the contents, it highlights the military characteristics, focusing on the military significance of the "typical contents", "typical events" and "typical regions"; in the design of the representation method, it focuses on the unity of art and coordination to allow people to enjoy its beauty. The compilation and publishing of the Officer Atlas represents the science and technology level of military surveying and mapping in China since the 1980s.

The atlas won the Excellent Military Scientific Research Product Award in 1992 and the Military Science and Technology Progress Award in 1999.

(8) ***China's City Atlas*** (Chief Editor: Ministry of Construction; preparation unit: Science and Technology Development Promotion Center, Ministry of Construction; publishing units: China Map Publishing House; publishing time: June 1994)

Compiled by the China Ministry of Construction, this atlas was the first large-scale comprehensive city atlas created in China. It incorporates all the planned cities of China from the founding of new China until the end of 1989, with a total of 445 cities (Taiwan, Hongkong and Macao were not yet included). This atlas comprehensively and systematically reflected the natural environment, the cultural and geographical characteristics, the urban development evolution and the long-range plan for Chinese cities; this atlas propagates the great achievements of city construction in the forty years of new China; the atlas is an all-around, multi-level atlas to show the style of the Chinese city. The compilation of the atlas began at the end of 1989, and after four years of efforts, it was published in 1994. It uses fair drawing and scribing, as well as the copy process method, with spot color printing and a wireless adhesive hardcover process. In addition, it will play an important guiding and promoting role in the modern construction and the scientific management and planning of cities.

(9) ***The National General Atlas of the People's Republic of China*** (Chief Editor: YU Cang; preparation unit: China Map Publishing House, Shanxi Provincial Bureau of Surveying and Mapping, Chinese Academy of Surveying and Mapping; publishing units: China Map Publishing House; publishing time: January 1995)

The National General Atlas is the first volume of the series of the National Atlas of the People's Republic of China and is the basis for the preparation of other sub volumes. The atlas includes four parts, such as introductory maps, regional maps, provincial maps, and the index of place names. The introductory maps consists of 18 nationwide thematic maps and 20 statistical charts, reflecting the macro distribution the regional mapping factors in the whole country. There are 20 maps in the regional map group, highlighting the natural and geographical landscapes in four regions of Northeast China, Southeast China, Southwest China and Northwest China, as well as the major residential areas, the traffic line distribution and the terrain of both sea and land, especially the topographic features of the sea bottom. The provincial map group is the main body of the atlas, the nation and provinces (autonomous regions and municipalities directly under the Central Government) each have a map, the 9 typical areas under the jurisdiction of provinces (autonomous regions) each have an enlarged map, and 108 major cities each have a map; the distribution and relationship of the six elements and related contents of the general geographic map are expressed in detail. They can be used as the important basis to understand the actual situation of one province and the natural geography and social economic ties of the surrounding provinces. The index of place names contains a total of more than 40000 place names of provinces and autonomous regions; each of them lists the names of Chinese characters, Roman alphabet spelling and the coordinates on the map for convenience of quick retrieval on provincial (autonomous regions) maps (the place name index is compiled according to the place name database established by the census data of the national geographic name).

The atlas is compiled using the latest measured data and new research results of geosciences surveys, and parts of them are updated using satellite images to ensure the novelty of the maps. There is innovation in the representation method. For example, the terrestrial maps of the regional map group use the new expression method of landscape tinting for the general map, changing the traditional concept of paying more attention to the land than the sea; it uses a light purple color system to express the Qinghai-Tibet Plateau, breaking through the usual practice of China's hypsometric contour. The altitude scale is divided into three kinds of colors according to three major ladders of China to better shows the terrain features of all regions.

(10) ***Atlas of Shenzhen City*** (Editor: LIU Jiasheng; deputy editor: HUANG Rentao, GUO Renzhong, HUANG Fulai; preparation units: Shenzhen Municipal Planning and Land Source Bureau, Wuhan Technical University of Surveying and Mapping; publishing units: Shenzhen Municipal Planning and Land Source Bureau (For internal use only); time of publication: April 1997)

This atlas is the representative work of the China urban atlas and is the first large-scale comprehensive atlas of China released simultaneously in both printed and electronic editions. It integrates the map language, pictures, video, graphics, text and other multimedia tools, and comprehensively and systematically reflects the resources, environment, population, social economy, and developmental planning of Shenzhen city, and this atlas is innovative in content structure and expression mode.

The atlas uses the latest map of Shenzhen city, 1:10,000 aero photogrammetric topographic maps, and the latest professional data provided by various departments as basic data; additionally, computer mapping and electronic publishing technology were used to compile this atlas. It is the earliest successful attempt to create a large engineered atlas by using computer mapping.

The atlas consists of five groups, such as introductory maps, regional detailed maps, social and economy maps, natural and environment maps, and development and planning maps. It is a standard 8 folio atlas, with 177 maps, 154 charts, 162 photos and 6 aero photos.

The representation method of the atlas is novel, the thematic information is prominent, the symbol design is scientific and reasonable, and the drawing is delicate. In terms of the map color design, the color is very exquisite, scientific and reasonable.

The design of the main tone focuses on the map content and makes full use of the contrast color to enhance the level of content expression and to enhance the clarity and beauty of the graphics expression.

The atlas won the Outstanding Cartography Achievement Award in the nineteenth general assembly of ICA (International Cartographic Association) in 1999. This is the first time that China won this award, which is the highest honor in International Cartography.

(11) ***Atlas of the Yangtze River Basin*** (Editor: WEN Fubo; deputy editor: YAO Chuguang, CAO Shengzhong, CHEN Bingquan; preparation unit: Yangtze River Water Conservancy Commission, Ministry of Water Resources; publishing units: China Map Publishing House; time of publication: August 1999)

This atlas is a large comprehensive river basin atlas that provides a comprehensive description of the trunk stream and branch of the Yangtze River. It focuses on water conservancy construction and also considers the natural resources, social and economic development, and the historical and cultural environment. It is also a large mapping project. From content to form, the atlas tightly encircles the theme of water resources and water conservancy construction, which is a highlighted subject. It takes the comprehensive utilization plan report of the Yangtze River Basin approved by the State Council (revised in 1990) as the basis, while absorbing and summarizing the achievements of various disciplines in the development of the Yangtze River. This atlas is based on maps and supplemented by text descriptions, photographs and tables to facilitate the readers understanding and knowledge of the Yangtze River intuitively, vividly and profoundly. The atlas is divided into 7 groups. The introductory group mainly includes the introduction and overview. The historical map group describes the historical evolution of the Yangtze River and its general situation, aiming to provide historical data for the development of the Yangtze River. The map group of natural conditions reflects the geology, topography, climate, soil, vegetation, mineral resources, hydrology, water quality and other natural geographic factors of the Yangtze River Basin; this river has a longer length, which is the most important part in the atlas. The social and economic map group mainly reflects the land use, industrial and agricultural, urban, transportation, tourism, economic development and other contents of the Yangtze River Basin. The map group of environmental protection mainly represents water quality evaluation, water quality pollution, water source protection, distribution and protection of rare animals and plants, fish resources and protection, distribution and control of schistosomiasis and endemic diseases, and other contents in the Yangtze River Basin. The map group of water resources development and use mainly reflects the comprehensive utilization plan of water resources, water conservancy and hydropower engineering, flood control, power generation, irrigation, water and soil conservation, river regulation of the middle and lower streams, urban water supply, the South-to-North Water Diversion Project, and so on in the Yangtze River Basin; this group highlights the social and economic benefits of the Three Gorges water conservancy project, which is the focus map group of the atlas. The map group of dry tributaries mainly reflects the relative position of the main tributaries of the Yangtze River, the basic situation of each river section of the mainstream, the general situation and development plan of the main tributaries and is also the focus map group of the atlas. The 7 groups are interrelated and complement each other, which basically reflects the whole picture of the Yangtze River Basin.

(12) ***Atlas and Gazetteer of Standard Geographical Names in the People's Republic of China by Administrative Division*** (Editor: LI Baoku, LI Zhiguang; preparation of units: Ministry of Civil Affairs of the People's Republic of China and the Bureau of Surveying and Mapping of the General Staff of the People's Liberation Army; publishing units: the Planet Map Publishing House; time of publication: September 1999)

This atlas is a thematic atlas that is based on the achievements of administrative divisions, geographical name management and mapping in the 50 years since the founding of the People's Republic of China. This atlas comprehensively and accurately reflects the standard name of China's administrative region and the distribution of the administrative areas, government residential locations and the historical evolution. It has more than fifty-three thousand administrative region standard names. The atlas was compiled by using digital mapping technology and is in standard 8 folio format. Then, after more than 1 year, the atlas was revised with updated data, and it was compiled and published in 16 folio format in January 2001.

The atlas consists of three major parts, including introductory maps, provincial administrative area maps and an appendix. The introductory maps include the China administrative division, relief, urban distribution, distribution of nationality autonomous districts, and evolution of territory and contemporary administrative regions; this atlas reflects the evolution of China's territory and the first-level administrative division since the Qin Dynasty, and it emphasizes the development situation of the provincial administrative regions of contemporary China. The provincial administrative area maps include sketch maps and detailed maps of administrative divisions, city maps and text descriptions. The appendices include the contrast between new and old names of some cities and counties, a table of the names and the setting time of Ethnic Autonomous Regions, a name list of city municipal districts and street offices, the national administrative region name index, and a list of the uncommon words in the administrative district names and their local pronunciations.

This is the first atlas with an administrative division standard name as the theme after the founding of the People's Republic of China. The administrative division names of the 4 levels of province, region, county and township in China are the standard names approved by the government at all levels. All of them are represented in the atlas, and the full names are given to reflect their categories and levels and to reveal the rich connotation of the administrative division standard name. This is one of the main features of the atlas. This atlas includes the map, table, text description and other means to illustrate the theme from multiple aspects and to show the evolution of the administrative division in each of the main historical periods over the past

2000 years. At the same time, it also highlights the main contents of the contemporary administrative divisions. The information is rich and clear with a prominent theme that is easy to read. The atlas has the characteristics of rich information, distinct levels, prominent theme and readability. It won the First Excellent Map Award issued by Chinese Society of Geodesy, Photogrammetry and Cartography (CSGPC) in 2004.

(13) ***Atlas of the Xinjiang Uygur Autonomous Region*** (Editor: LIU Geqing; deputy editor: LI Quanzhan, CHANG Gejun; preparation unit: Bureau of Surveying and mapping of Xinjiang Uygur Autonomous Region; publishing units: China Map Publishing House; publishing time: August 2004).

This is a general atlas that reflects the natural geographical features and main social economic phenomena in the autonomous region. There is also cartography and geographic information system engineering involved in multiple disciplines and multiple domains.

The atlas contains information according to the principles of accuracy and completeness, which includes scientific, unified, up-to-date data, as well as and safety data. The new technologies of an integrated GIS and a digital computer desktop mapping system are used to establish a database and compile the atlas, reflecting the use of an advanced level of modern mapping and GIS technology.

The atlas is composed of two parts: introductory maps and regional maps. The former part generally reflects the natural, economic and social development profile of the autonomous region (29 pages); the latter part reflects in detail the basic situation of each prefecture, county and city (278 pages). The atlas contains 169 maps, 13 charts and 20 photos, including approximately 100,000 words of text to comprehensively, systematically, intuitively and vividly show the basic situation in Xinjiang. The introductory map group has 14 mapping units, including administrative regions, geographical location, satellite image, terrain, climate, water system, vegetation, land use, mineral resources, transportation, tourism, Production and Construction Corps and others. The regional map group consists of 140 mapping units of three map types, including prefecture, county and city maps. In the aspects of layout arrangement, map decoration, symbol design, representation method, color scheme and so on, the atlas is fresh and natural, novel and unique, with a distinctive theme and clear level; it also contains supplements from before and after the echo, which are harmonious and united.

(14) ***Atlas of the Jiangsu Province*** (Editor: SHI Zhaoliang; preparation of units: Basic Geographic Information Center of the Jiangsu Province; publishing units: China Map Publishing House; time of publication: September 2004).

The atlas consists of five parts, including introductory maps, population, resource and environment maps, economic and societal maps, sustainable development maps and city, district and county maps. The atlas is in standard 8 folio format. There are 48 thematic maps, 150 general maps, 80 color photos, and approximately 200,000 words of text in a brief introduction. Therefore, it has the largest format and is the most fundamental masterpiece in the Jiangsu Province atlas (copies) series. The atlas is compiled based on the latest geospatial data of various-scale digital line maps and satellite remote sensing image data provided by the Basic Geographic Information Center of the Jiangsu Province, as well as the latest statistics from the *Statistical Yearbook of the Jiangsu Province* and various departments.

An advanced integrated computer mapping and publishing system was used to complete the atlas design, data processing, map drawing, prepress processing and plate making. The structural arrangement, graphic selection, symbol making and element generalization methods are novel and scientific, meeting the requirements of modern cartographic theory. The decorative layout and the use of color not only contribute to the intuitive and vivid display of contents, but overall, it reflects the distinctive professional characteristics, local characteristics and the characteristics of the times.

In 2006, the atlas won the gold award, "PEI Xiu Award for Excellent Map Works".

(15) ***Atlas of Chongqing City*** (Editor: JIANG Yong; preparation of units: Chongqing Investigation and Surveying Institute; publishing units: Xi'an Map Press; published: April 2007).

This atlas is a comprehensive provincial city atlas and is the first atlas created since Chongqing city as the municipality directly under the central government.

This atlas is in a standard 8 folio format. The content includes introductory maps, population, resource and environment maps, economic and societal maps, district and county maps and development plan maps. There are a total of 248 pages of maps. Among them, there are 51 pages of thematic maps, 84 pages of general maps, 7 pages of aerial pictures, 2 pages of satellite images and 76 pages of photos.

The atlas is designed and compiled by using a combination of multiple elements, such as illustrations, text and pictures and is completed with full digital mapping technology. It brings together many of the latest 1:10,000 topographic maps of Chongqing city, the 1:50,000 place names database of Chongqing city, the administrative maps of Chongqing city, and a variety of the latest yearbooks from Chongqing city and other information; thus, the atlas systematically and thoroughly presents the situation and development of Chongqing city.

The representation method of the thematic map is novel, the thematic information is prominent, and the symbol design is vivid and intuitive, with color that is harmonious and unified. The district and county maps use layered tinting coupled with the hill shading method to represent landforms; the symbol and color of urban areas is prominent and the level is clear. The geographical status and development characteristics of natural and social factors in different regions were well expressed.

In 2008, the atlas won the gold award, "PEI Xiu Award for Excellent Map Works".

(16) ***Atlas of Zhejiang Province*** (editor in chief CHEN Jianguo; preparation of units: The First Surveying and Mapping Institute of the Zhejiang Province; publishing units: China Map Publishing House; publication dates: January 2008)

The atlas is composed of six parts: the introductory map group, population, resource and environment map group, economic and societal map group, development plan map group, regional geographic map group and indexes. The atlas is in standard 8 folio format. There are 52 pages of thematic maps, 79 general maps, 74 urban maps, 350 color photos, and approximately 200,000 words of text in a brief introduction. The atlas mainly shows the Zhejiang Province's natural environment and resources, humanities, social and economic development and planning, the administrative divisions at or above the county level, township boundaries, topography, geomorphology, water systems, infrastructure, transportation and appendages, the distribution of residential areas and other conditions; the atlas provides basic geographic information and a scientific basis for the formulation and implementation of the sustainable development plan for all levels of government.

Based on the latest multi-scale digital line map, satellite remote sensing image data, a digital elevation model and other geographic spatial data provided by the Bureau of Surveying and Mapping of Zhejiang Province, the atlas is compiled and published by using digital mapping technology and with the support of computer technology, database technology, remote sensing technology and geographic information system technology. The thematic map is represented in a novel way; the, thematic information is prominent, and the symbol design is scientific and reasonable, with color that is harmonious and beautiful. The general maps use layered tinting coupled with the hill shading method to represent landforms and overlay various ground feature symbols, the symbols and colors of urban areas are prominent and the level is clear. The characteristics of the regional geographical environment are well expressed. The index contains nearly a thousand geographic names that appear in city and county maps, such as streets, islands, mountains, scenic areas, nature reserves, forest parks, the main tourist attractions and so on.

The atlas won the gold award, "PEI Xiu Award for Excellent Map Works" in 2008.

(17) ***Atlas of the Jiangxi Province*** (Editor: LIU Baohua; preparation unit: Third Surveying and Mapping Institute of the Jiangxi Province, Jiangxi Province Basic Geographic Information Center; publishing units: China Map Publishing House; publication date: May 2008).

The atlas is composed of five parts: the introductory map group, population, resource and environment map group, economic and societal map groups, sustainable development map group, and the map group of districts (cities and counties). The atlas is in standard 8 folio format. There are 166 pages of thematic maps (38 topics), 186 general maps, 167 color photos, and approximately 200,000 words of text in a brief introduction.

By using visual map language and supplementing with text, charts and color photos, the atlas comprehensively and systematically shows the basic geographic information and the economic, societal, resource and environmental and other basic situations of the Jiangxi Province.

Based on the latest multi-scale topographic map and satellite remote sensing image data and with the support of the latest professional data provided by various departments, the atlas is compiled and published using digital mapping technology. In terms of topics and content, the atlas includes a thematic map to highlight Jiangxi's features, such as the red revolutionary cradle (former red capital, the cradle of the People's Republic China), green homes and the Millennium Ceramic Capital. The symbols are of very fine quality; the color is exquisite, and the atlas is scientific and reasonable.

The atlas won the gold award, "PEI Xiu Award for Excellent Map Works" in 2008.

(18) ***Atlas of the New Beijing Great Olympics*** (Editor: ZHANG Hongnian; deputy editor: DENG Nan; compilation units: Beijing Municipal Planning Commission, Beijing Surveying and Mapping, Design and Research Institute; publishing units: China Map Publishing House; time of publication: July 2008).

This is a thematic atlas that introduces Beijing and its service to the Beijing Olympic Games. On the basis of the Beijing urban basic geographic information data platform, with "New Beijing, Great Olympics" as the main line, it uses a bright, innovative design style and intuitive and visual map language to show the long history of Beijing, the harmonious and livable environment, the dramatic development of the central city, and the thriving new towns and suburban counties; the atlas expresses the development course of the Olympic movement and the twenty-ninth Olympic torch relay route; and the atlas demonstrates the beautiful desire for "One world, one dream" and "To ignite passion, to pass the dream". Also, the atlas exhibits the Olympic competition venues, training venues, non-competition venues, contracted hotels, designated hospitals and other thematic information, as well as the cityscapes of the assisting host cities, such as Qingdao, Hongkong, Tianjin, Shanghai, Shenyang and Qinhuangdao.

The atlas includes an introduction to the Olympic Games, an overview of Beijing city, the central city of Beijing, Beijing's new towns, Beijing's suburban counties, an index and an electronic CD-ROM; the atlas is available in both Chinese and English. The data are from December 2007.

(19) ***Atlas Series of China's Provinces*** (Editor: XUE Guijiang, YAO Jie; deputy editor: ZHOU Ruixiang, LIU Hongnian; compilation units: Planet Map Publishing House; publishing units: Planet Map Publishing House; time of publication: June 2009).

This is a large provincial series atlas that was edited and published for the first time in the 60 years after the founding of the People's Republic of China. The atlas uses visual maps that are complemented by graphics and tables, texts and photos to vividly reflect the 60-year construction achievements of China's provinces (autonomous regions, municipalities directly under the central government) and special administrative regions in the administrative divisions, towns, transportation, tourism and other aspects.

The atlas is rich in content and novel materials, and the map compilation, layout and cover design are expertly planned. The complete set includes 34 volumes in 16 folios. The main color of the cover is gold, which conveys magnificent, elegant and brilliant character, and the artwork is pleasing to the eyes.

Each volume of the atlas is mainly composed of introductory maps, prefecture level maps, county level maps, and city maps, as well as text introductions and scenic pictures. The introductory maps offer an overview, which shows the provincial administrative divisions, topography, transportation, tourism and other comprehensive information. The prefecture level map acts as an administrative map to show the geographical information of all the county level administrative regions under the jurisdiction and plays a connecting role. As the main body of atlas, the county level map is a topographic map to show in detail the administrative division boundaries, residents in townships and parts of the villages, highways and ancillary facilities, as well as county and township roads, railways, railway stations, airports, ports, water conservancy facilities, scenic spots and so on. A single county's map uses the layering method to reflect the relief and topographic features of the county. These are the highlights and features of the atlas. The city map is mainly to show the city's periphery, main street, transit route, public facilities and so on. Through the above four map groups and the photos and texts, the reader can intuitively, systematically and easily understand the situation of a province or county and the regional geographic characteristics of the country, as well as experience the great changes that have occurred in new China in the past 60 years from a multiple perspectives.

The unique compilation method and layout design style gives the atlas a unique position in the current domestic map market.

(20) ***Atlas of the Jilin Province*** (Editor: ZHANG Limin; preparation unit: Geographic Information Engineering Institute of the Jilin Province; publishing units: China Map Publishing House; publication dates: September 2009).

This is the first large-scale comprehensive atlas compiled by the Jilin Province; it comprehensively reflects the natural and human geography and the economic and social development of the Jilin Province.

The atlas is in standard 8 folios and consists of five parts, including the introductory map group, population, resource and environment map group, economic and societal map group, sustainable development map group, and the map group of cities, districts, counties and development zones. The atlas is compiled by using geographic information system technology, remote sensing technology, and database technology. The atlas integrates feelings, science, artistry and practicality in one; it is rich in both pictures and text, with strong practicability. The atlas uses up-to-date thematic information and data, among which the geographical base map of the provincial thematic map contains the 1:500,000 *Jilin Province Map* compiled in 2007 as basic data. All kinds of thematic information are included based on the latest information provided by the professional departments and the *Jilin Statistical Yearbook* in 2008. The county maps are compiled using the latest versions of the 1:50,000 digital topographic maps from the National Basic Geographic Information Center. The city maps are updated using the 2.5 m resolution SPOT satellite image from the recent two years. The atlas is both current and accurate in its content.

(21) ***Atlas of the Anhui Province*** (Editor in chief: XU Tiejun; preparation unit: Fourth Surveying and Mapping Institute of the Anhui Province; publishing units: China Map Publishing House; publications date: January 2011).

This is the first large-scale comprehensive atlas compiled by the Anhui Province. It consists of five parts: the Anhui overview map group, population, resource and environment map group, economic and societal map group, scientific development map group, and the map group of cities, districts and counties. The atlas begins with an introduction to Anhui's Location in China and contains 49 pages of thematic maps, 113 general maps, 600 color photos, and approximately 200,000 words of text in a brief introduction. The atlas uses visual design language and is supplemented by charts and tables, text descriptions and color images, systematically reflecting the basic situations of the Anhui basic geographic information, economy, society, resources, environment, and so on. To exhibit the regional characteristics of Anhui, four maps, i.e., "Historical celebrities", "Ancient rhyme of Huizhou ", "The war flames in the Yangtze-Huaihe region" and "Beautiful Mount Huangshan" are set up to highlight the rich cultural heritage and beautiful natural scenery of Anhui; meanwhile, the layout design of the four maps is unique, which breaks through the centralized and unified layout style in the traditional map, which is a novel characteristic.

The atlas is in the standard 8 folios. It is compiled and published using an advanced integrated system based on the latest basic survey and mapping data, remote sensing image data, digital elevation models and other geospatial data. The perfect combination of science, cartography, color, aesthetics and printing allow the atlas to not only reflect distinctive professional characteristics but also the local style and features of the times.

References

Ai, T. (1998). Dynamic symbol and dynamic map. *Journal of Wuhan Technical University of Surveying and Mapping, 23*(1), 47–51.

Bai, Y. (2007). Research on the compilation of Atlas of Xinjiang Uygur Autonomous Region. *Bulletin of Surveying and Mapping,* (5), 61–62, 76.

Chen, Y. (1995). Mental map and it's role in map design. *Journal of the Institute of Surveying and Mapping, 12*(4), 290–293.

Chen, Y. (2000). A color matching experiment on electronic maps. *Journal of the Institute of Surveying and Mapping, 17*(1), 53–56.

Chen, Y., & Chen, Y. (1999). Application of visual perception theory on electronic map design. *Journal of the Institute of Surveying and Mapping, 16*(3), 64–67.
Ding, C. (1992). Some problems in the compilation of urban atlas. *Bulletin of Surveying and Mapping,* (4), 32–36.
Fu, S. (1991). The research and compilation of atlases in the information era. *Map,* (2), 3–6.
Fu, Y. (2007). Relationship between the two-dimensional map symbols and three-dimensional visualization. *Journal of Zhengzhou Institute of Surveying and Mapping, 24*(Suppl.), 73–75.
Gao, J. (1963). International map of the world in seventy years. *Bulletin of Surveying and Mapping* (2), 26–34.
Gao, J. (1993). Spatial cognitive basis of map design expert system (MADES); *Symposium on space technology, geodesy and geophysics.*
Gao, J. (1997). Talk about map at the turn of the century. *Geomatics Technology and Equipment,* (1), 1–15.
Gao, J. (1984). Experimental method of map feeling and map design. *Progress of Cartography.* Textbook for new technology lecture class of cartography of Chinese Society of Surveying and Mapping in summer of 1984.
Guo, L., & Li, L. (2004). Automatic implementation of color design in administrative map. *Editorial Board of Geomatics and Information Science of Wuhan University, 29*(11), 1025–1029.
Guo, Q. (1991). Map of the aesthetic characteristics. *Map,* (1), 57–58.
He, Z., & Du, Q. (1999). Study on "Atlas of Shenzhen city" production. *Bulletin of Surveying and Mapping,* (10), 28–31.
He, Z., & Zhang, H. (2003). The computer map designing and compiling for the historical atlas of shanxi province. *Bulletin of Surveying and Mapping,* (12), 42–44.
He, Z., & Zhu, G. (1991). A study of optical illusions used in map making. *Journal of Wuhan Technical University of Surveying and Mapping,* (3), 40–43.
Hong, J., & Zhou, H. (2006). Design and compilation of Hangzhou city atlas. *Bulletin of Surveying and Mapping,* (5), 45–47.
Hou, J. (2004). Some experience of color design in map aesthetics. *Printing Field,* (6), 45–47.
Huang, H., & Zhong, Y. (2006). Improving the mathematical definition of scale symbol, non-scale symbol and semi-scale symbol in the map. *Geomatics and Information Science of Wuhan University, 31*(3), 244–246.
Huang, M. (1964). Suggestion on the symbol of the desert symbol. *Bulletin of Surveying and Mapping,* (11), 36–38.
Huang, Y. (2002). Designing and mapping of the shanghai image-maps. *World Regional Studies, 11*(4), 106–112.
Jia, J., & Xiao, J. (2002). The overall design of Comprehensive Atlas—A "naval officer atlas" as an example. *Hydrographic Surveying and Charting, 22*(2), 45–48.
Jiang, B. (1991). A primary study on map colouring expert system. *Map,* (1), 49–52.
Li, Q. (1987). Preliminary discussion on compilation of "atlas" of the yangtze river basin. *Map,* (1), 47–50.
Li, S., Liu, H., & Wang, Y. (2002). The realization of relief representation method on the small scale wall map. *Journal of the Institute of Surveying and Mapping, 19*(4), 300–302.
Ling, S. (1997a). Aesthetic research on map design. *Map,* (1), 5–7.
Ling, S. (1997b). On the development trend of map symbol. *Map,* (4), 3–4.
Ling, S., Huang, S., & Liang, D. (2007). Research on design of lettering annotation on maps. *Journal of Anhui Normal University (Natural Science), 30*(5), 603–606.
Ling, S., & Wang, X. (2000). Study on the color of boundary in map. *Map,* (1), 1–2, 8.
Liu, D. (1963). The overall design of atlases and some technical problems. *Bulletin of Surveying and Mapping,* (2 Suppl.), 1–7.
Liu, D., Lu, Q., & Yu, C. (1958). Compilation of a new map of China. *Bulletin of Surveying and Mapping,* (10), 442–447.
Liu, F. (2003). Automatic map subdivision of electronic atlas. *Journal of the Institute of Surveying and Mapping, 20*(2), 148–150.

Liu, Z., Chen, Z., & Zhao, R. (1992). The general principles and methods for designing the legends of modern thematic maps. *Acta Geodaetica et Cartographica Sinica, 21*(2), 144–150.
Liu, Z. (1965). The use of shading in the political map. *Bulletin of Surveying and Mapping*, (3), 16–18.
Lu, H. (1988a). Thoughts on the relation between maps and modern design art. *Map*, (2), 15–17.
Lu, S. (1988b). Brief views for the development of atlases in China. *Map*, (1), 11–14.
Lu, Y. (1997). The National Atlas sequence map editor design features. *Map*, (4), 22–24.
Ma, Y. (1995). A study on the symbolic constitution elements and its designing models. *Acta Geodaetica et Cartographica Sinica, 24*(4), 309–315.
Ma, Y., & Yin, Z. (1994). "China City Atlas" technology design. *Bulletin of Surveying and Mapping*, (2), 31–32, 26.
Mao, Z., Yuan, D., & Fang, B. (1982). The characteristics of compilation of "Atlas of People's Republic of China". *Bulletin of Surveying and Mapping*, (4), 40–43, 37.
Min, B. (1957). Design principle of map schema (legend or special symbol). *Acta Geodaetica et Cartographica Sinica, 1*(2), 131–145.
Pan, J. (1989). Making use of color infrared aerial photography to compile the city current situation. *Map*, (1), 41–44.
Pang, X., E, D., & He, Z. (1999). *The progress of multimedia electronic map design features—On the electronic map of the Arctic, for example In geographic information system and electronic map technology* (pp. 127–133). Hunan Map Press.
Pang, Y. (2007). On the master design of atlas. *Bulletin of Surveying and Mapping*, (5), 63–64, 65.
Qi, Q., & Xiao, Z. (1991). Discussion on the characteristics of "Atlas of Xi'an City". *Bulletin of Surveying and Mapping*, (2), 23–28.
Qian, Z. (1985). The preliminary test of military value on the current cartographic symbolization of china and the improved opinioes. *Journal of the Institute of Surveying and Mapping, 2*(1), 92–98.
Qian, J. (1992). A study on the design and compliation of the "toursit atlas of beijing". *Acta Geodaetica et Cartographica Sinica, 21*(4), 299–306.
Shi, J., & Chen, C. (2008). Design and integration of the electronic atlas of municipal region. *Bulletin of Surveying and Mapping*, (9), 54–57.
Shi, Z. (1964). The production of "China shading map" of the new taste. *Acta Geodaetica et Cartographic Sinica, 7*(1), 58–72.
Song, A. (2002). Map design and visual perception. *Surveying and Mapping of Sichuan, 25*(1), 40–42.
Wang, G., Liu, F., & Liu, N. (2005). Related technology in electronic atlas design: Research and practice. *Bulletin of Surveying and Mapping*, (9), 47–50.
Wang, G., Yu, J., & Gu, S. (2008). Research on the representation of network electronic maps. *Bulletin of Surveying and Mapping*, (2), 66–69.
Wang, J. (1991). The designing characteristics of the atlas for officers. *Journal of the Institute of Surveying and Mapping, 8*(3), 33–42.
Wang, J. (1993). An introduction to the development of Modern Military Cartography. *Journal of Institute of Surveying and mapping, 10*(Suppl.), 36–66.
Wang, J., & Wang, G. (2007). The design and study of the photomap atlas of Suzhou. *Bulletin of Surveying and Mapping*, (2), 65–69.
Wang, J., Wang, H., & Chen, X. (2003). The symbology standards for the electronic cartography. *Geo-information Science, 5*(2), 16–19.
Wang, T. (2003). Wang Tao on map color (three): The color of the shock. *Map*, (4), 24–25.
Wang, T. (2004). New thinking of map symbol design (two): The principle of formal beauty. *Map*, (2), 59–61.
Wang, X. (2008). A study of comprehensive design of the situation atlas of Fujian province. *Bulletin of Surveying and Mapping*, (1), 59–61.
Wu, H. (1965). The principle and method of the choice of the height of the relief map. *Acta Geodaetica et Cartographic Sinica, 8*(2), 126–144.

Wu, S. (1988). The calculating shaded representation of the terrain surface. *Acta Geodaetica et Cartographic Sinica, 17*(1), 38–48.

Wu, Z. (1959). Department of surveying and mapping: Cartography—Part 3: method and comprehensive describes the elements of the map (Continued). *Bulletin of Surveying and Mapping*, (9), 31–33, 36.

Xu, D., Cai, Z., & Wu, G. (2003). Design and implementation of urban multimedia electronic atlas. *Bulletin of Surveying and Mapping*, (7), 46–48.

Xu, D., Du, Q., & Zhong, M. (2007). Techniques of the network map design. *Science of Surveying and Mapping, 32*(3), 153–155.

Xu, G., Gao, X., & Lu, Z. (2002). The design and compilation of the 1:4 000 000 administrative map of the People's Republic of China. *Bulletin of Surveying and Mapping*, (12), 52–54.

Yang, D. (1993). The encoding of cartographic symbols and orthers. *Journal of the Institute of Surveying and Mapping, 10*(2), 24–27.

Yang, K., & Ma, Y. (1992). An overall designing idea on compiling Xi'an Atlas. *Journal of Shaanxi Normal University (Natural Science Edition), 20*(1), 66–70.

Yang, X. (2007). Esthetics design of map color. *Geospatial Information, 5*(3), 112–114.

Yin, Z. (1992). Computer compilation of urban planning atlas. *Bulletin of Surveying and Mapping*, (5), 26–28.

Yin, G. (1997). Improvement of design of Chinese basic topographic maps. *Journal of Wuhan Technical University of Surveying and Mapping, 22*(1), 59–64, 71.

You, X. (1990). Aesthetics in the technology of cartography. *Map*, (1), 10–13.

You, X., & Wan, G. (1996). Visualization in map color design. *Journal of the Institute of Surveying and Mapping, 13*(3), 218–223.

Yu, L. (1989). The problem of Cartography in aesthetics. *Bulletin of Surveying and Mapping*, (3), 29–31.

Yu, L. (1990). Combining of the cartology and the aesthetics—Thought on the aesthetics of maps. *Acta Geodaetica et Cartographic Sinica, 19*(4), 307–313.

Yu, L. (1995). The properties and design requirements of applications. *Map*, (1), 6–9.

Yu, L. (1996). The nature discrimination of character symbol on map. *Journal of the Institute of Surveying and Mapping, 13*(1), 45–49.

Zhan, Q. (1993). Development of modern city maps from compiling Xi'an atlas. *Journal of Shaanxi Normal University (Natural Science Edition), 21*(1), 64–66.

Zhang, B. (1996). Electronic map (set) and its development. *Bulletin of Surveying and Mapping*, (6), 26–28.

Zhang, W. (1990). Cartography and expert system. *Map*, (4), 48–50.

Zhong, Y. (1989). The color rules on the color map. *Map*, (1), 49–52.

Zhang, X. (1966). On the evolution and discussion of layering method in map printing. *Acta Geodaetica et Cartographica Sinica*, (2), 57–67.

Zhong, Y., Wei, W., & Peng, Y. (2001). The mathematical definition of scale symbol, non-scale symbol and semi-scale symbol in the map. *Editoral Board of Geomatics and Information Science of Wuhan University, 26*(5), 465–468.

Zhou, L. (2006). Discussion on the colour application in map. *Geomatics & Spatial Information Technology, 29*(3), 98–99.

Zhou, W. (1965). The evaluation of general regional atlas map compilation process. *Bulletin of Surveying and Mapping*, (2), 30–34.

Zhu, G. (2006). Design of dynamic map symbol based on multi-transform assembly. *Geomatics and Information Science of Wuhan University, 31*(6), 548–551.

Zou, L. (1990). A study on the selection of contents of synthetic city atlases: The "Atlas of Shanghai city" as an example. *Map*, (3), 1–4.

Zou, J., & Sun, Q. (1998). Characteristics of digital mapping production based on Microstation: Taking water conservancy Atlas of Jiangsu Province as an example. *Bulletin of Surveying and Mapping*, (5), 29–32.

Chapter 5
Cartographic Generalization

Fang Wu and Jiayao Wang

5.1 Introduction

Cartographic generalization has always been one of the most challenging and creative fields in cartography. Automated generalization of spatial data under a digital environment is still one of the core issues that modern cartography faces. First, the automated cartographic generalization method must be used when producing smaller scale maps by means of large-scale digital map data; second, the automatic cartographic generalization method is the most effective way to solve the automatic derivation of a multiscale spatial database based on a large-scale basic spatial database and the integrated updating of a multiscale spatial database; third, to meet the requirements of the multiscale expression of geographic information system data, the automatic generalization method must be adopted and should be online; and fourth, in the construction of a spatial data warehouse, the automatic generalization method should be used to extract the required spatial data from heterogeneous databases in different places and departments according to the user theme (problem solving). It is because of the importance of the automatic generalization of spatial data that international academia has given full attention to research in this field for a long time, especially in the past 10 years, and a large number of valuable theoretical and technological achievements have been obtained. However, because of the complexity in the automatic generalization of spatial data, there is a certain gap between the present study results and a global solution to the problem, which is a practical application. This gap is not only because of the technical realization but also because of the ideological conception.

F. Wu (✉)
Information Engineering University, Zhengzhou, China

J. Wang
Henan University, Kaifeng, China

J. Wang and F. Wu (eds.), *Advances in Cartography and Geographic Information Engineering*, https://doi.org/10.1007/978-981-16-0614-4_5

5.2 Development Process

5.2.1 Cartographic Generalization: From a Subjective Process to a Scientific Objective Cartographic Method

According to the related information, in 1921, Eckert first discussed the concept of cartographic generalization. He believed that the essence of cartographic generalization is a choice, and the generalization of mapping objects and the main factors affecting its role is map use. This is an undoubted fact. However, Eckert also believed that generalization is a subjective process, as the rules cannot be ascertained from this process, and the generalization depends on only the drawing skills of the personnel. Eckert's view greatly influenced Europe until the 1960s.

In the 1940s, Salichtchev, a former Soviet Union cartographer, published the *Cartography Principle* and other works on the basis of summarizing the mapping production experience and scientific research achievements in cartography during the Second World War. He discussed cartographic generalization as an objective scientific method, and he systematically put forward the general principles, basic factors and expression (method) of cartographic generalization, believing that the scientific basis of cartographic generalization is the concept of dialectical materialism about the mutual connection, mutual restriction and development of natural and social phenomena. Under the guidance of this idea, the former Soviet Central Scientific Research Institute of Surveying and Mapping studied the cartographic generalization of all topographic map elements and summarized the design and compilation experience of large map works, such as the 1:2.5 million Soviet hypsometric map, Atlas of the World and Ocean Atlas. The principle and method of generalization for the topographic map was extended to the compilation of a small-scale general geographic map, and they edited and published the *Cartographic Generalization Principle for Small-Scale General Geographic Map* (from 1:1 million to 1:4 million), beginning to pay attention to the quantization of cartographic generalization. This reflects the principles of the unity of cartographic generalization theory and mapping practice, and the unity of the cartographic generalization method and regional geographic characteristics (Wang et al. 1992).

The viewpoints have ranged from Eckert's view that cartographic generalization is a subjective process without rules to Salichtchev's view that it is a scientific objective mapping method, which shows great progress. This progress reveals a truth that as a scientific mapping method, cartographic generalization has the characteristics of epistemology and methodology, and its regular use can follow.

5.2.2 *From Qualitative Description to Quantitative Description of Cartographic Generalization*

For a long time, the research and practice of cartographic generalization has generally involved a qualitative description, which affects the scientific nature of the cartographic generalization practice.

With the development of cartographic production and cartographic generalization theory, many domestic and foreign experts were dedicated to the applications of statistical methods in cartographic generalization. In foreign countries, cartographers from the former Soviet Union published the *Mathematical statistics method in mapping work*, which systematically uses mathematical statistics methods to study the distribution rule of geographical features and to determine the cartographic generalization index of some features. In approximately 1962, the Germany's Töpfer published a number of articles, which suggested the use of the square root of the ratio of the scale denominator between the source map and newly edited map as the basis of determining the selected number of objects, and a formula was developed for feature selection. In addition, in 1972 he published a monograph, *Cartographic Generalization*, which comprehensively described the applications of the square root selection law formula. In 1983, the French Franke published the results by using the graph theory method to study cartographic generalization. In China, some people begin to study the quotas of settlements selected for mapmaking by using the mathematical statistics method and graphical calculation method by the end of the 1950s and early 1960s. Since the 1970s, many people use the correlation analysis and regression analysis method to study the mathematical model of the habitation selection index, and a number of theoretical and practical achievements were obtained (Fan 1978; He 1986). In the middle of the 1980s, some scholars used fuzzy set theory and graph theory method research on a terrain structure selection model (Wang et al. (1985). At the beginning of the 1990s, the mathematical analysis method began to be widely used in cartographic generalization (Zhu et al. 1990; Wang et al. 1992).

The use of mathematical methods to determine the quantitative indicators of cartographic generalization is a large step forward from a qualitative description to a quantitative description. This change reveals the historical trajectory of cartographic generalization from the subjective process to a scientific cartographic method and from a qualitative description to a quantitative description of cartographic generalization, i.e., the measurement of cartographic generalization quotas. At the same time, this change indicates that mapping has adapted to digital map data and cartography technology innovation, and it will continue to deepen and be further systematized in the quantitative aspects, eventually leading to the modernization of cartographic technology.

5.2.3 From the Generalization Based on the Map Model to the Automatic Cartographic Generalization Based on the Model, Algorithm and Knowledge

From the qualitative description to the quantitative description, the quantifiable measure of cartographic generalization quotas had not been developed to use computer technology. With the arrival of digital map data, cartographic generalization is no longer manual mapping in the paper, it has many new characteristic (Wang et al. 1998a). Under the condition of digital map generalization, the complicated process of creative thinking is completed by cartographers, while the arduous and heavy operation process is realized by computer programming. Even the application of expert system technology, the computer can only imitate the domain experts' way of thinking to deal with problems in the process of generalization and to solve the problems that can only be solved by cartographic specialists. Generalization quality depends on the rationality, completeness and intelligence level of the model, algorithm and rules. The generalization decision is made by cartographers and subsequently carried out by a computer.

In twentieth century, from the 1960s to the end of the 1980s, domestic and foreign scholars have studied the map model and generalization model. In 1967, the British Scholar Board proposed the concept map as a model in the paper, *Map as a model.* An increasing number of people regarded the idea of map as a model. Cartography, especially cartographic generalization, entered the stage of a more rigorous theoretical model test. Many scholars believe that the map model can be described in mathematical form. They studied the mathematical logic expression methods of the map model. A logical mathematical description of the modeling elements and object identifier were studied (including the description of a concept-hierarchy, the logic mathematical description of an object's position in space and the spatial structure of an object, as well as the modeling description of the topological correlation between objects).

The concept of a map as a model has been a theoretical guide for cartographic generalization and has developed from manual to computer. The logic mathematical description of the map model elements helps to build a cartographic generalization transform model $M_{k1} \overset{G}{\to} GM_{k2}$ (G denotes the model transformation operator; M_{k1} is the model before the operator works, and M_{k2} is the model after the operator exerts an effect). As a logical mathematical description of an object identifier, the description of concept-hierarchy will help to build the directed graph of cartographic generalization, and information searches can be implemented in the mapping database with the help of a directed graph; the logical mathematical description of an object's position in space will decompose the object set into subsystems on a map, and according to the objects' spatial location measurement, the set will be divided as follows: three dimensions (body objects, such as landforms), two dimensions (surface objects, such as lakes) one dimension (line objects, such as rivers), and zero dimensions (point objects, such as control points). The logical mathematical description of an object's spatial structure will help to divide the previously introduced body objects, surface objects, line objects and point objects into several subsets. Further, the object's spatial structure and characteristics will be described in detail. Finally, the modeling description of the topological correlation between objects will contribute to the effective use of topological information between objects in cartographic generalization to solve problems, such as "merger" or "displacement".

In the digital environment, the foundation of automatic cartographic generalization is the model, algorithm and knowledge. Because the method is easy to write into a program (computer program and artificial intelligence program), a computer can perform the cartographic generalization operations well; the model, algorithm and knowledge is easy to program. Therefore, studying the model, algorithm and knowledge for cartographic generalization is the basic work of automatic generalization (Wang et al. 1998a, 1998b; Wu et al. 2001b; Wu 2004).

The so-called generalization model refers to mathematical expressions describing some relationships in cartographic generalization, that is, the mathematical formula of cartographic generalization rules based on the mathematical expression method. The model is usually used to manipulate map content transformation from a source map to a newly edited map (Wang 1989). In addition, the model mainly includes the quota (overall) selection model, which macroscopically controls the selection number and maintains the balance of the selection number between different areas, including the root mean square model, the regression model, and the fractal extended model of root mean square model. The structure (location) selection model can separate a more important part of the objects from the source map based on the structures among objects, which solves the problem of the "what" selection, including the common comprehensive evaluation model, the fuzzy comprehensive evaluation model, the graph theory model, and the fractal selecting model that considers the self-structural and distribution characteristics of a map spot set. The neural network selection model considers the configuration characteristics of the targets on the map, and finally, the quota-structure selection model or selecting process model simultaneous control the selection of the number and specific objects. It should be pointed out that the generalization models are the mathematical expressions used to describe some of the relationships in cartographic generalization. However, because these relationships are generally nondeterministic dependencies, they are mathematical descriptions of statistical rules and are not determined by a general functional relationship. Thus, the reliability of the cartographic generalization index calculated by these formulas is restricted by many factors, for example, the rationality of the number and distribution of statistical samples, the rationality of the determination of formula parameters, the completeness of evaluation factors, and so on.

The so-called cartographic generalization algorithm refers to the process used to finitely and mechanically determine (calculate) a certain type of cartographic generalization problem. By using only a finite number of instructions, the computer can execute a finite step calculation process according to the instructions and obtain the cartographic generalization result. Basically, the automatic cartographic generalization based on the algorithm is object oriented. The first proposed, classic algorithm is the Douglas-Peucker algorithm. Later, a fractal expansion was carried out by some scholars to make the algorithm applicable to the automatic generalization of contour lines in addition to the automated simplification of general curves (Wang et al. 1998c). The applications of a genetic algorithm, Agent and elastic mechanics in automatic cartographic generalization are new research fields, which have been studied for the past 10 years and have gained the most achievements. The genetic algorithm is a kind of bionic algorithm, which is abstracted in the process of the organism evolution; through a comprehensive simulation of natural selection and genetic mechanism, the algorithm uses coding space as an alternative to the parameter space in the problem, the fitness function is used as the evaluation basis, and the coding species is used as an evolutionary basis to establish an iterative process. In this process, the individual in the group is evolving, the function is close to the optimal solution, and finally, the goal of solving the problem is reached. This algorithm is mainly used for point cluster selection, line simplification, generalization of road networks, the selection of rivers and artificial network automatic generalization, automatic configuration of annotations, and so on. However, the genetic algorithm also has shortcomings, such as a low efficiency and slow convergence rate. The low efficiency of the genetic algorithm is because it is a global optimization algorithm. However, with the parallel computation of the genetic algorithm, each individual in the genetic operation is independent of each other and can be calculated at the same time, which is in full compliance with the requirements of distributed computing. For this reason, using the parallel genetic algorithm to solve the efficiency problem of the model is effective; resolving the convergence problem of the genetic algorithm has always been a theoretical issue in academic circles. It is impossible to fundamentally solve this problem with general convergence theory, the Markov chain model-based method and so on. At present, the research approach is to limit the number of iterations, for example, when the maximum fitness value and the average fitness value are close enough or the number of iterations reaches the specified number of times, the iterations will stop. Of course, this is not an ideal solution. The Agent technology originated from the field of distributed artificial intelligence. It is a well-packaged computing entity in a certain environment and a new method of calculation and problem solving. TIN technology is very powerful for geometric processing, but in the face of intelligent challenges, this technology still cannot meet the needs of automated cartographic generalization. The ABTM algorithm (Agent Based TIN Model—ABTM) can be constructed by combining the two technologies of Agent and TIN and is mainly used for aggregation of buildings, selection of point group elements and simplification of line elements. Elastic mechanics is the study of the principle of elastic body deformation under the action of force. That is, deformation will be caused when the elastic body is subjected to a certain external force, and the original shape can be recovered when the external force is removed. If only part of the elastic body is affected by the external force, the overall shape of the object is basically unchanged but local deformation occurs. This result is the desired effect for cartographic generalization, which is to maintain the overall invariance in an object's spatial relations. How to deal with the relationship of a cartographic feature is one of the cartographic generalization issues, while graphical symbol displacement is an important way to deal with the relation. An important constraint of the displacement operation is to correctly express the spatial relationship between elements. An important factor that caused the complexity of the displacement operation is that the displacement has the characteristic of propagation. Only by solving and controlling the spread of the displacement can we avoid the new conflict and maintain the correct spatial relationship. Back in 1970s, the University of Hannover, Germany, began to study the issue of displacement operations, and these studies have not been interrupted. Early studies used mechanical methods; the current study uses the optimization method. The displacement operation based on the principle of elastic mechanics is the most effective method to date (Hou et al. 2005). A depth study of the target collision detection method was conducted by Wu et al. (2005), and the two methods of translation and deformation were put forward on the basis of the object stress analysis. The translation method is adopted for objects with strong deformation tendencies, including the translation method based on plane space deformation and the translation method based on planar linkage. Compared with the latter two methods, the former method is more comprehensive in the determination of the spatial relationship, and the constraint conditions are more prominent, which is because it is a translation method based on the spatial deformation of planar triangulation, and correspondingly, the displacement operation should be limited to one area; the object group cannot be less than three, and it is not suitable for a linear distribution. The scope of the application is limited; the latter is to use the minimum spanning tree to describe the spatial relations among the objects, which is adaptable to describe almost all the spatial patterns. For objects with deformation tendencies that are not very strong and with complicated shapes, the object deformation method from the displacement operation is suitable. In other cases, the detection method and the stress situation of the deformation target is studied, the object deformation method is proposed, and the influence of the parameters of the stiffness element on the deformation is analyzed. Finally, the displacement model based on the principle of elastic mechanics is studied in detail, including the implementation strategy of the displacement method under various conditions. Taking roads, settlements and surface targets in their natural forms as examples, the use of the target displacement method is demonstrated, and the precision of the algorithm is analyzed. In addition, studies on automated generalization algorithms based on mathematical morphology, artificial neural networks and Circle have also yielded many results (Wang 2008a).

Not all cartographic generalization problems can be modeled and algorithmic. Automatic generalization based on knowledge refers to the formal description of some problems in cartographic generalization. At the beginning of the 1990s, cartographic generalization expert systems and knowledge inference attracted interest and attention of many scholars, and some exploratory results were obtained (Wang 2008a). However, their applications in automated cartographic generalization are low, as the intelligent development of cartographic generalization meets many difficulties (Wu 2004). Cartographic generalization is essentially a highly intelligent system. Intelligent activity cannot be independent of knowledge. Knowledge is the only way to achieve intelligence. Cartographic generalization is the combined process of knowledge representation and knowledge abstraction. Cartographic generalization knowledge is the foundation of the practice (Wang et al. 2006). For example, using the general knowledge of cartographic generalization to check the data before generalization, you can obtain the characteristics of the generalization region, comprehensive content, key generalization methods and other information. By using the general knowledge of cartographic generalization to check the data after generalization, you can determine whether the results meet the requirements. The knowledge of cartographic generalization can provide parameter support for algorithms and constraint conditions for the calculation results. The general knowledge is the main basis for process control. Based on this understanding, the knowledge classification of cartographic generalization, generalization knowledge acquisition and transformation, knowledge base construction of XML format, structured descriptions of generalization knowledge, cartographic generalization knowledge attributes, cartographic generalization knowledge management and organization, its applications to map generalization process control and reasoning, and other problems have been researched by Wang et al. (2006). It should be said that by far, these are the most innovative research results in the field.

From map model to generalization-based model, algorithm and knowledge, the theory, method and technology of cartographic generalization is continuously deepening and improving. Over the past 30 years, especially in the most recent 10 years, automatic map generalization research has been carried out around this theme. Practical applications have proven that automatic map generalization is a right way.

5.2.4 From the Goal of Full Automation to Human-Computer Collaboration

After a long history of traditional manual mapping, there are two understandings when the map data are digitized: the map production is quantified and the computer technology is used in map generalization. One understanding is to think that map generalization is a complete process of individual labor based on the mapping experience, and it is impossible for a computer to complete a process that is not yet clear. Even in the digital map environment, in practical map production, only the computer screen and mouse are used to complete map generalization instead of the traditional mapping tool for an analog map, which is essentially a manual mode. Another understanding is to exaggerate the role of the computer. According to this understanding, by writing a program, we can make use of the computer's strong processing ability and high processing speed to complete the cartographic generalization operation within a very short time, which requires much labor and time to complete by manual mapping. This process should be conducted by full automation. These two understandings are not scientific. Both of them lack in-depth analysis and research regarding the ability and characteristics of both human and computer information processing, and the relationship between the human and computer in the generalization process. How people and computers work together in the process of map generalization is not fully understood. In theory, these understandings developed because of the lack of correct understanding of the nature and characteristics of map generalization process.

In view of the above existing problems in the human-machine collaboration in the digital map generalization, the human way of thinking in the process of map generalization and the ability of a computer to simulate the thinking of a human were studied, and an optimal collaboration theory of automatic map generalization was presented. Considering the way a human thinks in the process of map generalization, the abstract thinking mode of map generalization (inductive reasoning based on connection, imaginary inference based on process, deductive reasoning based on rules), visual thinking mode (visual selective thinking, visual fixation thinking, visual structure associative thinking), inspirational thinking mode, and so on, are studied. Regarding the ability of a computer to simulate the thinking of a human in map generalization, research shows that the current computer simulation of abstract thinking is relatively easy to conduct, in particular, the map generalization expert system technology can effectively simulate the rule-based deductive reasoning. However, it is difficult or impossible for a computer to simulate visual thinking, especially the inspirational thinking during the process of map generalization; at the same time, by using a computer to simulate the way of thinking and solve the problem of map generalization, we must have three prerequisite conditions, including problem formalization, computability and reasonable complexity. In addition, a practical automatic generalization must be supported by a database, but the existing database is not built for map production, and even though some map generalization problems can be formalized and algorithmic, they cannot be solved without data for calculation. Even though some map generalization problems can be summarized into the rules and knowledge, no conclusion will be achieved from the rule-based reasoning if there is not prerequisite information to match the conclusion. Therefore, it is not yet practical for the data in the database to objectively and accurately reflect the thinking system of the human brain. This finding affects the effective simulation of human thinking in the process of map generalization. Since the computer cannot effectively simulate the whole way of human thinking in the process of map generalization, the irreplaceable role of the human in map generalization is obvious, and the automatic mapping system can only be a human-machine collaboration system. The human-machine collaboration in automatic map generalization should be based on the characteristics of the human and computer in the map information processing to achieve an optimal human-machine collaboration. That is, people's creativity is at the forefront, making full use of the ability of computer in processing map information, and the human takes the leading role in computer aid (support) in the process of automatic generalization.

According to synergetic theory, collaborative automatic generalization not only requires human-machine collaboration but also requires collaboration among the model, algorithm, knowledge inference and other technical means. Each area and each kind of map element has its own characteristics, and each technique has its own advantages and disadvantages. Their ability to solve the problem of map generalization is limited. However, if they are effectively combined to focus on their respective advantages and to make up for their deficiencies through the complementary advantages, then the ability of the whole system to solve the problem of cartographic generalization will be greatly enhanced. Obviously, such a system should be composed of many subsystems that are able to perform different tasks and should be an open system (Wu et al. 2002b).

5.2.5 From the Research of the Algorithms and the Automatic Generalization Test to the Process and Quality Control of Automatic Generalization for All Map Features

In a very long period of time, the studies of automatic map generalization mostly focused on the model and the algorithm, both of which were in an untidy state. The application of an expert system and knowledgeable inferences in automatic generalization are still low. Many people believe that automatic generalization is an ill-defined complex process, which is a significant problem in the field of cartography and GIS. Some scholars even think that it is a NP-complete problem (i.e., computer solutions cannot be designed). Although a number of software systems were launched in the 1990s, such as DynaGEN of the Intergraph Company, Change of the University of Hannover, Germany, STRATEGE and Carto 2001 based on the Agent of Institut Géographique National Français (IGN), Polygon Generalization System (PolyGon) of the University of Zurich and Clarity based on the Agent of Laser-Scan Company, the problem remains unsolved. Prospects are difficult to find for a comprehensive solution and overall application. The main reason is that there is no research that treats automatic map generalization as a whole (all geographical feature types, a whole process and complete control). The automatic generalization system lacks the support of knowledge and intelligence. Numerous generalization algorithms can only solve specific problems in the specific environment and lack overall coordination with each other. There is no spatial data model or data structure that can support the automatic generalization operation (Qian et al. 2006b).

The two problems of process control and design for quality must be solved in order to treat automatic map generalization as a whole (all geographical feature types, the whole process and complete control). For the problem of process control, some scholars proposed in the 1980s that the generalization process was a process of thinking, and the thinking process can be modeled as a control mode (Wang 1989). After a lapse of seventeen years, based on the analysis of the characteristics of cartographic generalization and referencing the research results in the field of artificial intelligence, Qian et al. (2006a) proposed a new method for the classification, acquisition and expression of generalization knowledge, as well as knowledge organization and management methods to support the model, algorithm, process control and quality evaluation. He also put forward a new classification method of cartographic generalization Agent based on the Agent technology of artificial intelligence, existence and communication mode and the structural description method of the Agent entity, which were studied in detail to support the framework design and development of the automatic generalization system. The map generalization algorithms based on ABTM and circle characters have been developed with the characteristics of strong graphics operation, intelligent detection capability and high computing speed. According to the idea and technology of the workflow in the field of industry, an automated generalization chain integrating model, algorithm, knowledge, evaluation and map generalization process-control model based on the generalization chain were put forward to support the control of the whole generalization process. Thus, a practical automatic map generalization software system was constructed. For the automated cartographical generalization quality issue concerns in academia and industry, some scholars think that the factors restricting the automated cartographical generalization quality originate from two aspects: first, the computer is convenient to process abstract thinking but cannot easily deal with image thinking and inspirational thinking in cartographic generalization. Second, the quality is decided by the rationality, completeness, intelligence degree and the evaluation model of the cartographical generalizations process model, the generalization algorithm and the knowledge (especially rules). In addition, the connotation of automatic map generalization quality problem should include three aspects: the framework of the quality control system, the quality evaluation strategy and the quality control process. According to the above information, the framework, quality management mechanism and mathematical description of the cartographic generalization model based on the design of quality were presented (Wu et al. 2007a). On the basis of analyzing the restraint condition of cartographic generalization, the knowledge representation of cartographic generalization for quality design based on the database has been extensively studied. By performing requirement analyses of the data model for generalization quality control, a data model for generalization quality control and the quality control process based on the model were presented. Considering the existing weakness in the research of topological consistency verification and evaluation for cartographic generalization, the theory and method of the topological consistency evaluation and preservation in cartographic generalization was studied; taking the generalization of a road network as an example, maintaining topological consistency in the process of cartographic generalization was studied and verified. Since the map generalization model based on the quality design requires the support of the multidimensional constraint space, a quality evaluation model based on the multidimensional constraint space and another quality evaluation model based on the most common used line element simplification algorithm were proposed; taking the simplification of a contour line as an example, the evaluation model of the line element simplification algorithm was verified and analyzed.

The intelligent control of the automatic cartographic generalization process and automatic cartographic generalization based on quality design are the most outstanding research achievements in the field in recent years.

5.3 Main Research Achievements

After the founding of new China and through 60 years of development, cartographic generalization, especially automatic cartographic generalization, has made considerable progress. As one of the important research fields of cartography at home and abroad, and as one of the difficult problems in cartography that remains to be solved, many map scientists and cartographers have carried out many studies and experiments in this field and have achieved very rich research results over a long period of time.

5.3.1 Theoretical Framework of Cartographic Generalization in the Digital Mapping

(1) **Basic theory of map generalization.**

The research of automatic map generalization has a close relationship with cognitive science. According to the map generalization theory based on cognitive science, cognition is the foundation of map generalization realization. You (1992) discussed the role of visual perception in cartographic generalization, considering that cartographic generalization is the product of the visual perception function, and the implementation of cartographic generalization should be subject to visual perception, while modeling of the mental image in mapping behavior is the main factor that causes an unreasonable generalization result.

Several basic theoretical problems and technical methods on cartographic generalization have been discussed and the academic viewpoint of structural generalization was proposed by WU Hehai. He believed that generalization is not unique to cartography and GIS but a common means of scientific cognition; thus, map generalization is a special application case of the philosophical abstract generalization principle in the processing of spatial data; as a result, he presented an information transform concept for map generalization and a DLM concept for the map generalization object; he considered the necessity of generalizing the generalization operators given a three level implementation model for structural generalization. Finally, taking river networks, topography and point clusters as an example, a comprehensive theoretical system for structural cartographic generalization based on graph theory, fractal theory and computational geometry was established (Wu 1996, 2000b).

Based on the essential characteristics of generalization objects and combined with its own law of map generalization, Guo (1999a) decomposed the problems of automatic map generalization, proposed a complete set of map generalization operators by analyzing the existing map generalization operators, and analyzed the relevance and orderliness of the operators. Based on the existing partitioning methods for generalization operators, Wu et al. (2001b) redivided the automatic generalization operators and proposed a collection of algorithms in the operator library to establish the algorithm library of automatic generalization operators.

Zhai et al. (2000) considered that only by the combination of the recognition, measurement and generalization method can we fully reflect the concept of digital chart generalization. Emphasizing the identification of graphical features and through introduction of the Douglas binary tree method, the recognition and measurement function of the graph feature is given, and the automatic synthesis of the linear feature of a digital chart was realized. Based on the idea of the mapping transformation of relational algebra, Ai (2003) divided the conceptual framework of map generalization into 2 types: spatial entity mapping and spatial relation mapping; he discussed the conceptual expression forms of the 2 types of maps in the map generalization process. The significant difference between this method and other traditional methods is that it emphasizes generalization of the relationship.

(2) **The viewpoint of human-machine collaboration for an automated map generalization system.**

Computers cannot simulate the whole way of human thinking in the process of map generalization, which determines a human's irreplaceable role in that process, and the automatic mapping system can only be a human-machine collaboration system. The way of thinking and the ability of a computer to simulate human thinking in the process of cartographic generalization were studied, and an optimal human-computer collaborative theory for cartographic generalization was proposed by Wang et al. (1992). WU Fang et al. (2001a, b) discussed the main reason that the research of automatic cartographic generalization has not made a breakthrough. The problem lies in their high dependence on the judgment of the human brain, that is, the fuzzy judgment of the human in the process of map generalization cannot be sorted and described clearly at present; this knowledge and these rules cannot be summarized into a formal description that can be accepted by a computer. Thus, based on the analysis of a human's way of thinking and characteristics, the role and position of a human in automatic cartographic generalization are discussed, and a collaborative model and generalization process of automatic map generalization was proposed. The model focuses on the unique advantages of human, computer, generalization algorithm and control strategy, and the model is a kind of automatic generalization, which can be effectively combined to enhance the ability of the system to solve the problem.

(3) **Multiscale modeling and representation in GIS.**

As one of the important application fields of cartographic generalization, multiscale modeling and multiscale representation in GIS is a research direction of cartographic generalization (Ai et al. 2005a). Qi et al. (1999) first put forward their view on the data consistency problem in a multirepresentation database. In addition, two application examples of database multirepresentation technology in the multiscale display of the electronic map and in hierarchical spatial reasoning were given.

WANG Yanhui, CHEN Jun, LI Xiaojuan, MENG Hao et al. have made many achievements in the aspects of multiscale modeling data model design, multiscale representation of ontological modeling and so on. As examples, Wang et al. (2006c, d) discussed the connotation of multiscale expression, reviewed the research history and current situation of the multiscale representation of geographical elements in GIS from the perspective of multi scale spatial database construction, and analyzed the academic problems from the aspects of the internal connections among objects, the maintenance of consistency, the dynamic query, and collaborative updating. The multiscale conceptual model of geographical features in GIS is studied, and the performance characteristics of spatial features at the geometric level, in the physical layer and at the attribute level were discussed. The basic conditions for design of the multiscale conceptual data model of elements were studied and a key strategy for constructing the concept model of multiscale spatial data was proposed. Taking urban map features as an example, the conceptual model design for the multiscale abstract expression of features was realized by using the extended E-R modeling method, which met the requirements of multiple expression modeling. The hierarchical interconnectivity among multiscale representations from GIS features was studied, a hierarchical semantic matching rule based on the aggregation relationship among multiscale representations was constructed, and the hierarchical interconnectivity among the abstract representations of the same element at different scales was realized based on the semantic, geometric, topological and attribute information of multiple expressions. An ontological analysis was conducted on the entity and relation from multiscale road network representations. Based on the "natural law", an ontology solution for the multiscale representation of road network features was proposed. An ontological analysis method of the multiscale modeling of geographical features was studied, and a design method of multiscale modeling based on ontology was proposed (Wang et al. 2003a, 2006c, d). Yin et al. (2005) discussed the basic concepts, such as the scale reference system, the scale event and so on, and proposed a multiscale representation model based on Petri Net.

The organization of multiscale spatial data is an important aspect of GIS data management. Based on the idea of the share scale, Wu (2008) studied the method of multiscale spatial data organization: first, he analyzed the changing characteristics of data with different scales for database building, and the hierarchical connectivity of multiscale data. Then, a multiscale spatial data organization method based on the idea of the share scale was proposed from the perspective of reducing data redundancy and facilitating data maintenance to solve the problem of the longitudinal inconsistency of partial spatial data. Because the existing R-tree cannot support the multiscale representation of spatial data, a new R-tree deformation index structure for the multiscale representation of spatial data was proposed, the spatial data multiscale query process of the index structure was analyzed, and the constraint condition, insertion algorithm and splitting algorithm from the index structure generation algorithms were emphatically discussed (Deng et al. 2009).

(4) **Map generalization in the mobile devices.**

Due to the small screen and other features of mobile devices, data for the screen display must often be processed with cartographic generalization. Ai et al. (2007b) applied variable scale visualization technology to the visualization of the navigation electronic map and achieved the "magic mirror" display, where the object is large when near and small when far. At the same time, the adaptive visualization of regional targets with different scales was studied. To overcome the disadvantage of a weak sense of distance in the visual map, the distance calculation method was improved according to the deformation characteristics of the variable scale map.

Yang et al. (2008) proposed the concept and model of the adaptive multiscale representation of the navigation electronic map and designed a specific implementation method. The method can automatically adjust the map scale in different areas of the display screen according to the density of the road network and the range of display, as well as realize the multiscale display of map data on the same screen; furthermore, the points of interest can be reasonably retained or discarded under some restriction rules.

SHEN Jie et al. proposed a basic idea, feature and key problem for the adaptive map generalization in the mobile devices based on an analysis of the current situation of the adaptive map generalization in the mobile devices. Taking the residents in a business district as an example, a user location adaptive simplified algorithm was presented for residential areas, the efficiency of part of the generalization algorithms was investigated in a mobile devices, and their usability was tested. By considering contextual information, such as user's location in a mobile devices and the user's current task, a real-time cartographic generalization algorithm centered on user location and the current task was designed (Shen et al. 2008; Wang et al. 2008).

5.3.2 *Basic Models and Algorithms of Automatic Generalization*

The cartographic generalization model is the mathematical expressions describing certain relationships in cartographic generalization, usually a mathematical description of a statistical law; the cartographic generalization algorithm is a finite, mechanical determination (calculation) process for a certain type of cartographic generalization problem. The model and algorithm research has always been one of the focuses of cartographic generalization.

(1) **Study on the cartographic generalization model.**

Using a mathematical model for the generalization of map features is a typical idea that is in accordance with the needs of computer mapping. The advantage is that the problem is clearly described, and the model is easy to realize by computer. The research of the automatic generalization method based on modeling is varied, and new theories and methods have been applied in the automatic generalization field with the continuous development of basic research and related disciplines.

Overviewing the research results during the 60 years since the founding of new China, most of the research results regarding the generalization model are mainly focused on the determination of a quota selection index. As examples, Zhang (1958) studied the choice criteria of the inhabitant areas on the map of China and used a mathematical formula and statistical method to solve the choice criteria problem of residential points in the atlas. Chen (1964) derived the formula of the density coefficient of landform cutting and applied it to the preparation of a general-purpose map. Lu et al. (1964) discussed the problem of measuring the river network and the density of lakes and ponds on the map, the research direction and treatment methods were put forward after practice. Chen (1981) studied the mathematical description of curve information content and subsequently provided the relative compression method. He (1986) established a mathematical model to calculate the selective index of an inhabitant area by using the multivariate statistical analysis method. Zhu (1989) inferred the degree of selecting a river on a map with a mathematical model, and taking the river as an example, he listed the mathematical models for inference and their application methods. He (1995) conducted an exhaustive and systematical analysis on the selection index for the main contents of the topographic map in the current compilation specification of the topographic map, and he put forward the idea of improving the selection index. Wang et al. (1996c) analyzed the existing problem of the square root model, using the fractal theory to study the change law of the quantitative index of relevant features with scales and then established a fractal transformation of the square root model considering the shape and distribution characteristics of the features. Wu (2007) discussed the mathematical foundation for control table generation in a selection model, determining that the mathematical model is essentially a geometric sequence of difference, and its internal structure law was revealed.

(2) **Study on the cartographic generalization algorithm.**

Many problems in the automatic map generalization process cannot be described by mathematical models. At this time, the use of a certain algorithm to describe the process will have a significant effect. Automatic map generalization is divided into two categories, one is the basic algorithm, and the other is the advanced algorithm. The basic algorithm refers to a simple implementation of the geometric transformation of the generalization operation. The advanced algorithm may be a composite algorithm composed of a few basic algorithms or an intelligent algorithm. Due to the complexity of the automatic map generalization problem, currently developed algorithms cannot solve all the problems, although they can achieve better results for a certain situation or situations. This is the primary reason why the new model and algorithm of automatic generalization are still undergoing continuous improvement and development.

The basic algorithm of automatic generalization mainly includes selection, simplification, aggregation, displacement, and so on. The basic algorithms listed here only contain the basic generalization and displacement of point, lines and polygon features; other algorithm results can be found in the section "cartographic generalization models and methods for map features" in this chapter.

A. **Selection of the point cluster**

Automatic selection of the point objects is an important aspect of automatic map generalization. In general, the selection of a point cluster includes two aspects: one is to calculate the selected quantity, and the other is to determine the selected objects. The main research results are now mostly shown in the latter. The key to maintain the distribution density of a point cluster and the distribution characteristics of the external contour.

As examples, Wu (1997) used the principle of the convex hull to realize the generalization of point cluster objects. In the basic principle, considering the aggregated distribution of a point cluster, a convex hull algorithm is used to form multilayer nesting and reflect the characteristics of a layered distribution. This process provides an integral distribution control method for the structured selection of point cluster objects. With the help of Voronoi, an additional quantitative basis is provided for the regional assessment of all objects. The whole process is divided into two sub processes, i.e., the aggregation of the convex shell and the generalization of the polyline vertices of a polygon. Qian et al. (2005a) proposed an algorithm of point cluster selection based on the circle characteristic transformation; the key steps of the algorithm are to find the center point holding maximal empty area, calculate the character space of the map objects, transform the Gauss coordinates to character space, create line clustering in character space and simplify the lines in character space. Liu et al. (2005) simulated the way of thinking of a person in cartographic generalization with the basic cartographic generalization principle of soundings on a chart, and an automatic generalization model for water depth annotation-expression, recognition and generalization was proposed. In addition, the practicability and effectiveness of the model are verified by experiments. Peng et al. (2005) proposed an optimum method for the automatic generalization of spatial point clusters, where spatial point clusters were divided into three subsets: boundary points, aggregated centers and interior points. For the boundary point subset, the Delaunay triangulation method is used to determine the boundary points and the choices of boundary points. For the aggregated centers subset, a fuzzy clustering analysis method was used to determine the composition of the center point subset. Finally, after determination of the boundary point subset and the aggregated centers subset, an optimization method for the automated simplification of the interior point subset was designed. Cai et al. (2007) improved the standard SOM (Self-Organizing Feature Map) algorithm based on the Kohonen network model and studied the selection and typification of the point set under the condition of preserving the original spatial distribution properties of the point set.

B. **Simplification of curves**

Line elements are abundant basic map elements on a map. As an important aspect of automatic map generalization, the simplification of a curve has attracted great attention. Since the Douglas-Peucker simplification algorithm was put forward in 1973, the research results in this field are continuously emerging.

Huang (1995) proposed a vector data compression algorithm for curves, and this algorithm has predictive function. Under the condition of satisfying the given accuracy tolerance, the method can be used to analyze the changes in the curve in real time; additionally, the reservation point sequence with a maximum compression ratio can be automatically extracted from the point group of the composition curve. Zhang et al. (2001) presented a curve generalization method based on the compound bending analysis according to the curve shape analysis method. Wang et al. (2002) put forward several parameters for describing tiny concave-convex shapes, and these parameters were taken as control parameters for curve simplification to establish a tiny concave-convex removal algorithm. Ying et al. (2003) applied commonly used curve simplification methods, such as the DP algorithm, BLG tree algorithm and so on, combining with the topological relations of point, line and surface to carry out line simplification based on feature constraint points (whether they are in or out of the curves), and put forward a consistent simplification method for all kinds of situations. Wu (2004) analyzed in detail the main problems of the curve generalization method based on the DP algorithm, and presented a bilateral offset method and a multitree structured curve generalization method considering the value of the equivalent offset. Qian et al. (2007a) proposed a new line simplifying algorithm with an oblique dividing curve. The approach divides curves into monotonous arcs with an oblique section and focuses on curve shapes on both sides of the line. At first, the method identifies whether each monotonous arc is a U-type or V-type arc, or a large or small arc, and then, different types of arcs are generalized with different details. During the course of generalization, if one arc is simplified, the approach will redivide the whole line into new arcs and simplify them again individually, or the next arc will be simplified. He deduced this process by analogy. Thus, the algorithm is a dynamic simplifying process.

C. Simplification and aggregation of polygonal objects

Simplification and aggregation of polygons is one of the commonly used operators for automatic map generalization. The generalization of residential land, vegetation and other features cannot be competed without this process. Qi et al. (1996) discussed the algorithms, such as automatic selection, simplification, exaggeration and displacement, as well as aggregation for the discontinuous area phenomenon (i.e., area patch). Lu et al. (2006) proposed the definition and construction algorithm of the layer-structure of speckle data based on the measure of information and structured selection of speckle data, as well as the spatial distributing and connection, and modeling of selection process was realized. After analyzing the capabilities to detect conflicts existing between polygons and in polygons of the CDT-based simplicial data structure (SDS) model, a unified solution for polygon simplification and merging is proposed by Zhang et al. (2006), and at the same time, an improvement of the conflict detection algorithm is applied to avoid the excessive increase in area and a loss of the characteristic vertex during polygon simplification and aggregation, which is caused by involving peaked triangles located at the edge of the triangulate mesh.

D. Displacement

Displacement is the most important way to solve the conflict of space occupancy and is an important aspect of the study of automated map generalization.

The theory, method and implementation processes of automatic map generalization, which are supported by the cartographic database, were discussed and studied by Wu et al. (1992). Focusing on the displacement of the line features overlapping on the map, the methods used to detect the feature interference and perform feature displacement were described; additionally, the methods of detecting interferences based on the map database were discussed and implementation examples were given (Wu 1997). Based on an analysis of using the finite element method to deal with the displacement of map objects and the propagation characteristics of spatial relation conflicts generated by the displacement of map objects, an algorithm for the maintenance of the map objects' spatial relations in their displacement was presented by Mao et al. (2003). The algorithm is an integration of the spatial reasoning method and the finite element method. WU Fang et al. discussed the method to solve the displacement according to the theory of elastic mechanics. Based on the finite element method, the conflicts of targets were explored, and the force bearing on the targets was analyzed. As an example, they aggregated the road network targets by regarding this aggregation as an elastomer. In addition, the size of the displacement was obtained with a plane pole model elastomer. Additionally, the point features with closer spatial relations and mutual independence were clustered to form a number of organic entities by using the method of spatial distance clustering and the minimum spanning tree. Further, the finite element method was used to implement the iterations of each entity to realize the displacement operation for the point cluster of residents (Wu et al. 2005; Hou et al. 2005). Liu (1994) studied the handling of features' relationships in the cartographic database graphic output process, putting forward a new method to make full use of the retrieval functions of the cartographic database (CDB), creating quantitative priority levels in the graphic output of features, defining the sizes of the feature graphs, and providing a detection method of the graphic conflict based on the CDB. In the treatment procedure, all relationships of the graphic output were divided into five classes, such as the conflict between lines, the conflict between point and line, the conflict between point and surface, the conflict between line and surface, and the conflict between points, which will be treated separately. All the algorithms used to process the relationships were introduced in detail. Finally, the preliminary experimental result was analyzed, and some other methods were put forward to solve this question.

Ai (2004) presented a field analysis based method to deal with the displacement of a building cluster, which is driven by street widening, and the field theory with an isoline representation model was introduced to describe the abovementioned displacement phenomenon. Based on the skeleton of the Delaunay triangulation, a geometric construction similar to the Voronoi diagram was created and the displacement field was built, in which the propagation force is related to the adjacency degree with respect to the street boundary. The study offers the computation of displacement direction and offset distance for the building displacement.

To solve the displacement problem, FEI Lifan and HE Jin put forward concrete methods to enable the computer to simulate human cartographers and have the functions of 'visual sense' and analysis. For this purpose, a special hybrid raster-vector data structure was designed, the graphic conflicts were solved by the cartographic displacement and reshaping of all the building symbols was constrained within the displacement zones. The mathematical methods involved in the constrained reshaping of the buildings were discussed to maintain the correct inner relationships of the buildings and the correct interrelationships between the street and nearby building symbols; when the multiple nearby street symbols are broadened for map compilation at smaller scales, there were no more places for ordinary displacement. The experimental results are shown and their evaluations were given from the aspects of visual and digital analyses. At the same time, a procedure was introduced based on raster detection and vector calculation to detect sequential conflicts after displacement of buildings, and a method for removing the sequential conflicts was described (Fei 2004; He et al. 2007).

Aiming at the specific conflict problems between streets and buildings in a map, Li et al. (2006b) analyzed the constraints of displacement and designed a scheme of displacement to solve the specific conflict problems between streets and buildings. Wu et al. (2008) used the Snake model to displace road features and eliminated the conflict. First, the theory of the Snake model was illustrated in detail. Then, according to the demand of cartography, the Snake model is improved as follows. First, the relationships between the Snake parameters and road shape character were built up to better maintain the shape of displaced road. Then, the force propagation extent in the Snake model was controlled to maintain the accuracy of the road position. Finally, the power attribute of the intersection point between roads is set to control the displacement of the intersection point and maintain the topological relation of roads. The general concept of disposing of the spatial conflict in road networks was brought forward by using the improved Snake model. In terms of the general concept, one research experiment showed that the roads are displaced and the spatial conflicts are solved.

Conflict detection is the precondition to solve the displacement. A novel method for the automatic detection of spatial conflict based on the plane scan algorithm was proposed by Liu et al. (2006). According to this method, the topological relationships between line segments were calculated in the plane sweep process. In addition, the line-line topological relationships were deduced based on the relationships of the line segments. Then, the spatial conflicts were detected according to the line-line relations, and the rules were defined.

5.3.3 Cartographic Generalization Models and Methods for Map Features

As the most frequent map elements occupying the maximum load, the generalization of residential land, road networks, rivers and contour lines is particularly important.

(1) **Generalization of road networks**

The difficulty in road network generalization is maintaining the typical network structure characteristics before and after road generalization.

With the characteristics of the simple road network topological structure, the road network was abstracted as the node set and edge set, and the adjacency matrix was established by Wang et al. (1985b). The data structure of the pointer table was used to store the data, and the strength value of the node was obtained by using the matrix iteration algorithm according to the connectivity principle of the graph. The results were used as a basis for selecting road networks. At the same time, the connectivity, road grade, road mesh size and other factors of the road were considered in the mathematical model of road selection.

Aimed at the problems of road object selection at any unknown scale, the existing methods of object selection were integrated and extended in the paper by Wang et al. (2007b), and a new object interpolation method was proposed, and this method reflected the inheritable and transferable characterization of related information among multiscale representation objects, as well as considering the attribute effects. Then, the interpolation basic idea, overall framework and technical flow were put forward, and the synthetic weight function of the interpolation method was defined and described. Aimed at the same problem, the key technologies of the interpolation model, including evaluation of the synthetic weight, calculation of the synthetic weight coefficient, and processing of the dimensionless weight factors, were further studied by Wang (2008b).

Hu et al. (2007) proposed a new approach involving the selective omission of roads based on mesh density. The density of roads is calculated using meshes as units and represents the local region density in a network. Usually, the density is different between maps with different scales. The density thresholds are acquired by theoretical analysis and empirical study for mesh density at different scale maps. For meshes with a density over threshold, the largest density meshes are taken out individually and one of the road segments is eliminated. Meanwhile, the obtained mesh is merged with the adjacent mesh one at a time. A road segment to be eliminated is the least important of all segments on the boundary of this mesh according to the parameters reflecting the importance of road segments. In this study, the meshes are classified depending on the road segment types. For different mesh types, the thresholds are different, which can preserve the density difference. The process of eliminating road segments and merging meshes ensures the road network connectivity. This proposed approach considers topological, geometric and semantic properties of the road network. At the same time, the selection method for road data based on mesh density and road matching method considering the levels analyses is proposed for this procedure.

This approach is applied to the update the data at the 1:50 000 scale based on map generalization, and the results shows the feasibility of this approach.

In the paper of Chen et al. (2008), a series of experiments were conducted where the road networks were autoselected by different methods using the graph theory. The results were analyzed and compared, and then, a better method was proposed for selecting roads. According to the network characteristics of the plane figure, the road network was divided into two types (rural and urban), and different methods for each type of road network were put forward. In the paper from Deng et al. (2008a), a new generalization model of road networks based on topological similarity was proposed. This model selected road generalization outputs by establishing evaluating figures of road network topological similarities, and furthermore, an assessment method depending on a road's shape was put forward for different road deletion schemes of the same topological similarity.

According to the constraint-based automatic map generalization theory, an automatic generalization method for the urban street network was presented by Tian et al. (2008). The method relies on the constraints and improved dynamic decision tree structure in the generalization of the street network, and the constraints are incorporated into the specific data structure to realize the progressive generalization of the street network.

(2) **Generalization of rivers**

Structural selection of rivers and the feature preservation of a river system is the focus river generalization, but it is also difficult. Wu (1994) discussed the method of the automated cartographic generalization of digital rivers, and the emphasis was placed on the automated elimination of rivers based on the non-structured river network. Wu et al. (2007b) studied automated canal selection. A method based on a grid to detect the network density was put forward and the result showed that this method was better than the other method. In addition, a canal network selection model was set up based on a spatial knowledge decision. This model integrated five canal selection factors, such as the attribute, length, link condition, spatial relation between canals and network density. Then, the multiple criteria decision were utilized to evaluate each canal by these five factors and make alternating selections to preserve the whole network character. An automated selection model of a ditch based on multiobjective optimization by genetic algorithm was developed by Zhai et al. (2008), where several key problems in the process of building model were discussed, and three factors influencing model selection were proposed, which are the selection sums of ditches, minimum space between ditches, and adding a few 'primary' ditches to the initial selection.

A new rule of ordering river entities based on counting their tributaries was proposed by Zhang (2006), which indicated the density differences in the river entities among different areas. A complex index that combined length, grade and level was designed for structured river selection. An approach was developed to compute the new grades and complex indexes of river entities based on structuring river segments into river entity trees. Thus, a new method to generalize dendritic drainages while maintaining the density differences between subdrainages was proposed by Zhang (2007), which is based on the layered division of the number of retained rivers in the river trees. The layered structure of dendritic drainages and their density differences was analyzed, and the density differences inside the drainage was found to originate from the differences in the number of tributaries of its subdrainages. A method to allocate the number of retained rivers was accordingly proposed, which divides the overall retained number of subdrainages according to the ratio of the number of tributaries. Finally, the structured drainage generalization considering the density differences among drainages was realized.

(3) **Generalization of habitation**

The research effort of habitation generalization has focused on two categories: one is the aggregation and simplification of the boundary expression of habitation; and the other is the selection of habitation with a circle-shaped symbol. There are many research results for both categories, especially the former. The research results of the latter address the selection of point clusters. While the graphic grade transformation from boundary graphics to circle symbols is relatively simple, the research results are also quite few.

Guo (1994) discussed the method and implementation process of the automatic compilation of 1:500,000 maps (including automatic selection, simplification and transformation) using the original data files of the residents in a 1:250,000 map database. The process has a large proportion of human-computer interaction. The map generalization knowledge for the elements of habitation was summarized and obtained by Wang (1996) using the knowledge representation method of production rule in the expert system. In addition, based on this knowledge, habitation generalization for generating a small-scale map based on medium scale map data was tested.

Guo et al. (1996) researched the automatic generalization of an urban settlement on a topographic map, the urban settlement was represented by a network structure, and the generalization was decomposed into two main components, i.e., handling of streets or blocks, and handling of other features within blocks. Wu (2000a) discussed the necessity of creating the geo-entity data model to support the automatic generalization of city maps, the logical hierarchies of generalization and the major algorithms for implementing the merging operation on building graphics.

Some simplification rules of the area feature boundary and the method of obtaining spatial knowledge were discussed in the paper by Guo (1999b), which focused on the progressive method of graphic simplification of an area feature boundary turning at right angle. Li et al. (2006a) proposed an algorithm for building typification based on data matching between multiscale databases. The algorithm is based on the 'divide and conquer' strategy. The whole map sheet is divided into several parts by road networks, and then, each part is divided into several clusters by means of spatial clustering. The cluster is the basic processing unit of building typification, which supports parallel computing.

Guo et al. (2000a) distinguished two kinds of spatial relations, which are the topological neighbor relation and the visual neighbor relation for polygon automatic generalization under a digital environment, and the corresponding aggregation methods were presented based on the vector and raster data structure. For the building shape simplification, the paper proposed a rectangle differential mode approach to simplify the building polygon by a series of rectangles with different hierarchical levels.

Du et al. (2004) proposed a polygon simplification method based on combinatorial optimization. The polygon boundary was decomposed into a series of curve features according to the feature points, the generalization was translated into a combinatorial optimization problem by geographic-feature analysis of the objects, and finally, a simulated annealing algorithm was applied to implement a polygon simplification based on combinatorial optimization. Based on the figure characteristics of a building polygon, two models of building polygon simplification and rectangularity in the map generalization were presented in the paper by Liu et al. (2008). First, the convex and concave distance threshold of the building polygon at different scales was set according to the visual 'natural principle', and then, all points of the building polygon based on the threshold were grouped, and finally, the least squares adjustment was employed to linearly fit all the points of each group and to simplify the building polygon. As for the rectangularity of a building polygon, an unequal weight rectangular adjustment model with a limiting condition was proposed to ensure that the overall displacement of the building polygon was minimal and to avoid conflict with the two neighboring edges of the building polygon.

Song et al. (2005) presented a method for attribute reduction and the structured selection of settlements in the automatic cartographical generalization based on a rough set. A rough set with a strong ability to mine knowledge is applied to quantitatively analyze the importance of settlement attributes and attribute reductions. In the precondition without transcendental knowledge, attributes of settlements can be filtered based on the rough set so that the conditional attribute sets can be optimized, and the significance value of each attribute can be extracted.

Wang et al. (2005a) presented an approach to simplify the building polygon (one aspect of the inhabited map area simplification), which is supported by a data model combining the raster and vector model and adopts the methods of mathematical morphology and pattern recognition with the assistance of a neural network. This approach integrates the knowledge of map generalization into the graph simplification. In the paper by Qian et al. (2005b), a new method based on the technique of Agent, TIN and clustering was put forward, which was expected to improve the operation speed, intelligence level and automation level of generalization. In addition, the results were quantified and analyzed. In the paper by Qian et al. (2007c), the dynamic geometric spatial relationship among larger scale streets and buildings was pointed out, which is that buildings and streets influence each other, and they fill the whole area. Then, a new method of block generalization based on street skeleton-line and building skeleton-line was put forward. The process of obtaining skeleton-lines was defined, and the characteristics among the building skeleton-line and street skeleton-line were analyzed. Based on above information, several algorithms, such as mergence, simplification and displacement of buildings were put forward based on the dimension reduction technique.

Based on analyzing and studying of the generalization process of an urban residential area, Yuan et al. (2005) put forward an algorithm and constrained conditions of generalization according to the least squares adjustment theory, and the example shows the validity of the algorithm. In addition, a generalization algorithm for the mergence of urban resident land was established. Based on the research of the generalization rules and the analysis of the geometrical characteristics of a residential area, Yuan (2007) presented a method of buildings segmentation related to scale to implement the building polygon simplification.

By introducing the concept of a "central place" and combined with the principal component analysis method, Su et al. (2007) comprehensively considered multiple indexes of geographical elements and put forward an algorithm for automatically accepting or rejecting linear geographic elements based on the central place conception in cartographic generalization.

Aiming at the automatic simplification of 3D building surface models, Wang et al. (2007a) reviewed the current simplification algorithms and stressed the disadvantages of application to the simplification of 3D building surface models. Finally, some new simplification strategies for 3D building models were proposed. Ge et al. (2009) studied the facet shift algorithm in the simplification of buildings with parallel structures. First, the 3D building data are structured and then, a data structure that benefits the generalization was proposed. Therefore, a shifting facet simplification algorithm based on minimal distance was realized in the simplification of a building with parallel structures. At the same time, the facet shift algorithm was realized based on the minimal feature. For the simplification methods of nonparallel buildings, according to the extent of complexity, nonparallel buildings were divided into two types, including simple buildings and special buildings. Based on an analysis of the structural characteristics of the two type buildings, algorithms were developed and tested to simplify them.

In Yan et al. (2008), an approach was presented for automated building grouping and generalization. Three principles in Gestalt theories and six parameters were selected to describe the spatial configurations, distributions and relations of buildings. Based on the rules and parameters, an approach to building grouping and generalization was derived using the Delaunay triangulation. The method is different from existing methods because it considers the direction relations among buildings. This approach is fully automatic and can be used in the generalization of buildings in blocks.

The thesis of Zhang et al. (2008) analyzed five common methods, i.e., the longest edge, weighted bisector, statistical weighting, the smallest minimum bounding rectangle and wall average. Particularly, their applicability and limitations were analyzed according to their basic principles. Through two experiments, the applicability of these methods and their practical problems were further illustrated. It was also found that the smallest MBR was more robust than others.

(4) Generalization of contour lines

Generalization of a single contour line obviously does not represent the landform morphology expression with a contour lines group. Group-generalization of contour lines is inevitable, but it is also very difficult.

Zhao et al. (1983) conducted research on the automatic generalization of contour lines in sets. The key points of the relief generalization based on valley lines were discussed. The algorithm of selecting skeletons by using distances between valley lines was proposed, and some key improvement measures were put forward, such as the approach of modifying the heights of interpolated points, the use of supplemental points in tracking, adjustment of skeletons, etc.

Tang (1987) studied the automated generalization of submarine relief for the automated compilation of a chart. The algorithm presented here for automated sounding selection is based on the area characteristics and distribution of soundings. The automatic generalization of the shape of the depth contours focused on solving the following three problems: to find the position of a curve and calculate its size, to determine the features with smaller bending, and to process different curves with different methods. The automatic simplification of depth contours depends on the feature of the curves, which is determined with the "positive and negative" method. The automatic selection of the quality of the sea bottom is also included in the paper.

Yang et al. (2005) used the lifting method to process DEM data with multiscale representation. They introduced the structure algorithm briefly and emphasized the improved structure algorithm in detail; finally, they gave the experimental method and process of extracting the characteristic line of terrain from the multiscale DEM data by using the improved structure algorithm.

Based on the analysis of the principle of the 2D Douglas-Peucker algorithm, a 3D Douglas-Peucker algorithm was proposed by Fei et al. (2006), and the method was applied to the automatic generalization of 3D discrete points, which is the basis of a DEM. Later, the algorithm was further improved (He et al. 2008).

Zhao et al. (2008) put forward a topologically consistent contour simplification algorithm based on the contour tree and strip-tree. This algorithm ensures that contours do not self-intersect or intersect with each other after simplification and the set of contour vertices after simplification is a subset of the set of contour vertices before simplification.

In the paper by Zhu et al. (2009), a novel generalization method of contour lines based on the furthest visibility condition was proposed. First, the area of contour line simplification was divided according to the furthest visibility principle, and then, in each dividing area, the bend nesting of the curve is analyzed and classified. Thus, curves were selected or deleted according to the bending level of the curve segment and the preset threshold value.

5.3.4 *Applications of Mathematical Methods and Artificial Intelligence Methods in Cartographic Generalization*

(1) Application of fractal theory in automatic generalization

Fractal theory is a new branch of modern mathematics. The basis of fractal theory is that it views the shape of an object as smaller copies of itself and the copies are similar to the whole: same shape but different size (self-similarity). Many of the elements on the map, such as water systems, contour lines and coastlines, have extremely typical fractal characteristics. As a result, the fractal theory provides a new research method for automatic cartographic generalization.

A thorough study on automatic cartographic generalization using fractal geometry was conducted by WANG Qiao et al. A new cartographic generalization-oriented method of fractal dimension estimation was proposed according to the fractal theory. As a tool, some difficult problems in automated cartographic generalization, such as choosing thresholds automatically, maintaining the shape features of a map, and segmenting the curve as its complex nature, were researched. The basic principle of fractal dimension estimation was combined with one method of linear element generalization, and a cartographic generalization-oriented method of fractal dimension estimation was proposed. The method provided a quantitative characterization of the cartographic lines and described the relation between the tolerance values and the features of cartographic lines when the scale changes, which then helps to realize the modeling and objectivity of the cartographic generalization process. According to the fractal theory, a new method of complicated contour automated generalization, which is based on quantizing the features of contour line shape, was proposed. By discussing the self-similarity and fractal dimension of cartographic lines, the features of the cartographic line in terms of shape were quantized and extracted, and a new automatic generalization method was proposed. This method has a good effect on maintaining the shape features, quantization and modeling of the procedure of cartographic generalization, and provides a solution to some problems that are difficult to solve with other methods in automated cartography. By applying the fractal theory, a fractal model of quantizing the polygonal shape features on the map was proposed. The meaning and formula of the model parameters, for example, the shape factor and fractal dimension, were also discussed. Thus, a new solution for measuring the shape features on map was given. Based on the fractal theory, a new method of automatic generalization of map polygons was proposed, which can automatically generalize the map polygons according to the distribution characters and shape features themselves (Wang 1995a, b; Wang et al. 1995a, b, c, 1996a, b, c).

Research on extended-fractal-based automatic map generalization was conducted by Wu (2001), which concentrated on the essence of the application of generalization, i.e., how much information it can provide for us, its limits, and how it can be combined with other methods in map generalization. Several issues related to the application of the Extended Fractal in automatic map generalization were expounded. He et al. (2002) considered that hydrographic features show the fractal characteristics and proposed a method to measure the fractional dimension of coastlines and rivers on the series scale topographic map. As an example, coastlines and rivers on the map are generalized using the change regularity and fractal geometry. While Li et al. 2006c) studied the catastrophe-scale in the course of map information attenuation and presented a new method for the determination of the catastrophe-scale based on the theory of the extended fractal dimension. They found that from large scale to small scale, the self-similarity of the map object descends gradually and is reflected as the change in the map information. As a result, the catastrophe-scale between two known scales can be determined by making use of the extended fractal model of the inverse "S".

(2) **Application of the wavelet analysis in automatic generalization**

Wavelet analysis is a new kind of time (space) domain-frequency domain analysis method; it has good localization properties in the time domain and frequency domain. Moreover, the sampling density in the airspace/time domain automatically regulates the different frequency components of signals, and all the details of a function (signal, image, etc.) can be observed and analyzed from an arbitrary scale. Thus, it is known as a "mathematical microscope". Multiresolution analysis in wavelet theory (MRA), which provides an effective way to analyze and express the information at different resolutions, was introduced into the automatic cartographic generalization.

In the paper by Wu et al. (2000c), based on the wavelet theory, the relation between multiresolution analysis and multiscale data expression in GIS was found, a new model of using the multiresolution analysis was put forward, and the application example for using the proposed model was given. Additionally, a multiscale representation model of a complex landform based on two-dimensional wavelet transformation was proposed. The new idea and principle of this model can completely eliminate the intersections of different contours, which commonly occurred in almost all models previously used.Thus, it could be said that by theory and method, the model can preserve the lines of the landform and maintain the coordinate relations among the contour lines. In addition, this method has adaptive property. Wu et al. (2001) proposed a scheme for a multiscale representation and generalization of scale-dependent relief based on the basic principle of the multiresolution analysis (MRA) of a wavelet. Handling of the multiscale automatic generalization of relief was based on the model. The normal ratio of the wavelet coefficients was proposed as a quantitative index to measure the detailed extent between the original data sets and their derived counterparts.

According to the characteristics of the wavelet, a contour data reduction method based on the B-spline wavelet and the contour data compression model based on a wavelet analysis was proposed by Wang et al. (2003b) and Zhu et al. (2004). Yu et al. (2008) combined the multiscale wavelet analysis with the square root model to simulate the generalization process of DEM data. In addition, the method was used to generate two new smaller scale DEM data based on the original DEM data of the 1:1,000,000 scale topographic maps.

(3) **Application of mathematical morphology in automatic generalization**

Mathematical morphology is a kind of theory and method for digital image processing and recognition and a discipline based on strict mathematics. The comprehensive application of its basic operation is the basis of its use in cartographic generalization.

Zhang et al. (2000) discussed the method of identification and generalization of clusters and regular structural units by mathematical morphology. Clusters were identified based on two criteria, i.e., proximity between features inside the clusters and the ratio between distances inside and outside the clusters. Furthermore, neighboring features with similar sizes, spacing and orientation were identified as regular structural units inside the clusters. After cluster identification, regular structural units were generalized in a way to maintain their characteristic pattern in terms of size, spacing and orientation. Other features inside the clusters were generalized in association with their context, so that the general characteristics and distribution of the clusters were maintained.

According to Wang et al. (2000), by researching and analyzing the correlative algorithms of mathematical morphology and integrating the traits of uniting blocks in residential areas, a new method using the image closing operation in mathematical morphology was presented for the map generalization of uniting blocks in residential areas.

(4) **Application of the artificial intelligence method in automatic generalization**

Intelligent and automated map generalization is an idea that cartographers have dreamed of for a long time. The search for many intelligent methods to solve the problem of automatic generalization is an academic area of study that cartographic scholars have explored and attempted for many years. Many intelligent methods, such as Agent, artificial neural network and genetic algorithm, which developed in the field of artificial intelligence, have been applied to automatic map generalization.

In view of the global and parallel search capacity of the genetic algorithm and its simple, fast and stable characteristics, DENG Hongyan, WU Fang and others applied it in the field of automatic map generalization and achieved many results. As examples, a model of point cluster selection based on genetic algorithms was designed. The basic principle of the point cluster selection model was to first divide the point cluster into several subclusters according to the density, and then, the preserved amount of every subcluster was calculated according to every subcluster amount and the total amount to be preserved; finally, the convex hull was combined with the genetic algorithms to make a selection. Based on analyzing the methods of line simplification devised by other cartographers and the basic principles of genetic algorithms, a model of automated simplification was provided by using genetic algorithms and the key issue in the automatic generalization of the line was discussed together with the detailed procedure and output. A new generalization model of road networks based on the genetic algorithm was proposed from a global perspective. Based on the structural characteristics of the river network and the principle of river selection, a new automated elimination model was established using the genetic multiobject optimization algorithm. The structured data model for the river system's spatial knowledge was first constructed, and then, the river was autoselected using the genetic multiobject optimization algorithm. The indicators including length, interval and importance of river, and so on, and the collectivity structural characteristic of the river network was considered when selecting the river. Considering the preservation of the graphical characteristics of man-made river networks before and after generalization, the length of the canal, density control of the man-made river network and so on, a man-made river network elimination model based on multiobjective optimization by genetic algorithms was developed according to the purpose and principle of man-made river network generalization (Deng et al. 2003, 2006a; Wu et al. 2003; Zhai et al. 2006, 2008).

With a selection of settlements as an example, Wang et al. (1985) quantitatively described the importance of cartographic objects using the fuzzy comprehensive estimation method, and the automated selection of settlements was realized. Xu (1986) also studied the selection of cartographic objects in a map by using the fuzzy synthetic judgment principle. In a typical example, determination of the subordinate degree of the main factors in the factor set was emphasized, and the basic way to select all the cartographic features in a map was introduced. In the paper by Wang et al. (1999), the design of the sounding generalization neural network was discussed, along with the analysis on the network's overall structure, the working methods, network factors and learning. To solve the problem of dealing with both the spatial and attributive factors at the same time to selected soundings, a new set of operations was designed based on a practical problem-solving method called "Hierarchical Information Structure". The experimental result was shown in a protocol system.

Qian et al. (2004) applied the technique of Agent to map generalization, and a map generalization monitoring model based on Agent was put forward; this model was used to monitor the generalization behavior. The behavior greatly affects the automated map generalization process all the time.

(5) **Application of computational geometry and other methods in automatic generalization**

Delaunay triangulation is a very important research field in computational geometry. Since the Delaunay triangular network can conveniently calculate the massive neighborhood relations existing among spatial objects, it is very widely used in cartography and GIS and also has been used in depth in automatic map generalization.

AI Tinghua et al. conducted in-depth research on automatic cartographic generalization based on Delaunay and achieved a series of results. For example, with respect to the generalization of polygonal objects, they presented a method of Delaunay TIN partitioning in two-dimensional space, which can be used to support such operations as finding neighbor polygons, identifying the bent parts of polygons, detecting conflict regions between polygons and merging polygon groups. They presented one method that applies the constrained Delaunay TIN built in street regions to retrieve centerlines and further generate a street network structure. Three kinds of triangles were distinguished and different corresponding centerline extraction methods were offered. The spatial relations among street edges, street junctions and street blocks were described by graph element edge, node and loop. The method also supports the matching neighbor between centerlines and street block arcs. The provided data model the unified the street network and block polygon. By applying the circum-circle principle and closest to equilateral properties of the Delaunay triangulation, the simplification and aggregation of polygonal objects supported by the Delaunay triangulation structure was presented. Based on the Delaunay triangulation and Voronoi diagram model, with consideration of the Gestalt principles in visual adjacency cognition, a method was presented to find the distribution range polygon by progressively stripping the outside triangles. In addition, the distribution density was represented by the Voronoi cell size and visualized gray image. By applying an image process method, the distribution center can be extracted from a gray image. As a result, a method of point cluster simplification was provided in a dynamic way based on the Voronoi polygon establishment. They also presented an automatic method to extract terrain landform features and organize drainage system into tree structures based on the bend assessment using the Delaunay triangulation model. This approach not only obtains the topological structure of a drainage system in a planar graph but also the valley distribution polygon range. Depending on the geometrical computation and judgment of the vector line and polygon, the structured properties in the drainage representation are enhanced, avoiding the case of noise disturbance in the DEM-based method (Ai et al. 2000, 2001a, 2002, 2003a).

In addition, Qian et al. (2001) introduced a merge operation for area objects based on the Delaunay triangle interpolation, and several application examples of this method were given in terms of the automated map generalization of area objects in a residential area, vegetation and soil properties. According to the spatial proximity characteristics of the edges of the triangles in the Delaunay triangulation, Ai et al. (2003b) presented a method to extract the medial axis of a double line stream using the Delaunay triangulation skeleton and studied the measurement of the distance between nearby rivers. A generic algorithm for point cluster generalization based on Voronoi diagrams was presented by Yan et al. (2005). The new algorithm employs two methods to correctly ensure that different kinds of information is transmitted: (1) determine the point number on resulting maps by the radical law so that the statistical (positional) information is correctly transmitted; and (2) recursively construct Voronoi diagrams of the retained points and delete points by comparing the significance value of a point with those of its Voronoi neighbors so that topological, thematic and geometrical information are correctly transmitted.

GUO Qingsheng and others put forward the concept and method of progressive generalization, and much work was completed, which is mainly reflected in the following text. The progressive graphic generalization approach of a linear feature was given. This approach can control the self-intersection of a line and makes the map scale continuous, not dispersed. A set of practical methods for the progressive graphic simplification of contours was established, and a specific algorithm was proposed to handle some special cases. The algorithm of buildings progressive typification based on the weighted mesh simplification was put forward. The algorithm considers the resolution distance and density between objects, and maps with arbitrary scales can be progressively obtained. In addition, the characteristics of the original buildings can be maintained from the aspects of geometry and semantics. At the same time, the density variations between the different partitions can be maintained before and after generalization. Based on the component abstraction transformation method, the abstraction rules of topological relations between two areas and progressive transformation methods of these topological relations were investigated. A corresponding diagram of progressive transformations was given. The cartographic design criteria of the schematic road network map were analyzed and summarized, a set of quantitative constraints were set up, and a new road network generalization method that includes progressive selection and displacement was proposed. In addition, the topological checking method for a road network was studied (Guo 1998; Guo et al. 2000b, 2005, 2007; Dong et al. 2007).

5.3.5 *Acquisition and Representation of Cartographic Generalization Knowledge*

The difficulty and complexity of cartographic generalization is concentrated in the high dependence on human thinking activities, while the thinking activity of a human being in cartographic generalization implementation has the characteristics of subjectivity, flexibility and fuzzy judgment criterion. Therefore, not all cartographic generalization problems can be modeled or be an algorithm. For a subjective human judgment to be accepted and formalized by a computer in the course of map generalization, the key is to establish an evaluation criteria system and knowledge rule for map generalization.

Automatic generalization based on knowledge refers to the normative description used to deal with some problems in the process of cartographic generalization. At the beginning of 1990s, the research of cartographic generalization expert system and knowledge reasoning had aroused the interest and attention of many scholars, and some exploratory achievements were obtained (Wang et al. 2005b).

Generally speaking, knowledge acquisition of cartographic generalization can be accomplished in 3 ways. The first way is to obtain knowledge from the experts. However, people soon discovered that many of the actions of experts may be understood but not able to be explained. The second approach is to obtain knowledge from the existing specifications or rules, but people soon found that specifications or rules are too difficult to understand to learn how to do it. The third approach is to obtain knowledge from an existing map. However, people soon discovered that differences among areas are large, and knowledge obtained from one region may not be useful in another. For these reasons, the research of the cartographic generalization expert system was a hotspot for a while in the early 1990s, and then, its application has tapered down. The development of intelligent cartographic generalization has met many difficulties (Wu 2000b).

However, cartographic generalization is essentially a highly intelligent system, which cannot do without knowledge. In addition, map generalization is a process that combines knowledge representation and knowledge abstraction. As a result, some exploratory work on knowledge acquisition, representation and processing of map generalization have been conducted by cartographers; the main results are reflected in the following aspects.

(1) **Classification system of map generalization knowledge**

Qi et al. (2001) researched the index system and knowledge rules for geographic-feature-oriented generalization. In the index system, the knowledge rules include geometric, structure and procedure knowledge. From the viewpoint of the operation process, knowledge rules can be divided into 5 categories, i.e., geographic-feature description, operator selection, and algorithm selection, as well as specific geo-object-oriented and specific region-oriented knowledge. In the knowledge base, these knowledge rules are organized by three dimensions of variables, including the generalization condition, generalization scope/level, and generalization actions. Qi et al. (1998) expounded upon the methods of intelligent generalization in the GIS environment. The key points were outlined as follows: by using the method of knowledge reasoning and the mathematical model, knowledge reasoning was used to guide and control the mathematical model; simultaneously, the geographic characteristics-oriented mechanism was used in generalization knowledge. Additionally, the flow chart of the mathematical model process driven by knowledge inference, as well as the structure of the generalization expert knowledge base were described. In the 'Theory and technique model of automated cartographic generalization chain', Qian et al. (2006b) believed that cartographic generalization knowledge is the main basis of the process control of automatic cartographic generalization. Based on this kind of awareness and understanding, the knowledge classification, acquisition and knowledge base construction of cartographic generalization, the structured description of generalization knowledge, the generalization knowledge attribute, the cartographic generalization knowledge management and organization and their applications in automatic cartographic generalization process control and reasoning have been explored. In 'Spatial data checking with generalization knowledge', Qian et al. (2006a) thought that data checking using generalization knowledge before generalization can provide valuable information about the generalization region, the key generalization content and the generalization method, while data checking using generalization knowledge after generalization can be used to determine whether the results meet the requirements. According to Wu et al. (2007c), the knowledge in automated river network generalization includes structural river network data, river properties and generalization rules. The methods of capturing and formalizing this knowledge were studied. A production rule base for the automated selection of a river network was established. Driven by the rules from the rule base, ordinal reasoning of the production rules system was applied in river selection. According to Deng et al. (2008b), a cartographic generalization knowledge base designed for database quality control was developed by analyzing the existing knowledge base of cartographic generalization, and thus, the basic structure of a table and the development, maintenance and realization of reasoning principles were developed. The most notable characteristic of the knowledge base was that the quality knowledge was efficiently managed, making the effect of cartographic generalization knowledge not focused in terms of choosing a simple generalization operation aimed at different objects at the beginning of generalization but also automated in terms of initiation and control of quality during each generalization process.

(2) **Representation methods of map generalization rules and knowledge**

It is extremely difficult to obtain map generalization knowledge and rules. The traditional knowledge acquisition method is based on the exchange of experts, group discussions, problem sets, survey forms, learning according to instruction or observation, and so on. At present, some experts and scholars at home and abroad have accomplished much fruitful work in the establishment of the generalization index system as well as establishing the rules and knowledge (Liu 1999; Qi et al. 2001). According to the degree of automation of knowledge acquisition, the acquisition methods can be divided into 3 categories: manual, semiautomatic and automatic knowledge acquisitions (Liu 1999).

The use of a variety of models and algorithms to obtain structured knowledge of spatial layers and spatial relationships is a key point of map generalization knowledge acquisition.

In terms of obtaining the structural knowledge of the spatial point cluster, a polygon cluster pattern mining based on the Gestalt principle was created by Ai et al. (2007a), in which knowledge of the spatial distribution of cartographic objects was obtained by the combination of spatial cognition principles, Gestalt principles and the spatial clustering method. In addition, according to Guo et al. (2008a), the hierarchical clustering method of a group of points based on the neighborhood graph has been studied, the unsupervised hierarchical approach of a group of points was improved, the clustering algorithm procedure was elaborated, and the feasibility of the unsupervised hierarchical approach was validated through a detailed example. They also studied the rules of spatial topological relations abstraction in the process of graph simplification, and proposed a method for the progressive deletion of local topological relations in the process of spatial abstraction.

The correct recognition of the shape of the curve structure is the premise of the curve generalization. A binary tree representation of the curve hierarchical structure was presented in depth by Ai et al. (2001a), the constrained Delaunay triangulation model was used, and through triangle stripping trace, the bend structure was recorded by the binary tree, which describes the bend hierarchical levels. In this method, the Gestalt principle is used to investigate the characteristics of the curve, which can be used to extract the real bend in the cognitive sense. Simultaneously, in the paper by Zhai et al. (2007), a structural method of the curve shape was proposed, that is, the constraint Delaunay triangulation network model was used, the Gestalt principle was considered, the structure of the curve shape was partitioned, the curve bends were divided into basic bends and complex bends, and the bend characteristics were identified. The method of detecting and expressing the spatial structural relationship among the bends was given. Finally, the bend and its structural relationship were class-encapsulated by the object oriented method.

Acquisition of the structured knowledge of a river network is very important to the selection of a river. In addition, there have been many achievements in this area. In the paper by Du (1988), the author proposed a kind of link-based structured river network and discussed the method of its automatic establishment. Finally, based on the structured river network, which contains these two indexes, several retrieval functions were provided as a way to extract structural information during structure processing in a. The advantage of this method lies in that river reaches can be used as a target and the basic relations among different targets can be defined. As a result, the river network in a cartographic database is in the form of a hierarchical structure. Zhang (2002, 2004) considered that there are many structural units on maps. The structural pattern reflects the characteristics of geographic phenomena and thus should be analyzed and represented on a map before map generalization. Based on dynamic segmentation model, they proposed a method to organize arcs into route systems and analyze their relative importance in relation to the linear network. Structural parameters are further derived for each route for network generalization. Based on these parameters, linear entities are selected as a whole in a structured way, so that the overall structure of the linear network may be maintained. Additionally, according to the characteristics of dendritic river systems, a method of constructing a river-tree was proposed. Starting from river sources, river entities were constructed from river segments, and the left and right tributaries of the mainstream were identified. Zhai et al. (2007) considered that a reasonable structured data model is helpful to the automatic generalization of rivers on the map. A structural river network data model oriented to automated generalization was developed. The model adequately considers the spatial relationship between reaches, reach and areal drainage features and a reach's hierarchal relationship of the river network. In addition, the structured knowledge of the river network was acquired. Ai et al. (2007c) tried to build the watershed hierarchical partitioning model based on the Delaunay triangulation. The hierarchical partitioning model can compute the parameters of distribution density, distance between neighboring rivers, and hierarchical watershed area, which are the preconditional parameters used to determine the importance of a river branch in the simplification and selection of a river network. In addition, Guo et al. (2008b) analyzed the spatial characteristics of different types of river networks and studied automatic reasoning on the main streams of tree river networks, and their algorithm can automatically infer the relationship between main streams and river branches.

The best way to conduct the contour line generalization is through group- generalization (generalization of a set of contour lines). The preconditional condition is to obtain the construction lines of the contour lines. In terms of the automatic extraction of the knowledge of construction lines, the automatic establishment of the contour tree is the key point of the research. Wu (1996) thought that establishing the contour tree involved establishing the inclusion relation between the parent layer of the contour lines and its sub layers, which is in the form of tree. For this purpose, he studied the establishment of a contour tree, which was used to generate high quality DEM and contour group-generalization. According to Guo et al. (2005), an improved algorithm for determining the feature points from the contour data was presented. Then, through the establishment of a mathematical model matching the feature points, the structural lines were generated. Simultaneously, the classification information of the structural line was also extracted.

Wang et al. (2007b) studied the hierarchical relationship among road multiscale representations. By using the multiscale extended E-R method, the hierarchical relations among features was given in detail, which includes abstract data types and attribute-value domains for geographical features at different scales, as well as the scale, semantics, mapping and so on. The aim was to extract knowledge from these aspects.

(3) **Expression methods of map generalization rules and knowledge**

In terms of the map generalization rules and knowledge representation, Qi et al. (2001) proposed an representation method of a productive rule knowledge base for geographical features. In the practical application of productive rules, first, we need to establish the rule base and inference database. Rules are stored in a rule base, while the data or facts related to problem solving as well as the new facts that are produced in the process of reasoning are stored in a deductive database. When the system is running, if the conditional part of a rule matches the fact, it indicated that the rule is available. This part is marked in the available database or extracted and stored in another database; for those that failed to find a match, the search for other matching rules is continued. After all the available rules are found, one of the most applicable rules is chosen through the comparison, and the behavior is performed. Cai et al. (2002) proposed a method of the six tuple representation of map generalization rules. The basic idea is that all of the map generalization rules are summarized in a six tuple. The six tuple structure is defined as follows: <layer code>, <operator>, <attribute code>, <index item>, <up limit> and <down limit>. Among them, <layer code> determines the ground object layer for which a rule is applicable; <operator> determines the rule that is suitable for the generalization operation, such as delete, merge or simplify; <attribute code> determines what kind of object in a certain level is applicable to the rule; <index item> determines the rule-based feature item, that is, what will be the basis for simplification, the length or the area; and <up limit> and <down limit> determine the upper and lower range of values of the index items, respectively.

5.3.6 *Automated Generalization Process Control Based on the Generalization Chain*

Map generalization is an intelligent process, which contains the experiences of many experts and much knowledge accumulation. Because the objects on the map are not isolated, the overall environment of the object groups should be considered in the cartographic generalization. Moreover, there are no automatic generalization algorithms that can meet all the requirements. Different cartographic generalization tasks can only be fulfilled by using different algorithms. More important, there are usually no satisfactory results that can be obtained at a given time, so it must go through repeated operations, such as generalization, revocation and regeneralization, to achieve a good result. Therefore, we need a comprehensive theory and method of cartographic generalization to control the overall environment, the algorithm and the workflow.

By making full use of all the algorithms, model, operator and knowledge, forming a scientific and comprehensive operation process, and implementing intelligent control of the process, the automatic generalization system performance can make a qualitative leap, and the generalization results can be improved. This process is called the process control of automatic generalization. In addition, it is an important means to improve the automation degree and quality of cartographic generalization from a global point of view. However, it is still rare to study the process control of automatic generalization in detail.

In China, Qian et al. (2006b) proposed a process-control model of cartographic generalization based on the automatic cartographic generalization chain, which includes the following steps.

(1) Data preparation: check the data sheet number, unit of data, data range and the region the data belong to; the data are transferred into the system; and the data are clustered according to the main roads and rivers.
(2) Data quality assessment of the original data: If the original data are in line with the requirements of the object map scale, the cartographic generalization does not need to be conducted, and the task ends;
(3) System initialization: the knowledge base, algorithm library, display module and symbol library are transferred, the default parameters are initialized, the system's public information pool is initialized, and so on;
(4) Settings and adjustments before generalization: the cartographic purpose, the source scale setting, the target scale and the regional characteristics of the data, the adjustment and setting of the generalization operators, generalization algorithms and generalization parameter database, and so on;
(5) Knowledge-based data checks;
(6) Extraction of the generalization task based on the data checks;
(7) Formation of the cartographic generalization chain;
(8) Implementation of the cartographic generalization chain;
(9) Quality evaluation of the data after cartographic generalization; the quality evaluation information of the generalized data is obtained, and the data are judged as to whether they meets the requirements of the generalized result. If this step is not satisfied, then the generalization operators, algorithms and parameters are reset and readjusted; then, repeat each step after implementation; if the requirements are fully met, then end;
(10) CASE storage of the generalization chain: the generated generalization chain is stored in the CASE library as a CASE, and the comparison with the generalization chain CASE generated in other cycles is accepted;

(11) If the system had executed a certain number of cycles and there were no automatic generalization result that meets the requirements of the knowledge base, then the optimal generalization chain in CASE library will be selected and executed and its result will be taken as the final result.
(12) The end.

The above steps are the main flow of the automatic cartographic generalization chain. In the course of the execution of the cartographic generalization chain, a monitoring model of cartographic generalization monitors each operation of the generalization chain in real time to ensure the correctness of the generalization operation.

The whole execution process of the generalization chain is automatically recorded by the system; the user can easily analyze the entire implementation process of the cartographic generalization chain. After manual editing, this record can also re-enter system to be re-executed in the form of a generalization chain. Simultaneously, this record can also be used as a basis for the multistage fallback operation of a system.

5.3.7 A Quality Evaluation and Control Model for Automated Generalization Based on the Quality Control Design

As an important and weak link in the research of automated cartographic generalization, the research of generalization quality control is of great significance to effectively improve the accuracy of the automated cartographic generalization system. Because the result of automated cartographic generalization is highly uncertain, the quality of the result of the automated cartographic generalization is not predictable, and most of the results can not completely meet the needs of users, from this point of view, the quality problem of automated cartographic generalization is the key to the automated cartographic generalization at present.

The research on the quality control of automated cartographic generalization is at the beginning stage in China and other countries. In China, the main research results are shown in the following aspects.

(1) **Uncertainty evaluation of the linear feature simplification**

The performance evaluation of the linear feature simplification algorithm was carried out by SHI Wenzhong et al. In the evaluation of the linear feature simplification algorithm for vector data generalization, the uncertainty in the curve position in the linear feature simplification is divided into two aspects. First, the uncertainty in the position of the original curve yields a transmission error; second, the curve displacement error is caused by simplification, and in the course of displacement, the shape of the curve will change. In addition, the position error model of the line simplification is also presented (Shi and Chui 2006).

(2) **Quality assessment and control of automated generalization**

WU Fang et al. from the Institute of Surveying and Mapping at the University of Information Engineering systematically studied and evaluated the automatic generalization algorithm, the resulting evaluation of the automatic generalization, the automatic generalization framework model based on the quality design and the data model, and so on. Their study is mainly reflected in the following aspects.

In terms of the quality analysis and evaluation criteria for automatic generalization, Wu et al. (2007a) studied the quality analysis and evaluation criteria of automated cartographic generalization and believed that most of the automatic generalization tools cannot control the generalization quality in addition to providing the basic parameter settings for the user. Therefore, they presented the quality analysis and evaluation standard for automatic map generalization and provided the main research contents and ideas required. Through analysis of the quality problem of automatic generalization, Wu et al. (2007d) established an automated generalization framework designed for quality control. Based on the model framework, the structure of the automatic cartographic generalization system was designed for quality control. Considering the combined use of the quality function deployment and failure analysis, the quality management mechanism of automatic generalization model was analyzed based on quality control.

In terms of the quality assessment of automatic generalization, Deng et al. (2006b) proposed a quality evaluation model for the map generalization results based on a multidimensional constraint space. First, each constraint value was processed by normalization and a dimensionless number, and then, the virtual multidimensional constraint space was built based on a single constraint coordinate axis, which meant that the generalization outputs were in the same scale range. Third, the quality of the cartographical generalization was analyzed and compared by extending the Euler distance to the multidimensional constraint space. Based on the analysis of the influence of an algorithm on the location accuracy and spatial relation of other neighboring elements in the process of line element simplification, a research method known as the error propagation of line element simplification algorithm was proposed by Zhu et al. (2007a), the transfer error model of the line element simplification algorithm was established, and the model was visualized by using the error ellipse. The quality assessment of the city building geometry generalization with the reducing-dimension technique was studied in the paper by Qian et al. (2007b). Considering that buildings and streets influence each other and join together to fill whole areas on large-scale maps, a new idea of assessing a city's building geometry quality by street was put forward. Furthermore, the skeleton-line was used to assess the building's quality and was evolved into the quality assessment system based on reducing-dimension processing. The steps and six specific operation rules of the quality assessment by street skeleton-line for building contour simplification, building mergence, building displacement, and conflicting processing were expatiated. Wu et al. (2008b) believed that the influence of the simplification algorithm on a linear features' accuracy can be divided into the geometric accuracy and attribute accuracy. Aiming at the influence of the curves' characteristics on the geometric character and changes in positions of points in the process of simplification, the geometric accuracy of the line simplification algorithm was evaluated, and a series of evaluation figures of geometric accuracy, such as the sinuosity of a curve, position error, etc., were proposed.

In terms of the quality control of automatic generalization, aiming at the quality control of the cartographical generalization not supported by R-tree, a transmutation data model of R-tree used for quality control of cartographical generalization was proposed by Deng et al. (2007). A quality control process of the model for cartographical generalization was analyzed, and the constraint condition, insert algorithm and divide algorithm among the created algorithms in the model were discussed. Deng et al. (2008b) studied the generalization knowledge base for quality control based on a database. On the basis of analyzing the present knowledge base of cartographic generalization, the cartographic generalization knowledge base designed for quality control based on the database was developed, and the basic structure of the table and the development, maintenance and realization of the reasoning principles were expatiated. The most notable characteristic of the knowledge base was that quality knowledge was managed efficiently, which makes the effect of cartographic generalization knowledge not only focused in terms of choosing a simple generalization operation aimed at different objects at the beginning of generalization but also automated the initiation and control of quality during each process of generalization; meanwhile, the functional configuration on the design for quality was considered in the design of the table structure, which can implement the automatic reasoning mechanism to play the role of quality control in the whole process of cartographic generalization.

In terms of spatial data checks, Qian et al. (2006a) studied the spatial data checks with generalization knowledge. A generalization knowledge structure was put forward. The internal structure of the knowledge in the structure was analyzed, and the particular target of the knowledge attribute was defined. After ordering the sequence of data layers by weightiness, a synergic data checking method based on fuzzy knowledge and an automatic data checking method based on accurate knowledge were created.

(3) **Research on the uncertainty of automated generalization**

A systematic and thorough study on the uncertainty of automatic generalization was carried out by TONG Xiaohua et al. at Tongji University. For example, aggregation, simplification and adjustment of building polygon objects in multiscale map generalization were studied (Tong et al. 2007b). Based on the constrained Delaunay triangulation principles, the neighbor relation of the building polygons was first formulated. Then, an improved algorithm was presented for the building polygon aggregation on the basis of the SDS model, and the principle of the rectangular process for aggregated polygons was proposed. The building polygon simplification was explored by means of the direct area generalization algorithm. Due to the uncertainty in the simplification process with the uncertainty of the threshold value and turning points, a simultaneous adjustment model based on the least squares for the building polygon aggregation and simplification was thus established. According to

Tong et al. (2007a), a new approach for map conflation based on a least squares adjustment was presented. The constraints in the map conflation were first derived, and the coordinate's displacements of the conflated objects were then obtained by iterative adjustment computation. In addition, according to Lei et al. (2006a, b), two schemes for a data integrated process and partition adjustment calculation according to section were provided. Many equations for the data process and partition process methods regarding the partition adjustment calculation were discussed in detail. An integrated data process method of linear feature generalization based on curve fit was studied.

The automatic evaluation model of map generalization in a digital environment was put forward by WU Jie, LI Lin and GUO Qingsheng and so on. In addition, the generality and evaluation criteria of the automatic map generalization were studied (Wu et al. 2002a; Guo 2000).

5.3.8 Development of the Automated Generalization System

Putting the research result of the automatic map generalization into application is the goal that the vast numbers of map scholars have been working on for many years. Therefore, there have been many experimental systems in the course of development of automatic generalization in the recent 50 years. Some of them have been applied to the practice of map production.

Zhang (1988) introduced the general structure of a developed map generalization expert system MAPGEN. The expert system is composed of 5 parts, such as a knowledge base, inference engine, explanation interface, deductive database and graph I/O. The link between the graph I/O and map database, the connection between PROLOG and dBASE, the natural language query of dBASE and the combination of the knowledge base, inference engine and dBASE were also described.

A theory and method for the production of charts with computer mapping was presented by Zhao et al. (1990). In addition, some related schemes on how to solve the problems of matching the digital cartographic material, the automatic compilation of irregular symbols, automatic updating of map data, the automatic scribing of line printing and coat removal plates satisfying the requirements of print, as well as automatic cartographical arts were put forward.

Wu (1991) discussed and studied the theory, method and implementation process of automatic map generalization supported by the data from the 1:500,000 military traffic cartographic database; the selection of map contents, simplification of linear elements and displacement of overlapping elements were emphatically investigated. Wang et al. (1998b) introduced the design idea, structure and function module, the test result and summary of a compilation system, which would be used for the automated compilation of a small-scale map based on a large-scale map database.

From the discussion of the basic method of mapping, the realization method for selecting, simplifying, generalizing and displacing a map generalization under the GIS vector and raster data structure was discussed by Sheng et al. (1995), and in particular, the generalizing approach of raster data features based on mathematical morphology was discussed.

In the paper by Cai et al. (2002), by using human and computer cooperation for map generalization, the complex procedure of map generalization was decomposed, an intelligent human-machine task allocation was made, and an interactive map generalization environment for a large-scale topographic map was then designed and realized. According to the map generalization software AutoMap, a circular framework supporting the model generalization and its implementation method was given in Ying et al. (2002), and the analysis of the model generalization based on a separation of the topological, metric and semantic aspects of the map objects was presented.

To meet the requirements of China's NSDI construction, Ai et al. (2005b) investigated the development and application issues of the cartographic generalization software. From the point of view of software development, the study explored the hierarchical partitioning of cartographic generalization functions and realization in the computer system, construction of the generalization environment, normal representation of generalization rules and design of generalization operators. Based on the case study of a 1:50,000 database construction from a 1:10,000 basic database, a technological method was presented to build the multiscale map database.

In the paper 'On Land-use database intelligent generalization' by Teng et al. (2007), based on the land-use database, an object-relational database was used to manage the GIS graphics and attribute data. Through the research on the multiscale spatial data model of land information system, the land database, which supports the land information multiple expressions, was established to realize the scientific management of the land management decision at different spatial scales.

5.4 Problems and Prospects

5.4.1 The Realization of Automated Map Generalization Follows an Objective Development Process from Simple to Complex, Local to Whole, and Digital to Intelligent, and This Process Is Far from Finished

In the era of traditional cartography, the field of cartographic generalization mainly focuses on the research, summary and refinement of the principles and methods of map production. The transformation of cartographic generalization from a subjective process to an objective scientific mapping method has been realized. In the age of digital cartography, cartographic generalization mainly focuses on map generalization quantification research as well as the mapping model and cartographic generalization model and algorithm. The transformation from a qualitative description to a quantitative description has been realized. Many modern applied mathematical methods have been introduced in the research on the cartographic generalization model and algorithm, and remarkable achievements have been obtained. The era of intelligent cartography mainly focuses on the research of cartographic generalization based on the model, algorithm and knowledge, as well as human and computer cooperation for cartographic generalization, in particular, the knowledge-based intelligent control of automatic cartographic generalization of the whole element and whole process, as well as the generalization quality evaluation of automatic generalization based on quality design. The transformation from map model to map generalization model, algorithm and knowledge has been realized, and the pursuit of full automation has changed to a human-computer collaboration. In addition, the research on the isolated and scattered model and algorithm and the automatic cartographic generalization test with a single factor have changed to process control and quality design of automatic generalization as a whole (whole element, whole process and controllable). Of course, due to the complexity of automated cartographic generalization, there is still much work to be done to ensure such a change, and there is still a ways to go.

5.4.2 The Development of Intelligent Cartographic Generalization: The Importance of the Process of Constant Renewal of Ideas

In reviewing the development course of cartographic generalization research, especially the study of automatic cartographic generalization, it has been proven that the renewal of ideas is the key to the cartographic generalization progress. Because map generalization is a technical method that reflects the quantitative and qualitative features, distribution and interrelation among mapping objects, specifically, the subjective must conform to the objective, the cartographic generalization change from a subjective process to an objective scientific mapping method can be achieved. In addition, because of the awareness that the application of digital cartographic data and computer-mapping technology will inevitably lead to the quantification, modeling and algorithm of cartographic generalization, the transformation of cartographic generalization from qualitative description to quantitative description can be achieved, and further, the transformation from the map model to the generalization model and algorithm can be achieved. Because cartographic generalization involves all geographical feature types, the whole process and is completely controllable, there is a development trend from isolated and scattered model and algorithm research and automatic cartographic generalization tests with a single factor to process control and quality design of automatic generalization as a whole. Hence, to fully realize the process control and quality design of automatic cartographic generalization with full elements and the whole process, as well as eventually using it in map production practice, new ideas are still needed. Any new concept must go through the process from imperfect to perfect. With familiar concepts, not actively exploring new and more vital ideas is very detrimental to scientific and technological innovation.

5.4.3 The Model and Algorithm of Cartographic Generalization Must Be Continuously Improved and Optimized Since Human Cognition Is Progressing and Science and Technology Are Developing

A cartographic generalization model is a mathematical description of the law of cartographic generalization. The algorithm of cartographic generalization is a description of its characteristics with a finite number of instructions. In addition, whether this description is perfect or not is related to the cognitive level and the level of scientific development. For the Douglas-Peucker algorithm, there is fractal expansion because of the problems found in the applications and the emergence of fractal mathematics (Wang et al. 1998c). For the object displacement algorithm, the automatic cartographic generalization displacement model based on the elastic mechanics principle was developed because of some problems have been found in the practical displacement operations of the existing mechanical algorithms, optimization algorithms, genetic algorithms and simulated annealing algorithms (Hou 2004). For the Li-Openshaw algorithm, the Li-Openshaw algorithm has been improved since some problems had been found in applications, such as the local maximum point being omitted, circles and curves having multiple intersections and other issues (Zhu et al. 2007b). There are many other similar examples that are not listed here. All of these examples show that with the increasing application of the cartographic generalization model and algorithm, the incompleteness and deficiency of the model and algorithm will be revealed. In addition, with the continuous improvement of people's cognitive level and the application of new mathematical methods, the cartographic generalization model and algorithm are continuously improved and optimized. This concept will endure as a law in development of things.

5.4.4 Improving the Intelligence Level of Automated Cartographic Generalization Is Still the Main Aspect of Cartographic Generalization Study

Map generalization is a highly intelligent labor process and very creative work, which requires the knowledge, experience and wisdom of cartographers. Cartographic generalization is not only a process of knowledge simplification for mapping objects but also a process of knowledge reorganization to generate new knowledge. The accumulation of knowledge and experience is of great significance. In the process of automatic cartographic generalization, knowledge-based reasoning requires the support of knowledge, human-machine coordination requires the guidance of knowledge, and the generation of the automatic cartographic generalization chain and operation of generalization process control requires activation by knowledge. There is a very close relationship between the cartographic generalization process and cartographic generalization knowledge. The urgent problems that need to be addressed in map generalization are outlined as follows: summarizing the knowledge, experience and wisdom that cartographers need to address the issue of cartographic generalization; implementing knowledge classification that adapts to cartographic generalization operation; expressing this knowledge and creating a structural description; building a cartographic generalization knowledge base and knowledge reasoning mechanism; managing and organizing the cartographic generalization knowledge, and so on. In these areas, although the current research has achieved some results, there is still much work to do. Cartographic generalization knowledge engineering should be taken as an important research task; more people should take part in the research in order to achieve more breakthroughs.

5.4.5 Engineering of the Process Control for Automated Cartographic Generalization Already Have Basic Conditions, but the Research Is Still Endless

All scientific research is ultimately for practical application, and the research of the process and quality control for automatic cartographic generalization is no exception. At present, the existing test system of the process and quality control of automatic cartographic generalization (with all feature types, the whole process and controllability) is capable of operation; some of these results have been used in the production of large and medium scale topographic maps. However, there is still more work needed before comprehensive application of map production practice, as there are many areas to be further improved and optimized. Although the engineering and industrialization of scientific research results already have the basic conditions, these areas are still a long way from actual implementation. Our ultimate goal is to develop a fully digital cartographical system with self-owned intellectual property through the engineering and industrialization of scientific research results, the core of which is the automatic map generalization subsystem. We also need to further study and develop a related cartographic compilation system with the functions of data input and color separation film output. In addition, the automated cartographic generalization system itself needs the improved generation of the automatic cartographic generalization chain and process control to make the system more intelligent and extensively adaptable. In terms of automatic map generalization quality (and algorithm) evaluation and control based on quality design, we must focus on the research of the automatic generalization algorithm and operator library, the effective control of the algorithm and operator, and the cartographic generalization quality assessment model. Future studies should not only include the management of semantic knowledge but also the treatment of the knowledge expressed by the network, as well as simulations of the human's thinking process in cartographic generalization.

The engineering and industrialization of the automatic cartographic generalization system should be combined with mapping production. It is only in the practical production that we can find the problem and aim to make the system perfect, stable and mature.

5.5 Representative Publications

(1) ***General map compilation*** (ZHU Guorui, YIN Gongbai. Surveying and Mapping Press, Beijing, Nov. 1982)

This book has a total of five parts and is divided into two volumes. Volume one includes chapters one, two and three; volume two includes chapters four and five. The first chapter introduces the basic knowledge of map compilation, where the purpose is to understand the map, establish the basic map compilation concept, and lay a foundation for further study. The second chapter introduces the technical methods of mapping, which mainly include map mathematical foundation establishment, the theory and method of map content transfer, the use of equipment, the basic requirements for each link, and so on. The third chapter introduces the theoretical basis of map generalization; the principle of map generalization and the application of mathematical statistics in map generalization are discussed for further discussion on map generalization. The fourth chapter describes map generalization of the elements on a map, and the basic principles, methods and procedures of the map generalization of the elements on a general map were discussed in detail, focusing on the study of quantitative indicators and implementation methods and paying attention to the combination of theory and practice. The fifth chapter describes the compilation and editing of a map according to the map production process, content, procedure, method and requirements during each stage of map design, compilation, preparation for publication, revision and so on.

(2) ***Cartographic generalization principle for the general map*** (WANG Jiayao, FAN Yiai, HAN Tongchun, MAO Longdian, WANG Guangxia. Surveying and Mapping Press, Beijing, Jun. 1993)

This book was written on the basis of summarizing the experience of long term teaching and map production practice and discusses the research results of domestic and foreign scholars in this field. The whole book is divided into 8 chapters. The first chapter provides an introduction, which mainly presents the basic concept of map generalization, the process and sequence of map generalization, and the influence of map generalization on map accuracy. The second chapter mainly introduces the conditions of map generalization and provides a detailed analysis of map use, map scale, regional geographic features, the effects of map load and the minimum size of symbols in map generalization. The third chapter is the mathematical method of calculating the index of map generalization; in this chapter, the principles and methods for establishing the mathematical model of map generalization are introduced, and detailed descriptions of the models for calculating the index of map generalization are provided, including the graphical calculation model, square root model, geometric sequence model and regression model. The fourth chapter discusses cartographical generalization methods and introduces the four commonly used methods, including the selection method, shape simplification method, generalization method base on quantitative and qualitative features, and the displacement method. From the fifth chapter to the seventh chapter, the generalization methods for natural, social, and economic features, as well as typical areas and the most important factors are introduced. The eighth chapter gives a brief introduction to the history, current situation and future tasks in the study of cartographical generalization.

(3) ***Principles and methods of digital map automatic generalization*** (WANG Jiayao, WU Fang. The People's Liberation Army Press, Beijing, Mar. 1998)

Beginning with the basic theory of automatic map generalization, the principle and method of the automatic generalization of the digital map are discussed systematically and comprehensively in the book, and the research results in this field at home and abroad are reflected as much as possible. The whole book is divided into 8 chapters. The first chapter is an introduction, which mainly analyzes the research background, discusses the concept of automatic map generalization, and introduces map generalization based on the model, algorithm and rule; additionally, the process of automatic map generalization is introduced. The second chapter presents the database of automatic map generalization, including the data content, data model, automatic splicing and retrieval function of the map database. The third chapter describes the human-machine interactive processing in automatic map generalization and introduces the function of human-computer interaction processing system and application in automatic map generalization. The fourth chapter discusses the basic method of digital map automated generalization, from the selection and simplification of digital map features to the generalization and displacement of quantitative and qualitative features and other aspects. The fifth chapter, considering the control point (height point) and independent features, water system, geomorphic contour, residents, road, vegetation, boundary, pipeline, short wall, fence and so on, introduces the automatic generalization of various kinds of elements in digital map and provides some examples; additionally, this chapter discusses the automatic map generalization requirements for the database. In the sixth chapter, the structure and application of the cartographical generalization expert system are introduced. In the seventh chapter, the significance of the integration of automated map production and its main research contents are discussed, and a map design and publishing system are introduced. In the eighth chapter, the research situation and trend of automatic cartographical generalization are analyzed in detail; and major institutions engaged in research on automatic map generalization in foreign countries are introduced.

(4) ***Study on the fractal description and automatic generalization of map information*** (WANG Qiao, WU Hehai. Wuhan Surveying and Mapping Science and Technology University Press, Wuhan, Aug. 1998)

This book reveals the fractal theory potential and applications in automatic map data processing from the aspects of fractal expansion, fractal dimension, fractal hierarchy, changes in graphic shape features, fractal analysis and so on. Simultaneously, according to different factors, a series of automatic generalization methods based on fractal analysis is proposed.

The whole book is divided into 5 chapters. In the introduction of this book, the concept, influencing factors of map generalization and its universality in spatial information processing are analyzed, the operators and basic model of map generalization is discussed, and the main methods of cartographic generalization, especially the generalization of river systems, landforms and residential areas are described. The first and second chapters mainly introduce the basic idea of the fractal theory and the fractal method research and applications. In the third chapter, the fractal analysis method and its several extensions in the processing of geo graphics data are discussed. In the fourth chapter, fractal modeling and its application in automatic map generalization are studied; the automatic generalization methods of linear and surface elements based on fractal analysis are introduced, with a focus on the automatic generalization of river networks and landforms. The fifth chapter is about function modules and their preliminary applications, in which the basic functions and applications of the developed prototype system are introduced.

(5) ***Theories and methods of automatic map generalization*** (GUO Qingsheng. Surveying and Mapping Press, Beijing, July. 2002)

The book is based on the author's doctoral dissertation. The dissertation primarily covers the following topics. The first and second chapters introduces the concept framework, solvability and problem decomposition strategy of automatic map generalization, as well as the basic operator analysis and establishment of the new system concept framework of map automatic generalization. The third chapter describes the classification of spatial relations of vector space objects and the equivalence and fuzziness of spatial relations in the automatic map generalization process. The fourth chapter introduces the classification and hybrid expression method of automatic map generalization knowledge and line and plane spatial feature acquisitions. The fifth chapter gives a progressive map generalization method for linear and planar features. The sixth chapter illustrates the spatial feature reasoning of the map elements and its applications, such as the formation and application of a point cluster, the establishment of the hierarchical relationship between contour lines and progressive map generalization; additionally, an analysis of the configuration characteristics of the river system, automatic recognition of their spatial features, and so on is also presented. The final chapter gives a summary and outlook.

(6) ***Multiscale representation and automatic map generalization of spatial data*** (WU Fang. The People's Liberation Army Press, Beijing, Mar. 2003)

The book consists of 4 parts and 14 chapters. In the first part, based on a deep analysis of the automated map generalization in the digital environment, the characteristics and research focus of the multiscale representation and automatic map generalization of spatial data are discussed, and a conceptual framework and implementation strategy for automatic map generalization is proposed. The second part is the basis of the automatic generalization method, which mainly expounds upon the spatial relation and means of establishing the data model in the automatic generalization process. The third part introduces the applications of various methods in automatic generalization, including the Delaunay triangulation method, Circle model, wavelet analysis, mathematical morphology, and fractal theory, as well as some intelligent methods, such as the genetic algorithm, artificial neural network, Agent method, etc. The last part introduces the design method and implementation steps of an automatic map generalization system through the description of the system functions.

(7) ***Research on the basic theory and technology of map generalization*** (WU Hehai. Surveying and Mapping Press, Beijing, Jan. 2004)

This book summarizes the author's research results over more than 40 years in the fields of map generalization, map databases, GIS and geographic information integration and processing; the book includes a collection of more than 50 papers published by the author since 1965.

The whole book is divided into 5 parts. The first part, which includes several analytical calculations for the characteristics of map content elements, discusses 8 papers on the determination of river selection criteria, the calculation of river length, the smooth interpolation of the oblique axial parabola, the automatic determination of the inflection point of a digital curve and its application, and so on. The second part supports the map database system for map information generalization and includes a collection of 11 papers regarding some technical problems in the map database system in relation to map information generalization, such as the establishment of a map database management system, map information topological retrieval, edge matching and a combined map database, design and organization of the multimedia data model, design and implementation of the map digitization system based on the map graphic workstation, and so on. The third part discusses the study of the structural generalization of map content and includes a collection of 20 papers on the related basic theory, specific content and solutions to the map information structural generalization, such as the automatic generalization of a point cluster, establishment and automatic generalization of contour lines tree, establishment and automatic generalization of a river system tree, automatic generalization of a city plane figure, establishment of a multiscale spatial database, and so on. The fourth part covers fractal geometry and extended fractal theory applications in the field of automatic map generalization, including the related theory of the extended fractal dimension mathematical model and its applications in the automatic generalization of map information, the DP fractal dimension estimation method and its applications in geomorphology, the application of fractal theory in the GIS multiscale data output and the automatic generalization of a map polygon group, as well as the determination of a scale-free zone of a fractal object, and so on; a total of 7 papers were included in this chapter. The fifth part is another collection of 7 papers regarding the integration of GIS and geographic information, including the integrated management technology of geographic information, the design and implementation of MCGIS, the design and application research of CCGIS and other GIS software.

(8) ***Multiscale representation model of spatial data and its visualization*** (LI Lin, WU Fan. Science Press, Beijing, Jan. 2005)

The whole book consists of seven chapters. Chapter one is the prolegomenon that mainly discusses the theory, principle and method of the multiscale representation of spatial data; this chapter expounds upon the scientific significance and necessity of studying this problem in terms of development of earth information science, and the current situation and existing problems are investigated through the analysis of research topics in today's problem domain. Chapter two discusses the nature and characteristics of spatial data scale, in which the conceptual and theoretical framework of generalized scales is proposed. Chapter three discusses the scale-dependent spatial data model. Chapter four covers the multiscale aggregation model for spatial data. Chapter five discusses re wavelet theory and its application principles in the multiscale representation of spatial data, in which the theory and method of wavelet analysis are used to divide the geographic space into a multiscale nested sequence space. Chapter six covers the multiscale expression model of spatial data based on the wavelet theory, in which the proposed model is realized by the specific spatial data types, such as linear data, field data and so on. Chapter seven discusses the expression of scale-dependent spatial data and its visualization.

(9) ***Spatial reasoning and progressive map generalization*** (GUO Qingsheng, HUANG Yuanlin, ZHENG Chunyan, CAI Yongxiang. Wuhan Surveying and Mapping Science and Technology University Press, Wuhan, Sept. 2007)

In the book, the basic methods of simple spatial relationship reasoning and geospatial structure reasoning are systematically described. In addition, the basic principles and methods for progressive map generalization, which is closely related, are also studied. The whole book is divided into 15 chapters. The first chapter introduces spatial reasoning, spatial abstraction, the meaning of spatial cognition and the interrelations among them. The second chapter reviews the definition and research progress of map generalization. The third, fourth and fifth chapters expound upon the expression and reasoning methods for spatial relations. The sixth, seventh and eighth chapters discuss the reasoning method or the calculation method of a spatial structure. The ninth, tenth and eleventh chapters cover the model and algorithm of automatic map generalization, the control strategy of the automatic map generalization process, the application method of automatic map generalization knowledge, the engineering design of the map generalization system, and so on. The twelfth chapter studies the basic law of human spatial abstraction and the basic theory of progressive map generalization. In the thirteenth chapter, the principle and method of spatial topological relation abstraction and the basic law of spatial relation abstraction in map generalization are explained. The fourteenth chapter involves simplified thinking and the basic algorithm of progressive map simplification for linear elements and buildings. The fifteenth chapter describes the progressive generalization method of several different types of spatial object groups.

(10) ***Intelligent spatial information processing for automatic map generalization*** (WU Fang, QIAN Haizhong, DENG Hongyan, WANG Huilian. Science Press, Beijing, Mar. 2008)

This book is a theoretical summary of the scientific research and engineering practice in the field of automatic map generalization, and this book represents a new exploration in this field. The whole book is divided into 12 chapters; the main contents are as follows. Chapter one is an introduction. Chapter two is the foundation of relevant subjects applied in automatic generalization. Chapter three is auxiliary information acquisition for automatic map generalization. Chapter four is the automated generalization processing model based on knowledge. Chapter five is the automated generalization processing model based on the genetic algorithm. Chapter six is the automated map generalization processing model based on the ABTM. Chapter seven is the displacement processing model of automated generalization based on the elastic mechanics principle. Chapter eight is the automated map generalization processing model based on mathematical morphology and neural networks. Chapter nine is the automated map generalization processing model based on Circle. Chapter ten is the automated map generalization model of road networks based on the graph theory. Chapter eleven is the automated generalization process-control model. Finally, chapter twelve is the technical realization of the automated map generalization system.

(11) ***Map cluster(group) object description and automated generalization*** (YAN Haowen, WANG Jiayao. Science Press, Beijing, Aug. 2009)

This book is divided into 7 chapters. Chapter one is a prolegomenon to the rest of the book. Chapter two is the object description and automatic generalization of the point cluster. Chapter three is the object description and automatic generalization of the line cluster. Chapter four is the object description and automatic generalization of the line network. Chapter five is the object description and automatic generalization of the discrete plane cluster. Chapter six is the object description and automatic generalization of the continuous plane cluster. Chapter seven is the conclusion.

This book is characterized by the research of the object description and automatic generalization algorithm of the map cluster (group). In arranging the content, the cluster (group) objects are divided into three categories: point, line and surface according to the geometric properties of the map object. In addition, the differences in spatial distributions between linear objects and planar objects are considered. The linear object is divided into a line cluster and line network, while the planar object is divided into discrete and continuous clusters. In this way, the object group of map space is divided into five categories: point cluster, line cluster, line network, discrete plane cluster and continuous plane cluster. For the content, the author abandons the traditional idea of "overvaluing the algorithm and undervaluing the description" and stresses the equal importance of the description methods and algorithms of the map objects. Therefore, the book well-describes map cluster (group) objects.

(12) ***Quality assessment model for automated map generalization*** (WU Fang, DENG Hongyan, QIAN Haizhong, ZHU Kunpeng. Science Press, Beijing, Dec. 2009)

This book consists of 15 chapters. Chapter 1 is a prolegomenon to the book that introduces the development of automatic map generalization research and analyzes the quality problems in automatic map generalization. Chapter 2 covers the automatic map generalization framework model based on quality design; the basic theory and method of quality design is introduced, a framework of the automatic cartographic generalization model based on DFQ is proposed, a mechanism of automatic cartographic generalization quality management based on DFQ is established, and the mathematical description of the automatic cartographic generalization model based on DFQ is given. Chapter 3 covers the DFQ map generalization knowledge expression based on the database; the constraint conditions of cartographic generalization are analyzed in terms of the existing knowledge representation, and a DFQ cartographic generalization knowledge representation method is proposed based on the database. Chapter 4 discusses the data models for map generalization quality control; the data model for cartographic generalization quality control is analyzed, a cartographic generalization region division method is proposed, and a data model for comprehensive quality control, i.e., DFQR tree, is presented. Chapter covers the evaluation and management of the automatic generalization algorithm; the meta information of the map generalization algorithm is designed and an example is given, and finally, a means to evaluate the generalization algorithm is proposed. Chapter 6 discusses the geometric quality assessment of a point cluster object based on polarization transformation; the basic idea is introduced, and the process of point cluster polarization transformation is analyzed; additionally, the geometric quality assessment of point cluster object selection is studied from the aspects of feature points and integrity. Chapter 7 covers the geometric quality assessment of building generalization based on dimension reduction technology; the basic ideas and technical basis of the dimension reduction technology are introduced, the geometric quality assessment of building generalization (profile reduction, merging, displacement, and conflict resolution) considering the characteristics of buildings is studied, and the inversion of the generalization operation evaluation process based on the building generalization results is investigated. Chapter 8 covers the quality evaluation strategy of the line element simplification algorithm based on constraint conditions; the commonly used algorithms of line element simplification and its existing problems are analyzed, the basis and guiding ideology for the evaluation of the line element simplification algorithm are proposed, and the evaluation criteria and key point of the line element simplification algorithm are established. Chapter 9 discusses the geometric accuracy evaluation of line element simplification algorithm; the effect of the simplification algorithm on the line element accuracy is analyzed, the error propagation model of the line element simplification algorithm is established, and an experiment and analysis are carried out. Chapter 10 discusses shape feature evaluations of line element simplification based on the fractal theory; the feasibility of using the fractal theory for evaluating the shape characteristics of line elements is analyzed, and a general method for evaluating the shape and structure of line elements after simplification is presented based on the fractal theory. Chapter 11 discusses other evaluation methods for the line element simplification algorithm; other evaluation criteria for the line element simplification algorithm are introduced, such as area difference, maximum corner ratio, change trend of simplification, consistency in results and auto correlation, and so on; additionally, single line simplification and simplification of spatial data are offered as examples, and the Douglas-Peucker, Long and vertical distance methods are evaluated. Chapter 12 discusses the evaluation of parameters and the application scope of the line element simplification algorithm is introduced. Chapter 13 covers the evaluation and maintenance of topological consistency in map generalization, in which topological consistency in the process of cartographic generalization is analyzed; line element simplification rules based on the topological consistency and the topological consistency maintenance during the generalization process are designed. Chapter 14 covers the quality evaluation of automatic generalization results based on the multidimension constraint space; the basic idea is put forward, and a comprehensive quality evaluation model based on multidimensional constraint space is established; finally, an experiment and analysis are carried out. Chapter 15 discusses the spatial data checks with generalization knowledge; the inductive method of cartographic generalization knowledge is introduced, the spatial data checking methods are proposed based on fuzzy and precise cartographic generalization knowledge, and an experiment and analysis are carried out.

Acknowledgements This chapter is sponsored by the Construction Project of Innovative Scientific and Technical Personnel team in the Henan Province.

Professor AI Tinghua, School of Resource and Environmental Science of Wuhan University, reviewed this chapter.

References

Ai, T. (2003). The conceptual schema of map generalization from spatial mapping perspective. *Acta Geodaetica et Cartographic Sinica, 32*(1), 87–92.

Ai, T. (2004). A displacement of building cluster based on field analysis. *Acta Geodaetica et Cartographic Sinica, 33*(1), 89–94.

Ai, T., & Cheng, J. (2005a). Key issues of multi-scale representation of spatial data. *Geomatics and Information Science of Wuhan University, 30*(5), 377–382.

Ai, T., Guo, B., & Huang, Y. (2005b). Construction of 1:50 000 map database by computer generalization method. *Editorial Board of Geomatics and Information Science of Wuhan University, 30*(4), 297–300.

Ai, T., & Guo, R. (2000). Extracting center-lines and building street network based on constrained Delaunay triangulation. *Acta Geodaetica et Cartographic Sinica, 29*(4), 348–354.

Ai, T., & Guo, R. (2007a). Polygon cluster pattern mining based on gestalt principles. *Acta Geodaetica et Cartographica Sinica, 36*(3), 302–308.

Ai, T., & Liang, R. (2007b). Variable-scale visualization in navigation electronic map. *Geomatics and Information Science of Wuhan University, 32*(2), 127–130.

Ai, T., Liu, Y., & Huang, Y. (2007c). The hierarchical watershed partitioning and generalization of river network. *Acta Geodaetica et Cartographica Sinica, 36*(2), 231–236, 243.

Ai, T., Guo, R., & Chen, X. (2001a). Simplification and aggregation of polygon object supported by delaunay triangulation structure. *Journal of Image and Graphics, 6*(7), 703–709.

Ai, T., Guo, R., & Liu, Y. (2001b). A binary tree representation of curve hierarchical structure in depth. *Acta Geodaetica et Cartographic Sinica, 30*(4), 343–348.

Ai, T., & Liu, Y. (2002). A method of point cluster simplification with spatial distribution properties preserved. *Acta Geodaetica et Cartographic Sinica, 31*(2), 175–181.

Ai, T., Zhu, G., & Zhang, G. (2003a). Extraction of landform features and organization of valley tree structure based on delaunay triangulation model. *Journal of Remote Sensing, 7*(4), 292–298.

Ai, Z., Wu, H., & Ai, T. (2003b). The application of delaunay triangulation in river net automatic generalization. *Geo-information Science, 5*(2), 39–42.

Cai, Y., & Guo, Q. (2007). Points group generalization based on Kohonen net. *Geomatics and Information Science of Wuhan University, 32*(7), 626–629.

Chen, B., Wu, F., & Qian, H. (2008). Study on road networks' auto-selection algorithms. *Journal of Image and Graphics, 13*(12), 2388–2393.

Chen, Q. (1981). Mathematical description and reduction of the information of curved lines in automatic cartography. *Acta Geodaetica et Cartographic Sinica, 10*(3), 239–254.

Cai, Z., Du, Q., & Wu, H. (2002). Construction of database platform for interactive generalization on large scale topographic map. *Editoral Board of Geomatics and Information Science of Wuhan University, 27*(3), 289–295, 305.

Chen, Z. (1964). Cut density coefficient formula and its application of landforms in the preparation of general chart of. *Acta Geodaetica et Cartographic Sinica, 7*(2), 139–154.

Deng, H., Wu, F., & Qian, H. (2003). A model of point cluster selection based on genetic algorithms. *Journal of Image and Graphics, 8*(8), 970–976.

Deng, H., Wu, F., & Zhai, R. (2006a). A generalization model of road networks based on genetic algorithm. *Geomatics and Information Science of Wuhan University, 31*(2), 164–167.

Deng, H., Wu, F., & Zhai, R. (2006b). Quality evaluation model of automated cartographical generalization based on a multi-dimensional constraint space. *Journal of China University of Mining & Technology, 35*(5), 667–672.

Deng, H., Wu, F., & Zhai, R. (2007). A DFQR tree for quality control of cartographical generalization. *Acta Geodaetica et Cartographica Sinica, 36*(2), 237–243.

Deng, H., Wu, F., & Wang, H. (2008a). A generalization of road networks based on topological similarity. *Journal of Geomatics Science and Technology, 25*(3), 183–187.

Deng, H., Wu, F., & Zhai, R. (2008b). Study of generalization knowledge base for quality control based on database. *Acta Geodaetica et Cartographica Sinica, 37*(1), 121–127, 134.

Deng, H., Wu, F., & Zhai, R. (2009). R-Tree index structure for multi-scale representation of spatial data. *Chinese Journal of Computers, 32*(1), 177–184.

Dong, W., Guo, Q., & Liu, J. (2007). Progressive generalization research of schematic road network maps. *Geomatics and Information Science of Wuhan University, 32*(9), 829–832, 837.

Du, Q. (1988). The automatic establishment of structured river network in cartographic database. *Journal of Wuhan Technical University of Surveying and Mapping, 13*(2), 70–76.

Du, W., Ai, T., & Xu, Z. (2004). A polygon simplification method based on combinatorial optimization. *Editorial Board of Geomatics and Information Science of Wuhan University, 29*(6), 548–550.

Fan, X. L. (1978). *Research report on the density and selection index of Chinese residents*. Institute of Surveying and Mapping.

Fei, L. (2004). Solving graphic conflicts between streets and buildings in map compilation by simulating human cartographers. *Geomatics and Information Science of Wuhan University, 29*(5), 426–432.

Fei, L., He, J., & Ma, C. (2006). Three dimensional Douglas-Peucker algorithm and the study of its application to automated generalization of DEM. *Acta Geodaetica et Cartographica Sinica, 35*(3), 278–284.

Ge, L., Wu, F., & Wang, P. (2009). Facet shift algorithm based on minimal feature in simplification of parallel building. *Journal of Geomatics Science and Technology, 26*(2),103–105, 109.

Guo, J. (1994). Automatic synthesis of underground tunnel with 1:25 million database. *Journal of Institute of Surveying and Mapping, 11*(2), 140–147.

Guo, M., & Wu, F. (2005). The process of automatically extracted terrain structure lines from contours and the solutions to several problems. *Geo-information Science, 7*(3), 113–119.

Guo, Q. (1998). Study on progressive approach to graphic generalization of linear feature. *Journal of Wuhan Technical University of Surveying and Mapping, 23*(1), 52–56.

Guo, Q. (1999a). The decomposement and operators set of automatic map generalization. *Journal of Wuhan Technical University of Surveying and Mapping, 24*(2), 149–153.

Guo, Q. (1999b). The method of graphic simplification of area feature boundary as right angle. *Journal of Wuhan Technical University of Surveying and Mapping, 24*(3), 255–258.

Guo, Q. (2000). Research on generalized and evaluation criteria of map automatic generalization. *Surveying and Mapping Information and Engineering,* (1), 19–21.

Guo, R., & Ai, T. (2000a). Simplification and aggregation of building polygon in automatic map generalization. *Journal of Wuhan Technical University of Surveying and Mapping, 25*(1), 25–30.

Guo, Q., Wu, H., & Li, P. (2000b). Spatial relation rules and progressive graphic simplification of contours. *Journal of Wuhan Technical University of Surveying and Mapping, 25*(1), 31–34.

Guo, Q., & Cai, Y. (2007). Study on topological relations progressive transformation in graphical simplification. *Science of Surveying and Mapping, 32*(1), 30–31, 39.

Guo, Q., Li, H., & Liu, J. (2005). Buildings progressive typification based on weighted mesh simplification. *Geomatics and Information Science of Wuhan University, 30*(4), 293–296.

Guo, Q., Lv, X., & Cai, Y. (2008a). Rule of the spatial topological relation abstraction in graphic simplification process. *Geomatics and Information Science of Wuhan University, 33*(5), 520–523.

Guo, Q., & Huang, Y. (2008b). Automatic reasoning on main streams of tree river networks. *Geomatics and Information Science of Wuhan University, 33*(9), 978–981.

Guo, R., & Wu, H. (1996). A research on the automatic generalization of urban settlement on a topographic map. *Journal of Wuhan Technical University of Surveying and Mapping, 18*(2), 15–21.

He, J., & Fei, L. (2007). Mathematical methods involved in constrained reshaping for solving graphic conflicts between streets and buildings—taking the relationship between roads and buildings as an example. *Geomatics and Information Science of Wuhan University, 32*(4), 326–330.

He, J., & Fei, L. (2008). Further study on three dimensional Douglas-Peucker algorithm and its application to generalization of DEM. *Geomatics and Information Science of Wuhan University, 33*(2), 160–163.

He, Z. (1986). Using the multivariate regression-analysis method to establish the mathematical model to determine the quotas of settlement selection. *Acta Geodaetica et Cartographic Sinica, 15*(1), 41–50.

He, Z. (1995). Application of GIS and the social background. *Bulletin of Surveying and Mapping* (1), 37–40, 46.

He, Z., Ruan, Y., & Yin, W. (2002). Cartographic generalization of hydrographic feature based on the fractal geometry. *Geomatics and Information Science of Wuhan University, 27*(4), 427–431.

Hou, X. (2004). *An integrated displacement model of drawing based on elastic mechanics*. Master's degree thesis of the Information Engineering University.

Hou, X., Wu, F., & Liu, F. (2005). A model for point cluster displacement in automated generalization. *Science of Surveying and Mapping, 30*(2), 44–47.

Hu, Y., Chen, J., & Li, Z. (2007). Selective mission of road features based on mesh density for digital map generalization. *Acta Geodaetica et Cartographica Sinica, 36*(3), 351–357.

Huang, P. (1995). Vector data compression with prediction function. *Acta Geodaetica et Cartographica Sinica, 24*(4), 316–320, 249.

Lei, W., Tong, X., & Liu, D. (2006a). Design of spatial line feature generalization data processing system. *Bulletin of Surveying and Mapping* (7), 22–25, 29.

Lei, W., Tong, X., & Liu, D. (2006b). Data process methods of line feature generalization based on curve fit. *Geomatics and Information Science of Wuhan University, 31*(10), 896–899.

Li, H., Guo, Q., & Liu, J. (2006a). Rapid algorithm of building typification based on data matching. *Geomatics and Information Science of Wuhan University, 31*(7), 636–639.

Li, N., Du, Q., & Wu, X. (2006b). Research on displacement between streets and buildings in cartographic generalization. *Science of Surveying and Mapping, 31*(4), 92–94.

Li, W., & Wu, H. (2006c). On catastrophe-scale in course of map-information attenuation. *Geomatics and Information Science of Wuhan University, 31*(12), 1116–1119.

Li, W., & Wu, H. (2005). Fractal attenuation analysis of cartographic object's self-similarity on cartographic generalization. *Geomatics and Information Science of Wuhan University, 30*(4), 309–312.

Liu, C. (1999). Based on "knowledge rule" of GIS water feature's automated generalization reasoning. *Bulletin of Surveying and Mapping,* (9), 21–24.

Liu, J. (1994). Handling of features' relationships in the process of the graphic output of cartographic data base. *Acta Geodaetica et Cartographic Sinica, 23*(3), 222–228.

Liu, P., Ai, T., & Deng, J. (2008). Simplification and rectangularity of building-polygon based on least squares adjustment. *Journal of China University of Mining & Technology, 37*(5), 699–704.

Liu, W., Chen, J., & Deng, K. (2006). Automatic detection of spatial conflicts between line objects. *Journal of China University of Mining & Technology, 35*(6), 767–771.

Liu, Y., Zhai, J., & Lu, Y. (2005). Automatic generalization of sounding in digital chart. *Acta Geodaetica et Cartographic Sinica, 34*(2), 179–184.

Lu, L., Wu, J., & Liu, Z. (2006). Selection of speckle-data based on layer-structure construction in cartographic generalization. *Geomatics and Information Science of Wuhan University, 31*(2), 156–159.

Lu, S., & Ju, X. (1964). The method of measuring the density of rivers and lakes on topographic maps. *Acta Geodaetica et Cartographic Sinica, 7*(4), 292–300.

Mao, J., & Guo, Q. (2003). Maintenance of spatial relation in map objects displacement. *Geomatics and Information Science of Wuhan University, 28*(4), 492–495.

Peng, H., Qi, Q., & Liu, Z. (2005). The optimum method for automated generalization of spatial point clusters. *Geo-information Science, 7*(4), 122–126.

Qi, Q., & Jiang, L. (2001). Research on the index system and knowledge rules for geographic-feature-oriented generalization. *Progress in Geography, 20*(Suppl.), 1–13.

Qi, Q., & Liu, Y. (1996). An automated solution to the graphic generalization of discontinuous area phenomena. *Geographical Research, 15*(1), 1–10.

Qi, Q., & Pan, A. (1998). A study on the method of intelligent generalization in GIS environment. *Progress in Geography, 17*(2), 15–22.

Qi, Q., & Zhang, A. (1999). A study on some problems of master database multiple representation in multi scale GIS. *Geographical Research, 18*(2), 161–170.

Qian, H., & Wu, F. (2001). A merge operation for area objects based on delaunay triangle-interpolating. *Journal of the Institute of Surveying and Mapping, 18*(3), 207–209.

Qian, H., & Wu, F. (2004). The framework of inspect and control agent used for automated map generalization. *Journal of Institute of Surveying and Mapping, 21*(3), 211–214.

Qian, H., Wu, F., & Deng, H. (2005a). A model of point cluster selection with circle characters. *Science of Surveying and Mapping, 30*(3), 83–85.

Qian, H., Wu, F., & Tan, X. (2005b). The algorithm for merging city buildings based on ABTM. *Journal of Image and Graphics, 10*(10), 1224–1233.

Qian, H., Wu, F., & Guo, J. (2006a). Spatial data checking with generalization knowledge. *Acta Geodaetica et Cartographica Sinica, 35*(2), 184–190.

Qian, H., Wu, F., & Wang, J. (2006b). Theory and technique model of automated cartographic generalization chain. *Acta Geodaetica et Cartographica Sinica, 35*(4), 400–407.

Qian, H., Wu, F., & Chen, B. (2007a). Simplifying line with oblique dividing curve method. *Acta Geodaetica et Cartographica Sinica, 36*(4), 443–449, 456.

Qian, H., Wu, F., & Ge, L. (2007b). Quality assessment of city-building geometry-generalization with reducing-dimension technique. *Journal of Image and Graphics, 12*(5), 927–934.

Qian, H., Wu, F., & Zhu, K. (2007c). A generalization method of street block based on dimension-reducing technique. *Acta Geodaetica et Cartographica Sinica, 36*(1), 102–107, 118.

Shen, J., Long, Y., & Wang, M. (2008). An elementary approach of adaptive generalization in mobile environment. *Journal of Geomatics Science and Technology, 25*(4), 245–248.

Sheng, Y., & Guo, D. (1995). Application of GIS and the social background. *Bulletin of Surveying and Mapping,* (3), 26–30, 34.

Shi, W., & Chui, K. (2006). Performance evaluation of line simplification algorithms for vector generalization. *The Cartographic Journal, 43*(1), 27–44.

Song, Y., He, Z., & Su, W. (2005). Attributes reduction and structured selection in automatic cartographical generalization based on rough set. *Geomatics and Information Science of Wuhan University, 30*(4), 329–332.

Su, H., Cui, X., & Peng, Y. (2007). Research on automatically accepting or rejecting linear geographic elements based on central place conception in cartographic generalization. *Science of Surveying and Mapping, 32*(1), 40–42.

Tang, Q. (1987). A study on automated generalization of submarine relief. *Acta Geodaetica et Cartographic Sinica, 16*(4), 314–320.

Teng, Y., Nan, Z., & Yan, H. (2007). On land-use databases intelligent generalization. *Journal of Lanzhou Jiaotong University, 26*(4), 102–104.

Tian, J., Ma, M., & Yang, X. (2008). An approach to constraint-based urban street network automated generalization. *Science of Surveying and Mapping, 33*(5), 158–160, 139.

Tong, X., Deng, S., & Shi, W. (2007a). A new least squares adjustment method for map conflation. *Geomatics and Information Science of Wuhan University, 32*(7), 621–625.

Tong, X., & Xiong, G. (2007b). Aggregation, simplification and adjustment of building polygon objects in multi-scale map generalization. *Journal of Tongji University (Natural Science), 35*(6), 824–829.

Wang, G. (1996). The expert system method for inhabited place automatic generalization. *Journal of the Institute of Surveying and Mapping, 13*(1), 55–59.

Wang, G., & Yang, P. (2000). Application of mathematic morphology in uniting blocks of residential area. *Journal of Institute of Surveying and Mapping, 17*(3), 201–203, 206.

Wang, H., Wu, F., & Zhang, L. (2005a). The application of mathematical morphology and pattern recognition to building polygon simplification. *Acta Geodaetica et Cartographic Sinica, 34*(3), 269–276.

Wang, J., & Deng, H. (2005b). A model of cartographical generalization based on genetic algorithm. *Geomatics and Information Science of Wuhan University, 30*(7), 565–569.

Wang, J. (1989). On the controllability of map information transmission. *Military Mapping Album, 26,* 12–14.

Wang, J., Cui, T., & Wang, G. (1985a). Application of graph theory in automatic selection of road network. *Journal of Institute of Surveying and Mapping, 2*(1), 79–86.

Wang, J., & Yao, J. (1985b). An application of fuzzy comprehensive estimation in cartographic generalization. *Journal of the Institute of Surveying and Mapping, 2*(2), 45–52.

Wang, J., Fan, Y., & Han, T. (1992). *General map generalization principle.* Surveying and Mapping Press.

Wang, J., & Qian, H. (2006). Cartographic-generalization-knowledge and its application. *Geomatics and Information Science of Wuhan University, 31*(5), 382–386, 439.

Wang, J., & Tian, Z. (1999). The cartographic generalization of soundings on chart by artificial neural network techniques. *Acta Geodaetica et Cartographic Sinica, 28*(4), 335–339.

Wang, J., & Wu, F. (1998a). *Synthesis principle and method of digital map automatic drawing.* The People's Liberation Army press.

Wang, J., Wu, F., & Wang, G. (1998b). Automatic map compilation system based on map database. *Bulletin of Surveying and Mapping,* (9), 25–27.
Wang, Q., & Mao, F. (1998c). Geographic information integration in GIS. *Journal of Remote Sensing, 2*(2), 155–160.
Wang, J., & Lv, G. (2002). Micro concave convex elimination algorithm based on curve vector data synthesis. *Bulletin of Surveying and Mapping,* (1), 17–18.
Wang, M., Shen, J., & Su, K. (2008). The study on generalization algorithms of settlement in the mobile environment. *Geo-Information Science, 10*(2), 171–176.
Wang, Q. (1995a). Self-similarity analysis of cartographic lines and the automated line generalization. *Journal of Wuhan Technical University of Surveying and Mapping, 20*(2), 123–128.
Wang, Q. (1995b). A research into some problems of cartographic generalization in digital environment. *Journal of Wuhan Technical University of Surveying and Mapping, 20*(3), 208–213.
Wang, Q., & Hu, Y. (1995a). A study on the automated cartographic generalization by using fractal analysis. *Acta Geodaetica et Cartographica Sinica, 24*(3), 211–216.
Wang, Q., & Wu, J. (1995b). Fractal dimension estimation method for automatic generalization of drawing lines and its application. *WTUSM Bulletin of Science and Technology,* (1), 1–5.
Wang, Q., & Wu, J. (1995c). The research on fractal proceeding of complicated contour automated generalization. *Journal of the Institute of Surveying and Mapping, 12*(2), 125–130.
Wang, Q., & Wu, H. (1996a). Research on fractal method for automatic synthesis of map patches. *Journal of Wuhan Technical University of Surveying and Mapping, 21*(1), 59–63.
Wang, Q., & Wu, J. (1996b). Automated cartographic generalization based on a new method of fractal dimension estimation. *Acta Geodaetica et Cartographica Sinica, 25*(1), 10–16.
Wang, Q., & Wu, J. (1996c). Fractal transformation of square root model in cartographic generalization. *Acta Geodaetica et Cartographica Sinica, 25*(2), 104–109.
Wang, Q., Zhu, Q., & Ai, T. (2007a). A probe into the simplification algorithms of 3D building surface model. *Science of Surveying and Mapping, 32*(2), 84–86, 132.
Wang, Y., & Chen, J. (2007b). Interpolation method of the road object selection. *Acta Geodaetica et Cartographica Sinica, 36*(1), 78–83.
Wang, J. (2008a). The progress and trend of automatic generalization of spatial data. *Journal of Geomatics Science and Technology, 25*(1), 1–7, 12.
Wang, Y. (2008b). Selection of road objects in multi-scale GIS databases. *Journal of China University of Mining & Technology, 37*(2), 180–185.
Wang, Y., Chen, J., & Jiang, J. (2003a). Multi-scale conceptual model for GIS geographical features. *Journal of China University of Mining & Technology, 32*(4), 376–382.
Wang, Y., Zhu, C., & You, X. (2003b). The contour data simplifying based on B-spline wavelet. *Science of Surveying and Mapping, 28*(2), 23–25.
Wang, Y., Li, X., & Gong, H. (2006c). The basic problem of multi scale representation of geographical elements. *Science in China (Series E: Technological Sciences), 36*(B07), 38–44.
Wang, Y., Li, X., & Gong, H. (2006d). Ontology analysis method of multi scale modeling of geographical elements. *Science in China (Series E: Technological Sciences), 36*(B07), 116–123.
Wu, H. (1996). Structured approach to implementing automatic cartographic generalization. *Journal of Wuhan Technical University of Surveying and Mapping, 21*(3), 272–285.
Wu, H. (1997). Principle of convex hull and its applications in generalization of grouped point objects. *Engineering of Surveying and Mapping, 6*(1), 1–6.
Wu, H. (2000a). Problems of city plan generalization in the GIS environment. *Journal of Wuhan Technical University of Surveying and Mapping, 25*(3), 196–202.
Wu, H. (2000b). Research on fundamental theory and technical approaches of automated map generalization. *Journal of Wuhan Technical University of Surveying and Mapping, 25*(5), 377–386.
Wu, J., & Wang, Q. (2000c). A study on automatic cartographic generalization using wavelet analysis in GIS. *Acta Geodaetica et Cartographic Sinica, 29*(1), 71–75.
Wu, H. (2001). Research on extended-fractal based automatic map generalization. *Progress in Geography, 20*(Suppl.), 14–28.

Wu, H. (2004). Multi-way tree structure based on curve generalization method. *Geomatics and Information Science of Wuhan University, 29*(6), 479–483.

Wu, H. (2007). On selection model algorithm for geometrical difference sequences. *Geomatics and Information Science of Wuhan University, 32*(11), 1016–1021.

Wu, J. (2008). Method of multi-scale spatial data organization based on share-scale idea. *Bulletin of Surveying and Mapping,* (8), 10–13.

Wu, J., & Li, L. (2002a). On automatic evaluation model of digital cartographic generalization. *Science of Surveying and Mapping, 27*(2), 34–39.

Wu, F., & Wang, J. (2002b). Analysis and research of concept frame of map automatic integration. *Engineering of Surveying and Mapping, 11*(2), 18–20, 48.

Wu, X., Du, Q., & Hu, Y. (2008a). Disposal of spatial conflict between the roads networks based on improved snake model. *Acta Geodaetica et Cartographica Sinica, 37*(2), 223–229.

Wu, F., & Zhu, K. (2008b). Geometric accuracy assessment of linear features' simplification algorithms. *Geomatics and Information Science of Wuhan University, 33*(6), 600–603.

Wu, F. (1991). Automated cartographic generalization in military traffic map databank. *Journal of the Institute of Surveying and Mapping, 8*(2), 52–59.

Wu, F. (1994). Automated cartographic generalization of digital river data. *Journal of the Institute of Surveying and Mapping, 11*(1), 38–42.

Wu, F., & Deng, H. (2003). Using genetic algorithms for solving problems in automated line simplification. *Acta Geodaetica et Cartographic Sinica, 32*(4), 349–355.

Wu, F., Hou, X., & Qian, H. (2005). A model for road network displacement in automated map generalization. *Acta Geodaetica et Cartographic Sinica, 34*(3), 262–268.

Wu, F., & Zhu, G. (2001a). Multi-scale representation and automatic generalization of relief based on wavelet analysis. *Geomatics and Information Science of Wuhan University, 26*(2), 170–176.

Wu, F., & Qian, H. (2001b). The operator analysis of automated generalization and the establishment of the operatorbase. *Journal of the Institute of Surveying and Mapping, 19*(1), 50–52.

Wu, F., Deng, H., & Zhu, K. (2007a). A study of automated generalization framework designed for quality control. *Bulletin of Surveying and Mapping,* (9), 43–46.

Wu, F., Tan, X., & Wang, H. (2007b). Study on automated canal selection. *Journal of Image and Graphics, 12*(6), 1103–1109.

Wu, F., Tan, X., & Zhai, R. (2007c). Automated river selection based on cartographic knowledge representation and reasoning. *Journal of Liaoning Technical University, 26*(2), 183–186.

Wu, F., Zhu, K., & Deng, H. (2007d). On quality analysis and evaluating criterion of automatic map generalization. *Journal of Zhengzhou Institute of Surveying and Mapping, 24*(3), 160–163.

Wu, F., & Wang, J. (1992). Automatic displacement of line features supported by the cartographic data bank of military traffic. *Acta Geodaetica et Cartographic Sinica, 21*(4), 291–298.

Xu, Z. (1986). A mathematical model for determining the selection of cartographic objects on the basis of fuzzy synthetic judgement principle. *Acta Geodaetica et Cartographic Sinica, 15*(1), 51–52.

Yan, H., & Wang, J. (2005). A generic algorithm for point cluster generalization based on Voronoi diagrams. *Journal of Image and Graphics, 10*(5), 633–636.

Yan, H., Ying, S., & Li, L. (2008). An approach for automated building grouping and generalization considering multiple parameters. *Geomatics and Information Science of Wuhan University, 33*(1), 51–54.

Yang, B., & Sun, L. (2008). Adaptive multi-scale visualizations of road network for navigation. *Geomatics and Information Science of Wuhan University, 33*(4), 134–137.

Yang, Z., Guo, Q., & Niu, J. (2005). A study on multi-scale DEM representation and topographic feature line extraction. *Acta Geodaetica et Cartographic Sinica, 34*(2), 18–21.

Yin, Z., Li, L., & Zhang, X. (2005). Study on a multi-scale representation model based on petri net. *Geography and Geo-Information Science, 21*(3), 5–9.

Ying, S., Guo, R., & Yan, H. (2002). Framework design and implementation of model-oriented cartographic generalization. *Acta Geodaetica et Cartographic Sinica, 31*(4), 344–349.

Ying, S., & Li, L. (2003). Consistent line simplification based on constraint points. *Geomatics and Information Science of Wuhan University, 28*(4), 488–491.

You, X. (1992). The action of visual sensation and perception on cartographic generalization. *Acta Geodaetica et Cartographic Sinica, 21*(3), 224–232.

Yuan, C., & Zhang, J. (2005). The study on amalgamation of urban residential area in GIS environment. *Journal of Taiyuan University of Technology, 36*(2), 126–129.

Yu, H., Yang, Q., & Zhang, X. (2008). Generalization of DEM data based on multi-scale wavelet analysis. *Science of Surveying and Mapping, 33*(3), 93–95, 115.

Yuan, C. (2007). An optimized scheme of simplification in building polygon automatic generalization. *Science of Surveying and Mapping, 32*(6), 84–85, 32.

Zhai, J., & Lu, Y. (2000). Recognition, measurement and generalization for line features in digital nautical chart. *Acta Geodaetica et Cartographic Sinica, 29*(3), 273–279.

Zhai, R., Wu, F., & Deng, H. (2006). Automated elimination of river based on multi-objective optimization using genetic algorithm. *Journal of China University of Mining & Technology, 35*(3), 403–408.

Zhai, R., Wu, F., & Deng, H. (2008). An automated selection model of ditch based on multi-objective optimization by genetic algorithm. *Acta Geodaetica et Cartographica Sinica, 32*(1), 108–113.

Zhai, R., & Xue, B. (2007). A structural river network data model for automated river generalization. *Journal of Zhengzhou Institute of Surveying and Mapping, 24*(4), 294–298, 302.

Zhang, J., Zhou, Y., & Liu, Y. (2006). An improved algorithm for SDS model based polygon simplification and aggregation. *Journal of Image and Graphics, 11*(7), 1010–1016.

Zhang, K. (1958). On the standard of the choice of residential spots on the map of China. *Bulletin of Surveying and Mapping,* (12), 546–550.

Zhang, Q. (2002). Identification and generalization of spatial structures on map. *Acta Scientiarum Naturalium Universitatis Sunyatseni, 41*(1), 87–90.

Zhang, Q. (2004). Line generalization based on dynamic segmentation. *Acta Scientiarum Naturalium Universitatis Sunyatseni, 43*(2), 104–107.

Zhang, Q. (2006). Generalization of drainage network with density differences. *Acta Geodaetica et Cartographica Sinica, 35*(2), 191–196.

Zhang, Q. (2007). Drainage generalization by layered division of the number of retained rivers in river trees. *Geographical Research, 26*(2), 222–228.

Zhang, Q., & Liao, K. (2001). Line generalization based on structure analysis. *Acta Scientiarum Naturalium Universitatis Sunyatseni, 40*(5), 118–121.

Zhang, Q., & Qin, J. (2000). Identification and generalization of area clusters by mathematical morphology. *Geographical Research, 19*(1), 93–100.

Zhang, W. (1988). The integration of multiple language tools in the expert system mapgen. *Journal of Wuhan Technical University of Surveying and Mapping, 13*(2), 86–94.

Zhang, Y., Deng, M., & Li, Z. (2008). Computational methods of building orientation and their applicability for map generalization. *Geomatics World, 6*(2), 58–64.

Zhao, D., & Sheng, Y. (2008). A topologically consistent contour simplification algorithm based on contour-tree and strip-tree. *Science of Surveying and Mapping, 33*(4), 91–93.

Zhao, X., Li, J., & Deng, L. (1983). Automatic recognition and tracking of skeletons and its application in automatic generalization of contour lines. *Journal of Wuhan Institute of Surveying and Mapping,* (2), 102–110.

Zhao, Y., Li, H., & Liu, X. (1990). Study on the automatic compilation of full-element marine chart and it's practice. *Acta Geodaetica et Cartographic Sinica, 19*(1), 1–9.

Zhu, C., Wang, Y., & Li, Q. (2004). A model to compress contour data based on wavelet analysis. *Journal of Image and Graphics, 9*(7), 841–845.

Zhu, G. (1989). To infer the selecting degree of rivers on the map with the mathematical model. *Journal of Wuhan Technical University of Surveying and Mapping, 14*(4), 47–51.

Zhu, G., & Xu, Z. (1990). *Mathematical methods in general cartography*. Publishing House of Surveying and Mapping.

Zhu, K., & Wu, F. (2007a). Error propagation model of linear features' simplification algorithms. *Geomatics and Information Science of Wuhan University, 32*(10), 932–935.

Zhu, K., Wu, F., & Wang, H. (2007b). Improvement and assessment of Li-Openshaw algorithm. *Acta Geodaetica et Cartographica Sinica, 36*(4), 450–456.

Zhu, Q., Wu, F., & Zhai, R. (2009). A novel of contour line simplification algorithm base on visibility. *Journal of Image and Graphics, 14*(2), 359–364.

Chapter 6
Thematic Cartography

Nan Jiang

6.1 Introduction

Thematic mapping, which has gradually developed since the 1970s, is a main branch of cartography and has an important position in the development of cartography in China. Thematic mapping focuses on the theory and method of collecting, analyzing, classifying and integrating, as well as expression and graphical representation of various kinds of thematic information, by combining theory with technology, integrating science with art, and mixing qualitative methods with quantitative methods. Since the thematic information that can be expressed is very extensive, which not only includes the terrain, hydrology, vegetation and other natural geographic information but also political, demographic, economic, cultural and other social information, thematic mapping applications are very extensive. Thus, the thematic map plays the important role of both "achievement expression" and "scientific research method", which is also one of the important research methods in data mining and knowledge discovery.

Since the establishment of the People's Republic of China, thematic mapping from theory to technology and applications and other aspects have made rapid developments. The development course of thematic mapping can be roughly divided into 4 stages: the initial stage of thematic cartography (from the 1950s to 1960s), the development stage of thematic cartography (from the 1970s to 1980s), the rapid development phase of thematic cartography (1990s) and the information development stage of thematic cartography (from the beginning of twenty-first century to present). This chapter will introduce the development course and main research contents of thematic mapping since the founding of the People's Republic of China. The existing achievements will be summarized, the existing problems will be analyzed, and the direction of further research will be put forward to promote the continuous deepening of the research field and expand the application scope of thematic maps.

N. Jiang (✉)
Information Engineering University, Zhengzhou, China

J. Wang and F. Wu (eds.), *Advances in Cartography and Geographic Information Engineering*, https://doi.org/10.1007/978-981-16-0614-4_6

6.2 Development Process

The development of the thematic map is synchronized with the history of human culture, and its development process is closely related to the development of the social economy, the changes in politics and the progress of science and technology. The development of thematic maps in China has a long history. According to the three color maps (town map, garrison map and topographic map) unearthed at Han Tomb at Mawangdui, Changsha, Hunan Province, it can be confirmed that China had a high level of thematic maps more than 2100 years ago. However, formation of the complete thematic cartography field occurred in the period after the founding of the People's Republic of China. Since the founding of new China, thematic mapping has rapidly developed from theory to technology and application. Thematic mapping underwent 4 main periods.

6.2.1 The Initial Stage of Thematic Cartography (from the 1950s to 1960s)

In the early days of the founding of the P.R.C., thematic mapping was mainly based on the compilation of thematic maps of natural phenomena, which had the obvious characteristic of practicality and strong suitability. A category of thematic maps was prepared based on the professional surveys and scientific investigations by production departments, including large-, medium- and small-scale thematic maps for the population census, industrial investigations, and land and resource surveys, such as the national soil survey and its complete sets of maps, as well as stock map and forest resource map compilations for the national forest survey, and so on. Among these maps are *China Geological Structure Map* (Huang Jiqing as editor-in-chief), *Tectonic Map of China and Its Adjacent Regions* (Zhang Wenyou as editor-in-chief), *Soil Erosion Map* (prepared by geographer Huang Bingwei) and so on. Another is the compilation of a complete set of physical geographical regionalization maps, for example, hundreds of earth science and biological experts and scholars were brought together at that time to compile a set of 1:1,000,000 scale topographic maps, soil maps, vegetation maps, land type maps, hydrological maps, animal maps and comprehensive natural zoning maps, all of which are linked and interrelated. Mapping methods, such as the overlapping method and correlation factor method, have been utilized scientifically in attempts to make a series of thematic maps in China.

6.2.2 The Development Stage of Thematic Cartography (from the 1970s to 1980s)

From the late 1950s to the end of the 1980s, the development of various disciplines and maps as the main research means in earth science has been widely recognized and applied, and the theories, technical methods and products of thematic mapping have rapidly developed. One of the most prominent developments is that the compilations of a comprehensive atlas and regional thematic map have achieved fruitful results. In addition, a considerable part of the atlas has reached the international advanced level, and some examples include "*The Atlas of Endemic Diseases and Their Environments in the People's Republic of China*", "*Atlas of Population of China*", "*Historical Atlas of China*" and "*Atlas of Natural Protection China*", and so on. These atlases cover a wide range of topics and are rich in content, detailed information, diversity of graphic design and high levels of printing, which fully reflects the breadth and depth of the latest research results in China in the fields of geography, biology, environmental science and space science. These atlases provide an important basis for national macro decision-making. In addition, a systematic study on the representation method of thematic mapping, integrated mapping theory and its unification and harmonization, applications of remote sensing mapping technology and so on were carried out earlier by LIAO Ke, ZHANG Kequan, and CHEN Yu, who are the representatives of China's cartographers. Additionally, a number of excellent works have been written and published, such as *Automation of Thematic Mapping* by Liu et al. (1979), *Thematic Map Compilation* by Zhang et al. (1980), *Compilation of the Thematic Map and Atlas* by Li (1984), *Hierarchical Representation of Statistical Maps* by Lu (1989), and so on. These are all the earlier works on thematic maps, which have laid a good theoretical foundation for the development of thematic mapping. Simultaneously, remote sensing technology and computer technology have injected new vitality into thematic mapping. For instance, the latest remote sensing technology was used to compile the unique style of the *Tengchong Aerial Remote Sensing Atlas* and the *Atlas of Geo-Science Analyses of Landsat Imagery in China.* In addition, computer-aided mapping technology was used to compile the *Atlas of the Population of China,* the *Atlas of Population of Jiangsu Province,* the *China Economic Atlas*, and so on. Thematic mapping data began to use up-to-date aerial and space remote sensing images, and thematic mapping methods began to transform from the traditional manual mode to the computer-aided method.

6.2.3 The Rapid Development and Deepening Stage of Thematic Cartography (1990s)

In the 1990s, on the one hand, the rapid development of various disciplines put forward more requirements on the map; on the other hand, along with the development of space remote sensing technology, computer technology, geographic information system (GIS) technology, the emergence of digital maps and the unceasing progress of surveying and mapping technology, there have been profound changes in terms of the production and use of maps, and the thematic map underwent a rapid development trend. This trend is mainly manifested by thematic mapping development in the directions of depth and standardization, while the thematic mapping technology developed in the automation direction. The contents of thematic mapping are becoming richer, the map variety is becoming more varied, and its application scope is becoming increasingly wider. For example, in terms of map content, thematic mapping has not only developed rapidly in the fields of environment, ocean, urban areas, population, economics, transportation, etc. but there have also been other celestial maps with the development of space technology and deep research on other planets. There are also new trends in terms of map types: from the current maps to dynamic change maps and then to forecasting maps; from resource distribution maps to comprehensive evaluation maps and then to planning and design maps; from qualitative analyses to quantitative thematic maps, and so on. These are signs of the in-depth development of thematic mapping. In particular, due to international cooperation mapping, the thematic map has further developed in the standardization direction. Many thematic maps are available worldwide, and standardization of the geological map has been realized; additionally, map standardizations in terms of landscape, vegetation, climate and other aspects have also made certain progress. The development of computer-aided mapping, remote sensing mapping and the emergence of the electronic map has enabled thematic mapping to enter a new development level. Thus, the scientific and practical values of thematic maps were further improved, and the needs of all aspects were met. In this period, the basic theoretical research on thematic mapping was carried out, including studies on thematic mapping architecture, quantitative thematic mapping, thematic map symbols and design of the representation method and data processing. In addition, research on professional mapping, such as hydrogeological mapping, tourism mapping, disease geographic mapping and environmental mapping, was carried out. Additionally, the research of thematic mapping technology and its application, such as computer-aided thematic mapping technology, remote sensing mapping technology, object-oriented thematic mapping technology, and expert system technology, was carried out.

6.2.4 The Information Development Stage of Thematic Cartography (Since the Beginning of the Twenty-First Century)

In the early twenty-first century, human society entered the era of information along with the rapid development of space technology, communication network technology, and information processing technology. In particular, the development and application of a series of modern information technologies, such as satellite remote sensing, global positioning systems, GISs, and digital transmission networks has had a profound influence on the development of the theory, technology and application of thematic cartography in the new century. New theories of spatial cognition, visualization and usability were introduced into thematic cartography, which means that the depth of the study and the breadth of application of the thematic mapping theory have experienced unprecedented expansion. New breakthroughs have been made, and a number of textbooks, works and papers have been published, such as *Compilation of the Thematic Map* prepared by Huang et al. (2003), *Digital Thematic Map* compiled by Liu et al. (2009), *Electronic Cartography* prepared by Long (2006), and so on. Additionally, Guo (2002) presented a paper, 'Visual search processes and cognitive efficiency of statistical maps', Luan et al. (2007) published 'Cognitive expression in the production of thematic maps' and other articles have also been published. The research of thematic mapping technology is advancing with the times, and the achievements are remarkable. Especially with the aid of remote sensing, GIS, digital mapping technology and image processing technology, great progress has been made in the aspects of data processing, symbol visualization and digital and intelligent methods of representation. GIS software, such as SuperMap and Atlas 2000, has launched a module with the function of thematic map creation, and this module has different features in the production of electronic thematic maps. Simultaneously, due to the progress of the theory, the development of technology and the renewal of ideas, the mapping content of thematic map has developed in the thematic, personalized direction; the expressions tend to be diverse and novel to reflect people's aesthetic views in the information age. There have been many new varieties of thematic maps, such as 3 dimensional virtual maps, multimedia interactive electronic maps, mobile maps, network maps, animated maps, and so on. These maps greatly enhance the value of using thematic maps, expanding the scope of thematic map applications, and therefore, thematic maps play a large role in social progress, economic construction and national defense modernization.

6.3 Main Research Achievements

6.3.1 Studies on the Content System of Thematic Map Representation

Since the founding of new China, the thematic map has changed greatly with the development of the national economy, increased social demand and progress in science and technology. These developments mainly manifested in 3 aspects: first, the content of the thematic map developed from a single natural factor to many other factors, such as human economy and national defense construction, which involves more extensive disciplines. Second, the content developed from the expression of general geographical factor distribution characteristics to the expression of thematic feature multidimensional characteristics, reflecting the depth of scientific research. Third, the content developed from a single map with a single purpose to a series of multipurpose thematic maps and comprehensive thematic atlases. In short, the content of thematic maps has developed toward comprehensiveness and topicality. A series of new thematic maps have emerged, including the tourist map, geological map, ocean map, environment map and military thematic map, which makes the system of thematic mapping more perfect.

1. **Development from the thematic mapping of a single natural factor toward the thematic mapping of many factors including the human economy and national defense construction**

In the early days of new China, thematic maps were mainly maps of natural phenomena. Most of these maps were thematic maps aimed to express the natural phenomena, including relief maps, soil maps, climate and hydrology maps, geological maps, vegetation maps, and so on, and some examples include larger scale regional geological surveys and geological map sets, national soil surveys and its complete set of maps, a large number of forest stock maps and forest resource maps compiled during national forest surveys, and so on. Among them, the *Soil Erosion Map* compiled by Huang Bingwei according to the comprehensive investigation results in the middle Yellow River reflected the current situation of soil erosion in the middle Yellow River. The *China Geological and Tectonic Map* edited by Huang Jiqing emphasized geological and sedimentological viewpoints. The *Tectonic Map of China and Its Adjacent Regions* edited by Zhang Wenyou applied deep fault theory to highlight geological mechanics. After the 1980s, with the development of science and technology, the thematic mapping content began covering increasingly wider areas of technology. Thematic mapping has developed rapidly in many fields, such as environment, ocean, cities, population, economics and transportation, and the related products, such as environment maps, tourist maps, ocean maps, and urban maps were developed. However, with the development of space technology and the deep study of other planets, there have been celestial maps, for example, the research and production of Mars, the moon and other celestial maps, which serve as preparation for space exploration. Thematic maps not only represent the research results of the national economy but have also been widely used in military command, training, support, and so on. Thematic maps not only express a concrete physical object or phenomenon but also abstract concepts or phenomena. In a word, the contents of thematic maps are increasingly abundant, ranging from nature to economy, from concrete to abstract, from current situation to forecast, from land to sea, from the earth to other planets, from civilian to military, and so on. Thematic maps have played an important role in scientific research, resource development, disaster monitoring, situation analysis, environmental protection, military command and so on, which has consequently promoted the development of thematic mapping.

2. **Development from the expression of the distribution characteristics of general geographical elements to the representation of the diversity of thematic elements**

In the early years following the founding of the P.R.C, the large number of thematic maps were compiled according to professional survey and scientific investigation data from the production department, and these maps mainly expressed the distribution characteristics of the general investigation elements, including census, industrial investigation and land and resource surveys, such as the national soil survey and its complete set of maps, the forest stock map and forest resource map compiled for the national forest survey, and so on. With the further development of science and technology and the increasing use of maps, the content of thematic maps has demonstrated in-depth thematic development. In addition, there are many maps that place emphasis on themes, such as the economic atlas, population atlas, environment atlas, military atlas, traffic atlas, and so on. By means of the expression of various characteristics (quality, quantity, current situation, prediction, dynamics, etc.) of the thematic elements, the theme of the application was highlighted, the scientific nature, comprehensiveness and practicability were enhanced, and the breadth and depth of scientific research were reflected (Liao 1979; Chen 1988); some examples include the *Atlas of Population in China*, *Atlas of Land and Economy in Hubei Province*, *Atlas of Economic in Hunan Province*, *Atlas of Population and Economic in Shandong Province*, and others. There are not only distribution maps of ethnic groups, distribution maps of main cities and towns, and distribution maps of plants and animals but also a large number of maps that show the quantitative characteristics, development trends and zoning plans; some examples include the Zoning Map of Soil Improvement, Map of Per Capita Income, Map of Iron and Steel Production Capacity, and Production Map of Main Industrial Products, and others. In addition, there is the *Heilongjiang Agricultural Atlas* and other excellent atlases that take agriculture as a theme.

3. **Development from a single map with a single factor to a series of thematic maps and comprehensive thematic atlases with integrated elements**

Due to the increasing abundance of content and features, thematic maps have developed from a single use map toward a series of multipurpose thematic maps and comprehensive thematic atlases. For instance, while expressing natural phenomena, these maps also represent some of the major cultural and economic factors; aerial photographs can be developed simultaneously as a series of thematic maps, including topography, geology, soil, land use, and so on. In addition, the compilation methods of a comprehensive series of thematic map were studied. Since the late 1950s, along with development of the comprehensive research of various departments and the interpenetration among each other, the content of the thematic map has also developed in a comprehensive direction. The regional atlas and comprehensive atlas emerged. An atlas was created for geology, climate, soil, agriculture and forestry, and the research and practice of atlas compilation for emerging fields included environmental ecology, natural protection, disease prevention, urban planning, ecological tourism, military geography, and emergency response; some examples include the *Hydrogeological Atlas of the People's Republic of China* (Wang et al. 1979), *Atlas of Natural Protection in China, Endemic Diseases and Environment Atlas of the People's Republic of China, Atlas of Population in China, National Economic Atlas of the People's Republic of China* (Liu et al. 1993), *Atlas of Shenzhen City* (Liu et al. 1997), *Atlas of Natural Disaster System in China* (Shi et al. 2003), and so on.

6.3.2 Study on the Unification and Harmonization of Thematic Cartography Based on Composite Mapping

During the initial stage of new China, the main task of thematic mapping was to prepare a thematic map of natural phenomenon with a single factor, such as a relief map, soil map, climate and hydrology map, geological map, vegetation map, and so on. With the further development of earth science research, a wide range of multifactor regional comprehensive thematic mapping emerged. Since the late 1950s, with the development of the comprehensive research of various subjects and interpretations, the content of the thematic map also developed toward a more integrated approach. The regional atlas and comprehensive atlas were rapidly developed. Simultaneously, Liao Ke and other cartographical experts carried out a study on the unification and harmonization of thematic cartography based on composite mapping. The so-called unification and harmonization of composite mapping refers to the study on the unified coordination of all aspects of thematic cartography, with the geography theory (such as the zonal distribution law, regional differentiation theory, landscape theory, comprehensive natural zoning theory, etc.) as the guide, to correctly reflect the formation, development, structure, quality and quantity characteristics, as well as the objective law of the interrelation and interaction among the elements and various departments in the earth system or geographical environment and human-earth system. The representative research works are listed as follows: the twelfth chapter of *Modern Cartography* written by Liao Ke in 2003, which systematically expounds the theoretical basis and method of comprehensive cartography, and the scientific bases and content of unification and coordination; in 2005, the paper 'remote sensing complex series mapping for the ecological environment' was published by Liao (2005). Pang et al. (2006) also published the paper 'Unification and harmonization in true color city image map'.

The study on unification and coordination of thematic maps based on comprehensive mapping mainly includes the theoretical basis, contents and forms, and the principles and methods of unification and coordination; the types and patterns of connections between elements and phenomena; the factors and symbols that need to be unified and coordinated between each map; the unification and coordination of comprehensive cartography under the computer-aided mapping and GIS environment; thematic mapping standardization; and so on. These factors are represented in the aspects of topic and structure, content and index, classification and grade, scheme and legend, contour and boundary, map generalization, representation method and map decoration, and so on. For unification and coordination between classification and legend, the selection of the classification and grading system is emphasized, including the classification index and classification method; first, based on the scale, the comparable classification units corresponding to the regional level and same grade table are determined; second, to reflect the common zone regularity and regional characteristics, the representative types that can reflect the zonal characteristics (including horizontal and vertical zones) are selected and the transitional type or non-zone type should match; third, unified legend, structure, arrangement principle and naming methods are adopted. The unification and coordination between the contour and boundary mainly includes the following steps: first, the common natural figure of all features and phenomena must be correctly expressed, which can be divided into the zone boundary, the regional speckle and the basic natural figure according to its scope and scale; second, the common transitional characteristics of the contour line and boundary must be correctly reflected, since there are two kinds of transitional boundaries that are obvious mutation and fuzzy gradient in nature; third, the common joining relationship between the boundary and contour line must be correctly distinguished; fourth, the incidence relations among contour lines must be correctly differentiated since there exist 3 kinds of situations, which are coincidence, partial coincidence and non-coincidence between the contour lines of each element.

Since the 1970s, global mapping has been active. For example, the International Union of Geological Sciences has organized and compiled a complete set that includes the following maps: European Geological Map and Tectonic Map of the World, One of Million International Population and Land Use Map, One of Five Million World Vegetation Map, One of Five Million Soil Map of the World and the World Climate Atlas, and so on. Therefore, standardization of the thematic map is important content in unification and coordination research. Only standardization and normalization can be the most effective way to realize the automation of mapping to compare and use the same map data from different regions and countries. Thus, standardization and normalization is one of the problems that professional workers and cartographers focus on and research worldwide. In China, the standardization research focuses on the following aspects: standardization of the classification system of thematic maps, standardization of thematic map contents and the geographical units of the thematic map, and standardization of thematic map symbols and representation methods of different kinds of maps. The standardization for geological maps has been basically realized (Wang 1991). In addition, some progress has been made in the map standardization of topography, vegetation, climate, tourism and so on. Standardization of the more complicated economic map was also discussed. A variety of design schemes for the unification of contents, indexes and symbols of the economic map were put forward. These have laid a good foundation for the sharing of thematic map information (Liao 1979).

6.3.3 Study on the Representation of the Thematic Map

In the early stage of the founding of new China, since the social demand was not high, the information was insufficient, the content was less, and the practical purpose was strong, there was less research on the theory of thematic cartography. This research mainly aimed at producing thematic maps based on the results of professional surveys and scientific research. During this period, the requirements of graph expression and legend design were not high; a set of mapping methods suitable for specific conditions in China was established, such as the overlay method, the correlation factor method and other simple representation methods for the thematic map; in addition, the graphic structure type, content index, legend design and expression method of the physiographic map in the physical map and physical atlas were summarized, and the naming and definition of the thematic map representation methods were systematically summarized, which laid a good foundation for the development of the theory of thematic cartography (Zhang 1963). By the 1980s, with increasing mapping content and the development of a series of maps and comprehensive atlases, higher requirements on the representation of the thematic map were put forward. Based on the long-term experience and summary, with the development of the theory of cartography, the thematic mapping theory gradually formed, the index selection, legend design principle, content choice and representation method about the thematic map were discussed and summarized, and ten basic representation methods were identified, such as the location symbol method, linear symbol method, quality base method, choroplethic method, isoline method, area method, dot method, arrowhead method, statistical chart method and regional diagram method. The representative works are the *Thematic Map Preparation* written by Zhang et al. (1980) and the Compilation of Thematic Map and Atlas edited by Li (1984). From the end of the1980s to the early 1990s, thematic cartography involved increasingly wider areas of contents. In addition, thematic mapping developed to become more thematic. Thematic mapping studies were carried out on many thematic subjects, such as population mapping, environment mapping, tourism mapping, ocean mapping, city mapping, celestial mapping, etc.; among them, as a particular area of focus, the applications of the basic representation method of the thematic map in all kinds of mapping were studied and the application characteristics of the commonly used methods of different thematic content were analyzed. For example, the book *Land Resource Information and Cartographic Representation Methods* written by Tong and Liang (1988), the studies on tourism mapping by Ma (1996), Chen (1999), Guo et al. (1999), the study on medical geography mapping by Lu (1991), and the studies on environment mapping by Su et al. (1991), Ou (1995), and so on.

For decades, the representation methods of the thematic map basically have not changed. However, there are many innovations in the form and means of expression that greatly improve the performance and usefulness of the thematic map. Along with the development of computer-aided thematic mapping technology and the application of multimedia technology, on the one hand, the representative symbol types were more abundant, i.e., there were the symbols of a single factor and single indicator as well as the combination of multiple elements and multiple indicators; there were linear vector symbols and image raster symbols; additionally, there were 2-D symbols vs. 3-D symbols, static symbols vs. animation symbols and general effect symbols vs. special effect symbols (gradient, projection, relief) (Mao 1989). On the other hand, the form of expression is more varied and flexible, and the representation methods developed from static to dynamic interactive expressions, for instance, a form of animation, such as isoline method and arrowhead method, makes the performance more vivid. In addition, the form of dynamic interactive expression, such as the choroplethic method, quality base method and area method makes the performance more flexible and straightforward. The form of statistical charts expression makes the shape diversify (the diversification and three-dimensional form of the basic shape), makes the color harmonious (rich color, bright), and makes the style outstanding, innovative and vivid. The typical works are the *Multimedia Atlas of Ecosystem in Fujian Province* and *Electronic Map of Wuwei City*. Additionally, there is a topological relation representation method, which indicates the relative geographic location and highlights quantitative comparison. In addition, research on the selection of intelligent representations of the thematic map was carried out (Tian et al. 2007).

6.3.4 Study on the Thematic Map Data Processing Model and Method

From the 1950s to 1960s, thematic maps were mostly qualitative and single factor maps. Data processing was relatively simple, with a symbolic design for the purpose. Therefore, it was mainly a manual calculation. Since the 1980s, with the development of science and technology, the content of the thematic map was becoming increasingly extensive. In addition, the cases of expressing the statistical analysis and dynamic prediction results were increasing. Thematic mapping developed in the quantitative direction, while synthetic maps, such as the statistical map, zoning map, types map, and prediction map appeared; data processing of the thematic map developed from a simple manual calculation to complex automatic processing with a computer. Mathematics has been used to describe the quantitative characteristics of things or phenomena, and a systematic study on the mathematical model of thematic mapping was carried out by cartographers. For example, in the paper 'Some investigations on the development of thematic maps' published by Zhang et al. (1987), several mathematical models were proposed to first scientifically and abstractly clarify the intrinsic nature of phenomena and to turn the analytical conclusions (classified and hierarchical results) into statistical maps. 'The principle of the fuzzy multilevel comprehensive evaluation and its application in the quality evaluation of thematic map compilation' was published by He (1989). A thorough and systematic study on the classification and representation of statistical thematic map data were carried out according to the book *Classification Representation Method for Statistical Map* written by Lu (1989), where by taking the relationship between a hierarchical visual variable simulation and data processing as an important principle, a general formula for the simulation of the hierarchical visual variables and the visual error correction formula were given; also by applying fuzzy mathematics to the statistical classification, a gradual clustering classification, a fuzzy clustering classification and a gradual fuzzy pattern recognition classification, as well as an optimal classification with mathematical rules were put forward. 'The fuzzy-grey situation decision in the classification of the multifactor thematic data' was published by Xu (1992). In the book *Modeling Solution of Data Processing in Cartography*, which was written by Wang and Zou (1992), data processing is considered to be the core of modern cartography; guided by the theory of the map model for the first time, mathematical models are combined together with the mapping models for deep and systematic research on the cartographic method; in terms of the data processing needs of thematic maps and with data preprocessing, establishment of the mathematical model, data processing, establishment and interpretation of the map model as a clue, the theory and method of modern applied mathematics, such as the graph theory, fuzzy set theory, gray system theory, etc., and their applications in cartographic data processing are introduced; the "interfacing" problem of the mathematical model and the map model are a focus of study. In addition, the paper 'A new classification method of thematic features and its application' by Sun et al. (1994) and the paper 'Statistical data classification evaluation model based on multiple attribute decision-making' by Jiang et al. (2007), and others studied the statistical data classification model in depth. In the meantime, combined with specific applications, the applications of fuzzy mathematics and gray prediction theory for the production of thematic maps of type zoning, thematic maps of forecast and predictions and trend surface analyses were studied intensively by many scholars (Wu 1993; Huang 1994). The combination of computer mapping and GIS promotes research on the establishment and application of the data processing system for thematic maps (Wang et al. 1991) and the mathematical model base of thematic data processing based on the meta algorithm (Zhang et al. 2008).

6.3.5 Research on the Technology System of Modern Thematic Cartography

Since the founding of new China, cartographic scientists have continued to improve the thematic cartography technologies and methods to bring the development of thematic mapping to a new level. These improvements are mainly demonstrated by the development from manual mapping to remote sensing and computer aided thematic mapping, which expands the information source of thematic mapping, increases the automation level of thematic mapping, reduces the mapping cycle, and improves the quality of map making, which changes the use mode of thematic maps. Studies on remote sensing-based thematic mapping, computer-aided thematic mapping, GIS technology-based thematic mapping and graphics and image technology-based thematic mapping have been carried out. The theory and technology system of thematic mapping has been improved.

1. **Research on thematic mapping based on remote sensing technology**

In the 1980s, the introduction of the new technology of remote sensing promoted the development of the thematic image map in China. A large number of high-qualified workers and skill-based talents in the field of remote sensing application were trained, and a large number of image maps and image atlases were compiled and published, such as the *Tengchong Atlas of Remote Sensing*, *Atlas of Typical Image Analysis in the Changchun Remote Sensing Experiment Area*, *Atlas of the Pseudo Color Image of China*, *Geo-Analytical Image Atlas of China*, and others. After entering the twenty-first century, with the increase and application of multiresolution, multi-spectral and multitemporal remote sensing information resources, the application of remote sensing information in cartography and GISs was completely changed. On the one hand, it was deeply recognized that remote sensing information is an important information source in thematic mapping, which has a strong advantage in maintaining the diversity and timeliness of the thematic map. In addition, the thematic maps produced by remote sensing data became new variety of map, which greatly improved the performance of the map. On another hand, studies on the theory and method of remote sensing mapping were conducted from the aspects of image interpretation, identification, classification and thematic map representation, and further research was conducted on the application of remote sensing mapping in the study of dynamic phenomena, such as plant growth, yield estimation, forest resources, ocean situation, disaster prediction and forecasting, and so on. Many developed cities have compiled and published high-level image atlases, including the *Photo Atlas of Shenzhen City*, *Satellite Image Atlas of the Nanjing Urban District* and the *Photomap Atlas of Suzhou City*, which provide an important reference for the city's economic development.

2. **Research on the technology of computer-aided thematic mapping**

The development of computer technology and digital mapping technology provides advanced technical means for thematic mapping automation. During the 1970s and

1980s, computer-aided mapping developed from the preliminary exploratory stage to the application stage and was used to compiled several atlases, such as the *Atlas of Drinking Water in China, Atlas of Endemic Diseases and Their Environments in the People's Republic of China* (general map of the whole country), *Atlas of Population in China, Atlas of Population of the Jiangsu Province, Chinese Economic Atlas*, and so on. In addition, research on the basic principle and process of the automatic drawing of map symbols and computer-aided production of thematic maps (atlases) was carried out. Mapping expert system technology developed during the 1980s and 1990s was introduced in the field of automated production of the thematic map. Much of the knowledge and experience in thematic mapping were summarized. Mapping knowledge has been converted into rules and algorithms for effectively guiding the design and production of the thematic map and examples include the 'Object-oriented expert system for quantitative mapping' published by Lu (1991) and 'Determining map content of the thematic map with expert system technology' published by Hua (1993). Additionally, the application of object-oriented technology in thematic mapping and the thematic mapping system was studied, i.e., the 'Practical thematic map production' published by Wang Tiejun (1999).

3. **Research on thematic mapping based on GIS technology and multimedia technology**

After entering twenty-first century, with the development of GIS technology and multimedia technology, thematic mapping was developed from the pursuit of a full automation system toward a human–computer collaboration. The progress of GIS technology has greatly promoted the development of thematic mapping technology. The functions of GIS, including statistical analysis, terrain analysis, spatial analysis and visualization of data processing results, are the major technical means for thematic mapping. For example, SuperMap, GIS and other MapGIS software all have thematic mapping function modules with distinctive features, including the selection of thematic data and an automatic processing function, the choice of representation methods and an adaptive symbolic function, and so on, and there is a certain degree of intelligence and automation. Many scholars like Guo Qingsheng (1993), Jia Chengquan (2003), and Sun et al. (1998) have made a number of new achievements in the research on establishment and application of the thematic symbol database, the representation method base, the template library, the thematic data processing model base, and so on. In addition, due to the powerful editing functions of multimedia technology, many thematic maps and atlases were prepared by using commercial graphic and image software platforms, such as the *Historical Atlas of the Shanxi Province* (Editorial Department of the Compilation Committee of the Historical Atlas of the Shanxi Province, 2000) and *Earthquake Atlas in the Wenchuan Area, Sichuan Province* (State Bureau of Surveying and Mapping, 2008). With the application of multimedia technology in thematic mapping, map, image, text, video, audio, pictures, animation, websites and other multimedia information will be used through flash, animation, three-dimensional display, virtual reality and other means to realistically reproduce the change process of certain phenomenon, preview the developing trend of events and enhance the performance of the thematic map, which has better information transmission and cognition. A large number of thematic electronic maps and multimedia thematic electronic atlases have been produced for browsing, such as the *China Electronic Atlas* and *Hongkong Electronic Atlas*, and for interactive querying, such as the *Electronic Atlas of Transportation and Tourism in China, Traffic Map of Beijing City* and *Ecological Environment Atlas of the Fujian Province.* Additionally, a number of multimedia electronic atlases, such as *Electronic Map of Wuwei City*, have interactive functions and novel animation-based performances.

6.4 Problems and Prospects

For over a half century, thematic cartography has made great progress in terms of theory, technology and products, and great changes have occurred. Thematic cartography has been widely used in all aspects of social civilization, national economy and national defense construction and had played an important role. However, with the development of science, technology and social economy, thematic mapping, as an important tool for understanding the geographical environment and an important means for geographical expression, still has room for development.

6.4.1 Thematic Cartography in the Information Era Calls for New Ideas, New Theories and New Technologies

On one hand, we need to deeply understand the growing complexity and diversity of society, and the economy has placed high demands on the thematic map, such as more variety, flexible use, novel forms of expression, superb graphics, and so on. On the other hand, the research in various disciplines related to geography has also developed from qualitative to quantitative, from static to dynamic, and from digital to intelligent. As a language and means of geography research, the function of thematic maps is becoming increasingly prominent, and the demand is increasingly higher. We must renew our ideas and change passive service to active service; we must update technology and change the system of making thematic maps based on specifications for a system of on-demand customizing of the personalized thematic map; and we must explore new theories, such as an electronic thematic map design theory instead of the paper thematic map design theory. The data acquisition–model calculation–visualization–output and sharing, as an integral chain of thematic mapping, should be deeply understood. Only in this way can we promote the development of thematic cartographic theory and technology.

6.4.2 Strengthening the Research on the Intelligent Thematic Map Characteristics and Promoting the Development of the Thematic Cartographic Theory

Great changes have taken place in the field of cartography in the information era, i.e., traditional cartography has been replaced by digital cartography and is transforming toward intelligent cartography services as a core. Many traditional concepts have been extended and updated, for example, the map scale changed from fixed to unfixed and from invariable to variable; there are not only static visual variables but also dynamic visual variables, as well as perceptual variables of multisensory stimulation; not only has graphics generalization occurred but also data generalization, percept generalization, on-demand generalization, and so on have occurred. All of these factors directly affect the development of thematic cartography, and its theory has developed toward a digital thematic map, electronic thematic map theory; the symbol system has developed from simple a flat, static symbol to a multidimensional, dynamic, visual expression; the visual perception has extended from single visual to multiple sensory perception; thematic map design has developed from a single graphic design to the integrated design of data, graph and function; and the types of thematic cartographic products have developed from a single form to a multimedia form. Emotional thematic maps, fast food thematic maps and other personalized thematic maps have appeared and presented new challenges to the concept of the thematic map, as well as representation method innovations, formulation of mapping standards, data processing model research, map use effects, and so on. Intelligent data processing for thematic mapping and an adaptive design for the representation method with strong interactions, dynamic multidimensions, and the most intuitive visualization would be the main research directions.

6.4.3 Strengthening the Research on Thematic Map Data Processing by Improving the Scientific Nature and Practicability of the Thematic Map

With the advent of the information era of thematic mapping, the process of making a thematic map is actually a data processing and symbolic representation process. Therefore, it is necessary to focus on studies of the use of thematic mapping data sources, the establishment and intelligent processing of the thematic mapping data model, the interface problem of symbolic representation, and so on. For example, through research on multivariate (multisource) thematic mapping data fusion technology, data mining and knowledge discovery methods, all kinds of GIS databases, remote sensing image interpretations, existing (paper) maps, downloaded internet data, and so on, can be relied upon; additionally, intelligent data processing based on the model, algorithm, knowledge and knowledge reasoning can be used to construct a technical system and evaluation system for thematic mapping data processing to improve the science and practicality of thematic mapping.

6.4.4 *To Strengthen Thematic Mapping Technology Through Research, the Intelligence Level of Thematic Mapping Must Be Improved*

The rapid development of GISs and multimedia technology has led to the development of automatic thematic mapping technology. However, the technical status is far from the actual demand, and the technical means cannot easily meet the needs of multilevel mapping. This can be illustrated as follows: the symbol type is singular and the symbol library structure is not suitable for the requirements of rapid mapping and personalized mapping, that is, the configuration is not flexible enough to form a combination of symbols; the expression form and expression method remain dull and poor and cannot meet the needs of people's aesthetic concepts in the information age; in other words, the dynamic representation of the thematic map is far from computer games and synthetic movies; the structure of the data processing model base is not reasonable, and it does not have an intelligent call function; although the technology functions in graphics editing, the expert knowledge of thematic map design is not present. As a result, the technology cannot provide the environment and expert knowledge support for rapid personalized design and it is difficult to quickly make scientific, reasonable, beautiful, practical, and strongly adaptive thematic maps. For these reasons, we need more in-depth studies of the thematic mapping technologies in the information age, including setting up a special data processing model and evaluation model base, enriching the thematic map symbol library, representation method library and thematic map color template library, as well as the display style template library, interactive legend design tools, and so on, in order to greatly improve the intelligence level and automation degree of thematic mapping, improve the drawing speed and quality and make personalized drawing a reality.

6.5 Representative Publications

(1) ***Automation of Thematic Map-making*** (Liu Yue, Liang Qizhang. Surveying and Mapping Press, Beijing, April 1979)

This is the first of the thematic mapping writings in China. The whole book consists of 13 chapters and is outlined as follows: overview of the thematic map; development and application of automatic map-making; the main equipment in automatic mapping; program design and mathematical methods; a variety of thematic mapping software, the compilation method and program designs of the contour map, symbol map, three-dimensional map, statistical chart, type map, and so on. The book provides more than 30 source codes written in FORTRAN language, which is very useful for automatic mapping.

(2) ***Compilation of Thematic Maps*** (Zhang Kequan, Huang Rentao. Surveying and Mapping Press, Beijing, May 1980)

The book is divided into 4 parts and a total of 10 chapters, with a focus on several basic problems in thematic map and atlas compilations, as well as the compilation features of the major map types. The first part introduces the basic theory and methods of thematic map compilation, including the types and uses of thematic maps, the representation of thematic content and the design and compilation of thematic maps. The second part introduces some new techniques and methods of thematic map compilation, including the application of satellite remote sensing data, the application of mathematical methods, and the automatic drawing of thematic features. The third part introduces the compilation features of the main types of thematic maps, including preparation of the physical map and social and economic map, and the contents and characteristics of charts, aerial maps, teaching maps and tourist maps. The fourth part introduces the editorial preparation of the atlas, its compilation methods, and so on. In addition, there is an appendix at the end of the book, which contains the basic color knowledge and the design method of hieroglyphic symbols and perspective symbols for use in thematic map design with a larger practical value.

(3) ***Compilation of Thematic Maps and Atlas*** (Li Haichen. Higher Education Press, Beijing, Jan 1984)

This book is divided into two parts: thematic map compilation and atlas preparation. The first part is the compilation of thematic maps, which describes the basic types of thematic maps, the basic representation methods, and the compilation methods of various physical maps, social economic maps and other thematic maps. The book focuses on expression of the basic contents of the major branches of natural geography and social economic geography in the form of maps and charts, such as the compilation features and methods of political area maps, population maps, urban maps, historical maps, economic maps, and so on. The book also briefly introduces thematic mapping automation and its applications, including automatic drawing of thematic maps and an introduction to GISs. The second part includes preparation of an atlas, which expounds upon the definition, classification and characteristics of the atlas; the second part describes the content of atlas design and its compilation characteristics and methods, introduces several major atlases at home and abroad, and summarizes the experiences and lessons in atlas preparation.

(4) ***Thematic Mapping*** (Wen Changen. Seismological Press, Beijing, April 1984)

This book briefly introduces the basic concepts, classification methods, compilation principles and methods for thematic maps. In the book, the overall design of a thematic map, preparation of a geographic base map and collection and processing of mapping data are discussed in depth; the book provides a detailed discussion on thematic map representation methods, techniques, patterns and mapping processes. The principles and methods of remote sensing mapping technology and the basic principle and process of automatic thematic mapping are also introduced. Finally, the preparation methods and characteristics of the thematic atlas are briefly introduced.

(5) ***Mathematical Models of Thematic Mapping*** (Zhang Kequan, Guo Renzhong. Surveying and Mapping Press, Beijing, Mar 1988)

This book focuses on the statistical analysis of intrinsic distribution rules and property-structure relationships for the phenomenon system features. The book mainly introduces the basic principles and methods of solving various mathematical models, and the basic mapping theories and techniques using computational results, including the basic concept of the mathematical mapping model, the characteristics and processing of phenomenon system element variables, and the characteristics of various models of the phenomenon system elements, such as the hierarchical model, the interrelation model, the spatial distribution characteristic structure model, the dynamic analysis and forecasting model, the internal relationship analysis and information simplification model, the structural analysis model of compound features, and so on; finally, the adjustment of the type division and the determination of regional boundaries are included. Additionally, the structure type of complex model is also introduced.

(6) ***Land Resources Information and Cartographic Representation*** (Tong Yangfen, Liang Qianchao. Guangdong Map Publishing House, Guangzhou, May 1988)

This book is divided into 3 parts with a total of 13 chapters. The first part introduces the general situation of land and resource information, including the concept of land planning and the content and classification of land information. The book introduces different types of land resources information in detail, including land resources, human resources, mineral resources, energy resources, tourism resources, agricultural regional planning resources, and so on, as well as their representative contents and indexes on a map. In the second part, the representation methods of the land resources map are introduced, the representation methods of land resources information, which are distributed in the patterns of point, line and plane, are introduced, and a universal representation method—arrowhead method is described. The representation methods are classified and the coordination of several commonly used charts and representation methods are discussed. Finally, the application of mathematical methods in land resources mapping is discussed, with a focus on cluster analysis application. In the third part, the types and mapping methods of specific land maps, such as natural resource maps and social economy maps, are studied, and in particular, several problems regarding the composition of the *Provincial Land Resources Atlas* are discussed.

(7) ***Hierarchical Representation Methods of Statistical Maps*** (Lu Xiaozhong. The People's Liberation Army Press, Beijing, Apr 1989).

This book is divided into 3 chapters, and the content is listed as follows: introduction, hierarchical data processing and hierarchical visual variable simulation. In terms of hierarchical data processing, the book provides a definition of the grading error and accuracy and considers the hierarchical matching degree as a quantitative criterion for determining grade. In terms of specific methods, an optimal segmentation algorithm is used to generate the optimal segmentation and hierarchy with minimal grading error under the condition of a certain number of grades. The stepwise clustering method and the fuzzy clustering method are also modified and applied to data grading, which resulted in the stepwise clustering grading method and fuzzy clustering grading method in the book. To improve the efficiency of data classification processing and reduce the computer memory occupation, the method of successive approximation is adopted and the fuzzy pattern recognition method is applied to data grading; thus, stepwise fuzzy pattern recognition grading is put forward. The book also presents an optimal grading method with mathematical rules.

(8) ***Mapping by Remote Sensing*** (Feng Jiwu, Pan Juting. Surveying and Mapping Press, Beijing, Jun 1991).

This book comprehensively introduces the basic knowledge, theory and methods of mapping with remote sensing. The whole book is divided into 5 parts. The first part is an introduction to remote sensing and the physical basis of remote sensing. The second part is the imaging principle of the remote sensing image and nature of the image, in which airborne remote sensing images, such as aerial photography, thermal infrared and radar, as well as space remote sensing images, such as MSS, TM, SPOT, and so on, are introduced. The third part is about the visual interpretation of remote sensing images and the interpretation map, which introduces the principle and method of interpretation, the optical enhancement of images, and image and thematic interpretation applications. The fourth part is about digital processing of remote sensing image and its mapping applications, in which the characteristics, restoration, enhancement, image classification and applications of digital images are introduced. The fifth part is the methods for mapping using remote sensing images, and the accuracy analysis of the digital image, map projection and its transformation, as well as the techniques of map compilation and printing process are introduced.

(9) ***Models and Methods of Data Processing in Cartography*** (Wang Jiayao, Zou Jianhua. The People's Liberation Army Press, Beijing, Dec. 1992).

The whole book consists of 10 chapters. The first chapter mainly introduces the research object and content, actual background, development status, applications and basic process of the model and method of cartographic data processing. The second and third chapters introduce the foundation of the mathematical model and the mapping model for cartographic data processing, respectively. For the former, the book emphasizes the fundamental role of mathematics in cartography and focuses on the multivariate statistical analysis, fuzzy set theory and graph theory. For the latter, the basic problem of constructing mapping models, such as the scaling method, map symbol type, visual variable selection, and so on, is emphasized. From chapter four to chapter nine, the principles, methods and applications of the hierarchical model of map data processing, the correlation model, spatial distribution trend model, prediction model, internal connection and information reduction model, as well as the partition model types are discussed, respectively. These chapters are the focus of the book, which is mainly aimed at thematic feature data processing in thematic cartography. The tenth chapter is an introduction to the model and method of cartographic generalization; the chapter introduces the regression model and root mean square model for the quota selection of mapping objects, as well as the structural selection model based on the graph theory and fuzzy comprehensive evaluation method.

(10) ***Introduction to Thematic Maps*** (Zhang Shitao. The People's Liberation Army Press, Beijing, Jan. 1995).

The book briefly describes the basic theories and methods of thematic cartography. It consists of 11 chapters, including the theoretical basis of thematic cartography and a variety of mapping methods. For the former, the development history of thematic cartography, thematic map projection, thematic map color, thematic map symbol, representation methods, and so on, are introduced. For the latter, the statistical surface mapping method, the statistical diagram mapping method, and so on, are introduced. In addition, thematic map data acquisition and processing, as well as thematic map design, compilation and printing are introduced in detail.

(11) ***Environment Mapping*** (Zhu Guangrong. Surveying and Mapping Press, Beijing, Jan 1996).

The book consists of 13 chapters, including the basic knowledge of environmental mapping, such as representation methods of an environment map and the creation process of an environment map. The compilation methods of a variety of different contents and different types of environmental maps, such as the map of environmental pollution status, map of environmental quality assessment, environmental prediction and planning map, ecological environment map, environmental series maps, and so on, are also included. The book also introduces the applications of remote sensing data and computer-aided mapping technology in environmental mapping, and the applications of environmental map analysis methods.

(12) ***Tourist Mapping*** (Ma Yaofeng. Xi'an Map Publishing House, Xi'an, Mar. 1996).

Beginning with the structure of the tourism map, the main contents of the book include the geographical basis of a tourist map, the expression method of tourism elements, the symbol design of a tourist map, the color design of a tourist map, the compilation of a tourist map, computer-aided tourist mapping, and so on. Among the contents, the tourist map design is the central content.

(13) ***Thematic Analysis of Remote Sensing and Geo-Information Tupu*** (Fu Suxing. Science Press, Beijing, January 2002).

The whole book is divided into 12 chapters; the contents can be classified into 3 parts. The first part discusses the theory and method of the spatial remote sensing information mechanism and thematic mapping, which mainly includes geo-science development, geo-information Tupu, functions and effects of remote sensing mapping, characteristics and mapping analysis of the image information source, parameter analysis and application of remote sensing image recognition and mapping, and pattern mapping of the earth information science system. The second part describes the multivariate analysis of the remote sensing image and the practice of thematic recognition and mapping, which mainly includes remote sensing analysis and mapping of natural agricultural conditions, computer-aided analysis and mapping of remote sensing statistical data, remote sensing analysis and mapping of biological resources, remote sensing analysis and mapping of water resources and lake and marine environments, and remote sensing monitoring and mapping of urban environments. The third part introduces the output of remote sensing mapping and the design printing process. Simultaneously, the book also gives a brief introduction to the earth information fusion science system, map innovation of geo-information Tupu, and holographic automatic digital mapping, data sharing of digital earth, and so on, with a focus on the application of remote sensing and GISs and other new technology.

(14) ***Design and Innovation of National Physical Atlas of China*** (Liao Ke, Chi Tianhe, Qi Qingwen et al. Meteorological Press, Beijing, September 2003).

The book consists of three parts. The first part is the introduction to the compilation of the *National Physical Atlas of the People's Republic of China* and the development of the *National Physical Atlas of China* (electronic version), including a preface and foreword written by academics CHEN Shupeng and SUN Honglie. The first part provides an explanation of atlas preparation, overall design of the atlas, a development outline of the electronic version, requirements and technical regulations for map compilation, and so on. The second part is about the main theories and key technologies in the compilation of the *National Physical Atlas of the People's Republic of China* and the development of the *National Physical Atlas of China* (electronic version). The third part includes the related appendixes, such as the catalogue of the *National Physical Atlas of the People's Republic of China, National Physical Atlas of China* (electronic version) and several foreign physical atlases; additionally, the master code table of the physical map database, expert comments on the *National Physical Atlas of the People's Republic of China,* and inserts of the color map are included. Major articles and foreign experts' comments are attached in the English version.

(15) ***Compilation of Thematic Maps*** (Huang Rentao, Pang Xiaoping, Ma Chenyan. Wuhan University Press, Wuhan, October 2003).

Based on the previous two editions, this book was completed through further modification, complement and perfection. The book is divided into 9 chapters. The first three chapters discuss the basic theory of the thematic map, including the concept, classification, and development of the thematic map, as well as the relationship with other disciplines, distribution characteristics and representation methods; additionally, the data types and data processing of thematic mapping features are included. The fourth through seventh chapters mainly discuss thematic mapping technologies, including thematic map design and preparation methods, and in particular, remote sensing mapping technology and computer mapping technology are discussed. The latter two chapters are about the specific practice of thematic mapping, in which the compilations of several commonly used thematic maps are introduced, including natural, cultural, economic, statistical analysis, statistical analysis and evaluation results, as well as the compilation features of a series of thematic maps and atlases.

(16) ***Principles and Methods of the Cartographic Data Processing Model*** (He Zongyi. Wuhan University Press, Wuhan, Feb. 2004).

This book consists of 13 chapters. The first chapter briefly introduces the development and application of the cartographic data processing model. The second chapter introduces the mathematical basis of the cartographic data processing model. The third, fourth, fifth, and thirteen chapters mainly introduce the mathematical model of geographic information generalization and multiscale representation. The sixth through twelfth chapters introduce the mathematical model of thematic map data processing, including the hierarchical model of map features, cartographic evaluation model, cartographic features correlation model, distribution trend model, prediction model, information simplified model, type partition model, and so on.

(17) ***Electronic Cartography*** (Long Yi, Wen Yongning, Sheng Yehua, et al. Science Press, Beijing, Aug. 2006).

This book is the first electronic cartography textbook in China. The whole book consists of 11 chapters. The first chapter discusses the basic theoretical knowledge and initially expounds upon the concept, research object, basic content and development of the electronic map. Technologies and methods of electronic cartography are introduced from the second to sixth chapters, including data organization of the electronic map, basic and thematic geographic data visualization, spatial analysis methods of the electronic map and the electronic map software system. From the seventh to tenth chapters, combined with related technologies, the applications of several types of electronic maps are described, including the concepts, characteristic functions and applications of the multimedia electronic map, internet electronic map, 3-D map and mobile navigation electronic map. In the eleventh chapter, the electronic maps commonly used in several professional fields are introduced, such as tourism, urban, maritime, disaster, archaeological, military electronic maps, and so on.

(18) ***Digital Thematic Cartography*** (Liu Wanqing, Liu Yongmei, Yuan Kansheng. Science Press, Beijing, Sep. 2009).

This book is divided into 3 parts, each containing 3 chapters and a total of 9 chapters. First, the book provides an overview of digital thematic cartography, focusing on the basic theory of thematic cartography, including the thematic cartography outline, common types and representation methods, map making, cartographic generalization, and so on. Second is the digital thematic mapping technology, which mainly includes the overall design of the thematic map, data processing, base map digitization, and the electronic map publishing system. Third is the digital thematic mapping practice, which introduces the compilation process of the commonly used thematic maps, including a brief introduction to the GIS thematic mapping software, physical map compilation practices and the human map.

(19) ***Hydrogeological Atlas of the People's Republic of China*** (Producers: State Administration of Geology, Institute of Hydrology, Geology, and Engineering Geology; publishers: China Map Publishing House, Beijing; Publication time: 1979).

This atlas is the first large-scale comprehensive professional hydrological atlas ever published in China; in this atlas, the hydrogeological data of China over the past 30 years are systematically summarized, and the hydrogeological conditions and characteristics of Chinese territories are fully reflected. The publication of the atlas is of great significance to China's planning, construction and layout, as well as scientific research, teaching, and international academic exchanges.

The atlas includes national, regional and provincial map groups and consists of 68 total maps.

(20) ***The Historical Atlas of China*** (Editor-in-chief: Tan Qixiang; Producers: Fudan University, Institute of History, Chinese Academy of Social Sciences, Institute of Archaeology, Institute of Modern History, Institute of Nationality Studies, Central University for Nationalities, Nanjing University, Yunnan University, and others. Publisher: China Map Publishing House, Beijing. Publication time: 1982).

This atlas is a general atlas aimed at showing the territories and administrative regions of past dynasties. In addition, this is the highest quality atlas with the most detailed content and fine printing among similar atlases both at home and abroad. Since 1974, an internal trial version was published once under the Chinese Map Society. In 1980, a decision was made to revise the atlas and release it publicly. From 1982 to 1988, it was published by the China Map Publishing House, and in 1992, the atlas was published in the traditional Chinese version by the Hong Kong Triple Bookstore.

The atlas is divided into 8 volumes according to different historical periods. The first volume covers primitive society, the Xia Dynasty, Shang Dynasty, Western Zhou Dynasty, Spring and Autumn Periods, and the Warring States time. The second volume covers the Qin Dynasty, Han Dynasty and Eastern Han Dynasty. The third volume covers the Three Kingdoms period and the Western Jin Dynasty. The fourth volume covers the sixteen kingdoms of Eastern Jin and the Northern and Southern Dynasties. The fifth volume covers the Sui Dynasty, Tang Dynasty, Five Dynasties and Ten States. The sixth volume covers the Song, Liao and Jin Dynasties. The seventh volume covers the Yuan and Ming Dynasties. The eighth volume covers the Qing Dynasty. The atlas is divided chronologically into 20 map groups for a total of 304 maps and 549 pages. The map contents include the distribution of known sites of the primitive society and other important sites; the regimes established by each nations and their territories, administrative areas, scope of activities, and ethnic distributions; all geographic names, which can be found in the records before the Qin Dynasty; all place names above the county level since the Qin Dynasty and the boundaries of the first and second level administrative regions; historical sites of the Great Wall, barriers and ferries, forts, valley roads, tombs, imperial courts (the capital), and so on; and the main rivers, lakes, mountains, coastline, islands, and so on. There is a total of more than 7 place names. All volumes have compilation regulations and place names indexes in both Chinese and English.

The atlas won grand prize of the outstanding achievement award in college and university philosophy and social sciences in Shanghai city (in 1984); grand prize of the outstanding achievement award in philosophy and social sciences in Shanghai city (in 1986); first prize for the first outstanding achievement in the humanities and social sciences research of national colleges and universities (in 1995); and a special award for an outstanding achievement by the Chinese Academy of Social Sciences.

(21) ***Atlas of Landforms in China*** (Editor-in-chief: Li Weineng; Editors: Wuhan Institute of Surveying and Mapping, Surveying and Mapping Press, Shanxi Provincial Bureau of Surveying and Mapping, Guangxi Zhuang Autonomous Region Bureau of Surveying and Mapping, Heilongjiang Provincial Bureau of Surveying and Mapping, Guangdong Provincial Bureau of Surveying and Mapping, Surveying and Mapping Corps of the Yellow River Water Resources Commission, Aerial Survey Group of the Ministry of the Coal Industry, Sichuan Provincial Bureau of Surveying and Mapping, and others. Publisher: Surveying and Mapping Press, Beijing. Publication time: Oct. 1985).

Under the leadership of the State Bureau of Surveying and Mapping, the Surveying and Mapping Press was responsible for organizing and editing the atlas. In addition, the atlas was compiled by authors based on a collection of a large number of domestic aerial photographs. Although the atlas is mainly used as a reference for large-scale (primarily 1:10,000 scale) landform mapping by aerial photogrammetrics, it also has a certain reference value for terrain representation on small-scale topographic maps.

This atlas consists of 9 parts. A total of 206 aerial photographic images, which reflect the main landform types in China, were collected. Among these images, 156 images of typical landform patterns or difficult to express landforms were selected and produced as topographic maps using aerial photogrammetry. To help the reader to more intuitively understand the landform content, 79 ground landscape photos are also provided in the atlas. Aerial photographs, topographic maps and photographs are accompanied by a brief text description, and the genesis, landform formation causes, and key points of surveying are introduced.

In terms of content selection and layout, the main considerations are outlined as follows: first, the content and layout must be clear to service large-scale topographic mapping and must focus on the landform morphological characteristics and the contour lines feature. Second, as much as possible, the basic types of landforms in China need to be fully reflected, and the readers should be shown the unusual and unique landscape types in China with aerial photographs and topographic maps. Third, in the content arrangement, the relationship between the genesis and shape of various landforms is emphasized; in terms of landform naming, consistency should be maintained with the name of the academic circle as much as possible. Fourth, since the difficulty levels of the expression of various landform types differs, the length of each section is not equal. Fifth, most of the topographic maps in the atlas are 1:10,000 scale; individual types, considering the actual situation, are at the 1:50,000 scale; for topographic maps, the coastal type, elevation points and contour annotations are not labeled.

(22) ***Atlas of Soil in China*** (Editor-in-chief: Xiong Yi; Producer: Nanjing Institute of Soil Science, Chinese Academy of Sciences. Publisher: China Map Publishing House, Beijing. Publication time: Jan. 1986).

This atlas attempts to summarize the research achievements of China's soil science over many years in the form of a map and systematically and comprehensively reflects the main types and distributions of soil in China, as well as the geographic characteristics of the soil basic properties and the soil distribution regions and utilizations.

Based on the national map and supplemented by the regional map, this atlas is focused on the combination of science and production. In addition to the topics reflecting natural laws, several practical maps are also compiled. In line with the principles of whole first and local second, basis first and derivative second, natural landscape first and amelioration second, the 32 maps in the atlas are divided into 4 parts. Part I is the guide map with a total of 5 sheets, including a map of the China administrative area, map of relief in China, map of annual average temperature and annual precipitation in China, and map of vegetation in China. Part II includes the soil maps, which are 9 in total, including the map of soil in China and soil maps of eight typical regions; the map of soil in China reflects the main types of soil and their wide area distributions and displays the soil resources in China; the soil maps of the eight typical regions show the regional characteristics of soil in China. Part II is the soil properties and soil parent material map, with a total of 14 pieces, including soil parent material maps, soil geochemical type maps, soil texture maps, clay mineral maps, soil organic substance maps, and a soil PH map, as well as maps of soil fertility factors, such as phosphorus, potassium and trace elements, and so on. Part IV is soil zoning map, with a total of 4 maps, including soil regionalization maps, soil salinity zoning maps, an erosion soil distribution and zoning map, and soil utilization status and zoning maps.

Considering attention to scientific rigor, the geographical distribution regularity of soil and its properties must be embodied in the various maps. In addition, there are some logical requirements for the division index and the representation method of each mapping unit. In addition, the representation method and map decoration technique are cooperatively used in the atlas design to achieve a legible and elegant atlas and to obtain good reading and application effects.

(23) ***Atlas of Endemic Diseases and Their Environments in the People's Republic of China*** (Editor-in-chief: Tan Jian-an; Producer and Publisher: Science Press, Beijing. Publication time: Jan. 1989).

A new comprehensive medical atlas is compiled in cooperation with the medical department to prevent and cure endemic diseases and to show the relationship between human diseases and the ecological environment. This is a macro, comprehensive, basic and intuitive atlas and the joint research results from multiple sectors and disciplines.

This atlas is numerous in terms of scales, map types, and map subdivision specifications. The map types are divided into area division type, categorical type, isometric line and dot types, statistical type, and so on. There are approximately 30 scales. There are full page, 1/2 page, 1/4 page, 1/8 page, and other map subdivision specifications. The plastic sheet (polyester film) imposition and plate-making process is applied to ensure the overprinting precision. According to the actual technical condition of the map printing factory, to achieve high quality, high speed, and fund savings without adding new equipment or new personnel, new printing technologies were created, and a multicolor mesh screen overprinting technique was realized.

The atlas contains 221 maps, 100 photos, and contains a variety of introductions to endemic disease research and statistical charts. The atlas is rich in content, diverse in expression, and the key points are prominent; this atlas is a nationwide and multidisciplinary comprehensive scientific work.

(24) ***Atlas of Economics in the Hunan Province*** (Editor-in-chief: Cheng Yunguang; Producer: the Compilation Committee of the Atlas of Economics in the Hunan Province. Publisher: Hunan Map Publishing House, Changsha. Publication time: Oct. 1989).

This atlas is composed of 98 pictures in the quarto dimension, with 97 pages and 8 groups, for a total of 477 maps at various scales. Among these maps, 6 social conditions group maps account for 6.12% of all maps, 16 natural conditions group maps account for 16.32% of all maps, 22 agriculture group maps account for 22.44% of all maps, 23 industry group maps account for 23.47% of all maps, 13 traffic information business services group maps account for 13.27% of all maps, 11 science, education, cultural, and health group maps account for 11.22% of all maps, and 4 regional economic group maps account for 3.06% of all maps. There are also 386 analytical charts and more than 33 words of back page text.

In the atlas, 9 scales are selected to form the following scale series: 1:2 million scale, 1:2.5 million scale, 1:3 million scale, 1:3.5 million scale, 1:4 million scale,

1:5 million scale, 1:6 million scale, 1:7 million scale, and 1:8 million scale. The map scales larger than the 1:4 million scale are interspersed with the 1:2.5 million scale and 1:3.5 million scale maps, with a good scale interval. Expression modes are set according to the contents, which are complemented by using a variety of statistical charts; the form of a suspended map is used, and there are no contour edge and no adjacent region elements by using the natural curves as the boundary. Legend design focuses on symbol line and size, and the base map has a clear hierarchy. The map arrangement is proceeded by a condition map (natural condition) as a guide, a sector economic map as the main body, and a general economic map at the end. The layout of tables also reflects the logic of the timing sequence, administrative areas, and so on, which agrees well with the characteristics of human cognition. The descriptions with attached back text are refined, summarized, and of very practical value. Two kinds of bookbindings are simultaneously adopted, stick-page hardcover and box-packed loose pages, which is very convenient and practical.

(25) ***National Agricultural Atlas of the People's Republic of China*** (Editor-in-chief: Zhou Lisan; Producers: Nanjing Institute of Geography, Chinese Academy of Sciences, Institute of Geography, Chinese Academy of Sciences. Publisher: China Map Publishing House, Beijing. Publication time: Dec. 1990).

As the first volume of the National Atlas of China, the preparation, completion and publication of the atlas is highly influential. By making full use of a large amount of information and a variety of mapping means, the atlas reflects the latest research results of China's agricultural situation and Chinese geo-science, agriculture and biology. Technically, remote sensing mapping, computer-aided mapping and traditional cartography were organically combined to form China's first comprehensive agricultural atlas, which is the combined product of science, practicality and artistry.

Atlas description: in the quarto dimension, 34 × 50.4 cm–sized; the basic scale for national maps is 1:10 million to 1:40 million, the map scale for typical areas is 1:50,000 to 1:200,000; a total of approximately 300 color maps, with a large number of agricultural economic statistical charts; and for reference, tables of major agricultural statistics in 1988 are listed in the end of the atlas. This atlas was published in 1990 and issued both domestically and to foreign countries.

The atlas consists of 5 parts: part I contains the introduction maps, which mainly reflect China's complete territory, administrative region, terrain, water system, agricultural population, nationalities, agricultural distribution of major dynasties, achievements in agricultural production, comprehensive agricultural regionalization, and so on; part II contains maps of natural agricultural conditions and resources, showing that the natural conditions and resources closely related to the agricultural production in China, and their spatial and temporal distribution patterns and characteristics are given; part III contains maps of social and economic conditions and the technical level of agriculture, including agricultural population, labor burden, arable land, agricultural output value per mu of cultivated land, income level and other maps; part IV contains maps of distribution characteristics and the production level of agricultural sectors, mainly reflecting the agricultural economic characteristics, and the present distribution, output and some varieties of agricultural sectors and crops; part V contains maps of agricultural land use, including the national agricultural land use map and agricultural land use maps of 32 typical regions, which reflects the rich and varied forms of agricultural land use in China from the macro and micro aspects.

(26) ***National Economic Atlas of the People's Republic of China*** (Editor-in-chief: Liu Yue; Producers: Institute of Geography, Chinese Academy of Sciences, State Planning Committee, State Economic Information Center, and Statistical Science Research Institute of the National Bureau of Statistics. Publisher: China Map Publishing House, Beijing. Publication time: Jun. 1993).

This atlas is an important part of the National Atlas, the first scientific reference atlas that comprehensively and systematically reflected the economic and social development at the time and its geographical distribution in China. The atlas is in quarto dimensions, including 265 maps and 250,000 words of description. The atlas is composed of 10 parts, i.e., introduction (3 maps), resources (28 maps), population (10 maps), economic layout (18 maps), agriculture (32 maps), industry (84 maps), transportation, post and telecommunications (16 maps), construction industry, urban construction and environmental protection (12 maps), business, foreign trade, tourism and finance (10 maps), education, scientific research, culture, sports and health (21 maps), and the comprehensive economies of provinces (31 maps).

In terms of representation method, different methods, such as the point location method, cartogram method, arrowhead method, iso-line method and dot method, are comprehensively applied to a variety of thematic maps to display the characteristics and geographical distribution of the country's economy and society at that time at multiple levels. In terms of symbol design, by making full use of the functions and characteristics of various visual variables, the objective reality of the mapping elements is described in a scientific way. Contour hypsometrics and hill shading are applied to landforms to shape the rich and varied terrain features in China. In terms of color planning, to reflect the characteristics of grave, generous, clear and harmonious, a vivid, harmonious and balanced color design is used to meet the requirements for legible and readable map contents, as well as a prominent and distinct theme.

The atlas was completed by simultaneously applying computer-aided mapping and manual mapping. That is, computer technology is used to establish the economic atlas database and complete a large number of mapping index calculations, statistical analyses and mapping data classifications; meanwhile, the graphical representation and editing of statistical data are completed by using the symbol library software. However, map compilation, map carving and map printing are completed by manual mapping. In map plate making, a computerized modern plate-making and printing apparatus, as well as test tools and methods are used to ensure the high quality of printing and publishing for the atlas.

(27) ***Climatic Atlas of the People's Republic of China*** (Editor-in-chief: Zhao Guocang; Producer: China Meteorological Administration. Publisher: China Map Publishing House, Beijing. Publication time: Oct. 1994).

This atlas uses the latest compiled climate data in China, summarizing the wisdom of many scientists and the long-term accumulation of scientific achievements. The atlas systematically represents the basic climatic conditions, climate resource distribution and climate change facts in China. This atlas has richer contents than the past climatic atlas. The map groups of physical climate, climate change and applied climate are newly added contents in the atlas. Through this comprehensive atlas, readers can find what they want according to different needs; meanwhile, decision makers can consult the atlas to obtain high-level information, which is helpful for the development of disaster prevention and reduction countermeasures, as well as comprehensive exploitation and utilization of climate resources.

The atlas comprises a total of 339 maps, including the following map groups: introduction, basic climate, physical climate, weather and climate, climate change, applied climate, and so on. Among these groups, the map group of basic climate includes temperature, precipitation, sunshine, humidity, cloud, wind, temperature, and so on, for a total of 162 maps; the map group of physical climate includes radiation balance, heat balance, water balance, and so on, for a total of 53 maps; the map group of weather and climate includes tropical storm, severe tropical storm, typhoon, cold wave and other 9 maps; the map group of climate change includes temperature, precipitation and so on, for a total of 12 maps; the map group of applied climate includes agricultural climate, engineering climate, and aviation climate, for a total of 99 maps. The conformal conic projection is applied to the professional contents of the geographical base maps, and the series scales of 1:8 million, 1:17 million, 1:2.5 million, 1:350 million, and 1:42 million are used for the atlas.

The atlas uses computer mapping technology for preparation and the electronic publishing system for plate making, with four color printings and wireless adhesive bindings.

(28) ***National Physical Atlas of the People's Republic of China*** (Editor-in-chief: Liao Ke; Producer: Chinese Academy of Sciences. Publisher: China Map Publishing House, Beijing. Publication date for second edition: March 1999).

This atlas is a large comprehensive scientific reference atlas; the first edition was published in 1967, and the second edition was published in 1999. The main introduction of the second edition is outlined as follows.

Compared with the first edition, the second addition innovatively contains basic scientific information and data with obvious extensions and updates in time and space, and there are more comprehensive illustrations of scientific discoveries in the twentieth century; additionally, the discipline arrangement is further deepened, and animal and plant maps increased significantly. New topics and contents were added to highlight the new progress and contributions in earth sciences in China over the past 30 years; there are innovations in the atlas design and compilation methods and technology through the introduction of advanced digital map compilation, plate-making equipment and software system, a full digital computer-aided design, and editing and plate-making technology was realized for the first time. The atlas is recognized as representing the highest level of the China atlas at that time, which is also the world's most advanced level of outstanding achievements.

The 4 folio atlas contains a total of 580 map pieces, 225 pages and 200,000 words of map descriptions. The atlas includes 5 parts, i.e., the introduction map, natural environment map, natural resources map, natural disasters map, natural utilization map and natural protection map. Among these maps, the natural environment is divided into 7 groups, that is, geological and geophysical maps, topographic maps, climate maps, land hydrology maps, soil maps, biological maps and marine maps; the natural resource maps includes 4 groups, namely, mineral resources, water, gas, soil resources, biological resources, and natural scenery tourism resources; the natural disasters maps include 5 groups, that is, geological earthquake disasters, climate disasters, soil erosion and soil degradation, harmful animals, life elements and endemic diseases.

In addition to the paper atlas, a multimedia CD-ROM atlas was published and a national physical atlas information system was established to adapt to the development trend of information technology. This atlas can be seen as a comprehensive electronic atlas that reflects China's natural, economic, transportation, ethnic and population statuses. This atlas uses modern high-tech means to organize a large number of maps, text, graphics, three-dimensional landscapes and other data, which comprehensively and intuitively reflect the basic national conditions of China. It is an important reference for design, planning, resource development, and transforming the deployment nature and production; simultaneously, it is a powerful tool for the public to understand the national conditions and enjoy the magnificent rivers and mountains of the motherland.

The first edition of the atlas won the National Major Scientific and Technological Achievement Award at the National Science Conference in 1978 and won second prize of the National Natural Science Award in 1987; the second edition won the excellent atlas prize at the International Cartographic Conference in 1999.

(29) ***Atlas of the Remote Sensing Survey of the Ecological Environment Situation in the Western Region of China*** (Editor-in-chief: Wang Qiao, Huang Jie; Producers: Information Center of the State Environmental Protection Administration, Chinese Academy of Surveying and Mapping, Environmental Monitoring of China, and China Environmental Science Research Institute. Publisher: China Map Publishing House, Beijing. Publication date: January 2002).

To cooperate with the implementation of the western development strategy and ensure the sustainable development of society, economy and environment in the western development, the State Environmental Protection Bureau and State Bureau of Surveying and Mapping jointly carried out a remote sensing investigation of the ecological environment in the western region of China, and a comprehensive database of the ecological environment in Western China was compiled. The survey data covered 12 provinces (autonomous regions and municipalities directly under the central government) in Western China. With the database, by analyzing, sorting and using computer mapping technology, compilation of the atlas was completed at the end of 2001.

The atlas consists of five parts, namely, China's ecological environmental background, ecological environment status and dynamic change, land degradation status, comprehensive evaluation of the ecological environment, and the ecological environments in typical areas. The atlas is a standard 8 folio atlas, containing more than 200 remote sensing images and thematic maps, and with a large number of statistical charts, photos and text descriptions. The atlas has a very good reference value for the understanding and study of the ecological environment in the western region.

The atlas was awarded the first prize in excellent map works by the Chinese Society of Geodesy, Photogrammetry and Cartography in 2004.

(30) ***Atlas of the Natural Disaster System in China*** (Editor-in-chief: Shi Peijun. Producer: Beijing Normal University. Publisher: Science Press, Beijing. Publication date: July 2003).

This atlas was compiled based on the regional disaster system theory; the purpose is to express the spatial–temporal differentiation rule of the natural disaster system in China through a map and to fully reveal the formation laws of major natural disasters, as well as the regional differentiation pattern and dynamic change process in China.

The atlas contains two parts of China's integrated natural disaster system map series and major natural disaster system map series. Part I includes disaster-expectant environments map group, carrying body map group, hazard factor map group, disaster situation map group and disaster relief map group. Part II includes the earthquake disaster map group, flood map group, typhoon disaster map group, snow disaster map group, sand dust storm disaster map group, hail disaster map group, and so on. This atlas includes a total of 431 maps and 50 statistical charts.

The basic geographical map is the China land map, which used the secant conic projection and the South China Sea Islands as illustrations. The map is composed of series scale maps of 1:12 million, 1:18 million, 1:2.5 million, and 1:36 million. In addition, there are normal cylindrical conformal projection base maps corresponding to the map of sea areas and the double standard parallel conformal conic projection base maps related to meteorological contents. To express the relationship between the types of disasters and their quantitative relationships, it has a strict system for the expression of thematic contents, map color system, symbol system design, map decoration design, and so on. Based on the disaster database and thematic statistical data and by using computer digital mapping technology and direct plate-making technology, the atlas was finished in only 7 months. As a representative work, it is a useful attempt for atlas production in the digital environment.

The atlas was awarded the first prize in excellent map works by the Chinese Society of Geodesy, Photogrammetry and Cartography in 2004.

(31) ***Map of the Whole Antarctic*** (Editor-in-chief: E Dongchen, Pang Xiaoping, Yu Liansheng. Producer: China Antarctic Survey Research Center, Wuhan University. Publisher: China Map Publishing House, Beijing. Publication date: 2003).

Based on more than 20 years of field investigations in Antarctica and an accumulation of abundant data, this atlas was completed to meet the needs of national polar scientific expeditions and research. This is an excellent map work that integrates ideology, science, artistry, timeliness, and practicability in one and is among the fine new map works produced by new technologies.

The scale of the map is 1:5 million, and the map uses the polar zenithal projection (standard latitude is 71 degrees south).

Map description: two full sheets of paper, large format; large amount of information, high visualization, available in English and Chinese; comprehensive and rich in content, map surface enrichment, contents of the main map and accompanying map echo and complement each other; harmonious and beautiful surface, bright in color, color design achieves the artistic effect of highlighting the theme, with distinct levels, natural, harmonious, rich and vivid; both the main map and the accompanying maps are carefully designed, the map surface configuration an symbol expression embody both scientific and artistic characteristics; the large format has 6 color printings with full paper sheets, and the accuracy and exquisite effects of the map are guaranteed.

The latest data at home and abroad has been collected for the map. The remote sensing image and traditional map are organically combined, and the visualization method of the 2-D graph or image is used to represent the rock surface topography, ice surface topography, bare rock distribution, Antarctic research stations, seabed terrain of the Southern Ocean and other elements. The remote sensing satellite images cooperate with the two contour line systems of the rock surface and ice surface, the terrain feature of the Antarctica continent is visually expressed, and the visual effect is good. In addition, the submarine topography in the Southern Ocean is expressed with the combined method of depth contours and color gradients. In the color design, by making full use of the visual effect of color, the color contrast and coordination is unified, the symbolic meaning of color is expressed, and so on, the color of the map is close to the natural color of the background, both lively and bright, and does not lose coordination.

The map won the excellent map works award at the map exhibition of the International Cartographic Conference in 2005, and the excellent map works award of Chinese Society of Geodesy, Photogrammetry and Cartography in 2004.

(32) ***Series Map of the Antarctic*** (Editor-in-chief: E Dongchen, Pang Xiaoping, Yu Liansheng. Producers: China Antarctic Survey Research Center, Wuhan University, Key Laboratory of the State Bureau of Surveying and Mapping of the Polar Surveying and Mapping Science. Publisher: Wuhan University Press, Wuhan. Publication date: August 2003).

This series is composed of 5 series scale maps, including 1:1000 and 1:2000 scale maps of the Chinese Antarctic Great Wall Station, 1:1000 and 1:2000 scale maps of the Chinese Antarctic Zhongshan Station, and a 1:20,000 scale map of Larsemann Hills. These maps are compiled and published to meet the needs of the Chinese polar expedition and polar scientific research. In addition, these maps are the important contents of the basic surveying and mapping of the Antarctic area during the period of the Tenth Five-year Plan of China and basic surveying and mapping support for the Antarctic survey.

The mapping area includes China's two scientific research stations in Antarctica and other thematic elements closely related to the Antarctic scientific expedition and research; the atlas is comprehensive and informative in content. The expression method is the hypsometric contour, coupled with colored hill shading and building perspective symbols, and therefore, the natural geographical and environmental characteristics in the Antarctic area are vividly and intuitively represented. The map surface and color design highlights the theme, has obvious hierarchical characteristics, and appears to be natural, harmonious, lively and rich. The annotation is in Chinese and English. The series map uses 4 color printings; the printing process uses advanced digital plate-making (Computer-to-plate, CTP) technology, the printing line is fine, the color is vivid and beautiful, and the effect is good.

The publication of the series map won honors for China in the area of international Antarctic science and cartography, which enhanced China's international status in the Antarctic research field. The atlas won the PEI Xiu Award of excellent map works from the Chinese Society of Geodesy, Photogrammetry and Cartography in 2006.

(33) ***Atlas of Chinese Major Natural Disasters and Society*** (Editor-in-chief: Ma Zongjin, Gao Qinghua. Producer: Disaster Comprehensive Research Group, Ministry of Science and Technology, State Planning Commission, and State Economic and Trade Commission. Publisher: Guangdong Science and Technology Press, Guangzhou. Publication date: February 2004).

Experts from the Ministry of Water Resources, China Meteorological Administration, China Earthquake Administration, State Oceanic Administration, the Ministry of Land and Resources, Ministry of Agriculture, and the State Forestry Bureau investigated and studied 7 major natural disasters, such as floods, weather, earthquake, ocean, geology, agriculture, forest and so on. Thus, by taking the nature and social nature of disasters as the key link, this atlas was compiled through unified standards, unified scales, and a comprehensive analysis.

The atlas is mainly composed of plane maps and sequence maps. It systematically reflects the spatial and temporal distribution of the hazardous and destructiveness of major natural disasters in China, natural disasters in various historical periods in China, the impact and destructiveness of natural disasters on the social economy in China, the current disaster reduction situation in China, the natural and social conditions and development tendency for natural disasters in China and forecasting of regional disaster risks. Then, considering the disaster situations and national conditions in China, combined with the need to enhance the ability of sustainable development in the country, comprehensive disaster reduction system engineering countermeasures were proposed.

(34) ***Tool Software for Multimedia Electronic Map-making-TeleMap Atlas*** (Developers: Du Qingyun and Cai Zhongliang; School of Resource and Environmental Science of Wuhan University).

This software adopts a unified system architecture and integrates many modern technologies, such as GIS technology, digital mapping technology, multimedia technology and virtual reality technology. This software has completely independent intellectual property rights. The electronic map product developed based on the software takes the visual digital map as the background, with text, photographs, graphics, sound, animation, video and other media as a means of expression. This map is a new type of spatial information product, which shows the complex appearance of cities, enterprises, tourist attractions and so on. It can be stored in external storage and can be supplied in the form of a read-only optical disk (CD-ROM) and network. This software can be run and used on a desktop computer, touch screen computers, PDA and other equipment. Based on the software, a series of application systems can be further produced, such as an optical disc electronic map, touch screen inquiry system, small desktop GIS, 3-D electronic map (including a shaded relief map, 3-D fly-through, 3-D information system, and so on), network electronic maps, PDA electronic maps, and so on.

So far, there have been dozens of electronic map products produced by this software, including the *Electronic Atlas of Shenzhen City* (available in an English edition), *Interactive Electronic Map of Hong Kong, Electronic Map Guide of Tainan City, Know-All in Wuhan City, Map Consulting System for the Shandong Provincial Governor, Atlas of the Hubei Province, Window of Hangzhou City, Know-All in Shiyan City, Window of Nanchang City, Electronic Map of Songyuan City, Atlas of the Rural Power Grid in the Hubei Province, "Words of the Dragon City—Tour of Taiyuan" Electronic Map, "The Sound of Traffic"—Traffic Inquiry System of the Hangzhou Traffic Radio Station, Electrical Supplying Marketing System of Changchun City, Square Multimedia Query System of the New Hangzhou Train Station, Electronic Map Touch Screen Inquiry System of Nanning City, Electronic Map of Jiaxing, China, Window of Hefei, Know-All* in Shanghai City, *"New Beijing Great Olympics" Electronic Map, MobileMap System, Network Electronic Map of the Xinjiang Water Supply Network, Network Electronic Map of Jiaxing City*, and so on.

According to the 2001 annual domestic GIS software evaluation, this software is the only software product from colleges and universities that received national commendation.

Acknowledgements Guo Qingsheng, Professor at the School of Resource and Environmental Science, Wuhan University, has reviewed this chapter and made recommendations for revisions.

References

Chen, F., & Xie, G. (1999). *Tourist map, electronic map and electronic tourist map// Development of geographic information system and electronic map technology* (pp. 105–109). Hunan Map Press.

Chen, Y. (1988). The development of system of our country since the founding of the thematic cartography. *Map,* (2), 8–10.

Guo, L., Bai, Y., & Wang, M. (1999). *Development of Chinese transportation and tourism electronic atlas// Development of geographic information system and electronic map technology* (pp. 180–183). Hunan Map Publishing House.

Guo, L., Li, L., & He, Z. (2002). Visual search processes and cognitive efficiency of statistical maps. *Editoral Board of Geomatics and Information Science of Wuhan University, 27*(6), 637–641.

Guo, Q. (1993). The design of thematic mapping decision support system. *Journal of Wuhan Technical University of Surveying and Mapping,* 18(Suppl.), 91–95.

He, Z. (1989). The principle of fuzzy mathematical multi layer synthetic evaluation and it's application of quality evaluation in thematic mapping. *Acta Geodaetica et Cartographic Sinica, 18*(4), 266–276.

Hua, Y. (1993). Determine map content of thematic map with expert system technology. *Journal of the Institute of Surveying and Mapping, 10*(1), 56–61.

Huang, R. (1994). Review and dynamic cartography for level of social development in china. *Journal of Wuhan Technical University of Surveying And Mapping, 19*(4), 339–345.

Huang, A., Jing, X., & Yu, H. (2003). The fuzzy optimization and modeling of the avion visual. *Journal of System Simulation, 15*(8), 1146–1148.

Ji, X., & Guo, Q. (2006). Design and implementation of statistical maping software in internet. *Journal of Geomatics, 31*(5), 47–49.

Jia, C. (2003). The summary of compilation of maps on special subjects. *Sci/Tech Information Development & Economy, 13*(9), 286.

Jiang, N., Bai, X., & Sun, J. (2007). Classification evaluation model of statistic data based on multi attribute decision-making. *Acta Geodaetica et Cartographica Sinica, 36*(2), 198–202.

Liao, K. (1979). The development characteristics and trends of the modern thematic map. *Bulletin of Surveying and Mapping,* (2), 1–4.

Liao, K. (2005). Mapping method of eco-environmental remote sensing complex series. *Acta Geographica Sinica, 60*(3), 479–486.

Liu, Y., et al. (1993). *National Economic Atlas of the People's Republic of China.* Beijing: China Map Publishing House.

Liu, J., et al. (1997). Atlas of Shenzhen City. Shenzhen:Shenzhen Municipal Planning and Land Source Bureau (For internal use only)

Liu, Y., & Liang, Q. (1979). Mathematical method of drawing automation. *Bulletin of Surveying and Mapping,* (1), 37–41.

Liu, H., Yang, Z., & Gu, H. (2007). Inquiry & appraisal on reasonable allocation of groundwater resources in Yinchuan plain by building space database. *Ningxia Engineering Technology, 6*(2), 187–189.

Long, Y., Wen, Y., Sheng, Y. (2006). *Electronic Cartography.* Beijing: Science Press.

Li, H. (1984). Preparation of Thematic Maps and Atlas. Beijing:Higher Education Press.

Liu, W., Liu,Y., Yuan, K. (2009). Digital Thematic Cartography. Beijing: Science Press.

Lu, Y. (1991). Development and prospect of diseace mapping in china. *Map,* (3), 21–23.

Lu, X. (1989). *Hierarchical Representation Methods of Statistical Maps.* Beijing: The People's Liberation Army Press.

Luan, X., Sun, Q., & Wei, D. (2007). Cognitive expression in thematic map making. *Bulletin of Surveying and Mapping,* (4), 69–72.

Ma, Y. (1993). On the classification of tourist maps. *Map,* (2), 56.

Mao, Z. (1989). Map symbol of thematic map. *Map,* (3), 21–23.

Ou, Z. (1995). Discussion on environmental map and its representation. *Map,* (2), 20–24.

Pang, X., Zhao, X., & Gao, F. (2006). Unification and harmonization in true color city image map. *Geomatics and Information Science of Wuhan University, 31*(6), 481–488.
Shi, P., et al. (2003). Atlas of the natural disaster system in China. Beijing:Science Press.
Su, S., & Su, S. (1991). Actual base and development of serial mapping research on resources and environment. *Yunnan Geographic Environment Research, 3*(2), 9–16.
Sun, Y., & Du, D. (1998). A pattern-plate-based technique for thematic mapping. *Journal of Wuhan Technical University of Surveying and Mapping, 23*(2), 171–174.
Sun, Y., & Wang, R. (1994). A new grading method and its application in grading of thematic elements. *Acta Geodaetica et Cartographic Sinica, 23*(1), 59–66.
Tian, J., Huang, R., & Guo, Q. (2007). Study on intelligent choice of representation methods in thematic map. *Science of Surveying and Mapping, 32*(5), 170–172.
Tong, Y., & Liang, Q. (1988). *Land resources information and cartographic representation.* Guangdong Map Publishing House.
Wang, M., et al. (1979). *Hydrogeological Atlas of the People's Republic of China.* Beijing:China Map Publishing House.
Wang, Y. (1991). An overview of remote sensing thematic computer-aided mapping. *Map, (2)*, 42–44.
Wang, J., Jiang, N., & Zhuo, J. (1991). The research on thematic map data processing system. *Journal of the Institute of Surveying and Mapping, 8*(2), 44–51.
Wang, M. (1986). Hydrogeological atlas of the people's Republic of China general drawing design. *Map,* (3), 23–27.
Wang, J., Zou, J. (1992). *Models and Methods of Data Processing in Cartography.* Beijing: The People's Liberation Army Press.
Wang, T. (1999). *Practical thematic map production system// Development of geographic information system and electronic map technology* (pp. 268–271). Hunan Map Press.
Wu, F. (1993). *Application of discriminant analysis method in comprehensive zoning map // China map academic year* (pp. 33–35). China Map Press.
Xu, Z. (1992). The fuzzy-grey situation decision in the classification of multi-factor thematic data. *Acta Geodaetica et Cartographic Sinica, 21*(3), 214–223.
Zhang, K. (1963). Preliminary discussion on various representation methods of special map. *Acta Geodaetica et Cartographic Sinica, 6*(3), 195–205.
Zhang, K., & Guo, R. (1987). Some discussion on the development of thematic maps. *Surveying and Mapping Information and Engineering,* (3), 52–54.
Zhang, L., Jiang, N., & Zhang, Y. (2008). Establishment and application of math-model base of thematic data-processing based on meta algorithm. *Journal of Zhengzhou Institute of Surveying and Mapping, 25*(1), 72–74.
Zhang, K., Huang, R. (1980). *Preparation of Thematic Maps.* Beijing: Surveying and Mapping Press.

Chapter 7
Geo-information Tupu

Qingwen Qi

7.1 Introduction

Geo-information Tupu is a group of digital maps, charts, graphs, curves, or images arranged in accordance with some regularities of index progressive changes or classification, which reflect the temporal-spatial information laws of Earth science. Tupu should consist of 3 parts: one part is a series of graphs (apparent or features, visual and abstract); the second part is descriptive parameters (numbers, code, etc.) and mathematical models; and the third part is the reorganization and virtualization of an object feature on the basis of modeling (Qi et al. 2001b).

Geo-information Tupu has the following characteristics:

(1) The difference between geo-information Tupu and classical "Earth science map (or pattern)" is outlined as follows: First, it is built on the basis of modern space technology and information science, so it is very rich in information; second, it is built on the basis of the Earth system science and geographic information system (GIS), and the generation process of the map is intelligent and automatic; and third, it is based upon the "cyber space" and "virtual reality", which is able to reproduce the past but also to provide a variety of future projections and possible solutions for decision-makers to make judgments (Chen 1998).

(2) Geo-information Tupu adopts the abstract generalization method in graphic thinking, and the data mining method in information analysis is a research achievement of the analysis and processing of a large amount of Earth science information (Zhou et al. 1998).

(3) Geo-information Tupu is a standardized framework for a certain region. For any individual phenomenon, the abstract mapping graph can always be found in Tupu (Qi et al. 2001b).

Q. Qi (✉)
Institute of Geographical Sciences and Natural Resource Research, Chinese Academy of Science, Beijing, China

J. Wang and F. Wu (eds.), *Advances in Cartography and Geographic Information Engineering*, https://doi.org/10.1007/978-981-16-0614-4_7

(4) Geo-informatics Tupu has all the functions of "spectrum or family tree", that is, the demonstration and explanation function, the classification and location function and the planning and guidance function for spatial–temporal information in the field of geoscience (Qi et al. 2001b).

The study of geo-informatics Tupu will have a certain role in promoting the development of geoscience, geo-information science and its branches, and it will be of great value in application. First, the study of geo-informatics Tupu can greatly deepen the summary, refinement, expression and application of the geoscience regularity; second, geo-informatics Tupu can provide a new way of thinking and method for geoscience and geo-information science; third, geo-information Tupu can create a new direction of development for modern cartography, which greatly enriches and improves the theory, method and technology of cartography (Qi et al. 2001b).

Geo-information Tupu is a new kind of methodology and high-level information product of geo-information science, and it has an important influence on many related subjects and fields. For example: from the point of view of surveying and mapping science, all aspects of accurate positioning and high precision digital mapping are the soul of geo-information Tupu. Therefore, the significance of geo-information mapping lies in the in-depth study and thorough understanding of the relationship between the decomposition and composition of geo-spatial elements. Further, a precise mapping "template" is established so that we can implement measurement and mapping in accordance with the template and specification, as well as flexibly decompose and reconstruct various elements. From the point of view of the development of information science, the role of geo-information Tupu is to further clarify and deepen the important role of information, namely, information transmission function, virtual simulation function, integrated control function, and so on. These are all tasks that can be accomplished by geo-information Tupu. From the point of view of geosciences (especially geography), for research convenience, the objective entities of geoscience must be decomposed according to the essential components to obtain a series of minimal elements that are similar to a biological gene, and these entities are called "geo-information Tupu units". Therefore, virtual reorganization is carried out to obtain research ideas on the integration issues of geoscience (Qi 2004).

7.2 Development Process

In the foreword of the book *Exploratory Study on Geo-Information Tupu* edited by Chen (2001), it was pointed out that the top-level design of geo-information Tupu was initially inspired by academician MA Junru. In the early stage of the planning and design of Chinese major applied basic research project (973) in 1997, academician MA Junru asked the following questions: why do you only locate geography in the "complex, open giant system", and can you use a simple expression for the complex geoscience problems? This concept is similar to the "gene map" in biological science research and the "periodic table of elements" in chemistry. That is, the "spectrum or family tree" will be studied in the field of Earth science.

Chen (2001) believed that a long time ago, there were ancient traditional research methods and modes for Tupu in the geoscience field, and he advocated for an excellent combination of maps and written words; these studies had some advantages and disadvantages, sometimes more pictures than text, and vice versa. In ancient China, there was a good tradition of using the vivid, visualized and abstract expression language of Tupu to describe the spatial distribution of natural and social phenomena. The representative works include the Jiuding Map, Pictures of Classic Mountains and Seas, the Fangma Beach Map, and the Mawangdui Map (including a topographic map, garrison forces map, and town maps). PEI Xiu's Six Basic Principles for Cartography is a summary of the theory and method of ancient Chinese mapping, which is the precedent of Chinese cartographic methodology. In the prosperous periods of the Tang, Song, Yuan, Ming and Qing dynasties, many maps were published. In modern times, the masters in the field of geosciences all have made great contributions to geo-information Tupu. Mr. LI Siguang's tectonic theory and Mr. ZHU Kezhen's climate, natural regionalization and climate change curve over five thousand years have guided China's geological prospecting and agricultural production for half of a century.

Along with the rapid development of information science, especially the spatial information science and technology, such as GIS, remote sensing (RS) and satellite navigation and positioning system (GPS), modern cartography professionals are wondering how to develop cartographic information Tupu in the information age. At this time, the proposition of cartographic information Tupu was put forward by academician CHEN Shupeng, and domestic experts and scholars were organized to discuss and experimental studies. In 1997, Shupeng was in charge of the Institute of Geography, Chinese Academy of Sciences; with geo-information Tupu as its theme, a state key development program for basic research in China (973 program) was submitted and the preliminary study was carried out jointly with Peking University and Zhejiang University; subsequently, a number of academic papers were published. At the meeting of the Fragrant Hills in 1999, Shupeng organized a free discussion on geo-information Tupu, and a certain consensus was reached; in 2000, he invited experts to discuss the matters involves in the startup phase. For example, based on the Chinese seismic database, the State Key Laboratory of Resources and Environmental Information System began to investigate the spatial combination pattern of China's earthquakes and to study the flow field characteristic pattern of tidal currents in the East China Sea. Since then, Qi et al. (2001a) studied the morphological information Tupu of landforms on the Loess Plateau, established samples for expressing six types of Tupu, including an individual unit morphological chart of gully land, composition unit morphological chart of gully land, pedigree chart of altitudinal belts, spatial distribution pattern chart of gully land, spatial distribution pattern chart of ravines, and restructuring chart of individuals/combinations of landform configuration. Liao (2001) studied the design of the comprehensive information Tupu of natural landscapes in China and constructed the integrated information Tupu system of natural landscapes in China. Zhang et al. (2003) proposed a landscape spatial structure change pattern based on the basic geographic information unit of a mountainous region, summarized 63 kinds of vertical belts in China for the first time, carried out a digital integration Tupu of 239 mountain vertical belts, and established a Tupu vertical belt information system. RUI Jianxun et al. studied the law of urban spatial expansion in the rapid urbanization process from the end of the 1980s to 2003, and the pattern and process of a single element landscape, such as urban green landscapes, urban thermal landscapes and high-rise building landscapes, were designed and developed into an urban landscape model system (ULMS) (Zhang et al. 2009).

Generally, the research on geo-information Tupu is still in the initial stage of exploration and experimentation.

7.3 Main Research Achievements

7.3.1 Research on the Theoretical Framework

1. **Theoretical basis of geo-information Tupu**

The first important basis of Tupu is the geoscience foundation. The Earth's surface is a complex, enormous system, where its multilayers ordered structure, flow law and mutual transformation law of material flow, energy flow and information flow comprise the most fundamental theoretical foundation of geo-information Tupu. Additionally, the spatial–temporal distribution pattern of geoscience phenomena and features is the most direct theoretical foundation of geo-information Tupu, which includes the spatial distribution pattern, temporal distribution pattern and spatial–temporal composite pattern. This pattern comprises the external appearance of the complicated law of the Earth's surface complex, enormous system, and the most suitable geoscience rules are represented with universal, abstract and standardized geo-information Tupu (Qi et al. 2001b).

The second theoretical basis of Tupu is the foundation of geoscience cognition. Geoscience cognition is the process of a human understanding and perceiving an objective entity and phenomenon and obtaining a disciplinarian conclusion and knowledge. The process of a human understanding a geographic object and the phenomenon is from the relative perceptual understanding of our senses to the formation of imagery and memory in the mind. Then, the formation of a rational understanding is exactly one or several stages of the induction and refinement process of geo-information Tupu. That is, the process moves from a large number of geo-phenomena perceptual observations to conclusions on the spatial–temporal pattern graph of certain kinds of phenomenon, then to the classification and formation of description parameters, and finally to modeling and virtualization. The result and model of geo-cognition, the scientific and correct inspection of geo-cognition and secondary cognition are all important parts of the theoretical basis of geo-information Tupu (Qi et al. 2001b).

The final theoretical basis of Tupu is the mechanism of geo-information. This mechanism is the theoretical information foundation of geo-information science, which includes the geo-information Tupu itself. This mechanism refers to the information transmission mechanism of the Earth's surface, and the mechanism for predicting and controlling the material flow, energy flow and population movement at the Earth's surface through analyses of information flow on the Earth's surface. In terms of geo-information Tupu, this mechanism is also described in three aspects: one is the formation mechanism of geo-information Tupu, i.e., how to form a pattern or Tupu similar to a gene spectrum; the second aspect is the process mechanism from macroscopic graphic thinking to the inversion of an intrinsic geo law and its macroscopic readjustment, i.e., how to inverse the inherent geoscience law and implement macro control by way of the appearance-based macroscopic graphic thinking; and the third aspect is the recognition mechanism of a map pattern and RS image pattern, i.e., how to use a pattern recognition method to automatically identify, summarize and extract the geo-information Tupu from the map and RS images, which is the only way for the study of geo-information Tupu during development toward the advanced stage (Qi et al. 2001b).

2. **Standardized pattern of geo-information Tupu**

The standard pattern of the geo-information Tupu is a series of continuous tables, similar to the tables in a database. Different from the general tables, Tupu not only has text description and numbers but more importantly, it has graphic objects in OLE file format. These graphical objects are connected to a certain object in a GIS graph database by means of mapping or hyperlinks. There are thousands of records in a Tupu table, and all records are in the same format, each of which corresponds to an instance of a ground feature unit in the field; we can query any records according to the record index and click to jump to the graphics library of the GIS. Therefore, the geo-information Tupu is a combination of a series of graphics, text, numbers and models extracted from a GIS database. The contents of the Tupu tables may vary with the type, subject and name of the Tupu, but in general, it should be composed of 4 components, namely, graphics and special symbols, descriptive digital parameters, mathematical models, information restructuring and a virtual model or scheme. These tables are described as follows by taking China's Water Network Information Tupu (Figs. 7.1 and 7.2) as an example (Qi et al. 2001b).

(1) Graphical series and special symbols: graphic series includes the orthographic projection (horizontal) contour of ground features (the plane conformation of watershed units), orthographic projection (horizontal) abstract pattern (the network topology configuration of watershed units and its pattern orientation), vertical profile (optional), and an abstract pattern of a vertical profile (optional). Special symbols are the symbols of geo-objects that cannot be represented by their specific forms.
(2) Numeric descriptions mainly reflect the relationship between the geo-objects; as examples, a first-order stream is 1, a second-order stream is 2, a third order stream is 3, and so on. The inner river is 11, an outflow river is 12, and so on.

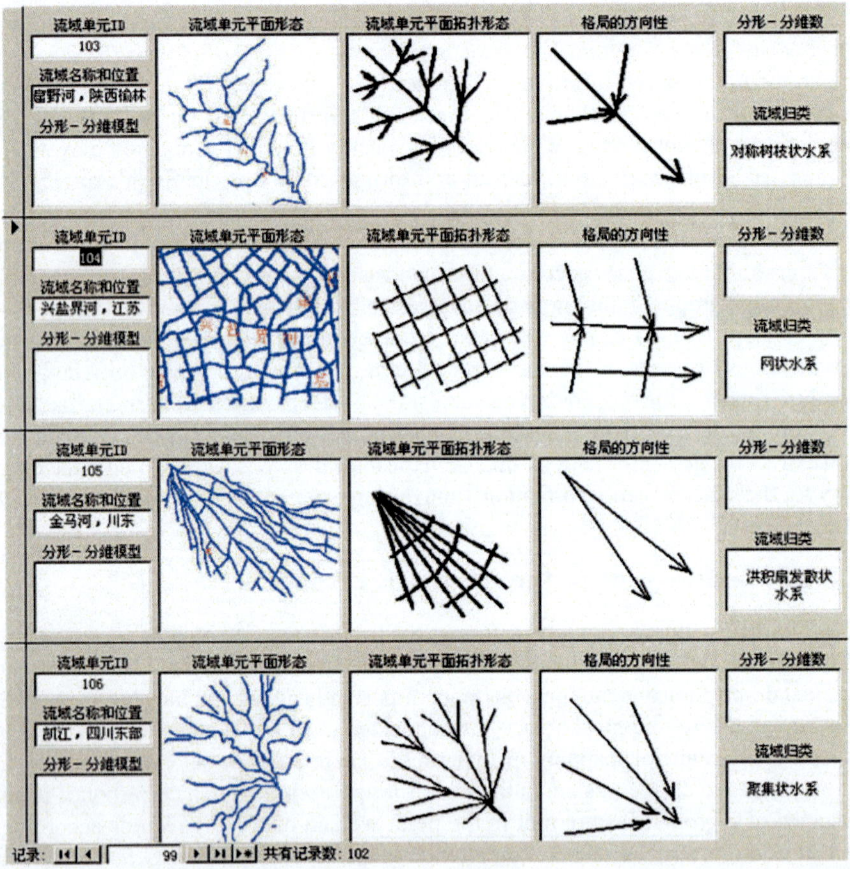

Fig. 7.1 The standard pattern of geo-info Tupu and compositional structure. Example 1: table of watershed units from China's Water Network Information Tupu, consisting of several fields such as the ID, name and position, horizontal shape, and topological shape of watershed unit, as well as the direction of pattern, fractals dimension estimation and model, and shape type, of watershed. The figure shows four records (rows) of watershed instances in the table

Additionally, the parameters reflect the spatial pattern of geo-objects, such as fractals and the fractal dimension.

(3) The mathematical model is composed of a variety of models reflecting the temporal-spatial distribution pattern of geo-features. For example, the fractal and fractal dimension models include the linear fractal dimension reflecting the distribution regularities of gullies and river systems, the area fractal dimension reflecting the distribution regularities of geomorphic characteristics of underlying surface of gullies, and the regression and trend surface model reflecting the relationships among average slope of river system, gully density and average downcutting depth, as well as the other models reflecting the contents of a river basin, such as aggregation intensity, distribution and aggregation area,

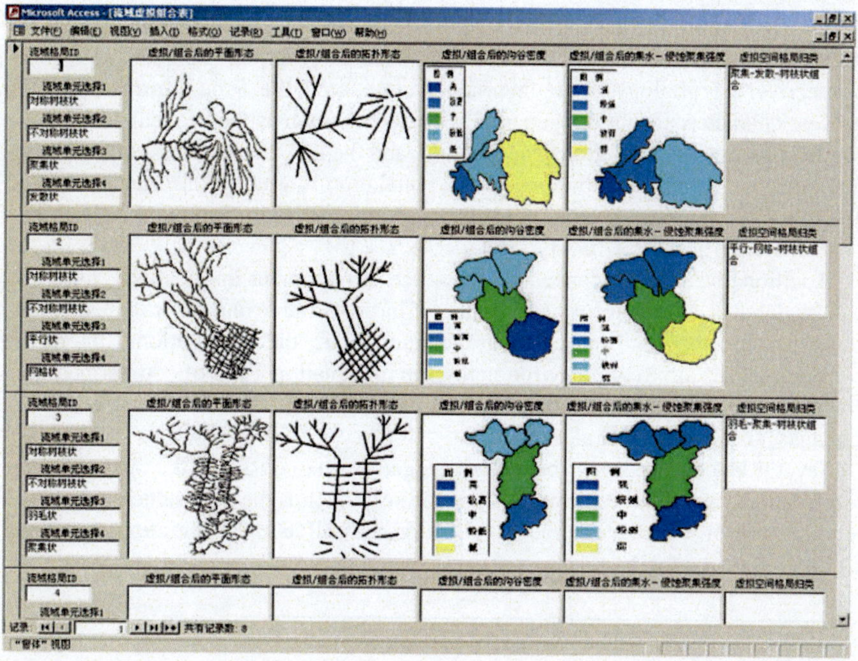

Fig. 7.2 The standard pattern of geo-info Tupu and compositional structure. Example 1: virtual combination table of a river basin from China's Water Network Information Tupu, which explains that a river basin could be virtually composed of three kinds of sub-watersheds, e.g., symmetrical dendritic, dissymmetric dendritic, concentrated pattern, and divergence pattern, therefore the new river basin would work with different 2D shape and topo-shape, different ditch density, different water catching density, and different soil erosion density pattern. The figure shows three records (rows) of different virtual combination choices in the table

aggregation range, spatial periodicity, measurement of the interaction between individuals and pattern orientation, and so on.

(4) Information reorganization and virtual model or scheme, which is the virtual model or scheme reflecting the morphological performance of the recombinant of individuals or combined forms of the river basin, and the ecological effects after the reorganization. A user can select a few forms (in this example, 4) from a large number of watershed units to carry out a virtual reorganization according to a certain rule, as well as to obtain the planar pattern, the topological morphology and the erosion water-harvesting/aggregation intensity pattern after the virtual reorganization, and then, the classification description of virtual reorganization is obtained.

Geo-information Tupu can be classified from different angles and according to different indicators.

For example, according to its functions, the geo-information Tupu can be divided into Omen Information Tupu, Diagnostic Information Tupu and Implementation

Tupu (Chen 2003); according to its geo-features, the Tupu can be divided into the information Tupu of geographical pattern (such as coastline, Loess Plateau landscape, river morphology, urban internal structure, etc.), the Tupu information of the process characteristics of the geographical process, such as the cyclical process, the sudden process, the evolution process, etc.), and the information Tupu of geographical behavior (spatial location selection, population migration, disaster prevention and mitigation measures, etc.), as well as the comprehensive Tupu of the composite characteristics of the above 3 types (Qi et al. 2001b).

Additionally, in accordance with resource environment applications, the Tupu can be divided into the information Tupu of basic resource and environment conditions, the information Tupu of geographic situations, the information Tupu of the evolution of resources and environments, the information Tupu of countermeasures for resources and environments, the information Tupu of interregional economic linkages, and so on (Qi et al. 2001b).

Several key factors must be acknowledged in the course of the study of geo-information Tupu. These factors include the scale factor, the hierarchical factor, the temporal factor, and the abstract level factor, as well as the integration of various factors (Qi et al. 2001b).

7.3.2 *The Research of Induction and Refining Methods*

For the study of geo-information Tupu, a variety of techniques (such as the exploration of the geoscience mechanism, the spatial–temporal analysis and data mining of geoscience information, computer pattern recognition, geo-information visualization, information reorganization, virtual reality, etc.) should be adopted to establish a geo-information Tupu system according to certain research, induction and extraction models and procedures; then, geo-information products with both basic theoretical research results of original innovation and practical application value should be produced (Qi et al. 2001b).

1. **General steps**

The induction and refinement of geo-information Tupu should be completed by several steps: first, in-depth study of geo-features (geoscience factors and phenomena, as well as geographical regions) should be conducted to master spatial and temporal patterns and rules; second, studies should be conducted to abstract basic components from the research objects, exhaustively describe the different geometrical configurations of these units one by one as much as possible, and form a series of pattern spectra; third, studies should address classification of the series pattern spectra and summarizing and extracting the abstract mapping graphs of Tupu and their standard types and grades; fourth, the mathematical parameter description of series graphs should provide the functions of quantification and formalization; fifth, the modeling of Tupu should be implemented to ensure it has the functions of computer pattern recognition and virtual reality; and finally, sixth, the reorganization and virtualization of information units should be implemented for practical Tupu applications to establish a control scheme for resources and environment problems and simulate predictive control results.

2. **Artificial visual induction and refining method**

In the early stage of geo-information Tupu research, there were the principal induction and refining methods. Because the concept and theory of geo-information Tupu has not been formed and a successful example has not been reported in public, the internal information mechanism of induction and refinement is not completely clear. The artificial visual induction and extraction method can enable the use of researchers' knowledge of geoscience and the thinking mode of the integrated sensibility and rationality. In addition, a large amount of geo-objects on a graph can be used to search for and discover information Tupu that represents fundamental laws.

3. **GIS/RS tool-aided induction and refinement**

This method is the only way for geo-information Tupu to further develop. Using GIS and RS tools to assist the induction and extraction of information Tupu involves using the powerful spatial analysis function of GIS for vector data and the powerful processing means of RS image processing system for raster image data, as well as to summarize and extract the information Tupu of geographic objects from the existing GIS database and RS images. At present, the potential methods include the spatial clustering method, neighborhood analysis method, map algebra method and mathematical morphology method. These induction and refining tools can help researchers to implement the regional division of geo-objects and the division of the landscape units, as well as helping to extract the standard form of the terrain units.

4. **Completely automated induction and extraction**

This method is the only way for geo-information Tupu to develop to an advanced stage. The method involves using computer pattern recognition and data mining tools to automatically summarize and extract the geo-information Tupu from the GIS database and RS images. This method involves two aspects: first, subjectively, human visual methods and all the knowledge, models and parameters of the GIS/RS tools are all input into the computer using an expert system and pattern recognition tool; second, objectively, computer artificial intelligence technology is highly developed, and the computer fully masters the thinking process of artificial induction and refining of information Tupu.

Funded by the national 863 project's frontier exploratory project in the information field, "Geo-information extraction, virtual reconstruction and their derivative technologies based on multisource spatial data" and by integrating pattern recognition, knowledge discovery in a database, GIS spatial analysis and RS image classification technology, we have proposed and successfully demonstrated a new method and technique for automatic or semiautomatic recognition and extraction of the spatial distribution patterns of geo-objects. That is, we use the algorithms of boundary extraction, conditional expansion, and connected unit area calculation in mathematical morphology, which is combined with a geo-object's texture recognition, clustering distance traversal calculation and other means. Simultaneously, with image grayscale enhancement, binarization, morphological filtering and other image processing algorithms, we identify geomorphic units, such as loess hill and loess ridge from RS images (Fig. 7.3) and refine the operation of the spatial distribution pattern. Finally, we use neighborhood search and structured assignment algorithm to identify gully and valley landforms from DEM data and refine the operation of the spatial distribution pattern (Qi 2004) (Fig. 7.4).

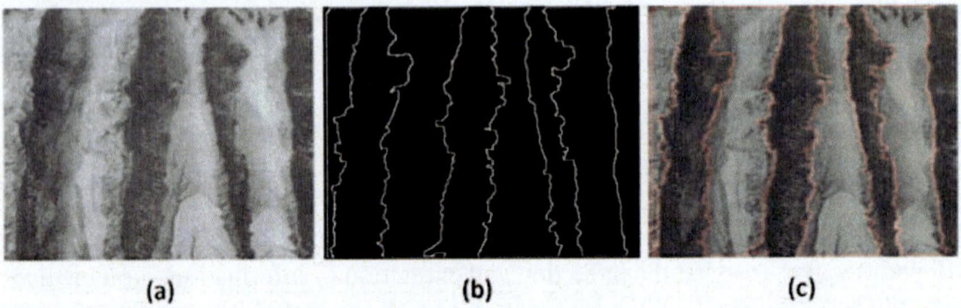

Fig. 7.3 Automatic/semiautomatic identification of Loess Plateau landform shape (beam) based on RS image, which shows the workflow from original RS image (**a**) to image edge extraction (**b**), and to the effect that extracted edge map overlaid on the image (**c**)

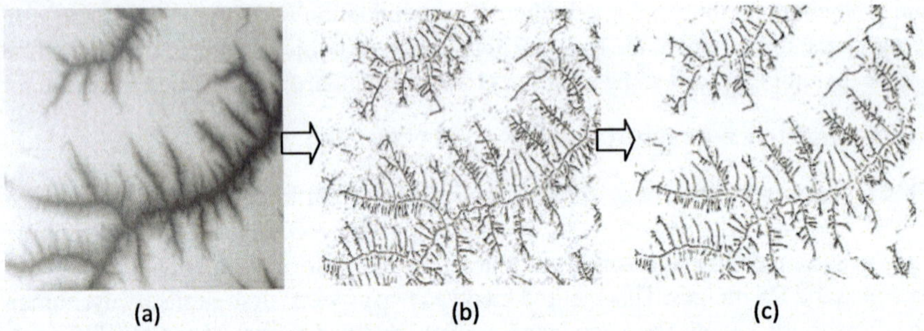

Fig. 7.4 Automatic/semiautomatic identification of Loess Plateau landform shape (beam) based on DEM, which shows the workflow from original DEM (**a**) to ditch extraction using morphological method (**b**), and to the effect of median filtered river map (**c**)

7.3.3 The Study of Virtual Reorganization and Derived Technologies of Geo-objects

Virtual reorganization of geo-objects is one of the main objectives of geo-information Tupu. The task is to carry out a virtual restructuring of the most basic components (which can be called geo genes) of geo-objects according to different schemes and to predict the ecological environment benefits after virtual reorganization in three-dimensional environments. Virtual reorganization uses the method of "first spatial modeling, then virtual reorganization" and is based on spatial–temporal distribution pattern recognition of ground objects and geo-information Tupu. In other words, this method draws on the function of the gene recombination of a biological gene map to extract a series of the most basic geo-object units (geo genes) and reorganizes them virtually in space. For example, by breaking the current space arrangement and combination pattern of geomorphic units on the Loess Plateau and rearranging the spatial combination relations among them, we can observe whether the ecological environment after reorganization develops in a benign or malignant direction. If a benign development occurs, we suggest that the Soil and Water Conservation Department transform the natural environment pattern according to this restructuring plan, and a variety of programs can be put forward for the production or planning department to choose from. This is the most powerful function that gives geo-information Tupu a powerful role (Qi 2004).

1. **Modeling**

Modeling is an important part of virtual reality. Based on the researcher's cognitive level and connection to the real world, the most essential characteristics from the ground objects or the objective entities (mainly appearance, anatomy, properties, etc.) can be extracted, and then, according to the categories used to establish a pattern or model with standardized features of physical shape, pattern, appearance and texture, modeling can be used to virtually display geo-objects.

Geo-information Tupu is a kind of information product, which is formed through the extraction, induction, and integration of shapes, patterns, and attributes of the objective entities on the Earth (including the Earth itself). Of course, geo-information Tupu is also a kind of methodology, that is, the process of induction and production of geo-information Tupu can be used to explore, mine and discover a large number of geo rules and phenomena. To carry out the virtual reconstruction of thousands of basic geo units or objects in geo-information Tupu, we must adopt the above modeling technology to build 3 dimensional models of the objects, and the techniques are described as follows:

(1) Extraction and classification of an object's appearance, anatomy and attributes;
(2) Based on the above features, the 3 dimensional models of various objects are constructed, including the construction of a regular entity shape algorithm based on CAD, the construction of a complex model based on a realistic terrain surface, and the construction of a virtual appearance model based on the unified shape and spirit;

(3) Based on the modeling of large geo-object classification, the parameter adjustment scheme and mechanism of individual variation can be determined so that we can easily and flexibly create a variety of geometric object models.

2. **Virtual reorganization**

The virtual reorganization of geo-objects involves selecting different types of objects (units) from the basic object features (units) of the thousands of established Earth science research fields and reassembling and arranging the selected units according to the design requirements of the virtual spatial–temporal pattern in the study area. Then, a new spatial pattern is created. Virtual reorganization can be designed and arranged in the direction of a spatial pattern; for example, a small second-order watershed containing different morphological types can be placed along the upper, middle and lower reaches of a first-order watershed. Further, different geo-objects can be arranged in the primary and secondary pattern directions to form a compound pattern consisting of the primary and secondary patterns and a lower level pattern in the region; as examples, compound barchan dunes as a sandstorm accumulation landform, composite beams in a gully landform on the Loess Plateau, and so on. According to the shapes of geo-objects formed by different development stages and in view of the differences in comprehensive regional natural landscape types, geo-objects can be placed forward, backward, lateral, oblique or arbitrarily to construct spatial distribution patterns (Ji et al. 2005). Of course, the premise of virtual restructuring is to follow an objective law; therefore, there may be a very absurd result if we casually reorganize geo-objects contrary to the objective law and the original purpose of the virtual reorganization.

The difficulty in the virtual reconstruction technology lies in the choice of the virtual reorganization scheme and the adoption of the virtual reality technology.

3. **Forecasting and assistant decision-making**

Ecological effect predictions and control of policy-making after the virtual reorganization are the highest goals of geo-information Tupu, which is also the most difficult content. The research and treatment process is outlined as follows:

(1) Understanding and mastering the law of the evolution process of the research objects derived from conventional research methods;
(2) Understanding and clarifying the evolution law of the research objects, which cannot be revealed by conventional means;
(3) Selecting an appropriate virtual reorganization scheme and implementing virtual reorganization; then, based on the evolution law derived from the conventional research methods, the results of the ecological evolution in the new state are predicted and evaluated, where the ecological environment is determined to be "improving", "worsening" or "not changing".
(4) Based on the conclusion of the above evaluation, a decision is made whether to adopt the current virtual restructuring scheme. If the conclusion is "yes", an ecological environment regulation and control scheme will be created. If the conclusion is "no", then the virtual reorganization will be reimplemented,

and predictions of the ecological environment will continue until an ideal conclusion is reached.

7.3.4 An Application Case Study

1. **Overland flow and a loess tableland, loess ridge and loess hill**

According to Qi et al. (2001a), with color infrared aerial photos (the scale is approximately 1:35,000) as the main source material and combined with 1:50,000-scale topographic map and DEM, automatic and semiautomatic extraction methods were used to induce and extract the landform shapes of single features, such as loess tablelands, loess ridges and loess hills over the whole Loess Plateau. Additionally, the overall pattern combination/virtual spectrum was obtained after reorganization, and the soil erosion, water and soil conservation measures were planned and other issues were forecasted on the Loess Plateau (Chen et al. 2004a, b).

The topographic information Tupu of the Loess Plateau is composed of 4 parts, that is, graphics and special symbols, descriptive numerical parameters, mathematical models, information restructuring and a virtual model or scheme. There are 9 kinds of Tupu expressions, which are described as follows:

(1) (Gully land) expression of an individual unit pattern, including horizontal shape contour, horizontal abstract contour, vertical profile shape, vertical profile abstract graph, three-dimensional graph, three-dimensional abstract graph, categorization, description parameter and location (latitude and longitude). Figure 7.5 shows the unit pattern expression for loess hills, and the legend shows the information Tupu contents of several typical loess hills, such as a gentle loess hill (pot cover shape), a turn hill, and a round hill (steamed bun shape). Figure 7.6 shows the unit pattern expression for loess ridges, and the legend shows the information Tupu contents of several typical loess ridges, such as a wide ridge, a feather ridge, a rib ridge, and a gentle ridge. All these features can be used in the dynamic localization, virtual reorganization and other functions of massive morphological units.
(2) Sectional view, 2-D and 3-D stereo shafting figures, categorization, description of location, parameters and classification.
(3) Altitudinal belt spectrum: the aspect graph, abstract graph of vertical profile, the 2-D and 3-D stereo graphs, and abstract graph.
(4) (Gully land) the table of spatial distribution patterns: aggregation intensity, distribution and aggregation area, aggregation range, spatial periodicity, measurement of the interaction between individuals, and pattern orientation.
(5) Individual gully table: horizontal form, vertical profile form, 3 dimensional graph, description parameter, and categorization.
(6) Spatial pattern table of valley: locations, planar shape of watershed unit, planar topology of watershed unit, valley density graph, downcutting depth graph,

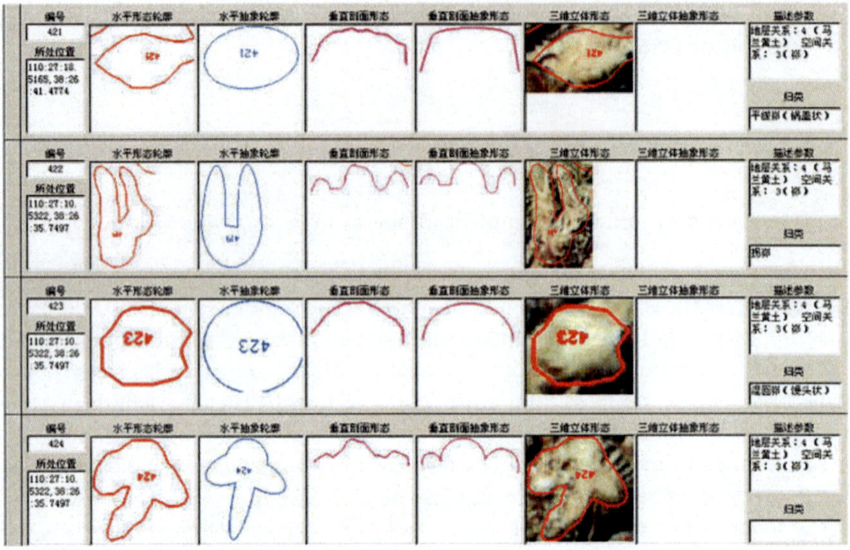

Fig. 7.5 Table for individual loess replat unit pattern from Loess Plateau landform Information Tupu, consisting of several fields such as the ID, name and position, horizontal shape, abstract horizontal shape, vertical shape, abstract vertical shape, 3D shape, abstract 3D shape, and shape type, as well as the descriptive parameter, of loess replat unit. The figure shows four records (rows) of loess replat instances in the table

water-harvesting/erosion aggregation intensity, pattern orientation, fractal dimension, and pattern classification.

Figure 7.7 shows several types of composite valleys and the planar topology shape, pattern direction, valley density, water-harvesting/erosion intensity and other information.

(7) Time series and pattern tables: temporal distribution configuration, time series, time interval position, and time identification code.

(8) Table of spatial and temporal patterns: spatial and temporal compound type, the temporal distribution pattern of both individual and combination spatial configurations.

Figure 7.8 shows the horizontal virtual configuration, vertical virtual configuration, stereo virtual configuration and other information Tupu contents in two cases (Record 1: tableland ID = 3, hill ID = 151, ridge ID = 51, watershed type = arborization; Record 2: tableland ID = 5, hill ID = 234, ridge ID = 122, watershed type = parallel shape), which is selected from the Tupu tables of loess tableland, ridge, hill and individual valley.

The landform types on the Loess Plateau are very complex, as the plateau contains gully landforms (including tableland, ridge, hill and loess plain and so on), valley landforms (including groove, rill, shallow trench, hanging gully, erosion ditch, dry ditch, graff, rivers, etc.), and loess gully slope landforms (loess slipping, collapse

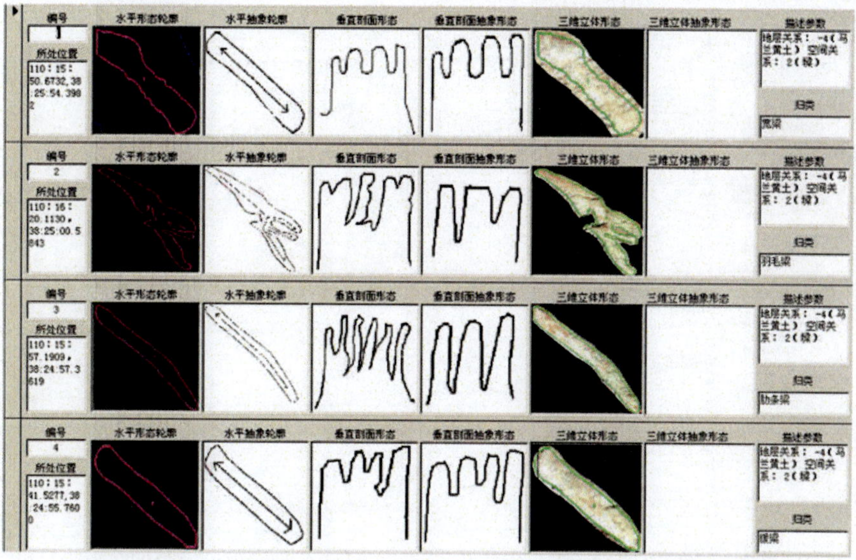

Fig. 7.6 Table for individual loess ridge unit pattern from Loess Plateau landform Information Tupu, consisting of several fields such as the ID, name and position, horizontal shape, abstract horizontal shape, vertical shape, abstract vertical shape, 3D shape, abstract 3D shape, and shape type, as well as the descriptive parameter, of loess ridge unit. The figure shows four records (rows) of loess ridge unit instances in the table

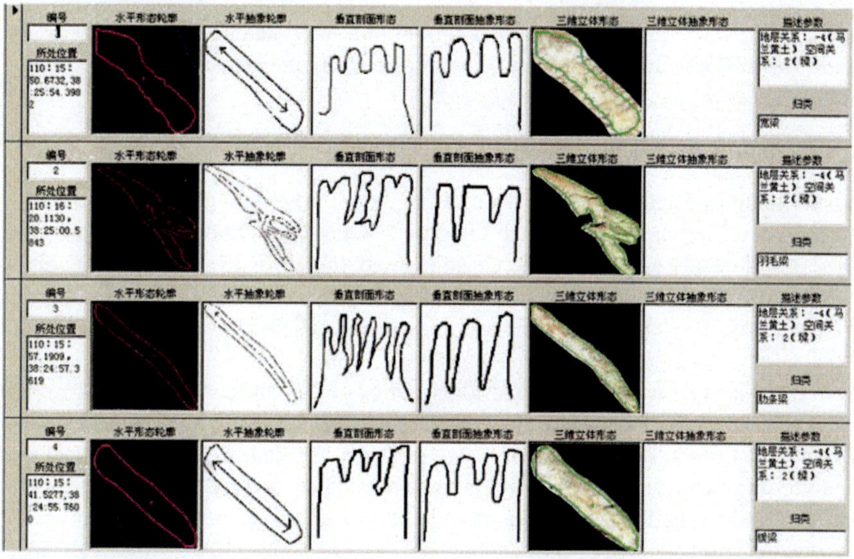

Fig. 7.7 Table for valley unit pattern from Loess Plateau landform Information Tupu, consisting of several fields such as the horizontal shape, horizontal topological shape, pattern direction, valley density, dinting depth, catching and soil erosion density pattern, of loess ridge unit. The figure shows three records (rows) of loess valley unit instances in the table

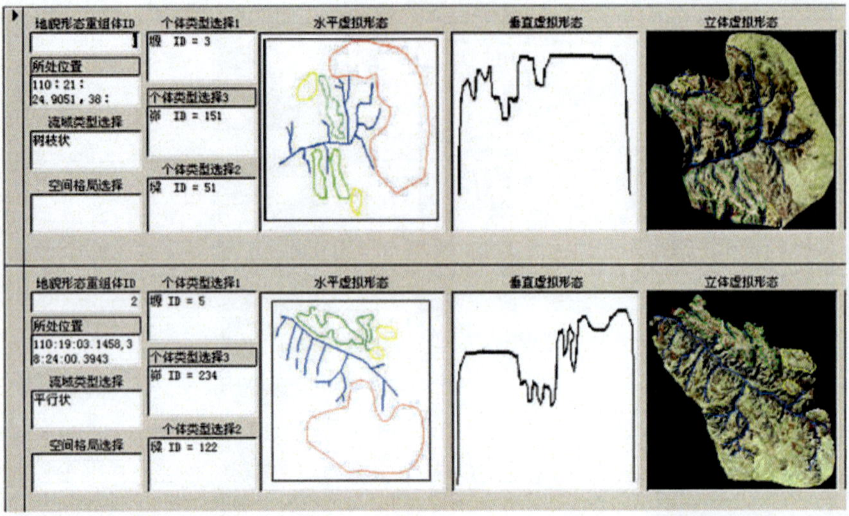

Fig. 7.8 Table for virtual reorganization of landform from individual/assemblage units from Loess Plateau landform Information Tupu, consisting of several fields such as the ID, name and position, loess landform units selection (from loess tableland, replat and ridge unit instances shown in the above tables), watershed type selection, pattern selection, horizontal virtual shape, vertical virtual shape, 3D virtual shape, of virtual loess reorganization unit. The figure shows two records (rows) of loess reorganization unit instances in the table

and landslide). In addition, in the background of above landform types, there are landforms of loess suffosion erosion, such as dish-like depressions on loess, loess sinking, loess bridges, loess columns, etc.

2. **Swinging melody of a riverbed**

When the middle and lower reaches of a large river flow into an alluvial plain, the riverbed development is less constrained by boundary conditions, but the general trend is still deeply affected by the Coriolis force; therefore, the local river sections swing freely among the nodes, similar to a group of plucked strings. This phenomenon is considered to be "the pendulum has swung back", where it moves back and forth, showing obvious regularity in the hydrological Tupu.

In LIN Yishan, according to the resultant action of the bend curvature, the riverbed in the alluvial plain of the middle and lower reaches of the river system in China was divided into three categories: the Yellow River type, the Yangtze River type, and the Heilongjiang type. The Yellow River is sandwiched between levees and with a wide and shallow riverbed and small longitudinal gradient, sediment pollution migrates rapidly, and the channel line frequently swings. LI Yizhi, QIAN Ning and other experts have conducted long-term research and put forward many control schemes of clearing sands with a clamping flow. During the 1985–1989 period, in cooperation with the Yellow River Conservancy Commission, ZHOU Chenghu, et al. (1998) used color infrared aerial RS images and the water conservancy's topographic map to analyze the geo-information Tupu of the riverbed and delta below the Sanmenxia Dam on the Yellow River (Fig. 7.9).

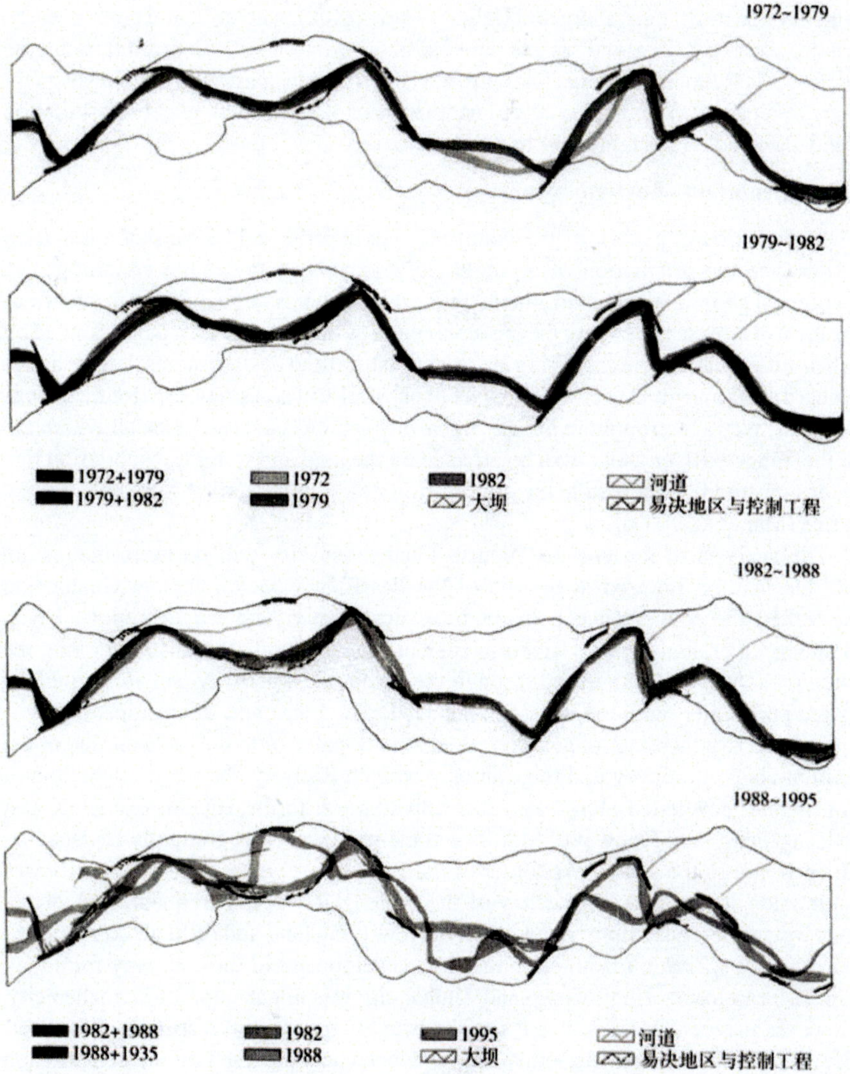

Fig. 7.9 Info-Tupu for the evolution of downstream channel of Yellow River (from Chen 1998), showing the swinging pattern of riverbed driven by Coriolis force in four periods of 1972–1979, 1979–1982, 1982–1988, 1988–1995, with different color representing different year, therefore display on the whole the spatial pattern between riverbeds and dams, and will be valuable for the control and adjustment projects in the downstream channel area of Yellow River

The string vibration rule is significantly reflected in the hydrological Tupu and contributes to traffic line optimization. In the case of traffic lines in Eurasia, during the age of voyages, humans had to bypass the Cape of Good Hope or break ice to travel from the Arctic to China; during the railway era, the Novosibirsk railway, which connects St. Petersburg and Dalian (Vladivostok), was built; and until recently, communication cables and oil gas pipeline transportation still occurred through the ancient Silk Road, the Caspian Sea and the Mediterranean coast, forming a convenient logistics channel (Zhou et al. 2001). All these examples reflect the chord oscillation trend and gradual straightening of routes.

3. **Orientation of urban expansion**

The urban growth pattern of the concentric circle proposed by Van Duren is used to describe the orientation of a single city's expansion under the condition of a theoretical geographical environment field, which extends outward from the inside in the form of a circle layer: with the business center as the core, the area is surrounded by urban areas, suburban areas, living areas, agricultural and pastoral areas, leisure areas, orchard areas and grain crop areas. Objectively, the ideal and single environment does not exist, even a metropolitan like Shanghai city, which is located in the alluvial plain of the Yangtze River Delta, will be affected by the land and water transportation line extension trend. As a result, the concentric circles are extended into amoeba-like forms (Mai et al. 1997).

The analysis of the growth Tupu of Guangzhou city will be more interesting and fascinating. Located at the top of the Pearl River Delta, ancient Guangzhou controlled the convergence point of three upstream rivers: the Dongjiang River, Beijiang River, and Xijiang River; in ancient times, the inland transport among the three rivers was communicated through the Sixianjiao waterway and Huangpu Port as sea port. Later, with the development of electric boats and steamships, the Sixianjiao waterway was no longer navigable. Additionally, with the construction of the Canton-Hankou railway and the Canton-Kowloon Railway, the main orientation of Guangzhou developed along with the railway north to the Huadu, Conghua, east to Zengcheng, and Dong Wu. Maritime trade in Macao was gradually replaced by Hong Kong. In the southwest corner of the Pearl River Delta, due to river tributary bifurcation in the Lower Reaches of the Xijiang River, the frequent river crossings were time-consuming; in addition, the low-lying land and difficult communication and transportation resulted in the slow development of the economy for a long time. Macao lost to Hong Kong, and Zhuhai city was not as good as Shenzhen city. However, in recent years, highways and railways in the southwestern part of the Pearl River Delta have flowed smoothly and have been accompanied by the construction of the Zhuhai airport and seaport, Macao's Return to China as the city-connecting Shunde city and Guangzhou city, the western movement of mulberry base agriculture and the cooperation between the Pearl River Delta and Xijiang Industrial belt. Thus, Guangzhou city has regained a full range of development opportunities (Fig. 7.10), playing the role of the core city of the Pearl River Delta (Chen 2001).

The layout of Beijing's inner and outer cities includes Tiananmen as the center, and the north–south axis and the East and West Chang'an Avenue form a chessboard-like

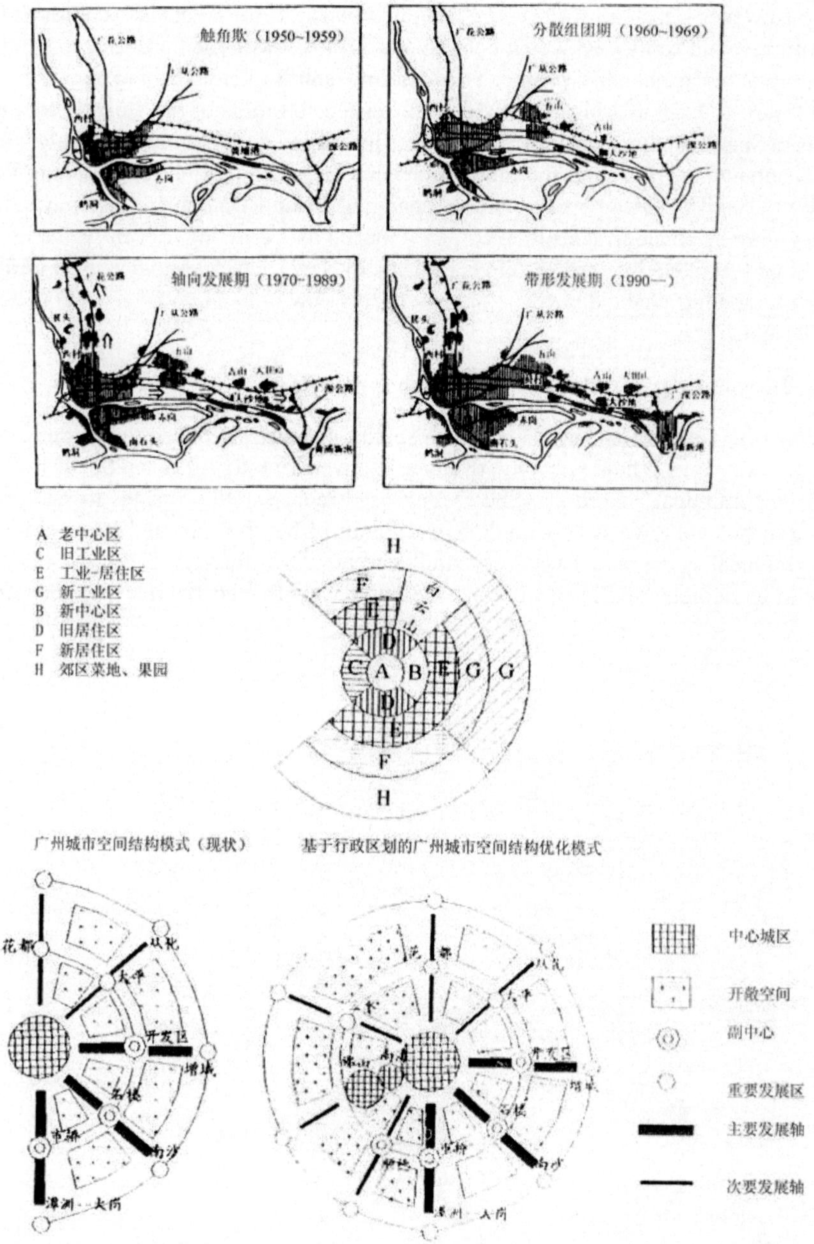

Fig. 7.10 Info-Tupu for historical development of Guangzhou City and its orientation change, showing the spatial pattern and orientation change in four periods of 1950–1959, 1960–1969, 1970–1989 and after 1990, as well as the present and optimized urban spatial structure mode of Guangzhou City. The Tupu is valuable for spatial optimization on the way of urbanization in the interested area

grid. However, considering the entire Beijing urban area, the layout also conforms to a northeast and southwest sketch contour and expands with the northwest to southeast water surface channel profile. The Ancient Central Plains Channel mainly runs northward to Lugou Bridge ferry along the eastern foothills of the Taihang Mountains in Shandong, moving into Yanjing and then to the northeast through Gubeikou or Shanhaiguan. However, the modern channel starts in Tianjin, moves along the Beijing Tianjin railway, and connects Zhangjiakou and Datong by highway. The two groups of channels constitute an "X" - shaped axis crossing. Recently, the joint development of Beijing and Tianjin in the Tianjin New District and convenient traffic construction has enabled the development of this new orientation (Fig. 7.11) (Chen 2001; Wu 2005).

4. **The point, axis and symmetry of the town system**

Towns originated from settlements. The spatial distances between the settlements originated from the field operation distance and were greatly restricted by the traffic and topographical conditions. SHAO Yun used radar satellite images to interpret the county-level town system on the Great Plains of North China and found that its geographical space layout was very similar to regular hexagonal cells, and exceptional amendments occurred only when the layout was affected by a river or mountain.

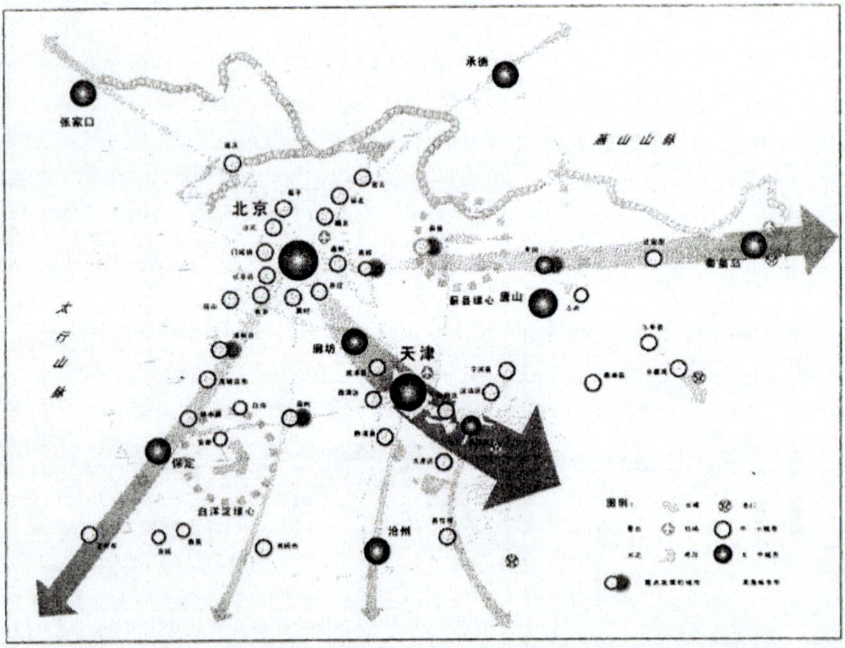

Fig. 7.11 Info-Tupu for the design of the development planning in Beijing-Tianjin-Hebei region, showing the spatial pattern and development orientation in the region, which is valuable for spatial development planning strategy in the interested area

Another study of towns in the Jiangsu Province also showed a geographical spatial distribution that was similar to the pattern of an equilateral triangle. These studies showed that for those regions similar to the North China plain, the pattern was quite similar (Chen 2001).

The concept of crystal structure was introduced into the analysis and research of urban systems by YE Danian, a famous mineralogist. Symmetric, anti-symmetric and oblique symmetric phenomena were revealed in many urban systems. He pointed out that in the Hunan and Jiangxi Provinces, a symmetrical distribution with the Wugong and Luoxiao Mountains as the central axis was formed due to the similarity of the lakes, basin and water system. His analysis of the Indian subcontinent resulted in similar conclusions. In addition, he also pointed out that interesting oblique symmetries existed in the Shanxi Province and Fujian Province (Chen 2001). These novel insights are inspiring creative ideas and represented a successful attempt to introduce natural scientific rules and methods into the humanities (Figs. 7.12 and 7.13).

5. **Differentiation of landscape**

A landscape represents the unity and harmony among nature and humanities and is the product of regional coordination and sustainable development. Since the 1930s, from the German Landscape School to the St. Petersburg Landscape School of the former Soviet Union, the research of landscape science has been debated, which has

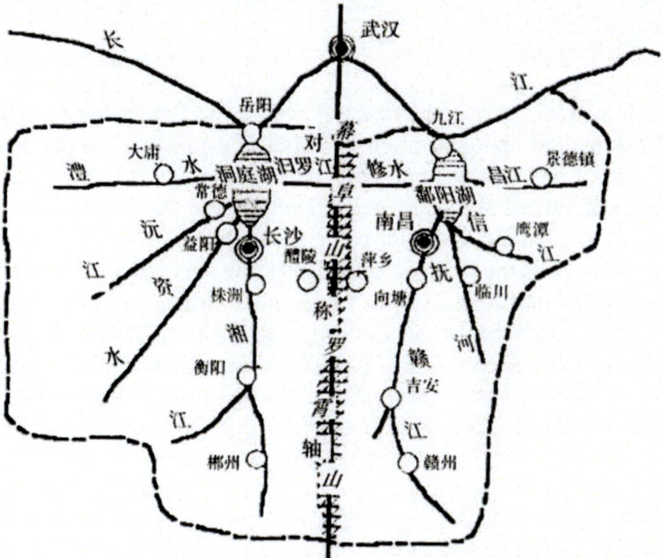

Fig. 7.12 Info-Tupu for the axial symmetry in the Hunan and Jiangxi Provinces (from YE Danian), showing the spatial axial symmetry pattern between Changsha and Nanchang, Yueyang and Jiujiang, Zhuzhou and Xiangtang, Lilin and Pingxiang, Hengyang and Ji'an, Chenzhou and Ganzhou, and so on, which is valuable for the spatial planning strategy of China's city clusters and belt

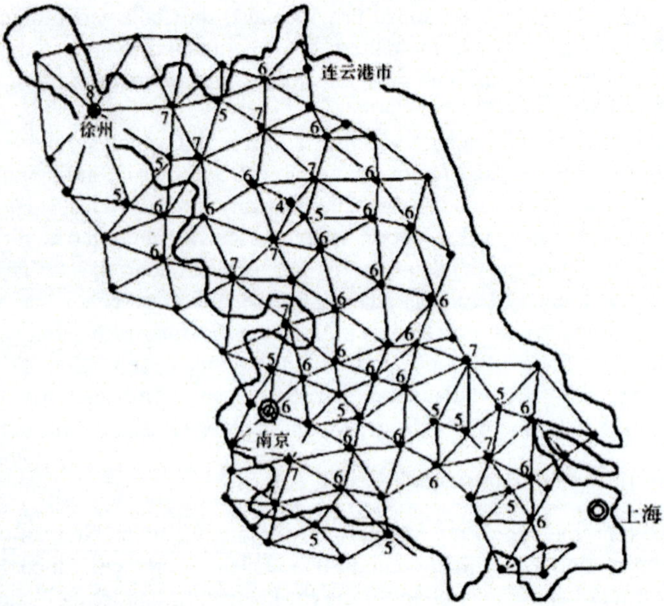

Fig. 7.13 Info-Tupu for the distribution of counties in the Jiangsu Province and coordination chart (from YE Danian), showing the approximate equilateral triangle spatial pattern of cities in Jiangsu Province, which is valuable for the spatial planning strategy of county system in northern China and related regions

influenced China, especially in the fields of ecology and soil science. In the 1950s, combined with natural regionalization in China, Chen (2001) led the organization of the multidiscipline landscape mapping method experiment in the East and West Dongting Hills of Taihu Lake. They pointed out that the landscape has an objective and multilevel form, but the landscape cannot be infinitely divided. Taking the material and energy cycle as the basic unit, the landscape structure of the local hills and valleys was divided into four level landscape units, which can be connected with the natural zones and physical geographic regionalization of China. An analysis was conducted on the gradual change law of the speckle pattern of plant and economic crops in the sub-tropical zone, which showed the development direction of Bi Luo Chun tea and the citrus industry (Fig. 7.14).

7.4 Problems and Prospects

Although there is a long history of traditional research methods and methods like geo-information Tupu in the field of geoscience, the geo-information Tupu in the information age has only been in use for just over ten years. During this period, the

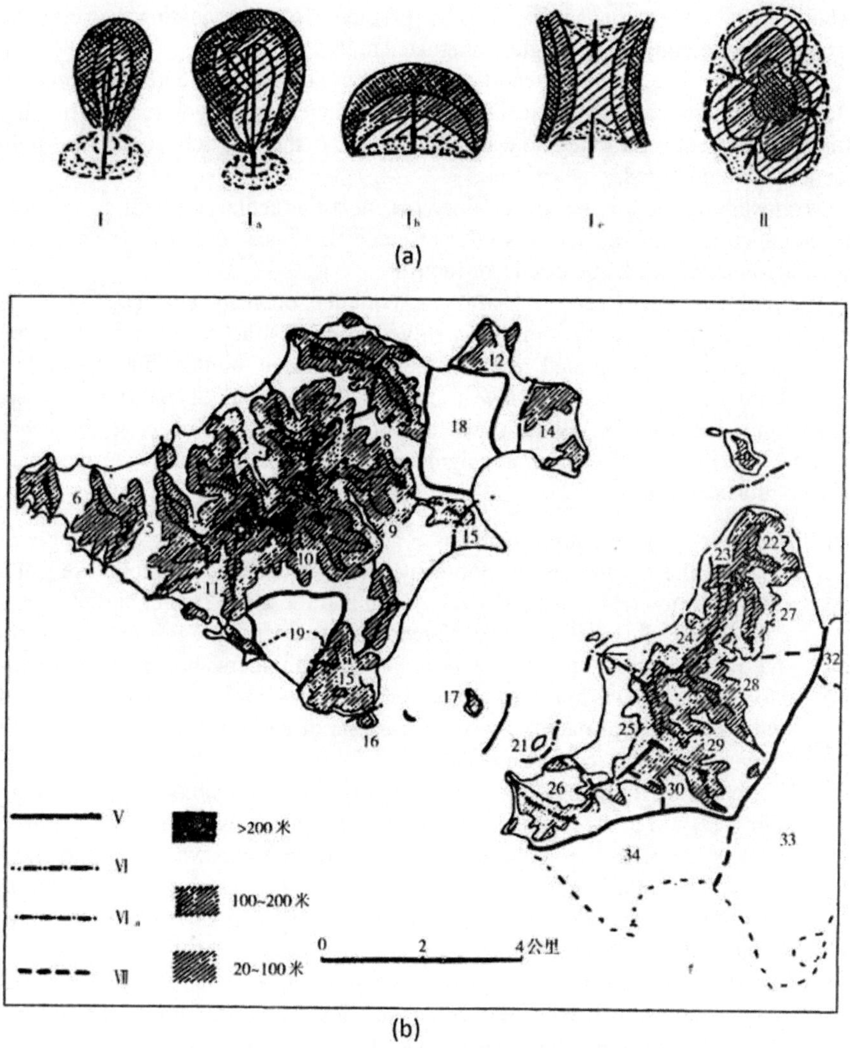

Fig. 7.14 Info-Tupu for the landscape zoning in the East and West Dongting Hills of Taihe Lake, showing the hill valley landscape structure models, e.g., single hill landscape model, single hill landscape asymmetrical complex model, crescent model, pathway model, isolated hill landscape model, and so on, which is valuable for the spatial landscape regionalization and the plantation mode in subtropical regions in China

research and experimental work has achieved many results, but overall, the field is still in its initial stage. The main problems are discussed as follows:

First, when the scientific proposition of "geo-information Tupu" was originally put forward, it immediately sparked heated discussions in academic circles, and some exploratory research experiments have been carried out by many experts and scholars. However, perhaps because of the problems in understanding the scientific essence and meaning, it has slightly stagnated at this point.

In addition, because of the short period of time of research and small amount of investment, the basic theoretical system and technical system of geo-information Tupu has not yet been formed, and strong guidance to the research work in this field cannot yet be provided.

Meanwhile, the current research does not meet the actual needs of sustainable development in Chinese society, so this research area does not drawn the attention of policy-makers, academic circles or industry.

The problem in geoscience is that it is a "complex, open and enormous system", and this problem cannot be completely solved by an abstract mathematical model. However, using the integrated research means of "shape, number and principle", we find order in chaos, and find law in disorder (Zhang et al. 2009). In this sense, geo-information Tupu is a kind of spatial data mining and knowledge discovery method. Therefore, it has important significance and broad prospects for the study of geo-information Tupu.

(1) Research on the scientific attributes of geo-information Tupu is very important to ensure the continuous development of this field. Some examples are listed as follows: the relationship between Tupu and the map (Tian et al. 2003; Yue at al. 2000), the relation and difference between the geo-information Tupu and traditional geoscience spectrum, gaining an understanding of what geo-information Tupu is in terms of scientific attributes or essence, why we should study the geo-information Tupu, the position of geo-information Tupu in the Earth system, and so on (Yu et al. 2005).
(2) The theoretical system of geo-information Tupu and its method and technology system should be emphatically studied. Some questions are listed as follows: What is the basic theory of geo-information Tupu and what is its practical theory? What methods and techniques can be used to construct geo-information Tupu? The answers to these questions will play an important role in deepening the study of geo-information Tupu.
(3) Geo-information Tupu should be closely combined with the scientific tasks of geoscience. Some reasons for this combination are listed as follows: to answer a series of questions related to global changes (Shi 2004; Li et al. 2008; Zhang et al. 2009); to develop global land cover change (Ren et al. 2004), wetland hydrological process, wetland growth and wetland landscape (Ye et al. 2003) and other related Tupus (Liao 2001; Chen et al. 2003; Chen et al. 2007); to produce urban land use, planting and heat island effects and other related Tupus (Zhou et al. 2002); to study the spatial–temporal differentiation Tupu of Earthquake disaster and situation (Li 2001); to make Tupus for the ecological evolution of river deltas and landform evolution (e.g., loess gully), and so on. Combined with practical research, Tupus may play a role in the sustainable development of social economy.

7.5 Representative Publications

(1) ***Exploratory Study on Geo-information Tupu*** (CHEN Shupeng. The Commercial Press, Beijing, December 2001)

In this book, the scientific connotation and main types of geo-information Tupu are described; combined with practical examples, the construction and application of geo-information Tupu based on a large spatial database are studied. The whole book is divided into 4 chapters. The first chapter discusses geo-information Tupu. Several traditional classification methods of geo-information maps are expounded upon, such as zonality Tupu, spatial pattern Tupu, process Tupu, chord Tupu and regional differentiation Tupu. Additionally, the concept, basic theory and research methods of geo-information Tupu are discussed, including spatial graphics thinking, fractal dimensions, geographic information units, and the inversion of geo-information Tupu, as well as application prospects, such as the relationship between geo-information Tupu and data generalization and compression, and the relationship between geo-information Tupu and data mining. The second chapter discusses hydrologic Tupu, where taking hydrological research as an example, the studies on the evolution Tupu of the Yangtze River channel, the watershed structure information Tupu, the delta information Tupu, the coastal zone information Tupu, data simulations of tidal waves and information Tupu, the Taiwan Strait flow field Tupu and others are discussed. The third chapter discusses the urban and town information Tupu, where taking urban research as an example, the symmetric distribution problem of the city is discussed, and then, urban system Tupu, small town Tupu, urban land use Tupu, urban traffic system Tupu and urban spatial evolution Tupu are introduced. The fourth chapter discusses the landscape Tupu, which involves the geochemical abundance Tupu, mountain altitudinal belt Tupu, landslide hazard Tupu, Tupu of sandy desertification in the west of Northeast China Plain, morphological analysis and Tupu expression of the Loess Plateau, Tupu of environmental characteristics of the transitional zone in Central China, Tupu of regional differences of supply and demand of water resources in western China, Tupu of environmental population load in western China, and so on.

(2) ***Thematic Analysis of RS and Geoscience Tupu*** (FU Suxing. Science Press, Beijing, January 2002)

This book was written on the basis of RS information science using the principle and method of comprehensive analysis of geoscience, which is combined with the author's long-term work in the field. This book is a monograph on thematic mapping of RS, which focuses on the connection between theory and practice. The whole book is divided into 12 chapters, and the overall content can be divided into three parts: first, the theory and method of the spatial RS information mechanism and thematic mapping are discussed (chapters one to five), including the development of cartography and geographic information Tupu, functions and effects of RS mapping, characteristics of image information source and mapping analysis, parameter analysis and application of RS image recognition and mapping, pattern mapping of the

Earth information science system, and so on. Second, the practices of multivariate analyses of RS images and thematic recognition and mapping are expounded upon (chapters six to eleven), including RS analysis and mapping of agricultural natural conditions, computer-aided analysis and mapping of thematic statistical data of RS, RS analysis and mapping of biological resources, RS analysis and mapping of water resources, lakes and marine environments, RS dynamic monitoring, mapping and RS surveys of ecological environments, mapping of urban environments, and so on. Third, the design of the output and printing process of RS mapping is discussed (chapter thirteen). Simultaneously, the book also gives a brief introduction to the Earth information fusion science system, map innovation of geo-information Tupu, holographic digital automatic mapping, digital Earth data sharing, and so on. The content is systematic and rich, with a focus on the application of RS and GISs, as well as other high-level technology.

References

Chen, S. (1998). An initial discussion on geo-information Tupu. *Geographical Research, 17*(suppl), 5–9.

Chen, S. (2001). *An exploration on geo-information Tupu.* The Mercial Press.

Chen, S. (2003). The exploration on geo-science. *Geographic Information Science, 6*. Science Press.

Chen, Y., Qi, Q., & Tang, G. (2004a). Research on slope-conversion-atlas in Loess Plateau. *Agricultural Research in the Arid Areas, 22*(3), 180–185.

Chen, Y., Tang, D., & Qi, Q. (2004b). Analysis of slope change atlas drawing from digital elevation models of different spatial scales. *Journal of Huaqiao University (Natural Science), 25*(1), 79–82.

Chen, Y., & Liao, K. (2003). Research on complex informatic Tupu of natural landscape in china. *Geo-Information Science, 5*(3), 97–102.

Chen, J., & Liao, K. (2007). Classification of the comprehensive informatic Tupu for the region's eco-environment: A case study on fujian province. *Geo-Information Science, 9*(2), 85–90.

Fu, S. (2002). *Thematic analysis of RS and geoscience Tupu.* Science Press.

Ji, C., Qi, Q., & Zhang, K. (2005). Indexes system of 3d symbols on the landscapes Tupu of loess plateau and its application. *Geo-Information Science, 7*(1), 47–52.

Li, J., Zhang, H., & Wang, R. (2008). Space-time change of oasis ecological landscape information spectrum in arid areas. *Agricultural Research in the Arid Areas, 26*(1), 106–111.

Li, J. (2001). Study on landslide information Tupu. *Geo-Information Science, 3*(3), 64–71.

Liao, K. (2001). Establishment and principles and methods of comprehensive information atlas of natural landscape in China. *Acta Geographica Sinica, 56*(Suppl.), 19–25.

Mai, A., & Cheng. Z. (1997). RS based study on the urban spatial dynamic expansion of Shanghai City in recent half century. *Shanghai Construction Science & Technology,* (2), 40–41.

Qi, Q. (2004). The latest development on geo-info-Tupu. *Science of Surveying and Mapping, 29*(6), 15–23.

Qi, Q., Chen, X., & Ji, C. (2001a). Study on the topographic information map of the Loess Plateau. *Acta Geographica Sinica, 56*(Suppl.), 32–37.

Qi, Q., & Chi, T. (2001b). Study on the theory and method of geo information map. *Acta Geographica Sinica, 56*(Suppl.), 8–18.

Ren, C., & Liu, X. (2004). Research on geo-informatic Tupu model of regional land use change. *Geography and Geo-Information Science, 20*(6), 13–17.

Shi, J. (2004). Preliminary approach to the informatic Tupu of the complex ecosystem. *Geo-Information Science, 6*(2), 109–114.

Tian, Y., & Yue, T. (2003). Discussions on geo-info Tupu and its development and application. *Geo-Information Science, 5*(3), 103–106.

Wu, L. (2005). *Enhance the conception of region and human settlement. Yearbook of Chinese Cities*. China City Press.

Ye, Q., Liu, G., & Yao, Y. (2003). Tupu analysis of land use changes in the new-born wetland of the yellow river delta (YRD). *Progress in Geography, 22*(2), 141–148.

Yu, M., Zhu, G., & Li, C. (2005). Approach to realizing mutual search and index between geo-informatic atlas. *Geomatics and Information Science of Wuhan University, 30*(4), 348–350, 354.

Yue, T., Chen, S., & Li, H. (2000). Study and application of geo information map. *Geographical Research, 19*(4), 337–343.

Zhang, B., Zhou, C., & Chen, S. (2003). The geo-info-spectrum of montane altitudinal belts in china. *Acta Geographica Sinica, 58*(2), 163–171.

Zhang, H., Wang, Q., & Lu, X. (2004). Perspectives on geo-informatic graphic methodology. *Advance in Earth Sciences, 19*(6), 997–1001.

Zhang, P., Dang, A., & Zhang, Y. (2009). Study on the method of constructing ecological sensitivity information map for digital urban master planning. *Geomatics World, 7*(1), 75–79.

Zhou, C., & Li, B. (1998). An initial probe into geo-spatial information Tupu. *Geographical Research, 17*(Suppl.), 10–16.

Zhou, J., Cui, G., & Zhang, J. (2001). A perliminary discussion on the study of transportation network's geo-informative Tupu of urban system. *Geographical Research, 20*(4), 397–406.

Zhou, J., & Xu, J. (2002). A preliminary discussion on town-information atlas. *Scientia Geographica Sinica, 22*(3), 324–330.

Chapter 8
Geospatial Database

Tiejun Cui and Jianzhong Guo

8.1 Introduction

Along with the development of database technology, many database systems that adapt to the special needs of some disciplines have been produced, which has promoted the application of specialized databases in these disciplines. Research areas that rely on large amounts of data, such as geological, geographical and other natural resource fields, are certainly users of database systems. Database systems have special data structures and processing requirements in the field of geospatial information and are mostly related to data with spatial distribution characteristics or special structures. Map database systems are one kind of data.

Map database technology is the core technology of digital surveying and mapping information engineering and is the foundation of digital mapping systems and geographic information systems. The map database was developed to meet the needs of information processing in the field of spatial data and to adapt to modern society through the provision of digital map products. The map database is based on the development of computer mapping technology and geographic information systems. Therefore, the initial map database is divided into two types: map databases in geographic information systems and map databases in computer-aided mapping. For map databases in geographic information systems, there is a wealth of thematic data in addition to the general map data; these databases are used for a variety of information retrievals, including the statistics and analysis of the data. This kind of map database later evolved into a geographic database or geographic spatial database. The map database in computer-aided mapping is also known as a mapping database, which mainly provides general map data and serves as a data source for the production of various thematic maps and general maps (Wang et al. 1998).

The basic issue in research on geographic spatial databases is the management of massive, orderly and unstructured geospatial data with linear structure computers.

T. Cui (✉) · J. Guo
Information Engineering University, Zhengzhou, China

J. Wang and F. Wu (eds.), *Advances in Cartography and Geographic Information Engineering*, https://doi.org/10.1007/978-981-16-0614-4_8

Computer database technology studies include the computer description of multi-dimensional spatial geographic phenomenon with time series characteristics, the indexing mechanism of massive data storage and visual operation language. Research and development of this technology can be understood from two aspects: the software system, that is, the geographic spatial database management system and the computer expression of the geographic information, which involves the data model, data acquisition, database design, data engineering, data standards, data quality and data application.

8.2 Development Process

In the 1960s, the introduction of computers into cartography resulted in computer cartography technology. People use computers to show map features and their mutual relations, as well as discretize spatial objects that exist in paper maps in both continuous and analog form so that the computer can recognize, store and process these objects. Early computer-aided mapping only used computers as a tool to accomplish the task of mapping, which freed people from heavy manual mapping and brought forth great economic and social benefits. The map data needs of the government, military departments and enterprises required considerable manpower and material resources to digitize maps of various scales and generate a large amount of map data, which became important resources for the country. Compared with other data, map data have a certain mathematical foundation, unstructured data structure and time characteristics of dynamic change, which makes data acquisition, processing and storage challenging. How to properly store and scientifically manage these map data has been a topic of great concern for a long time. Along with the development of computer data organization and storage technology, the maintenance, updating and management of map databases has undergone a development process from a low level to a high level. First, the file system is used, and then, the map database system (MDBS) is developed, which consists of map data, a map data management system, computer hardware equipment and map database management personnel.

In 1979, Prof. WU Hehai attended the Institute of Applied Geodetic Research (IfAG) in Frankfurt to study map generalization. He discovered the importance of map databases and turned to learning software design of map database management systems. Under the guidance of W. Weber, Prof. WU programmed the basic functions of the MDBS that they used, including two categories of retrieval: qualitative (layering) and window (positioning), as well as 30 kinds of Boolean logic combinations. After he returned to China in 1982, the software was further improved and applied to several departments, for instance, to establish a basic 1:500,000 scale military traffic map database. The map database management system won the State Science and Technology Award 3rd prize in 1985. Computer-aided drafting and thematic mapping technology, a key technology project of the State Eighth Five-year Plan, won the State Science and Technology Award 3rd prize in 1986. In 1983, Prof. WU opened a week of lectures on map databases for undergraduate students in the 1979 class at the Zhengzhou Institute of Surveying and Mapping. Based on the original lecture notes of "map database and map data processing", the book *Map Database System* was published in 1991, and the *Research on Basic Theory and Technology of Map Generalization* was issued by the Surveying and Mapping Press in 2004, in which the "cartographic database system as the core component of spatial information system" and the "topological retrieval of map information" were introduced in detail.

With the development of science and technology and the deepening of map data applications, especially with the successful development and practice of computer technology, database theory and information system theory, the application of geographic information was no longer confined to the single product of map; instead, all data with geographic coordinates and their applications were comprehensively used in the fields of resources, environment, economy and society. Compared with map data, this type of data mainly describes the quantity, quality and time characteristics of the geographic entities through the attribute data. These data are beyond the scope of map data in terms of the data content, acquisition method, representation method, and data organization. To distinguish these data from map data, they are called geographic information data. Map data and geo-information data are data with geographic coordinates and are two different representation methods of geospatial information, which are collectively referred to as geospatial data. Geo-information data acquisition, processing, management, analysis and its application have resulted in the emergence and development of geographic information systems. This system is a spatial information system used to collect, store, manage, analyze and apply the spatial information of whole or part of the Earth's surface by means of a computer and its peripheral equipment. The research object of this system is the whole geographical space, which provides a series of space operation and analysis methods for people to use in digital form and analyze realistic space. The core of the system is the geographic information database (GIDB), which combines the spatial geometric data in a certain area with thematic attribute data and realizes their collection, organization, management and renewal.

The database management system (DBMS) is the further development of file management systems. It plays a role as a bridge between user application programs and data files. The greatest advantage of this system is that it provides data independence between programs and files. Due to the unstructured characteristics of spatial data, the early relationship DBMS with fixed length records and no structured fields struggled to meet the requirements. Therefore, most of the early geographic spatial database systems used a mixed management mode (also referred to as binary management mode), namely, geometric graphical data uses the file system management and attribute data uses the commercial relational DBMS. The link between the two is matched by the spatial entity identifier or the internal connection code. In 1981, ESRI launched its first commercially available geographic information system, Arc/Info, which truly and effectively integrated geographic spatial information technology and database technology into a system. In China, based on Oracle and VAX780, CUI Tiejun et al. developed a 1:250,000 topographic map DBMS by using the binary management mode in 1989.

In this management mode, in addition to the use of the space entity identification code as a key field connection, the geometric graphical data and thematic attribute data are organized, managed and retrieved almost independently. This hybrid management model does not represent the true real essence of the geographic spatial DBMS. In this system, spatial data are stored in a special file structure and linked to the attribute data in the DBMS; although the solution is powerful, many shortcomings remain. For instance, there are some problems in terms of data security, consistency, integrity, concurrency control and data recovery.

With the continuous development of database technology, especially the use of binary big fields and the application of object-oriented thinking in database systems, traditional DBMSs can support access to spatial geometric data and multimedia data. However, in this storage mode, because the read/write efficiency of the binary block is much slower than the fixed length attribute field, especially when the object is nested, the speed is slower, and the efficiency is lower.

To improve the efficiency of data reading, some database product companies extended the management function of spatial data based on the relational DBMS, for example, Oracle Spatial and Structured Query Language (SQL) Server 2008 Express. At the same time, to facilitate development, spatial database engine (SDE) middleware was launched. The technical scheme of combining relational databases with spatial data engines had the merits of fast access and close application, which has advantages for certain applications, but since the spatial data engine is independent of database kernels, it is difficult to make full use of all kinds of mature data management and access technology in the database, which is a fatal weakness in its further development. In addition, other issues, such as not supporting extended SQL and difficulty achieving data sharing and interoperability have gradually been exposed.

Although the object-oriented spatial DBMS is the most suitable for spatial data representation and management, it not only supports variable length records but also supports object nesting, inheritance and aggregation; however, due to the lack of a good mathematical foundation and no major breakthrough in access speed, this system is difficult to develop and mature. It is estimated that object-oriented DBMSs are still unable to replace object-relational DBMSs over a long period of time.

Extended object-relational DBMSs will undoubtedly become the future development direction. Although the performances of Oracle Spatial, PostGIS and other geographic spatial databases still have certain gaps compared with traditional spatial data engine models, with the increasing emphasis on the spatial data management market, the further integration of structured data management and spatial data management, as well as the continuous improvement of data compression and transmission technology, the spatial expansion mode based on DBMSs will be continuously improved and become the mainstream technology in the management of spatial data in the future.

8.3 Main Research Achievements

The key technology of geospatial databases mainly includes 3 aspects: the first is the spatial data model, which is the core of the spatial database and the basis of other spatial database techniques; the second is the spatial data index, which is built based on the spatial data model and is the spatial data query clue; and the third is the spatial data query language, which provides the calling and operating method for spatial data. The combination of these three aspects extends traditional relational databases to spatial databases and makes it convenient and flexible to deal with complex spatial data. The main task is to study the data representation, data model and data storage structure in computers, establish a spatial data index, efficiently store and process spatial data with minimal cost, correctly maintain spatial data updates, and provide users with up-to-date, high accuracy, complete, open and easy-to-use geospatial data.

8.3.1 Research on the Computer Data Representation Method of Geographical Spatial Entities

Early map data mainly served map productions, which emphasize the visual features of the data. The method mainly involves "using graphics to express properties", and the quantity and quality characteristics of geographical objects are represented by a large number of auxiliary symbols, including line type, thickness, color, texture, notes, and dozens of others. The map data are based on the corresponding standard and specification of the map and still retain the characteristics of the map. To meet the requirements of map printing, according to the theory of the map overlaying process, map data are managed in the form of color layers instead of being classified according to the natural attributes of geographic objects. This method not only separates the organic connection between the geographical objects but also leads to repeated storage of the same object in different layers. For example, reinforced steep banks on both sides of the river imply the information of the waterside line. Affected by topographic mapping, map data are organized and managed according to the map sheet. Map sheets may break the integrity and continuity of geographic objects. For example, a boundary line will be divided into several records due to map subdivisions and stored in different map sheets. The national 1:250,000 map database developed by FAN Xiaolin and LIU Jiahao in 1989 is a typical representative of this period.

With the progress in the application of map data, the application of map information is no longer limited to a single map product. People need to comprehensively utilize all kinds of geospatial data to study and solve the problems of national defense and national economics, including all data with geographic coordinates in the fields of resources, environment, economy and society and others. Compared with map data, this type of data mainly describes the qualitative characteristics of geographic entities through attribute data and represents the location, quantity, quality and time features of spatial entities by numbers.

Wan et al. (2003) studied the division method of the morphological expression of spatial objects, in which spatial phenomena are abstracted to six kinds of objects: point object, line object, regular surface object, regular body object, irregular object and DTM. Based on the different combinations of six types of geometric elements (point, line, polygon, function construction surface, TIN surface, and regular body), the spatial phenomena are abstracted into five kinds of objects according to the spatial dimension, namely, 0 dimensional spatial object (point object), one-dimensional spatial object (line object), two-dimensional spatial object (surface object), three-dimensional spatial object (body object) and DTM; thus, the complex relationship description of geographic information is realized; by using faceted geometry elements, it is quite easy to implement an image texture mosaic on different sides of a surface spatial object or body spatial object. A 3-D realistic image texture can be created through three kinds of geometric elements: point, line segment and facets.

To improve the representation and processing of spatial data, spatial objects must be described accurately. Guo (1994) and Cui et al. (2000) provided the concepts of spatial variables and spatial attributes. Spatial variables are geographical phenomena that change with the extension of spatial entities, such as the depth, flow velocity and width of a river; conversely, the geographical phenomenon that does not change with the extension of a spatial entity is the attribute of the spatial entity, for example, name, length of a river, etc. The spatial variable is a partial description of the spatial entity as its definition domain, while the attribute of the spatial entity is a global description. Attributes are divided into two types: the former includes name, type, characteristics, and so on. The latter includes quantity and grade. This is generally an abstract concept that is obtained through classification, naming, measurement and statistics. In a paper about the representation of spatial data (Man et al. 2004), according to the four kinds of intrinsic properties of spatial objects, a spatial semantic rule that can accurately express spatial objects is designed by using spatial ontology to represent spatial data and a spatial classification method to classify spatial data. Past spatial data processing has been developed for knowledge discovery, and the precision of spatial information queries has been improved.

Spatial relations refer to some relationships among spatial entities that have spatial characteristics, such as topological relations, directional relations, and metric relations. Spatial topological relation refers to a topological invariance in a topological transformation, including the adjacency, connection and inclusion relations between entities; the direction relation describes a sort of order of entities in the geographical space, such as front and back, up and down, left and right, west, east, south and north, etc.; the metric relation uses the metric in a geographic space to describe the relationship between entities, such as the distance between entities, the area of a region, and so on (Chen et al. 1999). In the research of spatial relations and data models of map data, Cui et al. (2000) discussed map data and its spatial relations, the relationship between the geometry and attributes of map data, and the relationship between the establishment of spatial relations and the completeness of map features in detail; they proposed solving the problem of spatial relation segmentation caused by the hierarchical management of map features and the method for relation connections of multiple map sheets; they finally presented a data model based on the spatial relations of map data. In the research of spatial relations and their applications, Du et al. (2005) described the relationships between spatial objects in detail, including distance relations, direction relations, topological relations and similarities among spatial relations.

Multiscale processing and representation of spatial data is an important and frontier issue of theory and method research in the field of geographic information science. The research on spatial data multiscale representation is the foundation for realizing a scale-less geographical information systems (GIS). Sun et al. (2007) discussed the scale issues of geospatial data and its research progress, introduced the concept, the expression pattern and the effects of scale, and focused on the latest research results of scale problems in the field of information, including the scale theory of spatial data processing and the theory and method of multiscale spatial data mining. In a review of spatial data multiscale research, Li et al. (2007) introduced the basic theory and visual research of multiscale spatial data, discussed the research status of the multiscale representation of spatial data at home and abroad, and illustrated the specific application of the theory and method of spatial data multiscale representation using examples.

8.3.2 Research on the Geographic Spatial Data Model

The data model is a description of objects and their contacts. This description includes a description of data content and a description of the contacts among various entities, which is the basis for database design. Common data models cover hierarchical models, network models, relational models and object-oriented models.

The geospatial data model is the concept of spatial entities and their relationships in the real world, which provides basic methods for describing the organization of spatial data and designing spatial database schemas. It is the core and foundation of the spatial database, which is directly related to the algorithm implementation and operation efficiency of the whole DBMS. Because of the large amount of spatial data, spatial data management is much more complex and difficult than other database systems, thus forming a unique set of data models.

In China, data model research was mainly concentrated in the 1990s, and the main achievements in this field are outlined as follows: according to the characteristics and requirements of mining and considering the shortcomings of the traditional data model, a four-dimensional spatial-temporal data model for mining development was proposed; through a four-dimensional space linear coding technology, the required auxiliary operations for data storage were greatly reduced (Guo 1993); by aiming at the defects and lack of traditional relational data models when dealing with complex objects, an object-oriented spatial data model was studied to specify, organize and manage complex objects (Zhang et al. 1995); based on revealing the inner link between geographic spatial data, an integrated data structure with the characteristics of both vector and raster, as well as a target-oriented or object-oriented model that can express both graph and attribute data were proposed (Gong 1997).

The N1NF (non-first normal form) relational basis of the vector spatiotemporal data model was discussed (Huang et al. 1997); to express the temporal characteristics of spatial objects, an object-oriented spatiotemporal data model was studied, and the version information was labeled directly on the attributes to express the attribute values of multiple versions for the same objects at different times (Gong 1997). A temporal geographic data model based on a group of synchronous changing data items and a time stamp of segmented topology arc was proposed, and an object-oriented GIS data model based on a relational database was presented (Wang et al. 1998). Three integration solutions based on the characteristics of 3-D spatial data and data models were studied and given (LI Qingquan et al. 1998): one solution is the integration model based on TIN and CSG for urban 3-D GIS; the second solution is a hybrid model based on an octree and tetrahedral network (TEN) used for 3-D GIS in the fields of terrain, marine and other; and the third solution is a general 3-D spatial data model of vector and raster integration. A 3-D topological data model for spatial partitioning was proposed, and its data structure and spatial partitioning method for three-dimensional entities were presented (Chen et al. 1998); according to the characteristics of gray and dynamic changes in spatial objects in coal mining, a new GIS data model applicable for coal mines was proposed.

In "*Geographic Information Systems - Principles, Methods and Applications*", the type of spatial data model is divided into three categories, namely, object (feature)-based model, network model and field model, and the spatiotemporal data model and 3-D data model are considered to be at the forefront of future academic study. In this field, a hierarchical combined model for constructing building model was proposed, and then, the dynamic display, spatial analysis and interactive operation and other functions of the model were completed (Yang et al. 2000). A 3-D GIS object-oriented data model was proposed to solve key problems, such as the representation and organization of the complex relationship among the irregular entities (Wang et al. 2002). The reconstruction of 3-D complex geological bodies based on Tetrahedron Mesh was studied, and methods for reconstructing all kinds of complex geological bodies using simplified surface constraint conditions, as well as the expression of geological bodies with properties of gradually changing and mutating attributes were also put forward. The basic characteristics of geological objects and the basic requirements of computer 3-D geological modeling were discussed, an object-oriented 3-D topological data model based on components was proposed, and an object-oriented method was used to abstract geological objects as points, lines, surfaces, and bodies (covering four types, including the combined body, complex body, simple body and volume element); finally, 12 kinds of topological relations and corresponding data structures were designed for all object classes. Simultaneously, taking into account the temporal factors of GIS, an object-oriented spatiotemporal data model was proposed (Cao et al. 2002). A spatiotemporal data model based on graph theory was discussed and developed and used to describe the spatiotemporal geographical entity that spatially and temporally changed discontinuously; when the model was used to describe the polygon objects and their spatiotemporal relations, the vertices of the graph represented the new polygon objects formed by the spatial variation at the time point, and the edge of graph was connected to the vertex of spatial intersections and temporal encounters (Yin et al. 2003).

8.3.3 Research on the Geospatial Data Index

The geospatial data index is one of the key spatial database technologies. The multi-dimensional characteristics of spatial data bring great difficulties to computer data organization and data indexes. The overhead of spatial database queries is generally greater than that of transaction data management. Moreover, the one-dimensional index algorithm of relational databases cannot be extended to the spatial data index, and the efficiency is very low if the method of sequential scanning is used. Therefore, to improve the query efficiency, this technology has become one of the key technologies used to solve the contradiction between the linear structure of the computer and the nonlinear structure of spatial data.

As early as 1979, Prof. WU Hehai developed a map DBMS based on a file system, and grid retrieval (bit matrix) technology was used to realize two categories of retrieval: qualitative (layering) and window (positioning). This indexing technology is still the main geographic spatial database index technology, which has been affecting the application and development of GIS.

A new network based indexing mechanism for map data objects was proposed, and the algorithms and implementation techniques for searches, insertions, deletions and modifications were comprehensively discussed. A new spatial data structure for integrated raster and vector data was proposed, and the strategy of dividing tiles into three grades, as well as the representation technique of spatial occupancy enumeration were adopted, which converted spatial data into raster format. At the same time, to meet the data precision requirements and improve spatial query speed, a raster bit matrix for arcs and an adaptive spatial index structure for polygons were introduced (Tan 1995). Based on the analysis of spatial data index technology, the basic ideas of various techniques were studied, and a new and general spatial data indexing technique for high dimension feature vector retrieval was proposed; the traditional R-tree, which can only be built on the spatial coordinate data type, was extended to the R-tree of a class that can build an index on any data type (Yue et al. 2002). Xia et al. (2002) thought that R-tree was one of the important index structures that could be used to achieve fast spatial data processing; however, despite being studied for a long time, because of the complexity of its concurrency control, R-tree remains too small to be integrated into the commercial database. On the basis of reviewing the existing spatial database index technology, a QR-tree index method for large spatial databases was presented (Guo et al. 2003). The R-tree index structure for SDE was proposed, and the means of system use for the R-tree index was given; further, the divarication-demarcation method that used the R-tree index to achieve the nearest neighbor query was described, including the definition of a cost function and its upper and lower bound functions, as well as the pseudo code form of the algorithm (Zheng et al. 2004). The QR-tree, a spatial indexing structure based on R- tree and "Quad-tree" for space partition, was proposed, and its data structure and algorithms were also stated (Zhang et al. 2004). The MR-tree, an improved R-tree index structure for multiresolution spatial data, was presented and could effectively retrieve the multiscale organized spatial data, as well as conduct fast queries and display spatial objects at different resolution levels (Liu et al. 2004). The data structure of the spatial data index based on conventional QR-tree was analyzed and it is found that for the traditional Quad-tree, when the amount of data was particularly large, a particularly deep QR-tree, too much storage space and inefficient inquiries, as well as inflexible determinations of planar region segmentation limits would result; thus, an improved QR-tree data model was proposed to establish an efficient spatial data index; in this model, the node that a map element belongs to can be determined by checking the intersecting area horizontally and vertically, and the rapid retrieval of massive data was realized (Huang et al. 2005). By comparing R-tree with the R^+-tree and R^*-tree indexes, the R^*-tree index was considered to obtain a higher node storage and utilization, as well as avoiding the splitting of nodes (Xie et al. 2005). The CQRtree, a new spatial database index structure based on the R-tree index, was proposed, and the data structure of CQRtree, algorithms of insertion, deletion and queries, and a performance analysis and comparison were given (Lu et al. 2006). The construction method of the spatial data index based on D-Tree was studied, and based on the D-Tree index structure, a system flow framework for location-dependent information services was proposed; additionally, an approach of complementary void intersection in data areas was presented (Lai et al. 2006). A hybrid index structure for mobile spatial objects called Q^+R-tree was proposed, and the algorithm idea of the generation, updates and queries of the data structure were discussed (Xu et al. 2006). The distributed spatial data indexing mechanism was studied, and by using the distributed parallel technology, the key technology of the massive spatial data indexing mechanism in the network environment was established; additionally, the classic R-tree was improved; finally, based on the R-tree and Hash table, a distributed index tree structure (DR-H) for massive spatial data in a distributed environment was proposed, which made full use of the range querying of R-tree and the efficient single key querying of the Hash table (Chen et al. 2007).

Due to the complexity of the spatial data itself and the increasing requirement of fast queries of massive spatial data, at present, GIS are facing the challenge of large amounts of data storage and spatial data management. After comparing and analyzing the main techniques of spatial storage and spatial querying, a framework design of a spatial query system based on the R^* tree was proposed, which was realized in the application case of a geographic information system (Guo et al. 2002).

8.3.4 Research on Geospatial Data Manipulation and Query Language

Traditional relational DBMS query and manipulation language SQL is used to access data and query, update and manage the relational database system. SQL language is divided into 4 categories: data query language (DQL), data manipulation language (DML), data definition language (DDL), and data control language (DCL). Similarly, geospatial manipulation language is the foundation of spatial data management, which enables users to change the data in spatial databases. The main manipulation functions include spatial entity definition, spatial entity query, geometrical form of spatial entity, and manipulations between attribute and relation (surface to point, surface to line, surface to surface, line to point, and line to line), as well as additions, deletions, modifications, and so on. Geographical spatial DML is much more complex than SQL, so there is currently no complete geospatial database operation language.

An extended SQL for fuzzy querying was studied, and the method of transforming extended SQL into standard SQL was discussed. In view of the relational database SQL and its defects in the application of spatial data queries, the characteristic of the spatial database query language as studied by Li, and it was considered that the spatial database query language should have three features, namely, understanding 'spatial concepts', descriptive expression of queried results, and a declarative query process; finally, the main features of the three aspects in spatial database query languages and the key research topics in the query languages were outlined. An investigation on spatial information visualization based on the multimedia technique was carried out by Wang, where based on domain relational figure language QBE, a visual spatial information query language SIVQL (Visual Query Language on Spatial Information) was extended and constructed, and thus, spatial information could be queried directly by clicking a graph or chart, and the query results would be displayed in the forms of a graphic, chart, text, image, sound and others. According to Ju (1999), a visual spatial database query language CQL was introduced based on the three aspects of syntax, semantics and pragmatics, which allowed the end user to visually retrieve entities in a database by means of a card—the visual element, and a query sentence was expressed by the spatial combination of these cards. An extended data model oriented to a SDE and its manipulation language GSQL was presented, in which point, line, surface and other geometric data types are taken as the basic data types of the system, the manipulation language GSQL is based on the standard SQL language, and it supports the definition, retrieval, storage and management of both geometric data and attribute data (Zheng et al. 2002). In addition, considering the demand of multisource spatial data queries and the method for managing spatial data and attribute data separately, a data query algorithm that supports query decomposition and results screening was given, the optimization methods for spatial data queries were analyzed and illustrated, and a set of query techniques for supporting multisource spatial databases was formed (Liao et al. 2004); object-relational data models, OpenGIS specifications and SQL2 languages were analyzed. In view of the data manipulation system and its user interface, the architecture of spatial DML was studied by taking the intermediate layer as the medium; an achievable spatial DML architecture was proposed, and the whole manipulation process of spatial data by DML was illustrated (Xu Cuiling et al. 2007).

With the wide application of GIS in people's daily lives, research on natural language-oriented spatial database queries has received increasing attention. According to the basic spatial relations of geographical elements (metric relations, topological relations and direction relations), the basic query forms of various spatial relations in natural language and spatial semantics in query statements were discussed, which laid a foundation for the study of using spatial analysis and network analysis as a natural language query condition (Ma et al. 2003). Combined with the XML database technology and traditional spatial database technology, geographic markup language (GML) spatial data queries and indexes were deeply studied. Based on XML standard query language XQuery, the content of XQuery spatial expansion was proposed, and a GML spatial DQL was developed (Lan et al. 2006).

In view of the wide application requirements and practical situation of spatial data similarity queries, the concept of the Feature Based Spatial Data Similarity Query (FBSDQ) was proposed, and a formal definition was given. The characteristics of FBSDQ were analyzed, and a unified framework for FBSDQ processing and its key technologies were proposed. Taking the high dimensional index structure VP tree as an example, a high dimensional index technique based on distance metric space was discussed, which provided technical support for research of spatial data similarity queries (Xia et al. 2007). In the research of spatial data semantic queries based on ontology, an ontology-based spatial information query framework was designed, which effectively expressed the spatial information hierarchy and semantics and under the premise of guaranteeing semantic consistency, different sources of spatial data could be queried and accessed (Wang et al. 2007).

8.3.5 Research on Geospatial Data Engines

Geographic spatial data engines are another manifestation of geospatial database manipulation and DML. Under the influence of the object-oriented concept, complex space entity operations must be enveloped into a class, which simplifies the manipulation of the spatial entity. By means of mechanical terms for aircraft or vehicles, the class for spatial data operations in the database is called the SDE. Based on a specific spatial data model and specific data storage and management system, the SDE provides spatial data storage, retrieval and other operations for secondary development. SDE is composed of three parts: ① a pre-description pattern of storage, syntax, and semantics for the supported spatial data types; ② a spatial indexing mechanism; and ③ a set of operations and functions that perform the spatial query and management for the region of interest.

At present, spatial databases mainly use relational databases to organize and manage spatial data and attribute data, as well as to provide effective storage, query and analysis of these data and support the application of various geospatial data.

However, how can the relational database be used to store and manage complex spatial data in support of the functions of spatial relation operations and spatial analysis? How can users transparently access geospatial data without regard for the actual storage location, the method, the data structure and other practical issues? SDEs are a good way to solve these problems. SDEs can be used in traditional relational databases to manage and process spatial data and provide necessary functions of spatial relation operation and spatial analysis. Using the SDE to achieve the distribution of the client/server computing model, transparent access, sharing and interoperability of geospatial data can be realized to establish a real sense of the distributed geographic space database.

There are two ways to extend the relational DBMSs to manage spatial data. One way is the spatial data engine attached to the relational DBMS, for example, ESRI's Arc SDE, Spatial Ware of MapInfo company, and SDX+ from Hypergraph Geographic Information Technology Co. Ltd., as well as the spatial data engine of most domestic GIS software. Such systems are generally developed by GIS software vendors and have the merits of supporting the general relational DBMS. The spatial data are stored with BLOB and can be cross-platform operations and closely combined with a specific GIS platform. Their disadvantage is that the spatial operation and processing cannot be implemented in the database kernel, the data model is complex, the expansion of SQL is difficult, and it is not easy to achieve data sharing and interoperability. Another way is to directly expand the spatial database system of a general database, for example, the spatial expansion of databases such as Oracle Spatial, IBM DB2 Spatial Extender, and Informix Spatial DataBlade, as well as MySQL, PostgreSQL, DM4 and DIERAO. Such systems are typically developed by database vendors and have the advantages of integrating spatial data management into a general database system. Spatial data are accessed as objects, and their spatial operation and processing can be implemented in the kernel of the database. The extension of SQL is convenient, and data sharing and interoperability are easy to achieve. The disadvantage is that it is difficult to implement, its data compression is difficult, and compared with the first kind of system, a gap in function and performance still exists.

Zheng et al. (2004) used OCI (Oracle Call Interface) and OLE (Object Link Embed) to develop an Oracle Spatial database; the basic principle of spatial data storage with Oracle was studied, the two methods of developing spatial databases using OCI and OLE technology were explored and compared, and the problems of how to develop spatial databases and realize map visualization functions based on practical platforms were solved.

According to Liao et al. (2004), the key technologies of spatial data engines, such as the spatial data model, the efficiency of the engine, the data consistency of multiple user editing, and the support of multisource data, were analyzed using a three-tier C/S structure, which was compliant with OpenGIS interoperability specification. A multisource data SDE that supports data sharing and function sharing was developed, and this engine could be used for spatial database system integration while supporting the development of GIS software for the Internet.

Li et al. (2006) proposed an ideal geospatial data model based on the investigation of data models of common geospatial database engines, such as Geodatabase and Spatial Oracle, and according to the theory of data storage and query optimization of geographic spatial databases by applying an object-relational data model and a three-layer C/S framework, which met the high efficiency and seamless storage and management requirements of massive, multisource and multiresolution geospatial data, and the characteristics of the data model were described in detail. This model represented a paradigm shift from a system-centric GIS to a data-centric GIS.

Ni et al. (2006) believed that when using a large relational database to manage massive amounts of GIS data, the storage efficiency was low because of the indefinite length of the spatial data. In addition, since the existing SQL does not support spatial database retrieval, the data storage, reading and retrieval, management, and other functions between spatial data and DBMS must be solved. The concept, essence and characteristics of the SDE were discussed, and the arc SDE was studied.

In the paper by Lu et al. (2007), an object-oriented raster spatial data organization model was proposed, a distributed four-tier RasSDE architecture was designed, and the database service layer, application interface layer, application layer and other related technologies were analyzed. The architectures of the RasSDE core component OMSAC and an open, multisource geospatial data access component were discussed. The main goals, such as object-oriented raster spatial data organization, unified massive raster spatial data management and distributed raster spatial data computing, were realized. Tests such as the RasSDE and ArcSDE disk read and write and the comprehensive performance comparison of network transmission showed that the network transmission efficiency of RasSDE raster data was higher than that of ArcSDE, which met the requirements of mass raster spatial data transmission.

8.3.6 Research on Geospatial Database Architecture

Spatial database architecture is closely related to the development of spatial data models and spatial DBMSs. The architecture of the database is greatly affected by the operational environment of the database system, especially with the development of object-oriented technology, component technology, distributed computing technology and network technology. The architecture of the database system has also undergone great changes. The database system can be a centralized model and can also be a client/server model but, it can also be a distributed model across the entire geographic space.

In the 1980s, the map DBMS developed by Professor WU Hehai was a file system that provided a simple method of unified access and management data for users; users could establish their own logical file, which was generally divided into more than one file, each of which could be organized with the same or different logical files to complete the data storage, input, output and other processing techniques with the help of the file system.

In the paper 'Map DBMS design and implementation' (Cui 1990), the basic features of map spatial information were analyzed, and using the techniques and methods of managing attribute information with a relational DBMS and managing geometric information with special management software, a specific implementation plan to build a map DBMS based on an extended relational DBMS was proposed; thus, preliminary research and practice was carried out.

Zou (1999) pointed out that the map database based on client/server structure was a new type of MDBS. By analyzing the shortcomings of traditional single architecture map databases and the defects of simple map object databases, map complex object operations based on object-oriented technology were discussed, a concept model of map databases and a mode of map complex object management was designed, and finally, the distributed management technology of map databases based on C/S architectures was expounded.

In the paper 'Distribution management of map database based on client/server structure', the necessity of developing the map database of a client/server architecture was described in detail, and the concept, structure and network environment of map database distribution management based on client/server structure were discussed and designed.

In the paper 'Study on distributed processing of spatial data based on client/server architecture' (Yang et al. 2001), the development of distributed processing of spatial data was briefly analyzed. In a case study on cross regional cadastral management and application, a distributed processing scheme for spatial data ws designed by using the integration of spatial and nonspatial databases and distributed system technology based on client/server architecture; thus, a distributed design of cadastral data with spatial distribution characteristics was described, the data distribution and transparent distribution, integration of spatial databases and relational databases, distributed query processing, data updates in multiple places and data consistency were introduced in detail.

The distributed storage and management technology of multiscale spatial data provided technical support for the establishment of distributed network GIS and the implementation of integrated dissemination and application of multiscale, multiple data types of digital geographic information products. LI Hongwei and GUO Jianzhong (2003) discussed the spatial data distribution and storage model system, integrated storage and management method of spatial data in relational databases, SDE technology, and integrated management architecture of multiscale spatial information, and other issues.

8.3.7 Research on Geospatial Database Security Control

With data repositories ranging in size from gigabytes and terabytes to petabytes, the storage and management of spatial data can no longer follow the centralized storage mode of the traditional host GIS mode. The massive and regional distribution characteristics of spatial data make it more suitable for distributed storage in the network environment. Through the integration of multiple data resources distributed on the network, the distributed storage will form a single virtual data access, management and processing environment, shield the underlying heterogeneous physical resources for users, and construct an integrated data access, storage, transmission, management and service architecture for the distribution of massive data. Thus, this issue relates to the security problem of each storage node's data. The use of a geospatial data encryption/decryption algorithm to implement the spatial data storage encryption, even if the node data are accessed by illegal users, can also play a protective role.

In the 1990s, a self-copyright DBMS, called DM, was developed by FENG Yucai's team. This system has all the features of an enterprise edition database and focuses on strengthening its security characteristics, which adds the functions of tagging and mandatory access control, as well as using the security mechanism of executive role separation of the database administrator, database auditor and database security guard; thus, this system can provide multiple identity authentication services and offer transparent, translucent and other storage encryption modes, as well as audit control, various communication encryptions and other security means. The security level reaches the B1 level.

In the paper, 'Semantics for multilevel secure spatial database' (Zhu et al. 2001), a new multilevel relation-hierarchical model with element-level labeling was proposed, and this model was capable of processing multilevel spatial data. Based on the model, an upper-lower layer relational integrity was introduced; by absorbing ideas of belief-based and fact-based semantics, the format and semantics of the SELECT statement were modified, and semantics were extended from data-based semantics in MLR. To prevent the conversion channels introduced by our data model, the semantics and manipulation of other SQL statements in MLR were modified. Finally, the multilevel relation-hierarchical model was proven to be secure.

In the paper 'Research and application of dynamic audit policy for secure spatial database' (Li et al. 2006), a dynamic audit policy model based on secure spatial DBMS was proposed, and the model could fully express audit policies based on time and space constraints and monitor user activities in real time by using a constraint mechanism. Furthermore, fine-grained audit policies could be expressed by introducing coherent property expressions. Then, the algorithm was presented to trigger the rules. Finally, by embedding the audit policy module, constraint policy module and log module, the dynamic audit policy model based on secure spatial DBMS Sec-Vista was realized.

In view of the audit practice of information network systems, Li et al. (2006) proposed methods and measures of data security and maintenance in the construction of spatial data infrastructure. With the development of network technology, spatial information was shared by many users through the network, and the security problem of multi-user access spatial databases had to be considered. In the case of the independently developed Vista spatial DBMS, the MAC design method of Vista was discussed in detail by Shi et al. (2007) from the three angles of safety rules, security attributes and mandatory access control. Formal security rules were defined, and the algorithm flow and the corresponding security information data structure were given. The design scheme is also applicable to other similar object-relational database systems.

8.3.8 Research on Geospatial Data Sharing, Integration and Fusion

With the wide application of spatial information technology in the national economy and national defense construction, spatial data have been widely used in many disciplines and departments and have played an important role in resource management, environmental governance, disaster prevention, regional planning, urban management, scientific research, education and national defense and other areas. For the geospatial data production department, it is difficult to produce data that fully meet the needs of the user, so users have to make data products according to their specific application purpose, which leads to a wide variety of data sources. Due to the lack of uniform standards, spatial data often use a specific data model and a specific storage format, which causes a spatial object with the same scale in the same region to be repeatedly collected by different departments and leads to the multi-semantics, multiple spatiotemporal characteristics and multiscale characteristics of spatial data, as well as different storage formats, data models and storage structures. Multisource spatial data sharing, integration and fusion have become hot topics in geographic spatial database research. By using a variety of techniques and methods, multiple sources of spatial data must be integrated into a unified spatial database system to become a data form that can be identified by the database system. The core task is to shield the heterogeneity of the multisource data model and to integrate the heterogeneous data sources that are related to each other so that users can transparently access multisource heterogeneous spatial data.

To realize the sharing of spatial data, the concept of a formal spatial object model was proposed by Gong (1996), which uses a formal spatial object model to replace the exchange standard of spatial data and to realize the sharing of spatial data. Wu et al. (2002) proposed building a distributed multispatial database system to realize the interoperability of heterogeneous spatial data sources. According to the needs of spatial information sharing, based on the technical solution of combining spatial database technology with web technology, remote sensing technology and WebGIS technology, Wang et al. (2006) explored the spatial data sharing, analysis and thematic application of grassland resources under the WebGIS environment from three closely linked aspects, including user requirements, spatial data characteristics and information services. This system can provide end users with convenient and effective map-text inquiries and homogeneous, integrated and seamless spatiotemporal data services. Based on an analysis of the composition and characteristics of the distributed database, Li et al. (2006b) advanced an object-oriented expression method of spatial data, discussed the composition of object-oriented distributed spatial databases in detail, proposed a method for realizing the shared method of distributed spatial data, and designed an integration mode of the distributed spatial data based on the Oracle Spatial circumstance, which offered a concrete idea for the distributed management and sharing of spatial data in urban areas and land resources. Metadata are an efficient route for spatial data sharing, and for that reason, Liang et al. (2006) proposed a metadata updating mechanism based on a distributed spatial data sharing model. As an example, this metadata updating mechanism has been applied to the Information Integrated Management System of Digital Changping and has attained good results. Jia et al. (2007) analyzed four primary models of geospatial data sharing, pointing out that the picture-based quick browsing model was an efficient model being used at the time. Then, we introduce the quick browsing system of land and resource geospatial data to which the aforementioned model is applied.

Based on the research of current sharing and integration modes of the different formats and clues of UDA for sharing data among traditional heterogeneous data sources, Ma et al. (2002) proposed a data integration mode of the Universal Spatial Data Engine (USDE). The USDE is composed of a data engine manager and some data engines for different data formats. The former responds to calling the correct engine according to the data format and providing a unified data access interface; the latter is for accessing the data. Using COM technology, the author designed USDE for some popular GIS data formats and realized the integration of some different spatial data. Starting from the concept of a spatial information grid, Zhang (2006) introduced the framework and service mechanisms of data sharing and service of SIG-based geological maps and discussed the software component model for Internet-oriented SIG. The sharing of geological map data and web services was realized by using the registration, management and UDDI (unified description, discovery and integration protocol) of space information resources on websites. LI Kelu et al. (2006c) held that by sharing information in urban and rural planning, it was possible to increase the scientific tenability in the decision-making of urban and rural planning and avoid statistical contradictions in urban planning and other fields. The key to information sharing in urban and rural planning lies in the establishment of a unified technical standard: as far as data classification is concerned, the Categories of Urban Land Use and the Standard Index of Planned Urban Construction Land presently under operation should be used as the basis for the classification system; the original large categories are treated as medium categories and two large categories are added to the original categories. As far as the map standard is concerned, it is necessary to establish a relatively complete map system. As far as the data quality standard is concerned, the ISO geographical information standard should be used.

A spatial data accessing architecture based on the Seamless Integration of Multisource Spatial data (SIMS) technology was proposed by Song et al. (2000); it not only provides capabilities to access (read and write) spatial data in different formats but also gives users the ability to perform analyses among several kinds of spatial data formats. The authors described the SIMS technologic architecture and discussed its development and application. Li et al. (2006d) elaborated the production and performance of multisource spatial data as well as the hindrance of multiple data forms to GIS development and discussed multisource data integration methods. Combined with actual development experience, the authors summarized and proposed some multisource data integration methods and their applications in GIS application software development. Deng et al. (2006) proposed a spatial data integration service framework through analysis of the OGC uniform spatial data model and GIS user requirements. Based on framework, we discuss in detail the methods to transform multisource spatial data based on GML and to realize the data services based on soap, as well as the application model of virtual spatial data service in the grid environment. Finally, they constructed an experimental environment of multi-data sources and multi-type spatial data integration. Tian (2007) thought that the co-occurrence of multisource spatial data had existed since the beginning of GIS. The integration of multisource spatial data is a hot topic in the geo-information science field. Its significance and necessity relates to the information processing platform and character of the system. The article analyzed the concept, theory and research of multisource spatial data and provided an opinion on how to solve the problems with the technology.

A model of multisource heterogeneous spatial data integration based on GML was proposed by Liu et al. (2007), and this model solved the issues of transformation of the multisource heterogeneous spatial data to GML documents, the interaction between GML and spatial database servers, and the conversion of GML to SVG (scalable vector graphics); further, by using Web service technology, a multisource heterogeneous spatial data integration system was designed on the Microsoft.NET platform, which realizes spatial data integration based on GML and spatial data visualization based on SVG.

Wu et al. (2002) proposed implementing spatial data sharing by establishing a multispatial database system. First, the author discussed the basic concepts, the system structure and the key technology of distributed multispatial database systems; second, the author proposed the basic idea of the integration of the distributed multispatial database system, namely, "assimilating the heterogeneity and unifying the homogeneity", and established the layer model of integration.

For the three basic data integration models, a multisource spatial data integration model based on GML-data-interoperability was adopted by Jiang et al. (2007), and its running mechanism and key technology were analyzed by studying this data integration model. This model translates multisource spatial data formats into GML data files and then carries out the data integration. Depending on GML, this model will implement data integration and data sharing.

Guo et al. (2007) held that applications of GIS result in the generation of multisource data and make data integration and sharing more difficult. Through spatial data integration, the differences caused by multisource data, spatial data models, classifications and geometric positions were eliminated. Meanwhile, data integration can reduce the cost of the product, hasten the updates of geographical information and improve data quality. The principles of multisource geospatial data integration are discussed in the article, and their applications are discussed. According to the demand for multisource spatial data application, based on the analysis of the existing data fusion model, a basic model of spatial data fusion was established by Wang et al. (2008). Through the fusion of spatial data, the difference due to the spatial data model, the classification and grading of objects and the geometric location accuracy were eliminated. This fusion realized the sharing of spatial data, expanded the application scope of spatial data, reduced the cost of maintenance and increased the access to spatial data. Simultaneously, this article also explored a new way to produce spatial data to meet the requirements of a particular user.

Land data assimilation is a new means for multisource geospatial data integration. Li et al. (2004) introduced the conception and composition of the land data assimilation system (LDAS). Data assimilation is the method used to integrate observations with model simulation results. The observations can be both direct and indirect data from different sources and with variant resolutions. The output from an LDAS is datasets of land surface variables with spatiotemporal and physical consistency. Then, the author reviewed the major LDASs under development, which include the Global and North American Land Data Assimilation Systems developed by NASA (National Aeronautics and Space Administration), European Land Data Assimilation System, and Land Data Assimilation System for West China. They thought that the Land Data Information System derived from LDAS would play a very important role in the new generation GIS.

8.4 Construction of the Main Geospatial Database

Geospatial database construction is a difficult and complicated engineering task. The scientific management of geospatial database projects is a very important prerequisite and foundation for overall success. From the perspective of the life cycle of a project, the construction of a geospatial database project can be divided into five stages, including project initiation, project design, project implementation, project quality control, and project closeout; each stage has different management tasks and requirements.

8.4.1 The Basic Situation of Geospatial Database Construction

The research project of the national basic geographic database was commissioned in 1984, and the first phase of the development task was completed in 1994, in which the national 1:1,000,000 topographic database, geo-place names database, digital elevation model (DEM) database, 1:4,000,000 topographic database and experimental gravity database were completed; the construction of the national 1:250,000 topographic database, DEM and geographic database was completed in 1998; the 1:8,000 DEM (12 m grid spacing) database for key prevention areas of the seven major rivers and the 1:50,000 DEM (50 m grid spacing) database for the Three Gorges Reservoir Area of the Yangtze River were completed in 1999; and the national 1:50,000 digital line graphics database, digital ortho image database, digital elevation model database, and digital raster map database were finished in 2006.

Since 1999, according to the national development strategy and the requirements of national economic construction and development and in accordance with the overall planning of the key construction of the basic database for digital land engineering, the China Geological Survey systematically collected the results of a regional hydrogeological survey in China, with reference to the 1:200,000 regional hydrogeological maps, and built the most basic and representative hydrogeological result spatial database.

In the paper 'Development of the National Physical Atlas Database and its information products' (Chi et al. 2002), the authors first analyzed the background of the National Physical Atlas Database (NPAD) and its information products and concluded that NPAD represents China's scientific level and serves as the glue of its information products; they demonstrated an integration of the atlas' scientific value and social value. Second, the paper analyzed the project's scientific intention and concluded that it represents the transfer from a digital cartographic model (DCM) to a digital landscape model (DLM), as well as the integration of DCM and DLM. Third, the technical features of the project were related, i.e., they not only placed emphasis on database development but also paid attention to information sharing technologies in the project. Fourth, the paper summarized the total characteristics of the project, that is, the combination of dynamic, multiple-scale browsing of spatial-temporal features and static deep analysis and processing of geo-information. Finally, the authors predicted the prospect of China's National Atlas Series, i.e., to develop and construct a National Atlas Information System. Lei et al. (2007) discussed key questions about urban and rural cadastral spatial databases, analyzed the management mode of urban and rural cadasters, proposed the method of spatial and nonspatial information by spatial data engines, and studied the method from old cadaster types to new ones.

In Li et al. (2003b) classified and coded the programming information of land use planning based on the discussion of the demand of building spatial databases for land use planning in Guilin. On the basis of MAPGIS and Oracle 9i, the basic process and method of land use planning databases were proposed and studied, such as file collection, data input, data type design, coordinate integration, error correction, data audition, and index creation. The framework of the land use planning spatial database was designed for Guilin.

According to Wang et al. (2005), by using the GIS method, a wetland GIS spatial database of the Heilongjiang Province was built based on collecting a vast amount of wetland data. According to the database, we could not only inquire about the graphics and attribute information but also make conditional inquiries that wholly and systematically reflected the quantity, type, ecological characteristics, geographical distribution, animals, plants and natural reserves of wetlands in the Heilongjiang Province. This GIS technology could be applied to the protection and management of all wetlands in the Heilongjiang Province, which could greatly save labor, material resources and financial resources, as well as achieve unity among the ecological benefit, economic benefit and social benefit.

The basic geospatial database system is an important link in the construction of land and resource information and the geospatial framework of the 'digital province and digital city'. Zeng et al. (2004) introduced the content, system structure, development mode and function of the basic geographic database in the Guangdong Province and described the security strategy and characteristics in the DBMS. This study provided beneficial arguments and practices for the construction of provincial basic geographic database systems. The *'1:50,000 Spatial Database of the Shanxi Province'*, completed by the Remote Sensing Center of Shanxi Province, used remote sensing, databases, computers and other advanced technology to establish a 1:50,000 digital line graph (DLG) database, digital orthophoto database, DEM database, digital raster map database and meta database, which covered the whole Shanxi Province. This database had spatial data management, query, analysis, output and other functions. The article by Li et al. (2007) introduced the construction aim of the "Digital Wuhan Construction of Spatial Data Infrastructure", analyzed its system frame and the network infrastructure's construction; in their paper, the data classification was also studied, the data organization was discussed, and a model of data construction and services was introduced. Their paper also discussed the application of spatial data infrastructure in urban planning, territory planning, urban construction, traffic and public information services.

Location-based service (LBS) is one of the new application fields of GIS technology. Generally, geo-databases that are specifically created according to LBS or ITS (Intelligent Transportation System) requirements are called navigable databases. Compared with the ordinary geodatabase, the navigable database must follow some particular rules in data modeling, database creation and application, and some key technologies must be developed. After briefly introducing the concept of LBS, Jiang et al. (2003) discussed the characteristics of navigable databases. The procedure and some key technologies for database implementation and application were discussed. Jiang (2004) compared navigable data with general geographic information data in terms of the data model, accuracy and structure. The paper also briefly introduced the concept of a navigable database and discussed the source of the navigable data in the city transportation system and navigable data processing.

The article 'Construction of Spatial Database for Railway' (Liu et al. 2004b) noted that the spatial database for railways was the foundation of the Railway Geographic Information System (RGIS). After analyzing the development status of railway informatization and the construction situation of railway spatial databases, the paper proposed the structural composition of spatial databases according to actual railway requirements. The article further explored the construction and maintenance of spatial databases, which provided valuable references and ideas for the construction of spatial databases for railways in China.

Starting from forestry sharing information standards, Chen (2006) discussed the basic frame of forestry space database construction and data collection methods and proposed the forestry information systems integration platform architecture and technical route. Hong (2006) thought that the construction of a forest resource foundation geographical spatial database was the basic information content in the construction of industry information and expanded and enriched the national spatial information infrastructure in the field of industry information. This paper introduced the main content of forest resource foundation geographical spatial databases and spatial data production specifications.

To more rationally allocate groundwater resources in Yinchuan City and to make the survey and evaluation data visualization available for the whole society, it is necessary to further analyze, utilize and study the existing data for the sustainable use of groundwater, real-time update of monitoring data and dynamic prediction of groundwater resources. For this purpose, Liu et al. (2007) built a spatial database based on the platform of MAPGIS and collected and recorded abundant data. The functions of data acquisition, data entry, editing, management, browsing, querying and quality control were realized, which greatly improved the efficiency of data reuse in hydrogeological work.

8.4.2 Geospatial Database Design

According to the development time series, the construction of geographic spatial database projects is divided into 4 stages: requirement analysis, system design, system implementation, system maintenance and evaluation. Each stage will produce some documentation to ensure the smooth progress of the project construction. In the article 'Geospatial data engineering: a theoretical framework', the concept of spatial data engineering was proposed and defined. Meanwhile, the spatial data engineering meta-model and two children meta-model were described; thus, the theory framework was established, and finally, the spatial data engineering problems that needed to be solved and their research contents were expounded. In the article 'Database establishment of spatial data', the basic theory of spatial database was adopted and combined with the database construction practice of Beijing municipal land resources and housing management information system, the nature, principles, standards and procedures of the spatial database were studied, the organization of the spatial database management was discussed, and the key technologies of the seamless integration of spatial data, metadata database organization, spatial database incremental backups and incremental updating were analyzed.

Database design is a process used to abstract data that exist in certain ranges in the real world into a specific database structure. Specifically, for a given application environment, a logical design was provided to determine the optimal data model and process model, and a physical design was provided to determine the database storage structure and access mode, which aimed to establish a database that can reflect the real world geographic information and mutual contact; this design met user requirements, was accepted by a DBMS, and could effectively achieve the goal of the system and access data.

Zhu (1992) developed 1:50,000 system software for building a digital map database of the Hainan Province. In 'the research and practice of state 1:50,000 basic geographic database building', Wang (2006) introduced the 1:50,000 basis geographical database building in detail.

The national 1:250,000 database was built in 1998 and has been applied in many fields, which have achieved remarkable social and economic benefits. In the article 'The design and implementation of 1:250,000 map database', the construction scheme of the 1:250,000 map database was discussed in detail. In the article 'Research on the design and application of a national 1:250,000 database for national fundamental geographic information system' (Jiang 1999), the design, technical characteristics and application of the database were discussed. In 'The design and establishment of a 1:250,000 DEM of China', from the perspective of database design, the design idea, process flow and data accuracy analysis of the database, as well as its applications were introduced in detail. In the article 'Design and establishment of 1:250,000 toponym database of China', the data content, data organization and structure of the database and its process flow were described in detail.

Hu et al. (2002) discussed the design idea for 1:10,000, 1:50,000, 1:250,000 and 1:1,000,000 pyramid-like multi-layer and multiresolution scales of basic GIS in comparative detail, that is, the framework structural system based on the digital image digital raster graph (DRG), DEM and vector and raster integration (2-D or 3-D) of several layer digital line maps. A technical line for the construction of a basic GIS for the whole nation and various provinces for sustainable development was proposed. Some important GIS theoretical issues were discussed.

In the paper 'Construction and updating of the 1:1,000,000 geospatial database of China', summarized the several development stages of the 1:100,000,000 geospatial database since 1984, that is, the 1:100,000,000 topographic database, which was checked and accepted in 1994; additionally, the digital map database of China (1:100,000, international version) was published in 1996, and the China dataset of the global map was completed in 2000. The paper also described the general information, data sources, updating data source, product and character of the 1:1,000,000 geospatial database of the 2002 version in detail.

The spatial database is the core technique platform of digital city data infrastructure. Li et al. (2004) discussed the characteristics of spatial databases and determined that their functions should be data sharing and interoperability. The overall database mainly consists of a metadata database, remote sensing mage database, electron vector map, GPS database, geo-coding database and basic unit database of social economy. Finally, the authors designed and implemented the Spatial Database-based J2EE and Oracle Spatial for digital cities. The main technique includes the JPEG2000 compression arithmetic-based Jasper source code and a data updating schema.

Zhao et al. (2004) believed that to address the problem of water resources in the Heihe River Basin, the system of water resource configuration and management for the Heihe River Basin must be built by means of new and more advanced technology; in particular, the design and implementation of the spatial database must be the centerpiece for the system. Spatial database technology based on RDBMS has incomparable advantages in terms of storage, management, maintenance and inquiry of great capacity special data. This paper presented the design and implementation of a large spatial database. Practice indicated that the object-oriented data model was most effective for managing massive spatial data, based on middleware technology, and a large RDBMS was taken as the platform.

When researching the design of a nonpoint source pollution spatial database of the Three-Gorge Region based on Geodatabase, Guo et al. (2004) introduced a Geodatabase model and designed a nonpoint source pollution spatial database of the Three-Gorge Region based on Geodatabase with the object-oriented method and UML. The result indicated that the object-oriented method can perform well in designing the spatial database with the Geodatabase model.

With guidance on the theory of digital earth and data warehouses, Li et al. (2006) designed a public data model based on fields and objects according to the memory characteristics of spatial information and nonspatial information in geographical cells and space. Then, the spatial databases of different attributes was constructed. The authors first analyzed and designed the spatial data warehouse of land resources in the Songnen Plain, which faced application with the utilization of spatial data engines and operational objects. Currently, the research and exploration is in the preliminary phase.

The data type of the rural power enterprise distribution network is complicated, and the data quantity is large. As a result, reasonably organizing the data of the electrical distribution system and providing support for the high-level analysis function of the electrical distribution GIS system or enterprise-level decision-making is the key design of the distribution network GIS system. Fang et al. (2006) introduced the data organization of the distribution network GIS system, the data acquisition method and the construction method of the spatial database to meet the needs of the system.

Lin et al. (2006) introduced the design and construction of the Guangdong Forest Management Information System (GFSDMIS) and proposed the goal, database structure, system function structure model and construction steps of the GFSDMIS.

He et al. (2006) introduced the Spatial Database for Mining Area's Geological Environment Evaluation System of Sichuan Province, which adopted an object-oriented Geodatabase geographical data model and stored all the spatial data in the relational database Oracle 9i through the ArcSDE spatial data engine connection. By combining the basic geographical data of the mining area with the special topic data concerning the mine environment, geological hazards and basic geology, they constructed the spatial geographic information framework of the central mining area and a B/S- and C/S-based system and realized the efficient storage, management and distribution of a large amount of precious data obtained during the mining area investigation in different areas of the Sichuan Province.

According to Wang (2007), to process the multisource and spatial features of geological material and the prominent spatial characteristics and complex structural features of geological data, a design proposition to establish an urban geological spatial database was proposed based on spatial database technology and combined with the geological characteristics of Shanghai. Meanwhile, professional demands were also considered. The database system was accordingly divided into four levels: fundamental databases, professional databases, geological databases and 3-D modeling databases. How to establish urban geological spatial databases was also discussed at length.

In the heuristic design of spatial databases based on ontology orientation by Yang et al. (2007), the design of spatial databases was often challenged by opposing needs, such as expertise needs and corporate needs. The ontology-oriented heuristic design of spatial databases balanced these needs by exploiting a domain ontology model. Starting from the geographical cognition mode, a heuristic spatial database design model based on ontology was discussed and constructed, and a practical spatial database system was completed.

Spatial data can be expressed by vector data structure and raster data, which may be managed through ArcSDE. Regarding ArcSDE as the space data engine and using the large-scale relation database (RDBMS), Wu et al. (2007) set up a three-layer system structure that included a database, server and client, realizing the effective organization and management of spatial data, and successfully applied it in practice.

Zhang et al. (2007) held that it was very important for the development of information technology in the power industry to store and analyze power networks using GISs. Ordinary spatial and topological data were necessary for the power networks in GIS, and the data for describing the logical structure of the networks were also necessary. To store and analyze the logical structure data and meet the needs of the uniform management of data and development of distributed information systems, a set of hierarchical tables based on the relation data model was designed to match the characteristics of the power networks by improving the traditional space+ attribution data management mode in GIS. The tables represented the logical structure data of the macro power networks and the local electric networks and were convenient in the connectivity analysis and flexible modification of networks.

Effective methods for storing and managing data are key points in the GIS fields. Zhang et al. (2007) discussed SDE's basic principles and provided a plan for building spatial databases from client to server circles. Through effective data storage and database management design, the database system based on the GIS method and database technology had complete functions for spatial data storage and management, as well as designs and implementations of a functional client based on MapObjects components, and finally, a complete spatial database system was formed.

Tang et al. (2007) thought that using modern information technology, network technology, GIS, RS, GPS and traditional ground investigation techniques to monitor grasslands was an inevitable trend. Construction targets, system structure, system database, system function, the technology train of thought and key technology were designed all around. The technology basis for the construction of the Sichuan Province grassland monitoring information system was established by the authors.

Aiming at the demand of state environmental monitoring information construction, combined with the demand of state environmental monitoring operations, a state environmental monitoring spatial data infrastructure was developed by Shen et al. (2007). Based primarily on a large commercial relational database system and spatial engine (SDE), a platform management and operation of slate environmental monitoring spatial data infrastructure was designed and implemented.

In the paper by Zhou et al. (2007), questions regarding the content, organization method, construction and realization method of the spatial database system for the distribution of ChangQing oil gas resources and surveying work were discussed. The process of system development with MapObjects2.0 component object technology and technology through programming with VB was studied. Spatial data management and solutions for the application of digital oilfield and ComGIS were proposed.

Hu et al. (2007) studied the spatial data model of an underground pipe network, discussed the characteristics of the underground pipe network data, and designed the data organization model, data storage structure and corresponding spatial data model.

8.4.3 Geospatial Database Construction Project Management

Geographic database construction project management involves the owner side, the designer, the construction unit, the supervision and other involved parties and must deal with and coordinate the project cost, quality, progress, materials, and other aspects, including engineering requirements control, quality control, schedule control, risk control and other management technology. For large and complex geospatial database projects in China, such as the 1:50,000 basic geographic database, in the management process, especially in the construction process, the task is heavy, the construction period is short, and is concerning to many departments, industries, and cross-operations. Therefore, research on the construction project management of geospatial databases has become a hot topic in database construction.

In the article 'Present situation and prospects of construction and management of GIS projects in China', Cheng et al. (2004) introduced the present situation of the construction and management of geospatial database in China. In the book '*Geographic information system engineering design and management*', Guo et al. (2003) presented the design principles, methods and specifications of the software engineering and data integration projects of GIS. In the paper 'Construction project establishment and management of urban foundation GIS', Yu (2008) held that a project construction should include the integration of existing data, the incensement of the necessary data, the building of a data platform, the long-term maintenance and updating of data, as well as data management and distribution, data standards and data sharing policies, which was a complex social coordination project. Project construction should make a scientific and reasonable long-term plan and a phased implementation plan under the framework of overall guiding ideology. In particular, attention should be paid to choosing the basic framework of data construction and data sharing as a breakthrough, developing from the primary to advanced phase and gradually toward perfection.

8.4.4 Geospatial Data Standard Formulation

A standard is a document that prescribes common and repeated use rules, regulations or characteristics of activities or their results to obtain the best order in a certain scope.

This document is formulated by consensus and approved by an accredited institution. The standard should be based on the comprehensive results of science, technology and experience to promote the best social effect.

Jiang et al. (1996) introduced the progress of standardization work abroad, especially the standardization work by the Geographic Information/Earth Information Technology Committee of International Standards Organization since its establishment in 1994. Then, we expounded on China's participation in international standardization activities and countermeasures to strengthen the standardization work of geographic information in China. The basic policy of geographic information standardization in China was proposed to strengthen the development and implementation of geographic information standards and promote the sharing of geographic information resources. Accordingly, a number of specific policy recommendations were suggested.

Based on the practice of the production of basic geographic information products, Zhang et al. (2004) analyzed the problems arising from the implementation of standards in actual production and put forward several suggestions on standardizing the current technical standards.

The key problem of data sharing mainly includes data structure, data model and unification of data format; geographic information standardization; spatial data quality; data interoperability; network construction; and policies and regulations. The solution to these problems will help to create a geographic information sharing environment. Among them, geographic information standardization is one of the key problems in geographic data sharing. Du (2003) introduced the latest development of domestic and international geographical information standards and proposed countermeasures that should be taken in the research and development of geographic information standards.

The purpose of the National Spatial Data Infrastructure (NSDI) is to promote the sharing of geospatial data in a unified, reliable and economic way. The national standard for spatial data accuracy supports the NSDI by providing users with a unified approach to evaluate and compare the accuracy of map and digital geospatial data.

The problem of geographic information standardization will be further introduced in the eighteenth chapter.

8.4.5 Geospatial Database Quality Control

Geospatial data are an abstract expression of the phenomena and processes of the real world. Due to the complexity and fuzziness of the real world, as well as the limitations of human understanding and expression ability, to a certain extent, this abstract expression cannot be completely true and is only close to the true value. On the one hand, the problem of data quality is inevitable in geospatial database engineering; on the other hand, the processing of geospatial data also leads to some quality problems. Geospatial data quality is the degree of accuracy, consistency and integrity, as well as the spatial data unity that can be achieved among the three types.

Therefore, spatial data quality is a relative concept and is only feasible to a certain extent. Despite this fact, we can still put specific applications aside to evaluate and control the quality of spatial data according to the objective regularity of spatial data.

The research on spatial data quality control will be introduced in the seventeenth chapter of this book.

8.4.6 Representative Geospatial Database Projects

1. **National fundamental geographic information system (1:1,000,000 database construction)**

The national 1:1,000,000 database is the minimum scale database in the multiscale database of the national fundamental geographic information system. In the construction process and after completion, spatial information and high-quality service are provided for dozens of departments, scientific research and teaching units and enterprises in China, and great social and economic benefits are obtained. This database is the largest-scale geographic space database delivered to the public at home and abroad by the State Bureau of Surveying and Mapping, and this database is an indispensable geographical basis for various departments in the country. A continuously updated national 1:1,000,000 database is urgently needed by all circles of society.

A basic surveying and mapping project, that is, the national 1:1,000,000 database update (2002 Edition), was undertaken and completed in 2002 by the Chinese Academy of Surveying and Mapping. Thus, China's 1:1,000,000-scale digital map (2002 edition) was established. This project passed the acceptance check by a panel organized by the State Bureau of Surveying and Mapping on April 10, 2003, and the national 1:1,000,000 database was considered to be an important part of the multiscale database of the national fundamental geographic information system; additionally, updates to the database improved the timeliness of the national fundamental geographic information system, and the updating results better met the needs of national economic and social development on fundamental geographic information. As a basis of spatial location and important basic information for various kinds of professional GIS, this database played a greater role in national macroeconomic policy, national defense, and many other areas. The project's results have been applied to dozens of departments and enterprises, including the comprehensive national geographic information system of the State Council, the E-government project of the International Liaison Department of the Central Committee of the CPC, the Chinese Academy of Sciences, the Yellow River Reconnaissance Planning Design Ltd, the China National Publishing Industry Trading Corporation, and the Chinese People's Armed Police Force. Good social benefits have been achieved.

2. National fundamental geographic information system (1:250,000 database construction)

The national 1:250,000 database construction is another major project of the State Bureau of Surveying and Mapping that after completion of the 1:1,000,000 database project and is an important part of NSDI construction. This project consists of 3 parts, the terrain database, the DEM database and the geo-place name database, which covers the whole country. Before data collection began, the 1:250,000 topographic map published in the 1980s was updated. Renewal of the overall database occurred until the end of 1995, and data in some of provinces continued to be renewed until to 1997. In the construction process, the project developed complete technical regulations and a unified construction plan in strict accordance with the issued national standards or industry standards and the relevant technical requirements, which ensure the normalization and standardization of the database.

The national fundamental geographic information system (1:250,000 database construction) is an application-oriented large-scale industrial system engineering system that is organically combined with production, scientific research and management. It is an important part of the national fundamental geographic information system and the NSDI and is also the key project of the Ninth Five-Year Plan of the State Bureau of Surveying and Mapping. To meet the needs of national economic construction and adapt to the process of China's national economic information, as early as the eighth five-year plan, the bureau began the overall deployment of China's national land fundamental geographic information system and completed the national 1:1,000,000 topographic database, geo-place name database and DEM database. Therefore, to further strengthen basic surveying and mapping, meet the construction requirements of the General Provincial Situation GIS, and provide better services for government decision-making, after more than two years of technical preparation, the bureau began the construction of a 1:250,000 database in 1996. The completion of the database marks a new stage in the development of China's spatial data infrastructure. This database will enable surveying and mapping departments to provide a new level of basic geographic information services to national economic construction.

3. National fundamental geographic information system (1:50,000 database construction)

Thus far, the 1:50,000 database construction is the largest investment project with the maximum engineering volume, the most complex technology, the longest elapsed time, and the most widely applied prospects. The 1:50,000 database consists of 7 sub databases, including the digital raster map (DRG), DEM, DLG, geographical names (GN), land cover (LC), digital orthophoto map (DOM), the corresponding metadata (MD), and an integrated management system. The database covers the national land scope, with a total of 24,218 sheets of 1:50,000 topographic maps (among them, the data of a 2,000,000 km^2 area in the western region is replaced by a 1:1,000,000 topographic map) and a total of 5.3 TB of data, which is currently largest multisource basic geographic spatial information database with rich content and high precision.

The construction work was undertaken by the National Fundamental Geographic Information Center.

Based on the original 1:50,000 topographic map data, parts of elements in the 1:50,000 database were updated using aerial or space image data, traffic data and interprovincial survey results; both the plane precision and elevation precision of DLG data meet the requirements of 1:50,000 database technical specifications; the accuracies of the DEM and DRG are reliable; and the overlapping accuracy of the DOM, LC, GN data and DLG data is good. The integrity of the database system, data accuracy, timeliness and safety all meet the design requirements. Data for the seven subbases were collected in the construction, and the databases were independently built. However, the relationship among subbases is coordinated, and their contents and elements are integrated. Therefore, their data contents are complete, and the relationships are reasonable. The subbases are unified and managed by a database integration management system, which has the functions of database loading and unloading, view management, querying and retrieval, data output, map production, database maintenance, order management, security management, and so on, and these subases can meet the needs of data management and data supplementation and has reached designed requirements. The project was well organized and managed in a standardized way, where a total of 8 state and industry standards and 32 temporary technical regulations were formulated to ensure the progress and quality of the data collection and database construction. The project adopts the mode of extensive cooperative building, which greatly improves the authority and efficiency of the data and avoids repeated construction.

The project has been in service for thousands of units of the department of land, urban planning, water conservancy, agriculture, forestry, environmental protection, transportation, communications, electricity, defense, scientific research and education and other departments. The benefits are significant, with broad application prospects.

8.4.7 Representative Geospatial Database

The national fundamental geographic information database stores and manages a variety of basic geographic information scales, including topographic, river system, residential land, transportation, place name and others, throughout the country. This database consists of a raster map database, vector terrain feature database, DEM database, place name database and digital image database. National 1:1,000,000 topographic databases (including place name databases), DEM databases and 1:4,000,000 topographic databases were built in 1994 by the State Bureau of Surveying and Mapping; national 1:250,000 topographic databases, DEM databases and place name databases were built in 1999; 1:1,000,000 DEM databases and digital ortho image databases for key prevention areas of the seven major rivers were built in 1999; national 1:50,000 digital raster map databases were built in 2000; national 1:50,000 DEM databases were built in 2002; national 1:250,000 and 1:1,000,000 topographic databases were updated; and finally, 1:50,000 place name databases, LC databases and TM satellite image databases were built in 2003. Subsequently, a national 1:50,000 vector feature database, digital ortho image database and others were built. Provincial 1:10,000 topographic databases, DEM databases, ortho image databases and digital raster map databases were built in all provinces, and the design and experimental studies of provincial and municipal basic GIS and their databases were conducted.

1. **Topographic database**

The topographic database is a database that is constructed by collecting, editing and processing the data of spatial location, attribute information and spatial relationship of objects, and various types of elements (including river systems, boundaries, transportation, residential land, terrain, vegetation, etc.) are organized in a hierarchical structure on a national basic scale topographic map according to certain rules and a standard classification code. According to the overall design of the national basic geographic information system, the scale of the national topographic database is divided into three levels of 1:1,000,000, 1:250,000 and 1:50,000, while the scale of the provincial topographic database is divided into three levels of 1:250,000, 1:50,000 and 1:10,000.

(1) National 1:4,000,000 topographic database

The database is derived from a 1:1,000,000 topographic database through data selection and generalization. Data are divided into 6 layers, including the main river (level 5 and above), main road, all railways, residential areas (county and above), boundaries (county and above) and contour lines (contour interval is 1,000 m).

(2) National 1:1,000,000 topographic database

The content of the database mainly include survey control points, river systems, residential land, traffic, boundaries, terrain, vegetation, and so on. This database uses a 1:1,000,000 scale topographic map as the data source and implements the national standard of "*National land basic information data classification and coding*" (13923-92 GB/T).

(3) National 1:250,000 topographic database

The database contains a total of 819 maps and is divided into 14 data layers, including the river systems, residential land, railways, highways, boundaries, terrain, other elements, auxiliary elements, coordinate network and data quality. The database is stored in terms of the geographical coordinate system and Gauss-Krueger projection coordinate system. The data were mostly obtained by the end of 1995, and part of the region's data was obtained until the end of 1997.

(4) National 1:50,000 vector feature database

The database are composed of river systems, contour lines, boundaries, traffic, residents and other core terrain features, including the spatial relations among terrain features and related attributes. The database is based on the Gauss-Krueger projection, the 1980 Xi'an coordinate system, the 1985 national height datum, and the 6° zone division.

2. **Place name database**

The database is a spatial relation database. This database is built by inputting all kinds of geographical name annotations (including the names of residential areas, rivers, lakes, mountains, peaks, seas, islands, deserts, basins and natural protection areas; and their Chinese phonetic alphabet and attribute characters, such as the category, administrative code, ownership, grid number, traffic code, elevation, map number, name, publishing date and update date, X and Y coordinates, longitude and latitude) into the computer. This database is connected with a topographic database through a technical interface code, which can be accessed by each other and can also be used as separate relational databases.

(1) National 1:250,000 place name database. The main content of this database is all kinds of place name information and its related information in the 1:250,000 topographic map database, such as the Chinese phonetic alphabet, administrative divisions, coordinates, elevations and map information. The database uses six tables, including geographical name information, administrative division information, the relationship between the map and political areas, the categories of place name and the cross-references, and a comparison of administrative divisions and administrative region codes. The first 4 tables are the basic information tables; the latter 2 tables are the auxiliary information tables. This database contains a total of 805,431 place names.
(2) National 1:50,000 place name database. With the latest version of the 1:50,000 topographic map as a basis, place names were updated based on the combination of interior work and on-site verification; the national and provincial administrative division brochures provided by civil administration departments, gazetteers, census maps of place names and other geographical name data, as well as the latest achievements in surveying and mapping have been fully used. The verification and collection of the place name data above the village administrative system in the whole country was carried out, and a total of approximately 5,000,000 place name data were verified and acquired for a 1:50,000 topographic map; the amount of data is 1.2 GB, and nearly 1,400,000 place names were updated, which is approximately 26.4% of all place names. The time to completion of the place names database was the end of 2002, and renewal of the place names of street offices, towns, townships and villages was at the end of 2000. After the withdrawal of townships in 2001, the data in 9 provinces were updated.

3. Digital elevation model database

The DEM database is a spatial database. It is a regular grid elevation database that is built by collecting ground elevation (a set of points in the x- and y-planes or on an ideal ellipsoid) at regularly spaced intervals. The acquired vector topographic elements (contour line, elevation point or landform structure line) and some river system elements can be used as original data to derive interpolated point data and can also use the digital photogrammetric method to acquire data from aerial images directly. Among them, the grid value on land and islands represents the ground elevation; the grid value for ocean areas represents the depth.

(1) National 1:1,000,000 DEM database. More than ten thousand 1:50,000 and 1:100,000 topographic maps were used to collect the elevation value of the grid intersection point in accordance with the mesh spacing of 28.125 * 18.750 (difference in longitude * difference in latitude). After editing and processing, with a 1:500,000 map as a unit, the elevation values are stored in the database. The maximum allowable error of the height of the original data is 10–20 m.
(2) National 1:250,000 DEM database. The original data used to generate the DEM include contour line, elevation, bathymetric contour, soundings and some rivers, large lakes, reservoirs, and so on. Then, TIN interpolation is used to obtain the 1:250,000 DEM, which is stored in Gauss-Krueger projection coordinates and geographical coordinates. For the DEM in Gauss-Krueger projection coordinates, the grid size is 100 m * 100 m, with the map as a unit. The data of each map are delineated by a rectangle that contains the scope of the map; there are overlaps between adjacent map sheets. For the DEM in geographical coordinates, the grid size is 3" * 3"; the number of rows and columns for each map is 1201 * 1801; and the map is square shaped.
(3) National 1:50,000 DEM database. This database is produced by an all-digital method. Part of the database is based on data from the 1:50,000 database, that is, a 25 m * 25 m mesh DEM generated by using the TIN and GRID modules in Arc/Info software. The storage format is Arc/Info GRID, with a 6° zone Gauss-Krueger projection, 1980 Xi'an coordinate system and 1985 National Height Datum.

4. Digital raster map database

This database is a spatial database. It is a grid database built through the scanning, geometric correction, color correction and editing of published maps. This database can manage the DRG data directory and support data distribution. The smallest unit of storage and retrieval in a library is generally a map, which can be managed in a map/area.

The national 1:50,000 digital raster map database is the digital form of the existing 1:50,000 analog topographic map. The scan input resolution is 400–600 dpi and can be output with 4 m ground resolution. The data are stored based on a single 1:50,000 topographic map with a TIFF (LZW compression) format. The DRG database contains data from more than 19,000 1:50,000 topographic maps, covering approximately 70–80% of the entire territory.

5. **Digital ortho image database**

This database is a spatial database. It is an image database consisting of various airborne and space-borne remote sensing data, ortho images generated through the scanning of image data, radiometric corrections, geometric corrections and projection error corrections by using a DEM, as well as the occasional use of the main residential areas, place names, boundaries and other vector data. The image can be panchromatic and color, as well as multispectral. Image data can be stored in a compressed way to save storage space. The scale series is consistent with that of the topographic database.

The 1:50,000 digital orthophoto database is a spatial image database constructed based on the digital orthophoto dataset, which is generated through pixel-by-pixel geometric correction of the scanning digital image data of aerial photographs and cutting and mosaicking according to the standard scope of the 1:50,000 map. The image data are produced in accordance with the accuracy of the 1:25,000 topographic map, with a 1 m ground resolution, and therefore, the data have simultaneous map geometric resolution and image features.

6. **Land cover database**

The LC database consists of the Landsat satellite remote sensing images received before and after 2000, covering the land area of the whole country. A total of 752 1:250,000 sheets are included, and the amount of data is approximately 12 GB. LC is divided into 6 first-level classes and 24 second-level classes, with 6° zone Gauss-Krueger projection and includes raster and vector data formats. Database management uses ArcSDE and the ArcMap platform of 8i Oracle, which can meet the needs of retrieval, querying, browsing and distribution services.

7. **Airborne and space-born image database**

The database is a spatial image database that is designed and constructed by using various kinds of aerial and space-born remote sensing data or scanned image data as the data source, and the characteristics include multitemporal resolution, multispectral resolution, multispatial resolution, and multiple grayscale resolution.

(1) Aerial image database. This database includes a scanned aerial image database, aerial photo preview image database, aerial photo location database and the parameter database of aerial photographic documentation. The database contains the aerial photographic data of China since the 1950s, with a scanning accuracy of at least 4 μm. The database is currently under construction.

(2) Satellite image database. The spatial image database consists of the image data of the Earth observation satellite and its processing data. A TM satellite image database was built with data sources of 15 m-resolution panchromatic images and 30 m-resolution multispectral images obtained by the Landsat 7 satellite

ETM+ sensor. This database includes a total of 522 image scenes, covering the land area of the whole country. The data source of the SPOT satellite ortho image database is SPOT panchromatic data (10 m resolution) covering all national land areas (except for a few desert areas in Xinjiang and Tibet).

8. **1:4,000,000 soil spatial database**

Taking the China Soil Map (published in 2004) as a base map, according to the FAO soil classification system, a *1:4,000,000 Soil Type Distribution Map in China* was compiled. The data sources include the *Distribution Map of Soil Organic Carbon in China*, which is based on the soil species of the second soil survey of the Ministry of Agriculture and the results of previous work by soil scientists in the soil survey area, as well as the further compilation and calculation of the soil type and attribute data. The *Distribution Maps of Soil Organic Carbon in China* under different vegetation conditions, different vegetation and climatic conditions are compiled separately by overlapping the soil map, vegetation map and climate distribution map.

9. **1:4,000,000 China active tectonics spatial database**

With ArcGIS and MapInfo as software platforms, based on the China active tectonics data from Academician DENG Qidong's years of painstaking research and summary, using the unified coordinate system, the consistent projection method and the reasonable database logical structure, the China active tectonics spatial database is established. The effective management and system integration of the active faults, active basins, active folds and all kinds of related parameters, as well as mutual querying between graph data and attribute data, are realized. This database is convenient for use by each earthquake-related research department and engineering construction department.

8.5 Database Applications

Geographic spatial databases are widely used in national economies and national defense construction. In the construction of the national economy, the main application areas are listed as follows: urban and rural planning, dynamic monitoring and management, land use and dynamic supervision, monitoring of energy and mineral resources, mineral resources exploration and development management, hydrology and water resources dynamic monitoring, water resources development and optimal allocation management, water environment, water ecology monitoring and prediction, monitoring and forecasting of soil and water loss, weather forecast and major natural disaster monitoring, risk assessment, early warning and situation assessment, monitoring and evaluation of forest resources, wetland resources, wildlife resources and desertification, monitoring and management of forestry key construction and ecological protection projects, early warnings and forecasts, monitoring and evaluation of major forest disasters, atmospheric environmental monitoring and prediction, dynamic monitoring and evaluation of national ecological environments, coastal zone and marine resources exploration, development and management, marine ecological environment protection and management, marine disaster monitoring and early warnings, dynamic monitoring of agricultural and rural resource environments, emergency monitoring of sudden environmental pollution accidents, dust storm monitoring, environmental effects monitoring and evaluation of regional resource development, urban environmental monitoring, intelligent transportation and traffic information services, serving the public, electronic commerce, e-government, urban management and digital cities (Cui et al. 2007). As examples, in the article by Zhang et al. (2003), through the establishment of an ecological environmental background index system, using the method of spatial overlay analysis, a comprehensive evaluation of the ecological environment on a large scale was realized; by calculating the index of the ecological environmental background, the overall characteristics of China's ecological environment and regional differences were reflected more comprehensively and systematically with numerical values; these values can be directly used in regional comparative studies of national ecological environment backgrounds and their main influencing factors. They can also provide support for the monitoring of time series dynamic changes and the further research of regional characteristics.

Pan et al. (2003) analyzed the spatial data object, the spatial database, and the spatial database technology and its realization; the applications of cadastral management systems based on spatial database technology in spatial data sharing, massive data management, large range query retrieval, distributed spatial data processing and maintenance, cadastral mapping and data security control, and other aspects were studied. Based on the construction of a practical application system, the Cadastral Management Information System solution based on spatial database technology was proposed. By comparing the investigation results with those of the erosion statistics in 1987, it was concluded that the soil erosion area in Nayong County was reduced. This finding showed that soil and water loss management had achieved some results.

Cui et al. (2004) outlined the importance of the global mineral resources potential evaluation spatial database and considered it to be the basic database for evaluating the global geological and mineral resources potential, which involves a variety of information resources in the processes of economic, geographical, geological, mineral, physical and chemical exploration, remote sensing, exploration and production. They introduced the background, current situation and trends in global mineral resources database construction at home and abroad from the global, regional (or national) and local (or important metallogenic belts) levels, as well as discussed the research contents of the global mineral resources potential evaluation spatial database, and proposed a layer partitioning scheme to evaluate the potential of global mineral resources.

According to the requirements of information construction of national environmental monitoring, combined with the needs of environmental monitoring businesses, the construction of the national environmental monitoring spatial data platform was studied and discussed in Shen et al. (2007); based on the large-scale commercial relational database system and spatial data engine, the platform management and operation of the national environmental monitoring spatial data were designed and initially implemented.

In the article by Li et al. (2007), using Shaoguan city in the mountainous area of northern Guangdong as a research area, based on remote sensing and GIS technology, a set of evaluation index systems for sustainable land use in the study area was constructed, and a quantitative evaluation model for sustainable land use was proposed. Then, the evaluation index data were standardized and spatially processed. The land sustainable utilization evaluation based on raster spatial data was carried out in the study area, the land sustainable utilization situation in the study area was summarized, and feasible suggestions for sustainable land utilization in the area were put forward.

In the article by Qin et al. (2007), the integration modes of GIS and hydrological models were summarized, and the characteristics of distributed hydrological model databases and basic databases were compared and analyzed. According to the design steps for spatial databases, a distributed hydrological model database was designed from two aspects of function and structure and was subsequently established. Then, the spatial database was integrated into the distributed hydrological model database, and it played a key role in the model's operation, which improved the operating environment of the distributed hydrological model and improved the overall operation level.

According to Li et al. (2007), crop yield estimation has been an important topic for a long time in China. To this end, the characteristics and architecture of the spatial data warehouse were discussed, and the crop yield estimation methods were evaluated. In view of the complexity of the factors influencing crop yield estimation, the spatial data warehouse technology was introduced, and a multidimensional data model was established. Therefore, crop yield estimation was carried out, which provided a reference for decision-making by the Department of Agriculture.

8.6 Problems and Prospects

The research of spatial databases is based on the theory of geospatial information science, computer science, information science and database technology and involves a number of basic disciplines and application technologies. The research contents are comprehensive and multifaceted.

Considering the functions of geographical spatial databases, technologies including geospatial data acquisition and database building, spatial data representation, spatial data index, data querying and manipulation are the main technologies of these databases. At present, there are still some problems to be studied in geospatial databases.

(1) Abstract expression of geospatial phenomena

To efficiently extract data and organize the spatial data of different structures and the corresponding topological relations, it is necessary to study multiple expression modes of spatial data to meet the requirements of data consistency and accuracy, as well as the requirements of the data model, links, multi-agencies, and multi-scales of data.

(2) Organization of geospatial data

The organization of geospatial data includes the theory and method of the spatial data model, data structure, physical storage structure and spatial data index. Centering on the establishment of three-dimensional and multidimensional GIS, long-term research and further practice are also required for problems such as 3-D spatial data models, spatial and temporal data models, 3-D topological data structures, 3-D and spatiotemporal databases, and 3-D spatial queries and visualizations, as well as spatial and temporal data queries and visualizations. With the research and development of 3-D GIS, multimedia databases and spatial-temporal databases, the requirements of searching and updating functions for multidimensional spatial objects has become increasingly urgent. At present, the query efficiency of commonly used spatial indexing technology for three-dimensional or higher dimensional spatial data is low or even ineffective.

(3) Architecture of the spatial database

Issues to be studied include the distributed processing and application of the C/S mode, which will enable the spatial database to possess Internet/Intranet connection capability and can realize the functions of distributed transaction processing, transparent access, cross-platform application, heterogeneous network interconnection, and multi-protocol automatic conversion; the methods for realizing the data model, structure, algorithm and user interface of spatial database, the spatial query language, the logic-based query language, and the specific content and organization of metadata, as well as the data compression and encryption method also need to be researched.

(4) Geospatial database system

This system includes research on the SDE, query language and database architecture, geographic spatial data management system, geospatial database design and system engineering method.

Although commercial space DBMSs have their own characteristics and innovations in terms of storage models, process management, spatial queries, data cache and secondary development, these systems have poor portability, low flexibility, and poor scalability of application models. Therefore, these aspects are among the problems that needs to be solved to build a spatial data engine that is independent of a commercial database.

(5) Spatial relation language

This language is mainly studied based on the standardization of geospatial concepts, the use of natural language and mathematics methods to form the theory of spatial relationship expression, the calculation model of location expression, the acquisition and expression of spatial concepts, the definition of topological relations, the visualization of spatial information, and the user interface of spatial databases.

(6) Geospatial data sharing

The main studies of geospatial data sharing mainly include the theory, applicability and processing method of spatial data sharing, such as geospatial data specification, standards and metadata and the theory and method of multisource spatial data fusion, integration and interoperability; additionally, spatial data fusion, mutual operation and spatial data warehouses are studied to realize the integration and management of various distributed and independent spatial database systems; finally, according to user needs, studies are conducted through a variety of professional models to link a variety of thematic information, analyze them from a multidimensional perspective, and meet the needs of users' spatial decision-making support.

The spatial database based on the web provides the possibility of information sharing. However, it is still in the development stage, and many key issues remain to be addressed. Spatial data storage, retrieval and the establishment of indexing structures, as well as how to query and retrieve ultralarge amounts of spatial information rapidly and effectively based on the distributed architecture and the provision of efficient data transmission and display mechanisms, are problems yet to be solved.

(7) Geospatial data security

Spatial data sharing and spatial data security are a contradiction. To a certain extent, this contradiction limits the application and development of geospatial data. How can spatial data sharing and application be achieved to ensure the safety of geospatial data? This is a difficult problem facing the surveying and mapping management department.

The development and integration of computers and related fields have created unprecedented conditions for the development of spatial database systems. The advanced database system constructed by new technology and new methods will bring revolutionary changes to the spatial database system, which is reflected as follows:

a. The application of an object-oriented model makes the spatial database system more abundant in terms of semantic representation and has the ability to simulate and manipulate complex geographic objects.
b. The development of multimedia technology broadens the application field of geographic spatial database systems. Now, generalized geographic information not only includes graphics, images and attribute information but also includes audio, video, animation and other multimedia information.

c. Virtual reality technology promotes the visualization of geospatial data. Geospatial data are converted into a virtual environment, and people can enter the virtual environment, looking for the relationship between different datasets and feel the environment described by the data.

In the future, the ideal spatial database system should be a complex variable object, object-oriented, active, multimedia, visual and secure network database system.

8.7 Representative Publications

(1) ***Cartographic database system*** (WU Hehai. Surveying and Mapping Press, Beijing, November 1991)

In this book, the general principles of the database, spatial data structure, cartographic DBMS and the structured processing of cartographic data are introduced. The discussion proceeded step by step, from elementary to profound, from theory to practice, with a large number of illustrations and examples. The whole book is divided into 9 chapters and 1 appendix. The main contents include the introduction, data and file organization, data model, architecture of database, introduction of map data model, application of vector data model in general map data modeling, application of surface data model in thematic map data modeling, map DBMS and application of map database technology in map data processing. Basic mathematical knowledge related to database technology is introduced in the appendix.

(2) ***Navigation geographic database*** (JIANG Jie, HAN Gang, CHEN Jun. Science Press, Beijing, October 2003)

The geographic database that supports ITS and LBS is called the "navigation geographic database", known simply as the "navigation database". This book systematically discusses the technical characteristics, implementation methods and key technologies of navigation databases. The whole book is divided into 7 chapters. Chapter one briefly introduces the origin and development of ITS and LBS and expounds upon the function and implementation steps of the navigation database in ITS and LBS; chapter two discusses the basic concepts of the navigation database, including its definition, technical characteristics, key technical problems and so on; chapter three discusses the conceptual model of the navigation database in detail and addresses the classification and definition of spatial elements, the definition of spatial features' attributes, the definition of spatial features' relationships, and the representation method of spatial features; chapter four introduces the logical model of the navigation database based on the relational data model; chapter five explores the data organization and index method of the navigation database; chapter six introduces the production and service of navigation data; and finally, chapter seven presents the application of the navigation database.

(3) ***Spatial database indexing technology*** (GUO Wei, GUO Jing, HU Zhiyong. Shanghai Jiao Tong University press, Shanghai, April 2006)

This book consists of 13 chapters. The first chapter briefly describes the basic concepts of data objects and data organization, databases, spatial databases, database index structure and so on. The second chapter introduces the data file storage and organization. The third chapter introduces several commonly used data indexing techniques. The fourth chapter mainly introduces the related concepts of relational database retrieval and several representative database indexing techniques. In the fifth chapter, the characteristics of spatial databases and spatial retrieval are analyzed, and the requirements and classifications of spatial indices are discussed. The sixth through eleventh chapters focus on several typical spatial index structures, algorithms and performance analyses, including the spatial index based on binary trees, the spatial index based on quad trees, the spatial index based on B-trees, the grid index based on dynamic hash, the indexing based on spatial object sequencing and the spatial index based on QR-trees. The twelfth chapter focuses on the related concepts of spatiotemporal database index technology and several representative spatiotemporal index techniques. In the thirteenth chapter, the application and development trend of spatial index technology in several commercial databases are briefly introduced.

(4) ***Principles of the spatial database*** (CUI Tiejun. Science Press, Beijing, April 2007)

This book comprehensively and systematically discusses the basic concepts, principles and methods of geospatial databases, which involve the main contents of all aspects of geospatial databases. The whole book is divided into 11 chapters, which include the development process and trend of spatial data management and the computer representation of geospatial entities, as well as basic data structure, physical organization of geospatial data, spatial index method, spatial data model, database architecture, relational database interface technology, SDE, spatial DBMS, spatial query language, geospatial database design method, geospatial database construction method, data collection and processing, spatial data acquisition and quality evaluation, geospatial data warehouse, and metadata and spatial data interoperability.

(5) ***Spatial database*** (WU Xincai. Science Press, Beijing, May 2009)

The book systematically discusses the theory, practice and new development of spatial databases. In the theoretical section, we systematically elaborate on the basic concepts, basic principles and basic methods of spatial databases, as well as abstract expressions of spatial phenomena, spatial data models, spatial data index technology, spatial data organization and management, queries and access of spatial data, temporal spatial databases, and spatial database design. In the practical part of the book, we expound upon the idea, method and architecture of spatial database construction, taking the Jiangxia District of Wuhan city as an example, and we discuss the MapGIS 7 spatial database application in land use planning. In the new development section, distributed spatial database technology and spatial data warehouse technology are introduced.

(6) ***Research on the cadastral management database information system*** (ZHANG Xinchang, TANG Liming. Science Press, Beijing, June 2009)

This book consists of 7 chapters, each of which is related to each other, and each has a certain focus that combines theory and practice. In terms of theory and technology, this book expounds upon the connotation and function of cadasters, the structure and organization of cadastral spatial data, the management mode of cadastral spatial data, the design of cadastral management databases, and the technology and method of urban and rural Integrative Cadastral Management. In terms of practice, this book includes the quality inspection of cadastral data, cadastral management systems and software engineering. The book systematically discusses the basic concept of cadasters, modern cadastral property management and application of cadastral management information systems, as well as the latest theory and development trends and application examples in land planning and management. Presently, the main content and key technologies are the focused technical and methodological issues in the research and application of the GIS cadastral database.

References

Cao, Z., & Liu, Y. (2002). An object-oriented spatio-temporal data model. *Acta Geodaetica et Cartographic Sinica, 31*(1), 87–92.

Chen, J., & Guo, W. (1998). A 3D topological er model based on space partitioning. *Acta Geodaetica et Cartographic Sinica, 27*(4), 308–317.

Chen, J., & Zhao, R. (1999). Spatial relations in GIS: A survey on its key issues and research progress. *Acta Geodaetica et Cartographic Sinica, 28*(2), 95–102.

Chen, M. (2006). Forestry spatial database construction and information system integration. *Forestry Exploration and Design,* (1), 69–75.

Chen, Z., Wu, X., & Xie, Z. (2007). Study of distributed index mechanism of geospatial data. *Microelectronics & Computer, 24*(10), 54–57.

Cheng, C., & Guo, S. (2004). Present situation and prospect of construction and management of GIS project in China. *Geomatics World, 2*(5), 28–30, 35.

Chi, T., Qi, Q., & Liao, K. (2002). The construction of National natural map database and the development of information products. *Acta Geographica Sinica, 57*(B12), 18–26.

Cui, F., & Lian, C. (2004). Model of global mineral resources potentiality assessment spatial database. *Geological Science and Technology Information, 23*(2), 99–102.

Cui, T. (1990). Design and realization of MDBMS. *Journal of the Institute of Surveying and Mapping, 7*(1), 98–102.

Cui, T., & Dong, Y. (2000). Spatial relation of the map data and data models. *Journal of the Institute of Surveying and Mapping, 17*(3), 194–196.

Cui, T., & Guo, L. (2007). The study for multisource geospatial vector data integration and fusion. *Journal of Zhengzhou Institute of Surveying and Mapping, 24*(1), 1–4.

Deng, S., Wu, C., & Meng, L. (2006). Spatial data integration services for grid GIS. *Computer & Digital Engineering*, 34(2), 39–43.

Du, D. (2003). Development status of geographic information standards. *Geo-Information Science, 5*(2), 74–78.
Du, S., Qin, Q., & Wang, Q. (2005). Qualitative reasoning of topological relations between uncertain regions. *Acta Geodaetica et Cartographic Sinica, 34*(4), 349–354.
Fang, J., Zhang, C., & Yu, S. (2006). Space database design of the rural power enterprise distribution network GIS system. *Journal of Northeast Agricultural University, 37*(4), 508–511.
Gong, J. (1996). Formal spatial object model and its implementation. *Acta Geodaetica et Cartographica Sinica, 25*(4), 309–314.
Gong, J. (1997). An object oriented spatio-temporal data model in GIS. *Acta Geodaetica et Cartographic Sinica, 26*(4), 289–298.
Guo, J., Zhou, Z., & Guo, W. (2003). Research of indexing techniques for spatial databases. *Application Research of Computers, 20*(12), 12–14.
Guo, L., Cui, T., & Wang, Y. (2007). Study for multi-resources geospatial data integration. *Geomatics World, 5*(1), 62–66.
Guo, P., Liu, H., & Xie, D. (2004). Research on designing non-point source pollution spatial database of the three-gorge region based on geodatabase. *Computer and Modernization,* (8), 83–85.
Guo, Q. (1993). Establishment of complex object based on cartographic data base. *Journal of Wuhan Technical University of Surveying and Mapping, 18*(Suppl.), 59–62.
Guo, Q., & Wang, X. (2003).*Engineering design and management of geographic information system.* Wuhan University Press.
Guo, R. (1994). A philosophical study on spatial information. *Acta Geodaetica et Cartographic Sinica, 23*(3), 236–240.
Guo, Z., Wang, Y., & Yang, C. (2002). Research on index and query for spatial data and GIS application. *Computer Engineering and Applications, 38*(23), 176–178, 205.
He, Y., Gu, J., & Jiang, L. (2006). Design of the spatial database for mining area's geological environment evaluation system of sichuan province. *Land and Resources Informatization,* (6), 2–5.
Hu, D., & Li, J. (2007). Research on spatial data model of underground pipe network. *Sci-Tech Information Development & Economy, 17*(28), 224–226.
Hu, P., Hu, Y., & Yang, C. (2002). Design thinking, technical route and some theoretical issues on the geospatial digital framework. *Geomatics and Information Science of Wuhan University, 27*(3), 283–288.
Huang, M., & Chen, Z. (2005). Research on the spatial index based on improvement QR-tree. *Journal of Heilongjiang Institute of Technology, 19*(3), 18–20.
Huang, M., & Zhang, Z. (1997). N1NF relational basis of spatiotemporal data model. *Acta Geodaetica et Cartographic Sinica, 26*(1), 1–6.
Jia, W., & Zhou, Z. (2007). Geospatial data sharing models of land and resources. *Land and Resources Informatization,* (1), 22–25.
Jiang, J., Chen, C., & Han, G. (2003). *Navigating geographic database.* Science Press.
Jiang, J. (1999). The design and application research of national 1:25 million database of national basic geographic information system. *Remote Sensing Information,* (4), 14–18.
Jiang, J., & Liu, R. (1996). Progress of GIS standardization in foreign countries and China's Countermeasures. *Remate Sensing Information,* (3), 19–22.
Jiang, W., Xia, S., & Xia, Z. (2007). Study on GML-based data integration for multi-sources spatial data. *Computer Engineering and Design, 28*(14), 3310–3311, 3356.
Ju, S. (1999). Visual query language CQL for spatial database. *Chinese Journal of Computers, 22*(2), 205–211.
Lai, L., Li, D., & Xu, H. (2006). Research on spatial data index structure based on D-tree. *Computer Engineering, 32*(12), 102–103, 106.
Lan, X., Lv, G., & Liu, D. (2006). Research on query and index for GML spatial data. *Journal of Remote Sensing, 10*(6), 36–45.
Lei, H., Leiming. (2007). Study on key questions of urban and rural integrative cadstral sapatial database. *Science and Technology Square,* (9), 93–96.

Li, H., Peng, P., & Fang, L. (2007). Research of crop yield estimation based on spatial data warehouse. *Journal of Agricultural Mechanization Research*, (3), 162–164, 167.

Li, H., & Guo, J. (2003). Think about the distributed storage and management of multi-scale geographical spatial data. *Geo-Information Science, 5*(3), 56–59.

Li, J., & Hou, B. (2007). Sustainable land use evaluation in mountain area of north Guangdong based on the raster spatial data. *Hubei Agricultural Sciences, 46*(5), 719–722.

Li, J., Huang, J., & Chen, D. (2003b). Method of spatial database building in land use planning of Guilin. *Journal of Guilinin Statute of Technology, 23*(4), 449–452.

Li, B., Chang, D., & Liang, H. (2006). Research in data model of geographical database engine. *Journal of Zhengzhou Institute of Surveying and Mapping, 23*(5), 54–56.

Li, J., Tan, Q., & Yin, M. (2006b). Integration method of distributed spatial data based on oriented object. *Journal of Guilin University of Technology, 26*(4), 514–517.

Li, K., Chen, Y., & Jin, X. (2006c). Discussion on information sharing of urban and rural planning and some problems of spatial data standardization. *Planners, 22*(12), 13–15.

Li, Q., Chang, L., & Wang, L. (2004). The design and implementation of spatial database for digital city. *Computer Science, 31*(3), 89–91, 105.

Li, Q., & Li, D. (1998). Research on the conceptual frame of the integration of 3D spatial data model. *Acta Geodaetica et Cartographic Sinica, 27*(4), 325–330.

Li, W., Ding, J., & Miao, Q. (2007). Research summary of spatial data multi-scale. *Computer Knowledge and Technology (Academic Exchange)*, (13), 134–136.

Li, X., Qiu, J., & Peng, L. (2006d). Study on conditions of multi-source spatial data integration technologies and their application prospects. *Computer and Modernization*, (5), 39–42.

Li, X., & Huang, C. (2004). Data assimilation: A new means for multi-source geospatial data integration. *Science & Technology (Beijing)*, (12), 13–16.

Li, Z., Tang, J., & Sun, P. (2006). Spatial data warehouse's designing of land resource in songnen plain based on integration of multi sources data. *Application Research of Computers, 23*(9), 155–159.

Li, Z., Sun, Q., & Fan, G. (2006). Discussion on the security system of geographic spatial data construction. *Fujian Computer*, (6), 46–47, 53.

Li, Z., & Ju, S. (2006). Research and application of dynamic audit policy for secure spatial database. *Journal of Computer Applications, 26*(11), 2678–2681, 2684.

Li, Z., Peng, M., & Yao, C. (2007c). Construction and application of digital Wuhan spatial data infrastructure. *Digital Library Forum*, (6), 38–43.

Liang, C., Sun, Z., & Yao, Y. (2006). The core technique study and application of distributed spatial data sharing model. *Engineering of Surveying and Mapping, 15*(2), 57–59.

Liao, H., Wang, J., & Zheng, Y. (2004). Research on a multi-source spatial database engine. *Application Research of Computers, 21*(2), 138–140.

Lin, S., & Wei, S. (2006). Design and construction of the Guangdong forest management information system. *Central South Forest Inventory and Planning, 25*(3), 50–53.

Liu, C., Pei, J., & Sun, H. (2004). Multi-scale spatial index structure based on R-tree. *Shenzhen University Journal, 21*(3), 242–246.

Liu, H., Yang, Z., & Gu, H. (2007). Inquiry & appraisal on reasonable allocation of groundwater resources in Yinchuan plain by building space database. *Ningxia Engineering Technology, 6*(2), 187–189.

Liu, Y., Li, Y., & Liu, R. (2004b). Construction of spatial database for railway. *Journal of Northern Jiaotong University, 28*(2), 96–99.

Liu, Z., & Liu, H. (2007). Design of system of multi-source heterogeneous spatial data integration based on GML. *Computer Engineering and Design, 28*(8), 1962–1965.

Lu, L., Zhu, G., & Liu, R. (2007). Design and implementation of distributed raster spatial database engine. *Geo-Information Science, 9*(4), 64–69.

Lu, Y., Xiang, X., & Pan, P. (2006). The index structure and its implementation algorithm for the CQR-tree spatial database. *Computer Engineering & Science, 28*(6), 108–111.

Ma, L., & Gong, J. (2003). Research on spatial database query oriented natural language. *Computer Engineering and Applications, 39*(22), 16–19.

Ma, Z., Pan, M., & Lin, C. (2002). Research of sharing and integration mode for multi-source spatial data. *Computer Engineering and Applications, 38*(24), 31–34.

Man, J., Liu, Q., & Yang, D. (2004). Research on the representation method of spatial data. *Computer Applications, 24*(11), 97–99.

Ni, H., Qiu, X., & Cao, X. (2006). Research of spatial database engine. *Engineering of Surveying and Mapping, 15*(1), 17–19.

Pan, Y., Zhong, E., & Liang, J. (2003). A study on the cadastral management information system based on technology of spatial database. *Geographical Research, 22*(2), 237–244.

Qin, F., Tang, F., & Liu, F. (2007). Integrated application of spatial database in distributed hydrological model. *Journal of Anhui Agricultural Sciences, 35*(20), 6024–6025, 6028.

Shen, W., Wang, Q., & Wang, W. (2007). The design and realization of state environmental monitoring spatial data infrastructure. *Environmental Monitoring in China, 23*(5), 44–47.

Shi, H., Li, X., & Shao, X. (2007). Design of mandatory access control in spatial database management system VISTA. *Computer Science, 34*(10), 149–151.

Song, G., Zhong, E., & Liu, J. (2000). A study on seamless integration of multi-sources spatial-data. *Progress in Geography, 19*(2), 110–115.

Sun, Q., & Li, M., & Lu, J. (2007). Scale issue and its research progress of geospatial data. *Geography and Geo-Information Science, 23*(4), 53–56, 80.

Tan, G. (1998). Study of integrated spatial data structure and spatial index mechanism. *Acta Geodaetica et Cartographic Sinica, 27*(4), 293–299.

Tang, C., Su, Z., & Zhang, X. (2007). Technology design of sichuan province grassland monitoring geographic information system based on multiple source spatial data. *Remote Sensing Technology and Application, 22*(5), 661–667.

Tian, P. (2007). Analysis of multi-source spatial data integration. *Digital Library Forum,* (7), 15–16.

Wan, J., Li, G., & Wang, X. (2003). Spatial data models of 3D UGIS based on geometric primitives. *Application Research of Computers, 20*(3), 46–48.

Wang, C., Du, D., & Zhu, Z. (2008). Spatial data fusion model and method. *Marine Geology Letters,* (1), 37–41.

Wang, D. (2006). Research and implementation of construction of 1:50 000 scale topographic database in china. *Geomatics World, 4*(4), 4–5.

Wang, J., Yuan, Q., & Xu, Z. (2006). The solutions to the spatial data sharing of grassland resources based on WebGIS. *Chinese Journal of Grassland, 28*(2), 66–70.

Wang, J., Wu, F., & Wang, G. (1998). An automatic mapping system based on map database. *Bulletin Of Surveying And Mapping,* (9), 25–27.

Wang, L., & Zhou, Y. (2002). A study of 3D GIS object-oriented data model and its application. *Acta Geodaetica et Cartographic Sinica, 31*(3), 274–277.

Wang, Q., & Wang, J. (1998). Study and practice on GIS data model. *Journal of the Institute of Surveying and Mapping, 15*(4), 64–67, 70.

Wang, W., Li, G., & Wu, S. (2005). The establishment of wetlands GIS spatial database in heilongjiang province. *Journal of Northeast Forestry University, 33*(3), 53–55.

Wang, Y., Gong, J., & Dai, J. (2007). Spatial data semantic query based on ontology. *Journal of Geomatics, 32*(2), 32–34.

Wang, Y. (2007). Design of spatial database for urban three dimensional geology. *Computer Engineering and Design, 28*(5), 1176–1178.

Wu, H. (1983). Map database management system. *Miniature Computer System,* (1), 46–55.

Wu, L., & Zhang, Y. (2002). The integrated framework on distributed multi-spatial database system. *Geography and Territorial Research, 18*(1), 6–10.

Wu, M., Cui, W., & Mei, X. (2007). Design and construction of spatial database based on spatial database engine. *Computer Engineering, 33*(6), 48–50, 76.

Xia, Y., & Peng, D. (2002). Design and implementation of concurrency control of R-tree in spatial database system. *Computer Engineering, 28*(7), 242–244.

Xia, Y., Zhu, X., & Zhou, C. (2007). Research on feature based spatial data similarity query. *Computer Engineering and Applications, 43*(25), 15–17, 47.

Xie, G., & Su, J. (2005). Spatial data index technique and its application in GIS software. *Journal of Hainan Normal University (Natural Science), 18*(4), 372–376.

Xu, C., Cui, J., & Li, J. (2007). Research on realizable system structure for spatial data manipulation language. *Science of Surveying and Mapping, 32*(2), 166–168.

Xu, S., Wang, M., & Wang, W. (2006). A new index structure for moving object spatial database based on R tree and quan tree. *Computer & Digital Engineering, 34*(3), 54–57.

Yang, B., Li, Q., & Mei, B. (2000). Study of the visualization of 3D urban model. *Acta Geodaetica et Cartographic Sinica, 29*(2), 149–154.

Yang, C., Rong, F., & Peng, Z. (2001). Study on distributed processing of spatial dais based on client server architecture. *Computer Engineering and Applications, 37*(8), 87–89.

Yang, J., Li, Y., & Cai, G. (2007). Heuristics design of spatial database based on ontology-oriented. *Application Research of Computers, 24*(5), 24–26.

Yin, Z., Li, L., & Ai, Z. (2003). A study of spatio-temporal data model based on graph theory. *Acta Geodaetica et Cartographic Sinica, 32*(2), 168–172.

Yu, L. (2008). Construction project and project management of urban foundation GIS. *China Surveying and Mapping,* (5), 84–85.

Yue, X., Ju, S., & Li, Z. (2002). The spatial data indexes. *Application Research of Computers, 19*(2), 32–34, 41.

Zeng, Y. L. N., & Chen, Z. (2004). Construction of fundamental geographic database system in Guangdong Province. *Information of Land and Resources*, (1), 20-23.

Zhao, H., Jiang, Y., & Jiang, M. (2004). Design and implementation of the space database for the configuration and management system of water resources in Heihe River Basin. *Water Resources and Hydropower Engineering, 35*(4), 1–4, 8.

Zhang, Q., & Wang, Z. (2004). QR-tree: A kind of spatial index structure based on R-tree and quad-tree. *Computer Engineering and Applications, 40*(9), 100–103.

Zhang, Q. (2006). Spatial data sharing and service of geological map based on SIG. *Progress in Exploration Geophysics, 29*(1), 56–61.

Zhang, W., & Gong, J. (1995). Object-oriented data model for complex feature in GIS. *Acta Geodaetica et Cartographica Sinica, 24*(4), 293–300.

Zhang, X., Yuan, X., & Yuan, Y. (2007a). New method for designing the spatial database of power network. *Electric Power, 40*(2), 75–79.

Zhang, Z., Qin, Q., & Dong, K. (2007b). The design and implementation of spatial database based on ArcSDE. *Microcomputer Information,* (33), 133–135.

Zhang, Z., Yang, C., & Tian, G. (2003). Comprehensive assessment and analysis of china eco-environment based on spatial data. *Journal of Remote Sensing, 7*(1), 59–66, 87.

Zhang, Z., & Liang, M. (2004). Approach to related problems regarding standards of foundational geographical information. *Geomatics & Spatial Information Technology, 27*(2), 44–45, 48.

Zheng, J., & Chen, Z. (2004). Development of oracle spatial database by using OCI and OLE. *Computer Engineering and Design, 25*(6), 1007–1009.

Zheng, Y., & Liao, H. (2002). An extended data model for spatial database and its query language GSQL. *Computer Engineering and Applications, 38*(3), 123–125.

Zheng, Y., Liao, H., & Chen, Z. (2004). R-tree index in a spatial database engine. *Computer Engineering, 30*(5), 38–39, 97.

Zhou, Z., & Zhou, Y. (2007). Design and implementation of distribution of Changqing oil gas resources and the spatial database. *Science Technology and Engineering, 7*(21), 5652–5654.

Zhu, G. (1992). Hainan province 1:5 million digital map database system software. *Map,* (2), 24–26.

Zhu, H., & Feng, Y. (2001). Semantics for multilevel secure spatial database. *Mini-Micro Systems, 22*(9), 1086–1091.

Zou, Y. (1999). The map database system design based on client/server structure. *Acta Geodaetica et Cartographic Sinica, 28*(3), 251–256.

Chapter 9
Digital Cartography

Qun Sun and Haiyan Liu

9.1 Introduction

Digital cartography is a brand-new mapping technology that is developed and perfected with the extensive use of computer technology, graphics and image processing technology, network technology and spatial databases, as well as other related technologies in the field of cartography. The digital cartographic method has replaced the traditional production method of manual mapping, enhanced the scientific nature of the map production process, freed the map workers from heavy manual work, significantly shortened the period of mapping, improved the timeliness of the map, transformed the situation where the map product form is overly unitary, enriched the type of map products, changed the application mode of the map, and broadened its field applications, as well as lowered the professional threshold of mapping so that more people can participate in the production of the map, which deepens people's understanding of the map (Wang 2000). Cartographic theory has moved from ancient cartography theory and modern cartography theory to contemporary cartography theory, which is a fundamental change in cartography (Chen 1994).

Under traditional map production conditions, due to the long mapping cycle and difficulties updating and modifying maps, any changed content in the field is difficult to reflect on maps in a timely manner, which results in a lack of timeliness and inconvenience of use; dependence on an individual's production experience often affects the scientific nature of map works and the consistency among different maps. Therefore, computer technology was introduced into the field of cartography, which formed and developed into digital mapping technology. The first step in digital mapping technology is to enter maps into computers so that they become data and can be recognized and dealt with by computers. To ensure that the data have a wide range of uses, the map contents entered into computers are usually divided into four

Q. Sun (✉) · H. Liu
Information Engineering University, Zhengzhou, China

J. Wang and F. Wu (eds.), *Advances in Cartography and Geographic Information Engineering*, https://doi.org/10.1007/978-981-16-0614-4_9

types of information, that is, coordinate position, attribute code, and description of the relationship between the elements, and name information. The coordinate position provides the actual location of cartographic contents on the map; the attribute code assigns some regular digital codes to a category and grade according to the classification and grading system of map contents and links them with the map contents to facilitate computer identification and processing; a description of the relationship among elements is also called topological information, which describes the spatial structures and relationships of different or the same types of mapping objects in the form of a relational table; and finally, the name information is an important part of the map symbol system, mainly including the pronunciation of geographical names and descriptive texts, which for the convenience of processing and application are managed as a single type of data. The map contents can be stored, queried, analyzed and applied under the computer environment by means of this expression method. Therefore, through the study of map data acquisition, transformation, management, analysis and graphics output and other related issues, the digital mapping process can be realized (Zhu et al. 2007).

If we say that the formation and development of anything must include the internal conditions and external environment, then the external conditions of digital mapping technology formation are the development of computer software and hardware, as well as the continuous development of graphics and image processing technology research. The internal condition is that the mapping workers do their best to use new technologies to replace difficult manual labor, while the contents on maps are appropriately organized and expressed so that they can be created with computers.

9.2 Development Process

Research on digital mapping technology in the world began in the 1960s.

In 1958, a flatbed plotter was developed based on a numerical control machine by the Gerber Corporation of the United States, and an NC graph-plotter was successfully created by the Calcomp Company, forming the early automatic plotting system. In 1961, a professional committee on automatic cartography was established by the International Cartographic Association (ICA), and as a platform for international academic exchange, this committee was changed to the professional committee on computer-aided mapping in 1964. In 1963, the first set of human–machine interactive computer-aided drawing systems was developed by the Massachusetts Institute of Technology. In 1964, an Oxford automatic mapping system was established by the University of Oxford, which had drawn several map works by means of simulating manual mapping. Around the same time, the SYMAP system, a mapping system with a line printer as a graphic output device, was developed by the Computer Graphics Laboratory at Harvard University, USA. The Oxford automatic mapping system and SYMAP system have made pioneering contributions to the development of computer cartographic technology. By the end of the 1980s and the beginning of the 1990s, computer digital mapping technology has been fully used in the production of large-scale topographic maps, cadastral maps and middle-scale topographic maps, as well as aerial topographic maps and cadastral maps in western developed countries (Zeng 2014).

The start and development of digital mapping technology in China was approximately 10 years later than in the west, which can be traced back to the middle and late 1970s. Digital mapping technology has undergone a development process that started from scratch, from exploration to maturity; digital mapping technology is now widely used, and it contains the continuous efforts and hard work of the majority of mapping workers over nearly 40 years. The development of digital mapping technology in China in the past 40 years can be divided into three stages: the stage of experiments and explorations, the stage of development and improvement, and the stage of production, application and interdisciplinary and multi-industry development (Mo et al. 2005).

9.2.1 The Stage of Experiments and Explorations

This stage is the initial phase of the development of digital mapping technology in China, which is from the beginning of the 1960s to the end of the 1970s. This first stage mainly included the exploration of the basic theory, experiments on basic algorithms and development and improvement of the most basic drawing equipment to form the initial digital mapping system with a structure is composed of a manual tracking digitizer, a computer and a digital control plotter. By using a manual tracking digitizer and keyboard, map materials (mainly including paper maps and statistical data) were converted into digital mapping information and were stored in a card or paper tape in accordance with digitized order, and then, after computer processing, the map was drawn on a plotter (Fig. 9.1). Computer-aided mapping software basically includes a variety of plotting programs for drawing various kinds of lines and feature symbols. The lines drawn by this initial automatic mapping system are poor quality and cannot be applied to production practices except for principle experiments in the laboratory (Xu 2006).

In fact, China's cartographic workers began to think about how to realize the automation of mapping as early as the 1960s. At that time, academician CHEN Shupeng proposed that map production needed to draw on the achievements of

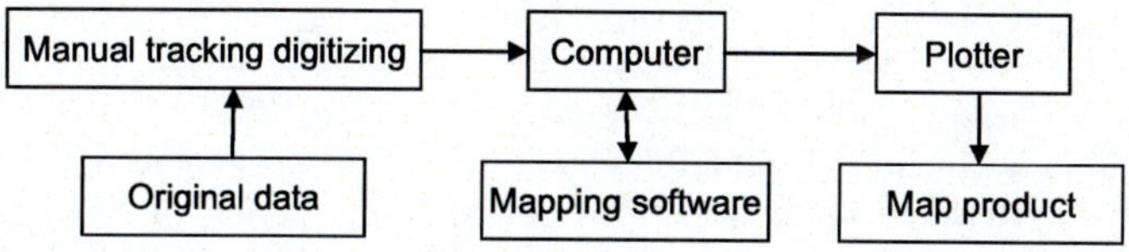

Fig. 9.1 Digital map system in the early stage

modern physics, chemistry and technology; with the introduction of electronic technology, the automatic calculation and simulation design in mapping production technical processes was a theoretical problem to be studied. However, over the following 10 years, due to the influence of the Great Cultural Revolution, the development of digital mapping technology in China was basically in a state of stagnation. At the beginning of the 1970s, under the guidance and organization of Professor WANG Jiayao, the Zhengzhou Institute of Surveying and Mapping began to cultivate and set up a teaching team for computer-aided mapping. In the mid-1970s, academician CHEN Shupeng began to lead the research of cartographic automation and put forward the idea and scheme of cartographic automation systems through investigation and survey, where based on the analysis of a geographical environment information series, the digitization of maps by means of automatic scanning and tracking was realized; additionally, through computer information processing, the automation of mapping and the semi-automation of map compilation were realized by automatically controlling the reproduction of maps according to the specified program (Fu 2009). At the end of the 1970s, the Zhengzhou Institute of Surveying and Mapping set up a computer-aided mapping laboratory and began to cultivate undergraduate students. Several other surveying and mapping, geographical research and teaching departments also set up their own research entities of computer-aided mapping and carried out research activities, including the automatic drawing of latitude and longitude lines, the automatic transformation of map projection in different regions, the acquisition of elevation data from scanned contour lines, vectorization and formation of digital elevation models, a digital software package for contour lines, and a program library for automatically drawing topographic map symbols, and some achievements were obtained (Yu 2002). In June 1977, the Geography Department of Nanjing University plotted a 1:50,000 scale map; in 1978, LIU Guangyun, a Professor of the Zhengzhou Institute of Surveying and Mapping, used a computer to plot whole topographic map contents, including automated topographic map border decoration, and he achieved success; LIU Yue, a researcher at the Institute of Geography at the Chinese Academy of Sciences created a variety of computer programs, including curve smoothing, contour map drawing, statistical graph and trend surface analysis through the experimental study of mapping automation; and in 1981, Professor YANG Qihe and Professor WU Zhongxing completed research in map projection transformation in an electronic computer-aided mapping environment.

9.2.2 *The Stage of Development and Improvement*

This stage was from the end of the 1970s to the end of the 1980s. This stage mainly served to implement further experiments, as well as improve and develop digital cartographic systems to perfection. This period was a period of rapid electronic technology development, when a variety of new computer-aided mapping equipment appeared, such as semiautomatic digitizers, scanning digitizers, drum plotters, optical plotters and human–machine interactive compilation equipment. Digital mapping

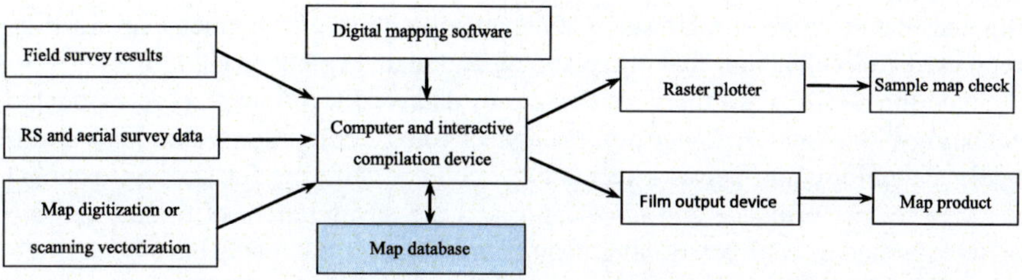

Fig. 9.2 Development and improvement of the digital cartographic system

software was further developed, a variety of graphics packages were created, and algorithms for various typical mapping problems were studied. The map database, place name database and geographic information database were established. The application results of the digital cartographic system in this period were mainly reflected in the digital mapping of various thematic maps; for example, the applications to geology, meteorology, population statistics, environmental protection and other aspects made high speed and quality breakthroughs, some of which are difficult to perform by hand. These advances have further promoted the development of digital cartographic technology for general maps (Xu 2006). Figure 9.2 illustrates the structure of a relatively mature digital cartographic system created in this period.

This was a period during which digital cartographic technology developed rapidly in China. Computer-aided cartographic systems were designed and developed or introduced in many scientific research and production departments, as well as applied to map compilation and data processing, and many atlases were produced. However, map publishing still used the traditional map printing method. The successful map works of computer-aided mapping include *the Population Atlas of the People's Republic of China*, *the National Economic Atlas of China*, the *Atlas of the Third Industry of the People's Republic of China*, and the *Environmental Quality Atlas of Tianjin City*.

9.2.3 The Stage of Production, Application and Interdisciplinary and Multi-industry Development

This stage occurred from the late 1980s to the present. Along with the development of computer and digital mapping software and hardware, the emergence of a variety of high-end computers, high precision graphics and image input/output devices, as well as the wide range of graphics workstation and computer network applications, coupled with the more powerful function of computer software, a very favorable environment was created for the development and application of digital cartography. Simultaneously, the theory and technology of digital cartography also began to develop to the depth of production and the breadth of application. During this period, various types of map databases and geographic information systems were established and widely used in automatic map production, environmental evaluation, trend prediction, decision-making, planning and management (Xu 2006).

Since the 1990s, with the development of database technology, object-oriented technology, graphics and image processing technology, animation technology, multimedia technology and network technology, coupled with the further improvement of hardware performance and price ratio, and the development of remote sensing technology and spatial positioning technology, broader prospects have been available for the development and application of digital mapping technology. On the one hand, whole digital map production technology has been extensively used for map production to gradually replace manual map production operations; on the other hand, the arrival of the information revolution has moved cartography in the direction of producing basic geographic information at the overlap of many subjects, and digital mapping technology began to penetrate multiple disciplines and fields, thus expanding its scope of application. Digital cartography has developed rapidly and is widely applied because it is not a simple combination of complex digital map processing equipment and traditional mapping methods but a great technological revolution in the field of cartography. The effect of this revolution is equivalent to or greater than the significant changes in mapping technology caused by the invention of the printing press and the emergence of photogrammetric technology (Liang 2006).

At this stage, the development and application of digital mapping technology is mainly reflected in the following aspects:

1. **Production and updating of large-scale basic geographic information data**

To meet the needs of the national economy, military and other areas, China has carried out production and updating of large-scale basic geographic information data. Basic geographic information products include digital line maps, digital elevation models, digital raster maps, and digital image maps (that is, 4D products), the most important of which is the production and updating of digital line maps. To unify the data standard of digital line maps and to provide unified basic geographic information for geographic information application, basic geographic information classification, coding, digital production standards, data quality checks and supervising systems, geographic information interchange data formats and other standards were developed by the state and the army. The core of this information is basic geographic information classification and coding. According to the demand of geographic information applications and the development of geographic information systems, all elements of series-scale topographic maps, charts and aviation maps were integrated, and based on statistics, analysis and categorization, the principle and method, feature coding and its definition for the classification and grading of basic geographic information elements were proposed. These features guarantee the correct expression and understanding of the vector data of the national basic scale map elements. The representative results from this period are listed as follows: national 1:250,000 map database (created in 1997, updated in 2005), national 1:50,000 elevation database (created in 2000), national 1:50,000 basic geographic information database (created in 2005), national 1:1,000,000 map database, world 1:5,000,000 map database, and world 1:4,000,000 map database.

2. **Using the map publishing system to realize whole digital map production**

For early digital mapping systems, after digital editing, modification and symbolization, the draft drawn by a printer or plotter was only a final manuscript (Kan et al. 2000). To obtain a film that meets the requirements of publication, printing and publishing of maps were required, including fair drawing, reproduction, color separation drawing, photocopying and other processes. The whole project process was complicated and had a long cycle, which became a bottleneck that restricted the development of digital mapping technology. With the continuous progress of printing technology, computer word processing and laser phototypesetting technology were widely used in the printing and publishing industries, which had a profound impact on cartography. The traditional map printing work may be finished by a prepress system. Through the combination of digital mapping technology and a desktop publishing system, a desktop map publishing system was formed. Map compilation results were output in a prepress data format, such as a PS/EPS file, and then output as a color film with high precision through a laser phototypesetting system for direct plate printing. These factors enabled map production to achieve practical applications and embark on the road to whole digital map production. In December 1996, based on Intergraph's map publishing system, the *Atlas of Water Conservancy of the Jiangsu Province* was compiled by the Zhengzhou Institute of Surveying and Mapping, which was the first thematic atlas in China that used a total digital production scheme (Zou et al. 1998). In January 1997, based on CorelDraw software, the *Atlas of Shenzhen City* was finished at Wuhan Technical University of Surveying and Mapping, which was the first comprehensive atlas created by total digital production. The *National Physical Atlas of China*, which was made by the Institute of Geography, Chinese Academy of Sciences, reached the international advanced level in the late 1990s. In addition, several other atlases were finished in this period, such as the *Educational Atlas of China, Atlas of Solid Fuel and Nonmetal Resources of China, Geological Atlas of China, Atlas of the Plague and Its Environment in China, Atlas of Sustainable Development in the Changjiang Economic Zone*, and *Atlas of Shanghai City*. The maps in the above atlases were generally produced by using vector graphics software, including CorelDraw, Illustrator, Freehand, MicroStation, AutoCAD, MapGIS, and EC Founder (Liao 1993).

3. **Creation of a multimedia electronic atlas**

The multimedia electronic atlas is another important representation of map results. There will be many books in the bookstore, and regardless of what type of books they are, they will be placed on the shelf, waiting for people to read. The *Atlas of the Ecological Environment in Beijing, Tianjin and Tangshan Are*a was published in 1989 and is the first electronic atlas in China. Since then, the research, design and production of electronic atlases have developed rapidly in China and have formed their own characteristics in electronic atlas map reading, retrieval and analysis. A number of electronic atlases have been published, including the *National Economic Atlas of China* (electronic edition), *National General Atlas of China* (electronic edition), *Electronic Atlas of Hong Kong*, *Electronic Atlas of Shenzhen City*, *Electronic Atlas of Beijing City*, *Electronic Atlas of Population and Environment of China*, *Electronic Atlas of Sustainable Development in the Changjiang Economic Zone*, *China Electronic Map*, and so on; the production technology of the electronic atlas and the independently developed software systems have gradually matured. Along with the application and popularization of the Internet, the dissemination of electronic atlases gradually shifted from CD discs to networks.

4. **Integrated production technology of paper maps and geographic information**

Once we achieved the construction results of the geographic information database, it was necessary to use the existing geographic information for the production of paper maps. The main technical means used in the construction of map databases involves carrying out digitization of the existing paper map; consequently, the obtained digital map has poor timeliness. Map production with these data must involve compiling and modifying the map contents and updating the map contents using the relevant data to obtain up-to-date map products while simultaneously updating the existing geographic information. Commonly used cartographic software has the functions of compilation, modification and output of map data, but this software cannot manage geographic attribute data and does not support updating, modifying, or saving geographic attribute data; therefore, it cannot be used for the production of geographic information. Geographic information system software (such as MAPINFO and ARCGIS) has relatively strong functions of geographic data management and geographic analysis, but it cannot satisfy the expression effect and design requirements in terms of spatial information expression and map output. Notably, the emergence of integrated digital mapping technology successfully solves the problem and enables updates to be included in geographic information during the production of paper maps. Figure 9.3 illustrates the integrated production process of the paper map and geographic information. Integrated production is a process of making maps while producing geographic information, that is, obtaining the two products of paper

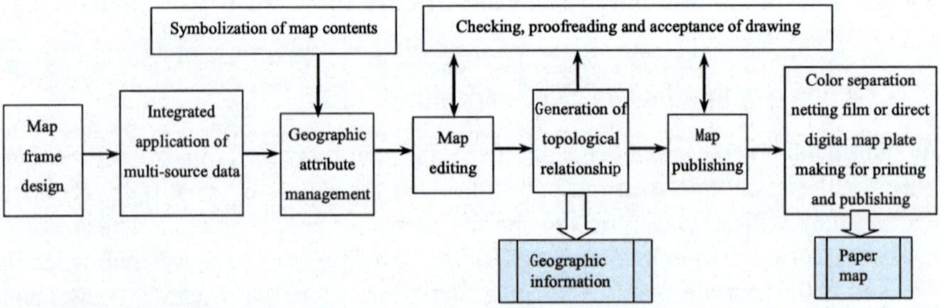

Fig. 9.3 Integrated production process of the paper map and geographic information

maps and geographic information in one production process, which saves considerable manpower and material resources, reduces duplication of labor, improves the quality of results, and ensures the consistency of paper maps and geographic information. It has practical significance in large-scale map production or renewal. The 1:250,000 digital mapping system is a successful example of the application of integrated technology, which not only completed the update of the national 1:250,000 map database but also produced 1:250,000 paper maps covering the whole country (Li et al. 2003a).

The superiority of digital cartography is obvious, which can be summarized with the following points: first, the results of digital mapping are easy to store and will keep non-deformation in data storage, thus improving the accuracy of the map; second, the application of digital mapping technology makes the contents of a map easy to edit, modify and update, and map contents can be added or modified according to the map user's request to increase the adaptability and practicality of the map and the universality of the user; third, the application of digital mapping technology not only reduces the labor intensity for the cartographer but also decrease the deviation caused by a cartographer's subjective casualness in the mapping process, paving a way for further standardization of mapping; fourth, using advanced laser film output (or digital direct plate making) technology and integrated digital mapping technology increases the speed of map-making, shortens the cycle of production, and improves the technological processes of mapping and printing; additionally, the map design, computer mapping, map printing and other contents are organically connected, which enhances the relationships among them and promotes the mutual infiltration of relevant contents; fifth, digital mapping technology can be used to produce multimedia electronic maps, as well as slope, aspect and visibility, three-dimensional maps, and so on, increasing the map varieties and broadening the service range; and finally, the use of digital mapping technology links the maintenance and updating of map database to the production process of new maps, the production of a map is combined with the acquisition and updating of geographic information, and it becomes possible to realize the mutual transformation between the three basic forms of a map (digital map, electronic map, and paper map).

9.3 Main Research Achievements

9.3.1 Research on the Related Concepts of the Digital Map and Digital Cartography

1. **Digital map**

A digital map is an abstract representation of geographic information in the real world and is a collection of geospatial data. The representation and storage form of a digital map in a computer is a set of data, which is composed of coordinate positions, attributes and a certain data structure, and a digital map can be reproduced as a symbolic map (called the screen map or electronic map) on a computer screen and output device through the processing of software and the application of symbolic methods. A digital map can also be output on film or other ways to obtain a paper map. According to the organizational form and characteristics of the data, digital maps are divided into four types: digital line graphs, digital elevation models, digital raster maps and digital ortho image maps, namely, 4D products.

2. **Digital line graph**

Digital line graph (DLG) is a vector dataset based on an existing topographic map, which also preserves spatial relationships among elements and related attribute information. The DLG product is the most important of the 4D products, but it is also the digital map product that has the largest investment and the heaviest workload, as well as the most complicated establishment and maintenance (update). At present, the following DLG products are available:

(1) World vector digital map data: these data were produced based on the digitalization of a 1:15,000,000 world map, including the world's coastlines, national boundaries, capitals, main rivers, and other main features; and the data are organized in accordance with the topological structure.
(2) National 1:4,000,000 topographic database: the data contents include major rivers (level 5 and above), main roads, all railways, residential areas (at or above the county level), boundary lines (at or above the county level), and contour lines (contour intervals equal 1000 m). The main features can be freely downloaded from the home page of the national basic geographic information system, which provides a variety of spatial data application software formats: MIF format of MapInfo; geospatial data transfer format; E00 format of Arc/Info; SHP format, and so on.
(3) National 1:1,000,000 topographic database (public release): The main contents include control points, river systems, residential land, traffic, boundaries, terrain, vegetation, and so on; E00 format in Arc/Info; data volume is 160 M; and the data deadline was June 2000, which was recently updated.
(4) 1:500,000 China traffic database: this database was produced based on the 1:500,000 topographic map and national traffic survey data. The main contents include railways, roads, residential areas, mountains, rivers, lakes, shorelines, place names, boundaries, and so on. The data were organized in accordance with topological structure; the database was completed in 1995 and updated in 2007.
(5) 1:250,000 topographic map database: this database was produced by the Bureau of Surveying and Mapping of General Staff, and the database is based on the digitalization of the 1:250,000 topographic map, which covers the whole country. The main elements are the control point, independent feature, residential land, transportation, pipelines, boundaries, river systems, terrain vegetation, soil, place names, etc. The data were organized in accordance with topological structure; the database was completed in 1997 and updated in 2005.

(6) 1:250,000 topographic map database: this database was produced by the State Bureau of Surveying and Mapping, and this database uses Arc/Info as the platform for data collection and database building; it was completed in 1997; the data volume is 8.2 GB; the data deadline was December 1995, and it was updated in 2005.
(7) 1:50,000 basic geographic information database: these data are mainly from aerial photographs and map digitalization, including the core elements of more than 24,000 pieces of 1:50,000 maps, which covers the whole country. The project was launched in 2005 and completed in 2001.
(8) Chart database: this database is completed by chart production departments; Arc/Info was used as the platform for data collection and management; this database is regularly updated; additionally, Notices to Mariners and electronic charts are released to the public.

3. **Digital elevation model**

The digital elevation model (DEM) is the digital expression of surface undulation in a computer, which intuitively reflects the landform situation of the real world in a computer. The main products are described as follows:

(1) 1:250,000 digital elevation model: a construction project simultaneously completed with the 1:250,000 vector map database, which is produced at intervals of 3 s or 100 m, with a 50 m elevation design accuracy.
(2) 1:50,000 digital elevation model: this DEM was produced based on 1:50,000 topographic maps and aerial photographs and was completed in 2000.
(3) National 1:1,000,000 digital elevation model: this DEM was produced based on more than 10 thousand pieces of 1:50,000 and 1:100,000 topographic maps; the DEMs consist of a sampled array of elevations for a number of ground positions at regularly spaced intervals of $28.125'' \times 18.750''$ (difference of latitude/longitude); there are approximately 25 million elevation points, stored in database with 1:500,000-scale.

4. **Digital raster graphic**

A digital raster graphic (DRG) is a digital product formed through scanning, correction, image processing and data compression of paper maps. It is a computer raster file that is completely consistent with the map. Now, series-scale topographic maps are all converted into digital raster maps.

5. **Digital orthophoto map**

A digital orthophoto map (DOM) is an image map formed through a series of processes, such as pixel correction and image mosaicking, with coordinate grid and border decoration, and line elements, text notes, etc., can also be added. Compared with traditional maps, orthophoto maps have the characteristics of abundant information and are easy to read; they also have short production cycles and are easy to update.

6. **Electronic map**

Electronic maps, also known as “screen maps”, are a new variety of maps formed by the use of digital map technology after the 1980s. It is a visual product of digital map data displayed by multimedia technology. Electronic maps can be stored on the hard disk, CD-ROM, DVD-ROM and other digital storage media. These maps can be operated interactively through the corresponding operation interface; the contents are dynamic, can be displayed on the computer screen, can also be printed on paper, and can perform displays, querying and a variety of statistical analyses. The connection and difference between the electronic map and digital map is described as follows: the digital map is the data foundation of the electronic map, and the electronic map is a symbolic map of the digital map on the computer screen.

7. **Cartographic database**

A cartographic database is a database based on a digital map, which is a collection of digital information files (the various elements of a map that are stored in a computer, such as the control point, landform, land type, resident land, hydrology, vegetation, traffic, boundary, etc.), database management systems and other software and hardware. Based on database technology, digital map technology and spatial information system technology, the map database was developed to meet the demands of spatial data in the field of information processing and to meet the needs of modern society for digital map products. Some developed countries in the world began to study cartographic database technology in the 1970s, and several typical databases were built in the middle of the 1980s. The exploration of cartographic databases in China began at the end of the 1970s, the establishment of cartographic databases for local areas began in the middle of the 1980s, and the widespread establishment and application of cartographic databases began in the 1990s.

8. **Digital cartography**

Based on computer science and cartographic theory, digital cartography is a theory and technique for studying and solving the problems of acquisition, transformation, storage, management and graphics output of cartographic spatial data, with the aid of certain hardware and related system software. The task of digital cartography is to use digital technology to implement map making and production, to establish a certain form of map database, and to provide basic geographic information for geographic information systems. Three major problems need to be solved in digital mapping (Qi et al. 2005): the first is the digitization of continuous map contents so that the computer can read and recognize the contents; the second problem, according to the requirements of production and application of geographic information, is processing a series of digital map data using a computer and forming a certain form of digital map products or creating a cartographic database; the third problem is converting the relevant contents of a digital map into a readable map graphic. The process of transformation from the graph to the digit and then from the digit to the graph is the essence of digital cartography.

9.3.2 Related Technologies of Digital Cartography

1. **Map information recognition and extraction technology**

Research on map information extraction in China began in the middle of the 1980s. The units that engaged in early research in this field include Peking University, Xi'an University of Electronic Science and Technology, Wuhan Technical University of Surveying and Mapping, and Zhengzhou Institute of Surveying and Mapping (Xie 2011). Since the 1990s, research on information extraction has received increasing attention. Many useful explorations of the theory and method of map pattern recognition have been conducted. Among them, Lv et al. (1995), Xi'an University of Electronic Science and Technology, studied the methods of color map recognition and linear feature tracking; LI Deren, Wuhan University, proposed an automatic raster to vector conversion method for CCD scanned contour map recognition, which applied the mathematical morphology method to the contour map processing; Lin et al. (1992) proposed a thematic map reading scheme based on computer vision and explored map symbol segmentation, neural network recognition and map color segmentation, as well as other issues in subsequent studies; Huang (1994) focused on the identification of map symbols, using neural networks, structural features and other methods to identify symbols on the map; Wang et al. (1999) investigated in detail the application of mathematical morphology in map recognition; Hao et al. (1997) focused on the automatic extraction of map elements, such as residential land, from the structural analysis; Ye et al. (1999) realized point symbol recognition based on knowledge and a fuzzy neural network; Sun et al. (2005) implemented point symbol recognition and semiautomatic extraction of color map information based on the wavelet theory. These achievements are widely employed in digital map production.

At present, the domestic software dedicated to map vectorization includes GeoScan of Beijing Geoway Technology Development Co., Ltd., FreeScan of Xi'an Institute of Surveying and mapping. In addition, all large GIS software platforms basically have a scanned vector function or module. Among them, the module in MapGIS, a basic GIS platform developed by Wuhan Zhongdi Digital Technology Co. provides two methods of fully automatic vectorization and semiautomatic vectorization. The above software uses vector tracking, manual adding, editing and other techniques to generate vector graphics, which realizes the semiautomatic extraction of map information. They have the functions of contour lines, roads, river system automatic (semiautomatic) line tracking, urban housing corner point extraction, and elevation values or text recognition.

2. **Multisource data integration and fusion technology**

Along with the practical application of surveying and mapping technologies and the diversification of geographic spatial information, a wide variety of data is available for map production, the sources have become more diverse, and the timeliness is getting better. In addition to the basic geographic information data, remote sensing images, GPS measurement data, place names, aerial photographs and geomagnetic data are all used for digital map production and updates. In the mapping process, reasonable and effective utilization of up-to-date material (data) may guarantee the quality of the map and improve the timeliness of the map, which can largely solve the problem that the newly mapped product is out of date. It is not easy to make full use of all kinds of information since the information and data are very complicated, for example, the use of different coding systems, the use of different spatial coordinate systems, the use of different map projections, and the heterogeneous data storage format. Its essence is the application of integration assimilation technology of multisource data in the field of digital map mapping. Liu (2006) studied the use of multisource spatial data to update and produce basic geographic information data; Cui et al. (2007) realized the integration and fusion of digital topographic map data and digital chart data; An (2008) used data integration and fusion technology to update and produce military digital maps and achieved good results.

3. **Map symbolization and map symbol database technology**

The map symbol is the language of the map, and it is the basis of geographic information visualization. At present, the research on map symbols is mainly focused on two aspects: one aspect is the basic theory of map symbols in the field of cartography, and the second aspect is the map symbol making and map symbol library establishment method within cartography and GIS systems. With the development of digital mapping technology, the method of drawing map symbols developed from the earliest programming method and the information block method to the interactive design method. The interactive design method, which uses the human–computer interaction method to design and draw map symbols, greatly facilitates the user, improves work efficiency, and has become the main development direction of symbol design and drawing. In recent years, with the development of visualization technology, traditional map symbols have become increasingly unable to meet the needs of map readers. Many scholars have carried out research on dynamic symbols and 3D symbols.

The map symbol is a graphical representation of geo environmental information, which can be divided into two levels: the first is to realize map symbolization in dynamic digital mapping, that is, symbols used in electronic maps and geographic information systems, which are schematic (about the object or location), without physical size, and suitable for screening; and the second level is to achieve full symbolization technology in static digital mapping to meet the requirements of map publication. The full symbolization also has two meanings: one is to symbolize all elements of a map, and the other is to symbolize the whole mapping process rather than a link and to meet the real-time symbolization demands of map element collecting, editing and modifying; additionally, the size, shape and color of the symbols defined on the screen are those of the symbol in the final published map, that is, 'what you see is what you get'.

4. **Map editing technology**

Map editing is the main environment for cartographic human–computer interaction. According to the operation method, map editing can be divided into three levels:

editing based on graphical symbols, map editing based on geographic attributes and map editing based on cartographic generalization. Graphical symbol editing is most commonly used in commercial graphics editing software. The map editing method based on geographic attributes directly uses the geographical attributes of map elements. For example, in the case of changing a county-level highway to a provincial highway, the map editing method can simultaneously update the attribute information and the map symbol of an element. The map editing method based on cartographic generalization mainly involves solving the interactive map compilation problems in the case of map generalization, and generally, this method is implemented by using a specific mapping generalization algorithm for specific elements. From the perspective of the development of map generalization technology, this method cannot do without the human–machine collaboration map editing environment to complete an integrated cartographic generalization product (Wu 2000). This type of editing is quite different from the first two, which is based on the previous two levels and forms a "Pyramid" map editing structure, as shown in Fig. 9.4. Among these tools, the most common are viewport tools and graphical editing tools. Viewport tools include viewport zoom in and zoom out, windowing, roaming, hierarchical display control, reference grid, etc.; graphical editing tools include graphics creation tools (for creating a variety of graphic elements, such as lines, primitives, curves, shapes, composite lines), graphic editing tools (including delete, move, copy, node deletion, node addition and node move), graphic attribute tools (for modifying layer, color, width, line, and filling type) and other auxiliary tools (fence, decomposition, measurement, and so on). These tools are the basic tools for the production of symbols and are also the tools for editing and modifying the elements that are independent of the attributes.

Viewport tools and graphics editing tools are the basic functions of various existing commercial cartographic software programs. However, there is a difference in their usages and effects. These tools will be enough for only the implementation of map symbols and publishing. If we want to realize the integration of paper maps and geographic information production, we must have editing and management tools based on geographic attributes. GIS-like software can accomplish the task of acquiring and updating geographic information data, but their graphics performance

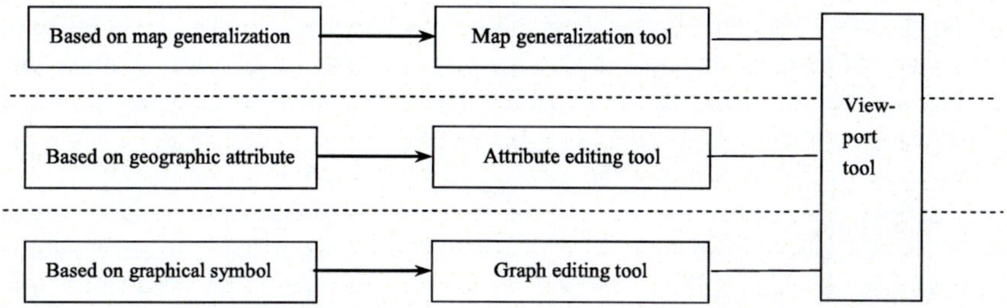

Fig. 9.4 The three levels of map editing

cannot meet the requirements of map publishing. Therefore, the key problem of integrated production is being able to edit and modify the attributes of map elements, that is, by making use of a full symbolic technique, when modifying a property, the symbol on the map is modified accordingly.

5. Quality control and check of the digital map technology

The production of geospatial data is a complicated process, and it will inevitably cause errors in the production process. Simultaneously, errors will gradually accumulate, accompanied by continuous production and may result in a mistake when they accumulate to a certain degree. Therefore, how to control the error within the specified range, to avoid and correct the error, and to ensure the quality of data are important issues directly related to the reliability of data and credibility of the system.

The main source of error in the mapping process includes three types: data source error, data acquisition error and system processing error (Liu 2007a). The data source error mainly refers to the error in the data used in the mapping process; for example, errors exist in digital base maps, and errors exist in remote sensing images. The data acquisition error is caused by the error operation of the operator in the process of human–computer interaction or low acquisition precision. System processing error refers to the error generated by the mapping system in data processing, such as coordinate system conversion, map projection transformation, and data format conversion.

Usually, the methods for error control and checking include map symbolization, implicit information display, and topology checking. Map symbolization is the most direct means used to visualize spatial data, which can be used to determine some obvious errors or mistakes in spatial data. The so-called implicit information display is to visualize some non-symbolic data, for example, to present the river code and place the name pointer on the map to determine correctness.

The map topology is mainly used to establish the relationships between points and lines and between lines and surfaces. Then, after topology calculation, significant signs are used to show the node graph and topological map, which can directly reflect the situation of the data, helping users modify the error in time.

6. Digital terrain representation method and technology

Among the various features of the geographical environment, landforms are the most basic. One of the main problems in map making is finding the best way to represent the roughness of the ground surface on a flat map, which is always the most difficult problem for the map maker. Therefore, how to express the undulating terrain on a flat map has always been one of the most challenging problems in cartography.

With the rapid development of digital mapping technology, manual drawing was gradually replaced by computer-aided mapping, and the map that is stored digitally in a computer storage medium completely changes and expands the production, use and expression of the map. At present, the traditional methods of representing terrain on a plane map, such as geomorphology sketching, contouring method, hypsometric tinting, and hill-shading method, can be implemented by computer software.

Due to progress in obtaining topographic data, there are a large number of high precision terrain data that provide a basis for the use of the geomorphology sketching method. It is convenient and rapid to obtain the real and objective terrain sketch map through 3D software, and according to the needs of mapping, landforms can be represented with a variety of different styles, such as perspective view, perspective axis view, serial section view, and sketch view.

Vector contour data can be obtained in two ways: one is through map scanning digitization, in which vector data of contour lines are extracted from contour images using semiautomatic digitalization software; and the other way is through the interpolation method to extract vector data from different scales of DEM data. The latter has become a mainstream approach for contour data acquisition because of its fast speed.

To enhance the three-dimensional effect, the illuminated contour method and thick and thin contour method have been applied in manual drawing; at present, they have been fully completed by computers through the preparation of algorithm software. Due to the lack of terrain information in coastal areas, this method can be applied to areas to obtain better effects.

Hypsometric tinting is also a commonly used terrain representation. Through the contour data obtained, it is convenient to obtain the color filling surface of different elevation belts in mapping software and then to represent terrain with the hypsometric tinting method.

The hill-shading method is widely used in terrain expression due to its strong, readable and realistic three-dimensional effect. In traditional mapping, hill shading is implemented manually, requiring staff to have high skill; it will take at least several months to finish the design and production of a standard-sized shading map and is troubled by staff shortages; therefore, it cannot meet the needs of a large number of users. Through research on the theory and technology of digital relief shading in a computer environment, the current developed map shading software is able to use DEM data to generate a hill-shading map with a multiscale and multiple-mathematical foundation and can be matched precisely with other geographical elements to obtain electronic maps, typed maps or printed paper maps, as well as other map products (Li et al. 2003b).

7. **Map publishing and prepress processing technology**

Paper maps are one of the most important products in digital mapping. Map publishing and prepress processing technology is used to convert digital maps to paper maps. In general, the production of digital maps and paper maps cannot be carried out at the same time; after the production of digital maps, a great deal of subsequent processing work must be carried out to meet the production requirements of the paper maps. This process mainly consists of two key steps: the first step is processing for publishing, including publication editing, map decoration, stream gradient, symbolization of area features, mask processing of map symbols, overlapping processing of map annotations, and overlapping relation adjustments to map annotations. Only the screen map after the above processing can conform to people's habit of map reading and map use; the second step is to convert the publishing formats of the screen map, usually with EPS or AI format, and to separate elements according to the printing color, as well as to generate the data with prepress data format, convert from screen color space to printing color space, implement image overlapping and other processes, and finally, the prepress system is used to output film or perform direct plate making, and the color paper map is printed through the printing machine.

9.3.3 Design and Research of Digital Cartographic Systems

A digital cartographic system is a computer system that is used to accomplish the task of digital mapping with the support of certain hardware and software. Among them, the leading role is the technological process of digital mapping. Although the different kinds of digital mapping software are different in the specific mapping operation, all of them aim to complete the task of digital map production and thus should follow the process of digital mapping.

1. **Process design of digital mapping**

Using a digital mapping system to produce a map generally requires the following process: cartographic data acquisition, symbolization of map content, editing, modifying and drawing checks, map publishing processing and separated film output. The processes and relationships are shown in Fig. 9.5.

If subdivided, the digital mapping process can be expressed in Fig. 9.6. The relevant contents of the flow chart shown in Fig. 9.6 are described below.

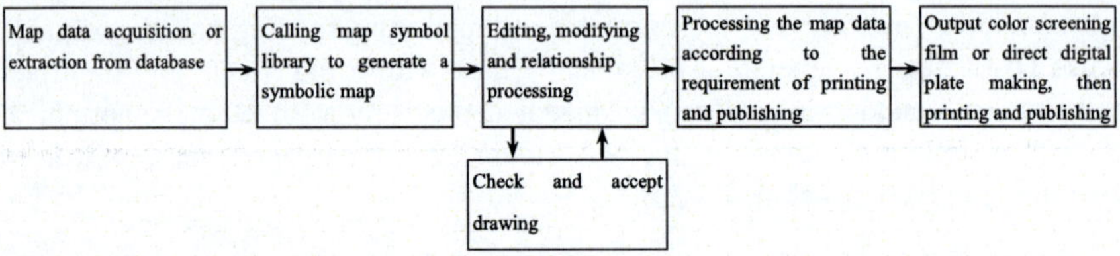

Fig. 9.5 General process of digital mapping

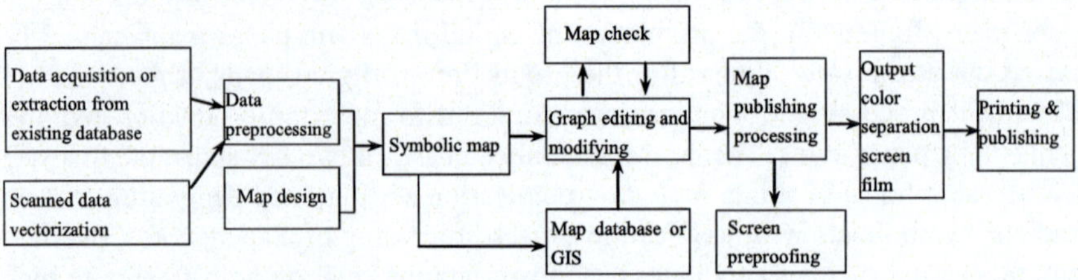

Fig. 9.6 The specific operation process of digital mapping

(1) Data acquisition or extraction from existing databases

In the case of no existing map data or no existing map data file, map data can only be acquired through manual tracking or scanned map digitization. The base map used for digitalization operations should be plotted with the latest changes in the field, and the quality and precision should be guaranteed. The quality of the digital base map, its contents and the level of detail have a great influence on the quality of the final map. If extracted from the map database or the existing data files, then it should be implemented by specific software according to the requirements of map production. If the scale of the final map is the same as that of the map database and the content of the final map is similar to that of the map database, then problems in this area should be less. Otherwise, he kind of content, how to extract, how to control selecting quota, how to supplement other contents and so on all need to be properly studied and sufficiently tested. At present, no software can take all user needs into account and solve all kinds of problems.

(2) Scanned data processing

For the map data that are well-made and updated, the scanning vectorization method can be used to obtain the map data. Raster data after scanning will be converted into the desired vector data through vector software. Satellite images and aerial image data can be directly used after image rectification for integration with vector data to generate the image map or used to extract the necessary information to supplement the map content.

(3) Data preprocessing

The work in this part mainly includes the map data format conversion, the transformation and correction of point coordinates, the necessary map data topological processing and sheet joining processing. The subsequent processing of the remaining problems cannot be solved by cartographic generalization and data extraction.

(4) Map design

The task in this part is mainly to design a map symbol system according to the needs of users, the use of map and the owned data, to determine the clipping range and map layout, to design the map color, to assign the printing color of each element and to determine the overlapping relationship. This part of the work should be considered before map production, and then, some adjustments and changes should be made according to the actual conditions.

(5) Symbolization of map content

This part of the work involves symbolizing the contents of the map according to the designed symbol system. To ensure the smooth progress of the work of graphic editing and map publishing, in the process of map content symbolization, different elements should be stored in different layers to carry out selective operations on different elements. To edit the elements in a certain layer, only that layer needs to be opened, which avoids mutual interference. All the layers are opened when you want to display all the elements. During the symbolic process, the size, color, thickness and the relationship among the symbols should reflect the actual situation after publication of the map, display and record, which is strictly according to the required size. Only in this way can the operator correctly handle the relationship among the elements on the map and solve problems such as overlapping, annotation placement and displacement, as well as edge-shared elements to guarantee the quality of the map. In most cases, an organic link should be established between graphic symbols and geographic information (the contents of the map database) to simultaneously implement the production of maps and update the map database contents to ensure the consistency between map products and map database contents (Liu et al. 2005).

(6) Graph editing, modification and drawing check

The map is represented by symbols and texts. The graphic symbols and geographic information are not exactly the same, so even if the symbolic software and the map data quality are very good, the formed symbolic map still needs to be modified by the method of graphic editing to correct defects. It is difficult to check the contents of a map on a computer screen, so the drawing method is mostly used for checking the contents of the map on the paper and then correcting the contents on a computer until they are all accurate.

(7) Map publishing processing

Map publication processing deals with the edited, revised map contents based on the demands of map publishing, the color separation plate data that can be received by a film camera are formed, and simultaneously, color proofing images are generated for content inspection.

Of course, in different circumstances, the above digital map production process may require some changes, but the overall approach will not change. For example, in the actual map production process, there will be problems to solve, such as the annotation input and the mixed typesetting of the map and chart, which need to be addressed on a case-by-case basis. In short, in accordance with the production process, with the support of the corresponding software and hardware, we will be able to complete the map production work in digital mapping mode. At present, the software used for map production is becoming increasingly practical and powerful, but its practicality is still a work in progress. Most software still needs requires step-by-step operation by an operator when dealing with mapping problems; the software itself is far from automation and intelligence. At present, some of the works in the map production process still need to be solved by the operator, such as the analysis and utilization of map data, map symbol design and map content selection. Software is still powerless to address these issues. Map data acquisition, map generalization, map content editing and other works, with high labor intensity still need to be implemented through human–computer interaction to a large extent. Therefore, we should focus on the superiority of computer-aided mapping, continue to carry out research on map design expert systems and intelligent cartographic generalization systems, improve the level of automation and intelligence of map production, and absorb the latest research results into map production software to free workers from boring and monotonous work.

2. **Representative digital cartographic system**

(1) Cartographic system based on graphic design software

The graphic design software commonly used in cartography includes CorelDraw and Illustrator. CorelDraw, a product of the Corel Corporation in Canada, is a graphics processing software based on vector graphics and is an advanced representative of graphics drawing and image processing software. CorelDraw has two powerful functions for the editing and publishing of maps: vector tracking of map compilation lines and typesetting of text, images and graphics. The representative map products completed by CorelDraw include the *Atlas of Shenzhen City* and the *Atlas of Shanghai City*. Developed by the Adobe Company, Illustrator provides a solution for desktop publishing and has been widely used in the field of digital publishing. Illustrators can achieve a variety of functions, such as graph-text mixed arrangement, layered processing and symbol library construction. Illustrator provides a perfect color management system for controlling the overprint effect between elements. The output format of Illustrator is EPS, which accords with the standard prepress system and forms a complete production process of desktop publishing. For this reason, Illustrator is also used to provide a final check for the EPS format map publishing products generated by other software before they are transferred to the prepress system.

(2) Domestic manufactured cartographic software MapGIS

MapGIS is a basic GIS platform with complete cartographic functions developed by Wuhan Zhongdi Digital Technology Co. Ltd. The functions of MapGIS include interactive, semiautomatic and other scanning vectorization inputs; powerful graphics editing and symbolic production; coordinate transformation and geometric rectification; drawing output and publishing EPS file output; and conversion among other vector data formats (DXF, DLG, ASCII, etc.). MapGIS is a commonly used mapping system in China (Wang 2000).

(3) 1:250,000 digital mapping system

This mapping system was developed by Zhengzhou Institute of Surveying and Mapping, and it is an integrated mapping and map database system. This system consists of symbol generation, data management, editing and modification, topology reorganization, a publishing process and data format conversion (see Fig. 9.7). Combined, map symbol making, map updating, computer-aided mapping and film output, with very strong practicability and pertinence, can adapt to the specific situation and special requirements of the production of collaborative maps. This system extracts the elements of the map from the map database and makes them symbolic; the symbolic contents are associated with the corresponding attribute information.

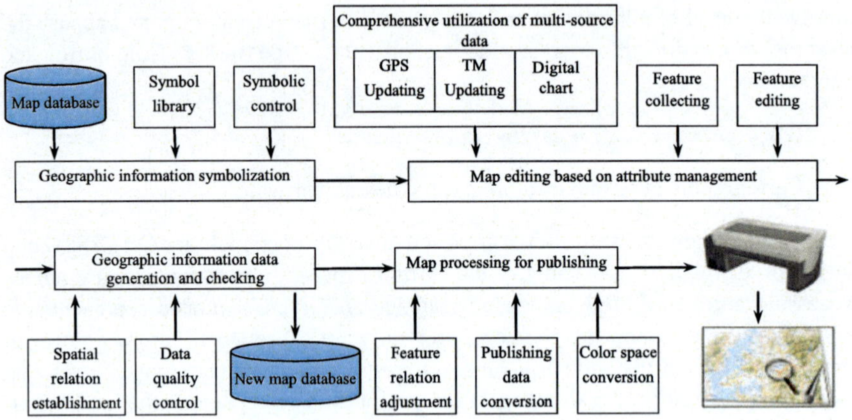

Fig. 9.7 Composition of the 1:250,000 digital mapping system

When the map elements are added and modified, the attribute information and the symbolic graph are both changed. When the editing and modification are completed, the map data extraction and topology processing are carried out, and then, the updated 1:250,000 map database data are obtained. Finally, the map contents are processed and the data are converted and output to obtain the paper map product.

9.4 Achievements of Major Construction Projects

(1) *Water Conservancy Atlas of the Jiangsu Province* and *Atlas of Shenzhen City*

The Water Conservancy Atlas of the Jiangsu Province was completed by the cartography department of the Zhengzhou Institute of Surveying and Mapping in December 1996. This is China's first thematic atlas produced by digital means, which is compiled by using MicroStation software and published by using Intergraph's map publishing system. This atlas comprehensively introduces the status of water conservancy facilities and the achievements of water conservancy construction in the Jiangsu Province and includes a total of 240 pages, 182 sheets of maps, and a page size of 16 K.

The main process of the digital production of an atlas is divided into three steps: first is to complete the preparation of the base manuscript according to the overall design requirements; second is the collection, symbolization and editing of map contents; and third, according to the printing requirements, is to process map data and output a color sample map and color separation film for the use of checking, modification and printing. It took a total of 9 months to finish the compilation, digitization, map content checking, printing and publishing of the atlas.

The *Atlas of Shenzhen City* is a large comprehensive atlas published by the Department of Land Information and Map Science of Wuhan Technical University of Surveying and Mapping (now the College of Resources and Environmental Science, Wuhan University) in January 1997. CorelDraw software was used as the processing and publishing platform for the atlas, and Photoshop was utilized for image processing. Through layout integration, the Adobe prepress data format was eventually formed and used for map publishing. The atlas has the page size of 8 K, is available in Chinese and English versions, includes five parts of introduction diagram, regional details, social economy, natural environment and development planning, and has 87 pages, a total of 177 maps, 154 charts, 6 aerial photographs and 162 color photos.

(2) *1:3 Million Geographic Map of the People's Republic of China* and the building of it's database

1:3 Million Geographic Map of the People's Republic of China is a large wall map consisting of $3 \times 3 = 9$ pieces of map. The paper size of a single map is 119.2 cm * 88 cm. The size of the inner border of the wall map is 300 cm * 216 cm. This map was published by the Zhengzhou Institute of Surveying and Mapping in 2001. The map uses a pseudo-azimuthal projection. The scope of the map extends as far as Guam in the east, Afghanistan in the west, Lake Baikal in the north and Indonesia in the south, covering the whole of China's territory and territorial sea. The production of this map adopts an integration mode of mapping and geographic information production, that is, the paper map is published and a 1:3 million China's geographic map database is established simultaneously. The map data of the domestic part are mainly acquired from the scanning vectorization of China's 1:2,500,000 traffic map, and the map data of foreign regions come from the comprehensive extraction of the International Map of the World on the Million Scale. The features of the sea and landforms are collected from digital 1:3 million charts and China's 1:2 million landform maps. Based on DEM data, a computer-aided relief shading method is used to obtain layered color images and monochrome shading images for the map. Then, these images are merged with the vector line EPS data of the corresponding color to output color separation film (8 colors) for final printing and publishing. The production of the map is highly difficult, involving a large volume of data processing, marking the maturity of digital cartographic technology.

(3) Construction project of the national 1:50,000 basic geographic information framework

The 1:50,000 basic geospatial information database is an important part of the national basic geospatial database. It consists of five different types of databases, including vector topographic databases, raster map databases, digital elevation model bases, digital ortho image databases, and place name databases. The 1:50,000 vector topographic database is the most difficult to construct, most widely used and most important database in the 1:500,000 basic geospatial database. The technical design started in 2002, trial production started in 2001, formal production started in 2003, the production task of the national 1:50,000 basic geographic information framework was completed in June 2005, more than 24,000 pieces of the 1:500,000 framework maps were produced, and the verification and acceptance of all data was completed in December 2005.

The construction project of the national 1:50,000 basic geographic information framework is based on a 1:50,000 topographic map, complemented by digital ortho images, change data of administrative divisions and boundaries, place name data, GPS measurement results of national and provincial roads, and other relevant updated information. The national 1:50,000 vector topographic map framework data are produced by using the methods of digital topographic map updating based on digital ortho images, map scanning vectorization, generalization of 1:25,000 topographic maps, digital mapping, and so on.

(4) Renewal project of the national 1:250,000 map database

The national 1:250,000 map database is organized by the State Bureau of Surveying and Mapping and was completed in 1998. The amount of data is 8.2 GB, and the deadline for updating data is December 1995. As one of the basic national geographic information data components, the database covers a wide range of data, including drainage system and landform information of natural landforms, boundaries, residential land, roads, railways, nature reserves and square km grids, for a total of 13 layers of data, with high data accuracy suitable for a wide range of planning, design and decision-making. To better utilize 1:250,000 map data as the basic geographic information and to meet the demands of all aspects of social construction for the accuracy and timeliness of the data, an overall update was implemented in 2002. Raw data for data updates include satellite image data, state backbone traffic network data, 1:50,000 place name data, the latest land boundary survey results and some updated reference information. The main contents of the update involve railway, boundary, provincial road and higher road levels; villages and towns, as well as higher levels of point residential land; county and higher levels of residential land in real shapes, grade five and the levels higher than grade five; and other important features. Since remote sensing images perform well and have real and reliable characteristics, they were used to identify changes in object features, such as changes in the waterside line of rivers and lakes, new reservoirs and highways, changes in the scope of resident lands, disappearance of lakes, alteration of road routes, migration of ditches and variations in the range of marshes. The specific attribute content of a feature was determined according to other updated data. Line data were updated based on image data, with high practicality and a fairly high positional accuracy, which is the main means of fast mapping and data updating. Since a 1:250,000 map database was built and updated, the data have been applied in many industries, such as transportation, communications, government departments and others, involving the fields of macro control, planning and design, assistant analysis and decision-making.

9.5 Problems and Prospects

Before the 1990s, the production of maps and atlases was completed by manual operations, with long periods, slow paces and complicated processes. With the application of computer technology, digital maps began to be put on the agenda, and experiments and research on the standards and production technology of digital maps was conducted. Over the subsequent 15 years, digital mapping technology was used to produce maps in the computer environment and to obtain the required digital maps and paper maps. The map production mode achieved a transformation, which was the first step in transitioning to digital cartography (Liu et al. 2007). Entering the twenty-first century, the integration of paper map-making and map database building solved the problem of the synchronous production of paper maps and map databases. A cartographic system that uses this technology can change the previously used single product mode into a multi-product mode to simultaneously obtain both paper maps and map databases during a single production process. This achievement can greatly reduce production costs, avoid duplication of work, improve the quality of products, and ensure consistent content and timeliness of both, bringing convenience to the application. In particular, paper map-making and map database building involving the whole country is a large project that requires a great deal of investment. Since the overall economic investment is still very limited, the use of more efficient, integrated production technology has more practical significance and application value. This new achievement in the new development stage of digital mapping is the second step.

The existing map digital production includes map data acquisition, data processing, map editing, mapping output and other technical links. Based on the practical applications of the past few years, there are some problems that need to be solved:

(1) The various types of software used for map data collection, map editing, mapping output, quality checks and other links are independent of each other, not supporting network collaboration or synchronization jobs, with poor emergency mapping ability and low production efficiency; thus, it is required to implement network map production.

(2) The existing digital mapping software is usually used for a single scale and a single map type and lacks general purpose software and a digital mapping system.

(3) It is necessary to further study how to apply the technology of multisource data integration and assimilation in map production. The function of the existing digital map drawing software is insufficient. For example, the processing function of multisource data is still relatively weak; there are not many correction and processing methods for image data, scanned map images and some existing vector data; the processing and application function for new data is weak; the transformation of different coordinate systems and projection systems is not flexible enough; and the import and export of different formats of data is not convenient.

(4) The existing digital cartographic software does not have the functions of automatic map design, generalization, and compilation and lacks data processing automation and intelligent technology. Map design and cartographic generalization have long relied on artificial approaches, even in computer environments, with slow operation speeds and nonstandard requirements, to solve the problems of intelligent map design and cartographic generalization.

Therefore, to improve the efficiency of map production and improve and strengthen the production system of digital mapping, we must develop a new generation of cartographic systems based on networks that have the integrated functions of data collection, data processing, map editing, publishing processing, mapping output and quality verification. That is, digital mapping technology needs to develop toward network map production, intelligent map design and generalization, integration of multisource information, and integrated paper, digital, electronic map production. Thus, digital mapping technology can realize emergency map production, shorten the map-making cycle, produce all kinds of digital maps and update maps more quickly and efficiently, allowing the production of a variety of maps and atlases to be more flexible and convenient, as well as meeting the needs of national defense and army construction for digital maps and paper maps.

Acknowledgements ZHOU Zhao, LIU Xingui, LI Shaomei, CHE Sen, XU Li and SUN Guangyu, who collected and provided information for this chapter.

References

An, X. (2008). *The study of theory and technology for basic geographic data fusion.* Master Dissertation of the Information Engineering University.

Chen, S. (1994). Challenges and opportunities cartography faces. *Acta Geographica Sinica, 49*(1), 1–8.

Cui, T., & Guo, L. (2007). The design and implementation for digital mapping system. *Journal of Zhengzhou Institute of Surveying and Mapping, 24*(1), 1–4.

Guo, L., Cui, T., & Wu, Z. (2002). Research on the technical problems of multi-source digital map fusion—taking the fusion of digital chart and topographic map as an example. *Hydrographic Surveying and Charting,* (2), 22–25.

Fu, S. (2009). The creation and application development of geo-information science—commemorative statement on the first anniversary of the death of Academician Peng and his 90th birthday. *Journal of Geo-Information Science, 11*(6), 695–697.

Hao, X., & Qian, Z. (1997). Automatic map orientation method in map scanning digitization. *Journal of Geomatics Science and Technology,* (2),109–114.

Huang, W. (1994). Map pattern recognition in cartography. *Bulletin of Surveying and Mapping,* (1), 16–19, 39.

Kan, Y., Liu, X., & Sun, Q. (2000). Processing and application of rarely used Chinese characters in digital mapping and GIS. *Bulletin of Surveying and Mapping,* (10), 12–14.

Leuzinger, L. H. H. (1996). The principle of map design and its influence on the production method of digital mapping (Q. Guo, & X. Wang, Trans.). *Wuhan Survey Translation,* (2), 43–47.

Li, D. (2005). Multilevel grid of spatial information and its function. *Geospatial Information,* (4), 1–3, 5.

Li, S., Sun, Q., & Xiao, Q. (2003a). Remote sensing image processing in 1:250,000 map database updating. *Journal of Geomatics, 28*(2), 20–21.

Li, S., & Sun, Q. (2003b). The current development situation of hill shading technology in digital mapping. *Bulletin of Surveying and Mapping,* (1), 18–21.

Liao, K. (1993). A review of the progress in the 80s and 90s cartography. *Map,* (2), 3–7.

Liang, H. (2002). *The theory and technology of integration of map production and spatial data capturing.* Doctoral Dissertation of the Information Engineering University.

Liang, L. (2006). *The map expression and achievement based on 1:10000 fundamental geographic information database.* Master Dissertation of the Liaoning Technical University.

Lin, Z., Lu, J., & Kai, D. (1992). Thematic map reading system based on computer vision. *Journal of Wuhan Technical University of Surveying and Mapping,* (2), 8–17.

Liu, H., Li, S., & Zhao, G. (2005). The method of map production based on the small scale geographic information database. *Bulletin of Surveying and Mapping,* (2), 58–60.

Liu, H., & Sun, Q. (2007a). The spatial data error classification and quality control based on the workflows. *Geomatics Technology and Equipment, 16*(4), 15–17.

Liu, H., Sun, Q., & Xiao, Q. (2007b). The integrative technology of the paper map production and the geographic information capturing. *Journal of Zhengzhou Institute of Surveying and Mapping, 24*(4), 239–243.

Liu, H., Sun, Q., & Xiao, Q. (2006). Multi-sources data's application methods in digital mapping. *Journal of Zhengzhou Institute of Surveying and Mapping, 23*(3), 161–164.

Lv, J., & Wang, B. (1995). A map automatic entry and identification system. *Journal of Xidian University,* (1), 64–67.

Mo, R., Liu, J., & Jin, Y. (2005). The development of the techniques for digital mapping. *Geomatics & Spatial Information Technology, 28*(6), 48–49.

Qi, Q., Liang, Y., & He, J. (2005). Discussion on the theory, method and technology system of digital map. *Science of Surveying and Mapping, 30*(6), 15–18.

Sun, Q., Fang, T., & Guo, D. (2005). Scale transformation in spatial data mining. *Computer Engineering and Applications, 41*(16), 17–19.

Wang, H. (2000). Preliminary study of the application of MapGIS to digital mapping. *Bulletin of Surveying and Mapping,* (6), 10–11, 19.

Wang, J. (2000). Cartography in information ERA. *Engineering of Surveying and Mapping, 9*(2), 2–5.

Wang, Q., Liu, Y., & Zhang, Z. (1999). Building and application of the integrated cartographic information system. *Geographical Research, 18*(1), 59–64.

Wu, F. (2000). *Research and implementation of collaborative automated map generalization.* Doctoral Dissertation of the Information Engineering University.

Xie, J. (2011). *On electronic sand table system from military topographic paper maps.* Doctoral Dissertation of the Northeastern University.

Xu, Q. (2006). *The development and implementation for the joint operations map production system of scale 1:5000.* Master Dissertation of the Information Engineering University.

Yang, Y., Zhu, C., Sun, Q., & Zhang, Z. (2006). Automatic extraction method of gully and steep cliff symbols in topographic map. *Journal of Xidian University* (Natural Science Edition), (3), 504–508.

Ye, J., Ren, L., & Ji, H. (1999). Automatic recognition of dot symbols based on knowledge. *Journal of Geomatics Science and Technology,* (3), 49–52.

Yu, C. (2002). Chinese cartography in the 20th century. *Maps,,* (2), 10–13.

Zeng, Y. (2014). *The research of cartography and application of digital map for middle school teaching.* Master Disseration of the Central China Normal University.

Zhai, J., & Wang, J. (1994). Morphological decomposition and representation of digital map images. *Journal of Geomatics Science and Technology,* (2), 122–129.

Zhu, R., Xiao, Q., & Zhao, G. (2007). The design and implementation for digital mapping system. *Journal of Institute of Surveying and Mapping, 24*(6), 454–457.

Zou, J., & Sun, Q. (1998). MICROSTATION digital mapping production characteristics based on "Jiangsu province water conservancy atlas" as an example. *Bulletin of Surveying and Mapping,* (5), 29–32.

Chapter 10
Map Printing and Publishing

Ruizhi Shi

10.1 Introduction

In the 28th year of Emperor Guangxu of the Qing Dynasty (1902), the Imperial Army School of Surveying and Mapping was established by the Qing government, and map drawing and plate printing classes were offered. This occurrence created a precedent for the government to use modern printing technology to print maps. After 1949, with the development of national economic construction and national defense construction, to meet the demands for maps from all aspects, several departments of map printing and publishing were successively set up. Some special map printing plants were founded in the Bureau of Military Surveying and Mapping and the greater military areas, which created a new era for China's map printing and publishing (Zhang et al. 2004).

Modern map printing is based on optics, color science, chemistry, polymer science, materials science, photography, electronics, information science and other disciplines, and map printing has its own characteristics and requirements, despite being part of publishing and printing. Map printing, through different printing processes and techniques, involves replicating a large number of maps based on an original manuscript (different types), which may be more beautiful and legible than the originals. Especially in the past 30 years, printing technology has changed greatly, from a photomechanical process to color electronic publishing technology; graphic and text information processing has transformed from analog to digital, and digital mapping technology and color electronic publishing technology have been combined to realize the integration of cartography and publishing technology. Map printing completely leaves behind the concept of 'lead' and 'fire', and the era of 'light' and 'electricity' has been entered. In the long development process of printing technology, every breakthrough is based on its own technological background, that is, the comprehensive application of multidisciplinary research achievements. The progress of science

R. Shi (✉)
Information Engineering University, Zhengzhou, China

J. Wang and F. Wu (eds.), *Advances in Cartography and Geographic Information Engineering*, https://doi.org/10.1007/978-981-16-0614-4_10

and technology has created conditions for the development of printing technology. On the other hand, the development of printing technology has also made important contributions to the progress of social civilization.

10.2 Development Process

10.2.1 Photographic Technique Development from Wet Plate and Dry Plate to Digital Photography

The printing technical process is divided into three stages: prepress, printing and postpress finishing.

Before the 1950s, only the printing plants in Shanghai, Tianjin, Beijing, Changchun and other major cities had camera equipment. Many printing plants do not have camera equipment, and even if they did, most of the cameras they used were small hand cameras with wooden structures, and the photosensitive material they used was silver iodide collodion.

What is important among the required materials is photosensitive material. Before the 1950s, the wet plate process was universally adopted; its photosensitive material was made as follows: taking a glass plate as a base, cotton gum solution containing iodide was evenly coated onto the glass plate and then put into a tray filled with silver nitrate solution. The iodide and silver nitrate in cotton gum reacted to produce silver iodide with photosensitive properties, and then, the glass plate containing silver nitrate was installed on the camera while it was still wet, immediately shooting and exposing the photo. The wet plate photographic process is a completely manual operation with a significant labor intensity, and chemicals do great harm to the health of operators.

In the early 1960s, the Shantou photosensitive chemical factory cooperated with the Beijing Printing Technology Research Institute to produce a plate-making film called ERA Brand, filling the gap in plate-making film in China; map publishing technology began to test the dry plate replication technique (Gao 1962). After the 1970s, the second film factory of the Ministry of Chemical Industry began to produce a Huaguang Brand plate-making film. By that time, domestic plate-making films basically met the needs of printing phototypesetting. After the 1980s, the domestic film completely replaced the use of more than 100 years of collodion wet plates, completing the transformation away from "dry instead of wet and soft instead of hard" in the history of China's photomechanical process (Zhang 1992).

Starting in the 1990s, with the development of digital imaging technology, film imaging has gradually been replaced by CCD digital imaging, and the image has changed from film and photo paper to a computer directly processing digital images. At present, digital photography almost completely replaces film photography.

For the lens imaging system, there is no fundamental difference between digital photography and traditional photography. However, the image recording materials and imaging principles are completely different. The image recording process of digital photography is outlined as follows: the optical signal on the imaging surface is converted into an electrical signal by a photoelectric conversion device (such as a CCD), and then, the electrical signal is converted into a digital signal by an analog/digital converter, and the data are stored in the "memory card" through a camera internal processor; thus, a digital photo shoot is completed. In the process of receiving the optical signal, by using color separation technology, the optical signal is divided into 3 channels of RGB and recorded. Therefore, the resulting digital image is an RGB (red, green, blue) color image that can be copied to the computer and used as a prepress manuscript.

Along with the coming of the 21st century, for map publications, a large number of image map manuscripts are directly derived from remote sensing digital images, with high image resolution that is informative with a rich tone, which provides a high-quality information source for image map publishing. Meanwhile, with the development of digital color separation technology and digital screening technology, as well as the improvement in the performance of printing equipment, the publication and printing of thematic maps based on digital images are also increasing rapidly and play an increasingly important role in the development of military and national economies.

10.2.2 The Color Separation Technique Has Developed from Manual Separation, Photographic Color Separation and Electronic Scanning Color Separation to Software Color Separation

The color separation technique for map publishing can be divided into two categories due to different types of maps: one is the color line map that used manual separation technology before the advent of digital publishing technology in the 1990s and developed into computer software color separation after the advent of digital publishing technology; another category is the color continuous tone map (such as an image map), which began to develop and gradually adopt photographic color separation in the 1960s (Gao 1962), moving to software color separation technology at the advent of digital publishing technology in the 1990s.

In approximately 1970, color separation for the original map copy was completed by adopting various structural (horizontal and vertical) plate-making cameras. That is, to apply the complementary color principle to photographic color separation and cleverly choose different color filters to achieve color decomposition when making the photographic color separation, only one kind of light is allowed to pass, and the rest are absorbed by the filter, leaving the image of a certain type of element in photosensitive materials.

In 1964, a C187 electronic scanning color separation machine was imported to the Beijing Xinhua printing factory and Beijing foreign language printing factory from the former Federal Republic of Germany, which created the history of applying electronic color separation technology in China. Electronic color separation, whether in terms of clarity or color correction, tones and color reproduction, has significantly improved in quality compared with photographic color separation. After the 1980s, the performance of the electronic color scanner was very good; a screening separated printing plate can be produced in just tens of minutes, which may have taken a few days in the past by using photographic color separation (Ye 1988; Jin 1988; Zhang 1989).

For the color continuous tone photomap, color filters of red, green and blue were used to take photographs to obtain the color separation filters of cyan, magenta and yellow, respectively. However, the color separation of the line map is relatively complex. Since the various elements of maps are represented by fine lines, it is not suitable to apply primary composites to achieve multicolor printing of line maps; instead, spot color is used in most cases. Therefore, photographic color separation cannot be realized through the RGB color filter and can only be achieved by the manual coating method (Hunan Xinhua Printing House, 1962). A transparent negative film with all elements will be obtained by photography, and more transparent negative films will be made through reproduction; next, only the elements of a color will be retained on a negative film, and other elements will be painted out with coating ink. Thus far, color separation for map elements has been completed. For elements, such as rivers, two pieces of negatives are first made for linear elements and area elements separately, and then, through screening and duplication, the elements of the two pieces of negatives are surprinted on a single piece of negative (Liu 1962; Yan et al. 1982). Figure 10.1 is a schematic diagram of the replication process of a line map, among which the steps in the dashed box represent the color separation process of the line map.

Due to the special requirements for the printing accuracy of maps and considering that the material deformation during replications should be as small as possible, the printing plate used for photography and reprinting is a glass plate. Especially for a large-format map, the operation is very difficult. Therefore, photographing and reprinting of maps at the time were labor-consuming activities (Xu 1962). Map printing technology using a glass plate as the graphic carrier remained in operation until the 1980s (Zhao 1988); later, along with the improvement in the performance of the polyester base, the film began to be used as a graphic carrier for the large-scale production of maps, which greatly reduced labor intensity.

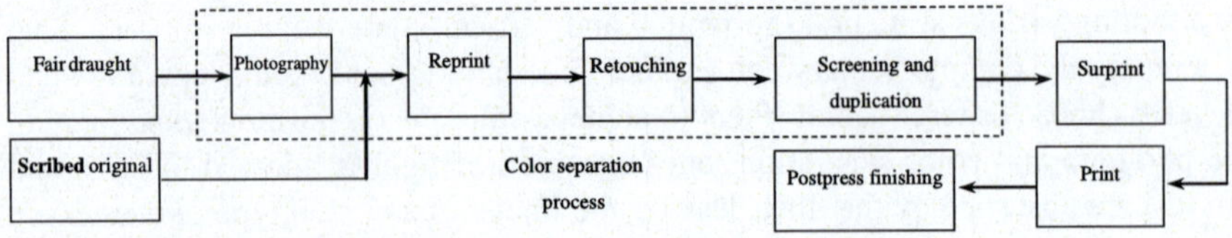

Fig. 10.1 The replication process for a line map

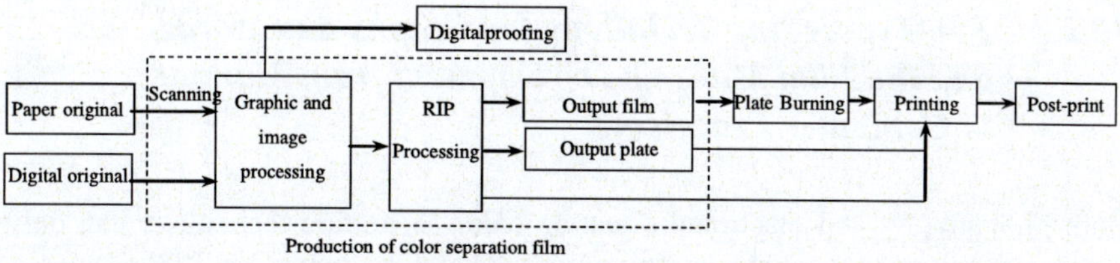

Fig. 10.2 Technique flow for digital publishing

With the popularization and application of computers, the Founder group of Peking University began to develop a color desktop publishing system and succeeded in 1994 and then came to market. The system uses a color electronic publishing system, in which the printing process, from scanning and color separation of the original, graphic and image processing, graph-text mixed arrangement and make up to the output screening, are all finished by computer. Color separation of the original is completed by scanner and computer software. The image after color separation is a digital image, and the tone can be adjusted by using image processing software. In the digital prepress stage, this digital image is displayed, processed and transmitted in the form of color, can be mixed with texts and graphics, and finally outputted to a film or printing plate in the form of color separation until all the page content is completed. Figure 10.2 gives a schematic diagram of the publishing process of the color electronic publishing system, in which the steps in the dashed box represent the processes involved in making a color separation film (plate).

Making a line or a continuous tone map in a digital mapping system or a color electronic publishing system is preceded by color separation from the beginning (the displayed color is the effect of the color composite by the software). The color setting mode is RGB (red, green and blue), CMYK (cyan, magenta and yellow, black) or spot color. When the color is in RGB mode, it will be converted into CMYK color or spot color through the software calculation (Shi 2002). With the development of color separation technology, color maps can not only be decomposed into four colors of CMYK but can also be decomposed into more than four colors of spot color to achieve high fidelity printing (Shi et al. 2007).

A color map depicted by CMYK color or spot color has been shown in the color form of the composite channel in the stage of digital prepress processing. When all the contents of digital prepress processing are completed, then a color separation film (or color separation printing plate) is made through the output device for printing.

10.2.3 The Typesetting Technique for Graphs and Words and the Film Make up Development from Manual to Computer Processing

Both photography and electronic scanning color separation technology can only handle images and cannot handle text directly. Therefore, before the 1980s, typesetting for graphs and words and film make up were conducted by hand. The image was pasted in the middle of the prearranged text with transparent glue to obtain the layout of the text and graph, and each page was pasted to the specified position of the imposition sheet with transparent adhesive to realize the film make up. There were traces of transparent glue paste on the positive film produced with this method, which needed to be treated after exposure. Simultaneously, all color printing plates were registered by hand, and their overlapping accuracy was limited. In this period, the sheet make up of standard products and the complex film make up of multicolor products often created a bottleneck for printing production. To achieve accurate registration accuracy, much time was spent on the correct arrangement of pages and their elements; nevertheless, they still could not completely avoid common mistakes of imposition.

In the early 1980s, a multi-duplicating machine was applied to this process, and particularly efficient imposing software was employed. Then, the full-page film and sheet film needed for the make up were automatically placed on the plate in the correct order. This part of the photosensitive plate was exposed. A multi-duplicating machine instead of manual typesetting can maintain a more accurate registration of each color separation film for the copy of a set of color separation films.

In the early 1990s, PageMaker typesetting software and large-format film printing and recording equipment became available, so the imposing software market has also grown. As an example, Heidelberg's SignaStation software was quickly promoted and applied. Typesetting software was generally used to implement a mixed layout of text and graphics, while imposing software was used for the make up. When a large page was fully assembled, color separation films would be produced by the output device, and the registration between each color separation film would be very accurate. Simultaneously, the film was clean and neat, with uniform density, which created good conditions for improving the print quality (Zou et al. 1996).

In the process of map printing and publishing, the content of the map was usually made with specially used cartographic software and then converted to EPS data format for use with the text and graphic layout as well as the film make up. The working platform for typesetting and collage was the same as that of general printing.

After entering the twenty-first century, the functional module of make up by computers was integrated into digital workflow software. In the digital workflow, the page data were standardized in a PDF format, imposing settings and page creation, and after the film make up was finished, the page data were interpreted and output as an RIP in the same system.

10.2.4 The Screening Technique Developed from AM and FM Screening to Hybrid Screening

In the past, a glass screen or contact screen was used as a screening method to generate printing dots; due to the limited number of screens, the screen line number was low, and the dot type was single (Wang et al. 1988). Along with the application of electronic scanning color separation, the printing dot was generated by the laser electronic screening mode of the digital dot generator. Although the quality of color separation screening was good, its electronic system was part of an analog special electronic machinery.

The dots generated by photographic screening and electronic color separation machine screening were called amplitude modulation dots (also referred to as AM screening). This kind of dot is characterized by a fixed distance between dots, that is, the frequency of dot appearance is invariant, depending on the size of the dots (different amplitude) to adjust the level of the image. Since there are disadvantages of turtle veins, rose stripes and tone jumps, AM screening cannot easily improve the printing quality or develop more than four-color printings.

Since the 1990s, with the development of color electronic publishing technology, frequency modulation (FM) screening theory and its application technology have emerged (Chen 1996). FM dots (referred to as FM screening), also known as random dots, are characterized by the same size of dots, and the density of the randomly arranged dots is used to represent the image level. There is no fixed dot angle to form turtle veins or rose stripes. However, due to the smaller FM dots that are easily lost in the process of printing down and screen printing, the requirement for the quality of printing down is high, and its use has a certain degree of difficulty. In recent years, both FM and AM screening technologies, known together as hybrid screening technology, have appeared, and the generated dot is called a hybrid dot. The screening technology can generate AM dots and FM dots in different parts in a timely manner according to the change in image color and level.

At present, the three kinds of screening technology exist simultaneously, and AM dots are still the main technology being used. With the use of direct plate-making technology and high-performance printing presses, FM dots and hybrid dots will be further popularized.

10.2.5 The Plate-Making Technique Was Developed from the Photographic Transfer Process and Film Printing to a Computer-to-Plate

In 60 years since the founding of new China, with the development and popularity of offset printing technology, lithographic technology has made great progress. It has experienced the development stages of stone plate, albumen plate, deep-etch plate, pre-sensitized plate and computer direct plate making.

The Lithoprint technique evolved from the stone plate process. Stone plate making is implemented by directly painting texts and graphics on the stone surface with fat ink or first painted texts and graphics on transfer paper and then transferred to the stone plate. Later, many printing plants gradually replaced stone plates with zinc plates, and their surface grains were treated by grinding. Initially, an albumen plate was used, for which photosensitive glue was prepared by using ammonium dichromate and albumin glue. Then, a gum flat gravure process was used, for which photosensitive glue was prepared using ammonium dichromate and Arabic gum. After the 1970s, Arabic gum was gradually replaced by domestic polyvinyl alcohol (Yan et al. 1982).

From albumen plate to flat gravure, there was a great improvement in both printing precision and the number of prints (press run). However, the photosensitive glues coated on these plates all contained dichromate, in which chromium is a heavy metal harmful to the human body. Second, this kind of photographic plate could only be produced when needed and could not be stored for a long time; it was affected by the environmental temperature and humidity and was unable to be used for industrial mass production. In the middle of the 1970s, the China Research Institute of Printing Science and Technology first successfully developed a positive-type pre-sensitized plate (PS plate), for which the synthetic high molecular photopolymer was coated on an aluminum substrate that was pretreated by surface roughening, then dried, packaged, and stored for a long time. Therefore, the production could be done in advance and ready to supply for printing use. Since the 1980s, map publishing technology gradually began to use PS print technology (Yan et al. 1982). The automation of the exposure and washing equipment was also increasing, which standardized and digitalized the plate-making process. It is currently the mainstream plate-making technology.

Before the end of the 1980s, print film was mainly obtained by means of photography (electronic color separation), copying, manual typesetting, etc. It required great effort to deal with the mess after plate burning, which brought some difficulty to the printing operation; from the end of the 1980s, the computer-to-film (CTF) technique appeared, and laser typesetter was used to directly output full pages of image text processed by computer, with neat and clean film. The print quality was significantly improved.

From the late 1990s, with the development of color electronic publishing technology, the computer-to-plate (CTP) technique appeared; page data processed by computers were received directly by plate-makers and exposed on photographic plates and then made into a printing plate after washing. The two processes of output film and plate burning in traditional plate making became one to avoid the problems of dot loss and film distortion in the process of film printing and one to improve the printing quality and production efficiency to a higher level (Wei et al. 2005).

The application of computer-to-plate technology not only improved map printing quality and production efficiency but also promoted the development of digital printing workflow technology and digital proofing.

10.2.6 Proofing Technique Developed from Artificial Overprint Proofing, Offset Press Proofing and Chemical Proofing to Digital Proofing

The proofing technique is an important part of printing reproduction technology.

In the early twentieth century, artificial overprint proofing was used, followed by ink debugged according to the printing oil and printing began to catch up to the effect of the specimen sheet. After the 1940s, the platform mechanical proofing technique was applied. Since the graphic transfer principle and the materials used were the same as those for printing, the preliminary sheets of mechanical proofing were coincident with printing proofs. However, mechanical proofing requires a printing plate with high cost and a long cycle.

Entering the 1980s, chemical proofing techniques arose (through chemical changes to show color effects), including Crow-Marin preproofing and multilayer color film proofing. Compared with mechanical proofing, although the speed was significantly improved, chemical proofing was still considered to be simulative proofing. In addition, the proof was quite different from printing color, with some restrictions on use. Therefore, for a very long period of time, the platform mechanical proofing technique has been the mainstream printing proofing technique.

After the appearance of a color electronic publishing system, a proofing technique with computer direct connected proofing equipment emerged, which was called digital proofing technology. This technology was divided into soft proofing and hardcopy proofing. Soft proofing involves observing the color on the screen; it is easy to use and modify the color, usually for internal control purposes. Hardcopy proofing techniques include color laser printing, ink-jet printing, sublimation printing, thermal transfer printing, etc. Color ink-jet printing is the most widely used method. In the early stage of digital proofing, the color difference between the proof of proofing and the proof of printing was large and could only be used to verify the layout, errors and negligence and was not used for color proofs. In recent years, with the development of color management technology, digital proofing and printing color matching have increasingly matured, and digital proofing has gradually become the mainstream of proofing technology. Upon entering the twenty-first century, the use of traditional platform mechanical proofing techniques ceased (Liu et al. 2007; Che 2007).

The most important contribution of digital proofing was the improvement in communications among mapping, printing processes and users. In the drawing and prepressing stage, we saw the printing effect, which not only saved cost but also enhanced communication and even made color separations while proofing. This was an important breakthrough in the printing process.

10.2.7 Printing Technique Development from Lithographic Printing and Offset Printing to Digital Printing

At the beginning of the 20th century, lithography was the mainstream technology of map printing. Large lithographic printing presses used electric power, manual paper feeding and automatically added water and ink. A sheet was printed in a round trip, which was called a duplicator. While a small lithographic printing press was completely operated by hand shaking, when printing, first, the surface of a stone plate was wetted with a wet towel, then the ink was loaded manually with an inking roller to complete printing; 70–80 sheets were printed per hour.

In the 1930s, more than 200 sets of manual-feeding folio paper flatbed offset presses were successively used by the Shanghai Amway Machine Factory. Since then, the stone plate has gradually been replaced by a zinc plate, and the duplicator has been replaced by an offset press; the heavy stone plate was completely removed, and both thin and light zinc sheets were used as printing plates. The printing plate device and the pressing device were both all roller types, which greatly improved the printing speed and could print 2500 pieces per hour.

After entering the 1960s, the offset press was equipped with an automatic paper feeder, which could print more than 4000 pieces per hour.

In the 1970s, China had the ability to manufacture automatic paper-feeding lithographic printing machines. The lithographic press machines in large and medium-sized printing plants of China were basically domestic monochrome, two-color machines. After the 1980s, when a multicolor lithographic press machine was introduced, a domestic multicolor offset printing machine was successfully developed and put on the market, which made the lithographic machine develop toward a multicolor mode with increasingly higher automation and significantly improved printing quality.

In the 1990s, because of the applications of computers, automatic control, lasers and other high-tech devices in the printing industry, digital printing machines were born; the map to be printed was directly transmitted to the printer by computers, thus eliminating the color separation, imposition, plate making and other processes. Digital printing machines made map printing one of the most effective processes: the entire process from the input to the output was controlled by a person to achieve one copy printing (Shi 2000). Digital printing implemented the concept of ‘first distribution and then printing’.

There is no doubt that the development of high-speed on-demand color printing technology will bring revolutionary changes to map printing. However, in the next few years, offset printing will still be a very active process. Today, almost all printers are equipped with a device that controls all printing functions. The commonly used manual work in the past, such as format adjustment, plate replacement, register verstellung and the cleaning of the blanket and roller, can be achieved now with the pressing of a button. The default amount of ink can be preset on the digital interface of the digital prepress system.

10.2.8 Publication Form Development from Print Publishing and Electronic Publishing to Online Publishing

From the birth of printing machines to the 1980s, printing was the only form of map publishing, and the form of publication was presswork. Beginning in the 1980s, with the development of computer graphics and image processing technology, image text, audio, video and other information could be stored on magnetic, optical and electronic media in a digital way and read and used through computers or equipment with similar functions. This form of publishing was called electronic publishing, and the publication form was an optical disk (or simply, a disk). With the advent of Internet technology, the spread of information using computer networks for information dissemination, recording, storage, reading, and information dissemination emerged as the times required. This publishing form used the web as media, known as network publishing, and publications were in the form of web pages (Deng 2005; Chen et al. 2007).

Printing publishing, electronic publishing and online publishing all have the same prepress process. The difference lies in the different output modes, namely, various sources of information can be simultaneously used in electronic publishing and network publishing through prepress processing; this phenomenon is called cross-media publishing, and the unified data processing stage is called the premedia stage. The three publishing methods are not mutually exclusive in nature but can promote each other, find many similarities and complement one another during development (Shi et al. 2008).

10.2.9 Print Quality Control and Process Management Development from Independent Process Control and Management to Whole Flow Digital Control and Management

Over the long development process for printing technology during the twentieth century, although the processing and techniques have undergone tremendous changes, the three processes of prepress, printing and postpress finishing were well-defined duties and functions, and only the hardcopy graphic information was transferred between the processes; thus, non-digital control of the whole process was formed (Zou 1988; Guan 1988; An 1988; Wang 1988).

Upon entering the twenty-first century, a hot topic in the field of print publishing was the "digital workflow"; this concept was introduced into map publishing, which is called "all-digital map publishing control technology". This technology is based on the digitization of map publishing information and integrates and controls the graphic-text information, production control information and production management information in the processes of prepress, printing and postpress finishing. The establishment of a digital workflow was based on the digitization of image text information and production control information.

In the field of printing, the meaningful digitalization process of image text information began with the rise in DTP (Desktop Publishing) in 1985, which covered text coding, graphic description, image scanning, image text layout information description, RIP interpretation, and record information generation until the output of the graphic-text record. The digitization of mage-text information in map printing gradually matured with the development of digital mapping technology, and with the development of digital cartographic and publishing software and hardware technology, digital cartography was seamlessly connected with publishing technology (Yang et al. 2003). China's first map publishing software was MAPCAD (upgraded to MapGIS), which was developed by the China University of Geosciences. The software integrated the functions of digital cartography and publishing and was put into use in the middle of the 1990s, which greatly promoted the development of digital publishing technology in China (Chen et al. 2004a; Wu et al. 2004; Deng et al. 2005a).

The digitalization of production control information was based on the establishment of the CIP3 international organization (International Cooperation for Integration of Prepress, Press and Postpress) in 1995. The organization became CIP4 (International Cooperation for Integration of Processes in Prepress, Press, and Postpress) in 2000, which expanded the scope of prepress, printing and postpress integration to the prepress, printing and postpress processes.

In the process of all-digital publishing, there are three kinds of information flow: first, the graphic-texture information flow, namely, the operation content entity; second, the production control information flow, that is, the information and instructions for automatic control of production order, which are automatically produced from each link in the production process and are applied to automatic control of related processes to coordinate the work; and third, production resources and production quality monitoring management information flow, which is used for the comprehensive monitoring and management of the personnel, equipment, materials, product quality control, and user information in a production unit. The integrated control and management of the three abovementioned kinds of information flow realizes a true sense of full digital map publishing, achieves the best quality, intelligent production management and production efficiency of map publishing.

The above nine changes show that map publishing technology has changed from manual to mechanical, from analog to digital, and from monochrome to multicolor after 60 years of development, and these changes can be fully reflected from various aspects, such as map product variety, quality, efficiency and so on.

For early maps, due to the restriction of color separation, plate-making and other printing technology conditions, only monochrome line drawings could be printed. After the 1970s, the application of a double-color printing machine and photographic reprint technology realized the printing of polychrome maps, layered maps and color hill-shading maps. However, due to the deformability of the photographic and copying material, with a low precision visual surprint, and the difficulty of double-color printing alignment, these factors can only be achieved by using a large number of spot colors instead of color overprints. After entering the 1980s, along with the popularization and application of the electronic color separation plate technique, color image map printing began. After the 1990s, with the application of digital publishing technology, the variety of map publishing grew, the amount of information was large, and the color was more abundant. In contrast, the number of color separation plates was increasingly less (due to the extensive use of color overprints), and the quality of the map product became increasingly higher (boutique maps were available). Moreover, the support ability of map printing increased rapidly, and the time required to print a sheet of maps was shortened from a few months to a few weeks or days and sometimes, " it was done while you waited".

In the 60 years of development of map printing and publishing technology, there are several important stages worth mentioning: first, the development of electronic color scanners from 1975 to 1985 provided the conditions for image map publishing; second, the appearance of digital prepress technology from 1985 to 1990 led the map publishing technology and mapping technology towards integration; third, the period from 1990 to 2000 was the heyday of the development of digital prepress technology, and digital publishing technology has been widely used; fourth, since 2000, CTP technology has been developing and maturing; and fifth, after 2005, the digital workflow entered the application period.

10.3 Main Research Achievements

In the past 60 years, map printing and publishing have achieved remarkable results in theoretical research, technique innovation, material and equipment manufacturing, software development and other aspects, and the research results have been unceasingly applied to map printing. Such great progress is inseparable from the cultivation of talent and scientific research.

10.3.1 Theoretical Research Results

1. **Color reproduction theory of map printing**

The study of color reproduction theory is accompanied by the development of color printing. Map printing generally uses two types of color: spot color and primary colors. The research on spot color printing focuses on the ink color matching theory, the subtractive color printing theory aimed at replacing spot color with three primary colors for printing to reduce the number of printing plates, and the high fidelity color reproduction theory to improve the performance of printing color. The study contents of the three-color process are mainly color separation theory and technology. Especially since the 1980s, the use of various color measuring instruments, such as density meters and spectrophotometers, has changed the control of map printing color from experience to data. The main research results include the spot color ink matching theory and model based on density or chrominance, the photographic color separation theory, the theoretical model of color space conversion and gamut compression, the high fidelity color separation algorithm, and so on. On the basis of theoretical research, several paper-based *map chromatographies* were produced, and a software platform for map digital chromatography was established to guide map color design and accurate color reproduction and to realize the consistency of color in each link of map design, map drawing and map printing.

Theoretical research and map printing and publishing technology are organically combined. 'Map color separation photography' uses red, green and blue filters for color decomposition of color images, replacing color separation by division of labor with photographic color separation technology (Gao 1962). 'A color map making-printing integration technique for multi-band satellite image' was proposed to simulate a natural color image map with a false color image by using photographic color selection method (Chen 1988); 'the main factors causing color cast and the mathematical model for color correction' analyzed the inherent law of color cast in map printing and deduced a mathematical model of color correction (Zhang 1989); 'a simple method to calculate the percentage of black dots in 4-colors printing' presented the mathematic model to calculate the gray component based on the 3 color Neugebauer equation, which solves the most difficult job in color space transformation (Liu et al. 2000a); in the paper 'research on color space conversion model of map electronic publishing system', the principle of the least square method was used to calculate the transformation matrix by multiple sample points to realize the conversion between the RGB value of computer display and CIE1931XYZ standard color space (Liu et al. 2000b); in 'the conversion of CIEL*a*b* and CMYK color space using the table look-up and interpolation', the concept of CMYK pseudo 4 dimensional space was presented, through GCR, the CMYK pseudo 4 dimensional space was transformed into CMY 3D space, and cubic interpolation, triangular prism interpolation and tetrahedron interpolation were used to realize the one-to-one correspondence between L*a*b* color space and CMYK color space coordinates (Shi 2002); 'a color separation model based on 7 color high fidelity color printing' proposed the partition theory of color separation and gave a color decomposition algorithm based on the Neugebauer equation (Shi et al. 2007); in 'research on cross-media color transmission based on color space conversion and gamut mapping', the problems in the process of cross-media color image transmission were analyzed, the implementation mechanism of color space conversion and gamut mapping were explored, the problem of color gamut boundary determination was solved, and the SLIN and CUSP algorithms were discussed and evaluated (Shi et al. 2008).

2. Map printing tone reproduction theory

This theory mainly delves into the printing tone transmission mechanism and the mathematical model of tone modulation transmission function of continuous tone maps, such as the hill-shading map and image map. The main research results include the photosensitivity of various photosensitive materials, the principle of the development process and its control mechanism, the principle of printing dot generation, the influence of dot distortion on tone reproduction, the standard tone reproduction curve, and the image evaluation theory.

The above theoretical studies combined map printing and publishing technology. 'The visual characteristics of copy screen used in map printing' studied the problem of how to make the screening area on a map accord with the visual characteristics, and several implementation methods were designed (Wang et al. 1988); in 'influence of halftone dot deformation on color reproduction', the deformation law of dots was analyzed by experiments, and it was pointed out that the edge part of the dot was uniformly expanded outward, and the dot was increased by the same proportion; additionally, the dot deformation increased with the increase in line density, and the increasing rate of dot for different percentages of line density was obtained (Du 1988); according to 'the investigation of the standard tone reproduction curve', the reproduction quality of image tone was accurately measured by making the standard tone reproduction curve according to the visual characteristics of human eyes (Jin 1988); 'design of a dot with gradual changed shape and its transmission characteristic test' presented a new algorithm model for the generation of a halftone dot with gradual changed shape, which effectively improved the tone discontinuity of dot distortion to achieve a better dot tone reproduction; in 'a kind of columnar halftone dot model suitable for lenticular screen stereo press image', the tone reproduction characteristics of lenticular screen stereo press were analyzed, and a halftone dot generation algorithm model was put forward to effectively improve the stereoscopy and sharpness of image (Shi et al. 2009).

3. **Digital map publishing theory**

This theory mainly studies the construction theory of digital publishing systems, the publishing and processing methods of multisource data, and the mathematical modeling of various process descriptions. The main research results include the theory and technology of digital cartography and publishing integration, the map publishing theory and technology based on the PostScript language, the theory and method of the generalization of map publishing data, and the theory and technology of map online publishing.

'The principle of map publishing system design based on PostScript' introduced the process flow of map publishing, discussed the concept and characteristics of PS language and RIP technology and its application in map publishing, and put forward the design idea of professional RIP (Chen et al. 2000); in 'Format conversion of digital map data to MapGIS data', the characteristics of two kinds of data formats of digital map and MapGIS were analyzed, and a digital sharing idea was put forward, that is, to convert digital map data into MapGIS plain code file and implement topological reconstruction at the same time, and then convert to MapGIS standard format file through the interface functions provided by MAP-GIS (Chen et al. 2004b); 'The system structure of cartographic and map publishing system' introduced a professional software system of cartography and publishing, which can be used to publish and distribute traditional maps and electronic maps, theoretically analyzed four aspects of the data model oriented to publishing, the process flow, the output interface, and the hierarchical model, and a feasible solution was put forward (Wu et al. 2004); in 'research on the ant aliasing algorithm suitable to the map publishing symbol', based on the analysis of the existing anti-aliasing algorithm, an ant aliasing algorithm suitable for the map publishing symbol was proposed, which was called the brush method, and a comprehensive assessment of the anti-aliasing effect and efficiency was conducted by experiment (Deng et al. 2005b); in 'Design of a four-view architecture for the map publishing system', a four-view architecture of the map publishing system was designed, in which the functional view and structural view were used to define the main business functions of the system and the system structure of "big grain", the control view was used to describe the flow control of data entry, editing, processing, and output, and the deployment view was used to describe the hardware and software configuration of the map publishing system (Wu et al. 2007); 'Research on map web publishing technology based on SVG' analyzed the advantages of applying SVG to map web publishing, discussed key technologies, such as SVG description and reference of map symbols, SVG coding and organization of map data, display control and web interaction of SVG map, and put forward a map network publishing system model based on SVG (Chen et al. 2007); 'Heterogeneous GIS data integration scheme and implementation based on PDF structure' proposed a method to rapidly reorganize GIS and other multisource map data as the prepress data (Wen et al. 2005); and finally, in 'A method of calculating the ink consumption of a printing machine', mathematical modeling was implemented based on the prepress film make up data and related printing conditions, then the amount of ink consumption in the ink-covered area of each color plate was calculated, and the prepress and the print data was organically connected (Shi et al. 2009).

4. Map publishing quality control theory

This theory mainly studies the quality control mechanism and principle in the map publishing process, the quality control model and so on, as well as the parameters and conditions, quality evaluation models and standards for controlling the quality of map printing; and construction of a process control with a node chain is proposed to realize the digitalization and standardization of the whole map publishing process.

The representative results are described as follows.

In the thesis 'Research on the quality control of whole-digital map publishing', color representation, dot transmission law and other factors affecting map publishing quality were tested and verified, the actual color performance of screen, digital proofs and printed pages, as well as the information transfer characteristics of digital pages—film—plate—printed sheets, were measured, a color separation model for map spot color was proposed, a subtractive color printing process was achieved, and the halftone dot model of image tone reproduction was optimized; 'Discussion on improving the quality of map printing' put forward nine measures and a detailed plan to improve the quality of map printing on the existing equipment and technology level, as well as recommending the addition of the necessary equipment to further test the new production techniques (such as scribing, color separation projection, multilayer metal deep-etch plate printing, and so on) that have been used in a certain way and gradually applying them to practice; simultaneously, it was pointed out that the fundamental measure to improve the quality of map printing was to continuously improve the technical level of map making and printing personnel (Yan 1962); 'Overview of map printing and the direction of our efforts' introduced some new technologies, such as new photosensitive material, non-negative plate making, laser computer-to-plate making, pre-coating plate making, preproofing and so on, as well as predicting the future development direction of map printing, that is, a simpler production process, standardized process technology, digitization and fast mapping, and the direction of our efforts was put forward to reform the unreasonable or outdated technology in the printing operation, to vigorously promote the use of the partial element scribing for publication original, and to develop digital map production. (Yan et al. 1982); in the paper 'Regression analysis of the readings of optical densitometer', in view of the difficulties encountered in the standardization of map printing engineering, nine regression models were used to analyze the correlation between the nine different types of densitometers, and the regression equation among them was established to make density measurements more accurate (Zhang 1986a, b); in 'Evaluation and standardization of printed images', the contents and principles of subjective and objective evaluation methods were analyzed, several problems to be paid attention to in the evaluation of map printing quality were put forward, and various methods for image quality controlling by using ladder rulers or test strips were introduced (Zou 1988); in 'Standardization of remote sensing image replication and its quadrant control system', based on a large number of research and experiments, with the aim to standardize remote sensing image reproduction, a quadrant reproduction system for assessment of remote sensing image quality was proposed (Zhang 1991); and finally, 'Printing quality detection and evaluation' expounded upon the principle of map printing quality inspection and evaluation, and presented two kinds of mathematical models for quality assessment, including fuzzy comprehensive evaluation method and extension comprehensive evaluation method (Sun et al. 2008).

10.3.2 Research Results of the Materials and Process Technique

1. **Map printing material**

In the field of map printing materials, plate-making photographic materials, non-silver-salt photographic materials, peel-coat films, and scribing films were developed. 'A new plate-making technology with domestic cherry gum instead of imported Arabic gum' (Yan 1962) greatly reduced the cost of plate making; 'Development of PS sensitive material with dual positive and negative purposes' promoted the development of map plate-making material from immediately coated chromate colloid to the pre-coated diazo photographic plate; the development and application of 'GS-II type positive diazo photosensitive strip mask' (Chen 1995) greatly simplified the production process of tone copy; 'Development of pre-coated photosensitive scribing film' and 'KM-300 map scribing film' had the advantages of non-pollution; the successful development of 'Dual positive-negative diazo duplicating film' (Chen et al. 2001) solved the major technical problems in implementing image replication with non-silver film instead of silver film and also solved the technical problem that the use of silver salt materials made it difficult achieve isomorphism image replication; and finally, 'Relief type photosensitive film for the purpose of map annotation' was a peelable photosensitive film suitable for map annotation, and its transfer printing technique could be applied to complement mistakes and omissions. In the development of map printing materials, a number of national invention patents were declared, such as 'exposure method for pre-staining diazo film', 'processing process of photosensitive peel-coat film', 'scribe coating and its preparation', 'positive and negative dual-purpose diazo photosensitive strip film', and so on.

2. Map printing process and technique

(1) Technical scheme for printing and mapping

In view of the types and features of the map publishing original, and according to the different equipment and technical conditions of each printing plant in different periods, the best process plan was designed to achieve high efficiency, high quality and low cost. Printing process schemes, such as fair plate drafting, separated drafting, separated scribing, regional color filling administrative maps and hypsometric tinting terrain maps, were developed. After entering the digital publishing period, research on the technical scheme of printing and mapping turned to the aspects of feature processing, color establishing and plate separating schemes, page make up schemes, hardware and software environments (Ren et al. 2007). These schemes have adhered to the principle of combining theory with practice and adapted to the needs of various kinds of map printing.

(2) Photographic operations

Replacing the wet collodion process with dry film and replacing glass plate (collotype) and pirated copies with a polyester film base, this kind of photographic technology, that is, "dry instead of warm, software instead of hardware", achieved vital innovations in map printing and publishing processes at the end of the 1980s; in the field of photographic color separation netting, the long-term indirect netting technique led to serious loss of tone level and greatly influenced the printing and publishing of high-quality satellite images. After the successful development of the direct netting technique, photographic color separation and netting could be performed at the same time, and the quality of the color separation filter was greatly improved (Chen 1988). In addition, significant achievements were obtained in the field of map microfilm technology.

(3) Plate making

A large number of technical studies and practices were conducted on map color separation techniques. For example, the use of chromated colloid dyeing retouching reduced the workload of manually coating negative film, which was particularly suitable for maps with fair drawing in two plates (Liu 1962); in the aspect of area element production, several processes for coat removing plates were developed; in the field of negative film reproduction and replica processes, negative and positive pirated copy processes were developed to reduce the number of pirated copies; and in the aspect of imposition and printing down, a step and repeat method was invented.

(4) Proofing and proofreading

During each period, research on proofing and checking was always a hot topic. In the period of manual coating, with the traditional mechanical proofing process, the main research content was the making of proofs, such as a monochromatic proof, double-color proof, red proof, panchromatic proof and printing proof. The role of content checking was different due to the different production methods of various proofs and cooperating with each other enabled map proofreading. To find a convenient and quick way around the cumbersome and high cost of mechanical proofing, new proofing technologies continued to be developed, for example, Kromah Lynn proofing and diazo film nested proofing, which played a role in sample map reviewing to a certain extent. After the application of digital publishing technology, research on proofing technology became more active. To adapt to digital publishing technology, the research focused on how to replace traditional proofing technology with digital proofing technology, and much research was conducted on digital proofing color management, color measurement and correction, simulation of digital proofing and printing effects, sample map checking of digital proofing and other aspects. Meanwhile, the software platform and samples were used for comprehensive review to complete the prepress proofreading work.

(5) Subtractive color printing

A notable feature of map printing color development was the use of a large number of spot colors. For early map printing, more than 20 colors would be used in a map, resulting in a greater number of overprints, poor precision, and long periods. Therefore, reducing the number of map printing colors to achieve subtractive color printing was the focus in industry. For map printing based on a double-color printing machine, there was a printing technology for large-scale topographic maps using brown and blue instead of black, brown, blue and green, which has not been widely used due to the large difference between color simulation effects and standard colors. After the application of four-color printing, research on the technology of printing maps with three primary colors (black, cyan, magenta and yellow) was carried out (Hong 1988; Wu 1988; Fan 1988); especially after the application of digital publishing systems, color management and data manipulation was used to accurately control the color ratio of various color plates to completely reproduce map design color with four-color printing.

(6) Special-use map products

In addition to the production of conventional paper maps, a special type of map product was developed, for example, maps printed on silk and other fabrics, as well as waterproof camouflage maps, which were very suitable for field use; a stereoscopic map printed according to the principle of lenticular screen imaging (Shi et al. 2009) was developed, in which stereoscopic effects were observed on one map without the need for stereoscopic glasses; a luminous map printed with special ink was used in the absence of light at night; and tactile maps printed with screen printing plates and foam ink was usable by blind people.

(7) Digital and standardized production technology

The slogan of digital and standardized production was put forward earlier, but it has gone through a long slow process. Since the 1980s, a proposal was given that implemented data detection and quality control using various kinds of signal fragments and punch-register devices in the printing process (Zhang 1987; Guan 1988; An 1988). Since the 1990s, according to the characteristics of map printing and publishing, the CW series of signal fragments for map printing were developed to strengthen the control of map printing, which plays an important role in the control of map printing data. Starting in 1999, the Digital Control System for Map Publishing and Printing and Quality Evaluation System were developed, which promoted the process of digital and standardized production of map printing and publishing. In 2008, Ink Consumption Calculation and Setting Software-InkSet was developed and put into production which also moved map printing and publishing technology toward a digital workflow. In the process of promoting the digitization and standardization of map printing and publishing, the national standard of topographic map color division and the standardization and quality control standard of image reproduction engineering were successively completed, the *Specification for Printing of Maps* (national standard and national military standard) were formulated and revised, and the Assessment Standard and Regulation of Map Printing was developed. At present, a relatively sound production management, quality management and business training performance evaluation system has been formed.

(8) Rapid replication

The main research goal of rapid map replication is to solve the problem of quick map supply. The first generation of rapid reproduction technology used wet process photography, chromated colloid plate burning, and monochrome offset printing, with slow response speeds; the second generation of rapid replication technology, from the exploration of new plate-making method, mainly used electrostatic plate-making technology, and the developed electrostatic plate-making machine was assembled on the field electrostatic printing vehicle; and the third generation of rapid replication technology used digital publishing technology, and the new developed type of map rapid replication system imported map data by scanner, processed prepress graphics and textures by computer, and directly printed maps by digital printing machine.

10.3.3 Development and Application of Map Publishing Software and Hardware Equipment

In terms of map publishing software, the 'Map Color Devising System', 'Map Publishing Data Format Conversion Software', 'Digital Detection and Control System for Map Printing and Publishing', 'Map Printing Quality Evaluation System', 'Map Prepress Data Analysis and Processing System', 'Database Management System for Digital Map Publishing Original', 'Ink Consumption Calculation and Setting Software (InkSet)' and others were developed.

In terms of map publishing hardware, the 'Multicolor Line Map Electronic Color Scanner', 'Multi-Functional Silverless Film Processing Equipment', 'TBJ-120 Silverless Coating Machine', 'Computer-To-Plate', and 'Map Rapid Replication Vehicle' were produced.

10.3.4 Map Publishing Standards and Specifications

1. ***Map color atlas***

To establish the standardized series for map color division in China, to provide map color standards for map design, map decoration and map making and printing workers, and to provide a basis for the establishment of an image color database of the Chinese surveying and mapping industry, a '*Map Color Atlas*' was developed by the National Institute of Surveying and Mapping according to the requirements of the State Bureau of Surveying and Mapping, and this atlas was published by Surveying and Mapping Press in July 1987. The main authors are ZHANG Qingpu, YE Taiqi, ZHAO Hongxia, AN Zhenzhen, Cao Tianjing, and JIN Lan. Based on the theory of colorimetry and color image reproduction, with standardization and digitalization as its main form, the color atlas carried out strict data detection in the process of developing, which has an important guiding role for the color design, printing and publishing of maps.

(1) Design principles. In practice, the focus is on both the characteristics of printing various types of maps and considering the technical requirements for color atlases in the general printing industry; in the content and layout of the color atlas, not only should the aim be to increase the number of color lumps but also to pay attention to the actual resolution of the color vision; in the selection and utilization of printing materials and printing processes, the advanced level

of foreign map color atlases should be reached and the possibility of domestic popularization and application should be considered.

(2) Structure of contents. This structure is composed of four parts, including double-color overprint, tricolor overprint, spot color printing and test sample, which contains 52 pages and 6768 color lumps that are detailed as follows: three-color overprints with yellow, magenta and cyan, a total of 6 pages and each page contains 144 color blocks, with a total of 864 color blocks; yellow, magenta, cyan and black double-color overprint, a total of 16 pages and each page contains 144 color blocks, with a total of 1584 color blocks; red, brown, yellow, orange, green, medium blue, purple, silvery gray, light blue double-color overprint, a total of 30 pages and each page contains 144 color blocks, with a total of 4320 color blocks; 5 color maps and test samples as an attachment to the map atlas; a group of colored foldouts was inserted between the foreword and catalog to express the Munsell color solid, and the basic concepts of additive color mixture, subtractive color mixture and FOGRA signal were also added.

(3) Printing process flow. Production of dot film master of 54 wire (CM)—copying and exposure of halftone dot film—cutting and imposition of halftone dot film—printing down—proofing—quality evaluation and binding.

2. ***Color atlas of military map***

According to the characteristics of the military map and the special requirements of color division, 18 basic colors and three kinds of dot films with different screen rulings are selected and divided into two parts: basic color overprint and spot color overprint, resulting in a total of 8348 color lumps. For ease of use and communication, the color number and percentage of dots of the color overprint of the color lump are attached to the top and bottom of each color lump. Readers can easily see the basic color and the proportion of dots in each color lump. At the bottom right of each page, there is a spot color sample of the basic color in the page as a reference for proofing and printing.

The color atlas mainly consists of four parts:

(1) Basic color. On the basis of the principle of three primary color mixtures and the technology of color image reproduction, two four-color combinations are separately designed. That is, for the dot of 60 wires (CM), yellow, magenta, sky blue and black series are combined into 4528 color lumps by using double-color overprint, tricolor overprint and four-color overprint, respectively, for a total of 36 pages and 6792 standard color lumps, suitable for satellite photo and color hill-shading, as well as a variety of fine color images, atlases and other special map reproductions.

(2) Spot color. This part consists of a total of 28 pages, with 1556 standard color lumps, in which two-color overprint series are listed according to the color characteristics of different kinds of maps: there are two kinds of dots with 54 wires (CM) and 40 wires (CM), and medium yellow, magenta, sky blue

and peacock blue, pale yellow, light yellow, reddish brown, relief brown and other spot colors are mixed by using double-color overprint and tricolor overprint standard color lumps that are comparable to one another and suitable for use with different types of maps, thematic maps and atlases. Another series of standard color lumps is designed for layered terrain maps, which can be combined into a variety of different operation schemes that are intuitive, practical and suitable for the color selection of aviation maps, atlases and large and medium-sized color wall maps.

(3) Samples. This part consists of 48 various types of maps.

(4) Textual descriptions. At the beginning of the color atlas, the content and application scope, the quoted standard, the printing and publishing conditions and the control parameters are given, which shows that when the color atlas is printed, strict quality control methods are adopted.

3. ***Color atlas of the thematic map***

This atlas was developed by the Institute of Geography of State Planning Commission, Chinese Academy of Sciences, and published by Surveying and Mapping Press in 1987 as a national matter standard of the P.R. of China (GSB: A26001-87), and this atlas is composed of two volumes of four-color parts and spot colors.

This atlas is mainly designed for maps, especially for color design and printing of thematic maps and atlases. In the design, the color design experience of China's map, especially the thematic atlas, is summarized, with reference to the characteristics of the foreign famous color atlas. Many color samples are given in the color atlas, including the color overprint system for four-color printing and the commonly used spot color overprint system for thematic maps, and a large number of continuous halftone tables and lines and symbol colors are collected and designed, which provides a color sample and basis for map color design and the standardization and digitization of the color image reproduction process. It is an indispensable utility for map design and decoration, image processing and reproduction, map making and printing, as well as quality control. This atlas was also an early work for the establishment of color databases and computer-aided design and has been of great reference value to scientific research, education, art design and other printing sectors.

Fourteen basic colors are used in the color atlas, and a total of 33,432 color lumps are overprinted, including the following 4 color overprint systems:

The double-color overprint, tricolor overprint and four-color overprint of medium yellow, magenta, cyan and black, with a total of 60 pages and 25,395 color lumps, which are designed based on the principle of three primary combinations and color image reproduction processes, are particularly applicable to satellite images, color shading maps and color images, photographs and the replication of art works.

The double-color overprint and tricolor overprint of spot color resulted in a total of 42 pages and 7215 color lumps. It is designed by considering the characteristics of thematic maps and the requirements of bright and vivid color. The overprint is divided into two parts:

(1) The tricolor overprint of lemon yellow, bright red and peacock blue has a total of 12 pages and 2067 color lumps. Application of this overprint system can be used make the commonly used orange series (lemon yellow plus bright red) and green series (lemon yellow plus peacock blue) brighter on maps.
(2) The double-color overprint of orange yellow, reddish brown, green, purple, lake blue and gray and the double-color overprint of lemon yellow, medium yellow, bright red, peacock blue and cyan include a total of 30 pages and 5148 color lumps, which can be used to meet the special needs of the color designs of various thematic maps.

The map continuous halftone table includes a total of 7 pages and 85 continuous halftone tables and 822 color lumps. Sections 1–36 are mainly designed based on contour maps, Sects. 37–74 are mainly designed based on statistical maps of grades, and Sects. 75–85 are mainly designed based on layered terrain. Of these sections, many of them directly use famous map color standards at home and abroad.

The color for the map line symbol includes a total of 2 pages and 128 types. It focuses on the color scheme of thematic maps but also considers that of topographic maps and general map line symbols, basically covering the color scheme of map line symbols, and with samples of linear, point and areal symbols, it is easy to use. These line symbol colors are partly printed in monochrome ink but mostly superimposed with different inks and different dot percentages for the convenience of the combination printing of symbols and solid colors. In addition to positive symbol colors, there are contrast symbol colors, which resulted from the development of the color scheme of thematic map symbols.

The color atlas uses 150 wires/inch square dot film, 157 g coated paper, ink from Tianjin printing ink factory, and was printed using domestic J2205 type two-color offset press. In the whole process of printing, the density gauge and signal bar are used for strict quality control.

4. **National standard—*Map Printing Specification***

The national standard of *Map Printing Specification* was revised in 2008 after the two versions of GBCHIV-301-82 and GB/T 14511-93. The new version is used to replace GB/T 14511-93 *Map Printing Specification,* GB/T 14510-1993 *Specification of Image Map Printing*, GB 14051-93 *Standard Colors for Topographic Map* and GB/T 15638-1995 *Specification for Optical Density Measurement of Map Printing.*

The standard specifies the basic principles of map printing, the requirements of process design, and the quality standards for prepress, printing and printing processes, as well as operation requirements. It is suitable for the offset lithography of national basic scale maps, general maps, image maps, thematic maps and atlases.

The main contents include terms and definitions, general rules, original artwork, receiving and inspection of accessories, process design, digital prepress processing, PS plate production, offset press proofing, proofreading, printing, sorting, and finishing.

10.4 Problems and Prospects

Map printing and publishing technology has transformed from analog to digital. However, in the present situation, there are still some links and factors that are not suitable for efficient digital publishing technology. For instance, film printing encounters a quality bottleneck in the printing process and will be replaced by direct plate making (CTP) technology in the future, and film will gradually withdraw from the stage of map printing. The data chain of digital publishing is still in prepress, and it is likely to be replaced by a digital printing system in a digital workflow environment. Each link in the process of digital map publishing is still in the state of operating by experience, and it will be replaced by the whole process digital control system. The post-processing of map printing is currently manual work, which is not suitable for digital publishing technology, and the digital operating level of post-printing quality inspection and classification will be improved in the future. Digital printing technology is not only the source of printing technology revolution but also the future development direction of printing technology, which has upgraded the printing industry from a handicraft industry (printing quality is heavily dependent on the experience of the operator) to a mechanical industry (the quality of printing products is completely monitored by computer programs).

At present, map printing is still based on the single machine operation or the LAN connected part of the equipment in the factory, with poor consistency and coordination of the work, inconvenient data transmission and finished goods transport. Driven by the development of network technology, prepress processing, printing and finishing equipment will be connected by using local area network technology in the future and can also be connected with the network system of remote equipment to realize printing on demand, that is, to print the required map contents at the required time, place and in the necessary quantity.

Currently, map printing is still mainly based on paper media, which can only express static contents, and once the map is made, the picture contents cannot be changed. In the future, with the development and application of a new type of information recording material, electronic paper and other media will be used for map publishing and will reach a content record form like a handheld computer and reading the new publishing carrier will be as comfortable as reading paper.

10.5 Representative Publications

(1) ***Process design of map printing*** (DONG Yuhui. Surveying and Mapping Press, Beijing, June 1993)

The book consists of 6 chapters. The first chapter is an introduction; the second chapter is the specification design of map printing; the third chapter is about the methods of map printing process design; the fourth chapter describes the design of the color table for printing; the fifth chapter introduces the design of the operation sequence for large-scale topographic maps; and the sixth chapter introduces the process design of small-scale topographic maps, in which the design principle and method of the work plan, the method and the process of program design are introduced in detail with examples of regional color administrative maps, layered color terrain maps and other types of maps.

(2) ***Applied colorimetry*** (JIANG Jiwang, WANG Xiuze, WANG Rong, MA Xuanwen. PLA Press, Beijing, November 1994)

This book systematically introduces the basic theory of colorimetry and the basic principle of color measurement technology and explains their important applications in many fields. Starting with explanations of the basic concepts of radiometry and photometry, the book discusses the origin, nature and basic laws of color vision step by step and introduces in detail the various color expression systems, color space, color difference formula recommended by the CIE, as well as the color measurement methods specified by the CIE, measurement standard, main color measuring instruments and the commonly used color atlas and color calibration methods in the world. Among them, OSTWALD, NCS, DIN, OSA, COLOROID and other color development and color expression systems, ANLAB color expression system, color rendering index measurement and calculation of light source, color metamerism index of light source color, color temperature measurement of light source, fluorescent surface color measurement, whiteness measurement, the color difference formula of recent years, and so on are all described. Important applications of modern chromaticity and color measurement technology in printing and dyeing, printing, coating, paint and color photography, color TV, computer color image matching, map color design and other fields are described in the application section.

(3) ***Map plate making and printing*** (YUAN Kansheng, CHEN Guangxue. Xi'an Map Publishing House, Xi'an, September 1995)

This book systematically discusses the process, theory and method of map plate making and printing. While considering the old process system, it focuses on new technology and new materials in the film-making process and discusses the quality inspection and control of map printing, as well as the subtractive printing technology of maps and other technical methods.

There are 11 chapters in the book. The first chapter is the introduction; the second chapter discusses original map printing; the third chapter introduces map photography and electronic color separation; the fourth chapter describes the reproduction of map negatives; the fifth chapter introduces map retouching and imposition; the sixth chapter presents the principle and method of lithographic plate making; the seventh chapter is about map proofing and proofreading; the eighth chapter is about offset lithography and atlas binding and layout; the ninth chapter introduces the specification and process design of map printing; the tenth chapter provides some simple methods for map reproduction; and the eleventh chapter introduces the new technology of map printing and its future development.

(4) ***Introduction to printing*** (LIU Zhen, GUO Chunxia. Printing Industry Press, Beijing, April 1996)

Beginning with a general outline of the printing process, the book describes the introduction of printing, the graphic and information processing of printing, plate making, postpress processing and the quality inspection and control of printing products. The book not only describes the basic principle of the printing process but also explains the inspection and quality control of printing equipment, printing materials and printing products. The main contents of the book are divided into 6 chapters: the first chapter is the introduction; the second chapter is the prepress graphic information processing; the third chapter is plate making; the fourth chapter is printing; the fifth chapter is postpress processing; and the sixth chapter is the quality inspection and control of printing products.

(5) ***Printing graphics*** (CUI Yuanri, WU Xinmin, LI Xiangting. PLA Press, Beijing, December 1997)

The book is divided into two parts. The first part is the foundation of printing graphics, which has a total of 6 chapters. Based on the brief introduction of the basic concepts of color vision and colorimetry, some problems and new progress in the research of color vision and color reproduction are introduced; and the basic concepts and methods of image function, optical image processing and digital image processing for printing graphic processing are introduced, which are the necessary basic knowledge of mathematics and physics for the color reproduction, tone reproduction, image clarity and granularity and noise control of digital color printing. The second part is about the application-color reproduction of printed images, which has a total of 6 chapters. Image input, output and processing are taken as the main line, then image density and definition, tone of printing image, and color reproduction and correction are discussed; color conversion between different color spaces and different media is discussed, the authors' latest achievements in the study of modified Neugebauer equation are introduced; and the basic principle and performance of the commercially available color scanning input devices and image recording output devices are summarized. In the last chapter, the digital color printing image processing system, the post-script page description language and the color management system are introduced.

(6) ***Map reproduction*** (LIU Zhen, CONG Wenzhuo. PLA Press, Beijing, August 1998)

The book consists of 9 chapters. The first chapter is the introduction, which mainly describes the characteristics and tasks of map reproduction, as well as the methods of map printing; the second chapter is about the preparation work of map publishing, and the main contents and process of map publishing preparation, original manuscript types and production methods are introduced; in the third chapter, the author introduces the principle and method of map photographic equipment and the imaging principle and photographic processing technology of several commonly used photosensitive materials; the fourth chapter is about the process of reprints and separate drawings, and the reprint function and its commonly used equipment, material and technique are introduced; the fifth chapter is about plate making, which describes three methods of plate making, such as the chrome protein negative process, the chrome polyvinyl alcohol positive process and the pre-sensitized plate-making method; in the sixth chapter, map printing, grading and packaging are presented, the types, structure and properties of map printing materials are introduced, and the map grading and packaging methods are described; the seventh chapter is the process and scheme design of multicolor map printing, which introduces the design of printing process, the design method and process of tint color table, the determination of the technical methods for each process and the drawing of the block diagram in the process of plate making and printing, as well as the preparation and description of the printing scheme; the eighth chapter is about electronic publishing system and the electronic publishing of map, which introduces the basic definition, development history and structure of electronic publishing system; and finally, in the ninth chapter, the authors introduce the replication technology of a small amount of copies and the principle and process of map reproduction with less need, as well as describing in detail the process of a variety of different replication methods.

(7) ***Printing information recording material*** (ZHAO Guochuan, Wuhan University of Science and Technology Press, Wuhan, October 1998, first edition)

The book is a rewriting and extension of *Photosensitive materials for photoengraving*. However, the recording materials are only within the scope of printing. Due to the rapid development of the printing industry, the related recording materials are also increasing. This book mainly involves two parts that discuss silver and non-silver recording materials. For silver, the imaging principle and technology are relatively mature; therefore, a systematic introduction is given. In addition, based on the basic principle, the classification method of conventional silver halide plate-making materials and unconventional silver halide plate-making materials is proposed, which is not common in the general teaching material.

In the non-silver halide imaging system, in addition to diazonium material, thermosensitive material and acid sensitive material, there are other materials, such as magnetic recording material. However, it is worth noting that the thermal imaging system may develop from the recording material. Additionally, the printed products using these thermosensitive materials may be plentiful and more attractive to customers because the required temperature of this kind of material is only 28–32 DEG C.

(8) ***Practical guide to the digital prepress*** (CONG Wenzhuo, SHI Ruizhi, LIU Shide. Planet Map Publishing House, Beijing, December 1999)

This book systematically and briefly introduces the knowledge of the electronic publishing system, which is helpful for students aiming to understand the electronic publishing system or for project managers engaged in the electronic publishing industry. The main contents of the book are divided into three parts: the configuration of the color electronic publishing system, the color electronic publishing software and the key technologies of color electronic publishing, for a total of 9 chapters. The first part, the configuration of color electronic publishing system, includes the first chapter of the color electronic publishing system and the second chapter of the configuration guide of color electronic publishing system; the second part, the color electronic publishing software, includes the third chapter of the Windows operating system, the fourth chapter of the Macintosh operating system, the fifth chapter of Photoshop 5.0, the sixth chapter of FreeHand 8.0, and the seventh chapter of PageMaker 6.5C; the third part, the key technologies of color electronic publishing, includes the eighth chapter of the principle of image publishing and the ninth chapter of related technologies of color electronic publishing.

(9) ***Digital publishing of the map*** (SHI Ruizhi, CONG Wenzhuo. Planet Map Publishing House, Beijing, December 1999)

For the first time, the concept of map digital publishing technology is introduced into teaching material, and with the support of computer hardware platforms and software environments, computer graphics and image processing technology are used to realize map publishing. The main content explains how to apply MapGIS software developed by the Zhongdi Cyber to the processing of map publishing.

The book is divided into 8 chapters, which mainly explain the basic concepts of digital publishing, digital map publishing techniques and processes, map data input methods, symbolization and editing technology, map databases and attribute database management, data format conversion, EPS file generation, printing and output.

(10) ***Principle and technology of digital prepress*** (LIU Zhen, SHI Ruizhi, WEI Bin, XU Dehe. PLA Press, Beijing, October 2005)

The book is based on the digital prepress process, discussing the latest technical achievements in the field of prepress technology, analyzing the basic principle of digital prepress technology, comprehensively introducing new techniques and methods involved in the process of data input and publishing output, and introducing the related knowledge of map publishing.

The book consists of 7 chapters. The first chapter is the basic concepts of digital prepress, the second chapter is the composition of the digital prepress system, the third chapter is the graphic and text processing technology of digital prepress, the fourth chapter is the principle and technology of digital prepress image processing, the fifth chapter is color space and color management technology, the sixth chapter is the publishing output technology, and the seventh chapter is the technical process of digital prepress.

(11) ***Prepress technology*** (SHI Ruizhi, LIU Shide. China Textile Press, Beijing, April 2006)

With the wide application of computer technology and laser phototypesetting technology in the printing field, the production technology of printed matter, especially prepress technology, has been fully digitized. Based on the prepress technical process, from the point of view of training practical operation ability, this paper expounds upon the main technical problems of each link in the prepress.

The book consists of 8 chapters: chapter one is the prepress technical process; chapter two is the original input technology; chapter three is the graphics preparation technology; chapter four is the image processing technology; chapter five is the typesetting and layout techniques; chapter six is the page output technology; chapter seven is the color management technology; and chapter eight is the plate-making and proofing technology.

Acknowledgements Professor YAN Mian and ZHANG Qingpu reviewed this chapter and put forward suggestions for improvement.

References

An, Z. (1988). *Using FOGTA signal to achieve the quality control of the printing of "chromatography"//Map printing papers* (pp. 94–100). Surveying and Mapping Press.

Che, S., & Wu, M. (2007). Realizing color management in map publishing system. *Geomatics and Information Science of Wuhan University, 32*(2), 184–187.

Chen, G. (1995). Study on the mechanism of the single bath washing of the positive heavy nitrogen film. *Map*, (4), 38–39.

Chen, G. (1996). Random hanging net technology and its application in map printing. *Surveying and Mapping Technology*, (1), 19–23.

Chen, G. (2001). The characteristics of modern cartography and its development and application. *Journal of Institute of Surveying and Mapping, 21*(1), 12–19.

Chen, G., Tang, J., & Zhang, D. (2001). Technical principle of dual positive-negative diazo duplicating film. *Image Technology*, (1), 8–10.

Chen, G., & Zhang, D. (2000). Design principle of map publishing system based on PostScrit language. *Journal of Beijing Institute of Printing, 7*(4), 36–41, 52.

Chen, Q. (1988). *Study on color map printing of multi band satellite images; A corpus of map making and printing* (pp. 140–146). Surveying and Mapping Press.

Chen, Y., Li, S., & Shi, R. (2007). Research of map web publishing technology based on SVG. *Journal of Zhengzhou Institute of Surveying and Mapping, 24*(2), 149–152.

Chen, Y., Liu, H., & Shi, R. (2004a). Transforming digital map data to MAPGIS data. *Journal of Institute of Surveying and Mapping, 21*(2), 154–157.

Chen, Y., Shi, R., & Liu, H. (2004b). Research on data format transformation software in map publishing. *Hydrographic Surveying and Charting, 24*(1), 47–49.

Deng, S. (2005). Discussion on web map design. *Journal of Henan Polytechnic University (Natural Science) General, 24*(1), 79–82.

Deng, S., Chen, S., & Liu, H. (2005a). The solution of displaying the kind of highroad symbols in map express system. *Journal of Institute of Surveying and Mapping, 22*(2), 134–136.

Deng, S., & Guo, J. (2005b). An anti-aliasing algorithm suitable to map publishing symbol. *Geomatics and Information Science of Wuhan University, 30*(12), 1120–1123.

Du, G. (1988). *The influence of dot deformation on color reproduction; A corpus of map making and printing* (pp. 39–44). Surveying and Mapping Press.

Fan, Y. (1988). *Geological map of color printing; A corpus of map making and printing* (pp. 124–127). Surveying and Mapping Press.

Gao, H. (1962). Map color separation photography//*Compilation of abstracts of annual academic papers on Cartography*, (p. 85).

Guan, D. (1988). *The feasibility of digitizing, standardization and quality management in map making process; A corpus of map making and printing* (pp. 14–19). Surveying and Mapping Press.

Hong, B. (1988). *On four-colour printing of map; A corpus of map making and printing* (pp. 49–52). Surveying and Mapping Press.

Hunan Province Xinhua Press. (1962).*How we print the "Hunan Province Comprehensive Atlas" and "Hunan Province General Atlas"//Abstracts of the Annual Conference of Cartography.* (pp. 86–87).

Jin, L. (1988). *Examination of standard order reproduction curve; A corpus of map making and printing* (pp. 119–123). Surveying and Mapping Press.

Liu, J. (1962). Chrome glue dyeing and coating method. *Compilation of abstracts of annual academic papers on Cartography*, (p. 86).

Liu, S., Sun, Q., & Wei, B. (2007). Thermal CTP and digital proofing technology. *Journal of Zhengzhou Institute of Surveying and Mapping, 24*(4), 274–277.

Liu, Z., & Cui, H. (2000a). A method to calculation the percentage of black dot in 4-color printing. *Journal of Institute of Surveying and Mapping, 17*(3), 210–212, 215.

Liu, Z., & Cui, H. (2000b). Research on color space transform model in map electronic publishing system. *Acta Geoadetica et Cartographica Sinica, 29*(4), 342–347.

Ren, X., Shi, R., & Yang, L. (2007). The publication technique mode and practice of photomap atlas under digital environment. *Bulletin of Surveying and Mapping*, (5), 42–45.

Shi, R. (2000). Digital press technology trends. *Printing Technology*, (4), 45–46.

Shi, R. (2002). Translation the CIEL*a*b*color space into the CMYK by looking up table interpolation. *Journal of Institute of Surveying and Mapping, 19*(3), 224–226.

Shi, R., An, J., & Wen, J. (2009). Suitable for stereo lenticular grating picture printing dot column model. *China Printing and Packaging Study*, (2), 26–32.

Shi, R., & Cao, Z. (2007). A color separation model based on 7-color Hi-Fi press. *Science of Surveying and Mapping, 32*(5), 58–60.

Shi, R., Cao, Z., & Ren, X. (2008). Cross-media color transmission research. *Journal of Geomatics Science and Technology, 25*(4), 79–82.

Shi, R., Ren, X., & Sun, D. (2009). Printing ink supply determination. *Journal of Geomatics Science and Technology, 26*(1), 1–4.

Sun, D., Shi, R., & Xie, G. (2008). Examination and evaluation for map printing quality. *Science of Surveying and Mapping, 33(suppl)*, 135–136.

Wang, F. (1988). *Opinion on quality management of map printing; A corpus of map making and printing* (pp. 156–158). Surveying and Mapping Press.

Wang, Y., & Zhu, Y. (1988). *Copy the screen visual characteristics used in map printing; A corpus of map making and printing* (pp. 101–108). Surveying and Mapping Press.

Wei, B., Liu, Z., & Zhai, Q. (2005). The research of sandwich plate silver film and output technology. *Journal of Institute of Surveying and Mapping, 22*(4), 275–277.

Wen J., Shi R., & Zhao X. (2005). Scheme and realization of different GIS data's unification based on PDF structure. *Bulletin of Surveying and Mapping*, (5), 62–64.

Wu, M., & Che, S. (2007). Design of a four-view architecture for the map publishing system. *Science of Surveying and Mapping, 32*(2), 92–94.

Wu, M., & Liu, Z. (2004). The system structure of cartographic and map publishing system. *Bulletin of Surveying and Mapping*, (7), 42–44, 50.

Wu, X. (1988). *Discussion on improving the overprint accuracy of 4-colour printing on map; A corpus of map making and printing* (pp. 53–59). Surveying and Mapping Press.

Xu, Q. (1962). The experience of color design and printing in Jiangsu map collection. *Compilation of abstracts of annual academic papers on Cartography*, (p. 87).

Yan, M. 1962. Discussion on improving the quality of map printing. *Compilation of abstracts of annual academic papers on Cartography*, (p. 84).

Yan, M., & Liu, J. (1982). *The development of map printing and the direction of our efforts. The collection of military surveying and mapping (selected papers of science and technology)*, (9), 58–62.

Yang, C., Zhang, Q., & Hong, L. (2003). The design and implementation of DBMS for digital publishing original. *Bulletin of Surveying and Mapping,* (9), 58–60.

Ye, T. (1988). *The influence of electronic color scanner on color reproduction; A corpus of map making and printing* (pp. 45–48). Surveying and Mapping Press.

Zhang, Q. (1986a). Regression analysis between reflection densitometer readings. *Acta Geodaetica et Cartographica Sinica, 15*(3), 25–28.

Zhang, Q. (1986b). The working principle and application of optical densitometer. *Acta Geodaetica et Cartographica Sinica,* (2), 19–21.

Zhang, Q. (1987). Digital detection and quality control in the production of map chromatography. *Bulletin of Surveying and Mapping,* (3), 23–25.

Zhang, Q. (1989). On the basic factors forming colour cast and investigation of mathematical models for colour correction. *Acta Geodaetica et Cartographica Sinica, 18*(3), 213–218.

Zhang, Q. (1991). Standardization of remote sensing image replication and its quadrant control system. *Remote Sensing Information,* (4), 25–29.

Zhang, Q. (1992). Research on remote sensing image and thematic map replication. *Bulletin of Surveying and Mapping,* (2), 25–26, 40.

Zhang S., Pang D., & Zheng R. (2004). *General history of Chinese printing.* The graphic communications fundation.

Zhao, G., & Li, Z. (1988). *Feasibility of dry film used in map making; A corpus of map making and printing* (pp. 63–70). Surveying and Mapping Press.

Zou, Y. (1988). *Evaluation and standardization of printed images; A corpus of map making and printing* (pp. 88–93). Surveying and Mapping Press.

Zou, Y., & Wang, Y. (1996). Map printing technology toward twenty-first century and its countermeasures. *Print in China, 14*(1), 38–40.

Chapter 11
Geographic Information Systems

Yixin Hua and Tao Feng

11.1 Introduction

In terms of defining geographic information systems (GISs), the definitions are different due to the different viewpoints, such as the viewpoint of maps, the viewpoint of databases and the viewpoint of spatial analysis. The *Introduction to Geographic Information System* written by Chen Shupeng (1999) defined GISs as follows: composed of computers, geographic data and users, through geographic data integration, storage, retrieval, manipulation and analysis, generates and outputs a variety of geographic information, provides new knowledge for land use, resource management, environmental monitoring, transportation, economic construction, urban planning and government administration, and serves engineering design, planning, management and decision-making.

There are two characteristics of a GIS: on the one hand, it is a computer system; on the other hand, it deals with data that are related to locations on the Earth's surface. There are many scientific sources of GISs, such as automated mapping, landscape architecture and environmental sensitivity planning, urban studies, population statistics, and so on. Now, it was generally believed that GISs originate from maps, especially computer cartography technology. However, due to the involvement of other disciplines and the maturity of GIS itself, the original relationship no longer seems accurate; conversely, cartography has become a basic subject of geographic information science and technology. In short, GIS is an interdisciplinary subject formed based on natural geography, surveying, cartography, remote sensing, graphics and images, and computer application technology (Li Deren 1997a).

GISs appeared in China between the late 1980s and early 1990s, approximately 15 years later than in western developed countries, and GISs were developed based on the introduction of western achievements and technologies. GISs emerged and

Y. Hua (✉) · T. Feng
Information Engineering University, Zhengzhou, China

J. Wang and F. Wu (eds.), *Advances in Cartography and Geographic Information Engineering*, https://doi.org/10.1007/978-981-16-0614-4_11

developed in China with a social background, international background and technical background.

In the late 1970s, with the strategic shift in the work focus of our country, China entered a historical period centered on economic construction, with great developments in culture, science and education.

'Outline of national science and technology development from 1978 to 1985' was formulated in China in December 1978; the report 'Economic construction must rely on science and technology, science and technology must be oriented to economic development, and maintain the coordinated development of science and technology, economy and society' was adopted by the National Science and Technology Conference in December 1980, and these publications provided the environment for the birth and development of GIS. In addition, since the reform and opening up, massive data have been generated during the country's massive infrastructure construction to meet energy demand and urban management needs, and others urgently needed to use GIS technology for data management and data analysis. Against this background, the concept of GIS technology was introduced to surveying and mapping, geography, computers and other industries, and with the support of related industries, it has developed, expanded and boomed in China.

The prosperity of the international GIS industry also offered a good opportunity for the development of GIS in China. After World War II, the third scientific and technological revolution centered on information science and energy pushed the development of human society to a higher stage, but with it came great changes to the ways of human life, as well as the global problems brought about by social development-environmental pollution, ecological destruction, and so on. None of these problems can be solved by any single country and require concerted international cooperation. As the container of geo information storage and management, GISs have become the tool and hope for people to solve this problem, which not only promoted the prosperity of international GIS but also put forward an urgent need for GIS to be spread and applied worldwide.

In addition, GIS is a discipline placing reliance on IT technology; the development of IT technology determines the depth and breadth of the development and application of GIS technology. Since 1978, China has been in a period of great development in terms of science, education and culture. One of the most important things is the rapid rise in the information industry under the background of the world's new science and technology revolution. In 1979, HDS-9, an integrated circuit computer with 5 million operations per second, was successfully developed in China; the first laser typesetter was invented by Wang Xuan; IBM entered the Chinese mainland; the State Council decided to set up the State Administration of the electronic computer industry; in 1980, the Yangtze River computer (Group) company successfully developed China's first DJS051 microcomputer using only the components made in China; in 1981, the Shenyang first machine tool plant started the computer-aided production management system project; in 1983, galaxy I, China's first supercomputer that computes at a speed of more than 100 million times per second, was successfully developed; and in 1989, Oracle landed in China. All these factors are related to the birth and development of GISs in China.

11.2 Development Process

The development of GISs in China over the last 30 years can be divided into the following 3 stages. Before the 1980s, GISs were in the embryonic and early stage; the 1990s were the stage of accelerated development and industrialization; the ten years after entering the twenty-first century was the stage of popularization and standardization.

11.2.1 Preparation and Initial Stage (from the 1970s to 1980s)

During this period, China took an important step in GIS technology, policy, and personnel training.

In 1974, a foreign Earth resource satellite image was imported to China for the first time; in 1977, the first map of all features was produced in China; in 1978, the first Nationwide Symposium on Database (space) was held by the State Planning Commission; in 1980, Chen Shupeng presided over the establishment of China's first geographic information research laboratory and opened a prelude to the research and application of GISs in China; an aerial remote sensing GIS construction experiment was implemented in the Ertan and Dukou regions of the Sichuan Province in 1981; in 1982, the Institute of Geography of the Chinese Academy of Sciences established the State's County Boundary Database; in 1984, the State Bureau of Surveying and Mapping began the construction of China's national basic GIS; in 1985, the State Key Laboratory of Resources and Environment Information System was established; and in 1988, the GIS major began to recruit undergraduate students in Wuhan.

Most notably, in 1983, under the support of the State Science and Technology Commission, academician Chen Shupeng carried out research on the National specifications and standards for China's Resources and Environmental Information and formed a report on the National Information System Specification for Resources and Environment (commonly known as the Blue Book), which guided China's GISs and their standardization. Subsequently, the GIS made a breakthrough and developed in the aspects of theoretical exploration, standard discussion, experimental technology, software development, system establishment, personnel training and regional testing. Some local governments started investing in the establishment of local GISs.

11.2.2 Accelerated Development and Industrialization Stage (1990s)

During this period, China perfected a relevant system of GIS development, and GIS discipline construction was continually improved and formed a multilevel personnel training mechanism. More important was the emergence of a number of influential GIS companies, and the pace of GIS industrialization significantly accelerated.

In terms of education, the first Chinese textbook *Introduction to Geographic Information Systems* (Huang Xingyuan 1990) was published by the Higher Education Press in 1990; the first GIS master's degree was set up in Wuhan in 1993; in 1997, 'Cartography and geographic information systems' and 'Cartography and geographic information engineering' were set up as the two secondary disciplines of the two first-level disciplines of geography and surveying and mapping in China and began to recruit doctoral students and master's students; in 1998, undergraduate majors reduced from 504 to 249 in China, but the GIS undergraduate science specialty was still available in the geography specialty.

In terms of national policy, on April 25, 1994, the GIS Association of China and the Education and Science Popularization Specialized Committee were established; the first annual evaluation of domestic GIS software held in March 1996 received the attention of the Ministry of Science and Technology of China, which provided specific guidance and instruction for the development of the GIS industry. The 'Resource and Environment Information and Digital Earth' conference was held by the China Academy of Sciences in November 1998. Thereafter, 'Recommendations on China's Digital Earth Development Strategy' was submitted to the State Council on January 20, 1999. The recommendation proposed to achieve sustainable economic and social development, to maintain a peaceful and stable international environment and to develop the capability of independent innovation of science and technology as three major strategic objectives for China in the 21st century; beginning with China's national goals, there is an urgent need for a digital Earth or digital China.

In terms of database building, a comprehensive national situation electronic map system supported by a 1:100 national basic topographic database was built in 1993; national 1:1,000,000, 1:250,000 and 1:50,000 basic geographic databases were built in 1994, 1998 and 2006, respectively; a distributed i comprehensive national situation GIS based on a local area network was built in 1997; and China's 1:250,000 basic geographic database was built in 1998.

In terms of software development and industrialization, the joint development project 'Comprehensive national conditions GIS' (9202 project for short) was initiated in 1992 by the Secretary Office of the State Council and the State Bureau of Surveying and Mapping; China's famous GIS software companies, such as MapGIS, GeoStar, SuperMap, CityStar, and EC Founder, appeared one after another in the 1990s, and a large number of GIS software was developed successfully and applied to all professions and trades.

11.2.3 Popularization and Standardization Stage (After Entering the Twenty-First Century)

After entering the twenty-first century, GISs began to spread in China and were mainly represented by three aspects: GIS education developed further, GIS services entered all aspects of people's lives through the network, and GIS standardization work was widely carried out.

In 2000, the Geo Windows software system based on the network environment was born; in July 2001, the Circular of the General Office of the State Council of the P.R. of China, upon transmitting and issuing several opinions of the State Planning Commission and 11 other departments to promote the construction and application of China's national spatial information facilities, pointed out that to further promote the sharing and application of geospatial information in China and demonstrate the role of geospatial information in China's national economic and social development and strategic adjustment of the economic structure, the construction of China's national spatial information infrastructure should occur quickly; additionally, geographical information standards and policies and regulations needed to be improved, and a perfect commonwealth, foundational geospatial information system and its exchange network system needed to be established to create conditions for the development of related industries. At present, the State Council has established a comprehensive national situation GIS, and most provinces and many prefecture-level cities have also established general provincial situation GISs and general city situation GISs.

The first domestic GIS magazine "GIS developer" started publication in August 2004; the first national university and college student GIS application and development contest ended successfully in March 2004; China's first network oriented large-scale GIS, SuperMap GIS 5, became available in September 2004, which marked the first successful implementation of industrialization of the national 863 program 'Network oriented large-scale GIS'; SuperMap, the first GIS software for China's secondary school geography education, was developed in 2005; Beijing Lingtu software company's 51ditu website first provided free and open access to its map service APIs in August 2005; for the first time, Beijing SuperMap Geographic Information Technology Co., Ltd. put forward an universal GIS software model based on philosophy on September 15, 2005; ArcGIS China Training Center and other units organized a Geographic Information System Day for primary and secondary school students at Capital Normal University on November 22, 2003; more than 700 million yuan was invested in the topographic database of national fundamental GISs, which was completed and accepted on February 25, 2006, and reached the international advanced level, providing accurate information at the 1:50,000 scale; in December 2006, Academician Wang Jiayao delivered a keynote speech titled 'Independent innovation and the development of GIS in China' at the GIS forum of China's universities and colleges, in which he analyzed the present situation of GIS development in China, and he believed that independent innovation was the foundation driving the rapid development of China's GIS; additionally, he put forward some technical breakthroughs for GIS independent innovation in China; on September 14, 2007, at the GIS independent innovation forum of the Chinese Academy of Sciences, Song Guanfu, President of Beijing SuperMap company, delivered a keynote speech of "Services GIS, a business agile GIS application" and presented the new concept of Services GIS (service-oriented GIS), and at the same time, he announced the launch of SuperMap 2008 to enable SOA.

For GIS standardization, China began to study geographic information standards in 1983 and published a research report on the national standards and specifications for resource and environment information systems (referred to as the blue book) in the following year, which was China's first book on the standardization of geographic information. The National Technical Committee for Geographic Information Standardization was founded in December 1997 and greatly accelerated the standardization of national geographic information. The committee completed the national standards for geographic information and identified and evaluated the formulation and revision of 83 items in 2004; more than 20 standards have been developed for the production and construction of foundational geographic information data; China's national geographic information standard framework, which includes 7 categories and 44 subcategories, was implemented in 2007. This framework can be divided into three levels: the general geographic information standard, the geographic information professional (interface) standard with close relation to related subjects, and the geographic information special standard for promoting various types of applications in information construction.

11.3 Main Research Achievements

11.3.1 Research in the Field of GIS Theory

Does GIS have a theory? The answer is yes. However, for a long period after the birth of GIS, most people, including experts, entrepreneurs and users, regarded GIS purely as technology and rarely considered its theory. Perhaps as Li Deren (1997b) said, because of the obvious application and technology-oriented characteristics of GIS, it has been widely used and vigorously developed before the formation of geographic information theory. However, the existing phenomenon of emphasizing technology and neglecting theory resulted in 'The development of theory cannot keep up with the demand of the development of technology, while theory has restricted the development of technology' (Chen 1997).

Since the 1990s, with the popularity of GIS in society, many well-known experts in geoscience and surveying and mapping in China have called for the theoretical construction of GISs and discussed the meaning and research contents of GIS theory.

Chen Shupeng is one of the most active promoters of geographic information theory. He initiated Earth information science, advocated Earth information science as the basis for studying GIS, and proposed the relevant theoretical system of GISs, including the basic theory, applied basic theory and basic technology theory (Chen

1997). Li Deren pointed out that the GIS discipline has many problems to be solved theoretically and proposed that the areas of science and technology need to be addressed to answer the questions mentioned above; he also discussed the establishment of a theory of geographic information (Li 1997b). Gao Jun and Wang Jiayao stressed that geographic information systems originated from maps, and related theories of cartography should be introduced in the GIS field. Wang Jiayao divided the theory of GISs into basic theory and specialized theory. Basic theory includes geographic information transmission theory, spatial cognition theory and geographical system theory. Specialized theory is derived from the basic theory, corresponding to the specific GIS technology, and is directly guided by GIS technology (Wang 1997). In addition, Cheng Jicheng proposed that geographic information theory, as the basic theory of geo information science, which had an important guiding significance for deepening the study of GIS theory; Yu Zhiqian's (1997) research on the relevance of geographic information provided theoretical guidance for the organization and utilization of geographic information in GIS. The five major sections of the GIS professional curriculum systems accepted by the Education and Science Popularization Professional Committee of the China GIS Association (2004) include knowledge of mathematics and physics, surveying and mapping, Earth sciences, computer science and GIS (including remote sensing), which can also be considered among the five theoretical bases of GIS.

Many domestic GIS experts and professors have conducted deep research on one aspect of GIS in their own fields, which enriched and perfected the theoretical system of GIS. The results of these studies are shown in their papers and writings. Chen Shupeng proposed the Earth information science and geographic information mechanism, geo information Tupu, and grid computing, as well as the book "*Geo information Tupu*"; Li Deren wrote '*On the definition, theory and key technology of RS, global positioning system (GPS) and GIS integration*' (1997a); Lu Xuejun wrote '*Analysis of the connotation of geographical cognition theory*' (1998); Wang Jinfeng wrote '*The theory system of geographic information spatial analysis*' (2000); Hu Peng wrote '*Foundational theory of GIS—the space view of map algebra*' (1998) and '*Discussion on the bottleneck of GIS development and its theoretical background*' (2000); Huang Maojun wrote '*Geographic ontology and its applications*' (2004); Lu Xuejun wrote '*Research on the spatial cognitive model*' (2004); Du Qingyun wrote '*Micro linguistic conceptual model of spatial information*' (2004); Wu Lixin wrote '*The principle and algorithm of the geographic information system*' (2003); Tang Guoan wrote '*The principle and method of the digital elevation model and terrain analysis*' (2002); Zhu Changqing et al. wrote '*The theory of spatial analysis modeling and theory*'; Wang Zegen et al. discussed the active service of spatial information (2006); Li Deren was the first put forward "knowledge discovery from GIS" and his research results are mainly embodied in the '*Theory and application of spatial data mining*' (2006); Wu Fang et al. wrote '*Intelligent processing of spatial information for automatic map generalization*', and so on.

Based on researching the above respects, Wang Jiayao concluded the methodology and epistemology of GISs from the viewpoint of philosophy, that is, geography lays the epistemological foundation for GISs, while cartography provides a methodological foundation for GISs.

11.3.2 Research on Spatial Data Organization and Management

In the early and mid-1980s, China first attempted to apply GIS technology in the field of resources and environment. At this point, the main achievement was the establishment of a number of databases, including a national 1:1,000,000 geographic database system and national land information system, a 1:4,000,000 national resources and environment information system, and a 1:250,000 soil and water conservation information system; additionally, special research and experiments on the Loess Plateau information system and flood disaster prediction and analysis system were carried out. Since these systems were built on the basis of the national geodetic and digital terrain models, the data were mostly grid data (including image data) and statistical data; the data storage management at that time was mostly in the form of data files, which had to be managed using a computer file management system.

The development and application of GIS systems began in China in the late 1980s. At that time, the GIS data structure was divided into vectors and rasters: the former was usually used for large-scale digital mapping in the fields of gas pipeline management, cadastral surveying, construction and other areas, while the latter was used in computer-aided mapping of land suitability, small- and medium-scale computer-aided mapping, land information systems, national or regional planning and special mapping. The early GIS system in China was mainly used in land planning, environmental monitoring and thematic mapping, so the data were mainly based on raster data. In addition, statistical data, attribute data and some vector data collected by the data acquisition module were stored. For instance, the general GIS software (PURSIS) developed by the Institute of Remote Sensing Technology, Peking University mainly included four data categories: grid data, remote sensing image and DEM data, and polygonal data; products of patch graphs (such as soil type maps, land use maps, administrative area maps, etc.) were gathered by the data acquisition module; the run length encoding data was compressed coding data for images with less gray level requirements, such as a variety of digital maps; attribute data was input by keyboard when the data were collected, and all kinds of nongeometric attribute information related to the image was stored. For raster data, run length encoding and quad tree methods were usually used for compression. The common approach to data organization at that time was to manipulate the picture data (including image and graphics) using the file management of a computer operating system and to manipulate attribute data by using DBASE II, DBASE III and other database compilation languages.

By the beginning of the 1990s, with the improvement of computer operation and storage capacity and the establishment and improvement of vector databases in many regions, a GIS system based on vector data structures with topological relationships began to be established in China, and a GIS system based on raster and vector hybrid data structures emerged. Simultaneously, research on the data structure began to receive attention. In his doctoral dissertation, Gong Jianya studied a unified data structure of the vector and the grid. The data structure and data model of GIS were introduced in detail in the book *Data Organization and Processing Method in GIS* written by Gong Jianya, Li Deren and others.

Since the middle of the 1990s, China's GIS industry has gradually caught up with the level of international development. With the rise in the object-oriented (OO) concept, data modeling based on the OO model has received attention and begun to be used in China's GISs. The object-oriented semantic data model and its application in spatial databases (Zhu Xinyan et al. 1993) preliminarily discussed the OO data model, and the object-oriented spatial data model (Zhang Wei et al. 1995) introduced the OO spatial data model in detail.

Meanwhile, since spatiotemporal GISs can be used to reflect the historical changes in a region and the track and state of a target, at this stage, many Chinese scholars began to study it. For example, Chen Jun (1995) and Huang Mingzhi studied a non-first normal form (NINF) spatiotemporal data model; Guo Dazhi established Mine GIS by using a spatiotemporal 4D data model; Du Daosheng et al. proposed a temporal geo data model with timestamps on the group of synchronously changing data items and the segmented topological arc; Gong Jianya (1999) presented an object-oriented spatiotemporal data model. In addition, many scholars, such as Lin Hui, Shu Hong, Ma Jinsong, Huang Xingyuan, Zhang Zuxun and Huang Mingzhi, Chen Shangchao, Le Yanfen and Du Daosheng, proposed or modified some models, which greatly promoted the application and development of TGIS theory and spatiotemporal data models in China (Wang et al. 2004).

With the application of Oracle and other large commercial databases in GISs, since the mid and late 1990s, great changes have taken place in the management of data, and many GIS systems began to use organization and management methods based on extended relational databases, that is, spatial data and attribute data were stored in relational databases through the establishment of a spatial database function extended module (usually known as a spatial data engine) on the relational database to achieve the organization and management of spatial data. During this period, China's well-known GIS software, such as MapGIS, GeoStar and SuperMap, adopted this method to manage graphics and attribute data simultaneously. After entering the twenty-first century, domestic GIS scholars set off an upsurge in the research of data organization and management based on spatial databases. This data organization mode was based on the spatial data model of Oracle Spatial, which directly managed the data with a spatial database system that stores and manages spatial data and attribute data.

11.3.3 Research on GIS Architecture

The rapid development of modern IT technology, which is based on computer technology, had a direct impact on the development of GIS software architecture. With the rapid development of computer operation speed, the more abundant and efficient data management methods, as well as the increasing network bandwidth, all these factors enabled GIS software architecture upgrades and perfection and correspondingly enhanced the effectiveness of GIS. The architecture of GISs experienced changes in host-based GIS, desktop GIS, WebGIS, distributed GIS and so on, which were subject to the constraints of computer hardware and software and adapted to different application requirements.

The host-based GIS uses a large mainframe as a GIS server and runs a database management system. All data and services are stored centrally in the host, and all terminals are connected to the host through the network to access data. In the early development of the GIS software system, structure is limited by computer performance and so on, mainly for high-end professional researchers. The advantage of this GIS is that the data and service are centralized, the security is good, and the data storage capacity is massive; the disadvantage is that software development is difficult and can only exchange data in the form of repeated backup or offline copy, with low efficiency. Moreover, the initial investment in the system is large; the maintenance cost is high; and it also needs to replace a more powerful host when the system has reached the capacity limit.

In the late 1980s, due to the application and popularization of PCs, we gradually moved toward the era of the desktop GIS. A desktop GIS combines the functions of user interface and interaction, geographic information processing and data management in a personal computer, mainly for individual users, and has the characteristics of low cost for GIS system development, flexible and convenient spatial information processing, convenient installation and deployment, and low price for a PC.

In the 1990s, with the development of database technology and enhancement of the spatial data management level, research on client/server GIS began to increase. For client/server GIS, GIS functions were deployed on both the client and server sides according to certain needs, and normal communication between the client and the server side was achieved by a certain protocol. Then, GIS functions were achieved by the cooperation of the two. In accordance with the structure, a spatial database was usually deployed on the server side with good computing and storage capabilities, while the GIS functions on the client side were customized according to the hardware and software conditions and actual needs. From the separated spatial data management of geometric data stored in the form of documents while attribute data are imported into the integrated database storage and management of geometry and attribute data, the client/server GIS made breakthrough progress in data management scale and level, GIS function and so on. This GIS developed rapidly in the 1990s, and there were some representative platform software systems, such as SuperMap, MapGIS, and MGIS (Military Geographic Information System). On this basis, a large number of client/server GIS application systems were developed. With the support of the architecture, the GIS was rich in functions that support the storage and management of massive spatial data and was suitable for domain-oriented multi-user applications within a department or an organization.

WebGIS is a kind of compatible computer information system that is stored, processed, analyzed, displayed and applied in Internet or Intranet network environments. The basic idea is to provide geographic information on the Internet, allowing users to access GIS data and functional services through the browser. In the 1990s, WebGIS and its related technology research and system development were carried out in China. After several years of development, the technology matured rapidly and moved toward application. By 2000, there were some influential WebGIS products, such as GeoBeans of National Remote Sensing Application Engineering Technology Research Center, GeoSurf of Wuhan Geo Information Engineering Technology Co Ltd., SuperMap IS of Beijing SuperMap Geographic Information Technology Co., Ltd., and others. WebGIS had the characteristics of a wide range of user groups, standalone client platforms and load balancing, mainly for public GISs.

Distributed GIS (DGIS) refers to a GIS ability that can transcend any number of organizations, can be distributed on any number and kind of platform, and can be accessed by any number of users. In DGIS, all the computing resources, GIS servers, database servers and geographic information are widely distributed on the Internet and users do not care about where to store the data in physics; they also do not know where the GIS service is, so users do not have to care where the data are physically stored, nor do they know where the GIS service is located; as long as you follow certain open principles (such as the OpenGIS specifications), any user can request geographic information and services on any server. Before and after 2000, there were many studies on the distributed GIS architecture and implementation methods. For example, Luo Yingwei (2001) proposed a dynamic distributed geographic information service architecture based on Agent; Wang Yuxiang (2002) explored the theoretical establishment of a distributed geographic information system framework based on CORBA technology; Wu Sheng (2002) performed deep research on the distributed GlS platform based on components; Chen Nengcheng (2003) proposed an architecture of distributed geographic information service based on JZEE; and Liu Nan et al. (2004) proposed a COM + -based distributed geographic information system architecture and its implementation. The distributed GIS architecture is an ideal architecture, which accords with the distributed characteristics and the diversified demands of geographic information sources, geographic information service providers and GIS users.

With the development of grid computing technology, after 2000, research on grid GIS (GridGIS) increased. According to the concept of grid service (Gridservice), GIS information and computing resources could be obtained on demand in a grid environment; they could be coordinated to form a "virtual" GIS (service) to achieve geographic information sharing. Since approximately 2003, the grid GIS architecture and key technologies have been discussed. Shen Zhanfeng et al. (2004) proposed a distributed architecture based on GridGML description language, Web Services technique and middleware; Wang Jinxin et al. (2005) presented an architecture model of grid GIS: the quadruple geospatial grid; Wang Jiayao (2006) proposed a grid GIS architecture based on Intelligent multiagent. Grid computing was based on the open grid service architecture, so GridGIS would be further upgraded to open the spatial information grid. From the current point of view, grid GIS may be the ideal solution for distributed GIS.

11.3.4 Research on Spatial Data Queries and Analyses

From the middle and late 1980s to the early 1990s, China's GIS industry developed preliminarily. In this stage, China's GIS system was based on raster data, supplemented by vector data. There were also certain attributes and statistical data; most of them were applied in urban planning, land evaluation and other aspects. The analysis and application methods included overlay analysis, land use analysis, urban planning decision analysis, and simple terrain analysis. The typical application model of raster overlay analysis was introduced in detail in 'Raster graphic data overlay analysis method and its application in urban planning' (Ding 1991). 'Urban general planning information engineering based on remote sensing and GIS technology' (Chen 1991) summarized the typical application of GIS in urban planning during this period as follows: investigation of the distribution, quantity and composition of urban land use; survey of all city density data; investigation and analysis of urban natural conditions and the geographical landscape; investigation and analysis of urban morphology evolution and urban land use planning implementation rate; investigation and analysis of foundational data of old city reconstruction; study of the distribution of pollution sources of the three wastes (waste gas, waste water and industrial residue) and comprehensive evaluation of environmental quality; prediction and analysis of urban social economic development index; comprehensive evaluation and analysis of the suitability of land construction and the sequence of construction and development in the planning area; and quantitative analysis and simulation optimization of urban comprehensive transportation planning.

In the field of terrain analysis, LV Guokai introduced in "Experimental study on a simple and practical geographic information system for county level" that building digital terrain model at the county level using the topographic map of a certain scale of county (1:10,000 ~ 1:50,000) provided map and terrain data (including absolute elevation, average elevation, slope and other features) for the comprehensive analysis of geography, economic regionalization and resource evaluation.

Conversion between raster and vector data is also a hotspot in GIS research. Many scholars have proposed the improved grid vector data conversion algorithm, and the quadtree data retrieval and its improved method are also used in data management. The analytical models provided by the typical GIS application platform at this stage included PURSIS with expert weight, overlay classification, expert proposition, etc., SPACE-MAN3.0 with spatial expansion model, overlay classification model and expert proposition model, logical classification model, discrete point fitting model, statistical graphical display and statistical analysis model, and fuzzy evaluation.

Since the late 1990s, with the development of IT technology, China's GIS industry entered a stage of rapid and comprehensive development; most GIS use vector data and many systems also use large commercial databases for data storage management, accelerating the speed of data query and retrieval.

In the twenty-first century, using data mining and ontology technology was developed to intelligently search and retrieve GIS data. 'Research on the theory and method of spatial data mining' (Yu et al. 2000) introduced data mining to the GIS; 'Review on the research of spatial data knowledge discovery' (Pei et al. 2001) provided a comprehensive description of GIS data mining; and 'Establishment and comparison of geographic ontology in GIS' (An et al. 2006) provided a detailed description of the ontology GIS.

11.3.5 Research on the Visualization of Spatial Data in GIS

Early spatial data visualization was not a digital display but an analog display or television display. It used a camera device to take pictures of a map, convert map information into a video signal, and then display it on a television screen.

According to the relevant information, as early as the middle of the 1980s, China's scholars began to pay attention to the problem of GIS spatial data visualization; however, the concept of spatial data visualization was formally put forward in early 1993. The sixteenth academic symposium of the International Cartographic Association (ICA) held in Cologne, Germany, in 1993 formally established a "Cartographic Visualization Committee", whose task was to regularly exchange the development and research focus of visualization technology in the field of cartography and to strengthen cooperation with the computer field.

In this period, virtual reality technology became a research hotspot in GIS. Chen Gang (1997) developed a virtual landscape simulation based on map data, that is, LOD technology, which is in line with the visual rules; and Gong Jianhua et al. (1999) attempted to establish an effective research framework for the visualization of geosciences. He believed that the visualization of geosciences included map visualization, visualization of geographic information systems and visualization in the field of professional application. Li Hongga et al. (1999) studied the visualization of spatiotemporal data in GISs. Gao Jun (2000) made a comprehensive summary of the background, application and model theory of visualization. He believed that visualization greatly expanded the family of maps, including electronic maps, dynamic maps, cyber maps (virtual environments) and other new styles of maps, as well as improved the relationship between GIS and cartography. Visualization has been applied in the fields of electronic maps and dynamic maps, virtual environments, knowledge discovery and data mining, spatial data quality detection and so on.

As of the twenty-first century, research on the spatial visualization of GIS has developed in a deeper and wider direction. During this period, the theoretical research has been further strengthened. The traditional visual variables have been expanded in the information age. Ling Yun et al. (2005) studied an adaptive user interface for map visualization systems based on user cognitive characteristics. Jiang Nan et al. (2006) proposed the concept of the electronic map display mode through the analysis and research of electronic map displays in GIS. With the rapid development of graphics hardware, there is a new way to develop large scene 3D visualization, that is, graphic display technology based on CPUs. GIS spatial data visualization extends from virtual reality to augmented reality and mixed reality, and the development of human-computer interaction theory and technology makes the research of GIS spatial data visualization focus on human-centered design and user-friendly design.

11.3.6 Research on Geospatial Data Sharing

In the age of information and networks, geographic information sharing refers to consideration of information security and data privacy; at the same time, through the computer network, geographic information sharing is convenient, fast, accurate, and safe with comprehensive queries, browsing, access, exchange, use and re-processing of digital information, which is directly or indirectly related to human survival, including the transparent use of some information resources, with particular emphasis on the sharing of foundational geospatial data and thematic data. In recent years, for their own needs, different GIS vendors and research units have their own geographic data models, data formats, data storage and data processing methods, and there is no uniform standard. Due to the development of GIS technology and the need for social application, interoperability and data sharing among different systems has been receiving increasing attention. Spatial data sharing has been a research hotspot in the field of GISs in recent years.

In the early stage of GIS development, data sharing was usually limited to research on the spatial data conversion of different data structures. The early method was to implement data sharing among different GIS systems by using intermediate data formats, such as Arc/Info E00 format and MapInfo Mif format. This data format was generally an ASCII code file; the user could directly read and write these external documents for their own use. In recent years, different countries and organizations have developed their own internal standards for their needs. Based on an integration model, Zhang Haitao et al. (2005) proposed a new method of data sharing for urban pipeline management departments and realized GIS information sharing in the Sihong and Lishui regions. In 1999, China developed the *Chinese National Spatial Data Transfer Format* (CNSDTF), which specified the exchange format of vector data, image data and grid data. Chang Yuan (2006) used CNSDTF as an original spatial data format to build GML Schema to describe the structure of CNSDTF and then used the XML file based on this Schema to store the contents of the CNSDTF file.

The data conversion method only considered spatial data sharing from the viewpoint of data and involved the integration of data, without regard for data processing, namely, functional interoperability. OGC developed a unified specification for data interoperability, which made it possible for a system to support different spatial data formats simultaneously. However, it required achieving the data access interface for the hosted software of each format according to the uniform specification, which was impractical for a certain period of time (Gong Jianya et al. 2006).

Seamless integration technology of multisource spatial data was a special data access mechanism. It not only provided the ability to access a variety of data formats directly but also made the GIS software function as a means of data analysis, which was an effective technology foundation to realize spatial data sharing (Song et al. 2000).

To realize the sharing and interoperability of geographic information in the network environment, the International Standardization Organization and OGC launched a series of standards, including simple object access protocol (SOAP), web service description language (WSDL) and universal description, and discovery and integration (UDDI), constituting the Web Service protocol stack, which made Web Service the primary means and development trend of geographic information sharing and interoperability.

China has accumulated some research foundations in the field of spatial data sharing. Yi et al. (2000) studied the sharing and interoperation of spatial information; Bao Shitai et al. (2000) explored the application of Web Service technology in geographic data sharing; Wang Peng et al. (2001) presented a new method of spatial information sharing and interoperability in XML-WebGIS; Chen Aijun et al. (2002) set up a hierarchical model of geospatial information sharing for digital Earth construction; Song Guomin et al. (2002) studied the mechanism of geospatial data sharing; Wu Lun et al. (2003) studied the distributed interoperability of GIS on the basis of Web Service; Gong Jianya (2003) studied spatial information sharing and interoperability technology; Wang Lingyun et al. (2004) summarized the adaptability of Web Service in spatial information sharing, interoperability and integration and pointed out that Web Service technology is an effective solution to realize a theoretical NG information sharing platform; Jia Yong et al. constructed a spatial data sharing platform based on Web Service from a theoretical point of view; and Song Guomin (2006) put forward a theoretical framework of geographic information sharing. All these studies promoted application research for Web Service technology in spatial data sharing to a certain extent.

In addition, Jia Peihong et al. (2007) used digital watermarking technology to hide and extract the identifying information of the data production unit and user's license in geospatial data to achieve order and security in geographic spatial data sharing and to ensure the technical security of geospatial data sharing.

11.4 GIS Applications

11.4.1 Applications in Various Fields

1. **In agriculture**

In China, since the middle of the 1980s, GIS technology has been applied in the field of agriculture, including land and resource decision-making management, agricultural resource information, regional agricultural planning, grain circulation management and food production decision-making.

Domestic research in this area is relatively extensive. Aerial remote sensing technology was widely used in the second national soil survey in China, which included fast mapping, high precision and cost-effectiveness, saving 3/4 of the funds compared with conventional methods; additionally, Liu Zhi designed a Land Use Evaluation System; Huang Xingyuan et al. (1993) studied the application of GIS in regional land use decision-making; Nanjing University developed the Parcel Land Price Evaluation System; the Beijing Urban and Rural Information Center developed the Beijing basic cropland land resource information management system; Beijing Normal University developed the regional land suitability evaluation system; and the Chinese Academy of Sciences developed the information system of soil and water loss on the Loess Plateau. All these systems have been put into practical use and have achieved good social and economic benefits.

2. **In forestry**

Currently, the applications of GIS in forestry include monitoring and management of the environment and forest disasters; forest surveys; analysis and evaluation of forest resources; forest structural adjustment; forest management; and wildlife monitoring and management.

From 1993 to 1997, the "Modernization of China's Forest Resources Survey Technology" project funded by the United Nations Development Program was successfully implemented. Under the support of ViewGIS, Ma Shengli (1999) established the Forest Resources Geographic Information System of Xixiang County, Shaanxi Province. According to the needs of modern resource and forest management, Li Fuhai et al. (2002) independently designed and developed a forest resource GIS. In the research of Chen Duanlv et al. (2002), GIS technology was combined with the optimal decision model of forest management, and spatial data and attribute data were combined for comprehensive analysis. Chang Yuan et al. (2004) applied ViewGIS to establish the forest resource management information system of Jiahechang village, Baiqizhai Township, Tieling County, Liaoning Province.

3. **In resource and environment management**

With the continuous development of technology, applications in land science include land evaluation, land use planning, classification and mapping of land use and land cover, and dynamic monitoring of land use and land cover.

In 1984, according to a unified deployment by the State Council, a survey of national land resources was carried out. GIS technology has played an important role in the compilation of land use maps, data calculation and aggregation, spatial analysis, and so on. Since 1996, the State Science and Technology Commission, the State Land Administration and the Ministry of Agriculture have implemented the Protection and Monitoring of Basic Farmland in China. Under the unified planning and organization of the Ministry of Land and Resources and according to the deployment and arrangement of the New Round Survey Outline and Implementation Plan of Land Resources, in September 1999, the construction project of county (city)-level land use database at the 1:10,000 scale was set up as one of the digital land projects and was officially launched in October 1999.

4. **In investigations, evaluations and predictions of mineral resources**

The application of GIS technology in ore prospecting is more extensive, and it has formed a series of new methods and ways of thinking in mineral prediction.

The MapGIS software developed by WU Xincai of the China University of Geosciences can be used to analyze the multisource geological information and to delineate a favorable position for ore formation. Hosted by the remote sensing center of the Ministry of Geology and Mineral Resources and conducted by the Changchun Institute of Geology, China University of Geosciences, and the Institute of Mineral Deposits of the Ministry of Geology and Mineral Resources, as well as other units, 'Research on the integrated image processing technology and application of remote sensing image and other geological data' was carried out in 1986. In April 1995, to improve the prediction level and the prospecting effect, an experimental study on the application of GIS in ore-forming prediction and ore-searching was conducted by the Geological Survey of China in the western margin of the Yangtze Platform in the Western Sichuan. The MORPAS system developed by Li Kun et al. (2009) adopted data warehouse and data mining technologies to realize the integrated management of multisource geo information and the extraction of latent patterns from data.

5. **In hydrology and water conservancy**

The use of GIS can effectively solve the problem of insufficient quantity and information in the study of hydrological models for a long time, as well as enrich the research contents of hydrogeology and improve the prediction and estimation accuracy of hydrological models.

The application research in this area in China is mainly dynamic monitoring information systems for small watersheds on the Loess Plateau, regional management and development information systems in the Sanchuanhe River Basin of the Loess Plateau, and flood disaster analysis information systems of the Yellow River Delta, as well as experimental studies on decision support systems for China's major lake information systems, which is funded by the National Natural Science Fund Committee and river course tests for the Digital Yangtze River. From 1987 to 1989, the Remote Sensing Technology Application Center of the Ministry of Water Resources initiated a collaborative effort with several other units, such as the Chinese Academy of Sciences, the State Bureau of Surveying and Mapping and the Air Force, to carry out flood control tests in the Yongding River, the Yellow River, the Jingjiang area, Dongting Lake and the Huaihe River and established a near real-time and all-weather monitoring system for flood disasters. After 5 years of efforts, a real-time transmission system for airborne remote sensing disaster monitoring was established and applied to flood monitoring in 1994, 1995 and 1996.

6. **In urban planning and management**

In recent years, GIS technology in China has been useful in urban management, and many cities, such as Beijing, Shanghai, and Tianjin, have established relatively perfect urban GIS management systems. The main applications are land planning and management; planning and management of resource utilization; environmental planning and management; urban planning and management; and traffic management and control.

In the sixth forum on an urban geographic information system held in Shenzhen in November 2003, the concept of "inter-city GIS" was proposed, and the application prospect and implementation scheme of GISs for urban control in densely populated areas were discussed.

7. **In disaster assessment and prediction**

GIS technology can be used for monitoring, forecasting, evaluating, disaster insurance, disaster relief, emergency rescue and post disaster recovery for a variety of natural or human-made disasters.

Severe flood disasters occurred in the Yangtze River and the Huaihe River area in 1991, and large flood disasters occurred in the Minjiang River and the Pearl River area in 1994. The floods of the two major disasters were monitored using a 1:250,000 scale GIS information system in the Taihu Basin and a 1:200,000 scale GIS land planning information database combined with remote sensing images to accurately assess flood disasters and mitigate losses. On the basis of full data analysis, a variety of analytical methods have been developed for the analysis of flood disaster data, such as the comprehensive analysis method of internal and external factors and multi-discipline (Li 1997a) and the analysis method combined with spatial information (Zhou 1993).

8. **In e-government affairs**

First class information enables the best decisions to be made. The GIS uses its own database through the construction and analysis of a series of decision models to analyze and compare various schemes proposed by policy makers and provide the basis for national macro decision-making. In 1993, the State Bureau of Surveying and Mapping utilized GIS to build a comprehensive national condition GIS for the State Council, which is a macro analysis decision system based on high-level, new technology for high-level leadership. At the beginning of 1999, China Telecom and the Economic Information Center of State Economic and Trade Commission, combined with more than 40 information authorities from departments, commissions, offices, and bureaus, jointly initiated the 'Government Online Project' and 'Two networks (Government intranet and Extranet), one website (Government Portal Website), four databases (population, legal entities, spatial geography and natural resources, macroeconomic and other four basic databases) and twelve golden systems (the business systems focusing on promoting office business resource systems)'; additionally, several comprehensive information systems were successively built, such as a comprehensive national condition GIS for the State Council, flood control weather information system for the State Council, spatial assistant decision information system for western development of the State Council, and the electronic government spatial assistant decision-making application project. These systems provided the central government with geospatial information, weather information for flood control, and information services for the development of the western region, as well as improved the level of comprehensive business processing and scientific decision-making.

9. **Applications of embedded GIS**

With the development of embedded hardware, embedded operation systems, navigation and positioning technology and mobile Internet technology, embedded GIS has been increasingly widely used in economic construction and national defense modernization. The main application fields include military and defense; field data acquisition; intelligent vehicles; intelligent transportation; information appliances; industrial control; environmental engineering and natural hydrological data real-time monitoring; and personal users.

At present, there are several kinds of embedded software in China: the SmartinHand software product for providing geographic information and location information services on mobile information devices, which was developed by Beijing Lingtu Technology Co. Ltd.; eSuperMAP, an embedded GIS development platform, was launched by Beijing hypergraph Geographic Information Technology Co., Ltd.; and TopMap CEGenius, a field geological data acquisition system that can be operated on Pocket PC for mobile mapping and GIS applications, was introduced by Beijing Huitu Technology Co., Ltd.

11.4.2 Applications in the Field of Public Service

Since 1999, the domestic GIS platform began to break into the Internet and gradually captured public attention. Its application field has also gradually transformed from the original provider of desktop geographic information applications to the provider of web map service applications and has achieved a number of functions, such as map search, public traffic transfer application, bus route and station querying, car navigation, electronic map, online map, mobile phone map, location-based service, map of China and city map.

1. **Map search**

Beginning in 2000, the combination of maps and search engines and the integration of map information and extended information of POI (point of interest) were achieved, and thus, search engines could be used to find geographic objects of concern. By 2007, map multimedia displays and search service functions appeared in combination with the network community (for example, MapABC's Cool map, http://diy.mapabc.com/). At the time, a user could show himself (herself) and communicate with others, just like being in the traditional network community; he (she) could label the visited location, favorite stores and other geographic information on the map for others to search and browse, which provided a user with an unprecedented experience in the network map. The most representative map search is the Google Chinese map (http://ditu.google.cn/).

2. **Public traffic query**

This application mainly provides query functions for city bus lines, bus stations, and transfer information for the public. For some cities, this application also provides real-time road information queries and browsing to aid in planning a trip. Websites, such as 58 bus networks (http://bus.58.com/), YAHOO Koubei networks (http://bus.koubei.com/) and others, have provided such services.

3. **Mobile phone map**

This application began in 2007, and the web map service functions were migrated from the internet to mobile terminals, which basically covered map searches, public bus inquiries, road information and merchant lists. Since 2008, with the wide application of GPS in mobile phones, mobile phone maps basically contained a navigation function, which made GIS applications closer to people's lives. Websites, such as MapABC mobile map (http://mobile.mapabc.com/), Lingtu Skywalker and others, have provided such services.

4. **True 3D realistic scene map**

This application began in 2004; the main purpose was to simulate the real geographical environment with three-dimensional images. For example, E-city (http://www.edushi.com) provides geographic information query application services based on true 3D scenes for a number of cities in China, and City Bar (http://www.city8.com), in another way, solved the application of geographic information to the world wide web, that is, provided map application services by integrating real photos and two-dimensional maps.

In addition to the direct service provided by domestic network map service providers, such as Sogou map (http://map.sogou.com), Mapbar (http://www.mapbar.com), MapABC (http://www.mapabc.com/), 51ditu (http://www.51ditu.com), etc., applications of GIS to the public are also involved in tourism life (such as the Aibang Network, http://www.aibang.com), house properties (such as Sina House Property http://map.house.sina.com.cn), city image advertisements (such as Yancheng City Information Port) and other aspects. The scope of its application is unprecedented.

11.5 Problems and Prospects

11.5.1 Problems

The development of GIS in China has a history of more than 30 years and has made brilliant achievements in the past 30 years; however, simultaneously, there are problems that cannot be ignored. Some of these problems are common problems in the GIS industry, and some are caused by China's current system and can be boiled down to the following issues.

1. **The problem of emphasizing technology and application while neglecting theory**

GIS is a technology-oriented applied discipline, and there has always been a tendency to emphasize technology and application while neglecting theory. In terms of theory, academician Li Deren (2003) believed that Earth information science included 7 theoretical issues. Academician Wang Jiayao believed that, from the viewpoint of philosophy, geography and cartography provide the epistemology and methodology for the development of GIS, respectively. However, at present, the existing research results are not sufficient to establish a complete theoretical system. It is necessary to strengthen the basic research in this field in the future.

2. **The problem of lacking standards in the GIS industry**

First is the lack of GIS engineering standards and specifications. Although the standardization work of GIS has been the focus of leaders and relevant standards have been actively developed by all interested parties, for users and GIS manufacturers, if there are no practical engineering standards and specifications, resulting in the lack of application engineering standards, the quality of the project cannot be guaranteed, which may damage the interests of users. Second is the lack of spatial data requirement specifications. Although data standards are being developed, there is no practical spatial data standard for enterprises. Currently, enterprises tend to use surveying and mapping standards, but sometimes, mapping, construction, transportation and other industry standards are contradictory or inconsistent, which makes the consistency of GIS spatial data in applications poor and difficult to share. Three is the lack of reference standards for market charges. There is no reference standard for the charges of undertaking GIS projects in the market. Although the cost of each enterprise is not the same, the charges can fluctuate, and because there is no reference basis, there is confusion in the market and even vicious competition, which not only affects the interests of users but also the development of enterprises.

3. **The problem of no perfect GIS industry service system**

First, there is no specific government agency available to manage and guide the GIS industry, and the GIS association is just a non-governmental institution that is unable to play a regulatory role in the market, which is one of the important causes of market disorder. For example, the software industry association is authorized by the Ministry of Information Industry and is responsible for the management of the software industry; software companies are required to provide annual reports for verification so that it can collect accurate data of software companies to facilitate and guide the development of the industry. Second, the restricted access to basic geographic data, to some extent, hinders the development of the GIS industry. Although the State Bureau of Surveying and Mapping keeps pace with the times, established a national database, and constantly modifies the relevant laws and regulations to facilitate the mapping of data to better serve the public, it is still far behind the needs of technology and market; whether to provide the people with the necessary basic geographic data services has become a fatal problem in the development of the geographic information system industry. Third, the high degree of confidentiality for high-resolution image data leads domestic GIS manufacturers to buy foreign commercial 1 m resolution remote sensing images. Fourth, the lack of a data sharing system resulted in many issues, such as the duplication of data acquisition, inability to interconnect application systems, and the waste of financial resources; there is also no mechanism for the use of basic geographic data, and it is only by providing a legal mechanism for using basic data that we can improve the illegal or poorly made map data situation.

4. **The problem that personnel training is out of line with the market**

At the fourth No. 3 conference of the Standing Committee of the GIS Association of China, which was held in 2008, Professor Tang Guoan was concerned that GIS education in China showed signs of cracking. This concern was because GIS education started late in China, the discipline at many colleges and universities was built too quickly, teachers were not consistent, and the education level was low. On the one hand, GIS specialty provision is relatively simple, lacking a GIS talent cultivation mode for different goals and social needs, and in particular, lacking interdisciplinary talents with technical knowledge and management knowledge; many students do not grasp the connotation of GIS after graduation or confuse learning GIS with learning computer programming. On the other hand, the low GIS threshold means that students from different industries perform GIS dissertations and applications, and then, the products are shoddy, which affects people's confidence in the application of GIS products.

11.5.2 *Prospects*

The development of GIS in China has undergone more than 30 years of glorious evolution and has made considerable progress; when looking forward to the future, the prospects are still bright. Some famous GIS leading figures, such as Chen Shupeng, Li Deren, Wang Jiayao, and Cheng Jicheng, have all discussed prospects for the future development of GIS in China. These prospects boil down to the following aspects:

Li Deren (2003, 2004, 2007) proposed that in the twenty-first century, in terms of theory, GIS is expected to form a more complete theoretical framework of geographic information science; technically, the acquisition of GIS spatiotemporal information will be space-ground integration and globalization; the handling and processing of spatiotemporal information will be automated, intelligent and available in real-time; the management and distribution of spatiotemporal information will be grid-enabled, the spatiotemporal data representation will tend to be multiscale, dynamic multidimensional and real-time 3D visualization; and finally, spatial and temporal information services will be popularized. Song Guanfu et al. (2007) proposed that service GIS- and SOA-oriented GIS architectures will be the main mode of future GIS development. In the ESRI 2008 User Conference in East China, HE Ning, the President of ESRI China (Beijing) Co., Ltd. proposed that enterprise GIS will be the future trend. Ye Jia-an (2008) believed that GIS in China will undergo substantial development in 3 areas: data sharing and opening up; industrialization and commercialization of GIS software development; and traffic GIS and network GIS. Wang Jiayao (2009) proposed that Web Service and Grid Service is the best way to solve the problem of geospatial information sharing, data interoperability, resource sharing and collaborative solutions (collaborative work) and will become the mainstream mode of geospatial information services. GIS will develop toward cross platforms, interoperability, resource sharing and collaborative solutions enabled by grid services.

Huang Xingyuan (1989), Gong Jianya (2002), and others also presented future development trends for GIS and generally considered the following aspects: grid GIS, spatial-temporal GIS, embedded GIS, 3D GIS, and Open GIS are the main directions of future GIS development and application; the application of GIS technology will take the road of integration, including "3S" integration, the integration of GIS and virtual reality technology, the integration of distributed technology, worldwide web and GIS, the integration of mobile communication technology and GIS, and the integration of GIS and decision support systems.

11.6 Representative Publications

(1) ***Introduction to geographic information systems*** (Huang Xingyuan Higher Education Press, Beijing, first edition in September 1990, second edition in December 2001)

This book was the first textbook on GISs in China, and for a long time, it was used as a teaching book by the surveying and mapping engineering specialty and the geographic information system specialty at many colleges and universities in China, with great influences in many aspects. This book systematically introduces the basic theory, technical system and application of GIS, and it is divided into 8 chapters. The main contents include the basic conception of GIS and the related basic theory, spatial data characteristics and data structure, GIS data model and spatial database, GIS spatial data processing and spatial analysis, application model and product output, GIS application system development and standardization research.

The original book was revised by the author in 2001. Compared with the 1990 edition, the revised edition provided a supplement, extension and expansion to the basic theory of GIS, the new development of GIS technology, GIS engineering and standardization, and the development and application of GIS.

(2) ***Data organization and processing method of integral GIS*** (Gong Jianya. Wuhan University of Science and Technology of Surveying and Mapping Publishing House, Wuhan, October 1993)

This book is further improved, revised and processed based on the author's doctoral dissertation. In the book, an integrated data structure with both vector and raster properties and an object-oriented data model that can express both graphical data and discrete data are presented. A data processing method based on the data structure and the data model is studied.

The book is divided into 5 chapters. The first chapter describes the development history and trend of GIS and the purpose of its study; the second chapter introduces the basic concepts and implementation methods of integrated vector and raster data structure in GIS; in the third chapter, the spatial query and spatial analysis algorithm based on the new data structure is systematically studied; in the fourth chapter, the object-oriented data model and its implementation are discussed; in the fifth chapter, the spatial database model integrating GIS and DEM with remote sensing data is derived based on an integral data structure and object-oriented data model.

(3) ***Geographic information system and its application in urban planning and management*** (Song Xiaodong, Ye Jia-an. Science Press, Beijing, October 1995)

This book comprehensively and systematically introduces the basic principle, application method, latest theory and development trend of GIS, as well its application in urban planning and management, providing some experiences and lessons for application. This book is divided into 12 chapters and 2 appendices. The first chapter is the introduction to the GIS; the second chapter is data structure and data management; the third chapter is data sources and inputs; the fourth chapter is the conversion and maintenance of spatial data; the fifth chapter is the query and analysis of spatial information; the sixth, seventh, eighth and ninth chapters are examples of applications, including land use planning and control, site selection, and urban construction management, as well as the combination of GIS, analysis and forecasting models; the tenth chapter introduces the method and process for establishing a GIS; the eleventh chapter describes the organization management in the implementation of GISs; and finally, the twelfth chapter summarizes the development, implementation and management of GISs and discusses its prospects. Appendix 1 is a brief introduction to computer knowledge related to geographic information systems; and appendix two is an English/Chinese comparison of commonly used professional terms.

(4) ***Principles and methods of the geographic information system*** (Bian Fuling. Surveying and Mapping Press, Beijing, August 1996)

This book consists of 14 chapters. The first chapter discusses the emergence and development of GISs. The second chapter is the introduction to GIS; the third chapter is the spatial characteristics of GISs; the fourth chapter is the concept of system engineering in GISs; the fifth chapter is the geographical basis of GISs; the sixth chapter is the data acquisition of GISs; the seventh chapter is the organization and structure of spatial data; the eighth chapter is spatial data management; the ninth chapter is the mathematical modeling of GISs; the tenth chapter is spatial analysis; the eleventh chapter is three-dimensional GIS technology; the twelfth chapter is network engineering of GISs; the thirteenth chapter is the cartographic output of GISs; and finally, the fourteenth chapter is the engineering conception of GISs. Appendix A describes the basic GIS software commonly used in China, appendix B is a brief introduction to the commercial database system, and appendix C describes the standard for spatial data conversion.

(5) ***The foundation and application of the geographical information system*** (Guo Dazhi, Sheng Yehua, Yu Zhaoping, Xie Chuhui. China Coal Industry Publishing House, Beijing, May 1997)

This book is divided into 8 chapters, plus an appendix. The first chapter introduces the background, development, hardware and software environment, function and application field of GISs; the second chapter focuses on the data structure of GISs; the third chapter focuses on the data model of GISs; the fourth chapter introduces the integrated spatial database; the fifth chapter introduces and reviews the commercialized GIS software, which mainly includes ARC/INFO, ArcView 2, GENAMAPGIS, Intergraph MGE GIS, City star and GeoStar; the sixth chapter introduces the processing and management of spatial data, including spatial data input and editing, the integration of remote sensing data and GIS data, spatial data management, and spatial data quality control and representation; the seventh chapter introduces the regional data analysis, data output and automatic mapping, and data sharing and standardization in GIS; and finally, the eighth chapter mainly introduces the application of GIS in engineering construction, national spatial data infrastructure, regional resources and the environment. The appendix lists the world's major GIS commodity software programs and their main features.

(6) ***Military geographical information system*** (Wang Jiayao, Hua Yixin. The PLA press, Beijing, August 1997)

China's first monograph systematically introduces the application of GISs in the military field. The book has 8 chapters and comprehensively introduces the theory, method and technology of military GISs. The first chapter is an introduction to the basic concept, composition and development of MGIS; the second chapter is about the data collection of MGIS, including the MGIS data source, the spatial data structure and its coding, the data acquisition software function, the data processing algorithm and so on; the third chapter is the MGIS geographic database, which introduces the concept of the geographic database, the traditional geographic data model, the object-oriented geographic data model, the geographic database system and its basic functions, the data quality of MGIS, etc. The fourth chapter is the spatial analysis of MGIS, which mainly introduces terrain analysis, spatial analysis based on vector and raster data, map animation and its application in spatial analysis; the fifth chapter is the MGIS analysis model, which introduces the concept of the model, the commonly used analysis model in MGIS, and the model base and its management; the sixth chapter is the application of MGIS in combat command; the seventh chapter is about mapping in MGIS; and finally, the eighth chapter is about the development and management of MGIS, which introduces the development and management of application-oriented MGIS and object-oriented MGIS tool software.

(7) ***Introduction to the geographical information system*** (Chen Shupeng, Lu Xuejun, Zhou Chenghu. Science Press, Beijing, May 1999)

This book is about the foundational theory of GISs, which comprehensively and systematically describes the technical system of GISs and focuses on the technologies and applications of GISs. The book is divided into 7 chapters. The first chapter is the introduction, which mainly presents the composition, function, development and present situation of GIS; in the second chapter, the author discusses the contents and theoretical problems of geographic space, spatial data and its model structure, quality and metadata; the third chapter introduces the contents and methods of spatial data processing in detail; the fourth chapter is the analysis of the spatial information model, focusing on the theory of spatial modeling, spatial model structure and its implementation; the fifth chapter deals with the problems in GIS engineering and standardization; the sixth chapter introduces the development and application of GISs with examples; and finally, in the seventh chapter, the author introduces the Open GIS and the prospect of geo information science.

(8) ***Applied geographic information system: Successful construction and management of geographic information systems*** (Chen Jun, Gong Peng. Science Press, Beijing, August 1999)

The book, from a practical point of view, explores and studies the theory and technology of GISs based on the author's years of experience in the development and management of GIS projects. The book has a total of 7 chapters. The first chapter is an introduction to the GIS, which first expounds upon the basic content and application scope of GISs; the second chapter is about the data of practical GISs; the third chapter is the design of the GIS database; the fourth chapter is the project management of the GIS, which describes the design, development and steps of the GIS database in detail, as well as the content, method and process of GIS project management; the fifth chapter is the application of computer network technology in the GIS; the sixth chapter introduces the standardization of the GIS; and the seventh chapter is the development trend and prospect of practical GIS. The last two chapters introduce the standardization organizations and standards related to GIS and forecast the development direction of GIS over the next 5 to 10 years.

(9) ***Urbanization and urban geographic information systems*** (Chen Shupeng. Science Press, Beijing, March 2000)

This book introduces the author's achievements in the research of urbanization and urban GISs in recent years. It consists of 9 chapters. In the first chapter, the author puts forward the main problems and urgent tasks in the research of urbanization and urban GISs; the second chapter briefly introduces the construction of urban GIS standards; the third chapter to the sixth chapter, focusing on city expansion and development, analyzes the evolution and regulation of urban and regional spatial patterns, the law of urban growth and land increment, the dynamic simulation of urban internal structure and spatial expansion process, and urban and regional development planning; the seventh chapter, focusing on urban traffic, introduces the organizing and processing of non-planar data of urban traffic networks based on features; the eighth and ninth chapters, focusing on the study of coastal zones and seaport cities, discusses how to control the environmental pollution of coastal zones, regulate the temporal and spatial patterns of coastal zone urbanization, and how to use GIS technology to evaluate the investment environment of seaport cities.

(10) ***Geographic Information Systems: Principles, methods and applications*** (Wu Lun, Liu Yu, Zhang Jing, Ma Xiujun, Wei Zhongya, Tian Yuan. Science Press, Beijing, February 2001)

The book introduces the basic theory, function implementation and application technology of GIS and includes 5 parts: the basic concepts and theories, the functions of the GIS, the application technologies of GISs, appendix one—computer foundation of GIS, and appendix two—introduction to GIS tool software. The basic concepts and theories include 5 chapters: introduction to the GIS, from the real world to the bit world, spatial data model, spatial reference system, and data in GIS. The

GIS function includes 6 chapters: spatial data acquisition and processing, spatial data management, spatial analysis, digital terrain model (DTM) and terrain analysis, spatial modeling and spatial decision support, spatial data representation and cartography. The GIS technology section includes 8 chapters: 3S integrated technology, network GISs, applications of the GIS, the organization and management of GIS application projects, the software engineering technology of GISs, the GIS standards, GIS and society, geo information science and digital Earth. Appendix one: the computer foundation of GIS includes 3 chapters: principles of computer organization, foundation of the database system, data structure and algorithms. Finally, in appendix two: introduction of GIS tool software, the current popular GIS software is introduced.

(11) ***Principles and techniques of the geographic information system*** (Hua Yixin, Wu Sheng, Zhao Junxi. The PLA Press, Beijing, February 2001)

The book has a total of 12 chapters, comprehensively introduces the principle and technology of GIS, and presents the latest developments in GIS. The structure of the book is outlined as follows: the first chapter, the basic concept of GIS, introduces the definition, development, and composition, application and related disciplines of GIS, and the types and characteristics of GIS data sources; the second chapter, the spatial data structure of GIS, introduces the vector data structure and raster data structure of GIS; the third chapter, the spatial database and data model of GIS, introduces the basic concepts of the spatial database and spatial data model, and the design and establishment of the spatial database; the fourth chapter introduces the mathematical basis of spatial data, and the commonly used map projections and map projection transformation methods; the fifth chapter is about the input of spatial data, including the coding of attribute data, the collection and verification of spatial data, and the standards for spatial data and spatial metadata; the sixth chapter is about the quality of spatial data and introduces the basic concepts, common methods and main theories of GIS data quality; the seventh chapter is about spatial data and introduces graphic editing, geometric correction, automatic topology, vector data compression, structural transfer between vector and raster data, and spatial data interpolation, processing of image data and other common spatial data processing algorithms; in the eighth chapter, spatial querying and spatial analysis are introduced, and the basic content of spatial data query in GIS is described; the ninth chapter is cartography in GIS; the tenth chapter is the development of the GIS; the eleventh chapter is the combination of RS, GPS and GIS; and finally, the twelfth chapter is the network GIS.

(12) ***Foundation of geographic information systems*** (Gong Jianya. Science Press, Beijing, February 2001)

This book is a comprehensive, systematic basic tutorial of GISs consisting of 8 chapters. The first chapter is the introduction, which briefly introduces the origin and development of the GIS; the second chapter, which covers the composition of the GIS, introduces the related hardware, software modules and functions of the system; the third chapter is about the acquisition of spatial data, including data collection, acquisition, conversion and quality problems; the fourth chapter is about the representation of spatial data, including the relationship between geographic phenomena and spatial objects, vector and raster representation, and spatial data model and data structure; the fifth chapter is the spatial data processing, which introduces a variety of spatial operation algorithms and implementation strategies; the sixth chapter is about spatial data management, including the database, database management system and model; in the seventh chapter, various methods of spatial querying and spatial analysis are introduced; the eighth chapter is about the visualization and mapping of spatial data, and the visualization and output of geographic information, such as the general map and thematic map, are introduced.

(13) ***Spatial information system theory*** (Wang Jiayao. Science Press, Beijing, May 2001)

A spatial information system (SIS) is a technology system of geospatial information science and an information system that uses computer technology and network communication technology to solve the problems of data acquisition, storage, transmission, management, analysis and application related to geospatial information science. The book is divided into 7 parts, for a total of 20 chapters. In the first part, the scientific basis of the spatial information system is discussed from 3 aspects: the concept of geo spatial information science, the basic theory of the spatial information system, and the structure and function of the spatial information system; the second part is the acquisition of spatial information, which includes cartographic data acquisition, remote sensing data acquisition, GPS data acquisition and 3S technology integration; the third part is the spatial data model, which mainly includes the spatial cognitive model and the geographic data model analysis, the object-oriented geographic data model and the spatiotemporal data model; the fourth part is the digital elevation model, which introduces the digital elevation model and its data structure, as well as the establishment of a TIN and regular grid DEM; the fifth part covers the spatial relationship and spatial analysis, and based on the discussion of the concepts, features and algebraic transformations of spatial graph algebra, the algebraic transformations, spatial relations and spatial analysis of various spatial graphs are introduced in detail; the sixth part is the multiscale representation of spatial information technology, and this part introduces the multiscale feature and automatic visualization of spatial data and spatial data visualization; and finally, the seventh part discusses the frontier technology of spatial information system, where the core technology of the modern spatial information system and new developments in spatial information systems are introduced.

(14) ***Tutorial of geographic information systems*** (Hu Peng, Huang Xingyuan, Hua Yixin. Wuhan University Press, Wuhan, February 2002)

The book consists of 8 chapters and 1 appendix. The first chapter, an introduction, discusses the concept, composition, function and development of GIS; the second chapter, the spatial data structure and database of the GIS, focuses on the spatial data structure and the GIS data model, and discusses the spatial database design; the third chapter is the acquisition and quality control of spatial data, including geographic coordinates and the control foundation of spatial data, geographic entity classification and data encoding, spatial data acquisition, GIS data quality and spatial data standards; the fourth chapter deals with spatial data processing, including the automatic establishment and graph editing of the topological relations of vector data, coordinate transformation, compression processing and structural transformation of spatial data; the fifth chapter, spatial query and spatial analysis, mainly discusses the statistical analysis, overlay analysis, buffer analysis, network analysis of spatial data, digital elevation model analysis, and Tyson polygon analysis; the sixth chapter, spatial information visualization, introduces the basic characteristics and forms of spatial information visualization, map language and symbols, visualization of spatial data, electronic maps, dynamic maps and virtual reality space; the seventh chapter, the application of GISs, summarizes the application of GISs, discusses the combination of GIS and remote sensing, GPSs and artificial intelligence, GIS applications in management, planning, and decision-making, and WebGIS is introduced; and finally, the eighth chapter, the development and evaluation of GISs, introduces the development method, development process, evaluation and application of GISs.

(15) ***Integration and implementation of a spatial information system*** (Li Deren, Guan Zequn. Wuhan University Press, Wuhan, May 2002)

The book expounds upon the geo spatial information science and technology system, as well as its core technology, 3S technology, and its theory and key technologies of integration. The book introduces the important methods or means of spatial information system integration and implementation in detail, such as spatial information fusion, spatial concept formation and association, spatial knowledge discovery and data mining. The integration problem in 3S is discussed, and one of the hot issues in spatial information integration, that is, integrated database technology based on spatial location technology, remote sensing technology and geographic information system technology, is described. The geo spatial data framework and its relationship with 3S are introduced, and the author's work in this field is also given.

(16) ***Design and implementation of geographic information systems*** (Wu Xincai. Electronics Industry Press, Beijing, June 2002)

This book is listed in the national '11th five-year' planning textbooks for general higher education. Starting from the recent development and practical needs of GIS design theory, mainly considering GIS design and application, as well as the latest research results of information technology and the author's years of development and design results, the characteristics, methods, processes and implementation techniques of GIS design are systematically introduced. The overall design, functional design, database design and several application examples of GIS are described in detail. The specific contents include system analysis, general design, system functional design, database design, GIS implementation and maintenance, GIS testing and evaluation, GIS project management and quality engineering, GIS standards, GIS design examples, and the development mode for a new generation of GIS.

(17) ***Applications of GIS in environmental science and engineering*** (Li Xuxiang. Electronics Industry Press, Beijing, January 2003)

This book is about the applications of GIS in environmental science and engineering. The book expounds upon the features, types, functions, data acquisition and database design of GIS and introduces the application, techniques and methods of GIS in environmental science and engineering. Consisting of 9 chapters, the book focuses on the composition and function of the GIS and its development status, the data characteristics, data sources and data quality of a practical GIS, and the main contents and methods of GIS database design; additionally, the book covers familiar GIS software at home and abroad, as well as their usages; further, the applications and development of GIS in ecological control, water pollution control, management and monitoring systems, as well as applications of GIS in air pollution control and other environmental control are discussed, and finally, the virtual environment and model establishment of the main factors affecting the environment are covered in this book.

(18) ***Engineering design and management for geographic information systems*** (Guo Qingsheng, Wang Xiaoyan. Wuhan University Press, Wuhan, February 2003)

This textbook is written for undergraduate teaching of cartography and GISs. The main contents include basic concepts and basic principles of geographic information; design principles, methods and specifications of GIS software engineering and data integration engineering; management methods of GIS engineering projects; principles, methods and legal issues of geographic information sharing; the establishment method of computer networks in GIS engineering and solutions for enterprise GISs, the basic concept and application method of Arc/lnfo; the design method of GIS engineering based on Arc/lnfo; and cases of professional GIS engineering design.

(19) ***Principles and methods of software engineering for geographic information systems*** (Bi Shuoben, Wang Qiao, Xu Xiuhua. Science Press, Beijing, July 2003)

The book systematically expounds upon the basic concepts, principles and methods in the field of GIS software engineering. The book consists of 15 chapters: Chapter 1 is an overview of GIS software engineering; Chapter 2 is the feasibility of GIS software engineering; chapter 3 is the system analysis of GIS software engineering; Chapter 4 is the analysis method of GIS software engineering; Chapter 5 is the overall design of GIS software engineering; Chapter 6 is the detailed design of GIS software engineering; Chapter 7 is the design method of GIS software engineering; Chapter 8 is GIS software data engineering; Chapter 9 is data quality control in GIS software engineering; Chapter 10 is the implementation and testing of GIS software engineering; this chapter is the GIS software maintenance project; Chapter 12 is the quality assurance of GIS software engineering; Chapter 13 is the standardization of GIS software engineering; Chapter 14 is the management of GIS software engineering; and finally, Chapter 15 is the GIS software engineering environment.

(20) ***Principles and methods for geographic information system integration*** (Lv Guonian, Zhang Shuliang, Gong Minxia. Science Press, Beijing, August 2003)

The book has a total of 9 chapters. Chapter 1 is an introduction that discusses the basic concepts of system integration and the integration and modelling of GIS; Chapter 2 introduces the integration of spatial data and attribute data; Chapter 3 introduces the integration of multisource data, including the integration model of multisource spatial data and the fusion of multisource spatial data, integration of multiscale spatial data, integration of spatiotemporal multiscale spatial data, and 3S integration; Chapter 4 introduces the integration of GISs based on spatial metadata; Chapter 5 introduces the integration of GIS with the application analysis model; Chapter 6 introduces the integration of GISs with knowledge and rule bases; Chapter 7 introduces the integration of hypermedia systems with GISs; Chapter 2 introduces the integration based on distributed computing; Chapter 9 introduces the integration of GIS application time platforms, reviews the significance of platform integration of geographic information application system, the framework of GIS application integration platforms, the integration of GISs with automation systems, management information systems and SCADA real-time systems, and integration between GIS application systems.

(21) ***Principles and methods of geographic information sharing*** (He Jianbang, Lv Guo Nian, Wu Pingsheng. Science Press, Beijing, August 2003)

This book systematically expounds upon the basic concepts, principles and methods of geographic information sharing. The book is divided into 8 chapters. The first chapter and the second chapter explain the general principle, system structure and main functions of geographic information sharing; the third chapter to the eighth chapter discuss the principles and methods for formulating and implementing the standards, policies and laws, which are important components of geographical information sharing, and the current status and trends of related work at home and abroad. The two appendices describe the basic format for the geographic information sharing standard and the external framework of "the law of geographic information sharing" (draft recommendation) in China.

(22) ***Principles and algorithms of the geographic information system*** (Wu Lixin, Shi Wenzhong. Science Press, Beijing, December 2003)

The achievements of GIS theory and core algorithms at home and abroad have been summarized systematically in this book, and they have been concluded, classified, compared, analyzed and refined. Taking the principle and algorithms as the key link, two parts of the GIS principle and GIS algorithm are formed. In the first part, the concepts of generalized geographical systems and generalized geographic objects are proposed based on disciplinary integration and intercrossing. Then, the basic principles of 2D GIS spatial data models, data structures, spatial relations, geo information classification and spatial coding, map projection and graph transformation are systematically explained, and recent progress in GIS theory, such as 3D GIS spatial data models, temporal GIS, temporal relations, and geographic metadata, is introduced. In the second part, based on the four aspects of spatial data processing, spatial graphics manipulation, spatial measurement and analysis, and visualization and data mining, the algorithms of spatial data compression, spatial data interpolation, spatial data conversion and spatial data analysis are systematically expounded upon and explained; further, the algorithms of automatic generation and clipping of polygons, construction of TIN, construction of Voronoi diagrams and spatial transformation, spatial metrics, digital terrain analysis, spatial statistics and spatial analysis, and the core algorithms of GIS visualization, spatial data mining and knowledge discovery are discussed.

(23) ***Environmental geographic information systems*** (Wang Qiao, Zhang Hong, Li Xuwen. Science Press, Beijing, August 2004)

On the basis of application and development practice, this book summarizes the implementation of environmental GISs. The book systematically discusses the general method of developing an information system software environment, including the technology standard and application mode of environmental GIS development, from data acquisition to database design, software development, project implementation, system security and maintenance. According to the environmental protection department, the book introduces the detailed steps and key methods of system construction, that is, a set of processes from requirements analysis to database design, functional design, detailed design, programming, and test installation, and the book focuses on the integration of GIS functions and environmental protection management business, integration of the application model and GIS functions, integration of the World Wide Web and GIS, and other hot issues in GIS applications. The book is divided into 7 chapters. Chapter 1 is the environmental GIS and its system analysis; Chapter 2 is the design and development of environmental GIS; Chapter 3 is GIS in environmental monitoring information management; Chapter 4 is GIS in pollution sources information management; Chapter 5 is GIS in environmental quality assessment and total quantity control; chapter 6 is GIS in environmental pollution accident emergency; and finally, chapter 7 is GIS in the simulation and prediction of environmental pollution.

(24) ***Urban planning and management information systems*** (Sun Yizhong, Zhang Jian, Zhou Sheng, Miu Hanshen. Science Press, Beijing, August 2004)

Closely combined with the demand for information construction in the current urban planning and management industry, this book systematically summarizes the contents of several urban planning and construction projects that the author has conducted and recently participated in, including urban planning and management information systems, urban integrated pipeline information system design, graphic and text database design, system development and construction, system running maintenance and database updates, and so on. At the same time, the book briefly introduces the construction of an urban comprehensive management information system, provides an example of GIS construction for the urban planning management industry, and provides a detailed technical route and implementation plan for the information construction of the city planning management industry. The book contains nine chapters. Chapter I is an introduction; chapter II is about urban planning and management information systems; chapter III introduces the design of the urban planning and management information system; chapter IV discusses the spatial database design of urban planning and management information systems; chapter V presents the design of the document database for urban planning and management information systems; chapter VI deals with the development, integration and testing of urban planning and management information systems; chapter VII expounds upon the operation, maintenance and management of urban planning and management information systems; chapter VIII gives the developing example of urban planning management information systems; and chapter ix introduces the construction of urban integrated pipeline information systems.

(25) ***Geographic information systems for government affairs*** (Zhang Qingpu, Liu Jiping. Science Press, Beijing, January 2005)

Based on the author's results from research in government GISs over a long time and the combination of theory and practice, the basic concept, theory and technology foundation, construction mode, operation mechanism, system design, database construction, software development, network communication of government GISs and their future trends have been comprehensively analyzed and theoretically summarized. This book is the first scientific monograph devoted to the construction and application of government GISs in China and abroad. The book consists of 9 chapters. Chapter I is the basic concept of government GISs; chapter II is the theoretical and technical foundation of government GISs; chapter III is the construction principle and operation mechanism of government GISs; chapter IV is the design of government GISs; chapter V is the database construction of government GISs; chapter VI is the research of software systems of government GISs; chapter VII is the network construction of government GISs; chapter VIII is the application of government GISs; and finally, chapter IX is the future development of government GISs.

(26) ***Geographic information system database*** (Zhang Xinchang, Ma Linbing, Zhang Qingnian. Science Press, Beijing, February 2005)

The book has a total of 7 chapters. The first chapter is the introduction, which summarizes the concept, formation and development of the GIS database, as well as the principle and relational model of the GIS database and DBMS; the second chapter introduces the expression and management of spatial data; the third chapter is the establishment and design of the GIS database, which mainly introduces the concept design of the GIS database, relational database design, object-oriented database design, geographic information metadata design, database design based on the geodatabase, and spatial data collection and database building; the fourth chapter introduces web GIS database technology; the fifth chapter is about the standardization of the GIS database; the sixth chapter introduces the basic applications of the GIS database, focusing on the topographic and cadastral database construction in urban and rural areas, the database construction of land and resources basic geographic information, the water resources planning and management database, and puts forward the application prospects of the GIS database; and finally, the seventh chapter discusses the forefront of GIS database development.

(27) ***Principles and methods of geographic modeling*** (Wei Yuchun, Chen Suozhong. Science Press, Beijing, February 2005)

According to the needs of geographic development and GIS specialty teaching, the book expounds upon the main models and methods of geographic modeling, covering the main parts of model building. The contents include the conceptual model, physical model, statistical correlation model, fuzzy mathematical model, dynamic data analysis model, spatial data analysis model, distributed mechanism process model, model construction and testing. At the same time, the model building and application of the model are discussed. The book contains nine chapters. Chapter 1 outlines the general concepts, features, classifications and uses of the model, modeling steps, and geographic modeling; the second chapter is about the conceptual model; the third chapter introduces the physical model; the fourth chapter introduces the statistical model; the fifth chapter is about the fuzzy mathematical model; the sixth chapter is the dynamic data analysis model; the seventh chapter is the spatial data analysis model; the eighth chapter deals with the distribution mechanism process model; and the ninth chapter discusses several problems in model building and model verification.

(28) ***Principles and techniques of web GIS*** (Meng Lingkui, Zhang Penglin, Shi Wenzhong. Science Press, Beijing, March 2005)

This book consists of 9 chapters. The first chapter mainly reviews the development process of GIS, looks forward to the trends in GIS, and briefly introduces the related technologies of web GIS; the second chapter introduces one of the foundations of web GIS, i.e., the basics of computer networks, so that readers can gain a basic understanding of the computer network; the third chapter focuses on the basic principles of web GIS, including the architecture, data organization and management, data sharing, and other basic knowledge; the fourth chapter introduces the data storage technology of web GIS and focuses on several network storage technologies, especially NAS and SAN, and introduces the application of network storage technology in web GIS through an example; the fifth chapter introduces the widely known web GIS technology, which is the most widely used and successful web GIS; the sixth chapter describes the rapid development of mobile GIS technology and its applications in location-based services; the seventh chapter discusses the principle and technology of grid GIS, which has great development potential and represents the important development direction of grid GIS; the eighth chapter introduces the engineering technology and engineering management of web GIS; and the ninth chapter briefly introduces several types of commonly used web GIS software, so that readers can understand the function and characteristics of the practical web GIS tool software.

(29) ***Principles and methods of digital elevation models and terrain analysis*** (Tang Guoan, Liu Xuejun, Lv Guonian. Science Press, Beijing, August 2005)

The book systematically introduces the basic concepts of DEMs, the data organization and management of DEMs, the establishment and processing of DEMs, the visual expression and accuracy analysis of DEMs, and other basic theories and key technologies; the basic theories and methods of DEM digital terrain analysis are expounded upon from four aspects, including slope terrain factor extraction, terrain feature extraction, terrain statistical analysis and geo model analysis; the study results of the DEM on the surface slope spectrum of the Loess Plateau, the characteristics of land and gullies, the analysis of climate features and the analysis of vegetation spatial differentiation are presented. The book consists of 15 chapters. The first chapter is an introduction; the second chapter is the DEM data organization and management; the third chapter is the acquisition method of DEM data; the fourth chapter is the establishment of the grid DEM; the fifth chapter is the establishment of triangulated irregular net (TIN); the sixth chapter is the visual expression of the DEM; the seventh chapter is the precision analysis of the DEM; the eighth chapter is the extraction of terrain factors; the ninth chapter is the extraction of feature terrain elements; the tenth chapter is the topographic statistical analysis of the DEM; the eleventh chapter is the geo model analysis of DEMs; the twelfth chapter is the analysis of surface slope spectrum in the Loess Plateau based on the DEM; the thirteenth chapter is the analysis of gully characteristics based on DEMs; the fourteenth chapter is the analysis of mountain climate characteristics based on DEMs; and the fifteenth chapter is the spatial differentiation analysis of vegetation based on DEMs.

(30) ***Land management information system*** (Sun Zaihong, Chen Huiming, Qiao Weifeng, Wu Changbin. Science Press, Beijing, September 2005)

The book systematically expounds upon all kinds of information systems related to land and their design ideas and development methods. The book is divided into 18 chapters: an introduction; the framework of land management information systems; the software engineering of land management information systems; basic database construction for land; land use investigation information systems; town cadastral management information systems; integrated urban and rural cadastral management information systems; information systems for revision of overall land use planning; urban land gradation and evaluation information systems; farmland gradation and evaluation information systems; land registration information systems; residential land examination and approval information systems; land market management information systems; land use planning and management information systems; land developing and neatening item management information systems; land statistical analysis and integrated affairs management information systems; the land information website; and land information service systems.

(31) ***Water resource management information systems*** (Chen Suozhong, Chang Benchun, Huang Jiazhu. Science Press, Beijing, February 2006)

There are 9 chapters in the book, namely, an introduction; demand analysis of water resource management information systems; overall analysis of water resource management information systems; development of water resource management information systems; database design of water resource management; water resource monitoring and management information systems; water drawing permitted management information systems; drain outlet management and water pollution simulation information systems; and groundwater resource evaluation systems.

(32) ***Geographic information systems for transportation*** (Liu Xuejun, Xu Peng. Science Press, Beijing, March 2006)

Based on GIS theory, the book systematically expounds upon the basic concepts, theories, techniques and methods of GISs for transportation. The book has 15 chapters, consisting of two parts. The first part introduces the basic theory, focusing on the concept of GISs for transportation, data acquisition, data models, organization, development of GISs for transportation, and so on. The second part describes an application case, focusing on the typical applications of GISs for transportation in the field of transportation, including transportation planning, highway engineering project construction management, highway maintenance, highway administration, highway graphic data, bus route planning, traffic facilities management, regional road and passenger integrated service information systems, driver and vehicle management, traffic control and dispatching.

(33) ***Tourism management information system*** (Zhou Chunlin, Liang Zhong, Yuan Ding. Science Press, Beijing, May 2006)

Based on national network infrastructure construction and social information environment development, the book introduces the development technology, application status and development trend of the tourism management information system under the information condition and expounds upon the requirements analysis, system design and implementation of the tourism management information system in the fields of tourism enterprises, tourism destinations, tourism resources, tourism planning, tourism e-commerce and tourism e-government. Combined with practical applications, several example cases of these management information systems are presented. There are 10 chapters in the book. The title of each chapter is listed as follows: overview of the tourism management information system; technical foundation of the tourism management information system; the hotel management information system; the travel agent's management information system; the tourist traffic information system; the tourism destination marketing system; the tourism resource management information system; the tourism planning information system; the tourism e-commerce system; and the tourism e-government system.

(34) ***Spatial analysis modeling and principle*** (Zhu Changqing, Shi Wenzhong. Science Press, Beijing, May 2006)

The book is written based on the author's research and teaching of basic theories, models and methods of spatial analysis. The principle and spatial analysis modeling are discussed in the book, including the basic concepts of spatial analysis, mathematical foundation, spatial overlay analysis model, buffer analysis model, statistical analysis model, network analysis model, digital elevation model and its precision analysis model, 3D terrain analysis model and wavelet analysis model. In the model discussion, the emphasis is placed on the idea of modeling, the exposition of principles, and the derivation of methods; meanwhile, the ideas and viewpoints of mathematics are applied to the models and methods for building spatial analysis.

(35) ***Fuzzy spatial object modeling theory and its applications*** (Tang Xinming. Surveying and Mapping Publishing House, Beijing, May 2006)

The book is a monograph on the theory and method of fuzzy spatial feature modeling. It consists of 9 chapters. The titles of each chapter are listed as follows: introduction; fuzzy set theory and fuzzy topology; topological relations in rigid and fuzzy topological spaces; topological relations in general fuzzy topological spaces; fuzzy spatial object modeling and its topological relation; the creation of fuzzy land cover objects; querying fuzzy spatial objects; variation reasoning of fuzzy land cover objects; and conclusion and discussion.

(36) ***Digital terrain analysis*** (Zhou Qiming, Liu Xuejun. Science Press, Beijing, May 2006)

Based on the digital features of curved terrain surfaces, the theories, methods and techniques of digital terrain analysis are expounded upon; the calculation and implementation of various terrain parameters and terrain features in DEMs are analyzed; the influence of digital terrain analysis methods on geological analysis is discussed; and algorithm evaluation and analysis methods for digital terrain analysis are established. The book consists of 10 chapters and 1 appendix, which are listed as follows: introduction; basic digit characteristics of terrain surfaces; digital elevation model and expression of ground morphology; parameter calculation of terrain and curved surface; terrain morphology feature analysis; terrain statistical feature analysis; compound terrain attributes; terrain visualization and analysis; error and accuracy of digital terrain analysis; and development and prospects of digital terrain analysis. The appendix includes a Chinese and English terminology comparison.

(37) ***Equipment and facilities management geographic information systems*** (Zhang Shuliang, Lv Guo Nian, Gong Minxia, Zhang Haitao. Science Press, Beijing, June 2006)

The book consists of 12 chapters. The first chapter outlines the AM/FM/GIS; the second chapter is about AM/FM/GIS domain engineering; the third chapter introduces the AM/FM/GIS data model; the fourth to twelfth chapters include the AM/FM/GIS software component environment, knowledge and rules in AM/FM/GIS, development and implementation of AM/FM/GIS software, AM/FM/GIS for city integrated pipelines, AM/FM/GIS for drainage systems, AM/FM/GIS for power distribution networks, AM/FM/GIS for gas, AM/FM/GIS for telecom, and AM/FM/GIS for street lights.

(38) ***Foundation of algorithms for geographic information systems*** (Zhang Hong, Wen Yongning, Liu Aili. Science Press, Beijing, June 2006)

This book comprehensively and systematically collects and sorts out the relevant materials in the field of GIS algorithms, aiming at the design and implementation of GISs; additionally, the book introduces the description, retrieval, storage and management of geospatial data, as well as the design and implementation of basic methods of geospatial information analysis. The book consists of 14 chapters. The first chapter outlines the design and analysis of algorithms; the second chapter introduces the computational geometric foundation of the GIS algorithm; the third to fourteenth chapters discuss the transformation algorithm, organization algorithm and space measurement algorithm of spatial data, as well as the spatial data index algorithm, spatial data interpolation algorithm, Delaunay triangulation and Voronoi diagram algorithm, buffer analysis algorithm, network analysis algorithm, terrain analysis algorithm, spatial data mining algorithms and data output algorithm.

(39) ***Foundational geographic information system*** (Zhou Wei, Sun Yizhong, Sheng Yehua. Science Press, Beijing, September 2006)

The book systematically expounds upon the principles, techniques and methods relevant to basic geographic data collection, processing and basic GIS design and construction. The book consists of 9 chapters. The first chapter introduces the basic concept, current situation and development trend of the basic GIS; the second chapter mainly discusses the mathematical foundation of basic geographic data; the third chapter describes some problems of basic geographic data collection; the fourth chapter discusses the system architecture design, function design and safety design; the fifth to ninth chapters discuss the basic geographic database design, basic geographic data processing and database building, basic geographic data update, basic geographic data standards and quality control, and basic geographic information system implementation; and finally, chapter ten provides the basic geographic information sharing technology.

(40) ***Geographic information sharing technology*** (Lv Guonian, Zhang Shuliang, Wang Yongjun, Tao Tao, Lan Xiaoji. Science Press, Beijing, January 2007)

This book expounds upon several key technologies of geographic information sharing from many aspects of GIS, and thus, this book analyzes the mechanism and method of urban geographic information sharing. The book is divided into 11 chapters, that is, the basic concept of geographic information sharing, and the structure and function of the geographic information sharing system; the traditional geographical data model; the unified geographic data model—SEDRIS; the geographic information sharing data model—GML; GIS data conversion technology; GIS interoperability; GIS map symbol sharing technology; spatial relation sharing and reconstruction technology; geospatial information sharing technology based on spatial services; geospatial information sharing technology based on spatial metadata; and the mechanism and method of urban geographic information sharing.

(41) ***Theory, method, standard and application of a new digital elevation model*** (Hu Peng, Yang Chuanyong, Wu Yanlan, Hu Hai. Surveying and Mapping Publishing House, Beijing, February 2007)

To establish an interpolation process in a DEM, the book discusses errors as numerical approximation errors, which differ from random errors generated with original mapping data. Thus, suggestions and test methods for developing DEM precision standards are proposed. In view of the 3 weak points of the DEM theory and technology in the past half century, error evaluations, high fidelity problems, generation methods and high precision have been discussed and studied in detail. The focus of this book is on the basic theoretical problem of the DEM—the problem of the DEM accuracy evaluation.

(42) ***Real estate management information system*** (Jiang Haiqin, Lv Guonian, Jiang Wenming. Science Press, Beijing, April 2007)

The book consists of 10 chapters: an introduction; the framework of real estate management information systems; the key technologies in the construction of real estate management information systems; the standard of property management information; database center design for real estate management information systems; data integration of real estate management information systems; real estate property registration and transaction management system; property mapping systems; property file systems; and housing demolition management systems.

(43) ***Geographic information system course*** (Tang Guoan, Liu Xuejun, Lv Guonian, Sheng Yehua, Wang Chun, Zhang Ting. Higher Education Press, Beijing, April 2007)

The author systematically summarizes the basic theories, practices and achievements in the existing GIS disciplines and summarizes the author's recent experience in GIS research and teaching at home and abroad. The book comprehensively and systematically explains the basic theory and application development of GIS; the contents include the basic concept of the GIS, mathematical foundation of geospatial and spatial data models, spatial data structure, spatial data organization and management, spatial data acquisition and processing, spatial data query and measurement, GIS spatial analysis, digital terrain analysis, spatial statistical analysis, geographic information visualization, and geographic information transmission. There are technical terms and questions for review in the book.

(44) ***Model and algorithm of the 3D spatial information system*** (Shi Wenzhong, Wu Lixin, Li Qingquan, Wang Yanbing, Yang Bisheng. Publishing House of Electronics Industry, Beijing, May 2007)

The book focuses on two aspects of 3-dimensional spatial data and visual modeling. The book consists of 9 chapters. Chapter 1 introduces the development status of 3-dimensional spatial data and visual modeling; in the second chapter, the method of acquiring 3D spatial data is discussed from three aspects of 2D space, land surface 3D space and underground 3D space; the third to seventh chapters discuss the latest research results in the aspects of the modeling and updating digital terrain, 3D spatial modeling, 3D geological spatial modeling, and the integrated true 3D modeling of ground and underground; additionally, 3D spatial modeling algorithms are described; the eighth chapter introduces a spatial data quality model; and finally, the ninth chapter introduces the experimental system and application examples developed by the author's research team and the main 3D modeling tools and typical systems used both at home and abroad.

(45) ***E-government system*** (Wang Weiguo, Lv Guonian, Wang Aiping. Science Press, Beijing, July 2007)

The book is written according to the construction needs of the e-government system for China's government and combined with the author's experience in developing the OA system, administrative examination and approval system, GIS, and other e-government systems over the years. The book consists of 10 chapters: an overview of e-government and e-government systems; analysis and framework design of e-government systems; network platform of e-government systems; database platform of e-government systems; workflow platform of e-government systems; GIS software platform of e-government systems; portal of e-government systems; e-government OA systems; GIS designs for e-government; and CA authentication and electronic seals in e-government systems.

(46) ***Logistics management information system*** (Sheng Yehua, Zhang Guiying. Science Press, Beijing, March 2008)

Combined with the practice of logistics management information system development, the author systematically expounds upon the analysis, design and development methods of logistics management information systems. The book consists of 12 chapters: logistics and logistics management information systems; object-oriented system analysis and design; analysis of logistics management information systems; design of logistics management information systems; database design of logistics management information systems; development and integration of logistics management information systems; storage management information systems; transportation management information systems; admeasure management information systems; supply chain management information systems; the logistics portal; and development of logistics management information systems.

(47) ***The principle and method of the location-based information service system*** (Sun Qinghui, Zhai Zhanqiang, Zhong Dawei, Wang Xiaoli, Shao Shixin. Xi'an Map Publishing House, Xi'an, February 2009)

The book systematically introduces the basic concepts and system architecture of location-based information service systems and the data engineering of system construction, system navigation and route planning, system information service, system positioning system, application of wireless communication technology in LBSIS systems, and so on. The book consists of 8 chapters: an introduction; system architecture; system data engineering; data organization and management; electronic map applications; route planning and navigation; system application technology; and system information services.

11.7 Typical GIS Software Products

11.7.1 Platform GIS System

1. **ViewGIS**

ViewGIS is a set of GIS software developed by Beijing Credit Electronic Technology Development Company, Chinese Academy of Forestry Science, and ViewGIS has the functions of geographic data collection, processing, management, analysis, query and map output. This software can adapt to the domestic geographic information management level, is easy to learn and use and meet the requirements of geographic data processing. ViewGIS fully considers the status of forestry production units at the grassroots level using a low-cost hardware configuration. All Chinese interface displays, with a large number of commonly used symbols and legends for forestry production, simple operation, and complete functions, meet the basic needs of forestry production.

2. **CityStar**

CityStar is a series of domestic GIS tool software programs developed by Peking University, and CityStar has an integrated vector/grid spatial querying and analysis subsystem, interactive editing subsystem, thematic mapping subsystem, digital terrain elevation subsystem, three-dimensional analysis subsystem, RS image processing subsystem, multimedia visualization developing platform and other modules, with powerful and intuitive functions. Citystar also has secondary developmental environments, and its GISOCX controls encapsulate TGISOCX visual controls, which are mainly used for layer display and query control. This software contains hundreds of function call interfaces (methods) so that users can use VB, VC and other common development languages to call these functions, as well as to obtain and display data and implement spatial analysis and query operations, and then, users can conveniently construct and build their own applications.

3. **MapGIS**

MapGIS is a GIS developed by Wuhan ZonDy Information Engineering Co., Ltd., Information Engineering Institute of the China University of Geosciences. MapGIS is an excellent desktop GIS software, a GIS platform with a vector data structure. The main advantages of MapGIS are digitization input, editing and topological integration of spatial data; powerful mapping capabilities, including the production of various thematic map symbols; and the basic functions and analysis functions of GIS. MapGIS is mainly used for the management of 2D or 3D geospatial data, which is managed by layer. MapGIS can organize geographic content as a separate layer according to feature data or combine several different types of feature data to form a layer, so it is convenient for data modification, extraction and updating. MapGIS uses a spatial database engine (MapGIS-SDE) to store and manage spatial and non-spatial data in a single way. MapGIS provides users with secondary development interfaces, as well as API functions, ActiveX controls, MFC class libraries and other development methods; additionally, VC, VB, Dephi and other development tools can be used with MapGIS.

4. **SuperMap**

SuperMap GIS is a large GIS platform software developed by Beijing SuperMap Geographic Information Technology Co., Ltd. SuperMap provides powerful functions, such as geographic information display, editing and geographic analysis, and it has a complete range of products, which can be used to develop fixed large-scale applications or mobile applications based on mobile terminals, and a combination can be used. SuperMap provides a wide range of products, including spatial data management, data acquisition, data processing, large-scale application system development, geospatial information publishing and mobile/embedded application development. SuperMap has many advantages, such as seamless integration of the system, diversified development language, good expansibility, powerful scalability and popular visual programming style, with strong vitality. In addition, SuperMap has powerful capabilities to support massive data management, and not only does SuperMap have the management mechanism of massive vector data but it also applies a large relational database to manage massive spatial data and has advanced massive image data management technology. SuperMap also has strong data compatibility and supports a variety of vector data formats, a variety of image data format conversions and the multisource seamless integration of spatial data. While SuperMap Objects is an all-component GIS development platform integrating mapping and GIS functions as one, it provides a powerful GIS function with ActiveX control, which is open and easy to integrate.

5. **TopMap**

TopMap is GIS software with full intellectual property rights, which was independently developed by Beijing Huitu Information Technology Co., Ltd. and includes all-component GIS development platforms, professional GIS products, and a GIS-related GPS. TopMap products are powerful, easy to learn and easy to use, so users can understand and grasp the product function in a short period of time and quickly invest in GIS project development and construction; it is an ideal choice for GIS application developers, spatial data providers and electronic map publishers. TopMap's series of products have a unified core structure and similar interface style, and the products complement each other, which can provide all kinds of GIS functions, such as business data collation, spatial database management, component development, spatial information publishing, and mobile and embedded GIS applications at the various stages of building GIS projects.

6. **VRMap**

VRMap is a three-dimensional GIS software platform developed by Beijing Lingtu Software Technology Co. Ltd., which provides a comprehensive solution from the underlying engine to the professional application in the field of 3D GISs and virtual reality, with many key technologies, such as massive amounts of data processing, advanced simulation, cross platform communication, database drivers and further development support. VRMap uses a new core technology to build a massive data engine and achieves the perfect unification of large-scale application and fine-grained application. At present, VRMap products have been applied in many fields of thematic analysis and simulation, including digital city, military operations command, electronic sand table and terrain simulation, intelligent building, real estate display, water conservancy and natural disasters, remote sensing mapping and land management, environmental protection, meteorology, geology, petrochemical, telecommunications base station management, and so on.

7. **GeoStar**

GeoStar is a component GIS launched by Wuhan GeoStar Company. It consists of an OLE control GeoMap and nearly 20 OLE automation objects. Among them, GeoStar control provides the methods and properties for GIS data manipulation and the visual interface of GIS data, while automation objects help users to achieve efficient organization and access to GIS data. The application systems developed based on GeoStar can achieve the following functions: basic operations, data organization and maintenance, visual operation, query, edit, image (library) overlay, print output, buffer analysis, and so on.

11.7.2 Thematic GIS System

1. **Digital Fujian**

"Digital Fujian" refers to information-based Fujian, that is, a system of digital, network, visual and intelligent information integration and application for the Fujian Province. This system processes the information from the province's various departments, industries and fields through digital means and computers and integrates and utilizes all kinds of information resources to the greatest extent to provide many information services quickly, completely and conveniently, as well as realizing the national economy and social informatization for the Fujian Province. The main means of information acquisitions are "3S" technologies (i.e., remote sensing, GPS and GIS). Through the application of GIS, the information related to the development of various industries is organized to form an easy-to-use information system that meets the needs of macro decision-making and overall supports decision-making.

2. **GridGIS Mobile**

GridGIS Mobile is an embedded GIS software platform launched by NanJing Creable Co., Ltd. GridGIS Mobile can provide embedded GIS development, project customization, program consulting and other services. GridGIS Mobile has many functions, superior performance, high stability, and high scalability, and it can meet the needs of various industry applications. GridGIS Mobile has been widely used in highway inspections, public security, firefighting, city emergencies, environmental protection, traffic, power, agriculture, GPS monitoring, urban management, data collection and other industries.

3. **Enterprise GPS vehicle command, scheduling and management platform**

The platform was developed by Beijing Legendsoft Information Technology Co., Ltd. It has many functions, such as real-time location tracking, historical track recording and replay, alarm target monitoring, vehicle sound monitoring, inner car image shooting, vehicle anti-theft setting, speed management, driving zone restrictions, command and schedule SMS, vehicle and motor checking, hands-free communication system settings in cars, custom tagging locations and routes, optional two point distance measurements and user information management. This platform can completely overcome the inconvenience caused by the C/S structure of the GPS system and standalone GPS system and further enhance the practical performance of the system. No matter where vehicles are distributed, this platform will know the vehicles' situation and issue instructions to vehicles directly. This system allows customers to be 'sitting within a command tent and devise strategies that will assure victory a thousand miles away' (quotation by SUN Tzi). The system can also provide a flexible configuration according to the actual needs and characteristics of users and realize purchasing on demand. The system has good scalability, reliability, security and stability.

4. **GPS vehicle locating and monitoring system**

The GPS vehicle locating and monitoring system was developed by Beijing Huitu Information Technology Co., Ltd. based on TopMap ActiveX components and TopGPS controls. Based on the Beijing city map, combined with the GPS positioning device, the system can accurately record and display the current driving vehicle location and driving direction angle, vehicle speed and travel distance. The system can draw the trajectory of the vehicle accurately and can record and display the distance and relative direction angle to a place (destination point) in real time. The system has basic GIS functions, which can control map layers, regularly maintain the database and save and print maps. The destination point can be generated by clicking on the mouse or by entering the coordinates of the input point directly. The system has the functions of vehicle track replay and playback time interval selection. During playback, you can see the actual running route, the speed, the position information and the distance of the vehicle. You can pause, continue, or display and hide tracks during playback.

5. **Guangzhou marine information management and control platform**

The platform was launched by the Guangzhou Marine Resources and Environment Monitoring Center (the first construction unit of 'China's Digital Ocean') and Guangdong Lantu Information Technology Co., Ltd. Developed based on SuperMap software, it aims to use information technology to support management mode innovation and effectively improve the management level of the Digital Ocean Construction project, help Guangzhou City Marine and Fisheries Bureau better monitor and control maritime space, realize the integration of resources and management collaboration, and give full play to the management and monitoring advantages of marine administrative authorities in the fields of illegal use of the sea, illegal reclamation, sea use dispute resolving, sea use planning and island management.

6. **Seismic hazard mitigation GIS**

This GIS is an information system for GIS query, analysis and management, which was developed by Beijing Huitu Information Technology Co., Ltd. based on TopMap ActiveX. The system will be able to provide a powerful network analysis function for urban earthquake prevention and disaster relief; it provides optimized disaster relief routes that are shown on the map. This GIS will mark the location of each seismic zone on the map and analyze the terrain in each area of the seismic zone, such as mountains and sea. It also analyzes the regional economic development, population distribution, building characteristics and seismic resistance capacity, industry and enterprise distribution, to find the weak link in the earthquake resistance behaviors of buildings and structures, thus developing earthquake countermeasures and disaster and disaster mitigation plans. By using the historical earthquake information stored in the database and combined with the recently measured geographic information data, the potential, location and magnitude of earthquakes are predicted; in conjunction with local economic conditions, a rapid and predictable assessment of possible disaster losses, such as possible economic losses and casualties, can be made. By using GIS technology, the accurate location of earthquakes will be determined quickly and accurately, and the range of influence will be analyzed. An overlay analysis based on local thematic map layers will be performed to determine the best route for relief supplies and rescue personnel to enter. Then, plans for post disaster reconstruction will be developed according to the degree of disaster and the distribution of the victims in various areas. In collaboration with the Lanzhou Seismological Research Institute of China Earthquake Administration, a joint study and analysis of earthquake disaster and water conservancy, flood control, disaster relief and dam failure analysis can be conducted.

7. **Municipal public geographic information integrated system**

The municipal public geographic information integrated system of Changzhou city was built and developed by the Changzhou City Construction Bureau and Wuhan ZonDy Information Engineering Co., Ltd. Based on a review by an expert team organized by the Ministry of Construction, the system was identified as "city planning developments, management and service engineering", which was funded by the National High-Tech Program for the Tenth Five-Year Plan period and a comprehensive demonstration project of urban digital demonstration engineering applications. The goal of the project is to establish four professional pipeline subsystems, including tap water, drainage, street lights and gas, and one integrated pipeline system, namely, the 4 + 1 mode. The integrated pipeline system is an organic integration of the four subsystems. In March 2005, the project was checked and accepted by the Ministry of Construction. In July 2006, the overall acceptance of the project was completed. After completion of the system, a municipal public GIS system covering the urban area (1864 km^2) will be built within 3–5 years, which mainly includes the integrated pipe network system of the Changzhou City Construction Bureau, pipe network data update subsystem, Changzhou water supply pipe network subsystem, Changzhou drainage pipe network subsystem, Changzhou street light management subsystem, and Changzhou gas pipeline network subsystem.

11.7.3 Public Application GIS

1. **MapABC full-featured map**

The MapABC full-featured map is an electronic map for internet applications, which allows users to embed electronic maps in their own websites with simple code without the need to write large amounts of complex code, such as the map API. An all-featured map lets users create electronic maps of their own websites by clicking the mouse. MapABC is simple and easy to use; you can generate map channels in your website in a few minutes; it is maintenance free, with automatic upgrades and an open source code; finally, MapABC is permanently free. It can be used to search all around for a location of interest, inquire about bus travel and transfer routes, inquire about self-driving travel routes, display current traffic information and load the user's own data, edit a user's own data and make queries, as well as other functions.

2. **Mapbar: a Map bar**

Mapbar was launched by Beijing Tuweixian Science and Technology Co, Ltd., a high-tech enterprise founded by Chinese overseas students and certified by the Zhongguancun Management Committee. Additionally, a famous American venture capital investor IDG has invested in Mapbar. The company focused on location-based services and adopted the most advanced map information engine technology with independent intellectual property rights to build China's largest map service portal, Mapbar provides the most comprehensive map yellow pages, driving routes, bus routes and other inquiry services. The map service covers more than 200 cities throughout the country. It has a characteristic map that displays proprietary interest point (POI) information (such as maps of banks in Beijing) and other community and personal services, such as map clubs, map blogs, and My maps). Mapbar map products have the characteristics of high-speed online browsing, high-speed queries, massive amounts of data, accurate routes, and so on. Mapbar is the best tool for mass travel, life information searches and enterprise address inquiries. At present, the company mainly provides map searching and content cooperation services for major portals throughout the country, such as Sina, NetEase, YAHOO 3721, 21cn, Tencent, eLong, Ctrip, Alibaba, and Chinavnet, as well as nearly one thousand other websites.

3. **Baidu maps**

Baidu maps (http://map.baidu.com) provide a powerful map search service. The Baidu map search has only one input box and a user-friendly interface, while the other map search has two. Baidu maps have distinctive Chinese characteristics; you can type pinyin or prescriptive code as search keywords, such as "Tiananmen", or enter "KFC" to represent Kentucky Fried Chicken. Even if you occasionally write the wrong words, Baidu maps will automatically search for the prompt. You can enter Chinese natural language, such as "from Xizhimen to Dongzhimen", to find your route. Baidu maps also provide a very detailed "driving route" for driver reference, and the information of left turn, right turn, and straight line, for example, is clearly marked, making these maps capable of providing correct navigational information for vehicle travel.

4. **SuperMap public service system**

Based on many years of experience in developing and constructing government GIS systems and public service systems, Beijing SuperMap Geographic Information Technology Co., Ltd. summarized a network electronic map public service solution with high feasibility, a network, expansibility and integration. Based on the SuperMap IS network GIS platform, GIS professional application middleware and related application development templates have been formed and widely used for the urban construction, landscape greening, public security, transportation, business travel and real estate information service industries, which provide rapid, embedded, integrated and high stability electronic map application services for end users and secondary development dealers. Major application areas of the system include urban integrated GIS, public security and police GIS, tourism GIS, real estate information inquiries and electronic map services.

References

An, Y., Bian, F., & Guan J. (2006). Design and comparison of geo-ontology in GIS. *Geomatics and Information Science of Wuhan University, 31*(12), 1108–1111.

Bao, S., & Yu, Y. (2000). Geo-data shared and interoperable technology. *Engineering of Surveying and Mapping, 9*(4), 32–36.

Bi, J., Wang, L., & Chi T. (2004). Research on the geographic services based on the web service. *Computer Science, 31*(1), 69–71.

Bian Fuling. (2004). Higher GIS education in China: Progress, characteristics and exploration. *Geomatics World, 2*(5), 16–18.

Bian Fuling. (2007). Higher GIS education in China: Issues, innovations and developments. *Geomatics World, 5*(2), 7–11.

Bian, X., Hu, R., & Yu, Z. (2000). Study on layered architecture model for distributed GIS. *Geomatics and Information Science of Wuhan University, 25*(5), 443–448.

Cao, Y., & Wu, W. (2001). Preliminary study on loess plateau with GIS. *Quaternary Sciences, 21*(2), 108–113.

Chang, Y. (2006). *Research and implementation on the interoperability of spatial data based on GML schema and CNSDTF*. Master Thesis of Nanjing University of Aeronautics & Astronautics.

Chen, A., & Li, Q. (2002). Geospatial information sharing methods. *Journal of Tsinghua Univesity (Sci &Tech), 42*(10), 1405–1409.

Chen, C. S., & He, J. (1999). Study of GIS conceptual model design for GIS data sharing. *Journal of Remote Sensing, 3*(3), 230–235.

Chen, C., & Zhang, C. (2000). The essence of GIS semantics sharing and its realization. *Science of Surveying and Mapping, 25*(1), 29–33.

Chen, D., & Chen, W. (2002). Forest management optimization and subsidiary decision-making system based on GIS. *Central South Forest Inventory and Planning, 21*(3), 44–47.

Chen, G. (1997). A virtual terrain simulation, based on the map data and according with the regular pattern of vision. *Acta Geodaetica et Cartographica Sinica, 26*(3), 241–246.

Chen, H., Yang, Y., & Li, W. (2006). Research and implementation of irid GIS service system.*Computer Engineering, 32*(13), 260–262.

Chen, J., Shi, P. (1995). The preliminary study of design of temporal GIS based on spatiotemporal data. *Remote Sensing of Environment China, 10*(2), 143–151.

Chen, J., Qiu, Z., & Sun, Y. (1991). Urban general planning information engineering based on remote sensing and GIS technology. *Acta Geodaetica et Cartographica Sinica, 20*(2), 125–130.

Chen, J., & Sun, Y. (1989). Experiment of using GIS technology to assist urban master planning. *Remote Sensing Information,* (3), 2–3.

Chen, N. (2003). *Distributed geographic information service based on J2EE*. Doctoral Dissertation of Wuhan University.

Chen, S. (2001). A new generation of geographic information system (Introduction). *Journal of Image and Graphics, 6*(9), 1–2.

Chen, S., & Chen, X. (2004). The cognition and practice of geo-information science. *Geo-Information Science, 6*(1), 4–10.
Chen, S., He, J., & Cheng, J. (1997). Basic research of geographic information system. *Geo-Information Science*, (3), 13–22.
Chen, S., Yue, T., & Li, H. (2000). Studies on geo-informatic Tupu and its application. *Geographical Research, 19*(4), 337–343.
Chen, W., Hua, Y., & Zhang, S. (2008). Tobacco precision fertilization formula information system based on GIS. *Journal of Zhengzhou Institute of Surveying and Mapping, 25*(6), 455–458.
Chen, X., Wan, Q., & Lan, H. (2004). Constitution and application of Jingjiang River flood forecasting information issuance system. *Journal of Yangtze River Scientific Research Institute, 21*(3), 65–66.
Cheng, J. (2008). The latest development trend of earth information science and technology. *Geo-Information Science, 10*(2), 280–281.
Chi, W., & Wu, X. (2008). Research on service-oriented architecture building development technology. *Application Research of Computers, 25*(6), 1717–1719.
Cui, H., & Hua, Y. (2008). Research on situation display system based on HLA. *Journal of Zhengzhou Institute of Surveying and Mapping, 25*(6), 417–419.
Ding, J. (1991). Overlay analysis method of raster graphics data and its application in town planning. *Journal of Wuhan Technical University of Surveying and Mapping, 16*(3), 59–69.
Du, Q. (2004). The micro-linguistic conceptual model of spatial information. *Geomatics World*, 2(6), 5–8.
Fang, Y. (2004). A decade of the development of china's GIS industry. *Geomatics World*, 2(5), 36–39.
Gao, J. (2000). Visualization in Geo-spatial data. *Engineering of Surveying and Mapping, 9*(3), 1–7.
Gong, J., & Lin, H. (1999). Perspective on Geo-Visualization. *Journal of Remote Sensing, 3*(3), 236–244.
Gong, J. (1992). Research on integrated data structure of vector raster in GIS. *Acta Geodaetica et Cartographica Sinica, 21*(4), 259–266.
Gong, J. (2002a). Development of GIS technology of china. *Engineering of Surveying and Mapping, 11*(2), 3–7.
Gong, J. (2002b). The development trends of the contemporary GIS. *Northeast Surveying and Mapping, 25*(4), 10–15.
Gong, J. (2003). Sharing and interoperation of spatial information resources. *Land and Resources Information* (5), 15–21.
Gong, J., & Gao, W. (2006). Sharing and interoperability of geospatial information. *Geomatics World, 4*(3), 18–27.
Gong, J., Li, D., & Bian, F. L. (1994). Data organization and processing method of GIS. *Bulletin of Surveying and Mapping*, (1), 28–37.
Gong, J., Yuan, X., & Chen, L. (1998). Cross platform distributed geographic information organization and processing. *Journal of Wuhan Technical University of Surveying and Mapping, 23*(4), 364–369.
Guo, X., Hua, Y., & Bai, X. (2008). The design and realization of public edition of china land boundary information system. *Bulletin of Surveying and Mapping*, (8), 56–59.
Guo, X., & Hua, Y. (2007). A GIS-based communication resources management application. *Science of Surveying and Mapping, 32*(1), 19–21.
He, L., & Gan, H. (1991). An improved algorithm for automatic conversion of missing data and raster data in microcomputer based Geographic Information System. *Map*, (3), 26–28.
Hu, P., Ai, Z., & Guo, Q. (1989). Geo information basic database system-GIS-the basic structure of dB3. *Journal of Wuhan Technical University of Surveying and Mapping, 14*(1), 48–56.
Hu, P., & Yang, C. (2002). Basic theoretical problem of GIS—space view of map algebra. *Geomatics and Information Science of Wuhan University, 27*(6), 616–621.

Hu, P., Yang, C., & Li, G. (2000). Bottle necks in GIS development, theories of GIS and the practice tool software, thee of million image GIS. *Journal of Wuhan Technical University of Surveying and Mapping, 25*(3), 212–215.

Hua, Y. (1997). An expert system for map design is established in GIS. *Map*, (1), 18–19.

Hua, Y. (2000). Boundary information system design. *Journal of Institute of Surveying and Mapping, 17*(1), 46–48.

Huang, X., et al. (1990). *Introduction to geographic information systems*. Beijing: Higher Education Press.

Huang, D., Yu, Q., & Le, J. (2008). WebGIS architecture based on mobile agent. *Computer Applications and Software, 25*(3), 56–58.

Huang, H., Xia, B., & Zhao, B. (2006). Research of generic architecture for GIS applications. *Journal of Geo-Information Science, 8*(4), 31–34.

Huang, M., Du, Q., & Wu, Y. (2004). A preliminary study on geographic ontology and its application. *Geography and Geo-Information Science, 20*(4), 1–5.

Huang, X., & Chen, Y. (1989). The tendencies of GIS development. *Acta Geographica Sinica, 44*(2), 230–236.

Huang, X., Ni, S., & Xu, S. (1993). The study of geographic information system supporting regional land use decision making. *Acta Geographica Sinica, 48*(2), 114–121.

Huang, Y., & Yang, J. (1999). Study on water saving irrigation regionalization based on GIS Technology. *China Rural Water and Hydropower*, (12), 10–12.

Jia, B., & Wang, X. (1990). Application of GIS technology to the evaluation of regional urban development land: A case study of Hainan Province. *Geographical Research, 9*(4), 86–93.

Jia, P., & Ma, J. (2007). Research on technical method of hiding watermark in GIS spatial data for sharing security. *Bulletin of Surveying and Mapping*, (2), 70–72.

Jia, Z., Liu, Y., & Zhang, T. (2004). Discussion on integration of GIS and visual interactive spatial data mining. *Journal of Nanjing Normal University (Engineering and Technology), 4*(4), 37–42.

Jiang, N., & Hua, Y. (1997). Application of remote sensing technology in military geographic information system. *Remote Sensing Information*, (2), 25–27.

Jiang, N., Xia, L., & Xue, B. (2006). Research on the multi-vision pattern of map in spatial data of GIS. *Journal of Geomatics Science and Technology, 23*(3), 157–160.

Jiang, Y., & Fang, Y. (2004). Spatial data sharing and integration platform based on web service. *Geography and Geo-Information Science, 20*(5), 1–5.

Lan, X., Hu, B., & Ma, L. (2003). Agent-based multi-layer distributed GIS. *Journal of Hohai University (Natural Sciences), 31*(5), 581–584.

Li, B., & Lin, J. (2000). Application of geographical information system and principal components analysis in analysis of long term climate monitoring data. *Advances in Earth Science, 15*(5), 509–515.

Li, D. (1997a). On the definition, theory and key technologies of RS, GPS and GIS integration. *Journal of Remote Sensing, 1*(1), 64–68.

Li, D. (1997b). Some thoughts about geographic information theory. *Journal of Wuhan Technical University of Surveying and Mapping, 22*(2), 93–95.

Li, D. (2003). The main trend of GIS Technology. *China Surveying and Mapping News*. The 003 Edition.

Li, D. (2004). On large scale surveying and mapping of earth and earth integration: geo-spatial Informatics. *Science of Surveying and Mapping, 29*(3), 1–2.

Li, D. (2006). Mobile mapping technology and its applications. *Geospatial Information, 4*(4), 1–5.

Li, D. (2007). On geomatics in multi-discipline integration. *Acta Geodaetica et Cartographica Sinica, 36*(4), 363–365.

Li, D., & Li, J. (2002). Establishment of the spatial data infrastructure of china. *Bulletin of Surveying and Mapping*, (11) & (12), 4–7, 1–4.

Li, D., Li, Q., & Yang, B. (2008). Techniques of GIS, GPS and RS for the development of intelligent transportation. *Geomatics and Information Science of Wuhan University, 33*(4), 331–336.

Li, F., & Liu, J. (2002). Primary discussion on GIS research and establishment of forest resource in Henan province. *Journal of Henan Forestry Science and Technology, 22*(3), 13–16.

Li, H., & Cui, W. (1999). Visualization of spatio-temporal multi-dimension data in GIS. *Journal of Remote Sensing, 3*(2), 157–163.

Li, H., Li, Q., & Duan, J. (2008). Research on GIS development methods based on ontology. *Computer Applications and Software, 25*(11), 69–70.

Li, J., Chen, W., & Bai, Z. (2003). Global scientific challenges: Earthquake prediction is likely to break through in China first. *Earth Science Frontiers, 10*(1), 89–91.

Li, K., Hu, G., & Duan, Q. (2009). Application of characteristic analysis in metallogenic prediction based on morpas: A case study of western gejiu tin-polymetallic deposit, yunnan province. *Geological Science and Technology Information, 28*(4), 65–70.

Li, X., & Wu, C. (2001). Component based architecture design for distributed geological and mineral resource point source information system. *Scientific and Technological Management of Land and Resources, 18*(6), 41–45.

Li, X. (2004). General review of GIS standardization. *Geomatics World, 2*(5), 11–15.

Li, Y. (2004). Review of market development of china's GIS. *Geomatics World, 2*(5), 40–42.

Li, Y., Hua, Y., & Wu, S. (2003). The design and realization of PowerComMap. *Journal of Geomatics Science and Technology, 20*(4), 297–300.

Liao, K. (2007). *Introduction to earth information science.* Science Press.

Ling, Y., Chen, Y., & Wang, Y. (2005). Cognitive basis of adaptive user interface for cartographic visualization system. *Acta Geodaetica et Cartographica Sinica, 34*(3), 277–282.

Liu, J., Zhong, E., & Zhuang, D. (2003). Development and application of national SARS disease controlling and pre-warning information system. *Journal of Remote Sensing, 7*(5), 337–344.

Liu, N., & Liu, R. (2004). Strategy of distributed WebGIS in server group and its implementation. *Journal of Image and Graphics, 9*(1), 99–104.

Liu, Q., & Cheng, B. (2007). Study on resource and environment spatial information grid for southwestern China. *Science of Surveying and Mapping, 32*(5), 138–140.

Liu, Z. (1993). The geographic information system based on the evaluation of land use and soil and water conservation planning: Farm in the North Island of New Zealand tapa card. *Journal of Soil and Water Conservation, 7*(2), 89–96.

Lu, F., & Li, X. (2001). Feature-based temporal-spatial data modeling: State of the art and problem discussion. *Journal of Image and Graphics, 6*(9), 830–835.

Lu, X. (2004). The research on spatial cognitive mode. *Geomatics World, 2*(6), 9–13.

Lu, X., & Cheng, J. (1998). Connotation analysis of geographical cognitive theory. *Acta Geographica Sinica, 53*(2), 132–140.

Lu, X., Li, H., & Chen, S. (2000). The initial research on the geographical spatial-temporal system spatial-temperal hierarchy. *Journal of Geo-Information Science* (1), 64–70.

Luo, Y. (2001). The study on agent-based distributed geographic information system. *Acta Geodaetica et Cartographica Sinica, 30*(4), 368–368.

Lv, G., Sun, Y., & Wang, W. (2004). Study of urban planning electronic government affairs system. *Modern Surveying and Mapping, 27*(5), 6–10.

Ma, S. (1999). Study on the application of county level Forest Resources Geographic Information System. *Shaanxi Forest Science and Technology* (1), 36–39.

Pei, T., & Zhou, C. (2001). Review on the proceedings of spatial data mining research. *Journal of Image and Graphics, 6*(9), 854–860.

Ren, F., Du, Q., & Cai, Z. (2003). The application of electronic map in urban public information service. *Urban Geotechnical Investigation & Surveying* (4), 42–45.

Ren, J., & Ren, F. (1990). Evaluation of regional environmental quality supported by GIS. *Environmental Science, 11*(6), 62–67.

Shan, G., Cheng, T., & Yang, P. (2007). Study on spatio-temporal data model for moving object trajectories. *Science of Surveying and Mapping, 32*(3), 33–35.

Shen, Z., Luo, J., & Chen, Q. (2004). System design of grid GIS and its application on remote sensing image processing. *Application Research of Computers, 21*(8), 122–125.

Shi, T. G., Zheng, Z., & Wang, Z. (2008). Land suitability evaluation of tea plant in rizhao southeast shandong province based on GIS. *Journal of Mountain Science, 26*(5), 560–564.

Shu, R. (1985). Discussion on the establishment of land base information system. *Bulletin of Surveying and Mapping* (3), 6–9.

Song, G., Zhong, E., & Liu, J. (2000). A study on seamless integration of multi-sources spatial-data (SIMS). *Progress in Geography, 19*(2), 110–115.

Song, G. (2006). Theoretical research framework of geographical information sharing. *Journal of Geomatics Science and Technology, 23*(6), 404–407.

Song, G., & Jia, F. (2002). The research to the method of geo-spatial data sharing. *Journal of Geomatics Science and Technology, 19*(2), 134–136.

Song, W., Wang, J., & Guo, J. (2006). An object-oriented spatial-temporal data model. *Journal of Geomatics Science and Technology, 23*(4), 235–238.

Su, B., Zhao, J., & Sun, Q. (2008). Study on the establishment for city land and resources information system. *Areal Research and Development, 27*(1), 105–107.

Su, F., Zhou, C., & Shao, Q. (2002). Development and application of marine fishery geographic information system. *Journal of Fisheries of China, 26*(2), 169–174.

Sun, Q., Luo, J., & Zhao, J. (2005). Strategy of data transmission in grid-GIS. *Journal of Geo-Information Science, 7*(1), 65–70.

Tang, G., & Qin, H. (2002). An application of using GIS Technology to return hillside cultivated land for forestry and grassland in loess plateau region. *Bulletin of Soil and Water Conservation, 22*(5), 46–50.

Wang, J. (1997). The theoretical basis and system framework of geographic information system science. *Journal of Geomatics Science and Technology, 14*(2), 122–127.

Wang, J. (2004). The achievement and mission of modern map science and geographic information engineering science and technology. *Journal of Geomatics Science and Technology, 21*(4), 235–240.

Wang, J. (2006). Independent innovation and the development of GIS in China. *China Surveying and Mapping News*, 2006.12.20.

Wang, J. (2007). Research on the progress of cartography and geographic information engineering in china. *Bulletin of Surveying and Mapping*, (5), 1–6.

Wang, J. (2009). Development of geographic information system and developing geographic information system. *Engineering Science, 11*(2), 10–16.

Wang, J., Wei, H., & Cheng, Y. (2004). The research and development of spatio-temporal GIS. *Hydrographic Surveying and Charting, 24*(5), 1–4.

Wang, J., Zhu, Y., & Wu, M. (2006). On grid and grid geographic information system. *Journal of Geomatics Science and Technology, 23*(1), 1–7.

Wang, J., Li, L., & Ge, Y. (2000). A theoretic framework for spatial analysis. *Acta Geographica Sinica, 55*(1), 92–103.

Wang, J., & Bian, F. (2005). An architecture model of grid GIS–the quadruple geo-spatial grids. *Geomatics and Information Science of Wuhan University, 30*(1), 73–76.

Wang, J., Zhang, C., & Liu, Z. (2007). Analysis on GIS architecture development driven by the computing models. *Geography and Geo-Information Science, 23*(5), 8–14.

Wang, L., Li, Q., & Yu, W. (2004). WebService and spatial geographic information interoperability. *Science of Surveying and Mapping, 29*(1), 38–41.

Wang, P., & Liu, L. (2001). XML new approach to share special information and co-operation in WebGIS. *Application Research of Computers, 18*(12), 142–144.

Wang, Y. (2002). Research on distributed network geographic information system. *Doctoral Dissertation of Institute of remote sensing applications, Chinese Academy of Sciences.*

Wang, Y., Ji, G., & Su, Q. (2005). Design and implementation in GIS of event-driven architecture. *Journal of Xiamen University (Natural Science), 44*(4), 472–474.

Wang, Z., & Hua, Y. (2006). Research on technology of active spatial information service. *Acta Geodaetica et Cartographica Sinica, 35*(4), 379–384.

Wang, Z., Zhang, P., & Hua, Y. (2005). Research and practice of ERP based on GIS. *Journal of Information Engineering University, 6*(1), 99–02.

Wen, X., Peng, S., & She, G. (2005). Current development and application of GIS in forestry of china. *Journal of Nanjing Forestry University (Natural Science Edition), 29*(2), 73–78.

Wu, M., Xie, G., & Yuan, C. (2006). A study on several key issues of GridGIS. *Journal of Geomatics Science and Technology, 23*(6), 415–417.

Wu, J., & Chu, S. (2006). The construction of hubei construction engineering laboratory resources management geographic information system based on GIS. *Resources Environment & Engineering, 20*(3), 301–305.

Wu, L., Tang, D. S., & Liu, Y. (2003). Interoperable and distributed geographical information systems based on web service. *Geography and Geo-Information Science, 19*(4), 28–32.

Wu, S. (2002). Research and practice of distributed GIS component platform. *Doctoral Dissertation of the Information Engineering University.*

Wu, S., & Wang, J. (2001). The evolvement of GIS developing approach. *Journal of Geomatics Science and Technology, 18*(2), 54–56.

Xia, S., & Han, Y. (2004). An improved implementation of shortest path algorithm in GIS. *Bulletin of Surveying and Mapping,* (9), 40–42.

Xing, X., & Xiang, N. (2007). Research of grid computing and grid GIS system structure. *Engineering of Surveying and Mapping, 16*(3), 27–30.

Xiong, S., & Fang, Z. (2002). Several applications of genetic algorithm in geographic information system. *Bulletin of Surveying and Mapping,* (1), 47–49.

Yang, C., & Wang, Y. (2001). Review of the main technologies of WebGIS. *Journal of Image and Graphics, 6*(9), 886–894.

Yang, K., & Chen, J. (1986). Urban development monitoring and municipal decision information system. *Wuhan Surveying and Mapping Technology,* (2), 22–26.

Yang, Q., & Lv, G. (1999). Research on the technology of geographic information data warehouse. *Journal of Image and Graphics, 4*(8), 621–626.

Ye, J. (2008). The development and trend of GIS in China. *China Surveying and Mapping News,* 13. May. 2008.

Yi, H., Cao, H., & Wang, Y. (2002). Data-mining and it's technology in GIS. *Engineering of Surveying and Mapping, 11*(3), 45–47.

Yi, S., Li, Q., & Cheng, J. (2000). Sharing and interoperability of spatial information. *Bulletin of Surveying and Mapping,* (8), 17–19.

Yin, X., Chen, X., & Luo, G. (2007). Research and realization of framework of WebGISVisA based on Web services. *Application Research of Computers, 24*(10), 298–301.

Yu, H., & Wu, L. (2004). OpenGIS reference model(ORM)and application schema for geographic information services. *Geography and Geo-Information Science, 20*(5), 15–19.

Yu, Y., & Bao, S. (2000). New GIS database—GeoDatabase. *Journal of Guangxi Teachers Education University(Natural Science Edition), 17*(4), 65–69.

Yu, Z. (1997). Research on geographic information relevancy (top, bottom): Construction of basic framework of Geographic Information Association. *Earth Information* (1), 24–32 & (2), 32–35.

Yuan, C., Yin, M., & Wang, S. (2004). Probing into establishing village-level management information system of forest resources. *Central South Forest Inventory and Planning, 23*(2), 49–52.

Zhang, H., & Lv, G. (2005). Urban pipeline GIS share based on integrated data model. *Bulletin of Surveying and Mapping,* (8), 57–60.

Zhang, L., Liu, G., & Jiang, T. (2003). The knowledge-based method of GIS shortest path searching. *Science of Surveying and Mapping, 28*(3), 27–29.

Zhang, Q. (2002). Electronic government and GIS. *Surveying and Mapping in Shanxi,* (3), 11–15.

Zhang, W., & Xu, Y. (1995). Object oriented spatial data model. *Journal of Wuhan Technical University of Surveying and Mapping, 20*(1), 18–22.

Zhang, Y., & Ma, Y. (2003). Design of universal compliance production system in GIS. *Journal of Geomatics Science and Technology, 20*(1), 76–78.

Zhang, Y., & Zhao, Y. (2003). The route-planning models in GIS for urban public traffic. *Journal of Geomatics, 28*(4), 29–31.

Zhang, T., Zhou, C., & Shao, Q. (2003). Marine GIS field data model. *Journal of Geo-Information Science, 5*(4), 25–29.

Zhao, B. (2007). Flexible GUI design in GIS. *Science of Surveying and Mapping, 32*(1), 150–152.

Zhao, H.,Fang, J., & Tang, Z. (2006). STL algorithms applied in GIS. *Computer Engineering, 32*(8), 70–72.

Zhao, J., & Gao, B. (2001). The latest representation of spatial information visualization. *Map,* (3), 7–10.

Zhong, E., & He, J. (1992). GIS technology and its application in disaster reduction. *Journal of Disaster Reduction in China, 2*(4), 24–28.

Zhou, C. (1993a). Research on flood disaster assessment information system. *Acta Geographica Sinica, 48*(1), 11–18.

Zhou, C. (1993b). Flood disaster surveillance and evaluation by remote sensing. *Geographical Research, 12*(2), 63–68.

Zhou, C., & Lu, X. (1998). Thinking about the information science of the earth. *Acta Geographica Sinica, 53*(4), 372–380.

Zhou, Y., & Chen, Z. A software architecture of distribute GIS. *Computer Engineering, 27*(9), 37–39.

Zhu, X., & Xu, Y. (1993). Object oriented semantic data model and its application in spatial database. *Journal of Wuhan Technical University of Surveying and Mapping, 18*(4), 76–81.

Chapter 12
Virtual Geographical Environment

Gang Wan, Hui Lin, Qing Zhu, and Yingzhen Liu

12.1 Introduction

The concept of the virtual geographic environment (VGE) was proposed just 10 years, but it has become one of the most dynamic and creative research fields in cartography and has drawn wide attention and discussion. First, the cognitive aspect of the virtual geographical environment is a three-dimensional, dynamic and interactive geographical space, the cognitive method is more in line with human cognitive law, as well as the extension and expansion of the map as a tool for human spatial cognition; second, the emergence and development of the virtual geographical environment has always been closely related to the latest computer technology and has been supported by the latest software and hardware in the field of information, and the development process has been advancing quickly; and third, the integration of various professional application models and geospatial environment models gives the virtual geographical environment vigorous vitality, which has gained a large number of valuable theoretical and technical achievements and has been widely used. However, due to the complexity of the real geographical space, the present research results remain some distance away from a complete solution to these practical problems.

Because of the importance and complexity of virtual geographical environment research, the research difficulties in this field have been the focus of academic circles at home and abroad, and a large number of valuable theoretical and technical achievements have been obtained. To date, more than 10 universities and research institutions in China have offered courses or enrolled postgraduates in this research direction. The State 863 Project began to support relevant studies during the Tenth Five Years

G. Wan (✉) · Y. Liu
Information Engineering University, Zhengzhou, China

H. Lin
The Chinese University of Hong Kong, Hong Kong, China

Q. Zhu
Southwest Jiaotong University, Sichuan, China

J. Wang and F. Wu (eds.), *Advances in Cartography and Geographic Information Engineering*, https://doi.org/10.1007/978-981-16-0614-4_12

Plan period. In 2006, the Ministry of Education approved the establishment of the Key Laboratory of Virtual Geographical Environment at Nanjing Normal University. The studies of Chinese scholars on the virtual geographical environment not only appeared in domestic academic journals but also gradually appeared at famous international academic conferences and in periodicals, such as PE&RS and IEEE. The symposium on the virtual geographical environment initiated and promoted by Chinese scholars has begun to influence the international academic community. Chinese scholars have been invited to communicate with relevant academic institutions in both European and American countries.

As more scholars join the research field of virtual geographic environments, many differences in the ways of understanding and solving problems are apparent, and many meaningful discussions have been aroused, which leads people to think from many aspects, such as the concept of virtual geographical environment, its difference from traditional maps and GIS, and the implementation methods, key technologies and applications of the virtual geographical environment. These studies have greatly accelerated the research progress of the virtual geographic environment.

In this chapter, the research progress in the virtual geographic environment will be systematically reviewed, the existing achievements will be summarized, the existing problems will be analyzed, and the direction of further research will be put forward to promote the deepening of research in this field.

12.2 Development Process

The virtual geographical environment is generated under the common influences of information science, geography science, Earth system science, computer science and practical application requirements; this environment is the result of scholars' active exploration and thinking in the fields of cartography and geography. The general background of the emergence of the virtual geographical environment involves the formation of the Earth system science view and the demand of the modern management science view: from the perspective of Earth system science, we should seek a platform of knowledge exchange (for recording and analyzing geoscience phenomena, simulating the Earth system and revealing the laws of macroscopic geoscience); modern management calls for public participation, and the establishment of contingency plans (which are needed to accumulate knowledge and optimize the exchange of programs for peacetime planning and training, as well as wartime and emergency response capacity-building) (Lin 2007).

From General MA Yuan's mountain valley made with rice (The History of the Late Han Dynasty, 25–220 AD) to the modern commander's sand table, there has been a long history of three-dimensional descriptions of the objective geographical environment for human beings. Maps based on traditional techniques are flat (commonly referred to as 2-dimensional maps) and still lack intuition and imagination when using these maps; in particular, a desire to make a stereoscopic map arises when a contour is used to represent topographic relief and is difficult to understand. Among them, the three-dimensional display of terrain has been a long-term hot topic in cartography and has been well explored, from the old hachuring and hill shading methods to the modern computer-aided terrain stereo representation, and there is a wealth of description and experience in the technical methods of stereoscopic drawing and the perception effect of stereoscopic vision. Such results are commonly referred to as 2.5-dimensional maps, which are produced by techniques that aid the reader to produce mental stereoscopic vision.

In the 1990s, computer 3-D image animation technology provided tools for making 3-D maps; it was not difficult to use binocular stereoscopic observation equipment to produce 3-D maps with physiological stereoscopic vision, and these maps were used in many places where maps were used. Thus, cartography has found a new growth point in the field of virtual reality technology.

12.2.1 The Phase of Emerging Technical Explorations

The initial exploration in China focused on the visualization and simulation of the geographical environment based on three-dimensional terrain, such as the virtual battlefield environment simulation proposed by Gao et al. (1996). *The Application of Virtual Reality in the Simulation of Terrain Environment* (Gao et al. 1999) was the first monograph in the field in China. Map experts believe that the simulation of the geographical environment involves the use of virtual reality and other simulation techniques to construct an accessible three-dimensional geographical environment, which is more in line with the characteristics of human spatial cognition, and is a natural and rational extension of the map as a spatial cognitive tool in the digital age, as well as a new spatial cognitive tool supported by digital maps. Simulation of the geographical environment creates a new kind of map that enables users to be immersive and go beyond reality and 'realize' a project ahead of schedule, that is, let the user experience, evaluate and understand by simulating the geographical environment.

In the development of GIS, 3-D GIS has naturally become a research focus. In Li et al. (1997), two kinds of spatial data structures in 3-D GIS, namely, a data structure based on surface representation and a data structure based on volumetric presentation, were studied, and the characteristics and applications of spatial data structures, such as boundary representation (BR), constructive solid geometry (CSG), nonuniform rational B-spline (UNRBS), octree and irregular tetrahedron network (TEN), were analyzed in detail. A 3-D data model based on vector and grid integration was put forward by Gong et al. (1997), where applications in mining and geology were taken as the background, the spatial objects and their links involved in 3-D spatial information systems were deeply analyzed, and several new spatial object types were proposed. In particular, it was pointed out that the vector and raster integrated data structure and object-oriented data model could be used to represent all kinds of 3-D spatial objects. A vector GIS topological relation and its dynamic construction algorithm were proposed by Li (1997), where in reference to the topologic relations of 2-D vector GIS and 3-D geometric construction and considering applications in mining and geology, five group relations were constructed to describe the topologic relations among five geometric elements of node, edge, cycle, surface and body. The above work laid a solid foundation for the design and establishment of a 3-D algorithm for GISs.

Furthermore, GIS technology was fully developed, Web GIS appeared after being combined with the Internet, 3-D GIS appeared to make people feel the stereoscopic effect when they observed terrain, and VR GIS appeared to enable users to simulate the evolution of the future or past. All these factors provided users with a virtual geographical environment, which provided more intense immersion in the real-world environment and enhanced users' perception of the objective geographical environment; finally, virtual, distributed and mobile geo-computing was also beginning to spring up.

12.2.2 The Phase of In-Depth Integration with Geographical Models

Researchers worked to make this new technique able to support the multidimensional representation of spatial environments and multichannel spatial visual, auditory and tactile perceptions, as well as multiuser collaborations; however, researchers were also concerned about another core problem, namely, the geographic application model.

The initial exploration was reflected in the combination of a geospatial database with a geosciences model. For example, Sun (1999) proposed the concept of the Resource and Environment Science Virtual Innovation Environment and established a virtual multidimensional information space generation system for geosciences. Lin (2000) studied tidal wave systems in the East China Sea. Li et al. (2000) proposed and studied digital cities, and other studies were conducted.

The virtual geographic environment (Gong et al. 2001) was the first writing in the study of virtual geographical environments. Based on a series of exploratory studies, the concept of virtual geographic environment was proposed, and the virtual geographical environment was considered to be a new generation of geographical language that evolved from maps and GIS.

Simultaneously, Lin et al. (2005) began to establish a virtual reality system of surveying and mapping; Zhang et al. (2006a), Gong et al. (2002a, b) carried out research on the application of virtual geographic environments in public health, digital watersheds, and collaborative virtual geographic laboratories; Tang et al. (2007), Yang et al. (2007), Zhu et al. (1998) carried out research on the theory of virtual spatiotemporal cognition, the rapid acquisition of spatial 3-D data and its integrated data model, the geographical analysis model and its integration with virtual geographical environment system, and the distributed group collaboration and spatial decision support system. It was the close connection between the virtual geographical environment technology and the geographical application model that showed great vitality and applicability in urban planning, geological surveys, military commands, space flight simulations, and so on.

12.2.3 The Phase of Key Technical Breakthroughs and High-Speed Development

At the beginning of the twenty-first century, although the demand for 3-D virtual environments in various applications was vigorous, and the pursuit of an immersive experience and realistic understanding became powerful driving forces for the development of virtual geographic environments, due to the expensive virtual reality equipment, the lack of software functions, and the shortage of high-resolution remote sensing image data sources and other factors, virtual geographic environment products were always "lonely products" and could not be widely used in various fields.

Until 2005, the United States Google company officially launched satellite image-based map software, known as Google Earth, and Microsoft Corp also launched the online products Live Local and Virtual Earth. A large number of mature virtual geographic environment-related software and systems have emerged both at home and abroad, such as ERDAS 's Virtual GIS, NASA's World Wind, Leica's Leica Virtual Explorer, Skyline's Terra Explorer, and ERSI's ArcGlobe; additionally, Supresoft Company's IMAGES, Geo Company's CCGIS and GeoGlobe, Lingtu Company's VRMap, Aladdin Company's E city and other software platforms became available in China. The emergence of these products has greatly changed the situation where people could not reach the product of the virtual geographical environment; since then, the concept of the virtual geographical environment and products has entered public life.

A series of new theories, techniques and methods for the study of virtual geographical environments have been embodied in these systems and products, which are listed as follows: integrated modeling and management of global 3-D geographic data based on multisource information acquisition and real-time processing; geospatial information grid technology based on distributed storage and computing and geographic information service technology in a wide area grid environment; spatial analysis, network analysis and thematic mapping technology based on the data structure of a complete 3D space model; multiple software and hardware platforms and diversified architectures based on a rich hardware platform and flexible architecture; flexible point of view control based on user demand and a three-dimensional symbol system; and so on.

In summary, the virtual geographical environment as a geographical language that evolved from maps and GIS that integrates multisource information acquisition, spatial data management, virtual reality, network, artificial intelligence and other technologies, and this virtual environment is the multimodal perception, expression, calculation and simulation of the geographical space environment, phenomena and laws. It can be widely used in comprehensive management, integrated publishing, human–computer interaction, collaborative research, design decisions, education, training, tourism, entertainment and so on and provides the methodological basis and technical support for the information construction of the geospatial information infrastructure (digital Earth, digital region, digital city and digital river basin).

Through statistical analyses of papers published in the virtual geographical environment field, we can see that the research institutions engaged in virtual geographical environment mainly include (according to the first author and their units) the Institute of Surveying and Mapping of Zhengzhou Information Engineering University, Wuhan University, Institute of Remote Sensing Applications of Chinese Academy of Sciences, National University of Defense Technology, Beijing University of Aeronautics and Astronautics, Peking University, Northwestern Polytechnical University, China University of Mining and Technology, Chinese Academy of Surveying and Mapping, Southwest Jiao Tong University, and Chinese University of Hong Kong. The main publications in which these studies are published include the Journal of System Simulation, Journal of Wuhan University (Information Science Edition), Journal of Image and Graphics, Journal of Surveying, Mapping Science and Technology, Acta Geodaetica et Cartographica Sinica, Science of Surveying and Mapping, and Geography and Geography Information Science. The peak period of development of the virtual geographic environment and its related technologies occurred in 2004 and 2005, and more papers were published.

12.3 Main Research Achievements

12.3.1 Relevant Theoretical Research

From the perspective of cartography, the virtual geographic environment is a new geographical language; it is closer to users of geographic information and more conducive to public participation in information exchange; it is also the extension and expansion of traditional map functions and a new spatial cognitive tool. The three functions of cartography (cognitive function, simulation function and information load and transmission function) have been inherited in the virtual geographic environments and continue to play a role. The basic theoretical problem of cartography has new features in virtual geographic environments and has made new progress.

1. **The theory of spatial cognition in the virtual geographical environment**

Spatial cognition research mainly includes spatial perception, spatial representation, conceptualization, mental representation of spatial knowledge and spatial reasoning, which involves spatial knowledge acquisition, storage and use. Based on the virtual geographical environment, how do readers complete a series of psychological processes, such as cognition, encoding, storage, memory and decoding for the geographic information? What is the difference and connection between the process and past spatial cognition based on maps? These are the research contents of the basic theory of the virtual geographical environment.

The virtual geographical environment has brought about profound changes in the way people think and work. Gao (2000) believes that visualization is generated in the background of digitization; its role is to exert the imaginal thinking of the human brain under the condition of a human–computer interaction. The map itself is a visual product, and a series of theories and methods have been formed in the course of development, which will naturally become the basis of the visualization technology of geospatial data. Cartography has gained new power because of the visualization method; GIS provides researchers with cognitive tools that combine logical thinking with imaginal thinking due to the support of visualization. According to Lin (2007), the characteristics of development and change are summarized as follows: from a single database core to a dual database core and model library, from abstract to image, from 2-D information to multidimensional information, from 2-D analysis to 3-D analysis, from single channel to multichannel interaction, the role that a human plays from the outside to the inside, and so on.

Gao et al. (2005) studied spatial cognition in the simulation of a virtual terrain environment and considered that the virtual terrain environment simulation, as a new spatial cognitive tool supported by digital maps, required new cognitive theories and modeling methods and required discussions from many aspects, such as spatial cognition, multisensory channels, human–computer interaction, sensory incongruity, performance and evaluation. Wang et al. (2005) argued that the study of geospatial cognition included the location of geographical objects in geographical space and the nature of geographical objects themselves, so cognitive science should be placed in the context of geographical science for special research. In Wan et al. (2008), a comprehensive summary on the formation of a cognitive map was created, the process framework of map spatial cognition was provided, and cognitive experiments based on paper maps, electronic maps, remote sensing images and virtual geographical environment were implemented.

Chen (2003) explored geo-environment simulations and cognitive verification: a case study at the Cudgewa catchment in Victoria, Australia, with the method of cognitive science, the validity of virtual environment technology, the authenticity of the virtual geographical environment and the reliability of environment cognition were tested, which provided a methodological foundation for geoscience research on the virtual geographical environment.

Chen et al. (2000) believed that the linguistic theory of cartography and GIS-semiotics was weakening in the virtual geographic environments. You (2002) proposed from the point of view of system simulation theory that similarity theory could be used to describe the consistency between the virtual geographical environment and the real environment. Lin et al. (2002) believed that the virtual geographic environment was designed to construct a virtual environment that reflects the real world, and whether it could obtain people's trust was dependent on whether it could realistically reflect reality. The above studies agreed that the virtual geographic environment did not seek consistency with reality in terms of appearance but could do its best to make it similar to reality. How can we achieve geometric similarity, temporal similarity, performance similarity, and feature similarity between virtual geographic environments and real environments? To what extent does similarity exist? These are the research areas of the similarity theory of the virtual geographic environment, that is, the question of fidelity and similarity. One of the purposes of virtual geographic environments is to promote simulation studies that are difficult to complete in reality, such as earthquakes, tsunamis and other natural disasters. Considering these geographical environment problems, we can make use of the virtual method, apply the similarity principle and system principle and build a model to study the relevant problems, which is the "virtual" aspect. However, some problems require more realistic input to achieve "reality", such as human behavior or the exact wind speed, temperature, humidity in certain locations, etc. In this case, the actual data can be directly input into the virtual geographical environment through the information acquisition sensor to achieve more realistic results, which is the "real" aspect.

Wan (2008) believed that the 3D image in the virtual terrain environment simulation gave users an immersive feel, with a strong visual impact, which was often regarded as a sign and style of new technology, and every effort was made to obtain it. It is necessary to clarify this kind of understanding to provide a scientific and objective definition of its application field and to study the standard of evaluation. The evaluation and test work should go through the whole process of research on the virtual geographical environment and should consider an evaluation of the sense of reality for the virtual geographical environment from the following four topics: the evaluation of the sense of reality should be carried out by combining the construction of model; the evaluation of the interactive feeling should be conducted by combining interactive techniques; the evaluation of immersion should be implemented by simulating the interactive behavior of human avatars in the virtual geographic environment and the suitability of human beings to the virtual environment; and the improvement and application of the virtual geographic environment system should be guided through the usability research of the virtual geographical environment.

2. The theory on virtual geographical environment construction

The construction theory of the virtual geographical environment was developed based on traditional cartographic theory, and it widely absorbs the latest theoretical results from virtual reality, network communication, visualization, human–computer interaction, and artificial intelligence. Among them, virtual reality theory is the key technological theory that guides construction of the virtual geographical environment; the network communication theory is used to solve the problems of speed and reliability of transmitting large quantities of geographic data in the virtual geographic environment; visualization theory is used to implement multidimensional visualization of scientific computing and results in the virtual geographic environment; human–computer interaction theory studies the interaction between the operator and the virtual environment in the virtual geographic environment; and to realize the self-governing behavior of an avatar in the virtual geographical environment, the artificial intelligence problem in the high dimensional space must be studied.

Mathematical foundations are the core of the construction of virtual geographical environments. In the early stage, due to the restriction of hardware and software algorithms, the construction of the virtual geographical environment was for the local part. To meet the needs of various application fields, it is necessary to study the construction techniques and schemes of the multilevel grid architecture for global geospatial information based on geocentric coordinate systems (and Earth ellipsoids) to break through the restrictions of traditional plane projections, scales and others. Many scholars at home and abroad have tried to build spherical grid systems based geographical coordinate systems. Interestingly, their studies have different emphases, which can be divided into two major categories: polyhedral mesh decomposition and geographical coordinate decomposition. According to Chen et al. (2002), Deng (2003), Du et al. (2005b), Ben et al. (2007), Han et al. (2008) and others, a variety of spherical geographic information grid division schemes were designed, and multi-level data management and efficient scheduling were implemented worldwide. Gong et al. (1999) proposed the research framework of geo visualization and considered that geoscience visualization includes map visualization, visualization of GISs and visualization in the field of professional application. Map visualization and GIS visualization are two basic parts of the theory and technology of geoscience visualization. You (2002) proposed the viewpoint that geographical environment simulation can be discussed from the two aspects of perceptual simulation and data simulation. Data simulation constructs different levels of geo-environment data models, providing computers with a "cognitive" space environment, which is mainly used for simulation and evaluation; the perceptual simulation constructs a geographical environment model that enables people to perceive, including the senses of hearing and touch, and is used for the cognitive environment of the human brain, especially for command and training. Wan (2005) proposed the concept of a schema cube for constructing a virtual geographic environment and considered that the modeling of the virtual geographical environment involved the combination of the four dimensions of map data scale, image spatial resolution, temporal resolution and application operation. Lin et al. (2006) considered that the spatial and temporal scales were endowed with new features in the virtual geographical environment. The continuous spatiotemporal model made the spatial and temporal scale of the virtual geographic space no longer discrete, and it could be sampled, computed, simulated and analyzed at any spatial and time intervals.

The design of the virtual geographic environment is closely related to modeling, which involves symbolic system design and dynamic visual variable analysis. Mo (2002) proposed the concept of textural simulation for virtual geographic environments and designed a different color scheme for different regions. Two-dimensional symbols are no longer the only language of expression due to the application of spatiotemporal multidimensional models to express geographical phenomena and their evolution. More intuitive three-dimensional symbolic representations that can describe spatial structure and shape evolution became the main expression method. Zhu et al. (2003) proposed a viewpoint of a 3-D symbol system; and Wu et al. (2008) proposed representation methods for various three-dimensional symbols. Wu et al. (2005) proposed 4 goal hierarchies of 3-D geoscience spatial modeling and compared and analyzed the functional requirements of the corresponding hierarchies. Zhong (2005) defined simulated map symbols and virtual map symbols according to the mapping relationship from the cartographic area to the cognition structure of a subject, discussed the evolution law from the cognition structure of a subject to a two-dimensional plane, and provided the mathematical model of the planning and prediction map and the mathematical definition of the geographical virtual space.

User immersion is an important feature of the virtual geographical environment, and it is also one of the important features that distinguishes it from traditional map products. The modeling of participant avatars and the modeling of equipment behavior are also part of the research contents of modeling. The avatar model in the virtual geographic environment represents the living individual (such as a human) in space, and the behavior of the avatar cannot be arbitrarily defined and should accord with the behavior of the person in reality. Behavioral theories in social science can be used to guide an avatar's virtual behavior; meanwhile, psychology, sociology, communication and other social science theories can also be used to guide avatar behavior in a virtual geographic environment. The behavior modeling of equipment and facilities in virtual geographic environments includes two categories: physical behavior and technical behavior. The purpose is to accurately describe the relationship and interaction between the equipment and geographical environment.

3. The theory of the virtual Earth system

The study of Earth systems is one of the most important application issues in the field of virtual geographic environments. The interactions among the atmosphere, lithosphere, hydrosphere, and biosphere of the Earth form a large and complex system; how can a relatively reasonable virtual Earth system be built? It must be in accordance with the actual conditions of nature, taking into consideration the various spheres and forming an organic system. The theory of virtual Earth systems needs to study how to map the Earth system into the virtual environment, form a virtual Earth system, and study its organization, structure, management and operation.

In 1998, former vice president Gore's speech on "Digital Earth" attracted the attention of political and academic circles from many countries. What he was is a virtual geographical environment with Earth as its goal; Guo et al. (2000) believed that the 3-D GIS will play an important role in the process of global information digitization based on the concept of digital Earth. Because of the complexity of spatial objects and 3-D GIS, in dealing with various spatial phenomena, such as the representation of geological phenomena and applications in mining, different data structures and visualization methods should be adopted, including 2-D, 2.5-D, 3-D and their combinations, to achieve good effects and benefits; on the frontier and developing trend of Earth information science, Li et al. (2004) believed that the focus in the field of Earth information engineering was information infrastructure and digital Earth, and Earth system science needed to study new relations between humans and Earth under the framework of the virtual innovation environment of geographical science, that is, computable human-Earth relations. Based on the new framework of the virtual innovation environment of geographic science and different spatial and temporal resolution data, the mass resources, environment, society, economy and population data of multiple resolutions related to matter and energy, as well as the data generated by data fusion, can be integrated, and then, multidimensional (such as 4 dimensional) dynamic simulations can be implemented from inner circle to outer circle, from lithosphere to biosphere and atmosphere, from local to whole, and from regional to global according to certain geographical coordinates. This framework provides a knowledge innovation platform and a technological development and theoretical research experimental base for solving the present situational description, historical inversion and future prediction of the large and complex system of geographical science. Based on the same understanding, different domain experts have carried out modeling research on different phenomena at different scales. Among them, You (2002) studied battlefield situation simulations; Lin (2000) studied tidal wave systems in the East China Sea; Li et al. (2003) developed a "digital city"; Liu et al. (2002) studied a digital Earth-oriented virtual reality system and designed and implemented a digital Earth-oriented prototype system based on the C/S structure, which can meet the modeling characteristics of digital Earth. Gu et al. (2004) studied the changes in the delta coastline. Li (2004) discussed the influence of digital Earth on the study of human-Earth relationships from the point of view of the research contents and research methods of Earth system science and geography science. Gong et al. (2008) carried out the study of small watersheds on the Loess Plateau.

12.3.2 Key Technology Research

As an application-oriented platform, the virtual geographic environment involves many key technologies, such as synthesis and integration. The goal of this research is to construct an integrated knowledge sharing geographic research and application environment through which researchers, government personnel (especially decision makers) and the general public can understand the issues of common concern, publish the research results (such as model publishing) and release their views. As the key technical support, the development of spatial environment information collection technology provides sufficient data support for the construction of virtual geographical environments; the development of geography and Earth system science provides an accurate model for virtual geographical environments; the development of computer science provides software and hardware support for the realization of virtual geographical environments; and the demand for applications, especially the demand for solutions to problems, provides a powerful impetus to the development of virtual geographic environments.

1. Geospatial multisource data acquisition and integration technology

With the rapid development of remote sensing technology, from single spectrum to high spectrum, from visible light to ultraviolet light, infrared ray and microwave, and from the use of passive reflectance spectrum information to the development of active laser, polarization and interference information resources, Earth observation can be progressively carried out all day long and in all weather conditions. The development of satellite navigation technology provides more accurate spatial positioning capability for the dynamic acquisition of geographical environment information. Research in astronomy, geology, atmospheric science, oceanography, meteorology, hydrology, topography, geodesy, mineralogy and other areas has resulted in a large amount of data, information, knowledge and models. These achievements provide valuable resources for the construction of virtual geographic environments. Integration of these multisource data is another difficult problem that must be solved in the construction of virtual geographical environments. At present, there are many studies on the integration of multisource data, but all are aimed at 2-D GIS. Whether these methods are suitable for the application of a virtual geographic environment with three-dimensional or higher dimensional features needs to be modified and checked. In particular, the data acquisition of the 3-D city model is the most typical.

Zhu (2004) pointed out that there are three main data sources of 3D urban models: remotely acquired data (satellite images, aerial images, space-borne laser scanning, etc.), near distance data (close range photography, close range laser scanning, manual measurement) and derived data. Different data sources correspond to different 3D urban modeling methods: 3D urban information acquisition methods based on remote sensing images and airborne laser scanning systems can be applied to quickly obtain ground models and building models in large urban areas and are one of the most promising methods for acquiring 3D model data in urban areas; on-board digital photogrammetric methods and airborne laser scanning methods are suitable for the modeling of corridor areas; ground photogrammetrics and near range laser scanning are suitable for modeling a single or small number of buildings; fast modeling methods based on conversion from existing 2-D GIS data to 2.5-D models have the advantages of low cost and high efficiency and have a wide range of applications in areas where fast 3-D urban models are needed; realistic CAD 3D solid modeling methods have low efficiency but fine effects; and by using image-based modeling and rendering (IBMR), we can also obtain highly realistic scene expressions in the absence of complex geometric models and can greatly simplify complex data processing work.

The image fusion technology for 3D terrain simulation is studied by Zhou et al. (2002), and the TM images and IKONOS images in the same region are fused; the result shows that the spectral information of TM images were retained, while the geometric features of IKONOS images were introduced. According to Shi et al. (2004), a 3-D urban model was reconstructed by using oblique photography images and vertical photographic images, and fast 3-D urban modeling based on various images was implemented. According to Du Wei et al. (2005a), by combining the streaming media technique and the virtual geographical environment, the structure and function modules of the virtual geographic environment system based on streaming media was designed, and a prototype test system was established by taking the virtual geographic research room as an example. To ensure spatiotemporal coherence in the distributed virtual geographic environment and solve the synchronization and consistency in the process of sharing virtual scenes, Zhou et al. (2005) proposed that a unified spatial benchmark and environment description was needed to ensure spatial consistency in the virtual geographic environment, and specific technical solutions for reducing data transmission delay were needed to ensure a real-time performance. Zhou et al. (2006) synchronously captured the 3-D point cloud and digital image of a building by using laser scanning and digital camera and realized the true color 3-D visualization of the point cloud model by giving the texture information of the image to the 3-D point cloud model.

2. **Construction and management technique of the virtual geographic environment model**

There are many kinds of models in virtual geographical environments. According to geographical factors, they can be divided into physical geography models, economic geography models, social culture models and traffic models; according to the terrain elements, they can be divided into geomorphic models and ground feature models; according to the geometric structure, they can be divided into point models, line models, surface models and body models; according to the description content, they can be divided into physical models, representation models, influence models and prediction models; and they can also be divided into static models and dynamic models according to space–time transformation and geometry models and behavior models according to functions. Different types of models differ greatly in building methods and management.

The key to the construction of geometric models in virtual geographic environments is to address the contradiction among the three aspects of model complexity, fidelity and real-time rendering.

Early terrain modeling is local, and researchers have attempted to improve the algorithm and display efficiency to adapt to the low performance of the early hardware. Chen (1997) proposed a virtual scene simulation based on map data, which conforms to the visual law, so that the virtual scene generated by the computer conforms to the visual law of people's "the closer you see, the clearer you'll find". Xia (1998) proposed a simplified algorithm of a 3D terrain model in a real-time battlefield environment simulation system. Zhang et al. (1997) studied the method of surface fractal dimension calculation and proposed a method of computing the land surface fractal dimension based on geographic data. Hao et al. (1997) was the first to propose a DEM improved quadtree for ground visualization and considered that the quadtree representation played an important role in the compression, storage and ground visualization of digital elevation models. Zhang et al. (1998) and Zhu et al. (1998) studied the reconstruction of urban 3-D house models based on triangulated irregular networks. Qi et al. (2000) studied the multiresolution 3-D terrain modeling method based on fractal technology and used the midpoint displacement method and triangle edge subdivision scheme of fractal technology to recursively subdivide terrain. Yang et al. (2001) studied the dynamic simplification of terrain based on the 3 viewpoint-dependent factors of image space error, intersection of view frustum and node orientation. Yang et al. (2003) studied and analyzed the mapping mechanism of a multiresolution texture model. Zhu et al. (2005) proposed a description method of LOD for 3-D urban models, including the planning of the LOD model, the determination of fineness, and the corresponding data production process. Du et al. (2006) proposed a new Mipmap texture generation technique and Mipmap texture compression algorithm to achieve the storage structure of Mipmap compressed textures in the global pyramid model.

To realize the unified management of multidimensional spatial data, combining the concrete application of 3D GIS in urban, geological, oceanic and other fields, Li et al. (1998) proposed three integrated 3D spatial data models: the integrated model based on TIN and CSG for urban 3D modeling; the hybrid model based on octree and tetrahedral mesh for geology, ocean and other fields; and the 3D spatial data model with integrated vector and grid data. Zhu (1998) proposed integrating a digital elevation model with a digital orthophoto map, regular vector data and various attribute information and established an integrated three-dimensional data input, operation and visualization mechanism. Thus, a spatial query and analysis model was proposed based on 2-D and 3-D hybrid representations. Xie et al. (2002a) introduced structured objects into virtual GIS data models to support the management of object-oriented geographic data in spatial databases and their expression in the geo virtual environment. Rui et al. (2002) divided the data of 3-D urban models into 5 types and established spatial databases, attribute databases and symbol databases.

Scholars have carried out studies on the management and allocation of massive global spatial data. Chen et al. (2002) proposed a dynamic triangulation algorithm for the RSG model of terrain, where by defining the constraints among vertices in the model, the inheritance of vertex errors and building the hierarchical structure of the vertex by quadtree, he effectively solved the problem of cracks between blocks with different resolutions, which was very good for real-time LOD control of terrain and for real-time rendering. Wu et al. (2005) presented an LOD-based algorithm of the TIN model, which was quite suitable for terrain rendering in large 3-D landscapes. The emphasis focused on issues of transition between TIN models with different resolutions and the generation of these models, such as how to design data structures and how to optimize algorithms when constructing a Delaunay triangulation. The above work laid a solid foundation for the implementation of "horizontal borderless roaming" and "vertical continuous scaling" for geospatial data display.

Chen et al. (2003) built the first global multiresolution virtual terrain environment in China. Aiming at the data characteristics and the rendering needs of the global multiresolution terrain environment, Du et al. (2005b) designed and implemented a global pyramid model based on the multiresolution LOD (MRLOD) model. On the basis of studying Google Earth, Virtual Earth, World Wind and ArcGlobe, several global geospatial data display platforms emerged in China. The emergence of geographic information grid technology provided new algorithms and ideas; the traditional method of data organization (i.e., map projection as a mathematical basis, data organized by scale and latitude and longitude, and subdivision based on a map sheet) was no longer suitable for the application of geospatial information under a global grid environment, and it was necessary to study the construction techniques and solutions of multigrid architecture based on the geocentric coordinate system (and Earth ellipsoid). Ben et al. (2007) and Han et al. (2008) proposed a variety of division schemes for global geospatial information grids. Liu et al. (2008) and Hui et al. (2009) implemented multiscale vector map data management, rapid allocation and display based on a global geographic grid.

Research on the 3-D spatial solid structure of underground pipe networks, mines and coal mines started late. Wu et al. (2007) proposed a generalized tri-prism (GTP)-based entity model for underground, real 3-D integral representation and revised and extended it in three aspects: modeling object, component element and topological description. In Wang et al. (2005), the roadway was divided into the roadway body and roadway nodes according to the data structure of the arc and node in the information system. Then, the 3-D model of the roadway body and nodes was established by using the 3-D vector surface wireframe model. In LI et al. (2007), the modeling and rendering techniques of 3-D solid clouds were studied by using fractal technology and particle system technology. In Tan et al. (2006), real-time shadow technology based on images was studied according to the requirements of real-time and realistic shadow effects in battlefield environment simulations.

Another type of modeling of the virtual geographic environment was the building of geographical process models (such as wind field, temperature field, pollution diffusion, water flow, etc.) and human behavior models (such as urban development models and crowd evacuation behaviors). This part would not be explained in detail in our article.

For the aspects of model accuracy analysis and evaluation, Shi (1998) proposed an error model for geometric features in 3-D GIS; according to the study of Wang et al. (2005), the features and terrain in the virtual geographical environment cannot be accurately fused, and taking the road as an example, a fusing algorithm of a road model and terrain model was put forward, and the accuracy of the fused road and terrain model was evaluated.

3. **The display and interaction technology of the virtual geographic environment**

Unlike traditional maps and GISs, the virtual geographic environment emphasizes the sense of immersion, and users become part of the system. The interaction between the human and virtual geographical environment can be called human virtual environment interaction (HVEI), and the main feature distinguishing it from human–computer interaction (HCI) is that the virtual environment has the ability to display interactive environments to users with multiple input and multiple output channels. This interaction is an important factor used to improve the efficiency of human tasks in a virtual world. Although traditional HCI devices (keyboard, mouse, joystick, etc.)

are also used in the virtual geographic environment, they are no longer the only interactive devices. The display system and HCI equipment in the field of virtual reality technology have been adopted in virtual geographical environments, and the virtual environment can be controlled through multiple perception and interaction channels by using data gloves, helmets, gestures, voices, etc.

Multiplatform virtual scene generation based on the DVENET system was studied by Chen et al. (1998), focusing on the construction of an environmental model during the generation of virtual scenes, the connection and combination of discrete models, the rendering of special effects, the application of segmentation and allocation technology and the unification of virtual visions on different developing platforms. A multichannel visual simulation based on a PC cluster was studied by Xia et al., which used a general PC as a hardware platform instead of the commonly used expensive dedicated graphics workstations. A geometric correction method for a multichannel surface projection system was studied by Zhang et al. (2006b). A geometric correction method based on texture mapping in a polar coordinate system was proposed for a curved multichannel large screen projection display system. By mapping an average angle-divided grid to the 3-D display surface, the algorithm could realize geometrical calibration for large-scale display systems with curved screens and oblique projectors. A multichannel synchronization technology in a ship handling simulation system was studied by Lv (2006), and multichannel synchronization was realized by using a multithread structure and dead reckoning technology.

Considering the problem of selecting observation information in a virtual environment simulation, a virtual observer model that can effectively help participants obtain multidimensional information in a virtual environment was proposed by Yang et al. (2001). In the paper of Wan et al. (2004), a spatial aural perception model and two models of aural perception were introduced, such as a physiological cognitive model and a psychological cognitive model. The mathematical model of realistic sound generation was discussed, and a variety of schemes for generating realistic sound were given. According to Liu et al. (2005a), movement data of the joints of the arm were obtained by using data gloves and a tracker. Then, the data were processed according to the binding between human joints, and a mathematical model of the virtual arm was established. The synchronized movement of the arm of an experimenter in the virtual arm model environment was realized with realistic effects. A new virtual reality system concept that adapted to human binocular stereo vision and could perform 3-D measurements accurately was proposed by Lin et al. (2005). A method of virtual human motion control for 3-D terrain was presented by Shi et al. (2006). People tend to lose their way when roaming in 3-D virtual scenes, and it takes time to find the target in the scenes; for these problems, a navigation map was introduced, and a method of building an intelligent navigation map was proposed by Li et al. (2008).

Wan Gang considered that the HCI was also an important factor affecting the efficiency of human beings in the virtual geographic environment and was the key to the construction of a humanoid virtual geographical environment. The general tasks of interactive technology included navigation, selection, manipulation, and system control. Viewpoint control techniques included four types of tasks: exploratory roaming, roaming determined by the dynamic relationship between the observer and the observed target, target-centered high-efficiency roaming and optimal viewpoint selection based on the target group.

4. **Distributed virtual geographic environment technology**

There are various problems based on the geographical environment that are complex and changeable, and it is difficult for a person or an organization to complete research on the problems, which requires multi-domain experts and multilevel personnel to participate in and solve together. Building a collaborative virtual geographic environment on the network is the direction of development. The collaboration of the virtual geographical environment is divided into several levels, namely, data, model, operation, visualization and decision. The establishment of a high-speed network has provided a very good network hardware foundation for the construction of a distributed virtual geographical environment. In software, more reasonable coding methods and data compression methods should be sought to reduce the transmission of network data. Simultaneously, corresponding measures should be taken to minimize the bit error rate of data and improve the quality of network data.

The DVENET (Distributed Virtual Environment Network) built by Zhao et al. (1998) was the earliest distributed virtual environment platform in China. It combined several simulators distributed across different regions and developed a virtual simulation platform by using virtual reality technology, which constituted a distributed virtual environment that could be used for tactical simulation exercises in different places. A multiuser collaboration and avatar technology objected country park management system developed by Gong (2000) realized the sharing of different places of knowledge with the aid of a network platform. Gong et al. (2001) discussed the concept and characteristics of distributed geo virtual environments and the relationship between GIS and virtual environments. The emergence of Virtual Reality Modeling Language (VRML) and Geo Virtual Reality Modeling Language (GeoVRML) provided a good way to solve the 3-D visualization of network geographic information. Gong et al. (2002a) designed and developed a virtual campus environment based on the Internet and established a prototype system of the Chung Chi College Campus of Chinese University Hong Kong. By using network virtual GIS design based on XML and CORBA combined with XML technology, the expression and transmission of many kinds of information in network virtual GIS was solved by Xie et al. (2002b), and the network structure of the network virtual GIS based on CORBA technology was designed. Real-time interactive browsing of 3-D massive terrain data on high-speed Internet was studied, and a transmission protocol suitable for real-time transmission of massive terrain data in the network was designed by Wang et al. (2002). A Web publishing system model for digital city 3-D Landscape based on the B/S structure was presented in Lan et al. (2003). The characteristics of five levels of user interaction in VGE were analyzed, and a method to combine the VGE system with a multiagent system was proposed in Qi et al. (2004). The characteristics of collaborative virtual environment navigation and the importance of spatial awareness were discussed in

Zhang et al. (2004). A 3-D WebGIS visualization technology based on GeoVRML was studied by Chen (2005). Xu et al. (2005) studied and designed the framework of a distributed virtual geographic environment system based on high-level architecture (HLA) and discussed its data stream and object model. Zhang Jianqin built a 3-D model of WEB based on VRML and Java3D, which provided users with a realistic, real-time and interactive 3-D virtual scene. Based on the traditional network structure and combined with P2P technology, a distributed network structure suitable for large-scale user participation in the virtual geographical environment was set up by Zhu et al. (2006). Research on peer-to-peer models suitable for distributed virtual geographic environment services was conducted in Bian and Tan (2007), and the adaptability of existing peer-to-peer routing structures was deeply analyzed and compared. Furthermore, a large-scale P2P network model suitable for distributed virtual geographical environment services was designed. According to Li et al. (2007), a mobile cooperative virtual geographic research room was discussed, and the specific model was designed and applied. Collaboration among multiple users in VGE based on genetic algorithms and multiagent systems was studied by Zhang et al. (2007). In Yang et al. (2007), the organization and management of spatial information and metadata in the context of virtual geographic environment was studied, and the architecture of spatial information sharing, which combines centralized spatial information object management and distributed spatial information service, was discussed. In Liu et al. (2007), based on the requirements and principles of dynamic terrain simulation, two key techniques of dynamic terrain simulation, i.e., temporal consistency and polymorphism consistency, were studied. Additionally, based on integrated environment data representation and exchange specification (SEDRIS) and HLA technology, a dynamic terrain simulation federation was designed and implemented.

The development of grid technology gives a great impetus to the research of distributed virtual geographical environments. In Wang et al. (2003), the application of grid computing in a distributed virtual environment was studied, a distributed virtual reality system built based on grid computing was discussed, and a virtual reality model based on grid computing was proposed. In Jin et al. (2004), 3D spatial data interoperability in the grid environment was discussed, and a 3D spatial data representation method based on SOAP (Simple Object Access Rotocol) and layers was proposed. Then, 2D vector data, attribute data, texture and model data in accordance with GML specification were obtainable by parsing SOAP messages, which facilitated data exchange and interoperability between software. Based on studying grid characteristics and the SF Express model, grid technology is used in virtual reality, and a virtual battlefield simulation model based on a grid is proposed by Ren et al. (2006). According to Wu et al. (2006), the architecture of the grid virtual geographic environment was studied by combining the grid technology with the virtual geographical environment, and by taking the service as the core, a virtual database was constructed, and a grid virtual geographic environment was formed, which realized the integration and sharing of heterogeneous data, the decomposition of tasks, the full use of idle resources, and the expansion of the virtual geographical environment services. A grid framework for 3D spatial information applications and its key technologies and system prototype were studied in Jin et al. (2006). A computing framework for the virtual geographic environment based on grid services was proposed, and the VGE data collaboration and computing service operation mechanism under the computing framework were mainly studied in Zhu et al. (2007). Ma et al. (2007) pointed out that the virtualization of remote sensing products was the development direction of future remote sensing data processing systems based on grid computing environments and proposed a software engineering methodology based on multiple agents to realize the virtualization of remote sensing products. Combining grid computing with 3D GIS roaming technology, a task allocation model for grid task-oriented task decomposition was discussed in Meng et al. (2007); then, complex 3D GIS computing tasks were decomposed and completed by several computers.

5. **Artificial intelligence technology in virtual geographic environments**

There are so many avatars in the virtual geographical environment that it is impossible to completely control them in reality. It is a reasonable scheme to use artificial intelligence technology to realize self-control of avatars in virtual environments. In addition to the control of avatars, artificial intelligence technology can be used in environmental settings to build intelligent environments. In Gong et al. (2003), individual-based spatiotemporal simulations of SARS transmissions in hospitals was studied. Based on epidemiological data from SARS, the characteristics and regularities of SARS spatiotemporal transmission in hospitals were analyzed by introducing the human behavior model and the factors influencing the environment in terms of SARS propagation. An intelligent Agent model for individual SARS spatiotemporal transmission was established by integrating Agent technology into a virtual geographic environment platform, and the application of the SARS transmission model in the platform was realized. Lin et al. (2004) proposed and discussed a research framework for content-based virtual geographic scene retrieval and intelligent navigation. In Zhu et al. (2007), the characteristics of multiuser interactions and the characteristics of multiagent systems (MASs) in the virtual geographical environment were analyzed, and the advantages of multiagent systems for multiuser cooperative control in virtual geographic environments were studied. On this basis, a hybrid cooperative control model based on Agent was proposed to realize cooperative control among users. A network virtual reality system architecture based on a multiagent system was proposed by Lin et al. (2007), and some key issues related to the implementation of prototype systems, such as Agent development platform selection, dynamic routing control policy, and user and shared resource management, were discussed.

12.3.3 Research on Virtual Geographic Environment Application

At present, the application fields of virtual geographic environments involve urban planning, land use, virtual transportation, virtual tourism, virtual education, military (virtual battlefield environment), environmental management, and so on, and obvious application results have been obtained.

1. **Applications in the field of digital Earth and environmental science**

A geo virtual innovation environment framework was proposed by the Digital Earth Laboratory of Henan University. It is believed that the digital Earth laboratory framework can provide a virtual innovation environment for the original innovation of geographical science. The virtual innovation environment of geographic science is based on the data of different spatial and temporal resolutions or based on the data generated by data fusion of massive matter and energy data or information at multiple resolutions (such as resources, environment, society, economy and population) and is used to implement integration, fusion and multidimensional (such as 4 dimensional) dynamic simulations of the Earth environment according to certain geographical coordinates; this virtual environment is from inner layer to outer layer, from lithosphere to biosphere and atmosphere, from local to worldwide, and from regional to global. It provides a platform for knowledge innovation and an experimental base for technological development and theoretical research to solve the present situation description, historical inversion and future prediction of the large and complex system of geographical science.

At Northwest Agriculture and Forestry University, research and application of landscape virtual reality technology for small watersheds was used for soil and water conservation and desertification control. At the China University of Mining and Technology, virtual reality technology was applied to environmental management, such as environmental planning, environmental governance, environmental decision-making, and environmental forecasting. Based on RS and GIS, Jilin University developed a three-dimensional virtual geographic environment modeling technique for the Songhua River basin (Jilin section), which has been used to predict the changes in and evolution of the water environment. A study on the construction of a virtual forest environment based on the stand growth law was carried out at Wuhan University to analyze the stand growth law and forest spatial data. Research on the visualization technology of the Hangzhou ecological planning decision support system was carried out at Zhejiang University, and the quantitative design and management of ecological planning was realized. The research and application of soil and water conservation supported by VGE was carried out by the Institute of Remote Sensing Applications, Chinese Academy of Sciences. Remote sensing research on cloudy and rainy areas was carried out at Chinese University Hong Kong. The technology flow and method of 3D terrain development for the Yellow River basin was carried out at Capital Normal University. At Henan University, the application of research to related technology for the virtual Yellow River was analyzed at a technical level for the main technical problems that needed to be solved in the construction of the virtual Yellow River. Computational visualization of the geo-process orientation was studied in Gong et al. (2002a), the objects in the dynamic changes in the geographical system were distinguished and defined, the necessity of building the visual data model was put forward according to the interaction requirements between human and model, as well as the real-time interaction requirements between humans and 3-D graphic images. A preliminary experiment was carried out by taking the flood routing simulation as an example. Li et al. (2003) analyzed the present situation of pipeline management, expounded the hierarchical model and organization method of pipeline data, and established a three-dimensional model of underground pipelines. Shu et al. (2004) analyzed the key problems of constructing a virtual forest scene in accordance with the rules of the stand structure, constructed a 3-D geometric tree model by using an interactive parametric modeling method, and finally built a virtual forest scene. Xu Zhiyong et al. (2005) studied the simulation of the climate environment in a virtual geographical environment and discussed the climate environment modeling method.

2. **Applications in urban transportation planning and exhibition**

In Zhu et al. (1998), the applications of virtual reality in environmental dynamic change research and urban planning were discussed. In Li et al. (2000), the concept, technical support and typical application of cyber city were systematically introduced, combined with the research and development of CCGIS software and its applications in the two cities of Shenzhen and Shanghai, and the key technical problems faced in the construction of cyber city were explained. In Wang et al. (2004), the 3-D abstract, description and expression of urban landscapes were studied, and the concepts and implementation methods of 3D geographical element systems, 3-D spatial description granularity, 3-D building partitioning and 3-D model libraries of urban landscapes were proposed. Tan et al. (2006) discussed the construction of a 3-D urban model and its application in the analysis of noise pollution.

The visualization and its system of urban 3-D pipe networks produced by Wuhan University provided a new way to solve the rapid visualization of complex 3-D pipe networks in urban areas. The use of virtual reality technology in urban planning and design, which was implemented by the Shenzhen Planning and Land Resources Information Center and Fujian University of Technology, provided perceptual knowledge for urban planning and design. The establishment of a 3D cadaster of urban underground space in Guangzhou illustrated the 3D cadastral data model and the integrated data management method of 2-D and 3-D cadasters. The planning and design of agricultural land consolidation based on the 3-D visualization model of the virtual geographical environment was completed by Huazhong Agricultural University, which was based on the current 3-D visualization model of the virtual geographical environment and better coordinated the works among different governmental departments involved in land consolidation. Based on highway design data, the method of establishing a 3-D model of terrain, highway and its structure was discussed in Zuo et al. (2004). According to Zhang (2005), based on the study of individual spatial behavior characteristics in urban traffic systems, a virtual experimental system for urban traffic management was developed, and a virtual experimental system of urban traffic management based on GIS and a multiagent system was implemented. In addition, there was the virtual environment railway location design system of Southwest Jiao Tong University, the 3-D highway and the 3-D digital channel of Wuhan University, and the traffic accident simulation and reconstruction of the Harbin Institute of Technology.

3. **Applications in the maritime field**

In Ma et al. (2002), virtual reality modeling technology was used to study coastal ocean tidal current visualization, a prototype system of virtual reality for coastal marine tidal current simulation was designed and implemented, and a comparative test was carried out in the numerical simulation test of tidal currents in the South Yellow Sea. Xiao et al. (2003) studied the multiresolution modeling and visualization of underwater terrain. In Zhu (2004), the particularity of the marine geographical environment and its dynamic process were analyzed; the theoretical framework of the virtual marine environment was presented by means of the theoretical description of objects, fields and field objects, and the technical emphases and difficulties of constructing virtual marine environments and their development trends were pointed out. Dalian Maritime University constructed a digital harbor and a virtual port. Du Ying researched ocean wave simulations using graphics processor unit (GPU) technology, proposed several different ocean wave models, and adopted a strategy of separate computation and display. Liu et al. (2005b) proposed a virtual implementation method for benthos terrain and current in the ocean environment based on scene simulation.

4. **Applications in the geology and mining fields**

Through the visualization method of 3-D seismic and topographic data, the original seismic and logging data were explained by Xi'an University of Science And Technology, and important information, such as the existence, location and reserve of mineral resources, was obtained. A 3-D stratum based on an analogical triangular prism was studied at Wuhan University, and the data model and basic algorithm for the 3-D modeling of stratum were studied based on the characteristics of geological prospecting engineering data. Through the research of GISs for open pit mines based on virtual reality technology, the multidimensional expression of geographic information of the geological structure of open pit mines was realized at the Wuhan University of Science and Technology.

A 3-D geosciences simulation and virtual mine system was studied by Wu et al. (2002), and a 3-D geosciences simulation method based on the analogical tri-prism (ATP) model was proposed. A method of 3D modeling based on quasi tri-prism volume (QTPV) was proposed by Cheng et al. (2004), and by taking the rock bedding band Laneway as the research objects, OTPV modeling methods and algorithms were designed, and a set of real borehole sample data, which come from Inner Mongolia China, were used to verify the developed prototype system. The 3-D numerical simulation and visualization of the groundwater flow field for the Changzhou-Wujin area were carried out, and the main causes of land depression in this area were revealed in Yan et al. (2004).

5. Applications in the military field

The virtual battlefield environment has many advantages for scientific, economical, antagonistic, intuitive, interactive and real-time military training exercise purposes. It provides an effective way for various countries to realize strategic, campaign and tactical exercises in the new period and can greatly improve the training quality of military operation simulations.

As early as 1996, the Institute of Surveying and Mapping of Information Engineering University proposed virtual reality and its application in military mapping and training simulations and considered virtual reality to be an artificial environment generated by computers that can interact with the operator and form a closed system to make people feel immersive. As a new means of surveying and mapping support, the virtual geographic environment extended the cognitive means and scope of commanders and changed the traditional techniques of the simulation method. You (2002) discussed the composition of the battlefield environment and the main contents of battlefield environment simulation and focused on the application and key technologies of virtual reality technology in battlefield environment awareness simulations. The research of battlefield visualization and application prototypes was proposed, especially in CGF, and was deeply studied. The mathematical model of ocean waves, the discretization process and the real-time problems in sea wave visualization were studied in Dong et al. (2003). Aiming at the visual simulation modeling of military aircraft, an optimized pretreatment method based on the fuzzy mathematical similarity principle was proposed in Huang et al. (2003). A conceptual framework for the synthetic environment generation technology based on the distributed virtual battlefield environment requirements was proposed in Hou et al. (2003). Some key technologies of situations mapping and representation in virtual battlefield environments was studied, a real-time conversion method for 2-D and 3-D plotting was proposed, and the interaction representation between the 2-D and 3-D situations was realized in Chen et al. (2005).

The Distributed Virtual Environment Network (DVENET) is a distributed virtual environment basic information platform that was developed by Beijing University of Aeronautics and Astronautics under the support of the national "863 Project", which combined several tactical simulators distributed across different regions to create a distributed virtual battlefield environment for collaborative and adversarial tactical simulation exercise across different areas. Simulation architecture and related problems of virtual globe war temporal-space were proposed and studied by the National Defense University. Special effects generation and entity model simplification technology in the virtual battlefield and terrain tracking matching technology of the weapon platform were studied by the National University of Defense Technology. The 3-D representation of radar electromagnetic information under the influence of topography, ocean and weather was also a key issue in the virtual battlefield environment. The composition and construction principle of the space battlefield virtual environment and the space battlefield virtual environment technology were studied by Northwestern Polytechnical University. The representation of the radar range under the influence of terrain in a virtual battlefield environment was analyzed by the National University of Defense Technology.

6. **Applications in the aeronautic and aerospace engineering fields**

The virtual geographic environment can be used for virtual control of future flight centers and air traffic. The visual simulation system of the flight simulator at Tianjin University focused on the multichannel visual simulation, HLA and nonlinear distortion correction. The real-time 3-D visualization simulation of space missions from the Institute of Surveying and Mapping of Information Engineering University realized the representation of spacecraft kinematic characteristics and the geometric modeling of spacecraft interior components.

7. **Applications in the culture and tourism fields**

Zhejiang University applied virtual reality technology to the development of tourism. This technology has an important influence on the investigation and evaluation of tourism resources, tourism resource development and planning, decision-making in tourism marketing and tourism guidance. Wuhan University applied GIS 3-D visualization to digital cultural heritage and constructed a typical virtual exhibition of the Tang Dynasty wooden architecture complex. Nanjing Forestry University realized the construction and application of the Virtual Reality GIS for the scenic spot of the Nanjing Treasure Dockyard Relics. Xu et al. (2001) proposed a virtual tour system scheme consisting of a Web GIS-based electronic map, panoramic image generation and network panoramic image browsing. By constructing the Virtual Olympic environment system, the temporal and spatial evolution of the north–south axis in Beijing was visualized, simulated, expressed and analyzed in Yi et al. (2004), which embodied the visual expression of the "People's Olympics" in the Virtual Olympic environment system. Through the analysis of digital cultural heritage, a design scheme of a 3-D visualization system of cultural heritage GIS based on a microcomputer platform was proposed by Zhu et al. (2006), and a typical virtual exhibition example of the Tang Dynasty wooden building complex was realized.

12.4 Problems and Prospects

12.4.1 Design, Construction and Evaluation of the Virtual Geographical Environment Guided by the Spatial Cognition Theory

In theory, we adhere to the basic principle that the virtual geographic environment is a natural and reasonable extension and expansion of maps in the digital age and is a new spatial cognitive tool supported by the digital map; adhering to this idea, the theory of map spatial cognition is the research core and basic theory of the virtual geographical environment. The mathematical basis, design principle, symbol system, development platform and service standard of the virtual geographical environment are unified, and the evaluation framework and engineering application environment of the virtual geographic environment are first set up.

12.4.2 Virtual Geographic Environment Service Using the Geographic Information Grid as a Platform

The importance of geospatial information grid service technology was fully realized in the departments of geographic information production and application, and the application mode of geospatial information experienced great changes from being data oriented to service oriented. That is, users no longer needed to store large amounts of geospatial data for use, but access to geospatial information services was available from the network platform without needing to know where the geospatial data were stored or who provided the geospatial information services. From the provision of data services to the provision of knowledge services, more data mining and knowledge services platforms will be available online to provide a variety of customized services for users. Under the support of mass storage and high-speed networks provided by cloud computing technology, ground observation technology, and virtual display software and hardware, virtual geographic environment technology will be fully adapted to this change and provide rich information services in virtual geographic environments.

12.4.3 The Intelligent Virtual Geographic Environment Supported by the Professional Geographic Model

As a technology, the further development and application of virtual geographic environments must rely on the development of Earth system science, the construction of various professional geographical models, and the deepening of theories, such as human-land relationships and human–machine-engineering relationships. The following work needs to be done to construct an intelligent virtual geographic environment: construction of various professional geographical models based on data mining and knowledge discovery; construction of geographical expert knowledge rule bases based on artificial intelligence; and construction of intelligent information service platforms based on automatic reasoning, task assignment and combinatorial optimization.

12.5 Representative Publications

(1) ***Application of virtual reality in terrain environment simulation*** (Gao Jun, Xia Yunjun, You Xiong, Shu Guang. The PLA Press, Beijing, May 1999)

This book discusses how to use virtual reality technology to build a virtual terrain environment. It introduces the basic concept of virtual reality and describes the relationship among battlefield digitization, visualization and virtualization; the book also introduces the classification, acquisition and processing methods of terrain data and its applications in modern warfare simulations, as well as describing contour-based digital terrain model modeling and its application in virtual reality systems. Additionally, the book presents the representation of 3D terrain planes and the principle and algorithm of terrain generation based on 3-D graphics, gives examples of using OpenGL to generate 3D terrain environments, and discusses the basic principle, main equipment and image generation algorithm of stereo vision simulations. This book highlights the applications in the military field, especially in modern warfare simulations; and as much as possible, the principles of virtual reality and terrain environment simulation are explained in detail. The book not only provides a reference for the research and design of military operation simulations and digital maps but is also suitable for scientific and technical personnel who are engaged in the study of geospatial data visualization.

(2) ***A study on the tidal wave system and coastal evolution simulation in the East China Sea*** (Lin Hui, Lv Guonian, Song Zhiyao. Science Press, Beijing, January 2000)

Supported by GIS and ocean hydrodynamic models, a systematic study of tidal dynamic systems in the East China Sea is carried out in this book. Through high-resolution numerical simulation, the basic characteristics, distribution law, formation mechanism and evolution process of tidal wave dynamics in the East China Sea are explored, and the relationship between the tidal wave dynamical system and coastal evolution is discussed. As a result, a marine tidal wave simulation support system based on the Internet geographic information system is formed.

(3) ***Virtual geographic environment—a geographical perspective on online virtual reality*** (Gong Jianhua, Lin Hui. Higher Education Press, Beijing, May 2001)

This book systematically introduces the features and significance of Internet-based web information space and online virtual reality from the standpoints of geography and philosophy. The concept of virtual community is defined, and its essence and characteristics are discussed from the standpoints of philosophy and aesthetics; the concept, structure, characteristics and evolution process of the virtual geographical environment are analyzed, and the concept, study objects and study content of virtual geography are also explained; the design of the system framework of the distributed geo virtual environment is discussed, and the system prototype is developed; the graphical model structure of computer geosciences and the framework of the graphical virtual environment resource system for geo knowledge analysis are established; finally, the relationship between the virtual community/virtual geographical environment and sustainable development of human beings is discussed.

(4) ***Construction of a virtual city—principles and methods*** (Liu Xiaoyan, Lin Hui, Zhang Hong. Science Press, Beijing, August 2003)

This book mainly introduces the background, theory and key technology of the virtual city system construction and some popular virtual system programming interface technology, the design methodology of the virtual environment and the conceptual model design method of the virtual environment manager, the modeling technology of the virtual environment and the establishment of the virtual environment scene database, and the interoperability of geographic information data and its implementation. Based on the above research, the author proposes the theory and method of the rapid construction of virtual cities, which takes virtual reality as the core and combines digital photogrammetric technology and GIS. It provides a typical demonstration for the research and application of virtual reality technology in geosciences.

(5) ***Cyber city geographic information system—a preliminary study of the 3-D city model in the virtual city environment*** (Zhu Qing, Lin Hui. Wuhan University Press, Wuhan, October 2004)

Because of its potential applications in the fields of city infrastructure management and city development decision-making support, 3-D digital cities with realistic scenes have become one of the hot issues of common concern in city information. How to change from 2-D geographic information systems to 3-D descriptions of urban environments is at the forefront of urban data management. From the standpoint of photogrammetric and remote sensing and geographic information engineering, this book systematically discusses the key issues of data representation, acquisition, management and analysis of 3-D city models for the first time and uses a cyber city GIS to describe them in a unified way.

(6) ***DVENET distributed virtual reality application system operation platform and development tools*** (Zhao Qinping. Science Press, Beijing, January 2005)

DVENET is a distributed virtual reality application system development and run-time infrastructure funded by the State 863 Project under the subject of computer software and hardware, which can support the development of virtual reality application systems in the whole process and full cycle, as well as stably and reliably supporting the operations of large-scale, cross route, distributed interactive simulations and distributed virtual reality application systems. In the book, the DVENET distributed interactive simulation platform, virtual environment platform, virtual entity modeling platform and related development tools are introduced from the aspects of system function, system design and system implementation; a batch of theories, methods and technological innovation achievements obtained in the development of DVENET are presented; and several virtual reality application system examples developed based on DVENET are given, which summarize the latest research progress of DVENET. The book focuses on the operating platform and development tools of the distributed virtual reality application system, as well as the related theories, techniques, and applications, which appear to be systematic and theoretical.

(7) ***Battlefield environment simulation*** (Guo Qisheng, Dong Zhiming. National Defense Industry Press, Beijing, April 2005)

This book is the first monograph that systematically introduces battlefield environment simulation technology. The contents of battlefield environmental simulation are systematically introduced, including basic concepts, basic computer graphics, distributed virtual environment standards, battlefield environment databases, terrain (including dynamic terrain) simulations, ocean wave simulations, simulations of special effects based on particle systems, infrared imaging simulations, sound simulations, electromagnetic simulations, and 3-D visual engines and tools. The book can be used as a teaching or reference book for undergraduates and graduate students and as technical reference books for scientific research personnel and engineering technicians.

(8) ***Virtual Geographic Environments*** (Lin Hui, Michael Batty. Science Press, Beijing, June 2009)

This book is based on the selected articles from the second International Conference on Virtual Geographical Environment, which introduces the latest development of GIS and visual integration, known as VGE. The main contents are as follows: online virtual environment and Web2.0 technology, virtual city and virtual landscape, user interface, public participation and geosciences visualization, mobile and network virtual geographic environment construction, and mobile and dynamic visualization. This book is mainly intended for use by scholars and students from Earth science and computer science, as well as governmental staff related to digital cities and company developers in the fields of engineering geology, construction, and others.

References

Ben, J., Tong, X., Zhang, Y., & Chen R. (2007). Research on generating algorithm and software model of discrete global grid systems. *Acta Geodaetica et Cartographica Sinica, 36*(2), 187–191.

Bian, F., Tan, X. (2007). P2P network model conformed to the service of distributed virtual geographic environment. *Geomatics and Information Science of Wuhan University, 32*(11), 1028–1033.

Chang, Y., & Shi, C. (2007). 3D Visualization and analysis of spatial information based on augmented reality. *Journal of System Simulation, 19*(9), 1991–1995, 1999.

Chen, A., Xu, G., & Shi, Y. (2002). Automatic 3D building modeling based on urban aerial stereo. *Acta Geodaetica et Cartographica Sinica, 31*(1), 54–59.

Chen, T. (2005) *Research and Implementation of 3D GIS Modeling Method*. Doctoral Dissertation of Jiangsu University.

Chen, G. (1997). A virtual terrain simulation based on map data and according with the regular pattern of vision. *Acta Geodaetica et Cartographica Sinica, 26*(3), 241–246.

Chen, G., Xia, Q., & Wan, G. (2002). An algorithm of dynamic triangulation for the RSG model of terrain. *Acta Geodaetica et Cartographica Sinica, 31*(1), 44–48.

Chen, G., Yang, M., & Wang, K. (2003). An algorithm of dynamic triangulation for the TIN model of terrain. *Journal of Institute of Surveying and Mapping, 20*(4), 286–289.

Chen, J., Wei, Y., & Wu, L. (2005). Situation plotting and displaying in virtual battle environment. *Journal of Computer Applications, 25*(7), 1701–1703.

Chen, J., Gong, J., & Zhu, X. (2007). Study on distributed emulation of virtual terrain scene. *Journal of System Simulation, 19*(5), 1057–1060, 1081.

Chen, X. (2003). Exploration on geo-environment simulation and cognitive verification: A case study at Cudgewa catchment in Victoria, Australia. *Geographical Research, 22*(2), 245–253.

Chen, X., He, H., & Duan, Z. (1998). Generation of virtual vision based on multiple platforms. *Computer Research & Development,* (12), 1079–1083.

Chen, Y., & Guo, D. (1999). The study on 3D-GIS vector data structure: A case study of mine. *Acta Geodaetica et Cartographica Sinica, 28*(1), 41–44.

Chen, Y., & Liu, H. (2000). Comparative analysis on some new conceptions of map. *Beijing Surveying and Mapping,* (1), 29–31.

Cheng, P., Gong, J., & Shi, W. (2004). Geological object modeling based on quasi tri-prism volume and its application. *Geomatics and Information Science of Wuhan University, 29*(7), 602–607.

Dai, C., & Zhang, Y. (1999). A rendering technology of virtual terrain based on digital photogrammetry. *Engineering of Surveying and Mapping,* (3), 56–60.

Deng, X. (2003). Research on service architecture and algorithms for grid spatial data. *Acta Geodaetica et Cartographica Sinica, 32*(4), 362–362.

Dong, Z., Guo, Q., & Song, J. (2003). Real time visual simulation of battlefield environment of foreshore. *Journal of System Simulation, 15*(11), 1524–1526, 1532.

Du, W., Gong, J., & Ding, G. (2005a). Study of virtual geographic environments based on streaming media. *Computer Simulation, 22*(1), 180–183.

Du, Y., Wu, Y., & Wang, X. (2005b). A LOD-based algorithm of TIN model for terrain in large 3D landscape. *Journal of Institute of Surveying and Mapping, 22*(2), 119–121, 124.

Du, Y., Wu, Y., & You, X. (2006). Research of Mipmap textures for virtual global terrain environment. *Journal of Geomatics Science and Technology, 23*(5), 355–358.

Gao, J. (2000). Visualization in Geo-Spatial data. *Engineering of Surveying and Mapping,* (3), 1–7.

Gao, J. (2004). Cartographic tetrahedron: Explanation of cartography in the digital era. *Acta Geodaetica et Cartographica Sinica, 33*(1), 6–11.

Gao, J., & You, X. (1996). Virtual reality and its application in military mapping and training simulation. *Journal of Institute of Surveying and Mapping, 13*(2), 133–137.

Gao, J., Xia Y., You X., & Shu G. (1999). *Application of virtual reality in terrain environment simulation.* The PLA Press.

Gao, Y., Zhu, Q., & Ying, S. (2005). Visual variable of 3D models for 3D GIS. *Science of Surveying and Mapping, 30*(3), 41–43.

Gong, J., Li, Y., & Wang, D. (2008). Characteristics and applications of knowledge maps in geographic knowledge visualization: A case study of spatial planning of dam systems in watersheds. *Journal of Remote Sensing, 12*(2), 355–361.

Gong, J., & Lin, H. (2001). Study on distributed virtual Geo-environments. *Journal of Image and Graphics, 6*(9), 879–885.

Gong, J., & Lin, H. (2001). *Virtual geographic environment—a geographical perspective on online virtual reality.* Higher Education Press.

Gong, J., Lin, H., & Tan, Q. (2002a). Designing and developing the virtual campus of the chinese university of hong kong. *Acta Geodaetica et Cartographica Sinica, 31*(2), 39–43.

Gong, J., Lin, H., & Zhang, J. (2002b). Research on Geo-Process based computation visualization: a case study of flood simulation. *Acta Geographica Sinica,* 57(Suppl.), 37–43.

Gong, J., Sun Z., Li X., Cao, C., Li, X., Qian, Z., & Zhou, J. (2003). Simulation and analysis of control of severe acute respiratory syndrome. *Journal of Remote Sensing, 7*(4), 260–265.

Gong, J., Lin, H., & Xiao, L. (1999). Perspective on Geo-Visualization. *Journal of Remote Sensing, 3*(3), 236–244.

Gong, J., & Xia, Z. (1997). A 3D data model integrated with vector and raster. *Geomatics and Information Science of Wuhan University, 22*(1), 7–15.

Gu, F., Cui, X., & Xie, C. (2004). Research on visual expression method of temporal information in virtual geographical environment. *Modern Surveying and Mapping, 27*(1), 11–13.

Guo, D., & Du, P. (2000). Research on digital earth and 3D geographic information system. *Acta Geodaetica et Cartographica Sinica, 29*(3), 250–256.

Han, Y., Wan, G., & Cao, X. (2008). The mix division scheme of global grid. *Engineering of Surveying and Mapping, 17*(3), 16–20.

Hao, P., & Huang, B. (1997). An improved quadtree of DEM for terrain visualization. *Journal of Image and Graphics, 2*(8), 578–584.

Hou, H., Shi, F., & Zhu, Y. (2003). The research of synthetic environment for distributed virtual battlefield. *Journal of System Simulation, 15*(4), 597–600.

Huang, A., Jing, X., & Yu, H. (2003). The fuzzy optimization and modeling of the avion visual. *Journal of System Simulation, 15*(8), 1146–1148.

Huang, K., Liu, B., & Huang, J. (2004). A survey of military simulation technologies. *Journal of System Simulation, 16*(9), 1887–1895.

Hui, L., & Michael, B. (2009). *Virtual geographic environments.* Science Press.

Jin, B., & Bian, F. (2004). Method for 3D spatial data interoperability based on grid. *Geomatics and Information Science of Wuhan University, 29*(12), 1075–1079.

Jin, B., & Bian, F. (2006). Grid architecture and application based on 3D geospatial information. *Geomatics and Information Science of Wuhan University, 31*(10), 920–923.

Lan, C., Chen, J., & Li, J. (2007). Real-time 3D visual simulation of space mission. *Journal of Geomatics Science and Technology, 24*(1), 47–56.

Lan, Z., Zhou, Y., & Xu, Q. (2003). Design and realization of a system model for the expression of cyber city 3D scenic on web. *Journal of Institute of Surveying and Mapping, 20*(3), 190–193, 197.

Li, D., & Li, Q. (1997). Study on a hybrid data structure in 3D GIS. *Acta Geodaetica et Cartographica Sinica, 26*(2), 128–133.

Li, D., Liu, Q., & Zhu, Q. (2003). An integrated representation of outdoor and indoor in cybercity GIS. *Geomatics and Information Science of Wuhan University, 28*(3), 253–258.

Li, D., Zhu, Q., & Li, X. (2000). CyberCity: concepts, technique support and typical applications. *Geomatics and Information Science of Wuhan University, 25*(4), 283–288, 311.

Li, F., Wan, G., & Ma, W. (2007). Cloud modeling and real time render research. *Journal of Geomatics Science and Technology, 24*(1), 76–78.

Li, Q., Zhuang, C., Wang, P., & Zhang, Y. (2008). Research on getting intelligent navigational map of vr scene created by 3D configuration software. *Computer Measurement & Control, 16*(12), 1969–1971, 1988.

Li, Q. (1997). 3D GIS topological relation and dynamic on struction. *Acta Geodaetica et Cartographica Sinica, 26*(3), 235–240.

Li, Q., Chang, Y., & Cao, D. (2002). Concept of "one surface with three layers" in 3D GIS topologic relation and its extension in 2D. *Acta Geodaetica et Cartographica Sinica, 31*(4), 350–356.

Li, Q., & Li, D. (1996). Data structure in 3D geographic information system. *Geomatics and Information Science of Wuhan University, 21*(2), 128–133.

Li, Q., & Li, D. (1998). Research on the conceptual frame of the integration of 3D spatial data model. *Acta Geodaetica et Cartographica Sinica, 27*(4), 325–330.

Li, Q., Yan, Y., & Yang, B. (2003). Research on 3D visualization of underground pipeline. *Geomatics and Information Science of Wuhan University, 28*(3), 277–282.

Li, S. (2004). Some considerations about digital earth and geographical sciences. *Journal of Geo-Information Science, 6*(4), 15–18.

Li, S., & Sun, J. (2005). Advancement of digital yellow river based on the virtual geographic environments. *Progress in Geography, 24*(3), 91–100.

Li, S., & Yao, J. (2005). The multi-dimension data model of virtual geographic environments and its expression in geographical processes. *Geography and Geo-Information Science, 21*(4), 1–5.

Li, W., Gong, J., & Zhang, L. (2007). Mobile oriented collaborative virtual geographic studio. *Geomatics and Information Science of Wuhan University, 32*(9), 817–820.

Lin, H., & Gong, J. (2002). On virtual geographic environments. *Acta Geodaetica et Cartographica Sinica, 31*(1), 2–6.

Lin, H., Lv, G.N., & Song, Z. (2000). *A study on the tidal wave system and coastal evolution simulation in the East China Sea.* Science Press.

Lin, H., Gong, J., & Shi, J. (2003). From maps to GIS and VGE--A discussion on the evolution of the geographic language. *Geography and Geo-Information Science, 19*(4), 18–23.

Lin, H, & Xu, B. (2007). Some thoughts on virtual geographic environments. *Geography and Geo-Information Science, 23*(2), 1–7.

Lin, H., Yang, Y., & Zhu, Q. (2004). Content-based virtual geography scene retrieval and intelligent navigation. *Geography and Geo-Information Science, 20*(1), 2–6.

Lin, Z., Liu, X., & Zhang, J. (2005). A stereo visual and precise measurable virtual reality system. *Science of Surveying and Mapping, 30*(2), 16–18.

Lin, H., Zhang, J., Yang, P., & Liu, J. (2006). Development on spatially integrated humanities and social science.*Geo-Information Science, 8*(2), 30–37.

Lin, Z. (2007). *Research on spatial data service and agent model in virtual geographic environment based on MAS.* Doctoral Dissertation of Wuhan University.

Liu, K., Wu, H., & Hu, J. (2007). The actuality and vista of applications of virtual reality technology in earth science in china. *Journal of Henan University (Natural Science), 37*(4), 371–376, 334.

Liu, Z., Yang, J., & Dai, C. (2008). Rendering of vector data over 3D scene. *Metal Mine,* (6), 94–96, 153.

Liu, K., Liu, J., & Feng, X. (2005). A kind of virtual implementation method of benthos terrain and current. *Journal of System Simulation, 17*(5), 1268–1271.

Liu, Y., Wan, G., You, X. (2005) Modeling of virtual hand and arm based on data glove and trackers. *Journal of Institute of Surveying and Mapping, 22*(4), 272–274.

Liu, Q., Li, D. (2004). Architectural 3D CAFM in Cyber-city. *Geomatics and Information Science of Wuhan University, 29*(1), 34–37.

Liu, X., Guo, G., & Qiu, X. (2007). Research of dynamic terrain simulation. *Journal of System Simulation, 19*(16), 3820–3823.

Liu, Z., Wang, H., & Wang, G. (2002). Research on key issues for digital earth oriented virtual reality system. *Journal of Image and Graphics, 7*(2), 160–164.

Lv, P. (2006). *Design and implementation of ship-handlings simulation system based on virtual reality.* Master Dissertation of Huazhong University of Science and Technology.

Ma, J., & Zhu, D. (2002). Visualization modeling and virtual reality for coastal ocean tidal current numerical simulation. *Acta Geodaetica et Cartographica Sinica, 31*(1), 49–53.

Ma, X., Du, W., & Yuan, X. (2001). Walkthrough in image-based virtual in-door environment. *Journal of Image and Graphics, 6*(1), 86–91.

Ma, H., Xu, H., Di, L., & Wu, J. (2007). On virtualization of remote sensing products under environment of geo-spatial grid. *Geomatics and Information Science of Wuhan University, 32*(9), 808–811, 816.

Meng, L. (2006). Some theoretical concerns along with the development of cartographic technologies. *Journal of Geomatics Science and Technology, 23*(2), 89–100.

Meng, L., Wu, C., & Zhou, X. (2007). 3D GIS grid model based on role oriented task assign. *Geomatics and Information Science of Wuhan University, 32*(10), 910–914.

Mo, F. (2002). The fracture of surface texture based on the map data. *Journal of Institute of Surveying and Mapping, 19*(1), 59–61.

Pan, M., Fang, Y., & Qu, H. (2007). Discussion on several foundational issues in 3D geological modeling. *Geography and Geo-Information Science, 23*(3), 1–5

Qi, M., Hao, C., & Tong, M. (2000). Research of modeling method to 3D multiresolution terrain based on fractal. *Journal of Image and Graphics, 5*(7), 568–572.

Qi, M., Shu, G., & Chi, T. (2004). Agent-based multi-user virtual geographical environment. *Journal of System Simulation, 16*(5), 1092–1095.

Ren, X., Wang, R., & Qi, Z. (2006). The simulation model of virtual battlefield based on grid. *Manufacturing Information Engineering of China, 35*(19), 88–91, 95.

Rui, C., Yang, C., & Wang, P. (2002). Study on the data acquisition & organization of three-dimension urban modeling. *Geography and Territorial Research, 18*(2), 11–14.

Shen, D., & Lin, H. (2005). Preliminary study on theory and methods for 3D simulation of dynamic soft geo-objects. *Geography and Geo-Information Science, 21*(1), 9–11.

Shi, M., Zhu, D., & Wang, Z. (2006). A method to generate character animation on 3D terrain. *Journal of System Simulation, 18*(5), 1247–1252, 1270.

Shi, W. (1998). Error models for geometric features in 3D GIS. *Geomatics and Information Science of Wuhan University, 23*(1), 18–20, 25.

Shi, W., Cao, H., & Zhang, J. (2004). 3D urban building modeling based on high resolution satellite images. *Geomatics and Information Science of Wuhan University, 29*(9), 783–787.

Shu, Y., Zhu, G., & Chen, C. (2004). Generation of virtual forest. *Geomatics and Information Science of Wuhan University, 29*(6), 540–543.

Su, L., Huang, Y., & Li, X. (2002). Generation of 3D real scene of remote sensing pixel. *Journal of Image and Graphics, 7*(6), 570–575.

Sun, J. (1999). An exploration of virtual recreation environment on resources and environment science. *Resources Science, 21*(1), 1–8.

Tan, R., Li, C., & Wang, H. (2006). Establishment and application in noise analysis of 3D city model. *Science of Surveying and Mapping, 31*(3), 101–103.

Tang, W., Lv, G., & Wen, Y. (2007). Study of visual geographic modeling framework for virtual geographic environment. *Journal of Geo-Information Science, 9*(2), 78–84.

Tao, C., & Wang, Q. (2002). GIServices-based 3D internet GIS: GeoEye 3D. *Acta Geodaetica et Cartographica Sinica, 31*(1), 17–21.

Wan, G., Cao, X., & Li, F. (2007). IBR real-time shadow technology in the battlefield environment simulation. *Journal of Geomatics Science and Technology, 24*(1), 18–21.

Wan, G., Chen, G., & You, X. (2002b). Research on geometry modeling of terrain features in virtual city. *Journal of Institute of Surveying and Mapping, 31*(1), 60–65.

Wan, G., Gao, J., & Liu, Y. (2008). Research on cognitive map formation based on reading experiments. *Journal of Remote Sensing, 12*(2), 339–346.

Wan, G., Gao, J., & You, X. (2005). Several spatial cognition problems in virtual terrain environment simulation. *Science of Surveying and Mapping, 30*(2), 48–50.

Wan, G., Wang, M., & Wu, Z. (2004). The creation of realistic sound in virtual environment. *Journal of Institute of Surveying and Mapping, 21*(1), 50–53.

Wan, G., & Xia, Q. (2002). The research of entity-behavior modeling in virtual scenic environment. *Journal of Institute of Surveying and Mapping, 19*(3), 214–216.

Wang, R., Yao, X., & Wang, H. (2003). Grid computing application in distributed virtual reality environment. *Journal of Chongqing University of Posts and Telecommunications (Natural Science Edition), 15*(3), 45–50.

Wang, B., Wei, Z. (2005). A research on 3D modeling of roadway in virtual environment for coalmine. *Journal of Institute of Surveying and Mapping, 22*(3), 204–206, 211.

Wang, C., Tang, X., & Chen, M. (2006). Adaptive road modeling method based on terrain-matching. *Journal of System Simulation, 18*(10), 2824–2826.

Wang, G., Wu, X., & Chen, B. (2002). The implementation of high-bandwidth network based real-time interactive browsing for 3D massive terrain data. *Acta Geodaetica et Cartographica Sinica, 31*(1), 34–38.

Wang, G., Liu, N., & Wan, G. (2005). A fusion algorithm and accuracy estimate between road and terrain in virtual terrain environment. *Acta Geodaetica et Cartographica Sinica, 34*(4): 337–342.

Wang, J., Li, C., & Lin, Z. (2004). Method for 3D abstraction and description of City. *Geomatics and Information Science of Wuhan University, 29*(8), 674–678.

Wang, J., Li, M., & Geng, W. (2002). Scene-planning system based on augmented reality technique. *Journal of Image and Graphics, 7*(4), 405–410.

Wang, X., Liu, Y., & Zhang, J. (2005). Geo-spatial cognition: an overview. *Geography and Geo-Information Science,* (6), 1–10.

Wang, X. (1999). Modeling and simulation: review and prospect. *Journal of System Simulation, 11*(5), 309–311.

Wang, Y., Wu, L., & Li, X. (2007). On 3D GIS spatial modeling. *Journal of Image and Graphics, 12*(8), 1430–1434.

Wang, Y. (2000). Overview of terrain visualization. *Journal of Image and Graphics, 5*(6), 449–456.

Wang, Z., Pan, M., & Qu, H. (2007). Triangulation visualization of 3D topological geology section. *Geography and Geo-Information Science, 23*(6), 42–45.

Wei, M., Yang, F., & Cao, Z. (2003a). Study on the animation matching virtual scene. *Journal of System Simulation, 15*(9), 1282–1284.

Wei, Q., Gao, L., & Gong, X. (2003b). The technique of browsing virtual world based-on cubic panorama. *Journal of Image and Graphics, 8*(9), 1061–1066.

Wu, J. (2005). Research on rapidly reconstructing texture for facades in 3D city modeling. *Acta Geodaetica et Cartographica Sinica, 34*(4), 318–323.

Wu, Y., Du, Y., Wang, X., & You, X. (2005). Lod-based algorithm of TIN model for large terrain simulation. *Journal of System Simulation, 17*(3), 665–669.

Wu, X., Li, S., Gong, J., Zhao, D., & Zhu, J. (2006). Research on the system architecture of grid-VGE. *Hydrographic Surveying and Charting, 26*(5), 71–74.

Wu, L., Chen, X., & Che, D. (2007). A GTP-based entity model for underground real 3D integral representation. *Geomatics and Information Science of Wuhan University, 32*(4), 332–335.

Wu, Z., You, X., Wan, G., & Li, F. (2008). Research of particle system applied in battlefield environment simulation. *Journal of Geomatics Science and Technology, 25*(1), 35–38.

Wu, L., & Shi, W. (2005). On 3D Geosciences spatial modeling. *Geography and Geo-Information Science, 21*(1), 2–4.

Wu, L., Zhang, R., & Qi, Y. (2002). 3D geoscience modelling and virtual mine system. *Acta Geodaetica et Cartographica Sinica, 31*(1), 28–33.

Xiao, B., & Xu, H. (2003). Underwater terrain construction and visualization. *Journal of System Simulation, 15*(3), 326–328.

Xiao, Y., Xu, Q., & Zhou, Y. (2006). Research and implement of shadow rendering algorithm in 3D city scene. *Journal of Geomatics Science and Technology, 23*(4), 308–309.

Xie, C., Meng, X., & Wan, H. (2002a). Data model of virtual reality GIS and its implementation based-on active object-oriented database system. *Journal of Geo-Information Science, 4*(4), 44–49.

Xie, C., Zhou, C., & Lin, H. (2002b). The design of web-oriented virtual GIS with XML and CORBA. *Acta Geodaetica et Cartographica Sinica, 31*(1), 66–70.
Xia, Q. (1998). Construction and application of 3D stereo terrain simulation. *National Conference of Graph and Image*, pp. 386–390.
Xu, B., Gong, J., & Lin, H. (2005). Architecture of distributed virtual geographic environment based on HLA. *Geomatics and Information Science of Wuhan University, 30*(12), 1097–1099.
Xu, H., & Shu, G. (2000). Research on 3D terrain model simplification technology in virtual terrain environment. *Journal of Computer Applications, 20*(S1), 261–264.
Xu, H., & Pan, M. (2005). The application study of volume rendering in the visualization of 3D GIS for geology. *Geography and Geo-Information Science, 21*(2), 7–9.
Xu, S., Wei, Z., & Yang, J. (2001). Application of virtual reality technology in virtual touring. *Geography and Geo-Information Science, 17*(3), 92–96.
Xu, Z., & Wu, X. (2006). Simulation climate environment in virtual geographic environments. *Journal of Geomatics, 31*(1), 39–40.
Xue, Y., Zhang, Z., & Zhang, J. (2003). Compression and restoration of texture images of walls for 3D visualization of urban buildings. *Geomatics and Information Science of Wuhan University, 28*(5), 526–528.
Yan, H., Ma, C., & Fei, L. (2004). The research on 3D numerical simulation of the groundwater flow field and its visualization: Take the Changzhou-Wujin district as an example. *Journal of Image and Graphics, 9*(12), 1491–1495.
Yan, H., Ma, C., & Zhu, G. (2005). The research and realization of 3D interactive techniques for the volume visualization of geo-information. *Acta Geodaetica et Cartographica Sinica, 34*(1), 65–70.
Yang, B., Li, Q., & Mei, B. (2000). Study of the visualization of 3D urban model. *Acta Geodaetica et Cartographica Sinica, 29*(2), 149–154.
Yang, B., Li, Q., & Shi, W. (2003). Research on multi-resolution texture model in 3D GIS. *Journal of Image and Graphics, 8*(3), 328–333.
Yang, C., Zhang, J., & Lin, Z. (2001). The real-time LOD algorithm for rendering the virtual terrain scene. *Acta Geodaetica et Cartographica Sinica, 30*(2), 134–139.
Yang, H., Wen, Y., & Lv, G. (2007). Research on VGE-oriented geospatial information sharing. *Journal of Geo-Information Science, 9*(4), 75–79.
Yang, W.,& Liu, Y. (1998). Integration of terrain data in a system of virtual battlefield environment. *Transactions of Beijing Institute of Technology, 18*(6), 748–752.
Yang, Z., Zhang, Y., & Wang, H. (2001). Virtual observer model in virtual environment. *Journal of System Simulation, 13*(4), 467–468, 500.
Yi, Y., Gong, J., & Ma, A. (2004). Construction of distributed virtual olympic environment system and visual representation of "Human Culture Olympic". *Geography and Geo-Information Science, 20*(5):32–35.
You, X. (2002). The battlefield environment simulation based on VR. *Acta Geodaetica et Cartographica Sinica, 31*(1), 8–11.
Zhang, J., He, S., & Su, G. (2005). Application of image texture reorganization to 3D model reconstruction. *Geomatics and Information Science of Wuhan University, 30*(2), 115–117, 186.
Zhang, J., & Liu, J. (1997). Generation of fractal terrain using terrain model fitting method. *National Conference on Remote Sensing, Telemetering and Remote control*, pp. 309–317.
Zhang, J., Zhang, Z., & Xu, F. (1998). Reconstruction of 3D landscape from images with large scale for urban region. *Geomatics and Information Science of Wuhan University, 23*(4), 356–358.
Zhang, J., Gong, J., & Lin, H. (2006a). A distributed virtual geographic environment system based on web service. *Chinese High Technology Letters, 16*(3), 296–300.
Zhang, H., Jia, Q., Sun, H., Gao, X., & Ruan, R. (2006b). New solution of geometrical calibration based on immersive multi-projector displays. *Journal of System Simulation, 18*(z2), 493–496.
Zhang, J., Liu, J., & Guo, Y. (2004). On collaborative navigation and spatial awareness in CVE. *Journal of Institute of Surveying and Mapping, 21*(3), 225–227, 230.

Zhang, M., Wen, J., & Zhang, Z. (2007). Modeling and virtual demo of multi-users' coordination in VGE. *Journal of System Simulation, 19*(24), 5726–5730, 5735.

Zhang, R. (2005). A virtual laboratory system for urban traffic management based on GIS and multi-agent system. *Geography and Geo-Information Science, 21*(3), 495–498.

Zhang, Z., Wu, J., & Zhang, J. (2003). A method for determining orientation parameters in image-based reconstruction of architectural scenes. *Geomatics and Information Science of Wuhan University, 28*(3), 265–271, 300.

Zhao, Q., Shen, X., & Xia, C. (1998). DVENET: A distributed virtual environment. *Computer Research & Development,* (12), 1064–1068.

Zheng, K., Zhu, L., & Wu, X. (2006). Study on spatial indexing techniques for 3D GIS. *Geography and Geo-Information Science, 22*(4), 36–39.

Zhong, Y. (2005). Mathematical definitions for imitative map symbol, virtual map symbol and geo-cyberspace. *Geomatics and Information Science of Wuhan University, 30*(6), 558–559.

Zhou, J., Gong, J., & Chen, Z. (2005). Study on collaborative virtual geographic environment system for multi-user interaction and communication. *Geography and Geo-Information Science, 21*(5), 34–37.

Zhou, K., Zhao, X., & Ding, Y. (2006). The 3D visualization's approach based on laser scanning. *Journal of Geomatics Science and Technology, 23*(1), 70–72.

Zhou, Y., Zhu, Q., & Feng, L. (2005). Research on spatial-temporal consistency in DVGE. *Geomatics World, 3*(5), 23–26.

Zhou, Y., Xu, Q., & Tan, B. (2002). The application of image fusion in 3D terrain simulation. *Journal of Image and Graphics, 7*(6), 592–595.

Zhu, J., Gong, J., & Lin, H. (2007a). A mobile agent computing model for DVGE. *Geomatics and Information Science of Wuhan University, 32*(4), 337–339.

Zhu, J., Gong, J., & Lin, H. (2007b). Study on VGE computing architecture and key technologies. *Journal of North University of China (Natural Science Edition), 28*(4), 326–331.

Zhu, J., Gong, J., & Zhang, J. (2006). Study on the network architecture of virtual geographic environment based on P2P. *Science and Technology & Innovation, 22*(12–1), 191–193, 242.

Zhu, Q. (1998). 3D dynamic interactive visualization model-3D representation and analysis for GIS. *Geomatics and Information Science of Wuhan University, 23*(2), 124–127.

Zhu, Q. (2004). A survey of 3D GIS technologies. *Geomatics World, 2*(3), 8–12.

Zhu, Q., Gao, Y., Wei, Y., & Huang, D. (2003). Design of 3D models for 3D GIS. *Geomatics and Information Science of Wuhan University, 28*(3), 283–287.

Zhu, Q., Gong, J., & Du, Z. (2005). LODs description of 3D city model. *Geomatics and Information Science of Wuhan University, 30*(11), 965–969.

Zhu, Q., Lu, D., & Zhang Y. (2006). Application of 3D GIS in digital cultural heritage. *Science of Surveying and Mapping, 31*(1), 55–57.

Zhu, X., & Lv, G. (1998). Virtual reality and its applications in geography. *Geography and Geo-Information Science, 14*(3), 60–63.

Zhu, Y., Zhang, Z., & Zhang, J. (1998). Research of 3D visualization method of urban buildings concerning terrain. *Geomatics and Information Science of Wuhan University, 23*(3), 199–203.

Zuo, X., Li Q., & Tang L. (2004). 3D modeling and data organization of highway. *Geomatics and Information Science of Wuhan University, 29*(2), 180–183, 161.

Chapter 13
Geographic Ontology

Hongwei Li

13.1 Introduction

Ontology is a concept that originated in the field of philosophy; ontology describes the nature of things and is generally understood to be the science of existence, or a philosophical study of the nature of being; ontology is an interpretation or explanation of an objective existence that is concerned with the abstract nature of objective reality, and it belongs to the branch of metaphysical theory. This concept, which has existed for more than 2000 years in western philosophy, is still a hot topic discussed by philosophers.

The understanding of ontology in academia is very different and far from consensus. Nevertheless, this does not prevent the continued popularity of the ontology concept and its research in various fields. In addition, ontological research in computational linguistics, artificial intelligence, knowledge engineering, databases and other fields is particularly hot and supported by the National Natural Science Foundation of China, the State "863" Project and many other research projects, and the field of ontology has achieved a series of research results, with much focus on this area in academic circles. The main reason for the rise in ontology studies in the field of science and technology is that semantic web, semantic modeling and semantic integration require ontology as the basic theory to support development.

According to the analyses and research of scholars, ontology first appeared in the field of science and technology in 1984 and was put forward and defined in 1991. Since then, the definition of ontology has been increasingly clarified. In general, the definition of ontology contains 4 main aspects: conceptualization of abstract models of phenomena in the objective world, which abstract the domain knowledge individually into a definite object; a clear definition, that is, the concepts of each object and the relation between them are reasonably defined; formalization, that is, accurate mathematical descriptions of concepts and their relationships are required and computer

H. Li (✉)
Information Engineering University, Zhengzhou, China

J. Wang and F. Wu (eds.), *Advances in Cartography and Geographic Information Engineering*, https://doi.org/10.1007/978-981-16-0614-4_13

readable levels should be reached; shareability, i.e., the knowledge reflected in the ontology is recognized by all its users (including experts in domains and general users). In short, the applications of ontological ideas in the field of science and technology involve using an object-oriented method to restore the existing knowledge, information and data into a reasonable semantic system so that they can be processed in computers and be shared by people.

The classification system of ontology varies due to the starting point and basis of classification. The most representative is the division of ontology based on the degree of detail and the degree of dependence (Guarino 1994), which divides ontology into three types of different abstract level ontologies: top-level ontology, domain ontology and task or application ontology. Top-level ontology studies generic concepts; domain ontology studies specific domain concepts; and application ontology is geared toward specific applications.

13.2 Development Process

The introduction of the ontology concept into the field of geographic information science occurred in recent years, reflecting a shift in the research focus from the previous emphasis on the formalization of computer models to the present focus on the spatial domain itself (Du 2001). When ontology was introduced into the field of geographic information science, the concept of geographic ontology (geo-ontology) emerged naturally. It is a special domain ontology, which has its own characteristics in addition to the basic features of general domain ontology. We define geographic ontology as studying the detailed connotation and hierarchical relationship of geospatial information concepts at different levels and in different application directions in the field of geographic information science, and we semantically identify the ontological concept. Geographic ontology abstracts knowledge, information, and data about geographic science into objects (or entities) of consensus and form a system in accordance with certain relationships; further, conceptual processing and explicit definitions are also given, and the formal representation of theories and methods are adopted.

A geographic ontology domain can be represented by a 5 tuple GO = (GCD, GRD, GFD, GID, GAD), where GO represents geographic ontology, GCD is the definition of a series of geographical concept or classes, GRD is the definition of a series of geographical relations, GFD is the definition of a series of geographical functions, GID is the definition of a series of geographical concept instances, and GAD is the definition of a series of geographical axioms.

A geographical concept or class has a wide range of meanings and can be used to refer to anything in the geographical world; it is a collection of geographical entities or geographical phenomena with the same attributes, which includes two categories: geographical entities and entity attributes. Mountains, rivers, seas, hills, plains, deserts, swamps, forests, crops, wind, rain, snow and frost are concrete examples of the concept of geographical entities, while altitude, sediment concentration, flow rate, sea surface temperature, area, forest volume, crop yield, wind speed, rainfall and so on are specific examples of the concept of geographical entity attributes.

Geographical relations represent the interaction of geographical concepts, formally defined as subsets of n-dimensional Cartesian products, including two classes of basic relations, namely, spatial and non-spatial relations. There are 3 basic spatial relations among geographical concepts, namely, topological relations, sequential relations and metric relations. Non-spatial relations include 4 basic types, namely, part-of, kind-of, instance-of and attribute-of. Topological spatial relations represent spatial relations among geographical entities, such as adjacency, inclusion and intersection; sequential spatial relations refer to the spatial arrangement of geographical entities, namely, the orientation relations between entities, such as front and back, left and right, east and west, north and south, etc.; metric spatial relations mainly refer to the distance relation between geographic entities; part-of represents the relation between a part and the whole of a geographical concept, for example, the Taiwan Province is a part of China; kind-of represents the inheritance relationship between geographic concepts, similar to the relationships between the parent class and subclasses in object-oriented relationships, for example, a perennial river and seasonal river are subclasses of a river; instance-of represents the relationship between geographical concept instances and geographic concepts, similar to the relationship between objects and classes in an object-oriented relationship, for example, the Beijing Shenyang Expressway is an instance of a highway; finally, attribute-of represents that a geographical concept is an attribute of another geographical concept, for example, the concept "flow" is a property of the concept "river".

A geographical function is a special kind of geographic relationship in which the preceding n-1 elements can uniquely determine the nth element. For example, on a given DTM surface, the geographical coordinates X and Y can uniquely determine the Z value (which can be an elevation or a certain concentration of chemical substances), and then, there exists a functional dependency between the geographical concept Z and X and Y.

In practical applications, it is not necessary to construct ontology in strict accordance with the above 5 tuples, and the relationship between concepts is not limited to the two basic categories and 7 basic relations. In addition, the instance-of in instances and non-spatial relations seem to have a possibility of repeated definitions, which should be paid attention to in the construction of geographical ontology.

With regard to ontology and its research in the field of geographic information science, overseas studies occurred earlier and more literature and achievements have been made; China's research in this field started in the early twenty-first century, which was less than 10 years, but the progress is unusually rapid, and some preliminary theories and application results have been obtained.

In search of the Wanfang database and VIP database, the first formal thesis related to geographical ontology was 'Resource and environment information classification and coding and their relationship with geographic ontology' written by He et al. (2003), in which the concept of geographic information ontology was discussed, and a suggestion to use information ontology technology to classify resource and environment information was proposed. In Ren (2003), the problem of ontology-based distributed information access and interoperation was studied, and an ontology-based framework for interoperable systems was constructed, including a source system directory, search engine, shared ontology service, mapping service, ontology interface, source level local ontology set, user query interface, etc. Cui (2004a) completed his first doctoral dissertation on geographic ontology, which was the research topic of the semantic integration and interoperation of geographic information system based on ontology. Sun et al. (2004) summarized the studies on geographical ontology. Wang et al. (2004) proposed an ontology-based middleware for geographic information systems (GISs), which was located between the GIS platform and the application program, and the semantic interpretation of spatial data was realized through an ontology library and an ontology server, and then, the tight integration of the GIS application system was realized. In An et al. (2004), an ontology-based method for describing and discovering network geographic information services was proposed, a concept layer and an application layer were used to describe geographic information services, and after the service was published, the degree of matching between a service and user was judged by measuring the input and output parameters in the service description and the function description in the user requirements; finally, the service discovery was implemented. In Cui (2004b), a new approach for interoperation between information systems using data mining and ontology was proposed, where semantic transformation between different spatial information systems was implemented through ontology-based logical reasoning, and the transformation set was determined from the candidate set by using data a mining technique; additionally, the result of the operation was returned to the user, and thus, the interoperability of the spatial information system was accomplished; finally, the robustness and flexibility of the interoperation of the spatial information system were improved. In Wang et al. (2005), the problem of semantic heterogeneity in the process of system integration was discussed, the possibility of using ontology to solve semantic heterogeneity was analyzed, and a method for building ontology in a data integration environment was described. In Wu et al. (2005), an information integration method based on hybrid ontology was studied, and through mapping between global ontology and local ontology, a unified interface for data accessing was provided to enable users to access semantically relevant data. In Tan et al. (2005), a geospatial information system interoperability based on standard spatial ontology and hybrid semantic ontology was proposed, and through ontology collaboration, heterogeneous spatial information interworking based on standard spatial ontology and spatial information reclassification, as well as semantic interoperability based on the hierarchical matching of heterogeneous semantic ontology were realized. In Jing et al. (2005), the meaning and levels of geospatial information sharing were analyzed, and referring to the research idea of semantic web, a framework for semantic sharing of geospatial information was established, and finally, ontology-based concepts and methods were proposed to formally define the domain knowledge and solve the problem of the semantic ambiguity of geographic entities in geospatial information research.

If the introduction of ontology to the artificial intelligence field is mainly based on the purpose of knowledge sharing and reuse, then the ontology of geographic information science has the dual meaning of both philosophical ontology and information ontology. Philosophical ontology focuses on the geographic object domain itself, mainly involving the research of geographical concepts, categories, relations and process, and the spatiotemporal ontology, uncertainty ontology and scale ontology are also important embodiments of philosophical ontology. Through the study of philosophical ontology, especially the design of geographical categories and entity types in the geographical object domain, the purpose is to produce a better understanding of the structure of the geographical world, provide a more rational conceptual model for the development of GIS, and finally avoid the large contrast between the existing data model and the spatial cognition mechanism of human beings.

Geo-ontology is a domain ontology, which acquires consistent knowledge in the domain of geography, and this knowledge is not private for an individual and can be accepted and shared by a community.

The purpose of building geo-ontology is to define a common vocabulary for researchers who need to share information in the geographic domain, which includes the definition of concepts in the geographical domain and the relationship between concepts; further, a common understanding of structured geographic information is shared among people or software Agents to support the reuse and analysis of geographic domain knowledge, as well as for other purposes.

First, geo-ontology can share the common understanding of structured geographic information among people or software Agents, which is one of our common goals in the field of geographic information science research. For example, suppose several different websites contain geospatial information or provide navigation and geographic information services; if these sites share and publish the same underlying ontology for all the terms they use, then computer Agents can extract and integrate information from different sites. These Agents can use this integrated information to answer the user's request or use this information as input for other applications.

Second, geo-ontology supports the reuse of geographic domain knowledge. Geographical information has the characteristics of socialization, accounting for approximately 80% of the Earth's information. The widespread use of geographic information is the driving force behind today's geo-ontology research. For example, many different domain models need the concept to express geographic space; if a researcher develops a geo-ontology that includes location information, directional information, and coordinate reference information and others can reuse it in their domain, then we can integrate the geo-ontology with other existing ontologies to describe each part of a wider range of applications. In addition, geo-ontology supports the reuse of expert knowledge and expands it to describe areas of interest for people.

Third, the creation of geo-ontology makes explicit assumptions about the geographic domain, which is the basis for the implementation of geo-ontology. If knowledge in the geographic domain changes, then geo-ontology will likely be able to adapt to these changes more easily, thereby changing these hypothetical expressions. Typically, at the programming language level of knowledge about the real world, the expression of hypothetical hard code not only makes these putative codes difficult to understand but also makes these assumptions difficult to change, and especially for those without programming experience, it is harder to make changes at the code level. Furthermore, explicit specification of geographical domain knowledge is useful for new users who have to understand the meaning of terminology in the geography domain.

Fourth, Geo-Ontology can separate domain knowledge from actionable knowledge, which is another common use of ontology. We can describe the configuration task of a GIS component based on the requirement specification of GIS development, and we can finish the configuration by using a program that has nothing to do with the product or its components.

Fifth, geo-ontology can help us analyze the knowledge in the field of geographic information. Once the specification of a term display can be exploited, it is possible to analyze domain knowledge in a geographic area. Formal analysis of the term set is of great value in the need to reuse existing ontology and extend them.

13.3 Development Process

The research of geographical ontology in China started in the early twenty-first century, and the related research results can be found in much of the literature since 2000. By summarizing the literature over the last 10 years, we find that achievements have been made in theoretical research and method exploration, but there is no practical system. Specific studies and results are primarily reflected in the following aspects.

13.3.1 Reviews of Studies on Geographical Ontology

The domestic research on geographical ontology started later than in other countries. Based on reading a large amount of literature and data, many scholars have created a general description of the international research progress of geographical ontology and played an active role in guiding Chinese scholars to study geographical ontology.

In the article "Geo-ontology, geographical information ontology" (Sun 2004), the ontology of geographic information was systematically explained, and the main research content, hotspots, research plans and research directions of geographic ontology were discussed in detail. In Wu et al. (2006), a human-Earth relationship discussion based on subject-human and geo-ontology was made through the analysis of geo-ontology and ontology, the relationship among geo-ontology, scientific ontology and philosophical ontology was discussed, the differences and relationships among geo-ontology and classes and metadata were analyzed, and then, a new model of "human-land-GIS", which considered geo-ontology to be the core, was constructed; the model was divided into two parts of cognition and reproduction and consisted of the core ontology model and human cognitive layer, GIS model layer and "geo learning" phenomenon and their relational layer; taking the subject-human cognition as the main line, the functions and relations of each module in the model were discussed, and the future development direction was pointed out.

In Chen et al. (2006a, b), the definition of ontology in the scientific domain was deeply analyzed, and on this basis, the definition of geographical ontology was given, the research significance of geographic ontology was discussed, the major international planning and academic conferences related to geographical ontology were cited, and the progress in the research and application of geographical ontology at home and abroad was analyzed in detail; finally, it was pointed out that geographic ontology was a new and developing field in geographic information science, and China should keep up with the research trends of international geographical ontology and actively carry out research on geographical ontology theory, methods and applications. In Xu et al. (2007), a new architecture of GIS was proposed for the application of ontology in GIS. In Li Hongwei et al. (2008), the software engineering method of GIS development based on ontology was discussed, and on the basis of analyzing the general methods of traditional information system development, the evolution from GIS modeling to ontology-based information system modeling was discussed, an ontology-based two-layer structure system for GIS development was constructed, the technical approach of geographic information searches based on ontology was discussed, and the mapping between ontology and the concept paradigm was analyzed and summarized.

13.3.2 Research on Geographic Ontology and Geographic Information Classification

Classification is the most basic reflection of a human's cognition and perception of geographical objects. When people face complex geographical environment objects and their relationships, the first thing they think of is the division of class, grade and attribution; thus, through the process of "cognition, practice, recognition and repractice", a basic geographic information classification system was summarized and perfected according to the characteristics and rules of geographical objects, and finally, the standardization of the geographic information classification was realized. The application of geographic ontology to geographic information classification is said to be the promotion of geographic information classification in cognitive methodology. In He (2003), the classification and coding of resources and environmental information and its association with geographical ontology were studied. It was pointed out that using geographical ontology to classify and encode resources and environmental information was prospective research work. In Li et al. (2004, 2005, 2006), based on an analysis of the existing classification methods of geographic information and according to cognitive classification and taxonomy, ontology classification methods were discussed in detail; taking drainage in natural geographical elements as an example, an experimental study of ontology classification was carried out; then, the concept of geographic information was divided into two categories: meta concept and compound concept; based on certain constraints and according to the meta attributes of the geographical information concept, the classification of basic geographic information was realized by way of formalization. In Li et al. (2007, 2009), the classification and hierarchical representation of land use data based on ontology were studied; a concept in the hierarchical system of geographic data described by OWL was used to navigate to its parent concepts or sub concepts, and by merging elements in a geographic element class associated with all the sub concepts of a geographic concept in a geographic ontology, the combination of geospatial data contents was realized; finally, by categorizing the elements in a geographic element that was associated with a geographical concept in the geographical ontology as the element classes that are associated with their sub concepts, the specialization of geospatial data contents was realized. It was pointed out that the classification and hierarchical representation research of geographic data based on the ontology method was a useful supplement and enriched the current visualization methods of geospatial data.

13.3.3 Research on Construction of Geographical Ontology

The construction of geographic ontology is the key step in applications of geographic ontology. The construction of geographic ontology is a very complicated task; it needs to be based on the cognition of geographical objects and not be done by a person or a geographic community alone; only those individuals, organizations and institutions engaged in the study of geographical ontology together are able to make breakthroughs. At the same time, the construction of geographic ontology is one of the key issues for semantic sharing. Domestic research on the construction of geographical ontology is very dispersed, the research results are mostly in experimental ontology in related papers and writings, which are relatively simple and ideal, and it is hard to find a large-scale geographical ontology project. In Huang et al. (2004), several simple geographic ontologies were constructed by using OWL language, and the way to construct the ontology for a geographic information service was explained. In An et al. (2006), geographic ontology was established based on GIS, the concept of geographic ontology was compared from two aspects of structure and semantics, and the similarity calculation was implemented and used to judge the relationships among various geographical concepts. In Han et al. (2006), the establishment of domain ontology for railway line selection was discussed, and a comprehensive solution for railway intelligent line selection was put forward. In Liu et al. (2007), a generalized geographic name and its ontology was studied, and the corresponding conceptual models and logical models were given; then, the generalized geographic name ontology was described from the aspects of the generation method of generalized place names, describing objects, geographic scope, information groups and ambiguity caused by context, and so on; finally, a logic model that was easy to implement in the information system was established by using UML language. In Li et al. (2008), a conceptual framework and a structure for place name ontology were studied, the concept design of place name ontology was carried out, a conceptual framework and design method for constructing geographical name ontology by comprehensively using place name dictionary and geographical thesaurus were proposed; it was pointed out that the place name ontology was composed of three kinds of geographical ontology: geographical entity, entity type, and spatial relation, and its structure design was introduced in detail. In Bian et al. (2008), the problem of how to construct location ontology for geographical knowledge base was studied, which laid the foundation for solving the spatial location, expression and retrieval of geographic knowledge in the geographic knowledge base. Taking the Liaodong Bay as an example, a bay geographic ontology was constructed by Du et al. (2008), and a conceptual model that reflects the hierarchy of the bay's geographic entities was established, which was used for gulf data organization and data management.

13.3.4 Research on Spatial Data Organization, Management and Expression Based on Geographic Ontology

Spatial data organization and management was the core content of GIS, which involved the expression of geographic objects, entity modeling, database design and so on. The introduction of geographic ontology into spatial data organization and management provided a richer choice of methodology. In view of the deficiency of OWL language to express geographic ontology, Huang et al. (2005) proposed to construct some formal axioms about spatial properties and spatial relations with the help of three theoretical tools, that is, mereology, location theory and topology, and then, they added these new axioms to the OWL axioms to build the geographic ontology, which represented its spatial properties. The tests proved that the formal representation mechanism was feasible and could effectively describe the spatial properties of geographic ontology. In Wang et al. (2006), the key factors affecting multiscale representations from road features were analyzed. It was considered that multiple abstraction and representation schemas were caused by inconsistent conceptual cognition, abstractions and representation rules, which was a natural ontological problem when ascertaining multiscale entities and relations from multiscale representations. Thus, an ontological solution based on the natural principle was provided to accommodate the inconsistent conceptual cognition, in which cartographical natural rules served as a basic and general guideline in modeling multiscale representations. Based on these strategies, multiscale objective abstractions and quantitative representations were implemented in the following steps, namely, smallest visible object (SVO) measurement, geometry type abstraction, and semantic relation investigation. Thus, it was followed by a detailed description of multiscale geometrical abstract type and entity hierarchy during road network multiscale conceptual views, as well as the hierarchical semantic relations at both the feature class level and representation level, which provided a unified and objective multiscale conceptual view for multiscale road network modeling. In Zheng et al. (2006), a formal approach based on DL ontology was presented to support multiple representation, which virtually implied the capability to model and access geographic information flexibly under different resolutions or different viewpoints. As an explicit specification of conceptualization, ontology was adept in semantic expression and played a key role in information sharing and integration. It was a promising means to construct a conceptual data model close to geographic reality based on formal ontology. To formalize the basic topological relationships between regions and the varying semantics due to different viewpoints or levels of abstraction, two extensions to the classical description logics about the concrete domain and context were created. With extensions to DLs and context, a framework for the multirepresentation geo-ontology (MRGO) was finally proposed. It provided a consolidated and logic-based foundation for the flexible representation of semantic multiplicity, as well as consistent modeling of geometric multiplicity. According to Yang et al. (2007), the design of spatial databases was often challenged by opposing needs, such as the expertise need or the corporate need, and the ontology-oriented heuristics design of the spatial database balanced these needs by exploiting a domain ontology model. Thus, a practical spatial database system was completed. Practical applications showed that the mechanism was feasible. According to Su et al. (2007), aiming at deficiency of building conceptual model of spatial database with the E-R model, the ontological ideology and methodology were introduced into the geographic information domain. The exact meaning and its formal represented language of geographic ontology were discussed. The characteristics of the spatial E-R model were clarified and the relationship between geographic ontology and the spatial E-R model was analyzed. The concept of designing the spatial E-R model based on geographic ontology as a semantic foundation was presented. In addition, a method of establishing conceptual model of a spatial database by designing a spatial E-R model from geographic ontology was introduced with an example of irrigation ontology represented by expanded OWL. According to Du et al. (2008), the bay was a complicated multisource, multidomain, multifeature and multihierarchical system, and its way of spatial data organization was crucial to building a digital bay. The paper used the idea of geo-ontology to represent the bay and put forward a new spatial data organization method based on geo-ontology based on the analysis of the layer-oriented and feature-oriented way of spatial data organization, as well as metadata technology. The steps in the method were as follows: first, the bay ontology was modeled, which included a conceptual model reflecting the hierarchy of the bay well. Then, a spatial database of the bay, which was based on geo-ontology, was reconstructed according to the mapping relationship between the bay concepts and existing multisource geo-datasets. Thus, the integration and management of the multisource spatial data for the sea area, land area and intertidal zone was realized. In the third part of the paper, the Liaodong Bay was taken as a case study, and its related multisource spatial data was well organized using the above method. The result showed that using the idea of geo-ontology to organize and manage the geospatial data was helpful for the understanding of users from different domains and was good for the semantic sharing and interoperation of the spatial information of the bay.

13.3.5 Research on Geographic Information Retrieval and Geographic Knowledge Query Based on Geographic Ontology

Geographic information query and retrieval is the most basic and common function of GISs. The introduction of geographical ontology theory and methods is not to overthrow the existing theories and technologies in GISs but to explore a new method for the effective organization, management and scheduling of geographic information and to provide more efficient geographic information services. Therefore, geographic information retrieval and geographic knowledge query are still the important contents of ontology-based GISs. Wang et al. (2004) believed that spatial information querying methods in traditional GISs had many limitations, their operation was complicated, the presentation of querying results was highly formalized, and their user interfaces were so complex that they were not suitable for most people or decision making; an ontology is a formal, explicit specification of a shared conceptualization, and as a tool for conceptual modeling, which can describe information system on a semantic and knowledge level, it is quite suitable for information retrieval on the semantic and knowledge level. In their paper, they developed a new spatial information query method, named ODSKQ, which means ontology-driven spatial knowledge query. The ODSKQ method was built directly for users who did not have special knowledge on GIS, and it used natural language to describe querying conditions. The result of querying are highly synthetically and knowledgeable, providing users direct answers while hiding the complicated calculation and reasoning processes; it is indeed a highly humanized, intelligent spatial querying method. At the end of the paper, they took spatial query of resources and environment domain in the coastal zone as an example to verify the possibility and usefulness of ODSKQ, and the result showed that the ODSKQ method was very useful and powerful in the process of solving spatial problems. Fu et al. (2006) described a design idea and process of the intelligent information retrieval based on semantic web technologies in detail and proposed a framework of an ontology-based information retrieval system. What is more, the paper also provided an analysis of the implementation technologies based on an application of the geographical information domain. Yu et al. (2006) built an Ontology-based Geographic Information Inquiry System (OGIIS). The process is outlined as follows: at first, the geographic ontology instance was created by using the Geo-Ontology Building Algorithm (GOBA) and was based on reference ontology; then, indexes were built and inference was done on the geographic ontology instances; through reasoning and semantic description, OGIIS performed precisely and intelligently. In his dissertation for the degree of master of science, Li (2007) discussed several methods that improved query efficiency, including optimized query, expanded query and presentation of query results-based geographic ontology, as well as designing and realizing an urban geographic information query prototype. This prototype system benefitted from the recall in the traditional information system to query spatial information (that is, input vocabularies or sentences) and satisfied people's custom queries and reduced difficulties for users in query spatial information. At the same time, the dissertation explored the problems of using ontology to optimize and expand queries and realized intelligent spatial information query knowledge and semantic levels.

13.3.6 Research on Geographic Information Sharing and Interoperation Based on Geographic Ontology

Geospatial information sharing is an important part of GIS research. The development of the GIS has a nearly 20-year history in China. Over the course of development, for various reasons, there exist some phenomena, such as different understanding of geographical objects, different data definitions, different data modeling methods, different geographic data organizations, different spatial data formats and so on, which form the so-called "information isolated islands", and this leads to difficulty in sharing geographic information. In this context, a great deal of research has been conducted on the sharing and interoperability of geographic information. Especially today, network technology is highly developed and increasingly popular, and the interoperability of geographic information has been increasingly conspicuous. How do we freely exchange all the information about the Earth? How do we collaboratively run the software that manipulates the information over the network to accomplish the task of communicating and cooperating with two or more than two entities in a heterogeneous environment and realize the blending and matching of each component of the information system freely? This is the goal of geographic information sharing and interoperability. Geographic ontology contains rich semantic information, which makes the definition and representation of geographical concepts clear. It brings new opportunities for the realization of geographic information sharing and interoperability. The research of spatial information sharing and interoperation based on geographical ontology has achieved fruitful results. On the basis of analyzing the ontology-based information system, Cui (2004a, b) introduced an interoperability mechanism, which used ontology and OIL as the foundation. Through integrating the ontologies that were linked to the sources of geographic information, multisource geographic information was integrated dynamicity and openly based on its meaning and the efficiency and accuracy of the interoperability of GIS was improved. In Jing et al. (2005) explored geospatial information semantic sharing based on ontology and Agent; they discussed the connotation and layers of geospatial information sharing, made use of semantic web research and presented the research framework of geospatial information sharing. To solve the ambiguous meaning of geospatial information research and application, the ontology concept and method were adopted and domain knowledge was formally defined. Furthermore, the authors discussed how to support semantic information sharing and interoperation among users; they presented query information and semantic mapping based on Agents to clarify the different understanding among geographical information communities and obtained a common meaning and realization of information. Li et al. (2005) applied ontology to achieve the semantic integration of the intelligent transport system. The thesis built the basic noumenon, the domanial noumenon and the applied noumenon oriented to the semantic integration of China, as well as introduced the noumenon classification, studied the expression methods and advanced the framework based on the noumenon according to the national intelligent transport system architecture. Tan et al. (2006) developed heterogeneous spatial information systematic interoperability based on Bayes data classification and ontology and presented a feasible scheme for ontology-based semantic interoperation between heterogeneous GISs. The concept matching of heterogeneous spatial information was obtained by the ontology, and the Bayesian classification technique was used to classify the geo-ontology objects. By using this technique, the semantic interoperation between the heterogeneous GISs was obtained. Wang et al. (2006) proposed that ontology should be introduced to GIS, such as construction of ontology, aimed at the difficulty in data exchange and data sharing for different GISs. The data and information sharing as well as the accomplishment of different functional modules in ODGIS could be solved by means of the interoperabilities among different ontologies. Cui et al. (2007) studied the semantic interoperability of Geo-Agent based on ontology to meet the semantic requests of GISs, and the concept of the ontology-based Geo-Agent (geography Agent) was defined. Using Geo-Agent, the geography data were encapsulated by geography ontology so that a conceptual interface was exhibited to other Agents. The semantic relationships among the GIS nodes were built by logical reasoning. A virtual environment was established after all GIS nodes were integrated into a global ontology by using the semantic relationships. An example was given; the example provides details on how the Geo-Agent could transparently and effectively access the information and serve as a semantic tool for the decision support system. In Wu et al. (2007), the research on ontology-based heterogeneous spatial data integration was carried out, the framework of spatial data integration based on ontology was improved, and integrated experimental research was conducted. Guo et al. (2008) studied the integration method of hydrologic feature data based on ontology. In the paper, the basic concepts, components and construction methods of geo-ontology were briefly introduced. The integration framework of geospatial data based on geo-ontology was presented. The geo-ontology integration was a hybrid architecture based on the common intension properties template. Finally, taking the class of dry shoal, which is an element in hydrologic feature, as a case study, the progress of geo-ontology integration and the data transition method were realized. The difficult problem of data sharing caused by semantic heterogeneity was solved in this way.

13.3.7 Research on Geographic Information Services and Their Compositions Based on Geographic Ontology

The basic purpose of GIS is to provide users with rich, practical and easy to use geographic information services. The form and content of providing geographic information services by GISs has also undergone profound changes, from data services to functional services, service composition, personalized services, etc., and its standardization degree is becoming increasingly higher. However, one problem that cannot be ignored is that while providing a variety of services, the GIS has been faced with the confusion of semantic problems. The introduction and application of geographical ontology theory and methods can make up for the lack of semantic information, which will help GISs to provide better geographic information services. Geographic information services and their combination based on geographic ontology has become the focus of research in recent years. Chen et al. (2006a) implemented spatial information web services based on ontology. To improve the interoperation of spatial information, a new solution to implementation of discovery and retrieval for spatial information web services based on ontology technology was proposed. After the knowledge expression of web services capacity using web ontology language for services (OWL-S) was introduced, an architecture for spatial information interaction was constructed, and a semantic matchmaker based on ontology was added into the traditional service registry center in the architecture. A practical process of service request under the framework of spatial information web services was given in the framework of standard open GIS consortium (OGC) web services. The results showed that the ontology technology could extend the semantics of web services, reuse the same or similar web services, and improve the efficiency of reasoning. In Le et al. (2006), geo-ontology instances search for QA-based mobile spatial information service was studied. Based on the technology of the common search engine, the paper defined several spatial semantic roles and instance expression patterns according to geographic ontology, and large quantities of geo-ontology instances were searched by means of spatial semantic annotation, semantic phrase recognition and pattern match. It also provided QA-based instance service for mobile user with the help of semantic web technology. The preliminary experiment showed good precision, but the recall rate needed to be further improved. Jiang et al. (2007) designed geo_DataType ontology and geo_ServiceType ontology. An approach for geospatial web service discovery was proposed, in which geospatial web services were described and advertised by OWL-S, and then, a matching degree was ranked according to the input/output and service category for requested and advertised service descriptions to implement web services discovered automatically. Semantic heterogeneity from various fields still could not be solved by the existing GridGIS resource sharing technique, and for that reason, the method of constructing distributed ontology was discussed, and a framework of GridGIS service discovery based on semantic similarity matching was proposed by Rao et al. (2007). The transmission mechanism of spatial information service was studied in Chen et al. (2007), the idea of setting up relevant ontology for each process to implement information conversion was put forward, and then, a solution for the intelligent service of spatial information was explored. Cheng et al. (2008) proposed a design method for the geographic information service discovery component based on ontology, which combined ontology-based metadata and ontology-based query to improve the query efficiency and accuracy of network geographical service discovery based on the OGC standard. In Li et al. (2008), the geographic information web service automatic composition based on task-ontology was discussed, the semantic relationships between specific task-ontology concepts were analyzed, and the technical routine of web service matching was described. The concept of service composition implementation matching rules was defined, and the steps of geographic information web service automatic composition algorithm were discussed. By this research, users could find all service composition projects that satisfied the requirements from the existing web services and chose the best service composition project by comparing service the composition implementation matching rules. The research played an important role in the computer explanation of semantics on geographic information web services. Wang et al. (2008) studied the key technologies in the automatic discovery and composition of geographic information services based on the semantic web, which are listed as follows: semantic geographic information service description of OWL-S, OWL-S geographic information service description supported by extended UDDI, geographic information service matching algorithm for similarity computation of geometric distance, and semantic geographic information service composition based on multi-Agents.

13.3.8 Research on Spatial–temporal Geographic Ontology and Its Reasoning

Spatial features are unique to GISs. Spatial features refer to the geometric features of space, such as location, shape and size, as well as the spatial relationship with adjacent objects. The spatial characteristics of topology, geometry, location and orientation have an important and even decisive influence on the construction of geographical ontology, and it is also the essence of geographic ontology, which is different from the general information ontology. Temporal features reflect the temporal evolution of geographical objects, and temporal ontology is an important aspect in the study of geographical ontology. Spatial–temporal geographical ontology comprehensively reflects the temporal and spatial features of geographic objects and is the basis of spatial reasoning. In the paper, Research Progress in Spatial–Temporal Reasoning, Liu et al. (2004) believed that temporal and spatial reasoning were two important parts of artificial intelligence and that they had important applications in the fields of GIS, spatial–temporal database, CAD/CAM, etc. The development of temporal reasoning and spatial reasoning was discussed from three aspects: ontology, representation model and reasoning methods, and the research progress of spatial–temporal reasoning was summarized. The problems in the current research were discussed and the future directions were pointed out. Xu et al. (2005) studied spatial–temporal ontology and its application in GISs. They analyzed the spatial–temporal characteristics of GIS, introduced the cellular automata, proposed a formal definition of spatial–temporal ontology and a method based on spatial–temporal semantic granularity, discussed an integrated GIS architecture based on ontology and semantics, preliminarily implemented the creation and browsing of the railway passenger traffic spatial–temporal ontology, and realized the sharing and reuse of ontology among the GIS subsystems by integrating bidirectionality in an experimental railway GIS project. Min et al. (2006) introduced the basics of geographical spatial–temporal ontology, which included the primitive, property and representation models of temporal ontology and spatial ontology. Then, the research progress in geographical spatial–temporal ontology was reviewed. The problems in the current research were discussed and the future directions were pointed out. Based on the background of GIS ontology theory, Wang et al. (2007) used the representation and reasoning solution based on ontology and SWRL to realize spatial relation representation and reasoning in the tour map of Xi-an based on ontology. The results showed that combining ontology description language and SWRL to represent the conceptions and rules in spatial relation representation and reasoning not only improved the interoperability between the rules and ontology but also helped to quickly and dynamically update the spatial relation information in the geographic information ontology base.

13.3.9 Research on the Geospatial Semantic Grid

In 2000, Berners-Lee, founder of the World Wide web, proposed the concept of the semantic web and its seven-layer architecture at the world XML conference, in which the function and value of ontology were fully embodied. He believed the semantic web would be the next generation Internet. The main purpose of the semantic web was to enlarge the current WWW, and the web semantics information would be understood by computers and enable easy communication and cooperation between human and computer. In his proposed semantic web architecture, Berners-Lee gave a detailed description of the hierarchical relationships in the semantic web, and on this basis, ontology and logic inference rules were constructed to implement semantic-based knowledge representation and reasoning. In the semantic web, all the information was added with clear semantic information so that an Agent (intelligent Agent) could processes the resources on the network automatically. In this seven-layer architecture, the fourth layer was about the ontological vocabulary, which was used to define shared knowledge. Ontology revealed more complex and abundant semantic information between the resource itself and among resources, and as a result, the structure and content of the information were separated, and the information was completely formalized so that the information on the Internet could be understood by computers. After the seven-layer architecture of semantic web was put forward, researchers engaged in geographic information services at home and abroad and transplanted it into the field of geographic information science; combined with the practical application requirements of geographic information services, the concepts and architectures of spatial information, semantic grid and geospatial semantic web were put forward, which promoted the development of geographic information science to a certain extent. According to Cui et al. (2005a), to meet the semantic request of the geography information systems, the concept of the semantic-based spatial information grid (SSIG) was defined and the method to build SSIG was advanced. Using this method, SSIG could be built as following three steps: the first was that GIS data were encapsulated by geography ontology so that a conceptual interface was exhibited to the grid management layer; the second was that the semantic relations of the GIS nodes were built by logic reasoning; and finally, all GIS nodes were integrated into SSIG by using the semantic relations. A virtual environment was established after SSIG was built. The virtual organization lightened the burdens of guests and the data access procedure was transparent to guests, that is, guests could access information from remote sites using the current disk. An example was provided in the paper, which showed a detailed way to design and build SSIG. In conclusion, SSIG could improve the ability of the integration and interoperability of GISs. In Cui et al. (2005b), to meet the semantic demand of the spatial information grid and expand the data types of computer grid, the ontology and LDAP (lightweight directory access protocol) were combined to create a resource management system for the spatial information grid. The LDAP could filter the nature of the local ontology systems to construct a virtual organization, which provided a global directory service. The result showed that the semantic grid of the spatial information system, which combined ontology and LDAP, could improve the integration and interoperability of the spatial information grid. In Li et al. (2006), to integrate spatial information systems, the geographic ontology and geographic grid were used, and the spatial information system grid (SISG) was built by using the geographic ontology and grid. The geographic ontology was used to describe the meaning of the geographic objects, and the geographic grid was used to manage, index and store spatial information. SIMG could integrate the spatial information system and improve the sharing and utilizing of spatial information. Li et al. (2008) discussed in detail the basic and core issues of geospatial semantic grid research-semantic analyses of the fundamental geographic information. By exemplifying inland hydrological categorization in the domain of the fundamental geographic information in China, a discussion was made on the ambiguous semantics of the normative definitions of categories in some specifications. Ontology was regarded as an effective means to overcome semantic barriers in the domain of geographic information. A paradigm of formal semantic analyses of geographic information was put forward in the paper in terms of the formal ontology. With conceptualization of a category, attributes of a category were used for expressing its semantics. By examining the given hydrological categories, the formal semantics of each geographic concept were explicitly specified by a set of the defined ontological properties over its attributes.

13.4 Problems and Prospects

At present, experts and scholars from the field of geography, cartography and GIS, database technology, computer technology and artificial intelligence have been aware of the enormous potential and application prospects of geographic ontology research; however, the research achievements of geographical ontology are still mostly theoretical research and exploration; several small-scale experimental systems have been designed and implemented, but a large geographical ontology project has not yet appeared; the application system of geographic information services based on geographic ontology is rare; compared with foreign research in the same field, there is a large gap. The main reason is that there is not enough research, and studies are scattered and lack coordination. The atmosphere of joint research is far from being achieved, and there is a lack of a corresponding mechanism. In view of the distribution of research power, at present, the research units studying the theory and method of ontology mainly include Wuhan University, Peking University, Institute of geography of Chinese Academy of Sciences, Institute of Remote Sensing Applications of Chinese Academy of Sciences, and School of Surveying and Mapping of Zhengzhou Information Engineering University.

In the future, the research of geographical ontology should focus on the following aspects.

1. **Combined research with the semantic web**

After the semantic Internet is proposed, the underlying domain ontology must be well defined so that the semantic web can be constructed smoothly. The foundation of the semantic Internet is ontology. To make formal ontology understandable to computers, it must be used to define the meaning of data and metadata in Internet resources. As an important resource of the semantic Internet, it is an important subject for geographic information science to integrate geographic information resources into the semantic web. Abstract and construction of geographical ontology is its important foundation. How do we make geographic data and information run on the global information grid? Realization of geospatial data sharing and interoperation needs to be supported by the ontology database to establish a unified semantic network for the semantic description and transformation of the same object in different specialized spatial databases.

2. **Semantic interoperability between GISs**

Realizing the sharing and interoperation of geographic information resources is an important development direction in GIS. From the perspective of information, to realize interoperability of two systems, their information models must be interoperable; and to realize interoperability of information models, the two systems must be syntactically and semantically interoperable. Syntax interoperability refers to the problem that information in two systems are flow and processed using the same structure, which has been basically solved. Semantic interoperability indicates that the two systems have the same semantic understanding of information flow and processing.

3. **Change from data construction to both data and geographical ontology construction.**

Geo-data is a digital record of geographic concepts or geographic entities, reflecting the spatial, attribute and temporal characteristics of the recorded object. Geographical ontology is a knowledge architecture of geographic concepts or geographic entities, which contains rich semantic information, including entity type ontology, spatial relationship ontology and entity element ontology, reflecting the hierarchical relationship, subordinate relationship, and co-location relationship of geographic concepts or geographic entities. The synchronous record and construction of data and geographic ontology embodies the unification of the formalization of geographic concepts or geographic entity data models and the formalized knowledge of the spatial object domain itself.

4. **Semantic modeling in Earth sciences.**

The modeling of the Earth system requires the aid of ontology to enable models in different domains to interact, be reused, and share with each other. In describing these processes, it is necessary to describe the behavior of the process, the temporal and spatial characteristics, and the relationships to those of other processes. The collection of process descriptions forms a process library and becomes the basis for the simulation framework. The use and expression of semantics in Earth system modeling will enhance people's ability to study and investigate the environment systematically. The semantic web provides a new approach for the various solutions we described, as well as platforms, models, and useful model results.

13.5 Representative Publications

(1) ***Research on the integration and sharing of coastal zones and offshore scientific data*** (Du Yunyan, Zhou Chenghu, Su Fenzhen et al. Maritime Press, Beijing, November 2005)

This book is divided into 9 chapters and discusses the integration of coastal and offshore data based on geographical ontology. Among them, Chapter 1 discusses the concept of geographic ontology and its research progress at home and abroad and puts forward the problem of constructing the integration and sharing platform of coastal zones and offshore scientific data based on geographical ontology. In the fifth chapter, the method of spatial data integration based on geographic ontology is studied, and a spatial data integration framework based on geographic ontology is proposed; taking land use ontology and data integration as an example, the method of spatial data integration and transformation based on geographical ontology is studied; in the sixth chapter, based on the establishment of geographical ontology, the conceptual model of coastal zone and offshore area is given, and the ontology-based spatial data integration, display and query methods for coastal zones and offshore are analyzed and discussed; in the fourth section of the seventh chapter, a prototype system of coastal zones and offshore data integration platform based on geographic ontology is implemented, including the function design and technical realization of the prototype system; Chap. 8 is an application example of the data integration platform that was established in the seventh chapter, which takes the Fujian province coastal zone as the research object, and the bay spatial database based on ontology is established.

(2) ***Key issues and applications in geographic ontology*** (Huang Maojun. Press of University of Science and Technology of China, Hefei, December 2006)

The whole book consists of 7 chapters. Chapter 1 introduces the current situation and progress of geographical ontology research at home and abroad; Chap. 2 is entitled "philosophy ontology, information ontology and space ontology", which puts forward the profound connotation of geographical ontology, and points out that geographical ontology covers the 3 aspects of philosophy ontology, information ontology and spatial ontology; Chap. 3 discusses the logical structure and establishment method of geographic ontology, that is, based on a comprehensive summary of existing ontology construction methods, a general method for building geographical ontology is summarized; Chap. 4 is the analysis of OWL's ability to represent geographic ontology semantics, pointing out that OWL is the ideal language to express the geographical ontology, but it needs to be extended; Chap. 5 is entitled "the formal expression mechanism of spatial features of geographical ontology", in which 3 theoretical tools are proposed, namely, integration, topology and position theory, and formal spatial features and spatial relation axioms are constructed and added to the modeling primitives of OWL so that spatial features can be expressed in the geographical ontology constructed by OWL; Chap. 6 is about the ontology-based map service and its improvement; the chapter proposes a solution based on geographic ontology and points out that network map services are facing outstanding problems and cannot provide effective query services, so a comprehensive utilization ontology and web mining technology are proposed to improve the QoS of the network map query; and finally, Chap. 7 discusses the future research direction of geographic ontology.

(3) ***Geographic ontology and geographic information service*** (Li Hongwei, Cheng Yi, Li Qinchao. Xi'an Map Publishing House, Xi'an, April 2008)

This book explains the exact meaning of geographic ontology and its role and significance in the field of geographic information science. The basic theoretical problem of geographic information services based on ontology-semantic inconsistency are emphatically discussed, that is, using the concept lattice and formal concept analysis (FCA) method, the problem of semantic inconsistency is studied deeply with examples; the conceptual design of geographic domain ontology is implemented, and a framework of three-layer ontology is proposed. Ontology-semantic relations, including semantic similarity, structural similarity and semantic relatedness, are comprehensively analyzed, which lays the foundation for solving the problem of geographic information web service matching and discovery; the semantic description of ontology web services is described; using the web service ontology OWL-S, which is based on OWL language; a prototype of land use information display and query service based on ontology is implemented; the ontology-based geographic information service composition is discussed; GIS engineering based on ontology is summarized; and finally, the future directions of geographic ontology are discussed. This book shows the research potential and value of ontology in the field of geographic information science.

References

An, Y., Bian, F., & Guan, J. (2004). Geo web services description and discovery based on ontology. *Geomatics and Information Science of Wuhan University, 29*(12), 1063–1066.

An, Y., Bian, F., & Guan, J. (2006). Design and comparison of geo-ontology in GIS. *Geomatics and Information Science of Wuhan University, 31*(12), 1108–1111.

Bian, F., Mei, K. (2008). Building location ontology for geographic knowledge based. *Geomatics World, 6*(4), 27–32, 39.

Chen, J., Yang, S., & Li, C. (2006a). Implementation of spatial information web services based on ontology. *Journal of Zhejiang University (Engineering Science), 40*(3), 376–380.

Chen, J., Zhou, C., & Wang, J. (2006b). Advances in the study of the geo-ontology. *Earth Science Frontiers, 13*(3), 81–90.

Chen, Y., Cui, T., & Guo, Li. (2007). Research of transmission mechanism of spatial information intelligent service based on ontology. *Science of Surveying and Mapping, 32*(6), 8–9, 13.

Cheng, G., Du, Q., & Cai, Z. (2008). Ontology-based component designed of geographic information discovery. *Journal of Geomatics, 33*(2), 31–33.

Cui, W. (2004a). *Using ontology to achieve the semantic integration an interoperation of the geography information system.* Doctoral Dissertation of Wuhan University.

Cui, W. (2004b). Ontology-based GIS interoperability. *Journal of Geomatics, 29*(1), 20–22.

Cui, W., & Jiang, T. (2005a). Method to build semantic-based spatial information grid. *Journal of Wuhan University of Technology (Transportation Science & Engineering), 29*(3), 421–424.

Cui, W., & Li, D. (2005b). Geography ontology and LDAP directory service. *Journal of Wuhan University of Technology (Transportation Science & Engineering), 29*(4), 491–494.

Cui, W., & Jiang, T. (2007). Semantic interoperability of Geoagent based on ontology. *Engineering Journal of Wuhan University, 40*(1), 126–129.

Cui, W., Jiang, T., & Zhang, D. (2004). Using data mining and ontology to realize interoperability geographic information systems. *Journal of Wuhan University of Technology (Transportation Science & Engineering), 28*(1), 118–121.

Du, Q. (2001). *Study on linguistic characteristics and automatic understanding mechanism of spatial information.* Doctoral Dissertation of Wuhan University.

Du, Y., Zhang, D., & Su, F. (2008). Geospatial data organization of the bay based on the Geo-ontology: A case study of the liaodong bay. *Journal of Geo-Information Science, 10*(1), 7–13.

Fu, Y., Jin, L., & Wang, K. (2006). Design and implementation of ontology-based information retrieval system. *Application Research of Computers, 23*(11), 155–157.

Guarino, N. (1994). The ontologic level. In R. Casati, B. Smith, & G. White (Eds.), *Philosophy and the Cognitive Science* (pp. 443–456). Vienna: Holder-Pichler-Tempsky.

Guo, L., Cui, T., & Liu, G. (2008). Geo-spatial information integration framework and realization on ontology. *Journal of Geomatics Science and Technology, 25*(6), 462–465.

Guo, L., Liu, C., & Chen, Y. (2008). Study on the integration method of hydrologic feature data based on ontology. *Geomatics World, 6*(2), 42–47.

Han, C., Yi, S., & Lv, X. (2006). GIS based railway location intelligent environment and domain ontology modeling method. *China Railway Science, 27*(6), 84–90.
He, J., Li, X., & Bi, J. (2003). Research on classification and coding of resources and environment information and its association with ontology. *Geomatics World, 1*(5), 6–11.
Huang, M., Du, Q., & Du, X. (2005). Formal representation of spatial properties of geographic ontology. *Geomatics and Information Science of Wuhan University, 30*(4), 337–340.
Huang, M., Du, Q., & Wu, Y. (2004). A preliminary study on geographic ontology and its application. *Geography and Geo-Information Science, 20*(4), 1–5.
Jiang, L., & Gong, J. (2007). Geospatial web service description and discovery based on OWL-S. *Geomatics & Spatial Information Technology, 30*(5), 19–22, 27.
Jing, D. (2005). *A research on Geo-spatial information semantic expression and service based on ontology*. Doctoral Dissertation of Institute of Remote Sensing Applications, Chinese Academy of Sciences.
Jing, D., & Wu, F. (2005). Geo-spatial information semantic share on base of ontology and agent. *Geomatics World, 3*(4), 35–38.
Le, X., Yang, C., & Liu, D. (2006). Geo-ontology instances search for QA-based mobile spatial information service. *Chinese High Technology Letters, 16*(11), 1107–1112.
Li, D., & Cui, W. (2004). Semantic grid of spatial information. *Geomatics and Information Science of Wuhan University, 29*(10), 847–851.
Li, D., & Cui, W. (2006). Geographic ontology and SIMG. *Acta Geodaetica et Cartographica Sinica, 35*(2), 143–148.
Li, H., Cai, C., & Li, Q. (2009). The application of Jena in spatial information query and reasoning based on geo-ontology. *Engineering of Surveying and Mapping, 18*(5), 5–9.
Li, H., & Li, Q. (2009). The classified and hierarchical presentation of land-use data based on ontology. *Areal Research and Development, 28*(1), 82–85.
Li, H., Li, Q., & Duan, J. (2008). Research on GIS development methods based on ontology. *Computer Applications and Software, 25*(11), 69–70, 127.
Li, H., Yuan, Y., & Meng, C. (2008). Research on geographic information web service composition based on task-ontology. *Computer Applications and Software, 25*(7), 165–166, 176.
Li, L., & Wang, H. (2006). Classification of fundamental geographic information based on formal ontology. *Geomatics and Information Science of Wuhan University, 31*(6), 523–526.
Li, L., Zhu, H., & Wang, H. (2008). Semantic analyses of the fundamental geographic information based on formal ontology: Exemplifying hydrological category. *Acta Geodaetica et Cartographica Sinica, 37*(2), 230–235, 242.
Li, Q. (2007). *Research on ontology based geographic information retrieval method*. Master thesis of the Information Engineering University.
Li, S., & Gao, J. (2008). Conceptual framework of the ontology of place and its components. *Journal of Geomatics Science and Technology, 25*(3), 164–167.
Li, Y., Zhai, J., & Chen, Y. (2005). Using ontology to achieve the semantic integration of the intelligent transport system. *Information Technology, 29*(6), 10–13.
Liang, R., Li, S., & Li, W. (2010). Research on data management and retrieval service based on toponym ontology. *Journal of Information Engineering University, 11*(2), 175–179.
Liu, D., Hu, P., & Wang, S. (2004). Research progress in spatio-temporal reasoning. *Journal of Software, 15*(8), 1141–1149.
Liu, Y., Zhang, Y., & Tian, Y. (2007). On general place names and the associated ontology. *Geography and Geo-Information Science, 23*(6), 1–7.
Min, M., Tan, C., & Jiang, L. (2006). Research progress in geographical spatio-temporal ontologies. *Journal of HuaZhong Normal University (Natural Sciences), 40*(1), 132–137.
Rao, Y., & Liu, P. (2007). Study on architecture of Grid GIS services discovery based on ontology. *Geomatics World, 5*(4), 45–48.
Ren, M. (2003). *Research and implementation on the key technologies of the distributed UGIS*. Doctoral Dissertation of East China Normal University.
Shi, Z. (2004). Ontology scientific translation. *China Terminology, 6*(4), 13–14.

Su, L., Zhu, Q., & Chen, Y. (2007). Conceptual modeling of spatial database based on geographic ontology. *Computer Engineering, 33*(12), 87–89.

Sun, M., Chen, X., & Zhang, F. (2004). Geo-ontology. *Geography and Geo-Information Science, 20*(3), 6–11, 39.

Tan, X., & Bian, F. (2005). Heterogeneous spatial information interoperability based on cooperative ontologies. *Geomatics and Information Science of Wuhan University, 30*(2), 178–181.

Tan, X., & Bian, F. (2006). Heterogeneous spatial information system semantic interoperability based on Bayes data classification and ontology. *Geomatics and Information Science of Wuhan University, 31*(8), 724–727.

Wang, A., & Li, D. (2006). Knowledge sharing in construction of ontology-based geographic information system. *Journal of Northeast Forestry University, 34*(6), 107–109.

Wang, G., & Bian, F. (2008). Research on automatic discovery and composition of geographic information service based on semantic Web. *Science & Technology Information,* (6), 102.

Wang, H., Li, L., & Wang, Z. (2005). Classification tiers of fundamental geographical information based on ontology. *Geomatics World, 3*(5), 27–30.

Wang, H., & Cao, H. (2007). A solution to spatial relation representation and reasoning based on ontology and SWRL. *Microelectronics & Computer, 24*(7), 166–168, 172.

Wang, J., Su, F., & Du, Y. (2004). Study on spatial knowledge querying methods based on ontology. *Journal of Geo-Information Science, 6*(4), 93–99.

Wang, S., & Liu, D. (2004). Ontology based geographical information system middleware. *Computer Engineering and Applications, 40*(22), 9–21.

Wang, Y., & Meng, H. (2006). Ontological analysis on entity and relation from multi-scale road representations. *Journal of China University of Mining & Technology, 35*(5), 689–694.

Wang, Y., Lu, Z., & Chen, Y. (2005). An ontology-based approach to data integration. *Computer Engineering and Science, 27*(6), 67–69.

Wu, H., & Xing, G. (2005). Research on technology of information integration based on ontology. *Journal of Computer Applications, 25*(2), 456–458.

Wu, L., Xu, L., & Chen, X. (2006). Human-earth-GIS relations discussion based on subject-human and Geo-ontology. *Geography and Geo-Information Science, 22*(1), 1–6.

Wu, M., Song, X., & Cui, W. (2007). On ontology-driven heterogeneous geographic data set integration. *Geomatics and Information Science of Wuhan University, 32*(10), 915–918.

Xu, J., & Zhao, Z. (2007). Ontology based geographical information system and its system structure research. *Microprocessors, 28*(4), 100–103.

Xu, W., Huang, H., & Qin, Y. (2005). Research on Spatio-temporal ontology and its application in Geographic information system. *Journal of the China Railway Society, 27*(4), 119–124.

Yang, J., Li, Y., & Cai, G. (2007). Heuristics design of spatial database based on ontology-oriented. *Application Research of Computers, 24*(5), 24–26.

Yu, W., Cao, J., & Chen, J. (2006). Distributed geographic information retrieval based on the geospatial semantic web. *Computer Engineering and Applications, 42*(30), 6–9.

Yu, W., Cao, J., & Chen, J. (2007). Geographic information retrieval and ranking based on ontology. *Computer Engineering, 33*(21), 157–159.

Zheng, M., Feng, X., & Jiang, Y. (2006). A formal approach for multiple representations in GIS based on DL ontologies. *Acta Geodaetica et Cartographica Sinica, 35*(3), 261–266.

Chapter 14
Spatial Data Mining and Knowledge Discovery

Rongqin Lan

14.1 Introduction

Spatial data mining and knowledge discovery refer to the theories, methods and techniques for extracting implicit, previously unknown knowledge and spatial relationships and the relationship between spatial and nonspatial data from spatial databases, as well as find the useful features and patterns of interest among them. Knowledge discovery (KDD) is an advanced processing method that extracts credible, novel, effective and understandable patterns from a large amount of spatial data; spatial data mining is a step in the generation of specific spatial patterns through specific algorithms in knowledge discovery. The two processes are difficult to separate and are used simultaneously. Spatial data mining and knowledge discovery can be used to extract interesting and regular knowledge from large spatial databases; in particular, it can be used to understand spatial data, discover the relationship between spatial data and nonspatial data, build a spatial knowledge base, optimize queries, and reorganize spatial databases.

The process of spatial data mining and knowledge discovery can be broadly divided into the following steps: data preparation, data selection, data preprocessing, data reduction or data transformation, determination of data mining goals, determination of knowledge discovery algorithms, data mining, model interpretation, knowledge evaluation, and so on. Obviously, data mining refers to a particular step in this process. However, for the sake of simplicity, spatial data mining is often used instead of spatial data mining and knowledge discovery.

The commonly used methods of spatial data mining are the probability-based approach, spatial analysis method, statistical analysis method, inductive learning method, spatial association rules mining approach, cluster analysis method, neural network method, decision tree method, rough set theory, neural network method, decision tree method, rough set theory, fuzzy set theory-based approach, spatial

R. Lan (✉)
Information Engineering University, Zhengzhou, China

J. Wang and F. Wu (eds.), *Advances in Cartography and Geographic Information Engineering*, https://doi.org/10.1007/978-981-16-0614-4_14

features and trend detection method, cloud theory-based approach, evidence theory-based approach, genetic algorithm, spatial online data mining, and data visualization method. The types of knowledge that can be found are geographical geometry knowledge, topological relation knowledge, spatial distribution knowledge, spatial feature knowledge, and regional differentiation knowledge. The main types of rules that can be found are spatial characteristic rules, spatial discriminate rules, spatial association rules, and spatial evolution rules.

Using the concepts, theories, methods and techniques of spatial data mining and knowledge discovery, as well as closely integrating GIS to make full use of the function of GIS storage and management of spatial data, large amounts of data in GIS can become an unlimited source of knowledge. This research is significant for effectively dealing with large amounts of geo data, improving the automation and intelligence levels of geo scientific analysis, and providing a powerful analytical tool for global change and regional sustainable development. Further, this research topic has social and practical significance.

14.2 Development Process

Data mining is a new research field that appeared in the 1980s and initially explored finding useful data in large datasets, such as relational databases, transactional databases, and time series databases. At present, the research and application of data mining and knowledge discovery have made great progress and gradually became the research hotspot in the field of computer information processing.

To address the "data rich but information (knowledge) poor" situation, in 1989, the concept of Knowledge Discovery in Databases (KDDs) was first proposed during the Eleventh International Conference on Artificial Intelligence held in Detroit, MI, USA. Subsequently, KDD symposiums were held in 1991, 1993 and 1994, which brought together researchers and application developers from various fields, with a focus on the issues of data statistics, mass data analysis algorithms, knowledge representation, and knowledge use. With the increasing number of participants, the KDD International Conference grew into an annual meeting. In 1995, the first International Conference on knowledge discovery and data mining was held in Canada; the data in the database were figuratively represented as ore deposits, and data mining (DM) emerged as a new discipline. According to Fayyad et al. (1996), KDD is defined as a nontrivial process of identifying valid, novel, potentially useful, and ultimately understandable patterns in data. Data mining is a step in the KDD process that consists of applying data analysis and discovery algorithms that, under acceptable computational efficiency limitations, produce a particular enumeration of patterns (or models) over the data. The classical algorithm of Apriori association rule mining was proposed by Agrawal et al. (1993) for mining the association patterns of data items (sets) in transactions. Since then, the concepts of data clustering, data classification and data cube have successively been put forward, which indicates the formation of a methodological system of data mining and knowledge discovery.

The research of spatial data mining started late. With the rapid development of remote sensing, geographic information systems (GISs), spatial information acquisition technology, network technology and the long-term accumulation of data, traditional methods were unable to meet the needs of spatial data processing, analysis and application. To solve the new difficulties caused by the large amount of spatial data, at the 1994 fifth GIS international academic conference held in Ottawa, Canada, academician LI Deren proposed the concept of knowledge discovery from GIS and suggested applying KDD methods to intelligent GISs.

With the deepening of research and applications, people's understanding of KDD and DM became increasingly comprehensive. There are many other terms with a similar or slightly different meaning as data mining, such as knowledge mining from data, knowledge extraction, data archaeology, and data dredging. Among them, DM and KDD are more commonly used. DM is mostly used in statistics, data analysis and information systems, while KDD is used in artificial intelligence and machine learning fields. To unify understanding, in the authoritative monograph "Data Mining: Concepts and Techniques", KDD is redefined as the automated or convenient extraction of patterns representing the knowledge implicitly stored or captured in large databases, data warehouses, the Web, other massive information repositories or data streams. Data mining is only one step in the entire KDD process, albeit an essential step since it uncovers hidden patterns for evaluation. DMKD is a product of the development of data acquisition technology, database technology, computer technology and management decision support technology. As a result, spatial data mining and knowledge discovery (SDMKD) emerged as a new discipline and became the branch of data mining and knowledge discovery for dealing with spatial (location) data.

Spatial data mining has developed into a new and promising subject. It involves the integration of techniques from multiple disciplines, such as database and data warehouse technology, statistics, information theory, machine learning (artificial intelligence), high-performance computing, pattern recognition, neural networks, data visualization, information retrieval, image and signal processing, and spatial or temporal data analysis. With the increasing influence and importance of large Earth spatial datasets, such as maps, remote sensing images, warehouses and digital cities, spatial data mining is becoming increasingly important.

Spatial data mining is being increasingly applied to many fields, such as financial data analysis and financial modeling, retail data mining and customer relationship management, telecommunication-related data mining, bioinformatics and biological data analysis, scientific data analysis, geographic data mining, knowledge discovery in science, intrusion detection, road traffic, navigation, robotics, and soil analysis (Han et al. 2001). At the same time, spatial data mining has also achieved many application results; for example, SKICAT has found 16 new extremely distant quasars; the POSS system improved the classification accuracy of star objects in sky images from 75% to 94%; the Magellan study system recognized volcanoes by analyzing approximately 30,000 high-resolution radar images of the surface of Venus; and the CONQUEST system used content-based spatial and temporal queries to discover sample knowledge of ozone hole formation in the atmosphere (Li et al. 2006).

Spatial data mining is an important research topic at many famous international academic conferences, for example, the internationally important biennial meetings of IJCAI, annual meetings of SIGIR, SIGKDD, KDD, ICDM, SDM, PKDD, PAKDD, and so on. The number of domestic conferences on data mining has also increased every year. According to preliminary statistics, the number of international conferences on spatial data mining held in China during 2003–2008 exceeded 20, involving rough sets, fuzzy sets, granular computing, spatial modeling, spatial reasoning, spatial analysis and data mining, knowledge discovery, machine learning, pattern recognition, artificial intelligence, Agent, and so on. There are three major KDD publications: Knowledge and data engineering of IEEE, Data Mining and Knowledge Discovery of Kluwer and Data and Knowledge Engineering of Elsevier. Looking at statistics from the Web of Knowledge, regarding the research results of data mining, the largest number is from American scholars, accounting for approximately 38% of the total, followed by Chinese scholars (including Taiwan), accounting for approximately 12% of the total; others, including Britain and Canada, account for almost 50% of the total. China Data Mining Research, which was founded in 2004, has 4 registered people, and daily visits have amounted to more than 8,000 people.

Although the research of data mining in China started late, it has developed synchronously in the field of spatial data mining. At present, the basic theory and application of data mining are studied in many domestic research institutions and colleges and universities, and the research projects in this field have been funded annually by the National Natural Science Foundation of China and the State 863 Project.

14.3 Main Research Achievements

14.3.1 Research on the Theories and Methods of Spatial Data Mining Based on the Framework of Earth Information Science

A solid theoretical foundation is the cornerstone of the research, development and evaluation of spatial data mining methods. After more than ten years of exploration, some important theoretical frameworks of spatial data mining have been formed and attracted a large number of researchers and developers for further work and development. Among them, the theoretical framework based on Earth information science can help us to exactly understand the concept and technical characteristics of data mining. Earth spatial information science has gone beyond all the contents of modern surveying and mapping science, reflecting the infiltration and integration of multiple disciplines, with particular emphasis on the application of computer technology; in spatial data, it emphasizes the whole process of information collection, processing, measurement, analysis, management, storage, and display and distribution of geospatial data and information. The core of the theoretical framework of

Earth spatial information science is the spatial information mechanism of the Earth, which, through the study of the information transmission process and physical mechanism of the Earth's sphere, reveals the laws of geometry, spatial distribution and variation in the Earth; the main research contents include datum, standards, temporal and spatial variations, cognition, uncertainty, interpretation and inversion, and representation and visualization of Earth spatial information. Earth spatial information science has laid a theoretical foundation for spatial data mining in the aspects of data processing, algorithm research, geoscience pattern extraction and geoscience rule interpretation.

Li et al. (2001) studied the meaning of spatial data mining and knowledge discovery, explained the knowledge of spatial association rules, feature rules, classification rules and clustering rules, and their relationships with data mining and knowledge discovery, machine learning, geo data analysis, spatial databases, spatial data warehouses, digital Earth and other related disciplines, summarized the emergence and development of SDMKD, and analyzed the application and development of SDMKD.

Wang (2001) systematically discussed the geospatial cognitive model and the geographical data model analysis, the object-oriented geographic data model, and the spatiotemporal data model, digital elevation model and its data structure, establishment of TIN and DEM, algebraic transformations of spatial graphs, spatial relations of graphs and spatial analysis, multiscale features of spatial data and automatic generalization, and so on, which laid the theoretical and technical foundation for spatial data mining methods and techniques.

According to cloud theory, rough set theory and inductive learning, Di (2003) systematically studied the theory and method of spatial data mining and knowledge discovery and used GIS intelligent analysis and automatic interpretation of remote sensing images.

Li et al. (2002) analyzed the connotation and extension of SMDKD; then, the theories and methods for SDMKD, such as probability theory, evidence theory, spatial statistics, rule induction, clustering analysis, spatial analysis, fuzzy sets, cloud theory, rough sets, neural networks, genetic algorithms, visualization, decision trees, spatial online data mining, and others, were studied. Finally, the development prospects of SDMKD were discussed.

Zhou et al. (2002) introduced the concept, architecture, commonly used methods, knowledge types and applications of spatial data mining.

Li et al. (2006) conducted early and continuous research on the theory and method of spatial data mining, which mainly focused on the concept and theoretical system of geo spatial information science, cloud models, data fields, geo-rough space and view angles of spatial data mining.

Hypergraph theory based on the graph theory and set has recently been a hotspot in the field of geospatial information. Hypergraphs can be used to visually represent the structure and relationship of knowledge contained in spatial data. It can not only represent the structure and composition of knowledge in an image style and make complex knowledge easy to understand but also combines object-oriented technology and visualization technology, which makes the spatial data mining process easy to implement. Therefore, this theory has certain development potential in the field of spatial data mining. In Sun et al. (2002), the theory and method of spatial data mining were proposed and applied to spatial data mining based on the hypergraph model. The hypergraph model combines the hypergraph theory, object-oriented technology and visualization technology and can be used to express the internal structure and mutual relationship of complex data, as well as represent the different attributes of objects and the degree of association between objects.

The development and application of geostatistics theory and methods promote their application in spatial data mining, such as exploratory spatial data analysis (ESDA), spatial autocorrelation, and kriging interpolation.

14.3.2 Research on Spatial Data Mining Technology and Algorithms

Spatial data have many characteristics different from relational databases. It contains topological and/or distance information, which is usually organized in complex multidimensional spatial index structures and accessed through spatial data access methods. Spatial data often needs to use spatial reasoning, geometric computation and spatial knowledge representation techniques. These characteristics make mining information in spatial data a challenging problem.

Spatial data mining uses existing techniques, such as spatial data structure, spatial reasoning and computational geometry, and many new and effective methods for spatial data mining have been put forward. Spatial data mining expands the traditional spatial analysis methods, focusing on the efficiency of implementation, integration with database systems, better interaction with users, and the discovery of new types of knowledge.

To complete all kinds of data mining tasks, people have borrowed many basic research methods and tools from the fields of statistics, artificial intelligence and databases, and many methods have been put forward. The commonly used main mining methods are generalization- and induction-based mining methods, spatial clustering methods, spatial association rule methods, spatial classification methods, spatial trend prediction methods, rough sets, and visualization. Among them, spatial classification, spatial clustering and spatial association rules are the most commonly used techniques and are also the focus of current research. In recent years, the research results have mainly focused on spatial association rules, spatial clustering, and spatial anomaly detection.

The main challenge of spatial data mining is the efficiency of spatial data mining algorithms, while the core of the research is to improve the function, performance and efficiency of spatial data mining algorithms. In this sense, the current work is still in its early stage, and we need to further apply the latest achievements of intelligent computing and soft computing to design a more effective mining algorithm.

1. **Spatial data mining method based on generalization**

The data and objects in a database contain detailed information at the primitive concept level, and a dataset containing large amounts of data often needs to be generalized and displayed at a higher concept level. Knowledge discovery based on generalization assumes that background knowledge exists in the form of a conceptual hierarchy. The concept hierarchy can be provided by experts or automatically generated by means of data analysis. Two types of concept hierarchies are defined in spatial databases: the nonspatial concept layer and the spatial concept layer. Lu et al. (1993) extended attribute-oriented induction to spatial databases and proposed two algorithms: spatial data dominant generalization and nonspatial data dominant generalization. Han and Fu (1996) proposed an efficient data generalization technique, attribute-oriented induction, which collects an interesting set of data as an initial data relation, performs the induction process on it attribute-by-attribute, including attribute removal, concept hierarchy ascension, generalization operator application, and so on; additionally, this technique eliminates duplications among generalized tuples (with counts accumulated) and derives generalized relations and rules. Generalization results can be expressed in the form of generalization relations or data cubes to perform further OLAP operations; they can also be mapped with generalized tables, graphs, or curves for visual representation, and characteristics and discriminant rules can be extracted from them. The spatial data mining mechanism based on generalization needs to be further developed to deal with multi-thematic maps and multilevel interactive generalization mining, as well as effectively combined with the spatial index, spatial access method and data warehouse technology.

In 1995, LI Deyi proposed a model for the formal expression and analysis of concepts, known as the cloud model (Li et al. 1995), which is widely used in knowledge representation, data mining and other fields (Li et al. 2005).

Hu et al. (2001) proposed that concept lattices and Hasse diagrams can be used as symbolic data analysis and knowledge discovery tools and can be used for visual description of data. In the process of data mining based on the concept lattice, the rule is described as the relationship between two intent sets, which is reflected in the inclusion relationship between the corresponding extent sets. The concept lattice, induced from a binary relation between objects and features, is a very useful formal tool and has been used in many fields. It realizes the unification of concept intension and concept extension, represents the association between objects and features, and reflects the relationship between generalization and the specialization among concepts, so it is useful for discovering the potential concept below the data (Xie 2001).

In view of the shortcomings of the existing numerical data concept generalization algorithm in the database, after introducing a data mining algorithm based on the conception hierarchy tree, Xiao et al. (2005) proposed an improved algorithm. Moreover, the results of two algorithm tests on actual data were provided in the paper.

In Li et al. (2005), a classification method based on the cloud model was proposed, which is called the cloud classification method. The method used cloud transform and concept jumping-up based on pan-concept-tree to discretize the numerical data and thus reduced the complexity of classification.

2. **Spatial data mining method based on clustering**

Spatial data clustering uses a distance measure to identify clusters in a large-scale multidimensional dataset to discover the entire distribution pattern in the dataset. Spatial clustering methods can be divided into partition clustering and hierarchical clustering, the boundaries of the classification are fuzzy, and a clustering algorithm can sometimes work on two or even three clustering patterns. For example, in terms of its implementation, Wave Cluster can be attributed to hierarchical clustering, as well as to grid-based clustering, but the DENCLUE method belongs to both the hierarchical clustering method and the partitioning clustering method (Wang et al. 2003).

Zhou et al. (1999) introduced the concept of information entropy into SDM and proposed a spatiotemporal integrated geoscience data segmentation and clustering model based on information entropy reduction. It is a hierarchical clustering algorithm. ZHOU Chenghu et al. carried out verification with the example of China's soil horizontal belt distribution pattern, considering that the segmentation model based on the entropy reduction concept has great application value in the analysis of geospatial data with many uncertain mechanisms since it focuses more on the detection of patterns and the understandability of results in data compared with the previous statistical models. Li et al. (1999) proposed an adaptive algorithm for the clustering of two-dimensional objects, which used a feature filter to obtain objects and their relationships with a linear feature. Then, it created curves and clusters based on the relationships. Experimental results showed that the presented method succeeded in the clustering task and was immune to noise. The method was classified as the segmentation + model approach. Li Cong et al. (1999) proposed setting up cluster centers in data-intensive areas and divided a circle of influence, which obtained a solution near the global optimum. Zhu et al. (2000) proposed a new clustering method (NCL) based on human principles used to identify clusters, i.e., the distance between any two objects in one cluster must be less than the distance between any two clusters. This new clustering method did not need to be given a cluster number K beforehand, and the complexity of its implementation algorithm was basically O (nm). Moreover, this method efficiently processed a large set of data (hundreds of thousands to millions objects). The experimental results demonstrated that this new clustering method had good performances in many aspects.

In Zhang et al. (2004), the K-harmonic means algorithm was proposed, which was insensitive to the selection of initial center points and effectively eliminated the local optimum; therefore, the effect was better than the K-means algorithm. In Su et al. (2001), an improved version of the K-Means method was proposed, and the concept of point symmetry distance was introduced into clustering, which extracted spatial categories with symmetric relationships. The use of fuzzy techniques made clustering algorithms robust against noise and missing values in the databases. In Chen et al. (2001), a fuzzy prototype algorithm integrating K-means and K-modes algorithms was presented and used in mixed databases. Experiments on several real databases demonstrated that the fuzzy algorithm obtained better results than the corresponding hard algorithm. A genetic algorithm with implicit parallelism and powerful global search capability searched for the global optimum within a very short period of time. In Li et al. (2002), a genetic algorithm was used to search the initial center for the K-Means algorithm. The concept of 'gene difference' was introduced to control the crossover operator and mutation operator in the genetic algorithm. Experiments on the standard database of UCI showed that the proposed method efficiently improved the clustering result.

In addition, under the condition of a large amount of data, multiple I/O operations were required since the data could not be read into memory at one time when the k-mean method runs. In view of this weakness, Tang et al. (2002) proposed an incremental multilevel CFK-mean algorithm for large-scale data. Because the centroid of the cluster was used as the clustering center, outliers away from data-intensive regions led to clustering centers deviating from the real data-intensive region, so the k-mean algorithm was very sensitive to outliers. In this regard, Wan et al. (2003) proposed the idea of separating the average data of clusters from the cluster centers to eliminate the negative effects of outliers. Chen et al. (2005) provided new visual hierarchical clustering based on the minimum spanning tree partitioning algorithm, which grouped and visualized cluster hierarchies consisting of both nonspatial and spatial attributes. Taking spatial clustering as an example, an overall function for the J2EE-based visual spatial clustering system was designed. Simultaneously, the key techniques were discussed. It provided a feasible solution to such an application system (SDM). Classic clustering algorithms ignored the fact that many constraints exist in the real world and affected the correctness of the clustering result. Zhang et al. (2007) discussed the problem of spatial clustering with obstacle constraints and proposed a novel genetic K-means spatial clustering with obstacle constraints based on the genetic algorithm and the K-means method. To analyze the spatial objects, a spatial cluster algorithm (TGD) was introduced by Kong et al. (2008), which gathered intensive regional components into categories, and the function of discovering intensive regions of arbitrary shape and satisfying constraints was realized.

3. **Spatial data mining method based on classification**

The spatial classification method is a rule used to classify objects based on the spatial or nonspatial characteristics of objects, and it is supervised; that is, it has prior knowledge, such as class numbers and typical characteristics of each class. This method is the opposite of the clustering method. The commonly used classification algorithms include the ID3 algorithm, decision tree algorithm, fuzzy c-means algorithm, and Ester's spatial classification algorithm based on adjacency graphs. Shi et al. (2003) argued that spatial classification methods still needed to find efficient spatial classification methods to address incomplete information. Zhao et al. (2004) considered that the spatial classification algorithm based on the adjacency graph considered the spatial characteristics of the object well, but it did not generalize or analyze the attributes, so the obtained classification rules were less satisfactory. Thus, an improved spatial classification algorithm was proposed. Hu et al. (2007) argued that since spatial data had the characteristic of spatial autocorrelation, spatial information needed to be added when using the fuzzy c-means algorithm for spatial clustering. Then, the fuzzy membership degree of each spatial object to all clusters was calculated, and the cluster that had the largest fuzzy membership was determined. Finally, the value of the dependent variable of the cluster center object was used as an estimate of the dependent variable of the spatial object for classification.

4. **Spatial data mining method based on spatial association**

Spatial association (or geoscience association) is the association of one or more spatial objects with other spatial objects. The concept of association rules was introduced by Agrawal et al. for mining large transactional databases. It was extended to spatial databases by Koperski et al. (1995). The form of spatial association rules is X → Y (s, c), in which X and Y are collections of spatial or nonspatial predicates, respectively, s is the support of rules, and c is the confidence of rules. There are 3 forms of spatial predicates: predicates that represent topological relationships, such as intersection, overlay, etc.; predicates that express spatial directions, such as east, west, left, right, etc.; and predicates that express distance, including proximity.

The task of spatial data mining involves extracting geoscience association rules. Research in the field of geoscience knowledge discovery appeared in the middle of the 1990s (Han and Fu 1996) and was focused on extracting object characteristic rules with geoscience associations, dependency mining or attribute-oriented generalization. These studies laid the basic theoretical and methodological foundations for geographic knowledge discovery.

The mining algorithm for transaction data was not suitable for spatial data, a progressive refinement algorithm was proposed by Lu et al. (2000), and an efficient spatial association rule mining algorithm was obtained step by step by the approximation and refinement. In Su et al. (2004), based on a review of the research background and development of the geo-association rule, the geo-association rule was categorized by its complexity as a spatial characteristic rule, spatial discriminate rule, spatial association rule and spatiotemporal association rule. With the definition of the spatiotemporal rule in mathematical language, the effect of the different indices on rule mining was analyzed. Furthermore, the presentation analyzed the data processed in rule mining with the level of generalization, data dimension, variable type and spatial topologic relationship. The geo-association rule can be categorized as a single-level rule or inter-level rule by the level of generalization, as a one-dimensional or multidimensional rule by data dimension, as a Boolean or numeric rule by the variable type, and as a local or focal rule by the topologic relationship. Finally, the presentation describes the flow to forecast the fishing ground using the geo-association rule. Lan et al. (2005) combined fuzzy association rule mining approaches with the methods of fuzzy concept hierarchy presentation, as well as the analysis of fuzzy spatial relation hierarchy (including topology, direction and distance) to develop a method for extracting fuzzy spatial association rules from geographical data and remote sensing images. Based on application examples, the mining algorithm was derived, and the computing of confidence and support was realized. Spatial association rule mining based on spatial objects was more difficult than relational association rule mining because the spatial object attributes were spatially autocorrelated, continuous and multidimensional. It was more difficult to define transactions in traditional association rule mining. Huang Tian et al. (2005) established a new spatial multidimensional association rules model (SMARM) for multidimensional spatial data mining, where spatial transactions were defined by a notion of impact zone based on spatial autocorrelation and replaced by a traditional transaction definition. A new mining algorithm (SMARBIA) was realized. The algorithm avoided enormous candidate items in the mining process by pruning techniques based on impact zones and spatial support. Finally, the experiment showed that it could decrease the number of candidate items. In Chen Jiangping (2006), the problem of building an automated spatial concept layer was settled by using concept lattice and ordered semi groups theory. The new method of simplifying the spatial concept layer relations by the Hasse graph with ordered semi groups theory was emphasized, and the spatial association rules in the simplified concept lattice were mined. The experiment proved that this technique was effective. Deng (2007) focused on studying the techniques of the process and methods of mining association rules with decision attributes in the GIS database. In the process of mining association rules, transaction data were first divided according to the indiscernibility relation of attributes. Thus, a discernibility matrix was used to calculate absolute reduction, and an absolute reduction algorithm was designed. A mining multilayer and multidimensional association rules algorithm with decision attributes was also designed based on the equivalence of the reduction attribute.

5. **Causal relationship discovery**

The analysis of causal relationships between spatial objects is a research topic in spatial data mining. A comparison between association rule data mining and causal discovery was made by He et al. (2005). On the basis of explaining the characteristics of association rules and causal rules, this paper conducted a theoretical analysis and comparison from three aspects: directionality, guiding significance of human behavior and how to link them. The analysis results showed that causal discovery found the inherent mechanism between objects and reasoned and tested the association rules accordingly. Finally, two data mining methods were applied to a demographic dataset, and the mining results were compared. In view of the related research results of GIS and spatial analysis and according to the requirements of fuzzy spatial queries, Lan et al. (2005) presented a spatial configuration rule mining and feature extraction method based on spatial association rule mining and fuzzy spatial relation reasoning. The principle and method, as well as application examples were introduced in detail. It was apparent that spatial association rule mining could be used to discover the spatial or nonspatial association relationship among spatial entities on maps. From frequent item sets with causal relationships, spatial configuration patterns can be tested and extracted, and meaningful and potential arrangement patterns can be discovered. It can be used to discover hidden geo-distribution patterns, correlation characteristics or interaction rules.

6. **Predictive data mining in spatiotemporal data**

Spatial and temporal relationships exist at all levels (scales) of spatial data models and may vary with scales. Therefore, spatiotemporal data mining is an important research area. With the development of spatiotemporal data models and spatiotemporal data mining methods, spatiotemporal anomaly detection methods have attracted increasing attention and have become one of the most important methods of spatial data mining.

The research in spatiotemporal data mining can be roughly divided into spatiotemporal association rules or pattern mining, spatiotemporal correlation analysis, spatiotemporal cluster analysis, spatiotemporal prediction, and temporal and spatial anomaly detection (Hsu et al. 2008).

As a powerful nonlinear axiomatic system, the dynamic recurrent neural network (DRNN) has stronger dynamic behavior and computational capability than the feedforward neural network and can directly reflect the characteristics of the dynamic process system, which makes the model more robust (Zhu et al. 1998). Therefore, an ISTIFF (Improved STIFF) model is proposed for the study of dynamic recurrent neural networks, which uses the dynamic recurrent neural network to predict the space instead of the static feedforward neural network in the STIFF model and has been used to predict the forest fire areas in Canada (Cheng et al. 2006; Wang et al. 2007).

14.3.3 Research on the Extension of Spatial Data Mining Objects and Applications

The spatial database was first used as the research object of spatial data mining. Then, spatial data warehouses and multidimensional cubes emerged and provided effective tools for the management and distribution of spatial information, and they were directly or indirectly used for spatial data mining. Thus, OLAP (exploratory spatial analysis, ESA) emerged. The development of WebGIS and GridGIS further enlarged the space for spatial data mining.

1. **Applications in topographic map analysis and cartographic generalization**

In the field of topographic map analysis and cartographic generalization, Wang et al. (1992) systematically studied cartographic data processing techniques using the model method, and a series of theories and methods were proposed, such as the spatial distribution features of mapping elements, the spatial distribution trend analysis of geographical environment factors and the prediction of geographical environment factors. Meng (1998) proposed the idea of applying knowledge discovery to cartographic generalization and discussed the strategy of cartographic generalization cognitive modeling. The technique used the fuzzy rules discovered from topographic maps to guide artificial generalization and to detect graphical conflicts. Based on the analysis of topographic maps, Zhang et al. (2008) designed a set of spatial data mining systems with perfect functions and analyzed and solved the difficult problems encountered in the implementation of the function. The system contained completed basic function modules, such as space calculation, terrain feature extraction, obstacle path analysis and spatial clustering analysis.

2. **Remote sensing image information mining**

Mining non-standardized and multimedia data is one of the trends in data mining. Digital images can be viewed as multimedia data. With the extensive use of digital imaging technology and sensor devices, large amounts of image data are generated every day, such as remote sensing satellite images. The need for automatic analysis of these images to obtain a large amount of useful knowledge is increasing. Among them, image information mining technology provides effective methods and techniques.

Due to the lack of a profound understanding of image information mining and some of its theoretical and technical obstacles, development is slow. Since the first multimedia mining conference held in 2000, image mining has not yet formed a complete theoretical framework and unified technical methods, and it is still in the exploratory stage. At present, image mining technology mainly includes image similarity search, image association rule mining, image classification, image clustering, and the relationship among different phenomena. The research of image information mining mainly lies in the establishment of mining systems and the finding of mining algorithms.

An information-driven model was proposed by Zhang (2001) according to different levels of image representation and different roles of information, and the image information can be divided into 4 levels: pixel level, consisting of the original image information, such as pixels, image features, such as color, and texture and shape features; object level, for processing objects and region information based on the original features of the pixel layer; semantic concept level, for generating high-level semantic concepts from the recognized objects and regions by combining with domain knowledge; and pattern and knowledge level, which is for discovering potential domain patterns and knowledge by combining with domain-related text digital information. Yan et al. (2003) proposed a mining algorithm for discovering association rules in 2D color images. The algorithm consisted of four steps: attribute extraction, object identification, aided creation of image and object mining. The algorithm focused on data mining of image content and did not involve other knowledge domains. In Qin (2004), the spatial relationship of remote sensing image objects was expressed as a binary relational transaction database, and the spatial association rules of grassland, forest land and water bodies were mined by using the concept lattice association rule mining method. The image knowledge discovery model IMDFSSM based on Hilbert space theory was introduced in You et al. (2005), and the model was used to quantitatively represent the knowledge representation and knowledge discovery process of image data. In Wu et al. (2006, 2007), the method of mining association rules based on concept lattice was used to mine the frequent patterns of image texture features, which were used as texture features for image segmentation experiments. Combined with the cloud model and concept lattice, a method of image texture feature data mining was proposed.

3. **Web data mining (distributed, heterogeneous data mining)**

With the rapid development of the Internet, the spatial data resources of networks are becoming increasingly abundant. As a large distributed and parallel information space, networks are more meaningful for knowledge discovery. To obtain a sufficient amount of high-quality knowledge in a complex network environment by making use of spatial data mining, it is necessary to study the spatial data mining technology and system adapting to the network environment.

Spatial data mining in the network environment must break through the partial restrictions on data. Based on the Internet/Intranet and other network environments, by using search engines, vector space models, distributed data mining and virtual database technology, we may develop distributed data mining systems oriented to multiple workstations, multiple databases and multiple types of data sources to achieve efficient distributed computing, seamless integration of data, reasonable knowledge representation, knowledge updating, and result visualization, as well as to support hyperlinks and media references between document hierarchies during data mining.

Web spatial data mining is generally based on multidimensional views. This type of mining emphasizes the efficiency of execution and the timely response to user commands. The direct data source is usually a spatial data warehouse.

Based on the analysis of the concepts and technical characteristics of data mining, spatial data mining and the Web Agent, a basic framework of spatial data mining on the Internet using the Web Agent was presented, and a verification was implemented in Wen et al. (2003). An Agent-based distributed spatial data mining model was proposed, and the implementation of communication, procedure call and code migration between the coordinator, data mining Agent, database and user was described in Zhu et al. (2006).

4. **Uncertain data mining**

To link uncertainty research with data mining, new theories and methods (such as rough set theory, cloud theory, etc.) have been adopted to establish a knowledge-based attribute uncertainty analysis model. Zadeh's fuzzy set extends the range of classical sets from {0, 1} to [0, 1] and develops the characteristic function into a fuzzy membership function; it solves the "specious" fuzziness in spatial data mining (Zadeh 1965).

To correctly reflect the association between fuzziness and randomness in uncertainty research, Li et al. (1995) proposed a cloud model, which established a transformation of uncertainty between the qualitative concept and the quantitative description, served as the basis for conceptual expression and fully reflected the uncertainty in the concept. Based on the normal cloud model, T clouds, triangle clouds, Cauchy clouds, spectrum clouds, geometry clouds, function clouds, power-law clouds and other derivative clouds were developed step by step, and floating clouds, composite clouds, integrated clouds, decomposition clouds, dynamic clouds and other cloud operation methods were realized. A series of techniques and methods, including cloud computing, cloud transformation and uncertainty reasoning, were studied (Li et al. 2005).

Shi et al. (2002) proposed spatial data mining based on uncertainty. Wang et al. (2003) studied the processing of fuzzy information in spatial data mining. Guo et al. (2004) discussed the uncertainty and propagation of spatial data, the main technical methods of SDM and KDD, and the uncertainty of the SDM and KDD processes and their results, such as algorithms, processing, propagation models and quality evaluation. To overcome the deficiencies in traditional spatial data association rule mining that lacked uncertainty processing and measurement, the uncertainties in spatial data and spatial data mining were properly combined, and an uncertainty processing model and measurement indexes for spatial data association rule mining were established. Four key problems were investigated and analyzed, including the uncertainty simulation of spatial data with Monte Carlo methods, measurement of spatial autocorrelation based on uncertain spatial positional data, discreteness of continuous data based on an uncertain spatial clustering algorithm and uncertainty measurement of association results. It was verified using environmental investigation data in a region of China.

5. **Multiscale data mining**

In view of the multiscale characteristics of spatial data, the technical idea of multiscale spatial data mining was proposed. In Sun et al. (2005), the concept of scale and scaling was discussed, the general theory and method of scaling in geosciences was summarized, the scaling problem in spatial data mining was put forward, and the solution was discussed. In Sun et al. (2008), an information reasoning model of the same geographical element at different scales was established, which was called the multiscale spatial data mining framework model. Taking the thickness of the coal seam as the study object and using cluster analysis as mining technology, the idea of applying multiscale analysis to spatial data mining was realized, and the multiscale phenomena in spatial data mining were confirmed. The research results enriched and aided in developing the basic theory of spatial data mining.

6. **Visual data mining**

It is not enough to only discover knowledge; you have to represent the results in a way that is understandable to the user. The most effective way to represent rules is by graphical visualization because humans are very good at interpreting visual data and landscapes, and this fact should be exploited in the data mining process.

The combination of spatial data mining and geographic visualization technology has spawned a new type of spatial data analysis technology - visual interactive spatial data mining. Based on a spatial database, Li (2000) studied a spatial data model and 3D visualization technology in 3D GIS. Based on the spatial database and VRML file, a 3D Web-GIS architecture was proposed. Jia et al. (2004) explored visual interactive spatial data mining technology.

7. **Deviation (small sample) data mining**

Data anomaly (also called outlier or isolated point) mining mainly studies the deviation and evolution law of data, also called spatiotemporal anomaly detection or small sample data mining. That is, we can find the evolution law of spatial features by summarizing the general characteristics of change data. For example, the areas where air pollution increased last year can be compared with areas where air quality improved. Anomalies can also be found and predicted in regional weather models, land use planning, and so on.

At present, there is less research on spatiotemporal anomaly detection, and the detection methods can be roughly divided into two categories: spatial and temporal divide-and-conquer methods (Cheng et al. 2005) and spatiotemporal synthetic analysis methods. The divide-and-conquer method uses the traditional anomaly discrimination method from time, space and adjacent domains to detect abnormal data both in time and space, that is, temporal and spatial anomaly data. The spatiotemporal synthetic analysis method considers time and space synthetically in the process of detection.

In recent years, the concept of temporal anomaly patterns was proposed, and the main research results focused on anomaly pattern detection methods of time series. In China, Huang et al. (2005) and Ma et al. (2006) studied the spatial anomaly detection method based on nonspatial attribute deviation, in which the metric function was used to calculate the deviation of the nonspatial properties of spatial objects from their spatial neighboring domains, and then, the largest spatial objects were selected from the list of deviations to form the spatial anomalous dataset. Wang et al. (2006) proposed an anomaly pattern detection method similar to the above literature; the difference was that their approach involved mapping the segmented subsequences into a feature space and then performing subsequence clustering in the feature space. Weng et al. (2007) proposed a method for identifying abnormal patterns in multivariate time series, in which a bottom-up segmentation algorithm was used to segment the multivariate time series into a set of subsequences, the extended Frobenius norm was used to compute the similarity of subsequences, and then, the k-mean clustering method was used to detect anomaly patterns.

8. **Personalized spatiotemporal data mining**

Personalization involves collecting and storing user usage information, and from this information, the user's specific interests and needs are analyzed; then, based on the analysis, the correct information about personalized requirements and personalized services is sent to each visitor at the correct time. Zhang (2005) believed that there were two main elements of personalization: one element was well-directed, that is, for different user needs and interests to provide different services; and the other element was automatic, that is, the personalization process was automatically accomplished by tracking and analyzing user access behavior. Based on context awareness, a meta-model MGFPU for personalized user requirement acquisition and formal processing was given in Wang (2007).

In practice, a personalized data mining system is designed based on multi-Agent technology (Meng et al. 2003), and the basic idea of constructing this system is first described. Then, the User Agent is designed with a detailed introduction of transition relations from one state to another of this Agent and their communication protocol, and an MAS called the personalized data mining system is established with the User Agent and Mining Agent. The algorithm on which the system is mainly based is given. The system is characterized by the fact that an Agent serves as a person in the data mining system. The application results show that personalized knowledge can be mined in the system and that knowledge that is truly useful for the user is given. The system preferably avoids repeated work and unwanted energy of the user. Simultaneously, mining irrelevant data seldom occurs; thus, the system can work efficiently.

9. **Research on urban complexity mining**

With the rapid development of cities, many studies related to spatial characteristics need to be deepened to provide decision support for better urban development and management. A city, as the research object of spatial data mining, not only gives full play to the technical advantages of spatial data mining but also has great practical value. Comprehensive research on urban spatial data mining is becoming an active research area. At present, many valuable urban development models have been proposed by scholars, and relevant simulation research work has been carried out, such as land use transformation and its effect model, system dynamics model, cellular automata model, metropolitan expansion model, multiagent model, and so on. To better reflect the dynamic feedback relationship between society, economy and nature, it has become a consensus among scholars both at home and abroad to explain the complexity of urban development through multidisciplinary, multiview angle, multilevel and multiscale analysis.

Li (2005) proposed a framework system of urban spatial data mining (USDM) and the entity information model of the combination of location and attribute, which provided the measurement of spatial distance as a basic rule of spatial computation. By extending the method of the spatial weighted matrix, the conception of the spatial entity association matrix and spatial entity state association matrix was proposed, and the method of their establishment offered new basic tools for SDM. From the aspect of urban spatial distribution DM, adopting the conception model of the combination of location and attribute, the author brought spatial coordinates, spatial relationships and attribute characteristics into the unified model of spatial computation, studied the city's land suitableness evaluation and the method of spatial clustering in the division of urban function districts and proposed the calculation for the classification layer.

14.3.4 Research on the Software System and Architecture of Spatial Data Mining

With the continuous development of data mining technology, the corresponding mining software also emerges in an endless stream. In the development of spatial data mining systems, the most famous universal SDM systems in the world have GeoMiner, Descartes and ArcView GIS S-PLUS interfaces, as well as GwiM and other systems. In addition, some commonly used statistical software, such as Clementine, SPSS/AnswerTree, SAS, and CART/MARS, all have spatial data mining modules. Some plug-ins and modules related to geostatistics have also considerably developed, such as the ESRI plug-in, Surfer, GeoDA, ArcGIS, etc., which have better performances in geostatistical analysis, spatial interpolation and exploratory analysis.

The development of spatial data mining software includes independent, embedded, attached, automatic and interactive methods, which can have their own spatial databases or spatial data warehouses. They can also integrate with GIS and use GIS databases and other external databases.

In terms of system research and development, the data mining prototype system, data mining component module and OLAP-based data mining system are three kinds of systems with different degrees of difficulty. In the early stage of research, it usually started with some well-directed component modules or building your own system using a more mature approach based on OLAP.

1. **Development of prototype system**

In China, prototype systems include the Multimedia Data Mining Prototype System developed by CAO Jiaheng from Wuhan University and GISMiner developed by the University of Science & Technology China and the Hefei Intelligent Machinery Research Institute of the Chinese Academy of Sciences. The research team of academician LI Deren has independently developed two software prototype systems, GISDBMiner and RSImageMiner, and has obtained the state's software copyright registration (Li et al. 2006). In Wang (2005), the basic theory of spatial data mining was systematically discussed, a spatial data mining system architecture in GIS and a variety of spatial data mining algorithms that can mine many kinds of spatial knowledge types were proposed, and a prototype system of spatial data mining in GIS-GIS_Miner was implemented. Zheng et al. (2006) considered that the development of the current prototype system had the following characteristics: paying attention to visualization technology, focusing on cross platform system development, joint function development, and dynamic data mining software. The development trend of spatial data mining software is outlined as follows: professional development languages emerged, better visualization methods emerged, cross platform and component software was further developed and improved, network data mining software was rapidly developed, all kinds of unstructured data mining was strengthened, interactive discovery, special data mining software was further development, powerful spatial data mining and dynamic spatial data mining software made great progress, and mainstream technology and mainstream software market became more obvious.

In the "Spatial data mining technology and application engineering" project hosted by Prof. LIU Yaolin, the methods of cleaning, integration and mining of multi-type and multiscale geospatial data and their implementation in spatial decision support application engineering were systematically studied; a number of spatiotemporal data mining models, such as the spatial clustering rules mining model for land use, spatial association rule mining model, spatial dependency rule mining model, and spatial evolution rule mining model, were constructed; a series of algorithms for multiple types of spatiotemporal data mining based on multivariate statistical analysis, computational intelligence, geostatistics analysis and computational geometry were developed; and a visual human-machine interactive data mining software system was developed, integrating data integration, integrated analysis, mining and spatial decision support. The project was awarded the first prize of the National Bureau of Surveying and Mapping for scientific and technological progress.

2. **Combination of SDM and GIS**

With the widespread application of GIS in all walks of life, an increasing number of people want to complete a variety of spatial analyses under a unified GIS software package interface; therefore, the combination of spatial data mining methods with GIS has attracted much attention.

At present, many departments have begun to combine spatial data mining with intelligent GIS methods and spatial database technology and to establish an intelligent GIS system with knowledge acquisition and decision support ability by integrating object-oriented spatial databases, spatiotemporal databases, statistical analyses, spatial reasoning and expert systems. In Jia (2008), a design idea of an intelligent agricultural land classification expert system supported by GIS, SDM and expert system integration technology was proposed, and the function, structure and workflow of the system were analyzed in detail. VC6.0 embedded GIS controls (MO2.0) were used to realize the development of a knowledge-based agricultural land classification expert information system.

3. **Modular mining system**

Generally, spatial data mining systems are rich in mining patterns, and the modular structure can simplify the data mining process and reduce the threshold of data mining. At present, the development of spatial data mining platforms mostly adopts modular structures and sets up modular data access interface layers. It allows users to visualize the creation of data streams, selectively perform some or all of the decomposition steps, and then examine the execution results through interactive views on the data and model. In Fan et al. (2006), an open "plug and play" data mining system was implemented, and through the integration of data mining technology and visualization technology, a set of visual spatial data mining theory frameworks, technical methods and prototype systems were formed. In Zhao et al. (2008), based on the MATLAB platform, a spatial data mining integrated information system was designed and developed to provide technical support and solutions for mass spatial data mining.

4. **Agent (multiple Agent)-based system**

The spatial data mining system based on Agent is generally composed of a number of soft Agents that can run independently and cooperate with each other. The introduction of Agent technology can effectively coordinate all the links in spatial data mining and realize knowledge discovery from spatial databases faster and more effectively.

In Li et al. (2004), Agent technology was used to build a four-layer distributed spatial data mining model. Spatial metadata were adopted to tackle heterogeneous data sources. Metadata management used a semantic network-based data source hierarchical registration mechanism. In the model, Agent-based data mining solvers were key modules. At the same time, domain knowledge bases were introduced to acquire and process the spatial information to strengthen the feasibility, validity and rationality of mined results. Xu et al. (2004) proposed the model of a personalized data mining system based on multi-mobile Agent technology and presented functions of every part and the method of Agents cooperating with each other to complete personalized data mining, as well as the corresponding algorithms. According to Liu et al. (2006), incremental data mining adopts large-scale dynamic data, reduces storage requirements and implements parallel processing; however, the presented incremental clustering methods have problems because there are comparatively many parameters and the results are not exact adequate. In a data mining structure with continuously changing sources, using a swarm of special Agents to modify the knowledge model incrementally, a clustering model based on swarm intelligence and an incremental maintenance algorithm was presented. Information entropy was applied to speed up clustering and modified existing clusters according to pheromone and incremental insertions and deletions in the database, which required a few parameters. Experiments proved that it obtained almost exact clusters. Ni et al. (2007) analyzed the general technique of spatial data mining based on Agent theory. After explaining the task principle of this structure, an Agent framework for spatial data mining was designed in the practice of digital agriculture spatial information platforms.

14.4 Problems and Prospects

The research of spatial data mining has been active for nearly 20 years, but there are many difficulties and problems. To summarize, in terms of geoscience spatiotemporal model identification and decomposition, intelligent spatiotemporal clustering and classification algorithms and so on, there has been some development, but it is not mature or perfect. The mining of spatiotemporal association rules, the high-dimensional visualization analysis of geographic data, especially the research of time series mining, are still in the initial stage in terms of the development of the data mining theory of spatial databases, the development of relevant software prototypes, the application of theoretical methods and so on. The application of causal relationship reasoning in the field of geosciences has not yet been carried out, and it is still immature in application modeling and uncertainty data management, although it has been involved in GIS and remote sensing image interpretation.

Future research on spatial data mining should focus on the following aspects:

(1) The theory and method of spatial data mining based on spatial cognitive theory and the framework of geo information science lay the theoretical and methodological foundations for spatial data mining. The theory and technology framework of spatial data mining and knowledge discovery involves knowledge expression, qualitative and quantitative conversion, intelligent reasoning, efficient search of spatial information, spatial data analysis and processing, spatial data 3D visualization and other theories and methods. These theories and methods are not isolated but systematically solve fundamental problems, such as knowledge expression, quantitative and qualitative transformations, and uncertainty reasoning, under the guidance of spatial cognitive theory and the framework of geo information science. Therefore, research on SDMKD basic theory is very important. Only by establishing a solid theoretical framework can the research of algorithms and technology be deepened.

(2) Intelligent and efficient spatial data mining algorithms, which need to be further implemented with the latest research results of other disciplines. Due to the complexity of spatial data, it is more difficult to extract knowledge from spatial databases than to extract knowledge from traditional databases. The main challenge of spatial data mining is the efficiency of spatial data mining algorithms. The core of the research is to improve the function, performance and efficiency of spatial data mining algorithms. In this sense, the current work is still in its infancy, and we need to apply the latest achievements of intelligent computing and soft computing to design a more effective mining algorithm.

(3) Combining spatial data mining with grid computing technology and high-performance computing technology. With the continual growth of the amount of data available in the network, data sources that need to be mined are often distributed and stored on each node of the grid. At the same time, with the in depth application of information technology, the object dataset is getting larger, which inevitably requires that machines performing the mining algorithms have fairly high performances, and it is usually difficult for one machine to accomplish this ultralarge number of computing tasks; correspondingly, machines on other distributed sites have ample computing resources but cannot be effectively utilized. Traditional data mining technology cannot be used to solve these two problems. Therefore, distributed data analysis and mining technology based on grids have emerged. Grid-based distributed spatial data analysis and mining can use grid computing resources to discover knowledge from distributed spatial data resources.

The research work of combining spatial data mining with grid computing technology and high-performance computing technology mainly involves two aspects: on the one hand, it must be able to perform joint analysis and mining of distributed spatial datasets across multiple grid nodes; on the other hand, it must be able to use grid computing resources for collaborative analysis mining to meet the large workload of computing needs.

(4) Methods and techniques for large-scale uncertainty in spatial data mining and personalized spatiotemporal data mining. Spatial data are characterized by spatial, temporal and multidimensional mass, complexity and uncertainty. With the rapid development of the network, spatial data resources of the network are becoming increasingly abundant. Cyber sources include network platform resources (e.g., network equipment, interface resources, computing resources, bandwidth resources, storage resources, network topology) and various resources supported by platforms, such as image data, text data, voice data, network software and application software. It is necessary to study the spatial data mining technology and system adaption to the network environment to acquire a large amount of high-quality knowledge in the complex network environment by using spatial data mining.

Spatial data mining in the network environment needs to break through the local restrictions of spatial data and extend from the local part of one person or department to the entire network and from mining local rules to global network topology rules. Based on the Internet/Intranet network environment and by using a search engine, vector space model, distributed data mining, virtual database and other technologies, a distributed data mining system oriented to multi-site machines, multiple databases and multiple types of data sources should be developed to achieve efficient distributed computing, seamless integration of data, reasonable knowledge representation, knowledge updating and result visualization, as well as to support the hyperlink and media reference among the document hierarchical structures (Li et al. 2002). Meanwhile, the spatial, temporal, and semantic inconsistencies in the spatial databases that are distributed across the network must be considered, as well as the differences in space-time references and semantic standards.

In addition, in the network environment, the number of users of spatiotemporal data is increasing constantly, and user demand is also personalized. It is an inevitable research topic to carry out spatial and temporal data mining of user personalized requirements. Personalized spatiotemporal data mining is a service process that mainly uses data mining tools to acquire, mine, express and visualize network spatial information resources and user resources to generate recommendation information to be submitted to the user according to the user's spatial information demand preferences. In recent years, the "My Maps" map service launched by Google and others attempted to personalize the map service, which demonstrates that the era of spatial information service personalization has arrived.

In summary, spatial data mining and knowledge discovery have been explored and developed over more than ten years, many important theoretical achievements have been obtained, and many practical applications have started. At present, the research in this field is becoming increasingly specialized, and further research has been conducted for some theories and methods; at the same time, the application fields of spatial data mining and knowledge discovery have attracted much attention, and some preliminary application systems have been constructed. Although there are still many technical bottlenecks to be overcome and the gap between theory and practice is relatively large, practice has proven that there is much promise and hope for this new discipline.

14.5 Representative Publications

(1) ***Spatial data mining and knowledge discovery*** (DI Kaichang. Wuhan University Press, Wuhan, December 2001)

Based on cloud theory, rough set theory and inductive learning and other techniques, taking GIS intelligent analysis and automatic interpretation of remote sensing images as the main application targets, this book systematically studies the theory and method of spatial data mining and knowledge discovery. Data mining and knowledge discovery is a frontier technology in the field of information technology that came about in the 1990s. It is generated in the context that data and databases increase dramatically and are far beyond people's abilities to process and understand data. It is also the result of the integration of database technology, artificial intelligence technology, statistical technology and visualization technology. The purpose is to extract implicit but potentially useful information and knowledge from a large number of incomplete, noisy, fuzzy data to provide intelligent and automated means for processing and understanding data and databases. Spatial data mining and knowledge discovery is the process of extracting implicit and interesting spatial and nonspatial patterns and universal features.

(2) ***Methods and applications of urban spatial data mining*** (LI Xinyun. Shandong University Press, Jinan, September 2005)

Aimed at the demand for spatial information in urban development decisions, guided by spatial information science, urban geography and regional economics and based on the research of spatial data mining, the theory, method, technology and application of urban spatial data mining are deeply and systematically studied in this book. The main contents include urban spatial data mining architecture, spatial basic computing model, urban spatial distribution (static) data mining, urban spatial dynamic prediction, urban spatial and temporal association rules extraction, urban agglomeration data mining, and so on. A general framework and some new mining methods are proposed, some existing methods of spatial data mining are improved, a large number of experimental studies are carried out, and an experimental system of urban spatial data mining is preliminarily established.

(3) ***Theories and applications of spatial data mining*** (LI Deren, WANG Shuliang, LI Deyi. Science Press, Beijing, October 2006)

The book mainly introduces some new technologies, such as cloud models, data fields, geo-rough space and a view-angle of spatial data mining, constructs a spatial data mining Pyramid, researches the data source of spatial data mining, derives the LI Deren method for spatial observation data cleaning, explores image data mining based on spatial statistics, proposes "data field-cloud" clustering, fuzzy synthetic clustering based on data fields and clustering algorithms based on mathematical morphology, and studies spatial data mining based on inductive learning, remote sensing image data mining based on concept lattice and GIS data control. Combined with application examples of landslide monitoring, bank revenue analysis and site evaluation, land use classification of remote sensing images, land resource evaluation and safety inspection of train movement, this book also systematically studies the feasibility of spatial data control. On this basis, the prototype systems of spatial data mining, GISDBMiner and RSImageMiner are independently developed.

(4) ***Spatial data mining of vector graphics features and their applications*** (BI Shuoben. Science Press, Beijing, January 2008)

The book systematically explains the basic concepts, basic processes and related methods in the field of spatial data mining of vector graphics features and its application in the first phase of the cultural settlement of the Jiang Zhai Ruins. The book consists of 12 chapters. The main contents include spatial analysis of the first phase settlement of Jiangzhai, profile data mining based on the GIS database, spatial classification mining of vector features, spatial clustering mining of vector features, association rule mining based on attribute databases, generalized mining based on attribute databases, spatial classification of neural networks based on vector features, spatial classification of genetic algorithms based on vector features, fuzzy spatial clustering of vector features, spatial knowledge reasoning, and fuzzy spatial reasoning of vector features.

References

Agrawal, R., Imieliński, T., & Swami, A. (1993). Mining association rules between sets of items in large databases. *ACM SIGMOD Record 22*(2), 207–216.

Chen, CC. et al. (2005). The study and application on visual spatial clustering. *Geo-information Science, 7*(2)

Chen, J., & Li, P. (2006). Spatial association rule mining algorithm based on ordered semigroup. In *Proceedings of Hubei Society of Surveying and Mapping 2006 Annual Conference on Science and Technology* (pp. 13–19).

Chen, N., Chen, A., & Zhou, L. (2001). Fuzzy K-prototypes algorithm for clustering mixed numeric and categorical valued data. *Journal of Software, 12*(8), 1107–1119.

Cheng, CC. et al (2005). The study and application on visual spatial clustering. *Geo-information Science, 7*(2)

Cheng, T., & Li, Z. (2006). A multi-scale approach for spatial-temporal outlier detection. *Transactions in GIS, 10*(2), 253–263.

Deng, Y. (2007). *Research on spatial association rules mining technology with decision attribute set in GIS database*. Master Thesis of Jiangxi Normal University.

Di, K. (2003). *Spatial data mining and knowledge discovery*. Wuhan University Press.

Fayyad, U., Piatetsky-Shapiro, G., & Smyth, P. (1996). From data mining to knowledge discovery in databases. *American Association for Artificial Intelligence (AAAI), AI Magazine, 17*(3), 37–54.

Fan, M., Tu, J., & Chi, T. (2006). Mining and visualization of spatial association rules based on web. *Application Research of Computers, 23*(2), 157–159.

Guo, DZ. (2004). The research on spatial data mining and its uncertainty. *Geomatics and Spatial Information System, 17*(5)..

Han, J., & Fu, Y. J. (1996). Exploration of the power of attribute-oriented induction in data mining. *Advances in Knowledge Discovery and Data Mining*, 399–421.

Han, J., & Kamber, M. (2001). *Conceptions and technologies of data mining* (Fan, M., Meng, X., et al., Trans.). Mechanic Industry Press.

He, B., Chen, C., & Fang, T. (2006). Uncertainty processing and measurement of spatial data association rules mining. *Geography and Geo-Information Science, 22*(6), 5–8.

He, W., Pan, Q., & Chen, Y. (2005). A comparison between association rule data mining and causal discovery. *Pattern Recognition and Artificial Intelligence, 18*(3), 328–333.

Hsu, W., Lee, M. L., & Wang, J. (2008). *Temporal and spatio-temporal data mining*. IGI publishing.

Hu, C. P., & Qin, X. L. (2007). A survey of spatial data mining research. *Computer Science, 34*(5), 14–19.

Hu, K., Lu, Y., & Shi, C. (2001). Advances in rough set theory and its appliations. *Journal of Tsinghua University (Science and Technology), 41*(1), 64–68.

Huang, T., Qin, X. L., & Ye, S. (2005). New spatial multi-dimensional association rule model and its algorithm. *Journal of Nanjing University of Aeronautics & Astronautics, 37*(3), 301–306.

Jia, Z. (2008). Expert system for farmland grading based on GIS and SDM. *Journal of Anhui Agricultural Sciences, 36*(14), 6035–6037.

Jia, Z., Liu, Y., & Zhang, T. (2004). The study on visual interactive spatial data mining. *Science of Surveying and Mapping, 29*(5), 34–37.

Kong, J., Zhou, C., & Jiang, J. (2008). Digital urban management using spatial cluster algorithm. *Information Technology & Informatization*, (2), 74–76.

Koperski, K., & Han., J. (1995). Discovery of spatial association rules in geographic information databases. In *Proceedings of the 4th International Symposium on Advances in Spatial Databases*, Springer-Verlag, London, pp. 47–66.

Lan, R., Liu, Z., & Yang, X. (2005). Methods of mining fuzzy spatial association rules. *Journal of Geomatics Science and Technology, 22*(1), 36–39.

Lan, R. Q., Shi, W. Z., Yang, X. M., et al. (2005, October 14–16). Mining fuzzy spatial configuration rules: methods and applications. In *ISPRS Workshop on Service and Application of Spatial Data Infrastructure* (Vol. XXXVI (4/W6), pp. 319–324).

Li, C., & Zhang, Y. (1999). A new clustering algorithm. *Pattern Recognition and Artificial Intelligence, 12*(2), 205–209.

Li, D., Wang, S., & Li, D. (2002). Theories and technologies of spatial data mining and knowledge discovery. *Geomatics and Information Science of Wuhan University, 27*(3), 221–233.

Li, D., Wang, S., & Li, D. (2006). *Theory and application of spatial data mining*. Science Press.

Li, D., Wang, S., & Shi, W. (2001). On spatial data mining and knowledge discovery (SDMKD). *Geomatics and Information Science of Wuhan University, 26*(6), 491–499.

Li, D., Meng, H., & Shi, X. (1995). Membership cloud and subordinate cloud generator. *Computer Research & Development, 32*(6), 15–20.

Li, F., Xue, B., & Huang, Y. (2002). K-means clustering algorithm with refined initial center. *Computer Science, 29*(7), 94–96.

Li, H., & Wu, C. (1999). An adaptive clustering algorithm for two-dimension objects. *Journal of Xidian University (Natural Science), 26*(3), 300–304.

Li, J. (2000). *Research on 3D GIS spatial data model and visualization technology.* Doctoral Dissertation of National Defense Science and Technology University.
Li, K., & Liu, Y. (2004). Research on agent based distributed spatial data mining model. *Computer Engineering and Applications* (Suppl.), 217–219.
Li, N., & Shen, J. (2002). An algorithm automatic generating fuzzy sets for quantitative attributes. *Computer Engineering and Applications, 38*(21), 10–11, 41.
Li, H., & Zhao, B. (2001). Numerical model of the circulation of the Bohai sea and Yellow sea and East China sea. *Marine Sciences, 25*(1), 28–32.
Li, DY., & Du., Y. (2005). *Artificial intelligence with uncertainty[M].* Beijing: National Defense Industry Press.
Li, X. (2005). *Method and application of urban spatial data mining.* Shandong University Press.
Liu, B., & Pan, J. (2006). Research of incremental data mining based on swarm intelligence. *Computer Engineering and Design, 27*(11), 1939–1942.
Liu, D., Wang, S., & Yu, Q. (2004). A multi-level spatial association rules mining algorithm based on qualitative spatial reasoning. *Computer Research & Development, 41*(4), 565–570.
Liu, W., Chen, J., & Deng, K. (2006). Automatic detection of spatial conflicts between line objects. *Journal of China University of Mining & Technology, 35*(6), 767–771.
Lu, W., & Jiawei, H. (1993). *Discovery of general knowledge in large spatial databases[C].* Proceedings of the Far East Workshop on Geographic Information Systems, Singapore (pp. 275–289).
Lu, Q., Liu, SJ., & Cai, QS. (2000). A progressive refinement algorithm for mining spatial association rules. *Application Research of Computers, 17*(6).
Ma, R., & He, Z. (2006). Fast mining of spatial outliers from GIS database. *Geomatics and Information Science of Wuhan University, 31*(8), 679–682.
Meng, L. Q. (1998). *Cognitive modeling of cartographic generalization.* Project report on "Strategies on Automatic Generalization of Geographic Data"—Stage 2, Högskolan i Gävle/Sandviken (pp. 1–31).
Meng, Z., & Cai, Z. (2003). Research on personalized data mining system based on Multi-Agent technology. *Journal of Central South University (Science and Technology), 34*(3): 290–293.
Ni, K., Chen, Q., & Ma, Q. (2007). Research on spatial data mining based on agent. *Microelectronics & Computer, 24*(11), 172–174.
Qin, K. (2004). *Research on image data mining based on formal concept analysis.* Doctoral Dissertation of Wuhan University.
Shi, Y.,Sun, Y., & Zuo, C. (2000). Spatial data classification based on rough set. *Journal of Software, 5*, 673–678.
Su, F., Du, Y., & Yang, X. (2004). Geo-association rule with spatiotemporal reasoning. *Journal of Geo-Information Science, 6*(4), 66–70.
Su, M. C., & Chou, C. H. (2001). A modified version of the K-means algorithm with a distance based on cluster symmetry. *IEEE Transactions on Pattern Analysis and Machine Intelligence, 23*(6), 674–680.
Sun, L., & Zhang, D. (2002). Spatial data mining based on hypergraph model. *Computer Engineering and Applications, 38*(11), 30–32, 52.
Sun, Q., Chen, Q., & Fang, T. (2008). Multi-scale spatial data mining model based on fuzzy clustering and its application in mine. *Journal of Shanghai Jiaotong University, 42*(2), 194–197, 201.
Sun, Q., Fang, T., & Guo, D. (2005). Scale transformation in spatial data mining. *Computer Engineering and Applications, 41*(16), 17–19.
Tang, C., & Jin, Y. (2002). The research on a large-scale text clustering algorithm based on clustering features. *Computer Science, 29*(9), 13–15.
Wan, X. J., Yang, J., & Chen, X. (2003). An improved k-means algorithm for documents clustering. *Computer Engineering, 29*(2), 102–103, 157.
Wang, G. (2001). *Rough set theory and knowledge acquisition.* Xi'an Jiao Tong University Press.
Wang, Y. (2007). *Research on gathering and formalizing of user requirements in personalized process.* Master Thesis of Shandong University.

Wang, J., & Cheng, T. (2007). Applications of spatio-temporal forecasting to forest fire prevention. *Acta Scientiarum Naturalium Universitatis Sunyatseni, 46*(2), 110–113, 116.

Wang, J., Zhang, X., & Zhou, H. (2006). A genetic k-means algorithm for spatial clustering. *Computer Engineering, 32*(3), 188–190.

Wang, J., & Zou, J. (1992). *Model method for map data processing*. People's Liberation Army press.

Wang, S., Shi, W., & Li, D. (2003). Cloud model based interpretations of some issues in spatial data mining. *Computer Engineering and Applications, 39*(24), 185–186, 217.

Wang, Z. (2005). *Research on key technologies of spatial data mining based on geographic information system*. Doctoral Dissertation of Zhejiang University.

Wen, J., Li, L., & Wu, H. (2003). Research on the web agents-based spatial mining framework & its implementation. *Computer Engineering and Applications, 39*(22), 185–186, 217.

Weng, X. Q., & Shen, J. Y. (2007). Identification of outlier patterns in multivariate time series. *Pattern Recognition and Artificial Intelligence, 20*(3), 336–342.

Wu, T., & Qin, K. (2006). The theory and method of image texture feature data mining. *Computer Era*, (8), 4–5.

Wu, T., Qin, K., & Xiao, Q. (2007). Uncertainty analysis of image texture feature data mining. *Computer Engineering and Design, 28*(12), 2905–2908.

Xiao, J., & Ye, F. (2005). Research on data mining algorithm based on conception hierarchy tree and its application. *Application Research of Computers, 22*(3), 61–63.

Xie, Z. (2001). *Research on knowledge discovery based on concept lattice model*. Doctoral Dissertation of Hefei University of Technology.

Xu, J., & Wu, Y. (2004). Study of personalized data mining system based on multi-mobile agent technology. *Journal of Computer Applications, 24*(7), 28–29, 32.

Yan, X., & Cai, Z. (2003). Research on an algorithm of discovering association rules based on image. *Computer Engineering and Applications, 39*(2), 209–211.

You, F., Su, Z., & Yang, B. (2005). Model of knowledge discovery state space and its application to image mining. *Application Research of Computers, 22*(2), 231–234.

Zadeh, L. A. (1965). Fuzzy sets. *Information and Control*, (8), 338–353.

Zhang, B. (2005). *Description, acquisition and inference of individual requirements: A case study*. Doctoral Dissertation of Graduate University of Chinese Academy of Sciences.

Zhang, L. (2001). *Concept, structure and application*. Doctoral Dissertation of Northwestern Polytechnical University.

Zhang, B., Hsu, M., & Dayal, U.K. (2004). Harmonic means-a data clustering algorithm[EB/OL]. http://www.hpl.hp.com/techrcports/1999/HPL-1999-124.oaf.2004-12-10.

Zhang, W., Wei, L., & Qi, J. (2005). Theory and method of attribute reduction of concept lattices. *Science in China Series E Information Sciences, 35*(6), 628–639.

Zhang, X., Ge, W., & Wang, P. (2008). The design and realization of spatial data mining system based on topographic map analysis. *Hydrographic Surveying and Charting, 28*(4), 60–62, 72.

Zhang, X., & Wang, J. (2007). Genetic k-medoids spatial clustering with obstacles constraints. *Computer Engineering, 33*(4), 168–170.

Zhang, X., Su, F. Z., & Shi, Y. (2007). Research on progress of spatial association rule mining. *Progress in Geography, 26*(6), 119–128.

Zhang, X. P., Zhang, Q., Fan, Z. S., et al. (2008). Clustering spatial data with obstacles using improved ant colony optimization and hybrid particle swarm optimization. *FSKD*(2), 424–428.

Zheng, X., & Liu, X. (2006). Discussion on spatial data mining method based on clementine decision tree—A case study of Ancheng Township in Pingyin County. In *2006 Chinese Association Annual Meeting Proceedings* (2), 67–73.

Zhao, Q., Wang, L., & Gu, Y. (2004). Improvement of spatial classification algorithm based on neighbor graph. *Application Research of Computers, 21*(9), 115–117.

Zhao, L., & Zheng, X. (2008). Design and implementation of spatial data mining system (m-sdm) based on MATLAB. http://www.paper.edu.cn.

Zhou, C., & Zhang, J. (1999). Entropy-based model for geo-data mining. *Journal of Image and Graphics, 4*(11), 946–951.

Zhou, H., Wang, J., & Wu, S. (2002). Spatial data mining and its applications. *Bulletin of Surveying and Mapping, 2,* 11–13.

Zhu, M., & Wang, J. (2000). A new approach of clustering. *Pattern Recognition and Artificial Intelligence, 13*(3), 262–265.

Zhu, Q., & Sun, F. (1998). Study on application of recurrent neural network. *Journal of Beijing University of Chemical Technology (Natural Science Edition), 25*(1), 86–90.

Chapter 15
Spatial Data Uncertainty and Quality Control

Changqing Zhu and Fanghua Wu

15.1 Introduction

Spatial data are the basic data of cartography and geographic information systems (GISs) and are the main research object of GIS. These data describe the location, shape, and size of a spatial entity, as well as the relationship between entities. The location of a spatial entity is usually represented by a three-dimensional or two-dimensional coordinate string. Spatial data can be divided into two types: geometric data and relational data. Geometric data can be divided into three basic types: point, line and area objects in terms of geometry; relational data are the information that describes the relationships (such as adjacency, association, inclusion, connectivity, and proximity) among different spatial entities. The establishment of relational data is helpful for various applications and spatial analyses.

Uncertainty indicates the ambiguity, indeterminacy, non-affirmative or undecided or unstable state of something. The objective world we are facing is a complex, changeable, open and nonlinear giant system that contains uncertainty. Therefore, spatial data are also uncertain.

The concept of uncertainty appeared early in the literature on electronic measurement and metrology in the 1970s; when the uncertainty is synonymous with error, it can be used arbitrarily or more by using this simple concept of error, but more often, the simple concept of error is used.

"Uncertainty" is a broader and more abstract concept than "error". It can be regarded as a generalized error, which includes both random and systematic errors, as well as gross errors; it also contains measurable and non-measurable errors, as well as numerical and conceptual errors. Generally, uncertainty refers to the degree of lack of knowledge regarding the object being measured. It is usually characterized

C. Zhu (✉)
Nanjing Normal University, Nanjing, China

F. Wu
Xi'an Institute of Surveying and Mapping, Xi'an, China

J. Wang and F. Wu (eds.), *Advances in Cartography and Geographic Information Engineering*, https://doi.org/10.1007/978-981-16-0614-4_15

by randomness and fuzziness. From a philosophical point of view, randomness is a break in the law of causation, and fuzziness is a flaw in the law of excluded middle (Shi 2005). From the point of view of information theory, uncertainty has many meanings. The errors of data, the fuzziness and incompleteness of data and concepts can be regarded as the contents of uncertainty.

The sources of spatial data uncertainty mainly include the instability of space phenomena themselves, the limitations and uncertainties of human cognitive processes, the errors in spatial data acquisition, and the uncertainty caused by spatial data processing and analysis.

15.2 Development Process

The research of spatial data uncertainty began in 1968 when Blakney published his two papers—the accuracy standards of topographic maps and the quality and use of soil maps, which are considered to be the earliest literature in the study of GIS uncertainty (Liu 1995). The development of this study area can be divided into three stages (Liu and Liu 2001, Tang et al. 2004).

The first phase was in the 1960 and 1970s when GIS was born, which marked the beginning of the study of uncertainty. When people developed GIS as a tool for many digital maps, they found that to achieve good business profits, they had to pay attention to the quality of GIS data. At that time, the research of GIS data quality mainly used the principle of mathematical statistics to analyze the quality and precision of mapping and graphic operations. According to calculation tools, Gong Jianwen (1982) systematically introduced, summarized and compared map area measurement methods, such as the hand graph number converter method, density slicer method, photoelectric scanning method, planimeter method, weighing method, templet method and graphic method.

The second phase occurred in the 1980s. The importance of the research on the quality and uncertainty of data received much attention from developers and users both at home and abroad. The National Center for Geographical Information and Analysis (NCGIA) was established in 1988 in the United States. In the theoretical research, the mathematical statistics principle was still the main area of research, some related error models of spatial data were established, and error analysis in the process of image classification and interpretation began. In 1982, Chrisman applied the "ε-band" to the spatial uncertainty of line elements in his doctoral thesis entitled "spatial analysis methods based on cartographic errors". After that, Blakemore (1984) developed the method into a "ε-band model" for describing the location uncertainty of linear elements.

The third phase has been since the 1990s. At this stage, research on the quality and uncertainty of data developed rapidly, and a series of achievements were achieved. Chinese scholars began to actively participate in the study area, which was mainly based on a theoretical model and experiments. In addition to applying the theory of probability and mathematical statistics, there were also fuzzy mathematics, information theory, rough set theory, cloud theory and evidence theory. The research of GIS uncertainty attracted extensive attention both at home and abroad. In the 12 subjects of GIS basic theory put forward by NCGIA in 1991, GIS spatial database accuracy and quality research were listed as the top priority of research topics. "The uncertainty of geographic data and analysis based on GIS" was proposed as one of the 10 key directions by the University Consortium for Geographic Information Science (UCGIS) in 1996. "Error Propagation in GIS" was chosen as the third item of 14 research subjects by the UK's Economic and Social Research Council (ESRC) in the 1990s. In formulating the 1990s research plan, the Nederlands Expertise Centruum Voor Ruimtelijke Informatiererwerkig (NEXPRI) presented four major research topics. In the first research topic, "spatial analysis theory", the first problem was "error propagation in spatial operations". From 1998 to 2004, the major EU countries, including France, Germany, the United Kingdom, Austria and Holland, conducted large-scale and high-level international studies on the uncertainty of GIS data in cooperation with Canada, which made it a joint effort in the subject of "Revision of the Uncertain Geographic Information". In 1997, academician LI Deren ranked the research on the theory of spatial data quality and uncertainty as one of the 12 urgently needed theoretical problems in China's GIS applications and future development. M. Goodchild, an international expert on GIS and an academician of the American Academy of Sciences, pointed out in 1998 that uncertainty was one of the four most important basic theories of geographic information science.

The academic exchange on the uncertainty theory of spatial data has been highly valued by the international GIS community. The "Kyoto space database seminar" held in California, USA, in December 1998 was considered to be a milestone in GIS uncertainty system research. Thereafter, the conference "Accuracy of Spatial Database" became a classical reference in this research field. In July 1997, the International Conference on spatial data quality was held in Hong Kong. The achievements and status of GIS spatial data quality and uncertainty were summarized at that time. The international series of thematic conferences on uncertainty in spatial data held in recent years included the International Symposium on Spatial Accuracy Assessment in Natural Resources and Environment Sciences—a biennial meeting that began in 1994; the International Symposium on Spatial Data Quality—a meeting initiated by Shi Wenzhong et al. in 1999; and GISC, which began in 2000. The uncertainty theory of spatial data also became one of the most important subjects in other important international conferences, such as SDH, ISPRS, Congress, ICA, and Congress.

In China, Zhengzhou Information Engineering University, Nanjing Normal University, Tongji University, Wuhan University, and the Xi'an Institute of Surveying and Mapping play leading roles in the research on spatial data uncertainty and have achieved a series of theoretical and practical results, published a number of books and trained a number of doctoral and postgraduate students.

At present, the research on spatial data uncertainty mainly focuses on the positional uncertainty and attribute uncertainty of spatial data and the uncertainty of digital elevation model (DEM) data; the research focus is changing from theoretical research to studies that combine theory with application. However, so far, the theoretical system of uncertainty research has not been completely established, there is not a complete theoretical system for dealing with the uncertainty in geosciences, and many problems need to be studied and solved.

15.3 Main Research Achievements

15.3.1 Research on the Positional Uncertainty of Spatial Data

Data that express basic geometric elements are the most critical part of GIS data, and the uncertainty of position data will lead to the uncertainty of attribute data, so research on the uncertainty of position data has important theoretical and practical value, which is good for the quality evaluation of GIS data. The uncertainty of positional data is the focus of spatial data research, and it is also the most active and fruitful part of the uncertainty research on spatial data. Chinese scholars have conducted in-depth studies in this field, and some of them have reached advanced levels internationally.

Points, lines and polygons are the basic elements of spatial data. Therefore, the uncertainty research of spatial data mainly focuses on the positional uncertainty of points, the positional uncertainty of line elements and the positional uncertainty of polygons.

1. **The positional uncertainty of points**

A point object is one of the most basic descriptive objects in GIS, and it is also the basic element of a linear object or polygon in GIS. Thus, the positional uncertainty of a point has an impact on the positional uncertainty of all spatial entities in GIS (Shi 2005). In fact, the positional precision of points is an important part of surveying and mapping discipline studies; it has developed well from the distribution law of point data error to the uncertainty processing model for point data (Zhang et al. 2005).

The distribution law of spatial point data error is usually described from two aspects: coordinate error distribution law and spatial point distribution law.

The coordinate error distribution is one-dimensional, which is usually described as a normal distribution by traditional error theory. However, some scholars, through experimental research on digital error, show that the coordinate error may also follow the Laplace distribution, NL distribution, P-norm distribution, Huber distribution, and Rayleigh distribution (Liu et al. 1999; Guo 2004). The P-norm distribution was studied in detail by Sun Haiyan (1998, 2001), and the commonly used degenerate distribution, Laplace distribution, normal distribution and uniform distribution are treated as the specific cases of the p-norm distribution.

The distribution of two-dimensional plane point errors is usually described by using a two-dimensional normal distribution; in addition, Shi Wenzhong (1997) gave a more general form of Rayleigh distribution. The Rayleigh distribution focuses on the analysis of errors in terms of distance and can be used to describe the error distribution of plane points.

The uncertainty index of a one-dimensional point is given by the error interval $[-k\sigma\ ,\ k\sigma]$, where k is constant and σ is the standard deviation; if k uses different values, a different error interval is formed, including probability error range, standard error interval, linear map precision, entropy uncertainty interval and limit error interval. The uncertainty index of the plane point is expressed by the root mean square error of coordinates σ_x , σ_y, root mean square error of point $\sigma_p = \sqrt{\sigma_x^2 + \sigma_y^2}$ and the error ellipse. The location precision of three-dimensional points is also described by the three-dimensional root mean square error of coordinates σ_x , σ_y , σ_z, root mean square error of a point $\sigma_p = \sqrt{\sigma_x^2 + \sigma_y^2 + \sigma_z^2}$ and the error ellipsoids (Liu et al. 1999; Guo et al. 2003). Shi Wenzhong (2005) suggested that the uncertainty of point objects could be described in a polygon, which could be either an error ellipse or an error rectangle. Under the assumption that the random point obeys the standard normal distribution, we find that the probability value of the observation point falls into the polygon.

The error ellipse is the most common method for describing a point error, but because the uncertainty index based on the error ellipse is the theoretical deviation of the measured point from the real point, the center of the error ellipse is in the true position of the point, whereas the true position coordinates of the point are usually unknown; therefore, the uncertainty index based on the error ellipse is inconvenient for application. To compensate for this deficiency, some scholars recently proposed using the confidence region as an index of location uncertainty. A confidence region is an area established at the center of the measured point, and the true values of points fall into this region with a certain probability. At present, there are mainly three kinds of confidence regions: circular confidence domains (Li Deren et al. 1995), rectangular confidence regions (Shi 1997) and elliptic confidence regions (Guo et al. 2003). Lan Yueming et al. (2004) also presented an error band method for describing the location uncertainty of line elements by point error.

2. **The positional uncertainty of lines**

In the field of classic surveying and mapping, the amount of studies on line elements and surface elements are relatively few. Among them, the research on line elements is the key because the line element is not only the foundation of the uncertainty of the polygon but is also the basic element of the GIS overlay analysis, buffer analysis and spatial address matching analysis.

Perakl proposed the ε-band to describe the error of the cartographic line in 1956, when GIS was not present. Chrisman (1982) and Blakemore (1984) discussed it further and proposed a "ε-band" model to measure the positional uncertainty of line elements. The ε-band consists of two semicircles located at the endpoint of the line element, with a radius of ε, and there are two parallel lines located in the middle of the line element, with 2ε as the width. Based on the ε-band model, Zhang and Tulip (1990) further studied an error band model, that is, "E-band" model. The model assumes that the errors at each end point are independent of each other and that the errors at both end points have the same variance and covariance. Based on this assumption, the "E-band" model holds that the error band is narrowest in the middle and wider at both ends, which is different from the ε-band model that considers the error band to be the same width everywhere.

The "G-band" proposed by Shi Wenzhong et al. (1998a) is the further development of the "E-band". According to the theory of stochastic processes, the "G-band" takes into account many situations, such as the error between any two points with different accuracies or even those related to each other; the method establishes the location uncertainty model of line elements and reveals the law in which the error of arbitrary points on the line element is isotropic and direction independent. The boundary line of the G-band is the envelope of the error ellipse family, which is composed of the error ellipse at any point of the line element. The "G-band" model overcomes the limitations of the "E-band" model hypothesis and can be used to establish the uncertainty model of the line object location when the location errors of two points are independent or correlated. According to the principle of solving the envelope of the curve family, Liu et al. (1998) derived an analytical expression for the boundary lines of the G-band and proved that the boundary line of the G-band for the plane line segment error band is a continuous and closed curve. The "G-band" is rigorous in theory, but it is inconvenient to apply since the boundary line formula is tedious and computation is complicated.

In 1998, a variance–covariance matrix of coordinates at any point of the line element was derived by Shi Wenzhong and Liu Wenbao by assuming that the coordinate errors at both ends of the line element are uncorrelated and their relevance is considered, while the median error at a certain point in the direction of the line element is taken as the error bandwidth (Tang et al. 2004). However, the median error in the direction perpendicular to the line segment is not necessarily the point with the largest distance from the error ellipse to the line segment; thus, Liu Dajie et al. proposed the ε_m model for the uncertainty model of the line segment. The ε_m model takes the maximum error at a point on the line element as the error bandwidth of the point. Liu Dajie and Shi Wen extended the uncertainty ε_σ model and ε_m model of plane straight lines to straight lines in three-dimensional space. Tong Xiaohua et al. (1999) studied the ε_σ model and ε_m model for the uncertainty of planar circular curves. Shi Wenzhong et al. (2000) and Shi Wenzhong (2005) studied the ε_σ model and ε_m model of general curve uncertainty. Zhang Guoqin et al. (2005) studied the ε_σ model of curve uncertainty in 3-D space. In terms of algebra, the ε_σ model and ε_m model of linear elements in plane and three-dimensional space are discussed in Zhu Changqing et al. (2007). The shape feature of the error band is strictly proven. It is concluded that the error band generally shows the shape characteristics of "larger extreme points and less middle" and shows the shape characteristics of "big at one end and small at another end" under special conditions; additionally, the location of the error band with a minimum bandwidth is obtained. Tong Xiaohua et al. (1999) extended the ε_σ model and ε_m model of the linear position uncertainty to the planar circle curve. Zhang Guoqin et al. (2005) extended the ε_σ model to the uncertainty model of circle curves in 3-D space.

To obtain the error band without changing with probability, Fan Aimin et al. (2001) established an uncertainty band model of error entropy (H-band) by using the information theory. The model can determine the bandwidth of the error band according to the error entropy at the endpoint of the line segment, which is the same, as it is equal in terms of width. Li Dajun et al. (2003) considered the uneven error distribution of the line segment and developed the 'H-band'; they proposed that the average information entropy of the marginal probability distribution of the whole line segment should be used as the basis for determining the error bandwidth to establish the mean entropy uncertainty band model of the line segment.

In Shi Wenzhong (1994), the uncertainty model of the confidence region of the planar line segment was proposed and proven for the first time worldwide. The real location of the line segment was included in the band of the confidence region with a probability not less than a given confidence level. In Shi Wenzhong (1997, 1998), the uncertainty model of the confidence region was extended from two-dimensional to three-dimensional space, and the uncertainty model of the confidence region for linear elements in N-D space was further developed. Guo Tongde et al. (2003) studied the confidence region uncertainty model of plane lines and curves in different ways.

3. **Visual representation of the positional uncertainty of vector data**

In the existing description index of point uncertainty, the error interval, coordinate MSE and point MSE are only numerical measures of uncertainty, but the error ellipse, point confidence region and error ellipsoid can be used to directly describe the positional uncertainty of a point with a visual graph. Among them, the error ellipse is often used to describe the location uncertainty of two-dimensional positional elements in the error theory of surveying science. Liu Wenbao (1995) proposed and defined several error ellipses as follows: standard error ellipse, probable error ellipse, map circle precision error ellipse, limit error ellipse, and root mean square error ellipse.

An error ellipse can show the positional uncertainty of any two-dimensional point element in a map so that the user can make a preliminary comparison of the spatial distribution of the uncertainty. However, there are some disadvantages. When a point and its corresponding error ellipse descriptors are displayed on a map at the same time, the map may become overwhelming, and some points may overlap with other uncertainty descriptors of spatial features. This weakness can be overcome by displaying only one or several error descriptors of spatial features at a time.

Zhang Baogang et al. (1998) proposed an uncertainty propagation model for vector data buffer operation based on the "ε-band" and "E-band". Dai Honglei et al. (1999) derived the analytical expressions of the boundary of the "G-band" model for describing the positional uncertainty of line elements, and then, the visual graphics could be obtained by computer programming. Tong Xiaohua et al. (1999) studied the adjustment model of GIS digital data and realized its use. Tang Zhongan et al. (2004) visually represented the equivalent probability density error model. Based on the uncertainty models of confidence bands of points, lines and planes, Mei Shiyuan et al. (2004) discussed the propagation of the uncertainty of point, line and plane buffers. Zhang Guoqin et al. (2005) carried out visual expressions of the "ε_σ-model" and "ε_m-model".

15.3.2 Research on Attribute Uncertainty in Spatial Data

Attribute uncertainty in spatial data is the description of attribute values (or attribute categories) of points, lines, polygons, or remote sensing images and can be divided into the two categories of quantitative data (such as land prices) and qualitative data (such as the use of land). The accuracy is also divided into two categories: quantitative data accuracy and qualitative data accuracy. Attribute data may be a discrete value (category value) or a continuous value (Shi 2005). Continuous attribute data are a continuous random variable, which is similar in nature to location data. The uncertainty can be determined by measuring the error, and its modeling can be done using a method similar to that of location uncertainty, such as the error propagation law and Monte Carlo simulation method. The assessment of non-continuous attribute data (categorical data) can be achieved through a set of classification results (Shi et al. 2001).

1. **Research domain of attribute uncertainty**

An attribute error is the difference between an attribute's observation and its true value. Attribute uncertainty is one of the indexes that describes the quality of spatial data, and its meaning is more extensive than that of an attribute error. The uncertainty of attribute data mainly comes from uncertainty in the data source, uncertainty in data modeling and the uncertainty introduced in the analysis process. Among them, the uncertainty of the data source comes from measurement, artificial judgment and hypotheses in the process of data acquisition (Liu et al. 2000). Attribute uncertainty involves a wide range of factors, and the main research areas are summarized in Shi Wenzhong (2001).

Shi Wenzhong (2005) summarized various methods that can be used for attribute uncertainty analysis. Among them, the target model and domain model are part of the classical data processing model; probability theory, probability vector, evidence theory and spatial statistics are of statistical significance; and rough sets, fuzzy sets, cloud theory, genetic algorithms, chaos theory and gray theory are widely used.

2. **Attribute uncertainty model of spatial data**

At present, the accuracy analysis of attribute data mainly focuses on the accuracy analysis of remote sensing attribute data classification and the introduction of some new methods, and the research in this area is relatively scarce.

In Zhang Jingxiong et al. (1999), based on the field concept and model, the field model of positional uncertainty and attribute uncertainty was studied, and both can be described and analyzed in a unified way. Zhang et al. (2002) studied attribute uncertainty based on a target model and domain model (also called a field model).

The target model has difficulty completely describing the attribute uncertainty, and the domain model is more suitable than the target model for discussing the uncertainty of its attributes.

The research on uncertainty in remote sensing data mainly focuses on the study of the uncertainty of attribute data classification. Based on probability vectors, Shi Wenzhong (1994) proposed four parameter indexes, namely, absolute uncertainty, relative uncertainty, pixel mixture degree and evidence integrity. Converting a probability vector into several parameters can reduce the amount of uncertainty in the descriptive data, thereby reducing storage space.

In the process of applying fuzzy sets to study the uncertainty in attributes, Wang Xinzhou et al. (1997) proposed a fuzzy comprehensive method that combines the fuzzy comprehensive evaluation with the fuzzy cluster analysis. Based on the fuzzy clustering analysis of the absolute mean distance, the uncertainty in land prices and grade attributes is evaluated. The uncertainty in the attributes of GIS spatial data are discussed by fuzzy analysis. Huang Youcai et al. (1995) assessed the uncertainty of areas based on the classification of fuzzy land vegetation. In Shi Wenzhong (1994), fuzzy membership was used to express pixel categories of uncertain adjacent boundaries in remote sensing images. In Liu Wenbao et al. (2000), based on the uncertainty of attribute region classification, boundary positional error and sampling error of quantitative attribute data in the region, the fuzzy logic uncertainty propagation model and arithmetic model were proposed.

Di Kaichang (1999) applied cloud theory to spatial generalized knowledge, association rule mining, spatial database uncertainty querying, remote sensing image interpretation and recognition, and so on. Wang Shuliang et al. (2002) used cloud theory to evaluate land prices and to mine monitoring data for the uncertain motion of landslides. Wang et al. (2004a) used cloud theory to combine qualitative analysis with quantitative calculation to address the randomness and fuzziness of GIS attribute uncertainty.

Rough set theory was developed in the 1970s by Professor Z. Pawlak of Warsaw Polytechnic University in Poland and other scientists based on studying the logical characteristics of information systems. Di Kaichang et al. (1999) comprehensively introduced it into the field of GIS and concluded a set of methods for attribute analysis and knowledge discovery in GIS by the rough set theory. According to the rough set method, all attribute values are treated as qualitative data. If quantitative data should be processed comprehensively with qualitative data, the quantitative data should be transformed into qualitative data. The rough set can be used to analyze the table consistency and attribute reliability of attribute tables in the GIS attributes database, to simplify attribute dependency and attribute tables and to generate minimal decision trees and classification algorithms. Wang Shuliang et al. (2002) proposed a concept of geo-rough space (which consists of rough entities, rough relationships and rough algorithms) and used it to study the uncertainty of attributes: a new mathematical explanation was provided for the uncertainty theory or model of existing attributes; the space domain could be universally divided to include other spatial theories; the certainty and uncertainty of spatial entities were taken into account, possessing authenticity; and the uncertainty of attribute observations were analyzed more comprehensively. In Hu Zhaoling et al. (2000), the accuracy of GIS attribute data classification was studied by using the rough set theory, and the accuracy of logical operation such as logical intersection, logical sum and logical complement were analyzed. In Deng Min et al. (2006), spatial objects were observed from the angle of spatial data granularity, and the uncertainty in locations, attributes and relationships of spatial targets were analyzed and expressed by the rough set method.

3. Measurement of attribute uncertainty

The measurement of spatial data attribute uncertainty is usually expressed by the rate of disfigurement. Briefly, the so-called rate of disfigurement is the number of defects contained in the data cell of the unit number. The number of defects is the test result of the quality of the attribute data. The number of defects in the sample can be obtained by the following methods: in accordance with the types of defects as specified in existing surveying and mapping data quality inspection standards, such as general defects, heavy defects and serious defects, for the sample quality of each attribute data to be examined, its inspection value is recorded in the form of counting defects; then, the number of sampled defects is obtained by weighted summing of the recorded test values of general defects, heavy defects and serious defects. In Liu et al. (2002), the rate of disfigurement is discussed to be used to measure the accuracy of the attribute data based on sampling inspection, which is a new method for researching the accuracy of attribute data in GIS.

In view of the quantitative properties of continuous changes in space or time according to a certain trend, the possibility of using information diffusion estimation to deal with this issue is discussed in Wang Qingguo et al. (2004b).

4. Combination of spatial data position and attribute uncertainty - S-band model

According to Shi Wenzhong (1998a), the indicator combining position with attribute uncertainty is defined as the integrated position and attribute uncertainty, i.e., PAT uncertainty. When describing a two-dimensional vector object in GIS, it can be divided into two regions: the fuzzy boundary region and the interior region. Generally, the uncertainty in the fuzzy boundary region is more complex than the uncertainty in the inner region. In the fuzzy boundary region, the spatial entity is affected by the location and the attribute error; in the inner region, the uncertainty mainly comes from the attribute error.

A theoretical model, the S-band model, was proposed by Shi Wenzhong (1998b) for the integrated attributes and location uncertainty. The basic idea of the "S-band" is that in the fuzzy boundary, the uncertainty of an object is expressed by combining the uncertainty of position, the uncertainty of attributes and their relationships, and its essence is the integration of the qualitative and quantitative nature of the two different variables. The "S-band" has opened up research on the integrated uncertainty theory of attributes and location uncertainty, and it has great practical significance.

15.3.3 Research on the Uncertainty of the DEM

The DEM is a discrete digital representation of the undulating topography of the Earth's surface. At present, the scientific measurement and accurate expression of the quality and uncertainty of DEM data has become a very important issue for the producers and users of DEM data.

The generation of DEM data errors is closely related to the DEM production process, which is divided into two stages: raw data acquisition and interpolation modeling. Therefore, in the research of DEM uncertainty, the main work is focused on DEM raw data error, surface model error and the theory and method of DEM uncertainty analysis. Chinese scholars, especially those from Nanjing Normal University, Zhengzhou Information Engineering University, Institute of Surveying and Mapping, Wuhan University and other institutions, have conducted productive work in these areas.

1. **DEM raw data error**

Zhu Zhiming (1997) evaluated the quantization method, spatial distribution and expression methods of the DEM error. In the paper by Tang Guoan et al. (2001), DEM errors are systematically classified according to different classification criteria; in particular, the paper proposes that the errors of DEM elevation can be divided into two basic types: DEM elevation sampling error and terrain description error according to their types of generation. Based on experiments in related areas, the generation conditions, spatial distribution characteristics and mathematical simulation methods of DEM terrain description errors are analyzed in detail. Ke Zhengyi et al. (1993) studied the error and error distribution of DEM raw data in their book.

2. **DEM propagation error**

The DEM propagation error is the error propagation of DEM surface measurement data in the interpolated data. Li Zhilin et al. (2000) studied the transmission errors of DEM linear and bilinear interpolation. The method was first to derive the transmission error of linear interpolation and then, based on interpolation theory, it was used to derive the transmission error of bilinear interpolation. Shi Wenzhong (2005) studied the transmission error of the DEM high-order interpolation function and considered that it had the same transmission error as the linear interpolation. However, since higher order interpolation has a higher model precision, DEM higher order interpolation has a higher precision in general. In addition, Zhu Changqing et al. (2005) carried out in-depth research on the transmission error. In Wang Yaoge et al. (2007), based on spline interpolation and Coons DEM, a different model for the propagation error of the grid DEM constructed from different grid point quantities was obtained. The model showed that the propagation errors of the spline DEM model and the Coons DEM model were different. The more grid points the grid DEM model was based on, the larger the propagation error of the DEM model was based on the same interpolation. When the number of grid points was the same, the propagation error of the DEM model decreased by weakening the weight of some adjoining grid points.

For the transmission error in TIN, the error estimation of linear interpolation based on TIN was studied by Zhu Changqing et al. (2005), the relation between the surface function and the original data point was established by the polar coordinate method, and the transmission error formula of DEM linear modeling based on TIN was theoretically derived. This formula represents the whole surface instead of the error of a single point, and it has important theoretical significance for studying the error precision of the DEM model based on TIN.

3. DEM model error

In the process of DEM production, no matter what interpolation algorithm is adopted, the elevation between interpolation points is always different from the actual measurement elevation, which is the main source of DEM error and is among the important research topics in photogrammetry and GIS. The interpolation error of the DEM is related to the mathematical model chosen for interpolation because the mathematical model is varied, can be continuous, can also be discontinuous, and can also be smooth or not smooth (linear surface). The commonly used DEM interpolation algorithms include bilinear interpolation, spline function interpolation, least square method, polyhedral function method, moving fitting method, etc., and fractal interpolation, Coons surface interpolation and interpolation based on morphology have appeared in recent years. Various interpolation methods have different errors in different geomorphic regions and different point sampling methods. Based on the Coons surface, Wang Yaoge et al. (2008) proposed a grid DEM model with good accuracy and proved the corresponding transmission error.

In Tang Guoan et al. (2001), the relationship between the DEM error and the spatial resolution and average profile curvature was studied by using the rectangular window analysis method, and six geomorphic types of experimental areas were selected, while the relationship between the DEM error and the spatial resolution and the average slope was derived, which was of great significance for improving the DEM accuracy and the range of application. This subject was been further analyzed and improved in Wang Guangxia et al. (2004c), where a new mathematical model of coefficient fitting was given, and the formula for the relationship between nonlinear DEM error and resolution, average profile curvature and slope was obtained, which provided a reliable theoretical basis for DEM application and error analysis.

The DEM error mainly originates from the transmission error of raw data in the process of modeling and the terrain simulation errors generated from using the interpolation model to represent the terrain surface. Among them, the transmission error is an accidental error, and the terrain simulation error is a model error. The main objective of the DEM analytical error model is to theoretically describe the two DEM errors. The transmission error is mainly estimated by the error propagation theorem. The terrain simulation error needs to know the morphology and statistical characteristics of the terrain in advance, and then, it is evaluated by a specific mathematical method, for example, based on the assumption that the error of the sampling point is zero, Tang Guoan et al. (2002) used the mathematical model of the difference between the elevation of the midpoint of the grid and the mean value of the four vertices of the grid to describe the error in the terrain simulation; Li Zhilin et al. (2000) used a statistical method to obtain the range of error probability for any distribution; and the truncation error in the interpolation process was theoretically deduced by the idea of function approximation in Hu Peng et al. (2006).

4. **DEM accuracy evaluation**

According to the application purpose and research content, the uncertainty of the DEM is often calculated by a numerical precision model. At present, the numerical precision models used are mainly the RMSE (root mean square error). The RMSE does not reflect the size of a single error but instead describes the dispersion degree of the terrain parameter and its true value in the whole sense; thus, RMSE provides the possible range of true values. RMSE depends on the location and number of detection points. For the same DEM data, different sampling sets yield different RMSEs; therefore, the DEM precision evaluation method based on RMSE has difficulty explaining the accuracy of the DEM well.

In Hu Peng et al. (2003, 2005), the DEM error was studied in depth, and it was proposed that the DEM interpolation error was an approximation error, so the concept of RMSE should not continue to be used for accuracy evaluation; instead, the approximation error should be adopted. Moreover, the error evaluation models of the two most commonly used DEM interpolation methods, i.e., the linear interpolation on the Delaunay triangle and the bilinear interpolation based on the rectangular grid, were given.

Wang Guangxia (2005) systematically studied the related theories and methods of DEM accuracy evaluation and proposed the content system of DEM accuracy evaluation, considering that DEM quality control should be a complete system integrating the theory system, precision control mathematical model and quality control verification system; and terrain simplification model and LOD accuracy models of different geomorphic types and different simplification methods were established. In addition, it was proposed that the accuracy of the DEM must include the fusion accuracy of the DEM and ground object model, while establishing the precision model and algorithm for integrating the DEM with the ground object model.

Based on the reconstructed contour line, Zhu Changqing et al. (2005, 2008) established a new DEM accuracy evaluation index and evaluation model by comparing it with the original contour, and it was used to evaluate the DEM accuracy. The reconstructed error was determined by the area error of the reconstructed contour and the length of the original contour; there was a definite number and they were unique for a given DEM, so the reconstructed error was unique.

Tang Guoan et al. (2007) studied the contour matching difference and its application in DEM quality evaluation, the concept of contour matching difference was proposed, and the automatic calculation and analysis method of matching difference was explained. Thus, the Loess Hilly and gully area with more complicated topography was taken as the experimental area, and the application conditions, application methods and effects of DEM evaluation on the quality of contours were studied.

15.3.4 Map Data Quality Control

The quality of the digital map data is the "life" of the basic geospatial database. It directly affects the correctness and reliability of all the application analyses and decision-making based on the map data and affects the success or failure of the system. The quality control of the map data and the quality verification of the map data have become very important problems in mapping digital information engineering. It has a profound influence on the quality evaluation of geospatial data, the improvement of spatial data processing methods, and the decrease in blindness in design and development.

The quality of map data involves not only the problem of control but also the question of how to evaluate it correctly. Data quality not only includes the positioning error but also includes the attribute precision, data integrity, logic consistency, time precision and other complex contents. In recent years, research on map data quality has focused on the error of location data and less on the uncertainty in attribute data. The research results are mainly focused on theoretical methods, and few practical systems are seen. In addition, data quality checks are mostly manual inspections; and automatic check methods are difficult to study. Automatic check techniques include template matching, implementation of standards and specifications, computer understanding of cartographic rules, artificial intelligence, pattern recognition, etc. It is necessary to design an automatic check and evaluation method of data quality that is reasonable and suitable for actual needs in the construction of foundational geographic spatial databases. It can greatly improve the efficiency of inspection, reduce the labor intensity of data quality inspection and acceptance, and improve the quality of the map data.

The study of quality control and verification of finished drawings started as early as the 1960s (Surveying office of State Administration of Surveying and Mapping 1960). Chen Zongxin (1962) analyzed the accuracy of 1:10,000 to 1:100,000 topographic maps. Min Buqiu (1963) discussed the precision and error of each process in detail from design to printing. Wang Quanrun (1964) discussed the inspection and acceptance process of topographic maps during production.

Since the 1980s, research on quality control theory and standardization of spatial data has been systematically launched in China (Kang 1988; Yu 1988). Using the principle of multilevel fuzzy comprehensive evaluation, a mathematical model for assessing thematic map compilation quality was established in He Zongyi (1989) and was used for the quality evaluation of map compilation for the *Hubei Provincial Atlas of Land and Economy*. The results agreed with the evaluation results of map experts. This quantified and standardized the process of map evaluation and analysis. However, many studies were carried out in the middle and late 1990s. The State Key Laboratory of Resource and Environmental Information Systems at the Institute of Geography of the Chinese Academy of Sciences, State Key Laboratory of Information Engineering in Surveying, Mapping and Remote Sensing of Wuhan University, National Basic Geographic Information Center and other institutions have carried out research in this area. The Xi'an Institute of Surveying and Mapping and Zhengzhou

Institute of Information Engineering University and other units have also performed much research work in this field.

Shi Wenzhong (2005) discussed the theory and methods for handling errors in spatial data. On the basis of simple random sampling principles, the statistical model of rate of disfigurement (RD) was proposed and described in detail. Based on the definition of simple random sampling for attribute data in GIS, the RD mean and variance were deduced as the characteristic statistical model value to explain the feasibility of the accuracy measurement of the attribute data in GIS using the RD. From the result of the deduced equation regarding the RD mean and variance, the RD mean was the measurement of the amount of defects, while the RD variance was the measurement of the scatteration of defects. Therefore, the RD mean was used to measure the attribute accuracy of the sampling and assure the data accuracy, while the RD variance measured the sampling confidence. Hua Hui et al. (1998) explored the statistical analysis of the location accuracy of digital maps from digital instruments, operators, the degree of difficulty of graphics and so on, as well as their influencing laws on digital accuracy. Wu Fanghua (1998) analyzed and discussed the quality check of spatial data in map databases. Based on the theory of color image reproduction, the calibration and adjustment of equipment that affects the quality of data transmission, the establishment of a standard system working curve and the printability value were discussed in Jin Lan et al. (1998). Tong Xiaohua et al. (1999) tested and processed the distribution of digital data errors in GIS. Wang Weian et al. (1999) described the quality characteristics of spatial data, the quality characteristics of the attribute data, the relationship between spatial entities, and the quality of the relationship between entities and attributes in GIS, as well as the content of quality control, and they explained the method of implementing quality control according to different links. Chen Neng et al. (2000) studied quality control in the production process of GIS digital graphic products. In Liu Chun's doctoral thesis, 'Accuracy Measurement and Sampling Techniques for Attribute Data in GIS', digital error analysis of GIS data and geometric correction of graphic elements, uncertainty model and quality control of linear elements were analyzed and studied.

The expectation and variance in the RD can be used as a measure of the quality of attribute data. Whether the rate of disfigurement is reliable depends on the selection of sampling techniques. The larger the sample size is, the higher the sampling accuracy is, but the higher the sampling cost will be. The sample size can be determined according to the principle of "limiting the maximum relative deviation" (Liu et al. 2002). In Shi Wenzhong (2005), the statistical model of the rate of disfigurement was given based on the general sampling principle, and the mean and variance in the two statistical eigenvalues of the RD were derived. In Liu Chun et al. (2003), the method of the rate of disfigurement was proposed to measure the accuracy of attribute data in GIS. Then, the stratified sampling technique was applied to measure the accuracy of GIS data that were stored in layers. A case study was provided to demonstrate the applications of the rate of disfigurement method for attribute accuracy assessment.

In Fan Hong et al. (2004), an evaluation model of high-quality map labeling was proposed to obtain a value representing the quality. There were four aspects relevant to the labeling quality, which we call the overlap for a label with a label, the overlap for a label with background features, the position optimal level and the definite attachment for a label to the feature, as represented in the evaluation model. The map labeling system MapLabel developed by the author used this quality model to define the fitness function of the Genetic Algorithm to guarantee the good quality of the labeling. Wu Fanghua et al. (2005) proposed that the data quality control system should mainly consider the quality control of the map element updates and the quality control of map element collection, as well as map data quality checks and acceptance and map data quality evaluation. In Hu Shengwu (2006), the fuzzy fault tree method was introduced into the reliability analysis of GIS products; additionally, the construction of fuzzy fault trees and the operation rules based on triangular fuzzy numbers and trapezoidal fuzzy numbers were studied, and considering the characteristics of GIS products, the steps of fuzzy fault trees for reliability analysis of GIS products were discussed. According to the characteristics of vector map data quality, Wu Fanghua et al. (2008) took the basic elements of quality as the starting point of research and by combining the actual situation of vector map data production, 6 kinds of vector map data quality evaluation models were proposed and analyzed.

Throughout the development situation at home and abroad, map data quality control has been a very active research focus in the cartography field, receiving increasing attention, and making great progress. However, because of the difficulty of map data quality control and evaluation, as well as the limitations of the breadth and depth of research, there are still many key issues that remain unsolved. For example, there are few studies on the error analysis of multisource data, quality control and error fusion, automatic inspection and evaluation of attribute precision, and there are no good technical methods yet available. The quality of map data is closely related to the adopted codes, specifications, data structures, collection rules and specific requirements, but there is no available data quality control or evaluation software for map spatial data. Therefore, it is necessary to explore new ways in terms of both theory and practice and further strengthen the theoretical research of spatial data quality control to establish reasonable data quality standards, data quality control procedures and regulations, as well as to further research on automatic checking methods of data quality and comprehensive quality evaluations of digital maps for fairer and more objective evaluations, to further strengthen the research of quality control systems and standards, and to develop special software for data quality control and evaluation according to the characteristics of map spatial data.

15.3.5 Representative Project Outcomes

1. **The uncertainty principle of geospatial data and spatial analysis** (Shi Wenzhong, Tong Xiaohua, Zhu Changqing, Wang Xinzhou).

The uncertainty theory of spatial data and analysis is one of the core theories of geospatial information science, and it is also on the frontier of international theory. Spatial data quality is the key problem in national spatial data infrastructure, digital earth and digital city construction. It directly affects the success and failure of spatial information application and decision-making.

This research has made important achievements in the field of spatial data and uncertainty analysis theory, as well as its applications.

(1) The theoretical defects of the classical epsilon band geographical error model were found and proven, which has been used for 40 years in this field. The spatial linear error model was proposed and established systematically from four aspects, the law of uncertainty was revealed, and the scientific problem was solved.
(2) The average error in the TIN surface was shown to have a definite quantitative relationship with the node error, the theoretical formula of the mean error of the TIN surface was proven theoretically, and the error distribution law of the irregular triangle mesh in the digital ground model was revealed, which provided the theory and method for the analysis of digital terrain and the application of military fields based on TIN.
(3) A method of uncertainty analysis and simulation for spatial analysis was proposed, and a general theoretical model of positional uncertainty for buffer space analysis was established. The static data uncertainty description theory was systematically extended to the field of dynamic spatial analysis.
(4) A line-based transformation model was proposed, a new theory of geometric accuracy control for satellite images under sparse control points was proposed, and a multilevel vector space data quality control theory based on geometric and attribute constraints was proposed. The theoretical development of vector space data from the error description to quality control was realized.

This study enriched and developed geospatial information science through theoretical innovation in spatial data and spatial analysis uncertainty. By expanding error processing objects and the related basic theories, the science of surveying and mapping was enriched and developed. A breakthrough was made in the theory of uncertainty in geospatial data and spatial analysis. As a representative achievement of China in this field, it led international theoretical development in this field, and the theoretical results have been successfully applied to the quality control of geospatial data in mainland China, especially Hong Kong, as well as in other countries.

2. **Research on the theory and application of spatial data quality uncertainty** (Zhu Changqing, Lv Xiaohua, Yan Xiaodong, Zhang Guoqin, Tao Daxin, Wang Zhiwei, Wang Yaoge, Wang Qisheng, Yang Chengsong).

In light of the lack of research on the theory and application of the uncertainty of spatial data quality, the model establishment, accuracy evaluation and application of spatial data quality uncertainty were systematically and deeply studied; theory was combined with application, which guided practice and further promoted the theory from practice. A series of pioneering theoretical achievements were made in the field of theory and application of spatial data quality uncertainty.

(1) The theory and application of geospatial data model establishment and accuracy evaluation were systematically and deeply studied, and the research ideas and methods played an important role in guiding the wide study of accuracy theory and application.
(2) A series of pioneering theoretical results have been obtained in the field of accuracy evaluation theory. In particular, the DEM transmission error formula of the triangulated irregular network, the location uncertainty index and model of buffer space analysis, and the DEM overall accuracy evaluation index are obtained. These leading international research achievements in this field have important theoretical and guiding significance for the development of accuracy assessments.
(3) A series of creative achievements have been made in the establishment of a spatial data model. In particular, the model for road extraction from high-resolution remote sensing images, the model of remote sensing image fusion and the model for cross-country channel analysis have expanded the spatial data model. The proposed modeling methods provide a new approach for the theory and application of spatial data modeling and precision evaluation and related fields.
(4) A series of innovative applications have been achieved in the application of spatial data modeling and precision evaluation. In particular, the proposed method of cross-country passage analysis, the contour conversion method from inch to metric systems, the data conversion method of terrain surface models in different coordinate systems, etc., solves many key technical problems in surveying and mapping and has important application value.
(5) The theoretical results are fruitful. Thirty papers and 2 books have been published. A number of application software systems have been developed. In particular, 8 SCI papers published in international core journals fully reflect the international advanced level of research results. The research achievements have had important and extensive applications in scientific research, teaching and production of surveying and mapping and other fields.

3. **Map data quality control technology** (Liu Pingzhi, Wu Fanghua, Zhang Baoming, Yang Yun, Jin Cheng, Xiong Shun, Guo Haitao, Huang Limin, Zhang Weizhu)

To change the situation in which map data quality control technology lags and the intensity of the work labor is too great to meet the needs of digital mapping production, as well as to greatly improve the efficiency of production operation and the quality of map data and to meet the urgent requirements of 1:50,000, 1:250,000 and other digital map production and database construction and their quality control, the research of "map data quality control technology" was started in 2001, and its technical authentication passed at the end of 2006. The technical authentication committee held that the content of the project is comprehensive, advanced in technology, highly automated in the production operation system, stable and reliable, and its technical level is in the leading position in China.

(1) The structure model of map data quality elements, the concept of map data disfigurement feature and its application model are proposed, the theoretical system of map data quality control has been developed and perfected, and the scientific, standardization and adaptability of map data quality control has been improved.
(2) Several technical methods, such as gross error automatic detection, attribute template matching, cartographic rules interpretation and logic relation judgment, are proposed, automatic verification of attribute accuracy, positional accuracy, topological relationship, consistency, and integrity has been achieved, and the efficiency of map data quality inspection has been improved by more than 6 times.
(3) A method of intelligent input of attributes based on knowledge and rule base is proposed, which can realize automatic control of attribute error and reduce manual operation error. The speed of data updating and the accuracy of data acquisition are greatly improved while reducing labor intensity.
(4) The data quality information is classified and categorized scientifically and rationally. Quality evaluation models, such as the multiple-level fuzzy comprehensive evaluation model and the defection subtraction score based on weight, are established, fundamental transformation from artificial qualitative evaluation to automatic quantitative evaluation has been realized, and the scientific nature, objectivity and accuracy of digital map operation quality evaluation has been improved.

The results have been widely used in map data production units and other relevant departments, and the quality control tasks of more than 40 thousand topographic maps, aerial maps and traffic maps have been completed. It has played a key role in the successful completion of major mapping projects, such as the database construction and updating of 1:50,000 and 1:250,000 topographic maps.

15.4 Problems and Prospects

The theoretical and practical aspects involved in the uncertainty of spatial data are very extensive, and many problems remain to be further studied.

15.4.1 Further Research on the Uncertainty Model of Spatial Data

The model that describes the positional uncertainty of spatial data mainly includes ε_σ models, ε_m models, ε_E models and so on. Each model has its advantages and disadvantages. It is of practical significance to study the characteristics of the model and provide an applicable scope.

Compared with the research of positional uncertainty, the research of attribute uncertainty is more complex and difficult. Because of its complexity and difficulty, there are few achievements in this field. Traditional data processing methods assume that spatial distribution can be expressed by a set of discrete points, lines and planes, and it is usually assumed that the attribute data has been checked. In fact, in many GIS applications, such as land evaluation, soil chemistry and environmental science, which are based on attribute data analysis, requirements for the quality of attribute data are even higher than positional accuracy. Therefore, it is of great significance to study the uncertainty of attributes.

15.4.2 Research on Uncertainty in Spatial Analysis

Spatial analysis is one of the indexes used to evaluate the function of a geographic information system. Spatial analysis is the basis of all kinds of comprehensive geoscience analysis models and provides a basic tool for building complex spatial application models. In the spatial analysis of data, spatial data are always affected by different types of uncertainty, and these uncertainties are further transmitted through spatial analysis, which leads to incorrect conclusions in spatial analyses. Overlay analysis and buffer analysis are two common forms of spatial analysis. The error propagation in the overlay analysis and in buffer analysis of GIS are worth further study.

15.4.3 Research on the Application of the Uncertainty Model

At present, research on the uncertainty of spatial data has made many achievements. A large number of models and algorithms have been accumulated, and while some uses have been applied, applications overall remain still insufficient. How to combine theory and application in practice and apply these achievements to solve practical problems is a problem that needs to be considered and solved at present.

The application research of the uncertainty model includes the existing theory, the application of the model, the simplified application of these models and the establishment of a new model. Currently, many ε_σ models, ε_m models, ε_E models and S-band models for 2-D and 3-D linear elements, as well as planar elements have been established, but in general, these models are complex because many parameters are involved. How to apply these models or simplify these models to meet the needs of the application is very important. In addition, it is also necessary to establish a new practical model according to the specific conditions. Practical models should be established according to specific problems, such as accuracy control and error distribution. We should combine research with specific problems, find problems from actual production, and then apply uncertainty theory and models to solve problems.

15.4.4 Research on DEM Accuracy Evaluation Criteria

DEM accuracy evaluation mainly uses contour data as the true value to evaluate the error in the DEM caused by interpolation. Generally, the maximum error and the mean error are used to evaluate this accuracy. These two indexes reflect the degree to which the elevation values of the grid are not inconsistent with the true values. Some scholars, such as Hu Peng, et al., by analyzing the difference between random error and approximation error, pointed out that the interpolation error of the digital elevation model belongs to approximation error, and the accuracy evaluation method should not continue to use the concept of root mean square error but should use the approximation error. In view of the non-uniqueness of the centralized error in different sampling points of the same data, ZHU Changqing proposed the concept of reconstructed error, which is the ratio of the area error between the reconstructed contour and the original contour to the length of the original contour and was used to evaluate the accuracy of digital elevation models. However, it is difficult to present a general evaluation criterion for the evaluation of DEM accuracy. Therefore, it is necessary to conduct theoretical and experimental research on the DEM accuracy evaluation standards and propose a more accurate and reasonable general standard.

15.5 Representative Publications

(1) ***Error analysis and processing of GIS spatial data*** (Huang Youcai, Liu Wenbao, Li Zonghua, Xiao Daogang. China University of Geosciences Press, Beijing, March 1999).

This book mainly introduces the determination of GIS spatial data error sources, the error measurement, the method of weakening the error influence, and the law of error propagation in GIS operation. The book comprehensively reflects the research direction and some present research results in the field of GIS spatial data error processing and precision analysis.

(2) ***Theory and method of spatial data error processing*** (Shi Wenzhong. Science Press, Beijing, July 2000).

The book comprehensively and systematically discusses the uncertainty of location and attributes, as well as their combination in spatial data modeling and processing. First, the related theories of uncertainty processing are systematically analyzed, and then, the existing uncertainty processing techniques, methods and models are reviewed. More importantly, the methods and models for dealing with uncertainty in the GIS developed by the authors are systematically explained; new concepts, such as the confidence interval of line segment in GIS, the S-band model combining uncertainty of location and attributes, and so on, are discussed in detail.

(3) ***Principle of uncertainty in spatial data and spatial analysis*** (Shi Wenzhong. Science Press, Beijing, June 2005).

The book introduces the uncertainty principle of spatial data and spatial analysis in terms of theoretical research. The contents include mathematical foundations of uncertainty, position error model of linear objects, uncertainty in digital terrain models, attribute uncertainty model, integrated model of location and attribute uncertainty, uncertainty theory of topological relations between spatial objects, uncertainty modeling for spatial overlay analysis, uncertainty model for spatial buffer analysis, visualization of uncertainty in spatial data and spatial analysis, metadata model, spatial data mining based on uncertainty, and quality control of cadastral data.

(4) ***Research on the uncertainty of DEM data based on fractals*** (Li Shuang, Yao Jing. Science Press, Beijing, September 2007).

The uncertainty of geospatial data is universal, and certainty is conditional and relative. In this book, the theory of fractal geometry is applied to study the uncertainty of the ground elevation data field expressed by the DEM. From the point of view of the fractal dimension, the spatial correlation and fractal dimension characteristics of the DEM data field with certain topographic relief are discussed to a certain extent, and the spatial correlation and fractal feature of section lines are mainly completed by semivariation function and texture analysis. At the same time, through the fractal geometry characteristics of the DEM data field obtained by semivariation function analysis, the spatial correlation and uncertainty in DEM data fields with different fractal dimensions under certain topographic conditions are investigated.

Referencess

Blakemore, M. (1984). Generalization and error in spatial databases. *Cartographica, 21*(23), 131–139.

Chen, N., & Liu, C. (2000). Quality procedure control of digital graphic products in GIS. *Bulletin of Surveying and Mapping,* (10), 6–8.

Chen, Z. (1962). Provides a discussion of the analysis of topographic map accuracy. *Bulletin of Surveying and Mapping,* (8), 21–27.

Chrisman, N. R. (1982). A theory of cartographic error and it's measurement in digital databases. In *Auto-carto 5: Fifth International Symposium on Computer-Assisted Cartography* (pp. 159–182). Crystal City, Virginia.

Dai, H., Xu, P., & Liu, W. (1999). The visualization model of positional uncertainties for random polyline in vector GIS. *Acta Geodaetica et Cartographica Sinica, 28*(3), 239–243.

Dai, H., Wu, S., & Chen, L. (1999). Geometric image display of positional uncertainty in vector GIS. *Journal of Image and Graphics, 4*(12), 1078–1082.

Deng, M., Li, Z., & Cheng, T. (2006). Rough-set representation of GIS data uncertainties with multiple granularities. *Acta Geodaetica et Cartographica Sinica, 35*(1), 64–70.

Di, K., Li, D., & Li, D. (1999). Cloud theory and its applications in spatial data mining and knowledge discovery. *Journal of Image and Graphics, 4*(11), 32–37.

Fan, A., & Guo, D. (2001). The uncertainty band model of error entropy. *Acta Geodaetica et Cartographica Sinica, 30*(1), 48–53.

Fan, H., Zhang, Z., & Du, D. (2004). The study of the evaluation model for map labeling. *Acta Geodaetica et Cartographica Sinica, 33*(4), 362–366.

Gong, J. (1982). The map area measurement methods at home and abroad. *Journal of Geomatics, 7*(2), 40–47.

Guo, T. (2004). The models and experiments for positional uncertainties of spatial data in GIS. Doctoral Dissertation of Zhengzhou University of Information Engineering.

Guo, T., Wang, J., & Wei, H. (2003). A study of confidence regions of the fundamental factors in GIS. *Acta Geodaetica et Cartographica Sinica, 32*(2), 164–167.

He, Z. (1989). The principle of fuzzy mathematical multi-layersy mthatic evaluation and it's application of quality evaluation in thematic mapping. *Acta Geodaetica et Cartographica Sinica, 18*(4), 266–276.

Hu, Z., & Guo, D. (2000). Application of rough set thoery to precision analysis of attribute data and logical operations in GIS. *Bulletin of Surveying and Mapping,* (5), 7–9.

Hu, P., XHuang, X., & Wu, Y. (2006). Some problems of DEM error theory. *Journal of Harbin Institute of Technology, 53*(12), 2143–2147.

Hu, P., Wu, Y., & Hu, H. (2003). Research on fundamental theory of assessing the accuracy of DEMs. *Journal of Geo-Information Science, 8*(9), 64–70.

Hu, P., Wu, Y., & Hu, H. (2005). A new research on fundamental theory of assessing the accuracy of DEMs. *Journal of Geo-Information Science, 10*(3), 28–33.

Hu, P., Huang, X., Wu, Y., & Liu, Y. (2006). Some problem of DEM error theory. *Journal of Harbin Institute of Technology, 38*(12), 2143–2147.

Hu, S., Yuan, Z., & Wang, X. (2006). GIS product reliability based on fuzzy fault tree. *Science of Surveying and Mapping, 31*(1), 49–50.

Hua, H., & Tong, X. (1998). Statistical analysis of location accuracy of digital map. *Bulletin of Surveying and Mapping,* (2), 31–33.

Huang, Y., & Liu W. (1995). Gross error detection and robust estimation for building error estimation models of digitizing. *Journal of Wuhan Technical University of Surverying and Mapping, 20*(2), 151–156.

Jin, L., & An, Z. (1998). Research on quality control method of digital cartographic production system. *Bulletin of Surveying and Mapping,* (11), 21–23.

Kang, L. (1988). Discussion on total quality management of surveying and mapping work. *Bulletin of Surveying and Mapping,* (2), 3–7 & (3), 1–4.

Ke, Z., He, J., & Chi, T. (1993). *Digital terrain models.* Science and Technology of China Press.

Lan, Y., & Tao, B. (2004). End points accuracy based error band method for determination of a line segment position uncertainty. *Acta Geodaetica et Cartographica Sinica, 33*(4), 289–292.

Liu, W., Deng, M., & Xia, Z. (2000). Analysis of uncertainties of attributes in vector GIS. *Acta Geodaetica et Cartographica Sinica, 29*(1), 76–81.

Li, D., Gong, J., & Xie, G. (2003). The model of error entropy of area unit in GIS. *Acta Geodaetica et Cartographica Sinica, 32*(1), 31–35.

Li, D., Peng, M., & Zhang, J. (1995). Modelling positional uncertainty of line primitives in GIS.*Geomatics and Information Science of Wuhan University, 20*(4), 283–288.

Li, Z., & Zhu, Q. (2000). *Digital elevation models.* Wuhan: Wuhan Technical University of Surveying and Mapping Press.

Liu, D., Shi, W., & Tong, X. (1999). *Accuracy analysis and quality control of GIS spatial data.* Shanghai Scientific and Technological Literature Press.

Liu, C., Liu, D., & Shi, W. (2002). Quality tolerance for attribute data in geographic information system based on rate of disfigurement. *Journal of Tongji University (Natural Science), 30*(11), 1355–1360.

Liu, W. (1995). *Uncertainty theory of GIS spatial data.* Doctoral Dissertation of Wuhan Technical University of Surveying and Mapping.

Liu, C., Shi, W., & Liu, D. (2003). Statistical model of rate of disfigurement for attribute data accuracy assessment in GIS. *Acta Geodaetica et Cartographica Sinica, 32*(1), 36–41.

Liu, D., & Hua, H. (1998). The more discussion to the modeling uncertainty of line primitives in GIS. *Acta Geodaetica et Cartographica Sinica, 27*(1), 45–49.

Liu, D., & Liu, C. (2001). The status of researching on uncertainty of spatial data and quality control in GIS. *Engineering of Surveying and Mapping, 10*(1), 6–10.

Mei, S., & Jiang, N. (2004). The confidence band model of uncertainty propagation on vector buffer in GIS. *Journal of Remote Sensing, 8*(4), 289–294.

Min, B. (1963). On the map compilation and printing precision (standard error) problem. *Bulletin of Surveying and Mapping*, (S1), 8–12.

Shi, W. (1994). *Modelling positional and thematic uncertainties in integration of remote sensing and geographic information systems.* ITC Publication.

Shi, W. (1997). A generic statistical approach for modeling errors of geometric features in GIS from 1 to N dimensions. *Acta Geodaetica et Cartographica Sinica, 26*(2), 68–75.

Shi, W. (1998). Error models for geometric features in three-dimensional GIS. *Geomatics and Information Science of Wuhan University, 23*(1), 18–20.

Shi, W. (2005).*Uncertainty theory of spatial data and spatial analyses.* Science Press.

Shi, W., & Liu, W. (1997). The statistical distribution of cross-point centering error on manual digitizing a map. *Journal of Wuhan Technical University of Surveying and Mapping, 22*(1), 57–60.

Shi, W.,& Liu, W. (1998a). The stochastic process model for handling positional uncertainty of line segments in GIS. *Acta Geodaetica et Cartographica Sinica, 27*(1), 37–44.

Shi, W., & Liu, W. (1998b). Error circle of location uncertainty of GIS median line. *Journal of Geomatics*, (2), 6–8.

Shi, W., Li, Q., & Zhu, C. (2005). Estimating the propagation error of DEM from higher-order interpolation algorithms. *International Journal of Remote Sensing, 26*(14), 3069–3084.

Shi, W., Tong, X., & Liu, D. (2000). An approach for modeling error of generic curve features in GIS. *Acta Geodaetica et Cartographica Sinica, 29*(1), 52–58.

Shi, W., Wang, S. (2001). State of the art of research on the attribute uncertainty in GIS data. *Journal of Image and Graphics, 6*(9), 918–924.

Sun, H. (2001). Approximate representation of the p-norm distribution. *Geomatics and Information Science of Wuhan University, 26*(3), 222–225.

Sun, H., & Yu, Z. (1998). The digital characters of the p-norm sample distribution. *Geomatics and Information Science of Wuhan University, 23*(3), 244–247

Tang, G.A., Gong, J.Y., Chen, Z.J., Cheng, Y.H., & Wang, Z.H. (2001). A simulation on the accuracy of DEM terrain representation. *Acta Geodaetica et Cartographica Sinica, 30*(4), 361–365.

Tang, G., Shi, W., & Zhao, M. (2002). Evaluation on the accuracy of hydrologic data derived from DEMS of different spatial resolution. *Accuracy 2002: 5th International Symposium on Spatial Accuracy Assessment in Natural Resources and Environmental Sciences.* Melbourne, Australia.

Tang, G., Gong, J., & Chen, Z. (2004). A simulation on the accuracy of DEM terrain representation. *Acta Geodaetica et Cartographica Sinica, 30*(4), 361–365.

Tang, G., Tao, Y., & Wang, C. (2007). Contour-matching-difference and its application in DEM quality assessment. *Bulletin of Surveying and Mapping*, (7), 65–67.

Tang, Z., Wang, X., & Chen, Z. (2004). Envelop of error model of 2D spatial curve in GIS. *Acta Geodaetica et Cartographica Sinica, 33*(2), 151–155.

Tong, X., & Shi, W. (1999). Uncertainty model of circular curve features in GIS. *Acta Geodaetica et Cartographica Sinica, 28*(74), 325–329.

Wang, S., Chen, G., Li, D., Li, D., & Yuan, H. (2004a). A try for handling uncertainties in spatial data mining. In *International Conference on Knowledge-Based and Intelligent Information and Engineering Systems* (pp. 513–520). Heidelberg.

Wang, Q., & Wang, X. (2004b). Research on processing attribute uncertainty in application of information diffusion estimation. *Journal of Wuhan Institute of Technology, 26*(1), 29–31.

Wang, G., Zhu, C., & Shi, W. (2004c). The further study on the accuracy of DEM terrain representation. *Acta Geodaetica et Cartographica Sinica, 33*(2), 168–173.

Wang, X., & Wang, S. (1997). Fuzzy comprehensive method and its application in land grading. *Geomatics and Information Science of Wuhan University*, (1), 44–48.

Wang, Q. (1964). Inspection and acceptance of topographic maps in drawing production. *Bulletin of Surveying and Mapping*, (3), 33–35.

Wang, S., Li, D., & Shi, W. (2002). Theory and application of Geo-rough space. *Geomatics and Information Science of Wuhan University*, *27*(3), 274–282.

Wang, W., & Sun, H. (1999). Quality control of geographic information system. *Bulletin of Surveying and Mapping*, (4), 32–34.

Wang, Y., Gao, J., & Zhu, C. (2007). The propagation error of grid DEM based on coons surface. *Journal of Geomatics Science and Technology*, *24*(5), 361–363.

Wang, G., & Zhu, C. (2005). LoD modeling and accuracy analysis of terrain information. *Acta Geodaetica et Cartographica Sinica*, *34*(3), 228–232.

Wang, Y., Zhu, C., & Wang, Z. (2008). A surface model of grid DEM based on coons surface. *Acta Geodaetica et Cartographica Sinica*, *37*(2), 217–222.

Wu, F. (1998). Quality check of spatial data in map database. *Geomatic Science and Engineering*, (1), 27–30.

Wu, F., Liu, P., & Ke, X. (2008). Analysis on the quality elements of vector map data. *Geomatic Science and Engineering*, (4), 6–9.

Wu, F., Liu, P., & Yang, Y. (2005). The quality control of vector map data. *Bulletin of Surveying and Mapping*, (5), 50–52, 64.

Yu, N. (1988). Several problems on strengthening the quality management of Surveying and mapping products. *Bulletin of Surveying and Mapping*, (3), 6–7.

Zhang, B., Zhu, L., & Zhu, G. (1998). The uncertainty propagation model of vector data on 'buffer' operation in GIS. *Acta Geodaetica et Cartographica Sinica*, *27*(3), 259–266, 281.

Zhang, G., & Tulip, J. (1990). An algorithms for the avoidance of sliver polygons and clusters of point in spatial overlays. In *4th International Symposium on Spatial Data Handing* (pp. 141–150). International Geographical Union IGU.

Zhang, G., Zhu, C., & Zhou, B. (2005). Uncertainty ε_σ model of spatial circular curve features in GIS. *Acta Geodaetica et Cartographica Sinica*, *34*(3), 233–238.

Zhang, J., & Du, D. (1999). Field-based models for positional and attribute uncertainty. *Acta Geodaetica et Cartographica Sinica*, *28*(3), 244–250.

Zhang, J., & Goodchild, M. F. (2002). *Uncertainty in geographical information*. New York: Taylor & Francis.

Zhu, Z. (1997). Discussion on DEM data accuracy checking method. *Surveying and Mapping Technology*, (2), 48–49.

Zhu, C., & Shi, W. (2006). *Spatial analysis models and theory*. Beijing: Science Press.

Zhu, C., Shi, W., & Li, Q., et al. (2005). Estimation of average DEM accuracy under linear interpolation considering random error at the nodes of TIN model. *International Journal of Remote Sensing*, *26*(24), 5509–5523.

Zhu, C., Wang, Z., & Liu, H. (2008). Accuracy evaluation model of DEM based on reconstructed contours. *Geomatics and Information Science of Wuhan University*, *33*(2), 153–156.

Zhu, C., Zhang, G., & Shi, W. (2007). Algebraic research on geometrical characteristic of error band for 2D linear segment uncertainty ε_σ model. *Acta Geodaetica et Cartographica Sinica*, *36*(4), 463–467.

Chapter 16
Standardization of Cartography and Geographic Information

Jingtong Jiang and Ruomei Liu

16.1 Introduction

To obtain the best order within a certain range, documents that specify common and repeated use rules, guidelines, or characteristics for activities or their results are referred to as standards. The document is established by universal agreement and approved by an authoritative standards organization. Standards should be based on the comprehensive results of science, technology and experience to promote the best social benefits.

For the purpose of gaining the best order within a certain range, activities that make common and repeated use rules for actual or potential problems are called standardization, which includes the process of developing, releasing and implementing standards. The importance of standardization is to improve the applicability of products, processes and services, to prevent trade barriers and to promote technological cooperation.

Standardization is a comprehensive discipline with both a technical and managerial nature. Standardization involves making a thing or concept repetitive to achieve reunification through the development, publication and implementation of standards in social practice, such as economics, science and management, to obtain the best order and social benefits.

The standardization process needs to solve such contradictions as development and stability, unity and disunity, production and use, general purpose and exclusive use, as well as high-level, low-level, complex and simple standardization and diversification. In short, the purpose of promoting standardization is to create standards in terms of using new technology, as well as to improve the overall optimum function of standards in production and products, to plan scientifically for production, product and other technical and quality levels and to establish a unified system for cross professional sectors that ensures optimal conditions to meet the objective needs of

J. Jiang (✉) · R. Liu
National Geomatics Center of China, Beijing, China

J. Wang and F. Wu (eds.), *Advances in Cartography and Geographic Information Engineering*, https://doi.org/10.1007/978-981-16-0614-4_16

complex coordination among the various departments of the national economy and to ensure the unified coordination and best overall effect of the standards at all levels.

Standardization is the technical basis for national economic and social development, as well as the bridge to transfer scientific and technological achievements into productive capacity and is an important condition for organizing modernization and intensive production. Standardization is an important technological foundation to promote technological progress, upgrade industries, and improve product quality, engineering quality and service quality, which will help speed up the modernization of China's agriculture and push China to complete its industrialization to move toward an information-based society.

The standard system is a scientific organic whole that is formed within a certain range of standards according to its internal connection. The development characteristics of modern high technology, such as high integration, industrial relevance and permeability, determine the systematic nature of high technology-related standards, which is the most prominent feature of modern standardization development. The study and formulation of a standard system table will help guide and control the standardization construction in the field to realize unified planning and organization and distinguish the priority of the standard in terms of hierarchy and content.

Like other areas of standardization, cartography and geographic information standardization are the same, and cartography and geographic information standardization are very important to solve the problems of mapping and compilation, the production of geographic information data and the construction of geographic information systems. Maps and geospatial information, as a kind of knowledge product, can only be used and accepted by users when they are in accordance with accepted standards and specifications. Therefore, the standardization of cartography and geographic information is the technical basis for the map to be accepted by readers and the sharing of geospatial information, as well as the guarantee of map and geographic information exchange and one of the necessary conditions for map and geographic information sharing. The standardization of map and geographic information is of great significance to promote surveying and mapping industry development.

According to nature, mapping and geographic information standards can be divided into three kinds: technical standards, management standards and working standards. Among them, technical standards refer to the standards developed for the technical matters that need to be harmonized in the field of cartography and geographic information; management standards refer to the standards developed for the management matters that need to be harmonized in the area, including management basic standards, technical management standards, economic management standards, administrative standards, and the production, operation and management standards; and work standards refers to the standards developed for work responsibility, rights, scope, quality requirements, procedures, effects, inspection methods and assessment methods. Work standards generally include departmental work standards and post (individual) work standards.

According to function, standards can be divided into basic standards, product standards, and method standards. Basic standards are within a certain range as the basis for other standards and are widely used with wide guiding significance; product

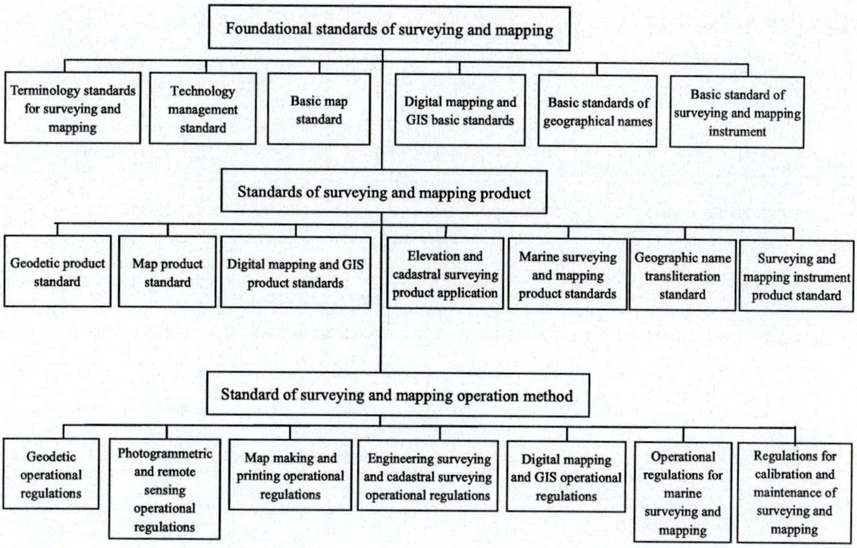

Fig. 16.1 Functional classification system of mapping standards (quoted from https://www.sbsm.gov.cn)

standards are technical specifications for product structure, specifications, quality and inspection methods; and method standards are the standards developed for test, inspection, analysis, sampling, statistics, calculation, measurement, operation and other methods or for producing reasonable, high-quality products and improving efficiency in production, operation, testing and business processes. Figure 16.1 illustrates a functional classification system for mapping standards.

According to hierarchy (or the scope of the role or the standard approval authority), the standards can be divided into international standards, regional standards, national standards, industry standards, local standards, and enterprise standards. Among them, national standards shall be planned and drafted by the administrative department for standardization under the State Council and shall be uniformly examined and approved, and serial numbers shall be issued. The code name for these standards is "GB". Industry standards refer to the technical requirements that need to be unified in the field of surveying and mapping and geographic information in the absence of national standards, which are mainly examined, approved, numbered, published and unified by State Bureau of Surveying and Mapping management. Some industry standards related to cartography and geographic information systems have also been developed, approved and published by other departments. The names and codes of the major industrial standards are shown in Table 16.1. Industry standards can be regarded as a supplement to national standards. Once the corresponding national standards are implemented, the industry standards are automatically abolished.

According to efficiency (or binding force or standard attribute), the standard can be divided into two kinds: mandatory standards and recommended standards. Among

Table 16.1 The names and codes for major industrial standards

Sequence number	Name of industry standard	Code of industry standard
1	Surveying and mapping	CH
2	Marine	HY
3	Geology and mineral resources	DZ
4	Land administration	TD
5	Town construction	CJ
6	Environmental protection	HJ
7	Traffic	JT
8	Railway transportation	TB
9	Water conservancy	SL
10	Forestry	LY

them, the standards for safeguarding human health, personal and property safety, or prescribed by laws and administrative regulations are considered to be mandatory standards, and others are considered to be recommended standards. The code of the mandatory national standard is "GB", and the code of the recommended national standard is "GB/T". Mandatory standards must be implemented. Cartographic and geographic information products that do not meet the mandatory standards are prohibited from being produced, sold or imported. In addition, the code of the national standardization guidance technical document is "GB/Z".

To guide and control the standardization construction of the surveying and mapping industry macroscopically, the standard system of surveying and mapping has been researched and developed many times over the years, but for various reasons, it was not until June 9, 2009 that China's first *Standard System of Surveying and Mapping* was issued and implemented by the State Bureau of Surveying and Mapping. It is the guiding document for the development and revision of national standards and industry standards of surveying and mapping in the current and future period. From now on, proposal and acceptance, project approval and standard review of mapping standards project shall be carried out in terms of the contents and requirements of this standard system.

In the *Standard System of Surveying and Mapping*, starting from the development demand of modern surveying and mapping technology, with the structure of standards under digital surveying and mapping technology systems as the main body, considering the actual needs of traditional surveying and mapping technology, the composition of surveying and mapping standards is described and constructed from a variety of perspectives. The standard system includes 5 categories (definition and description, acquisition and processing, calibration and test, results and service, and management) and 32 subcategories that contain a number of national standards or

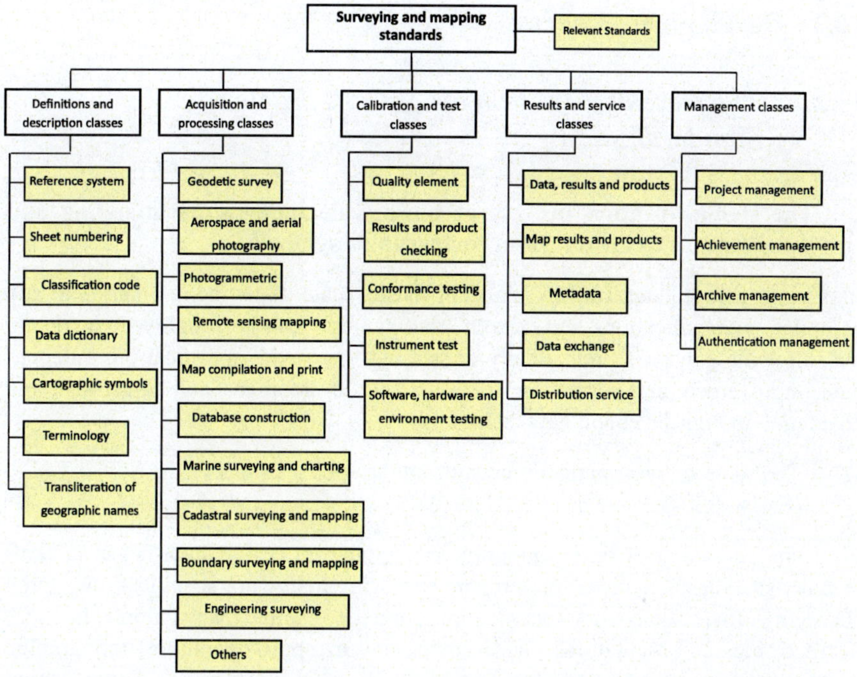

Fig. 16.2 Structure chart of the standard system of surveying and mapping (cited from the *Standard System of Surveying and Mapping*)

industry standards, for a total of more than 350 standards (see Fig. 16.2). There is a correlation between the 5 categories of standards; there are also correlations and interactions among subcategories in the 5 categories, as well as among the specific standards in the subcategories, thus forming a standard system covering the entire surveying and mapping field.

In the standard system, the current technical level and forecasts of future development needs are considered. On the basis of the main content and application scope of standardized objects and standards, the recommended names of specific standards are put forward, and all category standards, subcategory standards and specific standards are uniformly numbered in accordance with the *Principles and Requirements for the Compilation of Standard System Tables* (GB/T 13061-91).

The classification, scope, quantity, name of standards and so forth in that standard system are relative and expandable, and with the continuous progress and extensive application of surveying and mapping, geographic information and related science and technology, as well as the development of surveying and mapping standardization, they will be adjusted, supplemented and deleted, as well as gradually improved.

16.2 Development Process

16.2.1 The Development Process of Cartographic Standardization

1. **The standardization process of basic scale topographic mapping and compilation specification and cartographic symbols**

In China, the standardization work of basic scale topographic mapping and compilation specifications and cartographic symbols can be roughly divided into the following stages: systematically consulting the Soviet specification, independent improvement and perfection, recovery and development, and adjustment and adaptation to high-level and new technology.

(1) The stage of systematically consulting the specifications of the Soviet Union (Early 1950s)

From the beginning of basic topographic mapping in 1950, China began to draft technical standards, such as topographic surveying field work specifications, aerial surveying offices, field work specifications, and topographic map symbols. In 1953, detailed rules and regulations for surveying and mapping operations, topographic map symbols and regulations for surveying and mapping from the Soviet Union were systematically introduced.

(2) The stage of independent improvement and perfection (late 1950s to late 1960s)

Since the late 1950s, focus has been given to the improvement of the technical regulations of the basic surveying and mapping work and the perfection of various measuring datums to make the basic mapping technical standards better suited to China's national conditions to reflect the experience of China's surveying and mapping practice and to change the situation of copying foreign specifications and rules at the initial stage of the founding of the People's Republic of China. For this purpose, during 1957–1966, more than 40 kinds of standards were formulated and successively promulgated, including Topographic Map Symbols (scales from 1:500 to 1:50,000), Specifications for Topographic Mapping with the Plane Table (scales from 1:10,000 to 1:100,000), Fundamental Principles of Topographic Mapping (Draft) (scales from 1:10,000 to 1:100,000), Specification of Aerial Photogrammetric Field Work for Topographic Maps (from 1:10,000 to 1:100,000), Specification of Topographic Map for Aerial Photogrammetric Office Work (from 1:10,000 to 1:100,000), the 1:200,000 Scale Topographic Map Compilation Specification, and Mapping Symbols of the 1:10,000 Scale Topographic Map for Economic Construction.

By making use of the above technical standards and procedures, topographic mapping in most parts of China was completed in 1966, including a 1:50,000 scale topographic map of the eastern and central regions, a 1:100,000 scale topographic map of the western regions, and a 1:25,000 scale topographic map of the costal and

Northeast regions, except the Qinghai Tibet Plateau, Tarim and Tagrama Desert, Greater Khingan Range and Southwest Hengduan Mountains regions. During 1960–1966, 1:10,000 scale topographic mapping covered approximately 1000,000 km^2.

(3) The stage of recovery and development (late 1970s–early 1980s)

Considering the standardization of topographic map compilation, China formulated and promulgated the *Specification for 1:100,000 Topographic Map Compilation* in 1964 and 1975. On the basis of the existing work, the latter makes a new rule for the comprehensive selection of the elements of the topographic map so that the load and clarity of the map are greatly improved. A unified coordinate system, projection and map sheet number were adopted in the national 1:20,000 topographic map completed from 1957 to 1981. *Specifications and Symbols for the 1:1,000,000 Topographic Map Compilation* were formulated in 1979. *Basic Principles of Topographic Map Updating* and *Specification for Aerial Photography of Topographic Maps* (Scales of 1:5,000, 1:10,000, 1:25,000, 1:50,000, 1:100,000) were issued in 1980. Due to the small size of the map sheet, large amount of data and inconvenient use of the 1:200,000 scale of the topographic map, it was changed to a 1:250,000 scale topographic map in 1983, and the *Specification and Symbols for the 1:250,000 Scale Topographic Map* was issued in 1985 (GBCH IV-302-85).

By the end of 1975, China completed the mapping work of the first generation of 1:25,000–1:100,000 scale topographic maps. In 1973, the compilation and publication of national topographic maps at a scale of 1:500,000 were completed. During the period of 1976–1985, the mapping work of the second and third generations of 1:25,000–1:100,000 scale topographic maps were launched in succession. Compilations of 1:100,000–1:1,000,000 scale topographic maps were completed; among them, the nationwide 1:200,000 scale topographic map was finished in 1980, and the nationwide 1:250,000 scale topographic map was finished in the late 1980s. After completion of the first generation of the 1:100,000 scale topographic map in 1960, the compilation of the new version of the 1:100,000 scale map was completed in 1982.

(4) The stage of adjusting and adapting to high-level and new technology (mid 1980s–the end of the twentieth century)

Since the late 1980s, with the rapid development of computer-aided cartography, geographic information systems and digital photogrammetric technology, a number of standards related to topographic mapping and compilation have been revised to adapt to technological advances and the development of demand. Some of the new, revised and promulgated standards created during this stage are shown in Table 16.2.

2. **The standardization process of annotation of geographical names**

The standardization of geographical names in topographic maps is also an important component of cartographic standardization in China. Over the years, a great deal of research and standardization work has been carried out in this field in China.

Table 16.2 Standards of topographic mapping and compilation (partial)

Scales	The relation to the number of 1:1,000,000 topographic maps	The relation to the number of upper level scale topographic maps	Scope of map sheet	
			Longitude difference	Latitude difference
1:1,000,000	1		6°	4°
1:500,000	4	4	3°	2°
1:250,000	16	4	1°30′	1°
1:100,000	144	9	30′	20′
1:50,000	576	4	15′	10′
1:25,000	2304	4	7′30″	5′
1:10,000	9216	4	3′45″	2′30″
1:5,000	36,864	4	1′52.5″	1′15″

When new China was founded, the names of geographical place were cleared according to Instructions on Dealing with the Appellations, Place Names, Stone Tablets, Couplets and Inscriptions with Discrimination and Insults to Ethnic Minorities issued by the Government Administration Council in 1951 and Instructions on Changing Geographical Names. A geographical names review panel was established in 1964 by the State Council, and in the same year, a Unified Translation Committee for Names and Places was established in the Cultural and Educational Affairs Office of the State Council. Notice on Changing the Approval Authority of Names of Mountains, Rivers, Lakes, Bays, Straits and Islands was issued in 1965, and the *Scheme of Chinese Phonetic Alphabet Spelling for Chinese Geographic Names* was issued in 1974. In 1978, the State Council approved the establishment of the China geographical name commission. *Interim Provisions on the Nomenclature and Renaming of Geographical Names* and the *General Rule for the Translation of the Chinese Characters of Foreign Place Names* were issued in 1979. Reports using the Chinese Phonetic Alphabet Scheme as a Unified Standard for the Rome Letter Spelling of Chinese Names and Places and reports on speeding up the Census of Geographical Names of Islands and Reefs and Standardization of Geographical Names in China's Coastal Areas were approved by the State Council in 1978 and 1983, respectively. *The Spelling Rules of Chinese Geographical Place Names with the Chinese Phonetic Alphabet* (parts of Chinese geographic names) were issued in 1984. Administration Regulations for Geographical Names was issued in 1986. *China's Commission on Geographical Place Names* also formulated relevant regulations and standards, such as *Spelling of Chinese Place Names with Chinese Phonetic Alphabets, Transliteration and Writings of the Place Names of Ethnic Minorities with Chinese Phonetic Alphabets* (issued in 1976) and *Chinese Spelling Rules for Urban Streets* (Draft) (issued in 1979), all of which have standardized the annotations of place names on Chinese topographic maps.

To standardize the transliteration of place names in a minority language, the *Simplified Rule for Chinese Translation of Place Names of Mongolian Language*

(Draft) and *Simplified Rule for Chinese Translation of Place Names of Uygur Language* (Draft) were created in 1962. *General Rules for the Survey and Translation of Place Names in Minority Languages* was formulated in 1962, which requires that transliteration be a principle for *Chinese Character Translation of Place Names of Minority Language.* In the same year, *Transliteration Rules for Mongolian Place Names* (Draft) and *Transliteration Rules for Uyghur Place Names* (Revised) were published and finalized. *Transliteration Rules for Xishuangbanna Dai Place Names* (Draft) was published in 1965. *Transliteration Rules for Kirgiz Language Place Names (The first draft) and Transliteration and Writings of the Place Names of Ethnic Minorities with Chinese Phonetic Alphabets* (Draft), which were issued in 1965, were revised in 1979. Thereafter, rules for the transliteration of place names of many minority languages were sequentially formulated, including Mongolian, Uygur, Kazak, Kirgiz, Xishuangbanna Dai, Yi, Tibetan (Lhasa), Tibetan (Amdo dialect), Lahu, Hani, Tibetan (dege) Li, and Dehong Dai. These factors promoted the standardization of the translation of place names of minority languages with Chinese characters.

In addition, Specification of Aerial Photogrammetric Field Work for Topographic Maps stipulates that when place names are surveyed, the names of administrative divisions such as cities, counties, flags and so forth shall be subject to the official name promulgated by the State Council, the names of townships shall retain the official name of the government, and the names of villages shall be based on the natural name. The names of the mountains and drainage systems are clearly stipulated. These regulations improve the quality of field surveys for place names on topographic maps.

3. **The standardization process of marine charting**

In the early days of the founding of new China, the land part of a chart was made of old topographic maps, and the marine part was made of nautical chart data in foreign languages and supplemented with some new data. Therefore, the plane coordinate system and the elevation system were not exactly the same, and the accuracy was poor. In 1960, the 1954 Beijing Coordinate System was adopted as the plane control, and the Theoretical Lowest Low Water was adopted as the depth datum. The harbor chart was based on the Gaussian–Kruger projection, and the navigation chart was based on the Mercator projection. The scale series of different types of charts were unified; among them, the scales of the charts for military use were 1:1,000,000, 1:500,000, 1:200,000, and 1:100,000; the scales of the charts for civil use were 1:1,500,000, 1:750,000, 1:300,000, and 1:150,000; harbor and coastal charts were at larger scales, from 1:50,000 to 1:25,000 and more.

To promote the development of charting, during the 1960s, China formulated a series of relevant standards and specifications, mainly including the *Legend of Naval Hydrographic Chart, Symbols and Abbreviations of Charts, Regulation for Hydrographic Surveying Work Checking and Acceptance, Outline of Sea Area Information Investigation, Specifications for Hydrographic Survey* (control survey, bathymetry survey, terrain and shoreline surveys, and tidal control), *Publication Rules for Military Nautical Charts, and Publication Rules of Nautical Charts for Use by Domestic Shipping and Aquatic Vessels* (civil version).

The *Specification for Compilation of the Nautical Chart* was released in 1975, the *Specification for Compilation of the Nautical Chart in Foreign Seas* was released in 1979, and the *Specification for Nautical Chart Mapping in China Seas* was released in 1981. The new release standards reasonably stipulated the principle of compilation and mapping of the various elements on nautical charts and merged military and civil nautical charts into one edition.

To meet the needs of technological development and demand, since the 1990s, the standards of nautical charts began to consider electronic technology, and a number of charts, specifications and schemata were developed.

4. The standardization process of the thematic map

Compared with topographic maps, which have formed a systematic and complete standard system, the standardization of thematic maps is scattered and lacking. To date, only a small number of national standards and industry standards have been published, and the types of thematic maps involved are rare. The major standards issued by professional departments are described as follows.

Surveying and mapping departments: To provide geographical base maps for thematic maps and to standardize the mapping or compilation of maps other than topographical maps, the State Bureau of Surveying and Mapping issued an industry standard map, *Cartographic Symbols for Province, Prefecture and County* (CH/T 4004-93) in 1993, an industry standard, *Cartographic Symbols for Cadastral Maps* (CH 5003-94) in 1994, and they have laid an important foundation for the standardization of producing the two types of thematic maps in provinces, prefectures and counties.

Construction sectors: In view of the characteristics of urban mapping, the *Ministry of Construction formulated and promulgated the Urban Surveying Specification* (Draft) in 1959; the specification was revised in 1985 and officially issued as the construction industry standard, that is, the *Urban Surveying Specification* (CJJ 8-85) and was further revised and issued in 1999. The standard stipulates traditional urban mapping methods and technical requirements.

The Ministry of Geology issued *Specifications for Topographic Mapping of 1:1,000, 1:2,000 and 1:5,000* in 1966, which was suitable for geological exploration specialties. In 1995, the Ministry of Geology and Mineral Resources issued the *Specification for the Drawing of the Regional Geological and Mining Geological Map* (DZ/T 0156-95) and the *Specification for Compilation of the Geographical Base Map of the Geological Map of Provinces (City, District) at the 1:500,000 and 1:1,000,000 scales* (DZ/T 0159-95).

According to the characteristics of forestry thematic map compilation, the *Ministry of Forestry issued Legend Symbols of Forestry Maps* in 1959 and published *Cartographic Symbols of Forestry Maps* in 1982.

Some other departments have also formulated and promulgated relevant industry standards. To a certain extent, the formulation and implementation of these industry standards laid the foundation for the standardization of thematic maps.

5. **The standardization process of cartographic terms**

Over the years, China has attached great importance to the terminology of surveying and mapping, including the standardization of terms used in cartography, and a national standard *Basic Terms of Surveying and Mapping* (GB/T 14911-94) was issued in 1994, *Terms of Cartography* (GB/T 16816-1997) was issued in 1997, and *Chinese Terms in Surveying and Mapping* (ISBN-7-03-002105-3) was issued in 2002.

The *Terms of Cartography* defines the terms and definitions of cartography and map printing. *Basic Terms of Surveying and Mapping* and *Chinese Terms in Surveying and Mapping* define the disciplines of surveying and mapping, including the terms of mapping and printing, and the latter also gives definitions of terms.

16.2.2 The Development Process of Geographic Information Standardization

In the early 1980s, China had just researched and developed the geographic information system (GIS, then known as the information system of resources and environment), lessons were learned from the ignorance of standardization in some foreign countries, the standardization of geographic information was always taken as an important part of GIS development, and a great deal of work was done and some progress was made.

Over the past 20 years, the standardization work of China's geographic information can be roughly divided into two main stages: during the "seventh five-year plan" and "eighth five-year plan" periods, the standardization of data was at the core; during the "ninth five-year plan" and "tenth five-year plan" periods, the standardization of geographic information sharing was at the core. Several geographic information standards were studied, formulated and revised.

In 1984, the New Technology Bureau of the State Bureau of Science and Technology organized a group of experts to conduct in-depth research on the development of GIS technology at home and abroad and published a Research Report on National Standards and Specifications for Resources and Environment Information System.

During the "seventh five-year plan" (1986–1990), the research of geographic information standardization at national, provincial, city and county levels was included in the state key scientific and technological projects of the "seventh five-year plan". The New Technology Bureau of State Bureau of Science and Technology organized a National Standards and Specifications Research Group on Resources and Environment Information System, and dozens of experts and scientific and technical personnel were organized to tackle the problem. The four aspects of a unified geographic coordinate system, unified resource and environment information data classification system, unified geospatial data encoding system and unified geospatial data exchange format were studied in depth in combination with the research and development of the national resource and environment information system, and more than 20 standard proposals were put forward, which laid a good foundation for further formulating the corresponding national standards of geographic information.

During the "eighth five-year plan" (1990–1995), the national standards and specifications research group for resources and environmental information systems continued to work to refine the research results of geographic information standards that were completed, and efforts were made to place them at the level of formal national standards. The State Bureau of Surveying and Mapping was mainly responsible for the development of two national standards: *Geographical Grid* (GB 12409-90) and *Classify and Code for National Land Information* (GB13923-1992), while the Ministry of Forestry was responsible for the development of a national standard: *Classify and Code for Forestry Resources Data* (GB/T 14721.1-93). These standards were reviewed and approved by the State Bureau of Technical Supervision. The Code for China River Name, a national standard developed by the State Bureau of Surveying and Mapping and Ministry of Forestry, was completed as a draft.

At the same time, some departments and units have arranged much standard research and development work on their own. For example, the State Bureau of Surveying and mapping has set up more than 10 subjects and topics closely related to the standardization of geographical information in its scientific and technological projects of the "eighth five-year plan" and put forward a dozen draft standard, with a view to becoming corresponding national standards or industry standards. *Classification and Codes for the Features of 1:500, 1:1,000 and 1:2,000 Topographic Maps* (GB 14804-93), a standard formulated by State Bureau of Surveying and Mapping, was examined and approved by the State Bureau of Technical Supervision and subsequently promulgated.

During this period, with the rapid development of the urban GIS in China, developing the national standard for the urban GIS became very urgent and important. *Coding Rules for Urban Geographical Elements—Roads, Intersections, Neighbors and Municipal Pipelines* (GB/T 14395-93), a national standard formulated by the Beijing Municipal Planning and Design Institute, was approved, promulgated and implemented. In the middle of the "eighth five-year plan", the Social Development Division of the State Science and Technology Commission approved the establishment of a research project—the guidelines for standardization of the urban GIS, which was led by the China Institute of Standardization and Information Classification and Coding, National Geomatics Center of China and State Key Laboratory of Resource and Environment Information System, Institute of Geography, Chinese Academy of Sciences, and jointly studied and compiled with more than ten other units, such as Ministry of Construction Information Center, China Academy of Urban Planning and Design, Survey and Design Institute of Ministry of Construction, Information Center of State Land Administration Bureau, State Information Center, Wuhan University of Surveying and Mapping, and Peking University. It was published by the Science Press in 1998 and later reprinted. To a certain extent, the guide has a positive impact on the standardization of China's GIS.

Some of the national standards for information technology developed since the 1980s have also been directly adopted as geographical information standards, for example, the *Codes of the Administrative Divisions of the People's Republic of*

China (GB2260-1982), *Rules for the Code Representation of Administrative Divisions under Counties* (GB10114-88), *Naming and Coding for National Highway* (GB917.2-89) and so forth. They support the standardization of China's geographic information, and to a certain extent, they alleviate the contradiction of the lack of geographic information standards.

In the national Ninth Five technical key scientific projects, the standardization of geographic information once again became one of the focuses of attention; however, the focus of standardization shifted from the standardization of geographic information data classification and coding to geographic information sharing. For example, the Research on Key Technologies of National Resource Environment and Regional Economic Information System and Information Space Infrastructure was arranged by the State Planning Commission and undertaken by the State Information Center, National Geomatics Center of China and other units, as several geographic information standards were closely related to information sharing. The National Science and Technology Commission also arranged several standard research works related to information sharing in the Sharing Demonstration for Sustainable Development of China project (97-759). The State Bureau of Surveying and Mapping took the standardization of geographic information technology as an independent project to tackle key problems, which not only helped to promote the standardization of geographic information but also promoted the process of geographic information sharing in China.

During the "tenth five-year plan", especially in the new situation of economic globalization and China's accession to the WTO, the departments of the State Council and local governments at all levels recognized the importance of standardization from the overall and strategic level, and from the aspects of human power, material and financial resources, studies of standards, basic standards and method standards, as well as the important standards related to national security, protection of public health and the maintenance of national industries were greatly supported. Standardization work received increasing attention from all levels of government, enterprises, institutions, scientific research and education departments. This work also provided a rare opportunity for the standardization of geographic information. A large amount of money was spent on research of the standardization of geographic information in most major technology projects, such as the state key scientific and technological projects in the tenth five-year plan, national 863 projects, basic science and technology projects, major projects, fund projects, and "digital area" projects at various levels, including the digital industry, digital provinces, digital cities and digital communities. The standardization of geographic information in China showed an unprecedented excellent situation, and it was expected that the standardization of geographic information would soon leap to a new level and lay a good foundation for promoting the sharing of geographical information.

To meet the needs of national economic and social development for the standardization of geographic information as quickly as possible, as well as to alleviate the prominent contradictions of "lack of standards, duplication and overlapping, confused usages and difficult coordination", to better guide the standardization work of geographic information during the period of "11th Five-Year", to promote the development, sharing and utilization of geographical information resources and to improve the service support ability of geographic information to economic development, scientific and technological innovation, government administration and social life, the State Bureau of Surveying and Mapping and National Standardization Management Committee jointly formulated and promulgated the National Geographic Information Standardization in "11th Five-Year" Plan in December 2006. The main tasks of the plan included establishing a National Geographic Information Standard System, revising a number of urgent standards in key areas, transforming and adopting a number of applicable international standards, increasing the basic and preliminary research of standards, investigating a feasible method to establish a standard conformance test and evaluation system, strengthening efforts to publicize and implement standards. The proposed measures included establishing and improving standardized management and coordination mechanisms, raising funds for standardization work through multiple channels, training a team of qualified personnel for standardization, strengthening the guidance of standard revision projects, and actively participating in international standardization activities.

16.3 Main Research Achievements

16.3.1 Main Research Results of Basic Scale Topographic Standardization

To date, a set of complete and unified standards and regulations for topographic mapping technology has been established in China, and the standardization of topographic mapping has been achieved, which provides a basis and technical regulations for various types of basic topographic mapping and compilation and guarantees the consistency of mapping accuracy, specifications and requirements. They have the following remarkable features.

1. **Standardized series of scales of basic topographic maps**

The scale series of China's basic topographic map include 1:500, 1:1,000, 1:2,000, 1:5,000, 1:10,000, 1:25,000, 1:50,000, 1:100,000, 1:250,000 (1:200,000), 1:500,000 和 1:1,000,000. Among them, the scale of 1:200,000 was changed to 1:250,000 in 1983.

Uniform basic topographic map subdivision and numbering rules: According to the provisions of the specifications and technical indicators for basic topographic maps of large areas, the topographic maps of national or large areas are uniformly sheet divided according to the international map subdivision method, that is, 1:1 million maps should be divided in terms of the latitude difference of 4° and longitude difference of 6°; based on the subdivision of 1:1 million topographic maps, 1:500,000, 1:250,000, and 1:100,000 should be divided in terms of $2° \times 3°$, $1° \times 1°30'$, and $30' \times 45'$, respectively (longitude difference × latitude difference); 1:50,000 and 1:25,000 are quartered parts of 1:100,000 and 1:50,000 topographic maps, respectively; and 1:10,000 is one of 64 equal subdivisions of 1:100,000 topographic maps.

Based on the above uniform division rules, uniform topographic map numbering rules are formulated, that is, based on the number of 1:1 million scale topographic maps, the characters specified in suffixes at different scales are connected by hyphens. The greatest advantage of this rule is that it can clearly represent the scale of the map and the geographical location of the map coverage, and it is easy to judge the corresponding relation between the relevant topographic maps of different scales in the same region. The disadvantage is the complicated structure of numbering itself, and this problem is more distinguished for the numbering of larger scale topographic maps. Moreover, for a variety of reasons, this provision has undergone many local modifications, such as the division number of the 1:50,000 topographic map and suffix characters being added to the numbering of the 1:100,000 scale; the first map used Russian characters, such as А, Б, В, and Г, which changed to Chinese ordinal numbers, and later changed to English characters, such as A, B, C, and D. A certain degree of inconsistency occurred.

The national standard *Subdivision and Numbering for the National Primary Scale Topographic Maps* (GB/T 13989-1992) was issued in 1992, which specified the new division and numbering rules for 1:5,000–1:1 million scale topographic maps. The subdivision of various scales of 1:5,000 to 1:500,000 topographic maps were based on a 1:1 million topographic map; for specific divisions, see Table 16.2. The map number of the 1:1 million topographic map is the same as the original, without any changes. All other scales of topographic maps are numbered with new row and column numbering, that is, the 1:100 topographic map is divided into corresponding rows and columns according to the longitude and latitude differences in various scales of topographic maps. Then, the number of each topographic map is determined by the structure shown in Fig. 16.3. Among them, the scale codes for topographic maps are shown in Table 16.3.

The new map numbering has a uniform structure, and the number of each topographic map can be calculated by a formula. The latitude and longitude of the border

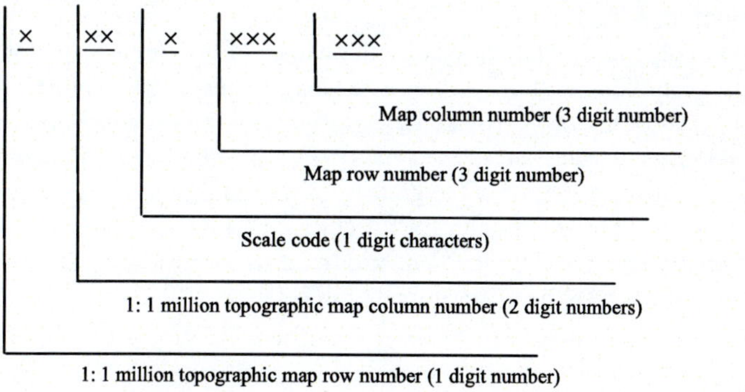

Fig. 16.3 Numbering of topographic maps

Table 16.3 Scale codes for topographic maps

Scales	Codes
1:1 million	A
1:500,000	B
1:250,000	C
1:100,000	D
1:50,000	E
1:25,000	F
1:10,000	G
1:5000	H

point at the southwest corner of the topographic map can be calculated according to the map number, and rows and columns of topographic maps at different scales can be converted within the number range of the same 1:1 million topographic map. However, the new number is up to 10 digits and characters, with poor readability, and it is prone to errors.

2. **Unified map content, symbol, color and decoration requirements**

Topographic map compilation specifications and symbols are divided into six groups according to their scales, including 1:500–1:2,000, 1:5,000–1:100,000, 1:25,000–1:100,000, 1:250,000, 1:500,000 and 1:1,000,000, and each volume was formulated, published and implemented according to the group.

Various scales of topographic map standards generally consist of two parts: topographic map compilation specifications and topographic map symbols. Of these standards, topographic map compilation specification defines the following contents of the topographic map: the nature, uses and requirements of a map; projection, subdivision and numbering of the map; contents and requirements of the editorial preparation; the contents of the map and compilation rules of each element; production of the final original; and publication of maps. Commonly used appendices include the size of the map margin, map face table, and illustrations of relief generalization in different landforms.

The topographic map symbols specify the names of the elements, color samples and color use of the compilation involved in the map content in tabular form. The color samples of symbols not only provide the symbols and their colors but also specify the size that makes up each part of the symbol. For different types of annotations on the map, it specifies the font type, the grading of the annotations and the font size at all levels, as well as the use of color for the compilation.

In addition, a number of technical standards relating to the printing and publication of topographic maps have been developed (Table 16.4).

Table 16.4 Technical standards relating to printing and publication of topographic maps

Standard numbers	Standard names
GB/T 2675-1981	Map paper
GB 14051-1993	Standard colors for topographic map
GB/T 14510-1993	Specification for printing of photomaps
GB/T 14511-1993	Specification for printing of maps
GB/T 15638-1995	Specification for optical densitometric measurements of printed maps
CH/T 4005-94	General rules of color separation map

3. **Unified spatial reference system and map projection**

To convert the geographic coordinates (longitude, latitude) of the ellipsoid into plane coordinates, Krasovsky ellipsoid and Gauss–Kruger projection (a traverse tangent cylindrical conformal projection) are used for all 1:5,000–1:500,000 topographic maps in China. Among them, zone division of 1:5,000 and 1:10,000 topographic maps is based on the 3° zone; others are based on the 6° zone. For the 1:1 million scale topographic map, a normal conic conformal projection in which the absolute values of side latitude and mid-latitude are equal are adopted alone according to the scope of the map. The two standard parallels of each map are 30′ away from both north and south side latitudes.

It was originally stipulated that various scales of topographic maps should adopt the 1954 Beijing Coordinate System and the 1956 Yellow Sea Elevation System. Since the 1980s, the 1980 Xi'an Coordinate System and the 1985 National Elevation Datum have begun to be used. Because the new generation of topographic maps that use the new coordinate system and elevation has not yet been covered nationwide, topographic maps using the new and old coordinate systems and elevation systems still coexistence and are being used simultaneously. China's geocentric coordinate system, the 2000 China Geodetic Coordinate System (CGCS2000), started being used on July 1, 2008.

4. **Standardization of place name placement**

On the map of China, place names with a nature of discriminating against and insulting ethnic minorities were deleted, and it was stipulated that the names of administrative divisions such as cities, counties, flags and so forth be subject to the official name promulgated by the State Council. The names of townships retained the official name of the government, and the names of villages were based on the usual natural name. The translation and Chinese phonetic alphabet spelling of place names were unified by the Provisional Regulation on Nomination and Renaming of Place Names, and the General Rules of Chinese Translations for Foreign Place Names was

Table 16.5 Rules for transliteration of place names of various minority languages

Standard number	Standard name
CH 4001-1991	Rules for transliteration of place names of Tibetan [Degg] language
CH 4002-1991	Rules for transliteration of place names of Li dialect
CH 4003-1993	Rules for transliteration of place names of Liangshan Yi dialect
CH/T 4006-1998	Rules for transliteration of place names of Dehong Dai dialect
CH/T 4007-1999	Rules for transliteration of place names of Mongolian
CH/T 4008-1999	Rules for transliteration of place names of Uighur
CH/T 4009-1999	Rules for transliteration of place names of Tibetan (Lhasa)
CH/T 4010-1999	Rules for transliteration of place names of Kazakh
CH/T 4012-1999	Rules for transliteration of place names of Kirgiz language
CH/T 4013-1999	Rules for transliteration of place names of Tibetan (Amdo dialect)
CH/T 4014-1999	Rules for transliteration of place names of Xishuangbanna Dai language

implemented. According to the regulations on the administration of geographical names, geographical names on the map have been strictly managed, and the transliteration of the Chinese phonetic alphabet of the place names of ethnic minorities was unified (see Table 16.5). As a result, geographical names on Chinese topographic maps have been standardized.

5. **Strict quality control**

A series of specifications, such as specifications for inspection and acceptance of surveying and mapping work, specifications for inspection and acceptance of surveying and mapping result plots, specifications for inspection and acceptance of surveying and mapping products, specifications for quality inspection and acceptance of surveying and mapping products, and so forth, were formulated to stipulate the requirements for quality inspection and acceptance of surveying and mapping products, the quality characteristics of surveying and mapping products and to verify the proportion, basic requirements and methods for quality evaluation of surveying and mapping products, as well as classification of defects per unit of product. According to these standards, the quality of the topographic map is strictly controlled, and the products provided to the users are all qualified products.

6. **Revision of technical standards for topographic (marine chart) mapping and compilation to adapt to technological advances**

To adapt to the technological advances for mapping and compilation of topographic maps (marine charts), the relevant technical standards have been revised since the late 1980s. These maps met the requirements of topographic mapping and compilation under the conditions of new technology (see Tables 16.6 and 16.7).

In Table 16.7, GB/T 15702-1995 "*Specifications for electronic charts*" stipulates the principles and methods of making electronic charts, the contents of electronic charts, and the basic requirements of using electronic charts for electronic chart

Table 16.6 Topographic mapping and compilation standards (1)

Standard number	Name of industry standard
GB 5791-86	Specifications for cartographic symbols for 1:5,000 and 1:10,000 topographic maps
GB 7930-1987	Specifications for aero photographic office operations for 1:500, 1:1,000, and 1:2,000 topographic maps
GB 7931-1987	Specifications for aero photographic field work for 1:500 1:1,000, and 1:2,000 topographic maps
ZBA79001-87	Compilation specifications and cartographic symbols for 1: 500,000 topographic maps
GB 12340-1990	Specifications for aero photographic office operations for 1:25,000, 1:50,000, and 1:100,000 topographic maps
GB 12341-1990	Specifications for aero photographic field work for 1:25,000, 1:50,000, and 1:10,000 topographic maps
GB 12342-1990	Specifications for cartographic symbols for 1:25,000, 1:50,000, and 1:10,000 topographic maps
GB 12343-1990	Compilation specifications for 1:25,000 and 1:50,000 topographic maps
GB 12344-1990	Compilation specifications for 1:100,000 topographic maps
GB/T 13977-1992	Specifications for aero photographic office operations for 1:5,000 and 1:10,000 topographic maps
GB/T 13990-1992	Specifications for aero photographic field work for 1:5,000 and 1:10,000 topographic maps
GB 14512-1993	Compilation specifications and cartographic symbols for 1:1,000,000 topographic maps
GB/T 14268-1993	Revision survey specifications for the national primary scale topographic maps
GB 5791-1993	Specifications for cartographic symbols for 1:5,000 and 1:10,000 topographic maps
GB/T 15661-1995	Specifications for aerial photography for 1:5,000,1:10,000, 1:25,000, 1:50,000, and 1:100,000 topographic maps
GB 15944-1995	Compilation specifications and cartographic symbols for 1: 250,000 topographic maps
GB 15967-1995	Specifications for aerial photogrammetric digital mapping for 1:500, 1:1,000 and 1:2,000 topographic maps
GB 15968-1995	Specifications for remote sensing image plan making
GB/T 7929-1995	Specifications for cartographic symbols for 1:500, 1:1,000, and 1:2,000 topographic maps
GB/T 16819-1997	Plane tabling specifications for 1:500, 1:1,000, and 1:2,000 topographic maps
CH/T 4011-1999	Compilation specifications and cartographic symbols for 1:500,000 topographic maps

(continued)

Table 16.6 (continued)

Standard number	Name of industry standard
CH/T 3001-1999	Operational procedures for the combination method with indoor interpretation and outdoor classification surveys on aerial photos for 1:5,000 and 1:10,000 topographical maps (Draft)
CH/T 3002-1999	Operational procedures for 1:10,000 and 1:25,000 scale orthophoto planimetric maps
CH/T 4011-1999	Compilation specifications and cartographic symbols for 1:500,000 topographic maps

Table 16.7 Topographic mapping and compilation standards (2)

Standard number	Name of standard
GB 12317-90	Symbols, abbreviations and terms used on charts
GB 12318-90	Specifications for Nautical Charts
GB 12319-90	Symbols, abbreviations and terms used on Chinese charts
GB 12316-90	Specifications for Chinese Nautical Charts
GB 15702-1995	Specifications for electronic charts
GB/T 17834-1999	Specifications for bathymetric charts
CH/T 7001-1999	Specifications for surveying and mapping for 1:5,000, 1:10,000 and 1:25,000 topographic maps of coastal zones
GB/T 14477-2008	Printing specifications for charts

application systems. This specification is suitable for the production and use of electronic charts.

GB/T 14477-2008 "*Printing specifications for charts*" was developed to replace GB/T 14477-1993 "*Printing specifications for charts*". The standard specifies the manufacturing process, quality standards and operating procedures for the prepress processing, plate making, proofing, printing, and finished product inspection in the chart printing process. It is suitable for printing general charts, nautical charts, special use charts, chart atlases, nautical books and tables.

16.3.2 Main Research Results of Geographic Information Standardization

In recent years, the standardization of China's geographic information has made much progress, mainly in the following areas:

1. **Initial establishment of the geographic information standard system**

The *Structural Framework for the National Geographic Information Standard* was issued and implemented in October 2007. On the basis of further refinement, the

National Geographic Information Standard System (Draft) was established, and in principle, it was examined and approved by the first plenary meeting of the third National Geographic Information Standardization Technical Committee held in December 2008.

The *National Geographic Information Standard System (Draft)* is composed of 7 categories, including general standards, data resources, application services, environment and tools, management, professional standards and profile standards, as well as 44 subcategories and other relevant standards. Among them, the first 5 categories are basic standards for geographic information, professional standards are the specific standards of the geographic information that are extended and revised based on the five types of standards mentioned above, which are tailored to various application fields; and profile standards are the geographic information profile standards used for special projects, which are formed on the basis of the above six standards.

The standard system will serve as a guiding document for the future national standards of geographic information, although there are still some deficiencies.

2. **Review and revision of several important national standards**

Since the founding of ISO/TC211, nearly 60 international standards of geographical information have been developed, and more than 40 of them have been issued and implemented so far. Several standards have already been revised. China has participated in the organization since the founding of ISO/TC 211 in 1994. In 1995, the State Bureau of Quality and Technical Supervision commissioned the State Bureau of Surveying and Mapping to act as the centralized management department of ISO/TC 211 in China, and an ISO/TC 211 technology centralized management office was set up in the National Geomatics Center of China. In 1998 and 2007, China hosted the seventh and twenty-fifth ISO/TC211 plenary meetings and work meetings in Beijing and Xi'an, respectively. From the beginning, China attached great importance to international standard activities, giving priority to participation in the development of international standards, closely tracking the standard drafting work, as well as standard discussion and test process, as well as time to convert international standards into Chinese national standards. To date, some of the standards have been converted, and some of them are in progress (Table 16.8).

3. **More international standards are converted into national standards**

A review was conducted on the national standard for geographical information, which reached and exceeded 5 years, and comments were submitted. Except for the part that continues to be used, some of the important standards were revised; the timeliness and practicability of these standards are improved to some extent.

For example, the *Codes for the Administrative Division of the People's Republic of China* (GB/T 2260-2007), since the first publication in 1980, have been revised every two or three years to meet the needs of all relevant departments and related fields. The standard uses administrative divisions at all levels as a spatial reference system based on geographical identifiers, and it has become the basis for the spatial positioning of fundamental information and thematic information in many fields in China. The latest revision was carried out in 2007.

Table 16.8 National standards for geographic information

National standard name and number	Corresponding international standard names and numbers	Degree of consistency	Release date
Geographic information—Conformance and testing (GB/T19333.5-2003)	Geographic information—Conformance and testing (ISO 19105: 2000)	Equivalent	2003–11–19
Geographic information—Metadata (GB/T 19710-2005)	Geographic information—Metadata (ISO 19115:2003)	Revision	2005–04–15
Geographic information—Metadata (GB/T 21337-2008)	Geographic information—Quality principles (ISO 19113:2002)	Equivalent	2008–01–09
Geographic information—Quality evaluation procedures (GB/T 21336-2008)	Geographic information—Quality evaluation procedures (ISO 19114:2003)	Revision	2008–01–09
Geographic information—Temporal schema (GB/T 22022-2008)	Geographic information—Temporal schema (ISO 19108:2002)	Equivalent	2008–06–20
Geographic information—Terminology (GB/T 17694-XXXX)	Geographic information—Terminology (ISO 19104:2008)	Equivalent	Submitted for approval
Geographic information—Spatial schema	Geographic information—Spatial schema (ISO 19107:2003)	Equivalent	Submitted for approval
Geographic information—Specific standard of the spatial schema	Geographic information—Core profile of the spatial schema (ISO/TS 19137:2006)	Equivalent	Submitted for approval
Geographic information—Web map server interface	Geographic information—Web map server interface (ISO 19128:2005)	Equivalent	Submitted for approval
Geographic information—Portrayal	Geographic information—Portrayal (ISO 19117:2005)	Equivalent	Submitted for approval
Geographic information-Metadata-XML schema implementation	Geographic information-Metadata-XML schema implementation (ISO/TS 19139:2007)	Revision	submitted for approval
Geographic information—Service	Geographic Information—Service (ISO/TS 19119:2005)	Equivalent	Under development
Geographic information—Methodology for feature cataloguing	Geographic information—Methodology for feature cataloguing (ISO 19110:2004)	Equivalent	Under development

(continued)

Table 16.8 (continued)

National standard name and number	Corresponding international standard names and numbers	Degree of consistency	Release date
Geographic information—Geography markup language (GML)	Geographic information—Geography markup language (GML) (ISO 19136:2007)	Revision	Under development

The *Classification and Codes for the National Land Information* (GB/T 13923-92), the first national standard for classification and coding of geographic information, was first promulgated in 1992. The standard adapts to the characteristics of basic geographic information in digital form and breaks through the division of scale for the first time. It adopts unified feature classification codes for the nationwide 1:1,000,000, 1:250,000, 1:50,000 and 1:10,000 databases in the national basic GIS. It has not only been widely used but also provided a useful reference for the development of many similar standards and played a significant role. The standard was revised in 2006 and renamed the *Specifications for Feature Classification and Codes of Fundamental Geographic Information* (GB/T 13923-2006).

The *Geospatial Data Transfer Format* (GB/T 17798-2007) was first formulated in 1999, but due to deficiencies in the standard itself, especially the weakness in the structure description of graphic performance attributes, it was relatively difficult to implement. The 2007 revised edition included many modifications and supplements, and the operability was improved.

The *Geographic Grid* (GB/T 12409) was promulgated in 1990 as the first national standard in the field of geographic information in China. It specifies the rules and codes for the division of geographical grid systems and designs 3 types of grid systems, namely, 10° * 10°, 4° * 6° and Cartesian coordinates. It is a geographical grid formed by dividing the Earth's surface into regular grids according to certain mathematical laws, which is used to identify various information related to geographical spatial distribution, especially thematic information that is distributed in planar patterns. For the first time, the standard set up a full geographic grid system for identifying all sorts of information about geographic spatial distribution, including basic information about land and offshore areas, marine information, meteorology, geophysics, engineering planning, design, construction, and so on. The standard was revised in 2008.

The *Urban Geographical Feature—City Road, Road Intersection, Block and Municipal Piping System Rules for Coding Structure* (GB/T 14395) was promulgated and implemented in 1993. It is the first national standard for the classification and coding of urban geographical features in China and played an important role in the design and construction of the early urban GIS. For example, in the early 1990s, the urban planning information system of Haikou city and the planning land information system of Shenzhen city introduced the rules established by this standard.

The standard was revised in 2008 to adapt to the development of new technologies and the development of cities in China.

4. **Many departments attached great importance to the formulation of geographic information national standards and industry standards**

The State Bureau of Surveying and Mapping always attached great importance to standardization work. In recent years, not only many urgently needed national standards have been formulated but also a number of industrial standards have been promulgated and implemented. For example,

(1) The national geodetic coordinate system 2000 was published

It is necessary to use a coordinate system with an origin at the center of the Earth's mass (referred to as the geocentric coordinate system) as the national geodetic coordinate system. It is conducive to the use of modern space technology for the maintenance and rapid updating of the coordinate system, as well as to determine the three-dimensional coordinates of high-precision ground control points and to improve the efficiency of the mapping work. On June 20, 2008, the national standard *Basic Specifications for National Geodesy* (GB 22021-2008) was issued. With the approval of the State Council, China's geocentric coordinate system, the China Geodetic Coordinate System 2000 (CGCS2000), began to be used on July 1, 2008. The transition and transformation of the old national geodetic coordinate system to the China Geodetic Coordinate System 2000 would occur over 8–10 years.

(2) Some national standards concerning the data classification, code, access, portrayal and quality control of basic geographic information were formulated and revised (see Table 16.9).
(3) A number of technical standards and standardized technical documents have been formulated and revised for the surveying and mapping industry; among them, the standards closely related to geographical information are shown in Table 16.10.

The Ministry of Housing and Urban–Rural Development has formulated, promulgated and implemented several industry standards related to geographical information, such as the *Code for Urban Survey* (CJJ 8-99), which is the most important technical standard for the production and processing of urban basic scale topographic data and is now being revised; for the first time, for the *Technical Specification for Urban Fundamental GIS* (CJJ100-2004), the basic geological data of the city is incorporated into the basic data of the urban fundamental GIS, and the urban 3D model data and the data related to the urban construction are incorporated into the urban spatial fundamental data system; and the *Standard for Urban Geospatial Framework Data* (CJJ 103-2004) defines the content and quality requirements of urban geospatial fundamental framework data and special-used framework data, as well as the methods and requirements of spatial registration of other data with urban geospatial framework data. In addition, many other standards have also been promulgated and implemented, such as the *Code for Classification of Urban Land Use and*

Table 16.9 Fundamental geographic information

Standard numbers	Standard names	Release date
GB/T 17160-2008	Specifications for digitizing 1:500, 1:1,000 and 1:2,000 topographic maps	2008–06–20
GB/T 20257-2007	Cartographic symbols for national fundamental scale maps Part I	2007–08–30
GB/T 20257-2006	Cartographic symbols for national fundamental scale maps Part II	2006–05–24
GB 21139-2007	Basic requirements for standard data of fundamental geographic information	2007–08–30
GB/T 21740-2008	Specifications for urban database construction of fundamental geographic information	2008–05–07
GB/T 13923-2006	Specifications for feature classification and codes of fundamental geographic information	2006–05–24
GB/T 20258-2006	Data dictionary for fundamental geographic information features	2006–05–24
GB/T 17158-2008	Record format for photogrammetric digital mapping	2008–06–20
GB/T 18316-2008	Specifications for inspection and acceptance of quality of digital surveying and mapping achievements	2008–07–02
GB/T 17941-2008	Quality requirement for digital surveying and mapping achievements	2008–06–20

Planning Standards of Development Land (GB50137-2011), *Technical Specification for Detecting and Surveying Underground Pipelines and Cables in Cities* (CJJ61-2003), *Technical Specification for Urban Surveying Using Global Positioning System* (CJJ/T 73-2010), *Specifications for Urban Database Construction of Fundamental Geographic Information* (GB/T 21740-2008), *General Software Standard for GIS of Engineering Construction* (JG/T 181-2005), *General Code for Measure and Evaluation of Application Software in the Field of Construction* (CJJ/T 116-2014), *Standards for Urban Construction Information Digitization and Urban Planning Data, Standards for Digital Community Applications, General Standard for Construction Information Platform Data, Standards for Scenic and Historic Areas Classification* (CJJT 121-2008), and *Standards for Classification of Urban Public Transportation* (CJJ/T 114-2007).

Recently, sponsored by the Ministry of Housing and Urban–Rural Development, a total of 7 industrial standards for the management of digital cities were compiled: *Urban Municipal Supervision and Management Information System—Rules for Basic Management Grid Division and Coding* (CJ/T 213-2005), *Urban Municipal Supervision and Management Information System—Classification and Coding for Urban Management Components and Events* (CJ/T 214-2005) (the standard was revised in 2007 and renamed *Urban Municipal Supervision and Management Information System- Classification, Coding and Data Requirements for Urban*

Table 16.10 Geographic information industry standards

Standard numbers	Standard names	Release date
CH/T 1011-2005	Digital products of fundamental geographic information for 1:10,000 and 1:50,000 digital line graphs	2005–12–07
CH/T 1012-2005	Digital products of fundamental geographic information land cover graphs	2005–12–07
CH/T 1013-2005	Digital products of fundamental geographic information for digital orthophoto topographic maps	2005–12–07
CH/T 1014-2006	Specification for managing and protecting data archives of fundamental geographic information	2006–08–21
CH/T 1015.1-2007	Technical rules for producing digital products of 1:10,000 and 1:50,000 fundamental geographic information, Part 1: digital line graphs (DLG)	2007–05–21
CH/T 1015.2-2007	Technical rules for producing digital products of 1:10,000 and 1:50,000 fundamental geographic information, Part 2: digital elevation models (DEM)	2007–05–21
CH/T 1015.3-2007	Technical rules for producing digital products of 1:10,000 and 1:50,000 fundamental geographic information, Part 3: digital orthophoto maps (DOM)	2007–05–21
CH/T 1015.4-2007	Technical rules for producing digital products of 1:10,000 and 1:50,000 fundamental geographic information, Part 4: digital raster graphs (DRG)	2007–05–21
CH/Z 1001-2007	Basic rules for the testing report of surveying and mapping product quality	2007–12–29
CH/Z 9001-2007	Construction requirements of the digital city geographic information common platform	2007–12–29
CH/Z 9002-2007	Rules of coding for addresses in the common platform for the geospatial information service of a digital city	2007–12–29
CH/T 9003-2009	Basic specifications for geospatial framework	2009–03–30
CH/T 9004-2009	Basic specifications for common platform of geographic information	2009–03–30
CH/T 9005-2009	Basic specifications for fundamental geographic information database	2009–03–30

Managed Components and Events (CJ/T 214-2007)), *Urban Municipal Supervision and Management Information System—Geocoding* (CJ/T 215-2005), *Technical Specification for Urban Municipal Supervision and Management Information System* (CJJ/T 106-2005), *Urban Municipal Supervision and Management Information System—Performance Evaluation* (CJ/T292-2008), *Urban Municipal Supervision and Management Information System—Mobile Device for Supervise Data Capture* (CJ/T 293-2008), *Urban Municipal Supervision and Management Information System—Supervision Case Register, and Treatment and Case Ended* (CJ/T 315-2009). These standards strongly guide and regulate the design, development, construction and application of digital city management systems in dozens of cities throughout the country and are standardized technical guidance documents to promote the new model of digital city management for the future.

To meet the needs of the informationization of land and resources and satisfy the Gold Soil Engineering, the Second National Land Survey, and Large-Scale Land and Resource Survey and other specific work, the Ministry of Land and Resources formulated a number of industry standards, including *Land Resource Information Classification Code and Data File Naming Rules*, *Basic Terminology for Standards of Land and Resources Information*, *Guide for Constituting Standard and Establishing Specification of Land and Resources Database*, *Specification for Fundamental Data Inspection and Acceptance of Land and Resources*, *Specification for Data Quality Inspection and Acceptance of Land and Resources Databases*, *Specification for Data Quality Inspection and Test Sampling of Land and Resources*, *Rules for Profiles of Land and Resources Information*, *Digital Archive of Land and Resources Management Standards*, *Technical Requirement for Integrating Land and Resource Database*, *Guideline for Data Management and Maintenance at Land and Resources Data Center*, *Classification and Coding Rules of Chinese Geological Domain Ontology*, *Inspection and Evaluation of Quality for Geological Data*, *Mineral Resources Programming Database Standard*, *Guide for Construction of Provincial Mineral Resources Planning Database*, *Cartographic Symbols for Land Thematic Maps*, *Standards of Urban land Classification Gradation and Valuation Database*, *Urban Cadaster Database Standard*, *Specifications for Data Quality Control of Land Use Planning Database*, *Database Standard for Remote Sensing Dynamic Monitoring of Land Use*, *Standards for Land Consolidation and Rehabilitation Projects Database*, *Standards for Land Consolidation and Rehabilitation Planning Database*, *Standards of Land Management Archives Database*, *Standards of Land Management Archives Catalogue Database*, *Standards for Agricultural Land Classification Database*, and *Standards for Prime Cultivated Land Database*.

In addition, many other departments have attached great importance to standardization work and formulated many national standards and industry standards related to geographic information, such as the Ministry of Industry and Information Technology, Ministry of Transportation, Ministry of Water Resources and State Forestry Administration.

5. **Batches of relevant national standards have been developed in many special application areas**

The most representative is a series of standards for location-based service navigation data and electronic map production.

Due to the urgent needs of the market for electronic maps, many units and companies in China have produced navigation electronic maps of hundreds of large and medium cities. To promote the healthy and orderly development of the city positioning information service and the promotion of the electronic map market, a number of urgently needed national standards have been issued, including the *Data Model and Data Exchange Format for Navigable Spatial Database* (GB/T 17911-2005), *Navigable Electronic Map—Basic Requirements of Security Processing Technology* (GB 20263-2006), *Specification for Collecting and Processing in-Car Navigable*

Geographic Data (GB/T 20268-2006), and *Criterion of Digital Map Production for Automotive Navigable System* (GB/T 20267-2006).

In the special project called the "standard research and development of satellite navigation applications", which was led by the Ministry of Industry and Information Technology, a series of national standards and industry standards for navigation electronic maps were developed, including the *Terms and Definition Related to Global Positioning System (GPS)* (GB/T 19391-2003), *Metadata for Navigation Electronic Map* (SJ/T 11419-2010), *Data Classification and Coding for Navigation Electronic Map* (GB/T 28442-2012), *Symbols of Navigation Electronic Map* (GB/T 28443-2012), *Application Storage Format for Navigation Electronic Map, Data Quality Specification for Navigation Electronic Map in Vehicle* (GB/T 28441-2012), *Data Quality Specification for Navigation Electronic Map in Personal Position* (GB/T 28445-2012), *Chinese Character Library for Navigation Electronic Map, and Network-based Data Updating Specification for Navigation Electronic Map* (GB/T 30289.1-2013). They involve the collection, processing, exchange and service of navigation electronic map data in the field of satellite navigation.

Research on e-government series standards is progressing rapidly. The standard is the basis for realizing the integration and sharing of e-government information among departments and regions. The standard system of natural resources and geospatial information database (referred to as information base) has been preliminarily formulated, including 70 standards, among them, the basic class standard has 7 items, the profile standard has 29 items, and the practical specification has 34 items. To ensure the quality and availability of standards, 31 standards with larger scale impacts are chosen from the standard system, which include 6 item basic class standards, 21 profile standards and 4 practical specifications. These standards are defined individually from the aspects of standard name, standard scope, application goal, compiling outline, relation with other standards, existing work basis and workload, status, and schedule. These standards, according to the contents, are divided into six categories: system standards, geographic information framework data content, data classification and coding, directory standards, information exchange service standards, database management, data products and quality standards. To date, standard formulation work has made periodic progress and effectively supported the design and construction of the fundamental information base of natural resources and geospatial information.

6. **More national standards work plans have been drawn up and are expected to begin**

Seven new international geographical information standards will be transferred during 2010–2011 according to the preliminary plan (Table 16.11). At the same time, a number of national standards will be formulated and revised. All these factors will help to further promote the sharing, service, application and industrial development of geographical information.

Table 16.11 Geographic information standards to be transferred into Chinese national standards

No	Standard names	Corresponding international standard
1	Geographic information—Profiles	ISO 19106:2004
2	Geographic information—Spatial referencing by coordinates	ISO 19111:2007
3	Geographic information—Metadata—Extensions for imagery and gridded data	ISO 19115-2:2009
4	Geographic information—Positioning services	ISO 19116:2004
5	Geographic information—Geodetic codes and parameters	ISO 19127:2005
6	Geographic information—Location-based services—Multimodal routing and navigation	ISO 19134:2007
7	Geographic information—Classification systems—Part 1: classification system structure	ISO 19144-1:2009

16.4 Problems and Prospects

16.4.1 Problems and Prospects of Mapping Standardization.

1. **Coordination between topographic maps of different scales**

As mentioned above, the scale series of fundamental topographic maps of China includes 11 levels. Compilation specifications and symbols for different scales of topographic maps are formulated, published and revised as a fascicle. They are divided into six groups according to scales: 1:500–1:2,000, 1:5,000–1:10,000, 1:25,000–1:100,000, 1:250,000, 1:500,000 and 1:1,000,000. The formulation and revision of the different periods and different times of the standard lead to disunity between the classification and the name of the elements between different cartographic symbols and scales of topographic maps, as well as inconsistencies in symbols on the map and their colors and sizes. It is necessary to coordinate the whole, determine the unified classification system, and formulate the unified cartographic symbol. Especially with the development of digital mapping and the gradual increase in its proportion, a series of comparative studies on traditional cartographic symbols are needed, and a set of compilation specifications and symbols for different scales of topographic maps must be designed.

2. **The focus of topographic map contents should be adjusted**

China's early topographic maps adopted a dual-use military and civilian version. The content of the map reflected military needs, especially large-scale topographic maps, which represented a large number of terrain and ground features that were meaningful to conventional operations. Currently, with the separation of military and civilian versions and the development of modern warfare technology, many topographic and ground features had little significance, and they needed to be carefully cleaned and adjusted. The military version of topographic maps should consider the characteristics of modern and high-tech warfare; the civilian version of topographic maps should be omitted from relevant military elements and supplemented with new contents and new elements that emerged with the progress of society and technological development; additionally, the civilian version of topographic maps should omit relevant military elements and supplement new contents and new elements arising from the progress of society and technological development, focusing on the required contents of economic, social, industrial and agricultural production, as well as the resource environment, transportation tourism, communication and public life, and so on.

3. **Parts of elements on topographic maps need to be coordinated with relevant professional fields**

Considering the classification of elements on the basic topographic maps that are mainly based on comprehensive characteristics and indicators, the grading and classification of some elements are different from those of related fields, which needs further coordination. For example, the classification of vegetation elements is not comparable with that of the fields of forestry, agriculture, land and other fields, the classification of highways is inconsistent with that of traffic departments, the classification of artificial water systems is different from the classification index of water conservancy departments, and the classification and names of special landforms and soil types are not in accordance with the professional classification and names of deserts, karst, glaciers and frozen soil. In addition, the mapping accuracy of large-scale topographic maps is different from that of cadastral surveys, which results in repeated mapping of terrain elements. All these need to be coordinated from a standardized perspective.

4. **The map subdivision and numbering scheme of topographic maps needs further consideration**

The division and numbering system stipulated by the national standard *Subdivision and Numbering for the National Primary Scale Topographic Maps* (GB/T 13989-1992) has its superiority; however, due to the unintuitive map sheet relationship between different scales and the long character number of up to 10 digits, it may not have obvious advantages compared with the old numbering scheme, it brings many inconveniences to the actual work and is still in the mixed use of the old and new subdivision and numbering system. It is necessary to further carry out scientific analysis and thinking on this provision, study the possibility and necessity of modification and improvement, and revise the standard in due course.

5. **Thematic maps are poorly standardized**

Compared with the fundamental scale topographic map, the standardization of China's thematic maps is quite low. Except for a handful of thematic maps, such as nautical charts, geological maps, and other various thematic maps so far, there are no unified specifications or standards of map classification, compilation and cartographic symbols, quality evaluation contents and methods, subdivision and numbering, or map border decoration.

6. **The standardization of the atlas is still unknown**

Over the years, China has compiled a number of atlases for different topics, different contents, and different versions and different regional coverages; however, the standardization of atlases has been ignored so far. Whether we need and how to implement standardization need to be further explored.

16.4.2 Problems and Prospects of Geographic Information Standardization

1. **Discuss the standardization of China's geographical information from the perspective of policy**

From the perspective of policy, the problems existing in the standardization of China's geographic information are mainly manifested in two aspects: the poor implementation of the relevant policies and the lack of standardized policies and regulations.

(1) There is a serious duplication among national standards, industry standards and other standards. For example, there are 3 national standards and more than 40 industry standards or project standards that involve "geographical information metadata". This not only wastes considerable manpower, financial resources and material resources but also impedes the sharing of information due to inconsistent standards. Further standardization is needed. This problem has improved in recent years, but the losses have been considerable.
(2) The published and implemented national standards for geographic information are basically recommended, with very few mandatory standards. Almost no technical regulations except Law of Surveying and Mapping of the People's Republic of China emphasized the importance and necessity of implementation of national standards. It leads directly to the high arbitrariness of the implementation of national standards, poor implementation and even neglect of standards.
(3) The standard revision cycle is too long, and it generally takes 4–5 years or even longer from the approval of standard project to release, and timeliness is poor.
(4) The standard age is too long, and the standard aging phenomenon is serious. The standard age analysis of nearly 100 national standards related to geographic information shows that most of the standard age has reached 5 years, of which 67% is over 5 years, approximately 30% is over 10 years, and the longest is 15–18 years.
(5) The review system under *Administrative Procedures on National Standards* has not been strictly enforced. Standard drafts were not widely consulted;

the review of the standard is simplistic; and there is no probation period for the release of the standard. These factors lead to uneven quality, incongruity between standards, and even contradictory phenomena.

(6) The work of formulating and transforming international standards for geographic information is lagging; in the work of formulating the international standard of geographic information, we are still in the role of learning and general participation, and so far, we have failed to undertake the formulation and revision of any international standards and failed to submit any proposals for new standards.

(7) The results of scientific research on standardization are rarely translated into productive forces. Government departments, experts and general scientific and technical personnel are not aware of the role of geographic information standardization as the technical basis for the industrialization of geographic information, and there is not enough guidance for enterprises; the enthusiasm and initiative of experts and scholars to participate in standardization work are not high. China's basic standards in the field of geographic information technology are also controlled by others, and the geographical information industry structure and product upgrading and updating are also subject to foreign enterprises.

(8) Severe imbalances in standard research and formulation. The investment in the development of each national standard is very limited, and the amount of investment in industry standards is relatively high, while there are more funds for standard research subjects. This imbalance has seriously affected the enthusiasm and standard quality of the development of national standards and has also induced anomalies that research achievements failed to translate into formal standards.

From the perspective of policy, the root of the problem lies in the failure to strictly implement the existing standard laws and regulations, as well as a lack of relevant planning, measures, coordination mechanisms and the standard of funding input quotas that are practical and operable for geographic information standardization. Moreover, the coordinating role played by existing institutions is limited. Therefore, to improve the overall level of the standardization of China's geographic information, it is urgent to strengthen the study and understanding of the standard laws and regulations and strengthen the implementation; at the same time, we should draw on the relevant policies and experience of international geographic information standardization, combine with China's national conditions, and research and formulate relevant policies and regulations to further promote the development of China's geographic information industry.

2. **Discuss the standardization of China's geographical information from a technical point of view**

From a technical point of view, there are still some problems in the development of China's standardization of geographic information, which need to be properly solved to achieve better standardization benefits. These problems are mainly manifested in the following ways:

(1) The standard system table of geographic information needs to be further improved and revised constantly. There are still some deficiencies in the established geographic information standard system table. For example, the complication and simplicity of national standard items listed in the system table are not uniform, category positioning for some standards needs to be adjusted, and the relationship between the professional and profile standards of geographic information and national standard needs to be defined more concretely. With the passage of time, the development of technology and the in-depth application, the system table itself also needs to be modified, supplemented, and dynamically improved to make it truly beneficial to the science, planning and orderliness of geographical information standardization.
(2) Translating international standards into national standards needs to be implemented to conform to international standards. To date, more than 10 international geographic information standards have been translated into China's national standards. However, it does not truly adopt these international standards for applications; instead, it simply converts international standard texts into national standard texts. For example, China has not yet established an effective conformance and testing mechanism, although the national standard *Geographic Information–Conformance and Testing* (GB/T19333.5-2003) has been completed for more than five years, and the relevant provisions of the conformance test are not contained in most national standard texts. In addition, metadata standards in many fields are not consistent with the national standard *Geographic Information—Metadata* (GB/T 19710-2005), which was completed in 2005.
(3) The quality of some standards needs to be further improved. With the increasing awareness of the necessity and importance of geographical information standardization, the overall quality of China's geographical information national standard has improved, the standard technical content has increased, and the standard practicability and operability have been strengthened. However, there are some standards with low standard quality, such as the technology adopted by the standard is not advanced enough or not in line with the technical development direction, and the scope of the standard and the object of its application are unclear; the technical specifications are vague and quantitative indicators are impractical, with weak practicability and operability; there is a lack of coordination and consistency with relevant standards, the language of standard content description is not standardized, and so on, meaning there is a need for further improvement.
(4) The relationship between special standards and national standards should be coordinated. In the standard system table, the positioning and attribute of the special standards are stated, but it is difficult to reflect the fundamental and special relationship between them and the corresponding national standards when formulating the special standard, and the phenomenon of copying

national standard texts occurs from time to time; how do we improve the pertinence, practicability and operability of the special standards for the special work demands and improve the quality of the standard? This requires practical research to avoid repetition.

(5) The quality of national standard revision should be further improved. The purpose of the standard revision should be to raise the level of technology and the level of application based on existing standards and not the other way around. However, the revised results of individual standards are questionable; for example, the revision of national standards *Specifications for Feature Classification and Codes of Fundamental Geographic Information* (GB/T 13923-2006), revised in 2006, abandoned the scale concept that desalted digital form basic geographic information, which was followed and proved effective by the national standard *Classify and Codes for the National Land Information* (GB/T 13923-92). More importantly, the international trend of development was the classification based on element ontology, and the number of categories is decreased. Through different attribute items to subdivide and meet different application needs, this represented the new technology development direction. The standard, however, was revised in the opposite direction. There were similar problems with the revision of individual standards. This is worth pondering, and the question of raising the quality of the standard revision needs to be emphasized.

(6) It is necessary to further explore the value of the standard and its implementation benefit. For example, due to the lack of unified data format conversion software tools and no support of large GIS software, and the standard data format was adopted as the main format of domestic GIS software, the existing GIS basic software and the commonly used foreign GIS basic software had the function of direct format conversion. As a result, the implementation of the national standard *Geospatial Data Transfer Format* (GB/T 17798-2007) was very difficult; to implement this standard, you had to develop your own transformation software, obtaining half the results with double the effort. From the point of view of technology development, few similar data format exchange standards were rarely seen in foreign countries, most of which were replaced by interoperability standards, such as ISO19118:2005 *Geographic Information—Encoding*. Therefore, it was necessary to explore the significance of this standard and to analyze its implementation benefits to reduce the cost of standard implementation, raise the level of technology, and explore more effective ways to facilitate industrial development.

3. Prospect of China's geographical information standardization development

To accelerate the pace of standardization, implement the standardization strategy and realize the leap-forward improvement of China's standardization level, the National Standards Committee defined the standardized development target for China over 5–10 years. That is,

In the first stage, to the end of the 11th Five-Year Plan, we basically built up a technical standard system with prominent focus and reasonable structure and adapted to the market so that the overall level of China's standards reached the level of moderately developed countries, that is, the formulation and revision of standards increased from 2000 items/year to 6000 items/year, the standard formulation cycle was reduced from 4.5 years to 2 years, the age of the national standard was reduced from 10.2 years to less than 5 years, and the proportion of using international standards and foreign advanced standards in related fields increased from 44 to 85%.

The goal of the second stage was that based on achieving the goal of the first stage, after 5 years' efforts, the overall level of China's standardization reached the level of developed countries in 2015; and the level of technical standards in parts of key areas reached the international advanced level. The specific objective was described as follows: China's independent innovation technology standards reached 5000 items; the international standards formulated based on China's standards or participation by Chinese technical personnel reached 2,000 items; China's share of the technical committee of the international standardization organization, the subtechnical committee or working group increased from 1.7 to 10%, the standard adoption rate in related areas reached 90%, and China became a permanent member of the international standardization organization; additionally, a mechanism by which China's leading enterprises actively participated in the formulation and revision of international standards was formed. This was a leap forward in terms of development goals, which strived for 10 years to cross the 30-year gap.

In the field of geographic information standardization, to achieve these goals, we need to rely on relevant domestic industries and enterprises, to give full play to the role of associations, intermediary organizations and scientific research institutions, to adhere to the principle of enterprise orientation, internationalization and independent innovation, to improve the content of independent intellectual property rights in the standard, to give full play to the standard's role as a bridge or bond link, and to improve the standard level, improve the applicability of standards, and enhance the comprehensive competitiveness of China's geographic information products.

To meet the needs of national economic and social development for standardization of geographic information, alleviate the prominent contradiction of "standard shortage, duplication, cross use, confusion and difficulty in coordination", to better guide the standardization of geographic information during the "11th Five-Year Plan" in China, to promote the development, sharing and utilization of geographical information resources and to improve the service support ability of geographic information to economic development, scientific and technological innovation, government administration and social life, the State Bureau of Surveying and Mapping and Standardization Administration of China jointly formulated the *National Geographic Information Standardization in "11th Five-Year Plan"* in December 2006. The main tasks of the plan include establishing a National Geographic Information Standard System, formulating and revising a number of urgently needed standards for key areas, transforming and adopting a number of applicable international standards, increasing the basic and preliminary research of standards, exploring and establishing standards conformance tests and evaluation systems, and strengthening efforts to publicize and implement standards. Proposed measures include establishing and improving standardization management and coordination mechanisms, raising funds for standardization work through multiple channels, training a team of qualified personnel for standardization, strengthening guidance on the formulation and revision of standards, and actively participating in international standardization activities.

Undoubtedly, as long as we follow the standardization development goals and the *National Geographic Information Standardization in the "11th Five-Year Plan"*, the standardization of China's geographic information will be effectively strengthened, and the development of the geographic information industry and the application, sharing and service of geographical information will be promoted.

16.5 Representative Publications

(1) ***Research reports on national standards of resource and environmental information systems*** (CHEN Shupeng, YAN Zhongyong, HE Jianbang, JIANG Kaifu, WU Funing, SHI Zhongzhi, ZHANG Jiaqing, ZHANG Jin, ZHOU Mingtao, ZHOU Hairong. Science Press, Beijing, January 1985)

Based on a summary of the situation at home and abroad, this report systematically discussed the nature, tasks and uses of resource and environmental information systems; the necessity, possibility, general idea and basic principles for the establishment of such a system in China; the basic technical regulations and requirements for the system; data sources, data management, and database systems; consideration of system hardware and software configuration; and the general situation of GIS abroad. The report aided each department to reduce blindness and avoid major mistakes, rework, repeated waste of labor and material resources among various departments to promote the circulation of information and comprehensive utilization. This study laid a foundation for follow-up work in this field in China. Soon afterward, the research group on the National Standards and Specifications of Resources and Environmental Information System of State Science and Technology Commission was established, and thus, the "research and development of national standards and specifications for resources and environment information", continued to develop and use geographic information standards at the national and provincial levels.

(2) ***The U.S. national digital mapping data standards*** *(7-5030-0362-6)* (translated by JIANG Jingtong et al. corrected by ZHU Changsheng. Surveying and Mapping Press, Beijing, November 1990)

The book was translated from the U.S. national standard, *A Draft Proposed Standard for Digital Cartographic Data* (STDS).

The standard was developed by the *National Committee for Digital Cartographic Data Standards* and is divided into four parts: Definitions and References, Spatial Data Exchange, Digital Cartographic Data Quality, and Cartographic Features.

The book was translated and published as an important reference for the design and establishment of various types of information systems in many departments, for the study of national and international standards on resource and environmental information, and for the research and construction of basic land information systems.

(3) ***Guidelines for standardization of urban GISs*** *(ISBN 7-03-006780-0)* (Editor in chief: YAN Zheng, deputy editor: JIANG Jingtong, HE Jianbang, ZHAO Yanhua, DU Daosheng. Science Press, Beijing, June 1998)

Since 1990, there has been an upsurge in the construction of urban GISs in China's special economic zones and southeast coastal cities, construction of urban GISs with application orientations was further defined, and a considerable number of urban GISs were completed or initially established, such as those for planning management, land management, underground pipe networks, traffic management and so on. City information queries, office automation, government departments, land planning, computer-aided real estate registration management, analysis evaluation and scientific decision-making of city problems, automation data and information updates, and so on, were realized to different extents, resulting in obvious social and economic benefits.

However, most of these systems were built independently and mainly considered the needs of their respective cities. The development and application of these systems lacked unified national standards and guidance, and all the cities and departments within the city were self-contained; it was difficult to realize the comprehensive utilization of information resources and information sharing; and the information for each city could not be used for statistical, comprehensive and comparative analyses by higher level authorities. The research and implementation of the standardization of urban GISs was comprehensive, inter-departmental and basic work. It had strategic and practical significance for the healthy development of urban GISs in China.

This book played a guiding and controlling role macroscopically in the design, construction and application of urban GISs and promoted the standardization of urban GISs, making the content of the guide as substantial as possible so that the reader could derive something useful from it. Authors attempted to use plain language, making it easy to read. In addition to the text, the appendices were also written in some chapters and sections, which provided a more detailed explanation of some specific problems related to urban GISs or a standard scheme for the reader's reference.

The guidelines covered many aspects, including the concepts and basic terminology of the urban GIS, standard system and standard formulation and standardized management, system design, data space positioning, information content and encoding, spatial data structure, hardware and software environment, data quality control, system implementation and maintenance, communication network and system interconnection, system security and confidentiality.

(4) ***Research on the metadata standard of geographic information in China*** *(ISBN 7-03-007487-4)* (Editor in chief: JIANG Jingtong, deputy editor: WANG Qiming, ZENG Lan, LIU Ruomei, XU Feng, CHENG Jicheng, ZHANG Jianzhong. Science Press, Beijing, June 1999)

In this book, the content, hierarchy, structure, classification, nature and characteristics of information sharing metadata standards were presented, the principles and methods of metadata extension were proposed, the basic model and operation tools of metadata standard implementation, metadata collection and meta-database creation and methods for linking with databases, metadata maintenance and updates, quality assurance, and so on were studied.

(5) ***Study on the sharing law of geographic information*** *(ISBN 7-03-008992-8)* (HE Jianbang, LV Guonian, WU Pingsheng, LI Xintong. Science Press, Beijing, December 2000)

The first 5 chapters of the book systematically described the general principle of geographic information sharing law, including resource theory, commodity economy theory, social public function theory and information technology theory of geographic information sharing law, which constituted the theoretical basis of geographical information sharing law. The subsequent two chapters discussed the legal theory of geographic information sharing law and compared it with property law and creditor's rights law, which provided a theoretical basis for the construction of the physical framework structure of geographic information sharing law. In the last chapter, based on analyses of many theories and practical information in China and foreign countries during both ancient and modern times, some suggestions on the current legislation of geographic information sharing in China were put forward systematically.

(6) ***Terminology of surveying and mapping*** *(ISBN-7-03-009896-X)* (Editor in chief: YANG Kai, deputy editor: CHEN Junyong, LI Deren, NING Jinsheng, GAO Jun. Science Press, Beijing, May 2002)

The standardization of nouns and their definitions are important basic work for the development of surveying and mapping science and technology. Approved by the China National Committee for Terms in Sciences and Technologies (CNCTST), the Chinese Society of Geodesy, Photogrammetry and Cartography (CSGPC) established an Examination and Approval Committee of Surveying and Mapping Terms in March 1987 to undertake the examination and approval of surveying and mapping terms. In March 1989, the first batch of 2146 basic surveying and mapping terms were examined, approved and officially released by CNCTST in 1990.

A total of 2,077 surveying and mapping terms and annotations were released; each Chinese noun corresponded to an English name, and definitions or comments were assigned to every Chinese noun for accuracy and authority. The book is divided into 7 parts: total class, 91 nouns; geodesy, 350 nouns; photogrammetry and remote sensing, 455 nouns; cartography, 365 nouns; engineering survey, 283 nouns; ocean mapping, 311 nouns; and surveying and mapping instruments, 222 nouns. The text is accompanied by an English-Chinese index and a Chinese-English index; the former is arranged in alphabetical order, and the latter is arranged in Chinese Pinyin order.

In terms of examination and annotations, the book strictly complies with the relevant provisions of the national standards GB1.6 and GB10112 and the *Principles and*

Methods of the Examination and Approval of Scientific and Technical Terms established by CNCTST. These nouns are the standard terms of surveying and mapping that should be used by scientific research, teaching, production and management departments. In 2002, they were formally published by CNCTST.

(7) ***Terminology of GISs*** *(ISBN-7-03-010828-0)* (Advisor: CHEN Shupeng; Editor in chief: XU Guanhua, deputy editor: LI Deren, JIANG Jintong, LI Jing. Science Press, Beijing, December 2002)

Approved by the CNCTST, the GIS Terminology Committee was established in 1998 to carry out the selection and standardization of geographical information system terms. Based on GIS terms, the principle of selection is that the terminologies of remote sensing, global positioning systems, computers, mapping and other related fields are appropriately absorbed. Through extensive collection of relevant information at home and abroad, as well as careful study, analysis and revision, the standardization work was completed.

The basic geographical information system nouns and neologisms and the corresponding English words and abbreviations used in international practice were selected. Chinese nouns are arranged in three categories: basic concepts; technology and applications; and international, major national organizations, institutions and others. The text is accompanied by an English-Chinese index and a Chinese-English index; the former is arranged in alphabetical order, and the latter is arranged in Chinese Pinyin order.

These nouns served as the standard GIS terms to be used in accordance with the regulations, totaling more than 1800, and these nouns were officially released by CNCTST in 2002.

(8) ***Collection of surveying and mapping standards*** *(ISBN-9787506628341)* (State Bureau of Surveying and Mapping, Institute of Surveying and Mapping Standardization of State Bureau of Surveying and Mapping. Standards Press of China, Beijing, published separately by volume during 2001–2006)

The collection includes the national standards and industry standards commonly used in the surveying and mapping industry, which was issued at the end of March 2002, and each volume was formed according to the classification of the discipline. All the standards in the collection are current and valid. However, due to the timeliness of the standard, they have been or will be revised or reenacted.

The collection consists of 7 volumes. The comprehensive volume contains 27 national standards, 5 industry standards, the Measures for the Supervision and Administration of Surveying and Mapping Quality, and the Provisions on the Quality Control of Surveying and Mapping Production, which was published in March 2003. Geodetic Surveying, Cadastral Surveying and Mapping Volume contains 13 national standards and 7 industry standards and was published in March 2001. The Engineering Survey Volume is further divided into four separate volumes. The Basic Volume of Engineering Survey Volume includes 7 national standards and 2 industry standards and was published in April 2003. The Water and Electricity Volume of

Engineering Survey Volume includes 1 national standard and 10 industry standards and was published in June 2003. The City Volume of Engineering Survey Volume includes 3 national standards and 5 industry standards and was published in March 2003. The Highway Volume of Engineering Survey Volume includes 7 national standards and 7 industry standards and was published in April 2003. Photogrammetry and Remote Sensing Volume includes 16 national standards and 1 standard of the geological mining industry and was published in March 2003. The Cartographic and Printing Volume includes 15 national standards, 3 industry standards, and the Map Compilation and Publication Management Regulations and was published in August 2003. Book one of Marine Surveying and Mapping Volume includes 13 national standards and book two of Marine Surveying and Mapping Volume includes 11 national standards and 6 industry standards, which was published in December 2006. The instrument and meter volume includes 2 national standards and 18 industry standards and was published in March 2003.

(9) ***Principles and methods of geographic information sharing*** *(ISBN 7-03-011571-6)* (HE Jianbang, LV Guonian, WU Pingsheng. Science Press, Beijing, August 2003)

The book systematically expounds the basic concepts, principles and methods of geographic information sharing, including the general principles of geographic information sharing, architecture and main functions, principles and methods for the formulation and implementation of standards, policies and regulations as an important component of geographic information sharing, as well as the current situation and development trend of related work at home and abroad. The appendix provides the basic format of the sharing standard of geographic information and the external framework of the *Geographic Information Sharing Law in China* (draft recommendation).

(10) ***Handbook of international standard of geographic information*** *(ISBN 7-5066-3201-2)* (Editor in chief: JIANG Jingtong, HE Jianbang; deputy editor: JIANG Zuoqin, DU Daosheng, LIU Ruomei, SU Shanwu. Standards Press of China, Beijing, February 2004)

The manual contains the text of the first batch of 19 international standards for geographic information (i.e., ISO 19101, ISO 19103–ISO 19120), which was developed by the ISO/TC 211 Project, and the manual is arranged in numerical order. These 19 standards are independent and have strong internal relations. They are structured and can be divided into 5 aspects, namely, framework and reference model standards, geographic information data models and operator standards, geographic information data management standards, geographic information service standards, and special standards. To date, the above 19 standards have officially been promulgated and implemented as international standards. Most of them have been or are about to be transformed into Chinese national standards.

To help readers understand the background, content, and progress of other standards being developed by ISO/TC 211, two articles, i.e., the status and progress of the development of the international standards of geographic information and the interpretation of the international standard of geographic information (Volume 1), were written and published in the manual.

The text and normative appendices of each standard were all translated; however, informative appendices were generally not translated unless it was necessary. The class name and attribute name, the unified modeling language (UML) diagram, the referenced program code, and the references in the text and appendix were not translated. Some of the untranslatable or nonunified terms from the original were attached to the back of the corresponding Chinese noun and enclosed in parentheses.

(11) ***National standard handbook of geographic information*** *(ISBN 7-5066-3623-9)* (Editor in chief: JIANG Jingtong, HE Jianbang, FANG Qing; deputy editor: LIU Ruomei, CAO Yanrong, LI Xiaolin. Standards Press of China, Beijing, November 2004)

The handbook comprises 5 parts.

The first part included an introduction to the major national standards of geographic information and related fields. More than 70 national standards closely related to geographical information and their scope, purpose, core content, appendices and so on were introduced individually. Among them, the national standards with a larger application scope or higher frequency were introduced in more detail; the rest were only about the time of release and implementation and how to obtain the standard text. A small section of the introduction also analyzed the shortcomings of the standard from the aspects of project, content and operability.

The second part included an introduction to some of the drafted national standards of geographical information. The drafts of some national standards on geographic information that were approved before 2002 were introduced. Through the introduction of these draft standards, the readers could participate in the discussion of the draft standards and make timely proposals to contribute to the development of these standards.

The third part included an overview of the standardization of some specialized fields related to geographic information. From a technical point of view, the tasks related to the standardization of geographic information in the fields of digital surveying and mapping, land, construction, transportation, railways, forestry, water conservancy, marine, environmental protection, agriculture, meteorology and so on, were summarized. In addition, as much as possible, the industry standard catalogue related to the geographic information was enclosed.

The fourth part included an introduction to some publications on geographic information standardization. Guide books and terminology related to the standardization of geographic information were chosen and separately introduced.

The fifth part, the appendix, mainly introduced the text of the national standard of geographic information. The texts of national standards for geographic information with a larger application scope or higher frequency, as mentioned in the first part, were introduced in detail.

The handbook also served as a companion to the *Handbook of International Standard of Geographic Information*, which formed a contrast between the geographic information standardization fields at home and abroad and complemented each other.

(12) ***Geographical terms (Second Edition)*** *(ISBN 7-03-001152-X)* (Editor in chief: ZHENG Du; deputy editor: LU Dadao, LIU Jiyuan, XU Xueqiang, CAI Yunlong, BAO Haosheng. Science Press, Beijing, April 2007)

According to the discipline, this book is divided into 21 categories, including general theory of geography, natural geography, geomorphology, climatology, hydrography, biogeography, soil geography, chemical geography, medical geography, environmental geography, glaciology, cryopedology, mire science, human geography, historical geography, economic geography, cartography, GIS, remote sensing, toponymy and quantitative geography. Chinese nouns are arranged according to the related concepts of the subject, and the English names corresponding to the concepts of the words are given.

Since the book was published earlier, it wasn't easily adapted to the current development requirements of the geography field, the new terminology and terminology in the new methods and new techniques as well as the newly established branch and edge disciplines in the field of geography were not involved, and the geographical terms that were enumerated in the book did not include descriptions of Chinese definitions; thus, in the process of scientific and technological communication, misunderstandings often occurred due to different understandings of the concept of nouns. Thus, the third Geographical Terminology Committee was organized by the geographical society of China, and the term standardization began in August 2008. Based on the original 1428 geography terms, some of the nouns that were not used in recent years were removed, many new geographical names were added, the terms being reviewed totaled nearly 4,000, the noun category was increased to 29 parts from the original 21, and each word was defined or annotated. They were released by the CNCTST in 2006.

(13) ***Guidance and practice of surveying and mapping and geographic information standardization*** *(ISBN 9787503018794)* (State Bureau of Surveying and Mapping. Surveying and Mapping Press, Beijing, December 2008)

In this book, the basic knowledge, system and planning of surveying and mapping and geographic information standardization are systematically introduced, the current standards of surveying and mapping and geographic information are summarized, and the policies and regulations commonly used in the standardization of surveying and mapping are selected and integrated. This book is the first to comprehensively introduce the standardization of China's surveying and mapping and geographic information and can be used as a reference or tool for decision makers, managers and technicians working in surveying and mapping standardization.

The book is composed of three parts and appendices. The first part is basic knowledge of surveying and mapping standardization, including a standardized overview, standard formulation procedures, structure and rules for standard writing, adoption of international standards, and standard quality requirements; the second part is surveying and mapping and geographic information standardization, including the overview of surveying and mapping information standardization, the goals and tasks, and the development of standard system framework; the third part is about the current standards of surveying and mapping and geographic information, including an overview of current surveying and mapping and geographic information standards and practical guidance for surveying and mapping and geographic information standards. Finally, the appendix includes the policies and regulations related to standardization.

Chapter 17
History of Cartography

Min An

17.1 Introduction

The history of cartography includes the study of the basic concepts of maps, mapping techniques, and the long-term impact of maps on human society and cultural development. From the human perspective, the history of cartography is also a medium with which to study how maps create interrelationships between the inner world of the human brain and the external world. A map is a basic tool that helps people perceive the universe in which they live at different scales. Undoubtedly, maps are among the oldest and most basic tools for spatial information transmission. Maps have been used for centuries as an instrument in analogical thinking.

The importance of maps is clearly defined in terms of their properties, the essentials of production and transmission, and their roles in society. To this end, the original hypothesis was that maps built a special graphical language, which was a tool for information transmission that affected human behavioral characteristics and social life. Maps are often used as repositories for spatial data, as well as memory banks for societies without using words. For centuries, scholars have believed in the power of persuasion and the formidable force of maps, which can transcend the barriers of ordinary language. As some historians have argued, maps form a common language that people from all nationalities and regions can use to express their social relationships and geographical environments. The function of the map was originally intended to tell others about the location or space experienced by the map maker. That is, from a historical perspective, a map is not only a summary of technical processes or artifacts in its production process but also a static image of a historical moment. In fact, the history of maps is composed of a series of complex interactions, including the use and production of maps. Cartographic research is about the knowledge of the real world (i.e., what to draw), the knowledge of the seeker or observer, or the knowledge of the art maker (i.e., map author), the knowledge of the map itself

M. An (✉)
Information Engineering University, Zhengzhou, China

J. Wang and F. Wu (eds.), *Advances in Cartography and Geographic Information Engineering*, https://doi.org/10.1007/978-981-16-0614-4_17

as a natural object, user knowledge, and so on. The history of cartography is also be concerned with the process of maps being created and used as graphic symbols.

17.2 Development Process

Since the founding of the People's Republic of China, the history of cartography as a discipline that integrates archaeology, psychology, philosophy, geography and bibliography has made great achievements in China. Archaeology, in particular, has seen amazing discoveries since the 1970s. For example, in 1973, three silk maps were unearthed at the Mawangdui tumulus in Changsha, Hunan Province, which were traced back to the Western Han Dynasty in the 2nd century BC. In 1978, Zhao yu tu (Graveyard Plan) of the Chinese Warring States period was found at the grave of King Zhongshan in Pingshan County of the Hebei Province. In 1986, seven ancient Chinese maps were found in an archeological excavation of a Qin State tomb in what is now Fangmatan, in the vicinity of Tianshui city in the Gansu Province, and a large number of prehistoric people's life ruins was discovered. These findings enrich the contents of research on the history of cartography, and temporally, these findings advance the physical maps studied in the history of cartography. Second, due to the involvement of surveying, mapping, geographic, psychological and philosophical workers in the study of maps and the introduction of theories, such as computer technology, cognitive science and information transmission, in terms of research contents and achievements, the history of cartography has created a single historical research approach that is focused on chronicling the past and shows multiperspective and multidirectional research results. In China, the study of the history of cartography has undergone four stages: collection and arrangement of ancient maps, analysis of ancient map contents and representation methods, new excavation and research of ancient maps, and the systematic study of the history of cartography.

17.2.1 The Stage of Collection and Arrangement of Ancient Maps

Maps have always been administered as state secrets in China, and most ordinary scholars have no chance to see maps except for special managers and users. In particular, maps of each dynasty are classified as secret materials and stored in the state agencies, so it is very difficult to study ancient maps in China. Mr. Wang used his opportunity to work at the Nanjing library and the Beijing library to read a large collection of maps of territory and census registers, especially the rare ancient books and maps. He also compiled, studied and summarized many of the original map materials he collected and then completed the book "*The outline of the Chinese history of maps*" in 1958. This book is a well-documented and illustrated book with a large number of ancient maps. It is the first book in China to study the history of maps and ancient maps, which laid a foundation for the study of China's map history. In 1983, led by Mr. CAO Wanru of the Institute of History of Natural Science (IHNS), Chinese Academy of Sciences (CAS), the official collection and sorting of ancient Chinese maps began. After 14 years, three volumes of the *Chinese Ancient Atlas* were completed; the first volume was published in 1990, the second was published in 1994 and the third was published in 1997. This project was a large project and a major achievement in the collection, collation and research of ancient Chinese maps, as well as a great achievement in the study of the history of science and technology in ancient China. These three volumes each contain eight folios. In the first volume, there are 205 maps from the Warring States period to the Yuan Dynasty; in the second volume, there are 248 maps from the Ming Dynasty, and the third volume contains 212 maps from the Qing dynasty; each volume has 120, 145 and 180 pages, respectively, which are known as the "macro volume". These ancient maps represent the crystallization of ancient Chinese science, technology and culture. These maps are treasures of cultural relics. After this atlas was completed, the collection and arrangement stage of ancient Chinese maps was basically complete. Subsequently, some studies were conducted based on time (such as the study of ancient maps according to dynasty), including the ancient maps of the Song and Yuan Dynasties and the ancient maps of the Ming and Qing dynasties; some studies were conducted based on geographical regions, such as ancient maps of the Hubei Province and the study of ancient city maps; and others studies were conducted according to the map color of ancient maps, such as painted ancient maps.

17.2.2 The Stage of Analysis of Ancient Map Contents and Representation Methods

Before the 1970s, there was no map of physical objects, except for the Yu ji tu (stone Maps of the Chinese Water Systems in the Song Dynasty) and the Hua yi tu (Map of both Chinese and Barbarian Peoples, Stone Map of the Han People and Minorities of the Song Dynasty), as well as the description of the text. Researchers could only guess about ancient maps based on these records. Therefore, before the 1970s, cartographers analyzed the contents and presentation methods of ancient maps according to the Yu ji tu and the Hua yi tu and the written records of PEI Xiu's Six Principles of Geographic Description and Map Making. As one of the earliest researchers, in the paper "Survey of the Compilation of the National and Provincial Atlas in the Ming and Qing Dynasties", Gao (1962) believed that maps began in the stage of confusion of pictures and words, when pictographic symbols were used to represent the phenomena and contents of the ground, such as mountains, water, villages and so on. The presentation of legend symbols on the map occurred later than the general content. However, as the map content became increasingly detailed, the characters on the maps could not meet the needs of researchers. Legend symbols appeared at this time, so the legend symbols represented the formation of the concept of "cartographic generalization". In the paper "On the evolution characteristics of mathematical elements and presentation methods of Chinese maps", Gao (1963) argued that "Drawing the Square Grid with Chinese Unit of Distance 'Li'" was a traditional method of drawing ancient Chinese maps. In the long process of evolution, it gradually played the role of transferring map contents and controlling map elements and objectively produced the effect of "grid projection".

In 1973, 3 silk maps from the Han Dynasty were unearthed in Mawangdui, Changsha, which opened a new stage of the study of ancient Chinese maps. From the 1970s to the 1990s, a large number of scholars conducted studies on the scale, geographical names, map contents and rendering methods for the Garrison (military) map and topographic map unearthed in the Han tomb in Mawangdui, and they considered that they are mapped under the guidance of certain mapping theories because the main contents in the two maps were described in detail, the less important elements were described briefly, the water system was expressed in great detail and shown with coarse lines, and the mountains and settlements used fixed-use symbols. Since 1986, a large number of scholars began to study the maps of the Qin dynasty after they were discovered in Fangmatan, Tianshui, Gansu. Through the discussion on the map plotting time, map combination, map layout direction, map properties, map scale, map annotation types, mapping language and system, map content and place names and a series of specific issues, they believed that the maps formed a unified schematic style and basic scale concept, which took the water system as a basic framework with higher accuracy. The method of map making was similar to that of the silk maps of the Han tomb in Changsha, which was a typical example of emphasizing river hydrology in ancient China, and showed that the mapping of ancient Chinese maps had basic standards and continuity.

After the 1980s, the research of map content and presentation methods was further broadened, and the study of the mathematical foundation of ancient maps began. In the paper "Regarding the History of Ancient Chinese Quantity Cartography", Yu (1991) explained the importance of cartographers, as represented by SHEN Kuo, to extend the theory of "Six Principles of Geographic Description and Map Making" and the significance of the twenty-four directions proposed by SHEN Kuo and discussed the mathematical foundation of the maps of the Song Dynasty.

In ancient Chinese painting maps, most of the drawing methods were similar to those of Chinese painting, which adopted image painting methods, such as the *Scenic Spot of the Yangtze River*, *Nine Frontiers Map* (Jiubian tu) of the Ming Dynasty, (Chengde) *Mountain Resort*, *Map of the Yellow River* (Huanghe tu) and *Mount Wutai Scenic Map* of the Qing Dynasty. In these drawings, the painting methods were used to represent the natural mountains, rivers, and cities; there were place names, the relative positions between the objects on the map were accurate, but the actual mileage was inaccurate, specific numbers were labeled between two objects, and the ground objects on the map were exaggerated. This kind of image painting was one of the drawing methods for ancient Chinese maps.

In short, in the past 60 years, Chinese cartographers have been able to determine the production technology and production theory of ancient maps of China according to the unearthed physical map. In the ancient Chinese maps, with the water system as the main controlling factor, the "Drawing Square Grid with Chinese Unit of Distance 'Li'" as the main mathematical basis, and supplemented by two symbols representing legend and pictograph methods, various required maps were accurately created.

17.2.3 The Stage of New Excavation and Research of Ancient Maps

The 3 silk maps of the Western Han Dynasty, the seven maps of the Qin State, and other ancient maps enriched the study of ancient maps and filled the blanks in terms of physical maps in ancient China. Therefore, many scholars performed much research on these maps, such as map excavation, the scope of map representation, the map scale, the content of map representation and the method of map making. It is generally believed that in these ancient maps, water system elements (rivers) were expressed in detail and were the controlling elements of the whole map. The Garrison (military) map and topographic map were unearthed in the Han tomb in Mawangdui, Changsha, and the map was painted in three colors. On the map, you can also distinguish a fine drawing area, a sketch area and a decorative area. There are relatively uniform symbols on the map to represent residents. In addition, resident annotations and transportation networks were analyzed.

In 1986, seven maps on wooden blocks were found at a Qin State tomb in Fangmatan, Tianshui, Gansu Province. These maps were analyzed from 4 aspects, including their annotations, the relationship between the maps, the area covered by the maps and the drawing techniques of scholars. The water systems are also the controlling factors on these maps, there are fixed symbols for residents, and the annotations are also marked on the maps. These maps are similar to the maps of the Han Dynasty that were unearthed in the Mawangdui of Changsha, indicating the inheritance of map-making technology in China. The maps were drawn between 323 B.C. and 310 B.C., with a history of 2,300 years.

In addition to the study of the map content itself, scholars reached beyond map making to study maps at this stage, for example, they studied the origin of the map from the graphic records in the Rites of the *Zhou Dynasty* (Zhou li), the rule of imperial power from the direction of ancient Chinese maps, and the relationship between the map and national culture from the content expressed on the map. Some scholars studied the different views of spatial understanding for different nationalities from ancient Chinese and Western maps. In short, with the in-depth study of ancient maps, more people saw that maps reflect the different views of spatial understanding for different nationalities, which represents the national culture.

17.2.4 The Stage of Systematic Study of the History of Cartography

After the 1980s, with the opening of historical map materials and the end of collection and arrangement work, a systematic study of the history of cartography began. The main manifestation of this stage was not only studying land maps but also studying charts, focusing on the research results of both domestic scholars and those abroad, paying attention to the views and achievements of foreign counterparts in terms of map history, and studying ancient maps and cartographic progress in the period of the Republic of China, the liberated areas, and the People's Republic of China, especially over the last 30 years. The research not only covered the production technology of the map itself but further extended to the relationship between map and culture and the comparative study of Chinese and Western cartography. Two books published in 1984, the *History of Chinese Cartography* and the *Story of Chinese Maps*, summarize land maps. The *Study of Modern Historical Charts*, which was published in 1992, provides a systematic study of Chinese charts. The *History of Chinese Cartography*, published in 2006 by the University of Chicago Press, is the Chinese part of the *History of Cartography* (Volume Two, Book Two). The *History of Chinese Modern and Contemporary Cartography*, published in 2008, is a summary of the development of Chinese cartography in the period of the Republic of China, the liberated areas and the founding of the People's Republic of China, especially over the past 30 years. Moreover, the *History of Chinese Cartography*, published in 2009, described the development course of China's maps from ancient to present, and it serves a comprehensive summary of the development of the history of Chinese cartography.

17.3 Main Research Achievements

In general, the study of the history of cartography is divided into four aspects. One aspect is the study of the outline of map history, which is to arrange maps into different categories according to their timeframes; the second aspect is the research of map theory, that is, map making technology, material base, map applications and so on; the third aspect is spatial cognition studies, namely, using the map as an expression of human spatial understanding, starting with the graphical representation of the map, to study prehistoric map expressions and analyze the spatial cognitive model of early humans; and the fourth aspect is the study of sociology, that is, with the map as an expression of the social force, political power and way of thinking, where the map represents the social form. In China, the research on the first two aspects is relatively mature, and the latter two studies have recently quickly developed. The following section includes discussions of 6 aspects.

17.3.1 The Study of the Outline of Map History

The *Outline of Chinese Map History*, published by Joint Publishing in 1958, is the first monograph to systematically discuss the development of Chinese cartography. The author, WANG Yong (1900–1956), also known as Yi zhong, came from Wuxi, Jiangsu Province. His major works include the *History of Chinese Geography* and *Investigation on Chinese Geography Books*. The *Outline of Chinese Map History* is divided into 11 chapters in chronological order, beginning with the original map of China and moving to the surveying and mapping of modern Chinese maps. By working in the Nanjing library and Beijing library, the author read a large collection of books, especially the ancient rare edition books or the only existing copies, and he sorted, studied and summarized many of the original materials about maps and then wrote his book. Therefore, the whole book is informative and excellent in terms of both historical materials and argument points and has high academic value. In particular, the collection of historical materials is extremely valuable and laid a solid foundation for further research. WANG Yong is the founder of the history of Chinese cartography.

Not long after the publishing of *Outline of Chinese Map History*, Joseph Needham, a famous British scientific historian, discussed the development of Chinese geography and cartography in Volume 3, Chap. 22 of the *Science and Civilization in China.* In the book, for the first time, the history of Chinese cartography was studied as part of the history of world cartography, the development of cartography in the east and west and the interaction of cartographic knowledge between different cultural regions was compared, which has greatly contributed to the history of Chinese cartography and the history of world cartography.

The *History of Chinese Cartography*, written by LU Liangzhi, was published in Surveying and Mapping Press in 1984 and provided a detailed introduction to the development and management of China's map science in various historical periods, and guided by dialectical materialism, summarized some of the experiences and lessons of the scientific development of maps in China in terms of the relationships between maps and military, politics, and production, as well as the relationships between cartography and other disciplines. Later, the author decomposed the contents of the book into papers of different topics and published them in various magazines. In 1984, the *Historical Narrative of the Chinese Map*, compiled by JIN Yingchun, was published in Science Press. Based on the detailed historical materials, the book systematically introduced the treasures of the Chinese maps, including Jiu ding tu (Map of Nine Tripod Cauldrons)—an original map in the legend, the earliest extant color silk maps in the world, unearthed in Mawangdui, and existing precious stone maps of Tang and Song Dynasties, such as Hua yi tu (Map of China and Barbarian Countries), Yu ji tu (Map of Tracks of Yu), Map of Nine Governing Districts, Map of Pingjiang City, geographical maps, the Sailing Charts of ZHENG He in the Ming Dynasty, and the Complete Imperial Map in the Qing Dynasty; the book also introduced the life and scientific achievements of Chinese ancient cartographic icon PEI Xiu and other famous cartographers, as well as the scientists who were closely related to mapping, such as JIA Dan, YI Xing, SHEN Kuo, ZHU Siben, LUO Hongxian, GUO Shoujing, ZHENG He, HE Guozong, MING Antu, WEI Yuan, YANG Shoujing, and others. The book also explained surveying and mapping technology, the theory of ancient maps and the evolution of map surveying and mapping mechanisms, along with some of the chart data of valuable ancient maps. These two books culminated in the study of ancient maps, and most of the books from later studies on ancient maps were out of reach.

Since the 1990s, with the increased availability of map historical data and the renewal of historical research perspectives, not only the nautical chart but also the map of the Republic of China was added to the historical data. One of the most representative works includes the *Study of Modern Historical Charts*, which was written by WANG Jiajun and published by the Surveying and Mapping Press in 1992. The book systematically discussed the research direction, contents and methods of modern history charts and their mathematical systems, as well as the investigation of modern historical charts and the transformation of mathematical foundations, and the preparation and application of modern historical charts. This book represents an early study of the historical chart. The second most representative work is the *Map Surveying and Mapping in Ancient China* written by Professor GE Jianxiong and published by the Commercial Press in 1998. In the book, the original map, the legend of Jiu ding tu, the ancient maps of Mawangdui and Fangmatan, and the maps of the Tang and Song dynasties, as well as the Ming and Qing dynasties, are all discussed in detail.

After entering the 21st century, the study of map history took another step forward, and the map development history following the founding of new China was added. One of the most representative works is the Modern and Contemporary History of Chinese Cartography written by LIAO Ke and YU Cang and published by Shandong Education Publishing House in 1998. This book comprehensively discusses the development process of modern and contemporary cartography in China, the main achievements, and the development background and historical enlightenment from the Ming and Qing Dynasties to the Republic of China and to the founding of new China. The book particularly focuses on the early development of modern Chinese cartography in the Ming and Qing Dynasties, continued development in the Republic of China, the preliminary formation of postmodern cartography in new China, and the further development of modern Chinese cartography since the 1970s. The other representative work is the *History of Chinese Cartography* written by YU Cang and LIAO Ke and published by Surveying and Mapping Press in October 2009. The book systematically introduces the history of the map and cartography in China. It elaborates on the historical trajectories of the development of ancient Chinese maps from the pre-Qin period to the Yuan dynasty, indicating that the development and achievement of ancient Chinese maps plays an important role in the history of the world map. The book illustrates the development history of modern Chinese mapping and cartography in the Ming dynasty, the Qing dynasty and the Republic of China and Liberated Areas. It comprehensively discusses map creation in new China and the development of modern cartography, especially the rapid development of remote sensing mapping, computer mapping and publishing systems, multimedia electronic atlases and Internet maps in China since the 1970s. There are many color map illustrations that reflect the development and level of modern maps, and a list of senior cartographers who contributed to the development of the map of new China and the development of modern cartography is attached at the end; further, a chronology of key map events (from ancient times to 2005 A.D.) are included.

17.3.2 The Collection of Ancient Map Works

In the past 60 years, the major achievements from the collection, arrangement and research of ancient maps in China have mainly been manifested in the publication of various large ancient atlases. Among them, the most representative is the *Ancient Atlas of China* compiled by CAO Wanru and ZHENG Xihuang et al., and this atlas is divided into 3 volumes. The first, second and third volumes, published in 1990, 1994 and 1997, respectively, are major achievements in the collection, arrangement and research of ancient Chinese maps. The atlas contains hundreds of ancient maps of various types, including Zhao yu tu (Graveyard Plan) of copper plate, which was found at the grave of King Zhongshan from the Chinese Warring States period; the topographic map, military map and prefecture map drawn on silk, which were unearthed in the tombs of the Western Han Dynasty in Changsha; the maps painted on the tomb wall and portrait bricks of a tomb of the Eastern Han Dynasty; the maps painted on the wall of Dun huang grottoes; and 15 ancient maps carved on stone tablets. In terms of map content, there are nationwide administrative maps and topographic maps; there are local administrative maps, border and frontier maps, sea defense maps, river way maps, etc.; there are traffic maps, such as the road map of the silk road drawn by the Qing Dynasty; there are also frontier maps, Western Xia maps, Khitan maps, mountain maps, urban plans, temple building plans, mineral area maps and a wide variety of ancient maps. The atlas also included a circular map drawn by Mahmud kashgari, a minority scholar born in Sashgar, Xinjiang in the 11th century, which reflects the scientific and technological achievements and geographical understanding of ethnic minorities. In addition, there are maps drawn by foreign missionaries in China in the Ming and Qing Dynasties, reflecting the exchange of science, technology and culture between China and foreign countries. The ancient maps in the atlas reflect the achievements and characteristics of geography and the related science and technologies in various periods. These include the investigation of the Yellow River source in the Yuan Dynasty, control of the Yellow River and maritime traffic in the Ming Dynasty, numerous large-scale national latitude and longitude measurements in the Qing Dynasty, and national maps drawn based on measurements, including the *Huang Yu Quan Lan Tu* (Map of China in the Emperor Kangxi Reign of the Qing Dynasty) and the *Qianlong imperial Atlas*. In the Qing Dynasty, great achievements were obtained in the study of the changes of the territory and political areas in Chinese history, especially the *Geographical Maps Through the Ages* drawn by YANG Shoujing, which is considered to be a milestone of historical map making.

In August 1996, the *Descriptive Catalogue of Pre-1900 Chinese Maps Seen in Europe*, which was compiled by LI Xiaocong, was published by the international cultural publishing company. Since 1991, the author visited the Netherlands, Sweden, Denmark, Germany, Austria, France, Belgium, Britain, Italy, the Vatican and other major European libraries and museums and looked at thousands of Chinese maps; he collected and arranged a batch of Chinese map information. The book is a joint bibliography and description of ancient maps that were deposited in Europe in Chinese and English. The book contains approximately 300 ancient maps, and the era of map compilation continued until 1900. Attached at the end of the book is the English translation of the commonly used names of local administration organization systems in the Ming and Qing Dynasties, names and abbreviation of collection units, main references and indexes, and 32 copies of reproduced ancient maps. In the book, the ancient maps are divided into 11 categories: world map, foreign map, mountain map, river and lake map, coastal map, traffic map, urban map, historical map, celestial chart, map of the whole Chinese territory/composite map of multiprovinces/atlas, and regional map. The same types of maps are arranged in terms of the time of drawing. The descriptive items include map name, author, drawing time, printing place, medium form, plate type, size, scale, description, collection place, reference, etc., and the code is composed of the classification number, the serial number of the catalogue and the age of the drawing. The book describes the location of the map collection, the serial number of collections, and published studies of Chinese and foreign references, especially for readers to read and explore.

Selection of Territory Maps: The Catalogue of 6827 Chinese and Foreign Languages Ancient Maps Deposited in the Beijing Library, compiled by the territory map group rare book department of the Beijing Library, was published by the Beijing Library Press in 1997. The collection of maps by the Beijing library started at the same time as the establishment of the capital library in 1909. At present, in the rare book department of the Beijing Library, there are more than one hundred kinds of painted maps of the Ming and Qing Dynasties, ancient maps that have been unearthed in the past, various kinds of territory maps made by public and private organizations and individuals since the late Qing Dynasty, and parts of foreign language maps. In the 1930s, the *Chinese Catalogue of Territory Maps in the National Peking Library* was arranged and published, and after more than half a century, the holdings of ancient geographical maps became richer. Thus, the *Selection of Territory Maps* was compiled through the efforts of the rare book department of the Beijing Library. This was an important disclosure of the collection documents in the Beijing library, which made it easy for people at home and abroad to study books and materials. It was also one of the important subject catalogues published in recent years.

The Selected Works of Chinese Ancient Maps were published by Harbin Press in January 1998. It was compiled by YAN Ping, JIN Yingchun, ZHOU Rong, HAN Beisha (Chinese Academy of Surveying and Mapping), Sun Guoqing (the original Beijing library), Yin Junke (Beijing Academy of Social Sciences), and QIU Fuke (Sino Maps Press), who visited major libraries, archives and local cultural relics departments in the country and reproduced, registered, and studied many ancient maps. This atlas contains 166 maps of China from 475 B.C. to 1911, reflecting China's long history of cartography and excellent technology; it is a high-grade ancient map with extreme collection and research value.

Compilations of Chinese Ancient Maps: Heyday in Kangxi Yongzheng and Qianlong Reign, edited by WANG Ziqiang, was published by Planet Earth Press in June 2003. With the one hundred years' reign of Emperors Kangxi, Yongzheng and Qianlong as the timeframe, the book includes a collection of approximately 1,700 maps drawn during this period, such as general maps, border maps, city maps, water system maps, mountain maps, coastal maps and a few landscape maps. All the maps that were collected without any modifications enable the reader to understand the true face of that time period. These maps were collected from *Da Qing Yi Tong Zhi* (Comprehensive Geography of the Great Qing Dynasty in the Qianlong reign), general annals of all provinces compiled during the Kangxi Yongzheng and Qianlong reigns, annals of each prefecture, state and county, and other ancient geographical books, such as *Huang Yu Xi Yu Tu Zhi* (Map of the Western Regions in the Qianlong Reign) and *Guang Dong Yu Tu* (Map of the Guangdong Province). The book is divided into three volumes, eight folios, and is in hardcover; each page is divided into two columns with one picture in each column; there is a directory in the front of the book and an index at the end. The first volume contains all 98 maps attached to *Da Qing Yi Tong Zhi*; in terms of sorting, there is a general map, provincial maps (frontier maps), and prefecture maps (Zhili State) or regional maps; the core of the second and third volumes is the maps attached to the general annals of all provinces compiled during the Kangxi, Yongzheng and Qianlong Reigns, and the *Shengjing Annals* and *Huang Yu Xi Yu Tu Zhi* compiled in Qianlong Reign, as well as other maps from the annals of each prefecture, state and county, and some ancient geographical books. In the second and third volumes, the order of the maps of 18 provinces is the general chart of the province, prefecture (Zhili State), sea, lakes, mountains and rivers, the cultural landscape, and finally the county (state) maps; if there is a city map, it is placed after the map of the administrative region of its jurisdiction. This book has three characteristics: originality, practicability and richness. First, many maps were collected in ancient Chinese books, but no one has ever developed and utilized them extensively. This book is the first to systematically collate and integrate the maps of ancient books and compile an atlas, so it is groundbreaking. Second, a map contains a large amount of natural, social and cultural information, describes the spatial distribution of the information in a specific time period, and has an important function of practicality; ancient maps are no exception. These maps are materials used to study ancient history and culture. This book started with the practicality of the map, focused on the maps produced during the period of the Kangxi, Yongzheng and Qianlong's reigns, and the collected maps were arranged according to the political area of the day so that the maps have inherent connections in time and space and can be referred to each other, greatly improving the use value of these maps, which were previously scattered in various ancient books. Finally, the richness of the book is reflected in its extensive coverage of ancient maps, almost all of the more than 200 administrative regions and most of the 1000 counties in the country have special maps at that time, and more than 200 city maps are also included in this book.

There is also the *Atlas of Ancient Maps of China·Urban Map Atlas*, which was published by the Xi'an Press in December 2004. This is a large atlas that was compiled by ZHENG Xihuang of the Institute of Natural Science, Chinese Academy of Sciences; in the fourth part of *Atlas of Ancient Maps of China*, a thematic map atlas, which brings together the city map of ancient China, provides a comprehensive expression of ancient Chinese city maps. *Huang Yu Xia Lan* (following extensive reading of imperial maps from the Qing Dynasty—the painted maps of the Qing Dynasty at the Peking University Library), was compiled by the Peking University Library and published by China Renmin University Press in September 2008. The drawing time lasted from the beginning to the end of the Qing Dynasty. In terms of type, there are national maps, provincial, prefecture, state and county maps, river and water conservancy maps, sea border defense and frontier defense maps, road mileage maps, land survey maps, salt industry and product maps, and other kinds of thematic maps. In terms of style, some of them are beautifully painted, such as the famous paintings of mountains and rivers; some reflect historical events, even the characters in the meantime, just like a historical scroll; some of the maps are illustrated with text and pictures and have a unique style; some of them are brightly colored but sketchy with only a synopsis. In general, the number of ancient painted maps in the Qing Dynasty is not large, but it is rich in type and diverse in style. It can best reflect the tradition and characteristics of Chinese map making and has unique research value and aesthetic value. These are the most important parts of the maps from the Qing dynasty.

Similar publications include the *Assembly of Ancient Maps* (chief editor: WANG Qianjin), which originated from the ancient maps at the Dalian Library, the *Assembly of Ancient Maps of Macao* (compiled by the First Chinese Historical Archive), and parts of ancient maps in the *Historical Atlas of Wuhan*, which has high historical value.

17.3.3 The Study of the Origin of Ancient Maps

There are few papers about the origin of ancient maps of China in the study of Chinese cartographic history. The problem is rarely discussed in the books related to cartography history. Therefore, the origin of ancient maps has become an outstanding question in the study of cartographic history, and some scholars have performed research in this area.

In the paper "Inscription of Yi Hou Shi Gui (food utensils of early Western Zhou Dynasty)—Maps of King Wu's and King Cheng's conquest over Yin Dynasty, East State Map, and comprehensive analysis of the map of the Western Zhou Dynasty" (Wang 1990), the Yi Hou Shi Gui inscription was unearthed in Yan Dun Hill, Dantu County in the Jiangsu Province in June 1954, which was considered to be bronze ware of Zhoukang Wang at the beginning of the Western Zhou Dynasty, and it is nearly

3,000 years old. The inscription shows that on April Dingwei (the forty-fourth day of the cycle of sixty in the Chinese calendar), King Zhoukang viewed the Maps of King Wu's and King Cheng's conquest over the Yin Dynasty and East State Map and then came to the place of YI, held a ceremony for appointment and was awarded feudal lord. It also pointed out that if the "Maps of King Wu's and King Cheng's conquest over Yin Dynasty" was actually mapped in 1027 BC, it was seven years ahead of the legendary "*Luo Yi Di Tu*" (Luo Yi refers to Luoyang in the Henan Province, which is the capital of the Eastern Zhou Dynasty).

In the paper "Talking about the Maps in the Rites of Zhou" (Yao 1990), the cartographic theory and application technology of the maps described in the Rites of Zhou was discussed. The maps described in the Rites of Zhou were discussed from four aspects. First, the map characteristics described by the Rites of Zhou occupy a special place in the history of the map. According to the Rites of Zhou, at that time, the number and type of maps greatly exceeded that of the past and played a major role in politics, military affairs and production, and the map began to shift toward science and practicality. Second, the map reflected the political system and social economy. In the Western Zhou Dynasty, to strengthen the control of the national territory, King Zhou implemented the system of enfeoffment, where the map guaranteed the enfeoffment system, and at the same time, it was the basis of taxation. Third, the Western Zhou Dynasty is the embryonic period of Chinese cartographic theory. Fourth, the Zhou Dynasty developed the map management system, which guarantees the development of cartographic causes.

In the paper "On the origin and development of Chinese maps" (Huang 1997), it was pointed out that the most primitive map was an object picture made up of the figurative painting of objects, and the Xinglongwa Site (approximately 8000 years ago) in Inner Mongolian, which was excavated in the 1990s, was used as evidence. Archaeological excavations showed that there was a fence in the cultural settlement of the Xinglongwa Site, and there were orderly and almost uniformly arranged buildings within the walls.

In the paper "Discussion on the cartographic description of the Rites of Zhou and its related problems" (Liu and Yang 1997), the cartographic description of the Rites of Zhou was proposed to reflect the development level of Chinese cartography in the pre-Qin period, which showed the application of the map in the society of the Zhou Dynasty, people's understanding of cartography and the management system of maps and documents at that time. The details of its contents, the rich pictures, the wide varieties, and the incisive discourse are unmatched by other documents and classic maps. Many scholars have conducted research on the use of maps in the Rites of Zhou, explaining the cartographic achievements of the Rites of Zhou, and attempting to answer some questions in the history of the development of various disciplines starting from their respective majors, such as maps and architectural drawings. However, previous work was seldom done from the perspective of the history of science and the literature to comprehensively investigate the Rites of Zhou.

In the paper "A preliminary exploration of map making in prehistoric times" (Wang 1998), by quoting a large number of ancient books, it was shown that as early as in the times of the Divine Husbandman and Xuanyuan Emperor, China already conducted map surveying; in the portrait of the Han Dynasty, Fu Xi and Nu Wa, the Chinese humanist ancestors, always held a rule and a moment. The rule and the moment (namely, the compass and the ruler) are the basic tools for measuring and drawing. According to folklore, the earliest river maps and Luo Book refer to the relief map of the Yellow River and the terrain map of the Luo shui River.

According to the paper "The probe of the most primitive maps in China" (Lu 2002), there are pictures of natural phenomena in primitive society, and these pictures appear before hieroglyphs; although there are pictures describing the natural landscape, they are not truly maps but only the paintings of ground objects. In the late Xia Dynasty and early Shang Dynasty, after the emergence of hieroglyphs, there were no obvious differences between the object paintings of the primitive society and the hieroglyphs; thus, the development stage of the map was formed due to the integration of object paintings and pictographs. With the fall of the Shang dynasty and the rise of the Zhou Dynasty, owing to the needs of production and campaigns, a large number of maps with obvious themes and practical use emerged.

Based on the above analysis, nearly all of the studies on the origin of the ancient maps focused on the research of ancient books and materials, without physical maps, which makes it difficult to study the origin of ancient maps. Compared with the amount of studies of prehistoric rock paintings by a large number of scholars abroad, the amount of studies of ancient maps is not adequate.

17.3.4 The Study of the Map-Making Theory of Ancient Maps

1. **Determination of the germination stage of ancient map-making theory**

It is generally believed that the theory of ancient map making in China was the "Six Principles of Geographic Description and Map Making", which originated from the Qin and Han Dynasties to the Wei and Jin Dynasties. However, according to the map unearthed in Fangmatan, Tianshui, the ancient map-making theory of China was created earlier. The main basis is described as follows.

In the paper "Several problems concerning the maps unearthed in the Qin tomb in Fangmatan, Tianshui" (Cao 1989), the author analyzes maps from four aspects, including annotations, relationships among maps, the region covered in a map, and mapping techniques. It is believed that the maps unearthed at the tomb of Qin Dynasty included rivers, watersheds (or mountain ranges), residential areas, roads, river names, valley names, forest distribution, and mileage, as well as the north and south directions on the map. According to these contents, the maps can be drawn in a certain direction, and the upper map is north. The digit of mileage that appears in many places shows that the principle of distance is considered.

In the paper "A masterpiece on the study of the history of Chinese cartography", Wu (2004) evaluates the *Study of the Maps on Wooden Blocks Unearthed from Fangmatan in Tianshui*, a book written by YONG Jichun. In his opinion, the book identified the time for the plot of the maps, which was dated from 323 to 310 B.C., being 2300 years old. Through discussions of the mapping time, map combination, map layout direction, map property, map scale, map annotation type, map drawing language and system, map content and place names, and other specific problems, he believed that the maps formed a unified schema and a basic scale concept. The maps take the hydrographic net as the basic frame, with higher accuracy. It was a typical example of ancient China's attention to rivers and water systems. It created the foundation of basic norms of map making in ancient China, which marked the formation of the ancient Chinese mapping theory as early as 2300 years ago.

In terms of the accuracy of the water system depicted in the topographic map unearthed from Mawangdui in Changsha and the basic framework for *Yu Ji Tu* and *Hua Yi Tu*, the application of cartographic theory in the maps unearthed from Qin tomb of Fangmatan in Tianshui is appropriate for the long-term.

2. **The study of ancient map-making theory**

The paper "On the mathematical elements of Chinese maps and the evolution characteristics of presentation methods" (Gao 1963) showed that Drawing Square Grid with Chinese Unit of Distance 'Li'" is a traditional drawing method of ancient Chinese maps, and in the process of long-term evolution, it gradually played a role of transferring map contents and controlling map elements and objectively produced the effect of "grid projection". After the 16th century, a gradual shift occurred to using the map projection method. In "A preliminary study on the theories and methods of ancient Chinese maps", Cao (1983) elaborates on the methods and tools used in ancient China's surveying and mapping and stated that the mathematical elements of ancient Chinese maps are scale, direction and distance, and geographical elements can be divided into human geographical symbols, natural geographical symbols and text annotations. Among them, human geographical symbols include residents, boundary lines, roads, bridges and symbols of the Great Wall; natural geographical symbols include mountains, rivers, lakes, oceans, forests and deserts; and text annotations include place names, mountain names, river names, lake names and other important notes. In two articles, that is, "The outstanding achievement of quantitative cartography in the Song Dynasty" represented by Shen Kuo and "On the problem of the history of quantitative cartography in ancient China" (Yu 1987, 1991), the contribution of the cartographer, represented by SHEN Kuo, to the expansion of the theory of "Six Principles of Geographic Description and Map Making", and the significance of the "Twenty-four directions" proposed by SHEN Kuo, are discussed, and the mathematical foundation of the map of the Song Dynasty is explained. In the map of SHEN Kuo, the twenty-four equal angular azimuth system is consistent with the compass of the Song Dynasty. This finding indicates that in the Song Dynasty, SHEN Kuo combined the map with the compass, and a geographical position was indicated with the magnetic azimuth marked by "Twenty-four directions". This is of great significance in the quantitative history of the world.

"The charm of Chinese traditional map drawing" (Yin 2004) put forward that traditional Chinese mapping is divided into two major branches. One is to draw the terrain, ground objects and other elements into a map using vivid and realistic methods, which is called a vivid drawing map; another is to draw a square map, which emphasizes the mathematical positioning of the contents of the map. The combination of vivid painting and square drawing constitutes a unique style of Chinese maps. There are many maps with vivid painting, such as the *scenic map of the Yangtze River*, which depicts the scenic spots, cultural relics and other scenic spots on both sides of the Yangtze River by means of an image. This map is a famous tourist map, with the Yangtze River as the central line, and the map is harmonious in color, comely and elegant. It is a fine work of art. There are also maps that use the same drawing method, such as the *Nine Frontiers Map* (Jiubian tu) of the Ming Dynasty and the (Chengde) *Mountain Resort*, *Map of the Yellow River* (Huanghe tu) and *Mount Wutai Scenic Map* of the Qing Dynasty. The mapping method of "Drawing Square Grid with Chinese Unit of Distance 'Li'" is the specific practice of "Six Principles of Geographic Description and Map Making" by PEI Xiu, and the main works include *Yu Ji Tu* and *Hua Yi Tu*. In "The mapping of ancient Chinese maps" (Ding 2001), the methods and characteristics of China's map making before the Qing Dynasty and the status in the world were introduced.

3. The development of the legend symbols of ancient maps

The use of symbols on maps has a long history. Map symbols were used in the maps from Fangmatan of Tianshui and the maps from the Han tombs from Mawangdui of Changsha. Gao (1962, 1963) thought that maps first emerged in the stage of confusion of pictures and words, when hieroglyphic symbols were used on maps to represent the phenomena and contents of the ground, such as mountains, rivers, villages, etc. The appearance of legend symbols on the map was later than the general content. However, as the map contents became more detailed, words were not enough to satisfy needs, and thus, legend symbols appeared. Therefore, legend symbols were a signal that people created "cartographic generalization". It is a great leap forward in the history of map development. The use of map symbols in the Ming and Qing Dynasties was studied in the above articles. Usually, cartographers believed that the first use of map symbols was in *Guang Yu Tu* (Enlarged Terrestrial Atlas), a map compiled by LUO Hongxian. However, through the analysis of "*Guang Yu Jiang Li Tu*" (a Mongol map of China drawn by a Buddhist monk named QING jun) in the *Shui Dong Diary* (a book written by YE Sheng in the Ming Dynasty), Zhang (1992) believed that the use of map symbols in China should have been mature in the early Ming Dynasty. This concept formed 80 years earlier than the common concept of Guang Yu Tu.

4. Determination of direction and dimension on ancient maps

In the paper "The inspiration from ancient Chinese maps" (Li 1997), it was believed that the different orientations used in Chinese maps were the orientation view of Chinese cartographic craftsmen for the purpose of their use. For example, traditional Chinese coastal maps typically have two visual directions, rather than the geographical coordinates commonly used in modern cartography. The ancient map of China is usually practical and seldom has decorative patterns. Although the type, style, pattern and content of Chinese maps are very different from western maps, they reflect the unique characteristics of Chinese culture. In the paper "Directions on ancient maps" (Yu 2003), the directions of ancient maps are discussed. It is believed that the directions of most ancient Chinese maps are "top south and bottom north", which are determined for two reasons: one reason is the domination and influence of the kingship, and the other reason is practicality. The determination of the direction of Chinese ancient maps as "top north and bottom south" is based on 4 main reasons: one reason is the great success of ZHANG Heng (AD 78–139) in the theory of "Hun Tian" (theory of sphere-heavens) as a starting point, which can be seen from "*Shi Jing Tu*" (marketplace map) and other urban maps of the Eastern Han Dynasty; the second reason is the "Six Principles of Geographic Description and Map Making" by PEI Xiu (AD 223 to 271) of the Western Jin, which can be seen as a starting point, although the 18 "Regional Maps of Yu Gong" produced by PEI Xiu and the *Hai Nei Hua Yi Map* (Jia Dan's Map) produced by Jia Dan based on PEI Xiu's six principles have been lost; however, the stone "*Hua Yi Tu*" of the Song Dynasty was based on Jia Dan's Map of the Tang Dynasty, which has directions of upper north and lower south; the third reason is the western mapping method introduced by Matteo Ricci in the late Ming Dynasty, which is also a starting point, and the direction of "*Knuyu Wanguo Quantu*" (Great Universal Geographic Map) was "up north, down south"; and finally, the fourth reason is the "*Huang Yu Quan Lan Tu*" (Map of China in the Kangxi Reign) measured during the period from 1654 to 1772, which can also be regarded as a starting point, and the direction of the map is "up north, down south".

Through the above four aspects, the map-making technology in China is improving, and the elements on maps have become increasingly accurate and more abstract. All these results indicate the continuous improvement of the spatial cognition level of human beings, especially the appearance of map legends, which shows the high generalization and abstraction of ancient Chinese people in the classification of object space.

17.3.5 The Study of Representative Ancient Maps

Due to the attention to unearthed cultural relics, the study of unearthed maps is quite fruitful. For example, silk maps of the Western Han Dynasty were unearthed at Mawangdui tumulus in Changsha, Hunan Province in 1973; Zhao yu tu (Graveyard Plan) of the Chinese Warring States period was found at the grave of King Zhongshan in Pingshan County of Hebei Province in 1978; and in 1986, seven ancient Chinese maps were found in an archeological excavation of a Qin State tomb in Fangmatan, Tianshui City, Gansu Province. A large number of papers have been published on these maps. For example, in the paper "The arrangement of maps unearthed at the 3rd Mawangdui Han Tomb in Changsha, Hunan" (Working Group of Silk Books from a Tomb at Mawangdui 1975), the arrangement of the first topographic map was described in detail, and the content, scale and so on of this topographic map were analyzed. Although the map of the early Han Dynasty is more than two thousand years old, except for the boundary line, soil, and vegetation, the map contents are as detailed as the basic elements of modern topographic maps, including landscape, drainage, residential areas, and transportation lines. Therefore, this map is a rather rich topographic map of ancient times. In view of the elements of the map, the drawing technique has also reached a level of proficiency. For example, the coarse–fine control and the natural curve of the river are vividly represented. There is no inverted flow in the river mouth, which is usually an error in drawing. The road was drawn with a single stroke; there was no sign of a joint wire. The roundness of the circular symbols for depicting residential areas is good, which shows that the drawing technique was at a high level.

Jin (1983) discussed in detail the two maps found in the 3rd Han tomb in Mawangdui, Changsha, i.e., "*Topographic map*" and "*Military map*". These two maps are the oldest maps ever found in the world. They are rich in content and represent the four major elements of the water system, geomorphology, residential areas and roads. The maps shocked the world cartographic community because of its use of scale and three colors. The two silk maps were produced by a military conflict between two southern vassal states in Changsha and Guangdong in the early Western Han Dynasty. The mathematics and astronomy in the Western Han and Eastern Han Dynasties reached a high level, and the Pythagorean Theorem and Cosine were applied in the actual measurements. Goniometer "Si Nan" (prototype of the compass), rangefinder "moment" and other measuring tools were developed and widely used to measure the heights and lengths of islands, the areas of town-settlements, the widths of rivers, the heights of mountains and the depths of valleys. The historical background and scientific and technical conditions of the two maps of the two Han dynasties were elaborated.

Zhang (1986) pointed out that the command center in the *Military Map* was located on the East Bank of the Anning River estuary, and the main area was equivalent to today's Tuo Shui watershed area of Jiang Hua Ma city in the Hunan Province, with an area of approximately 900 km^2 and a scale of 1:45,000 to 1:50,000. The military map is drawn based on field surveys, and the map is divided into the drawing area, the sketch area and the decoration area, depending on the military needs and the structure of the layout; for the drawing characteristics, the design core of the sheet is first determined, and then, the key areas are highlighted, taking care of key parts as much as possible and concentrating on the relevant elements. Finally, local adjustments are made to the main area. Wu (1990) examined the age of painting of Mawangdui's map and found that the range of the silk map spans three counties of the Qin Dynasty, including Changsha, Guilin, and Nanhai. On the map, the front-line forces of the drainage garrison map are deployed on the north side of the ridges along the border of South Vietnam. Under these circumstances, he judged that the map is only for Tu Wei's expedition to South Vietnam, and the upper limit of the mapping time should be 221 B.C. From the content of the map, the northern Xiaoshui River basin, Nanling Mountains, Jiuyi Mountains and its vicinity are detailed, while the roads, residential areas and drainages of adjacent areas are relatively poor; various elements of the South Vietnamese region were sketchy and did not consider the integrity of the administrative district. From these analyses, it is estimated that in the early stages of the expedition, the map was drawn based on the field survey and investigation data and in terms of the route of the ship's army. The lower limit for the mapping time is a minimum of thirty-three years for Emperor Qin Shi Huang (214 BC) before the Qin Dynasty conquered South Vietnam and Xi Zhen and established the three counties of Guilin, Xiang and Nanhai.

Wang (1992) made a comparative analysis of the relevant traffic elements described in the *Military Map* and Topographic Map, in contrast with the relevant traffic elements described in the history books of the Han Dynasty. Through further analysis of the Chinese map, the content and mapping time of the *Military Map* and *Topographic Map* of Mawangdui, Zhou (1993) believed that the two maps comprised the map of the early Han Dynasty, instead of the River map of the Qin Dynasty. Cao (1994b) further explained the annotations of residential areas on the *Military Map* and the *Topographic Map*. He believed that in these two maps, the residential areas below the county and township were generally noted as "Li", such as "Bo Li"; beyond that, "Jun" and "Bu" were the names of the administrative divisions of townships or were equivalent to the level of "Li", which was set up in ethnic minority areas during the Han Dynasty and was nationally autonomous.

In 1986, seven ancient Chinese maps were found in an archeological excavation of a Qin State tomb in Fangmatan, Tianshui city, Gansu Province. For these maps, there are also some relevant studies. For example, Cao (1989) analyzed the maps unearthed from the Qin tomb in Fangmatan, Tianshui from four aspects, including the annotations, the relationship between the maps, the area drawn by the map and the mapping technology. Zhang (1991) concluded that the specific drawing time of the No. 1 and No. 2 maps of Fangmatan were approximately 320 B.C., the first year to seventh year of Houyuan of King Huiwen, 150 years earlier than the map of Mawangdui, Changsha. Wu (2004) appraised Yong Jichun's book "*Study of the Maps on Wooden Blocks Unearthed from Fangmatan in Tianshui*" and believed that the book had set the date for the plotting of the map of Fangmatan, which dated from 323 to 310 B.C. and has a history of 2,300 years.

All these studies have enriched the research of ancient map entities and verified the correctness of the map application records in Chinese historical works. In addition to the unearthed ancient map, stone-engraved maps, painted maps and block-printed maps were also studied. For example, Lu (1982) discussed the change in the material base of ancient Chinese maps, from pottery in the later period of primitive society to a metal matrix in approximately 2000 B.C., then the silk base in the Han Dynasty, the paper map in the Sui Dynasty, and of course, there were some stone-engraving maps or wood-engraved maps. From the change in the map material base, we also observed the development of science and technology at that time. Wang (1989) introduced in detail the treasures of ancient Chinese maps preserved in Beilin of Xi'an city: "*Yu Ji Tu*" and "*Hua Yi Tu*" of the Song Dynasty, which are the earliest extant stone-engraving maps in China. The two maps are engraved together on the front and back of a stone tablet. The front is *Yu Ji Tu*, and the back is *Hua Yi Tu*. Zhu (1990) introduced *New Zheng He's Nautical Chart* from the aspects of the content structure and main features. Through the practice of atlas compilation, three suggestions were put forward for the study of ancient maps in China: first, it was necessary to delve into ancient Chinese maps because it was an important manifestation of Chinese culture; second, the study of ancient maps had to be carried out with a variety of disciplines, which required the collaboration of natural science and social science workers; third, the map academic community needed to establish an academic group on the history of maps and promote the development of the subject. Cao (1994a) believed that "*Yan Zhou Tu Jing*" (Book of Map and Record of Yanzhou) of the Southern Song dynasty is the earliest existing map book. Although there are only three volumes left, the accompanying map is in the frontispiece, which is a valuable historical material used to explore the relationship between scripture and maps. According to the description of these three volumes, compared with the accompanying map, it can be concluded that the text is not the description of the map but the situation of the area painted on the map. Li (1995) studied "*Huang Yu Quan Lan Tu*" (Map of China in the Emperor Kangxi Reign of the Qing Dynasty) from 3 aspects. First, the appellation of "*Huang Yu Quan Lan Tu*", which was believed to have different appellations for historical reasons, such as "*Huangyu Quantu*" (Complete Imperial Map), "*Qianlong Neifu Yutu*" (Emperor Qianlong Imperial Map), "*Shi San Pai Tu*" (Thirteen rows map, a row equals five degrees of latitude), "*Qing Neifu Yitong Mitu*" (Secret comprehensive geographic map of the Great Qing Dynasty), and "*Man Han He Bi Qing Neifu Yitong Mitu*" (Secret comprehensive geographic map of the Great Qing Dynasty combining Manchu with Han), all refer to one map, namely, "*Huang Yu Quan Lan Tu*". The second aspect is to determine the assumed time for the measurement of the map, that is, tests were conducted near the capital in 1684 and the end of 1707, and a nationwide large-scale survey began in 1708. In the end, it was confirmed that the people responsible for mapping were western missionaries, mainly French. In the paper "CHEN Zushou and his *Huangming Zhifang Ditu*", the life of the Chen group and the significance of the *Huangming Zhifang Ditu* (Territory map of the Ming Dynasty) were introduced. It is a comprehensive atlas compiled by CHEN Zushou and other people, which shows the development process and technical level of mapping in the Ming Dynasty and has important reference value for the study of politics, economy and military in the Ming Dynasty. Yang (2006) focused on the studied of ZHENG He's nautical charts at home and abroad, especially the latest research results and summarized the generation, characteristics and academic value of the book "*ZHENG He's Nautical Charts*".

17.3.6 The Study of Ancient Cartographers

For ancient Chinese map makers, additional studies include PEI Xiu, JIA Dan, SHEN Kuo, ZHU Siben and LUO Hongxian. In addition, there have also been in-depth studies on these people, including JING Xiangfan, who assisted PEI Xiu to produce maps, ZHENG Ruoceng, author of "*Wan Li Hai Fang Tu*" (Thousands of miles of coastal defense map) in the Ming Dynasty, Mongolian scholar MING Antu, who was responsible for surveying and mapping western Xinjiang for Emperor Qianlong in the Qing Dynasty, as well as ZOU Boqi, YANG Shoujing, ZOU Daidiao and others in the late Qing Dynasty and early Ming Dynasty, all of which contributed to map making. In the paper "Historical honor of ancient Chinese scientists in the history of surveying and mapping" (Chen 1995), first, the "Zhi Tu Liu Ti" (Six Principles of Geographic Description and Map Making) of PEI Xiu, an ancient cartographer in China, and the measurement plan of the nation's latitude launched by GUO Shoujing in the Yuan Dynasty were introduced; second, the application and significance of the four tools of compass, gauge and moment, paper and printing plate were introduced. Because of the invention of the compass, SHEN Kuo discovered magnetic declination and suggested that the azimuth be divided into 24 directions. Because of the gauge and moment, the gear for the "southward pointing cart" (A.D. 110) and "li-drum-hodometer" (around AD 400) was invented. These two kinds of cars were used to locate and measure distances. Because of the invention of paper and printing plates, the reproduction of maps became easy and widely applicable. Finally, the applications of mathematical results in measurement were introduced, such as "Zhou Bi Suan Jing" (the Arithmetic Classic of the Gnomon and Circular Paths of Heaven). In the paper "PEI Xiu, an outstanding cartographic scientist in ancient China" (Mao 1984), the life, achievements and main reasons for his achievements were discussed. In the first part, a brief introduction to Pei's life is given; it is pointed out that because PEI Xiu was in the imperial court as a "Sikong" (a prime minister) and responsible for managing state maps and household registration, this created the conditions for PEI to show his talent in the field of cartography. In the second part, PEI's contribution to the history of cartography was introduced: first, he created "Zhi Tu Liu Ti" (six principles of cartography) based on his predecessors' experience, which was the earliest cartographic guideline in the world; second, he compiled "Di Xing Fang Zhang Tu" (terrain map in one square zhang) based on the old "Tian Xia Da Tu" (a world large map), and in the process, he actually examined the six principles, and in the preface of 18 articles of "Yugong Diyu map" (an ancient historical atlas), he clearly explained the six principles and the relationships between the principles. In the third part, the historical reasons for the success of PEI Xiu, i.e., the progress of society, the experiences of predecessors and the emphasis on field investigation and research were analyzed. In "The outstanding achievement of quantitative cartography in the Song Dynasty represented by Shen Kuo" (Yu 1987), it was believed that SHEN Kuo was the first cartographer in China with the viewpoint of a spherical Earth surface, and this was evidenced by historical data. According to his own view of a spherical Earth, he first created a new kind of amplitude coordinate to reflect a spherical surface map on a plane. The author also noted that maps and compasses were combined by SHEN Kuo to indicate geographical location according to the twenty-four directions of magnetic azimuth. This had profound significance in the history of quantitative mapping in the world. In "The cartographer JIA Dan of the Tang Dynasty and his achievements in cartography" (Zhu 1999b), it was believed that JIA inherited and developed the cartographic theory of PEI Xiu, compiled a number of maps and books, and made outstanding contributions to the development of cartography. JIA's life and his works were introduced in detail. In "ZHU Siben—the founder of reaching the peak of China's traditional cartographic achievement" (Zhu 1999c), ZHU Siben, a distinguished geographer and cartographer of the Yuan Dynasty, was introduced.

"*Yu Ditu*" (a territory map) was plotted successfully, and the subsequent maps that originated from his cartographic method and were revised on the basis of his map formed ZHU's cartographic system and was representative of China's cartographic achievement during the period. In "On Taoist master ZHU Siben's cartographic science thought" (Yi 2008), based on the study of the cartographic science thoughts of Taoist master ZHU Siben in the Yuan Dynasty, the author pointed out that there were five features and innovations that resulted from his methods of geographic scientific thinking: paying attention to on-the-spot observations and field surveys; emphasizing the way of combining geographic literature knowledge with field investigations; complying with the scientific standard and strict attitude of "putting quality before quantity"; for the method of map drawing, inheriting the traditional Chinese "Ji Li Hua Fang" (Drawing Square Grid with Chinese Unit of Distance 'Li') and skillfully using the national map mapping method of "Drawing your maps anywhere, and then merging them into a single whole"; and in terms of the map legend, developing a geometric symbol notation with simple, clear and systematic notation.

17.3.7 The Study of Other Aspects

The study of the history of cartography also includes some other aspects, such as the changes in place names on maps, the relationship between maps and culture, and comparative studies of ancient Chinese and western maps. In "A study of ancient place names from the history of Chinese cartography" (Niu 1986), the relationship between place names and the history of cartography was studied from three aspects. In the "Comparative study on the history of Chinese and western cartography" (Lv 1994), a chronology of the history of Chinese and western cartography was put forward. Before the 17th century, Chinese cartography was influenced by Japan, north Korea and other countries, and since then, it has been greatly influenced by the western world. The migration of cartographic centers in China was consistent with the migration of political centers, while the west was subject to religious influence and economic development. Before the exchanges between China and the west, each area developed in its own unique way, forming two different cartographic systems. As for the history of cartography, the study of place names was divided into three stages, namely, ancient, modern and contemporary. The ancient stage was divided into two parts: one part was from the remote antiquity to the Western Jin Dynasty, when there were various kinds of maps, but the cartographic theory was not mature, and the study of geographical names was mainly aimed at the transition from the mythological narrative to the scientific record; and the other part was from the Western Jin to the late Ming Dynasty, when PEI Xiu's "six principles of cartography" laid the theoretical foundation of cartography, and the study of place names focused on toponym evolutions due to the appearance of the historical atlas. The modern stage was also divided into two parts: first, from the late Ming Dynasty to the early Qing Dynasty, due to large-scale measurements throughout the country, the study of geographical names was promoted in frontier areas; and second, from the end of the Qing Dynasty to 1950, the compilation of the atlas promoted the continuous emergence of geographical name dictionaries, indicating the maturity of the study of geographical names. The contemporary phase was from the 1950s to present, especially in the 1980s, which was when the establishment of the national geographical name database opened up new fields for the study of geographical names. In "On the relationship between maps and Chinese traditional culture" (Zhu 1999a), it was believed that map culture included map thinking, map language, map art, map technology, map works, map specification and so on. The research of map culture was mainly for understanding map culture, guiding the creation of maps, absorbing the essence of traditional culture, and finally creating map works with Chinese cultural charm. The relationship between maps and traditional culture was discussed in four aspects, including the glimpse of the development of map culture, the relationship between the map and traditional Chinese culture, and the relationship between the map and traditional Chinese cultural elements. Regarding the relationship between the map and traditional Chinese cultural elements, the relationship between the map and philosophy, the relationship between the map and art, the relationship between the map and religion, and the relationship among the map and the military, politics and the economy were also studied. In "Ancient maps and the history of thoughts" (Ge 2002), it was believed that the picture was often influenced by ideas, and even the maps that appeared to be fairly accurate were not always the result of actual measurements. In fact, there were always ideas and imagination in them. Maps drawn by people of different cultural backgrounds, faith backgrounds and political intentions, behind the scope of the space, the arrangement of the position, the size of the scale, and the similarities and differences of colors, there were interesting ideas and imagination that played a dominant role. For example, the imagination embodied in ancient maps, such as "land under heaven", "China, a country in the center of the world" and "Siyi, four barbarian tribes on the borders", the arrangement of "center" and "borders", and the arrangement of the proportions of various geographical regions, was the information from intellectual history.

Since the 1990s, no major archaeological discoveries have been made, so some reviews on the history of cartography emerged. "A review of the history of Chinese cartography in nearly forty years" (Cao 1990) summarized the development characteristics of Chinese cartography and discussed the research results over the past four decades. It is divided into four aspects, namely, book writing, discussion of theories and methods, research on unearthed maps, study of stone engravings, colored drawings and wood engraving maps, research on cartography and compilation of ancient atlases of China, and so on. "The history of Chinese cartography in the past 90 years" (Jiang 1997) discussed the studies by mainland scholars on the history of Chinese cartography in the last nine decades. According to the article, the more frequently discussed topics include the general history of the development of Chinese cartography, the study of individual ancient maps, the theories and methods of cartography, the study of ancient navigation charts, the cataloguing of the history of cartography and the arrangement of historical materials. Most books are more descriptive than analytical, with relatively few theoretical studies; there is hardly any research from the perspective of the world's macro architecture, nor are there any comparative studies.

17.4 Problems and Prospects

In the past 60 years, the research results of Chinese cartographic history have made great progress, but there are still some shortcomings that need further study.

(1) A map is the most direct and vivid graphical representation of the human brain's response to the outside world and should be studied as a tool for human understanding of space.

Maps are a very effective tool for human spatial information transmission. The map is a graphical representation of the human brain's understanding of the external world. The spatial perception of human beings includes the internal/external, central/edge, boundary, division, basic direction, connectivity and other topological relationships (no dimension) in space and includes the formation of the concept of primary classification of spatial objects, for example, the classification of topographic elements including rivers, mountains, roads, settlements, and vegetation and others, and the classification of persons and animals. However, these are the results of modern psychological research, and one of the ways to prove it is to find evidence. Prehistoric paintings were an instinctive description of their living environment by primitive peoples, so their paintings reflected their perception of space. For example, in the *History of World Map* published by the University of Chicago, a large collection of prehistoric rock paintings in Africa, Europe and other places was collected and studied. The work done on this aspect of China is still very scarce, so map studies should be focused on the graphic expression of spatial cognition. Through the depiction of an ancient's environment and the scene of production and life in pictures, the foundation and evolutionary order of human spatial consciousness can be determined.

(2) The map reflects different ethnic cultures and thinking, and a comparative study of ancient Chinese and Western map cultures should be strengthened to understand the essential features of different cultures.

Language and text are a linear means of transmitting information, and only pictures transmit in parallel. This is consistent with human understanding. Nature presents itself to humans as a picture, but in real life, people's perception of the outside world is multidimensional and multidirectional. The information we see will be reflected in the brain at the same time, forming a mental map of our own, and the information is drawn on the map according to our own experience and needs. Therefore, different ethnic groups formed different ancient maps because of their different environments, which were reflected in ancient maps. For example, the map of ancient Chinese farming culture is dominated by the river system "Yugong Ditu" (an ancient historical atlas) and its classification of soil; the navigation channel chart was made of sticks by the primitive inhabitants of the Pacific islands; the world map drawn by the ancient Greeks was based on nautical experience in the Mediterranean, and the description of the wind and zone divisions of human settlements were included on the map. These maps reflect the wisdom of human ancestors in the process of adapting to nature. In addition, human beings have been exploring both the real space and the overhead space, and many star maps have been drawn. In terms of exploring unknown worlds (spiritual space, religious space), the drawings of many of religious maps include rich materials for the study of the origin and development of different cultures in China and the western world.

(3) A series of ancient maps superimposed together is a dynamic reflection of the history of spatial understanding. Therefore, a comprehensive analysis of an ancient map series can strength the spatial understanding of humans.

At present, the focus of the study of cartographic history is to collect and compile all kinds of ancient maps and atlases or to arrange the outline or catalogue of ancient maps; most of them remain in the descriptive stage, and less analytical work has been done. For example, there have been less comprehensive analyses on the representation of symbols on these maps, such as the Fangmatan map of the Qin Dynasty, the silk map of the Han Dynasty, and Huayi Tu and Yuji Tu of the Tang and Song dynasties. There are few papers on the mathematical basis of ancient Chinese maps, including descriptive annotations, symbol system, rectangular mesh, PEI's six principles of cartography and SHEN Kuo's twenty-four directions, which is mainly because most of the experts currently studying the history of cartography are archaeologists and historians, but few cartographic experts specialize in this field. As people pay increasing attention to map culture, these problems will eventually be improved.

(4) A precise understanding of space requires technological advancements, comparative studies on the development of the measured topographic maps in each country should be strengthened, and the interrelationship between technological development and map development should be explored.

The measured topographic map is a perfect combination of surveying science and mapping art. The rise of measured topographic maps in the latter half of the fifteenth century mainly resulted from two reasons: first, the external expansion of emerging capitalist countries required an accurate understanding of the world; and second, various advanced measuring instruments were improved. These two factors later became the main driving forces of the development of cartography. The measured topographic map completely changed the method of mapping in the past. Instead of being completed by individual manual operations, specialized agencies were required to obtain measurements and drawings. Therefore, since the 15th century, corresponding measured topographic map organizations were established in each country, special personnel were organized for mapping, and the measurement level and presentation method of topographic maps was continuously improved. For example, the accuracy of map data was increasing, the unification of map projection and scale standardized the management of map data, and the cartographic symbols transitioned from pictographs to geometrical symbols, and a modern topographic cartographic system was formed. In the process, each country formed its own standard system according to its actual situation. By studying the historical background and applicability of these systems, the law and process of map development were discovered.

(5) Technology promotion and driving demand are two major factors of cartography development, and the study on the development law of cartography in the information age should be strengthened.

The cartography in the information age is quite different from that in the traditional era, which underwent major changes from data sources and map production to map making and map distribution. Map workers in the information age were only providers of data, and these data were derived from field measurements, but today, the data are mostly derived from satellite remote sensing systems. Map data are released to the Internet as a kind of information. Map users make their own maps according to their own needs. Map updates are no longer a special right for map data providers, and a large number of volunteers have made real-time updates to the maps on the Internet after field trips. The production and use of maps seems to be back to the era when everyone could make maps according to their own needs. Where will the map be developed? This is an important issue that needs to be studied in the future.

17.5 Representative Publications

(1) ***The story of maps*** (CHEN Shupeng. China Youth Press, Beijing, May 1955)

The book tells the story of the map's making and development. It answers the following questions: how did the ancient map keep improving with the development of society? What influence did the development of surveying and mapping science have on the map? What changes have taken place in the outline of the world map after navigation, exploration, and geographical discovery? The book also informs us about the creation and contribution of ancient Chinese maps, surveying and mapping of modern Chinese territory, and the important role that maps play in future socialist construction. This book is divided into 8 topics, in which 6 topics are related to the emergence and development of maps. The first topic is the map; the second topic is the germination of maps; the third topic is the foundation of modern cartographic science; the sixth topic is the scientific creation of maps in ancient China; the seventh topic is the development of modern Chinese maps; and the eighth topic is thematic maps that serve the socialist construction of the motherland.

(2) ***Chinese history of geography*** (WANG Yong. Commercial Press, Beijing, November 1956)

The book is divided into three chapters. The first chapter is the original geographic map and its change and historical development; the second chapter is the history of the map; and the third chapter is the history of chorography. Among them, the first chapter is divided into 5 sections: the first section is "Shan Hai Jing" (the Classic of Mountains and Rivers) and "Yugong" (the tribute to Yu), with Mu Tianzi Zhuan (Biography of King Mu); the second section is "Jiuding Tu" and the map of "Shan Hai Jing"; the third section is "Zhigong Tu" (Portraits of Periodical Offering) and Miao Yao Feng Su Tu (Miao Yao customs map); the fourth section is the Map for Foreign Countries; and the fifth section is the topographic map and terrain model. The second chapter is divided into 12 sections; the first section is doubtable maps; the second section is about the management, production and delivery of territory maps; the third section is the territory maps of the Qin and Han dynasties; the fourth section introduces the map making by Pei Xiu; the fifth section is about "Shidao Tu" (the map of ten administrative units of the Tang Dynasty); the sixth section introduces JIA Dan's maps, including "Longyou Shannan Tu" and "Hainei Huayi Tu"; the seventh section is the map of chorography and other general drawing; the eighth section is the general territory map of the Song Dynasty; the ninth section is the border map of the Song Dynasty; the tenth section describes ZHU Siben's map and its influence; the eleventh section explains the Ricci's Map of the World and its influence; and the twelfth section discusses map making in the early Qing Dynasty and its influence. The third chapter is divided into 3 sections. The first section is the development of chorography from the Han Dynasty to the Sui Dynasty; the second section is a general topographic map of the Tang and Song Dynasties; and the third section is the development of local chronicles in the Song Dynasty. The book is accompanied by 4 maps, including the Huayi Tu, Yuji Tu, and Wang zhiyuan geographical maps (stone-engraving maps by Suzhou prefectural school) and the Garrison Map of Shaanxi Province painted in the Ming Dynasty (one of the residual pages, Beijing library collection).

(3) ***Textual Research about Chinese geographic maps and books*** (WANG Yong. Commercial Press, Beijing, October 1957)

book consists of two parts. Part I is divided into 3 chapters, and they aim at the textual research and collection of Chinese geographic maps and books. Part I focuses the maps of the Ming dynasty, including the textual research of the general topographic map of the Ming Dynasty, the collection of northern border defense maps of the Ming Dynasty, and the collection of coastal defense maps of the Ming Dynasty. Part II is divided into 13 sections, which supplement the history of Chinese geography. They are Beiji Chu Gaodi (the Arctic and its uplands), south Vietnamese tzu, Xinding Jiuyu Zhi (the national chorography during the reign of Yuanfong, newly compiled), Jiu Zhu and Jiu Zhou (poetic names for China), Shanjing (part of Shan Hai Jing), mapping methods, map of the Qin Dynasty, topographic maps, map of the Western Regions in the Han Dynasty, Guo Zhao Tu and Xuzhi Tu Jing (taking the map as a mirror), Sui Shanchuan Tu (mountains and rivers of the Sui Dynasty), map coloring, and Shui Jing (part of Shan Hai Jing). It also discusses the relationship between historical maps and military affairs in China and historical land boundary maps in China.

(4) ***Outline of Chinese map history*** (WANG Yong. Joint Publishing, Beijing, May 1958)

The book consists of 11 chapters. The first chapter is about the original map of China and its qualitative changes; the second chapter introduces the ancient map of China and its military and political functions; the third chapter deals with Pei Xiu's cartography and its status in the history of Chinese maps; the fourth chapter is about the maps of the city landscape, the maps of states and counties, and their qualitative changes and collection; the fifth chapter introduces map making, production and "Shidao Tu" (the map of ten administrative units in the Tang Dynasty); the sixth chapter is the map of local chronicles and Jia Dan's map; the seventh chapter is about "Shiba Lu Tu" (map of the eighteen administrative units) and the border map; the eighth chapter discusses the influence of Zhu Siben's map (Yu Ditu) and Luo Hongxian's map (Guangyu Tu); the ninth chapter introduces latitude measurements and Ricci's Map of the World; the tenth chapter investigates the creation of the first entirely national map; and the eleventh chapter describes mapping in modern China. The book also includes the rubbings of 10 pieces of ancient maps, such as Huayi Tu, Yuji Tu, Dili Tu (Geographical map), Guangyu Tu (Enlarged Terrestrial Atlas), Huang Ming Zhi Fang Di Tu (Territory map of the Ming Dynasty), Knuyu Wanguo Quantu (Great Universal Geographic Map), "Kangxi Neifu Yutu" (Emperor Kangxi Imperial Map), "Qianlong Neifu Yutu" (Emperor Qianlong Imperial Map), Daqing Yitong Yutu part I (Atlas of Qing Dynasty), and Daqing Yitong Yutu part II (Atlas of Qing Dynasty).

(5) ***History of Chinese cartography*** (LU Liangzhi. Surveying and Mapping Press, Beijing, February 1984)

The book consists of 13 chapters. The first chapter introduces primitive maps and their applications in the pre-Qin period, including the origination of the map and the development and changes in the original map; the second chapter describes the prosperity of military maps during the Qin and Han Dynasties, including maps of the Qin, the western region maps of the Han, and the drawing of military maps; the third chapter discusses the maps during the Wei and Jin Dynasties and Pei Xiu's achievements in cartography, including map making in the Three Kingdoms period, Pei Xiu's "six principles of cartography" and its contribution in cartography, as well as the havoc caused by the map during the Northern and Southern Dynasties (420–589); the fourth chapter is about the prosperity of records and maps in the Sui and Tang Dynasties, including the integration of the administrative district map, record of the Sui dynasty, the compilation of the map, record of the Western Regions of China, *Dongdu Tuji* (Geographic map of Luoyang in Song Dyn), and *Yuanhe Junxian Tuzhi* (General Geography of the Yuanhe Reign). The fifth chapter is the revival and development of Pei Xiu's mapping in the Tang dynasty, including Jia Dan's contribution to cartography, *Shidao Tu* (the map of ten administrative units of the Tang Dynasty) and *Jiuzhou Shexian Tu* (a map of the success and failure of ancient and modern military operations all over the country), as well as the integration of cartography and astronomy; the sixth chapter is about the development of the map in the Song Dynasty and the advent of the woodblock printing map, including drawing of the nationwide general map, drawing of *Shiba Lu Tu* (the map of eighteen administrative units of the country), drawing of the outland map and borderland defense map, and the application of the woodblock printing map; the seventh chapter is about the development of the peak of traditional cartography in the Yuan and Ming dynasties, including the three pillars of Zhu Siben's cartographic system, other maps in Zhu Siben's cartographic system, and other cartographic systems, as well as the making of the military defense map in the Ming dynasty; the eighth chapter introduces the achievements of measurement technology, including the technical means of measuring terrain and ground features, the mathematical basis of measurement, and the practical application of measurement technology; the ninth chapter is about the three-dimensional terrain model and the urban map, including the stereoscopic terrain model map, the globe, and the urban map; the tenth chapter presents the river system map and navigation chart, including *Yuji Tu* and other water system maps, the drawing of *Huanghe tu* (the map of the Yellow River painted in the Kangxi period of the Qing Dynasty), and *Zheng He's Nautical Charts*; the eleventh chapter explains the input of the Western cartography at the end of the Ming dynasty and the surveying and mapping of the national map in the early Qing dynasty, including the spread of Ricci's maps in China, China's first measured provincial atlas, *Huangyu Quanlan Tu* (Map of China in the Kangxi Reign), and *Qianlong Neifu Yutu* (Emperor Qianlong Imperial Map) marked the final completion of the national measured longitude and latitude map; the twelfth chapter introduces the compilation of the atlas at the end of the Qing dynasty, including *Daqing Huidian Yutu* (map of the Qing Code in the Guang-xu period of the Qing dynasty), China's first World Atlas with new techniques—*Haiguo Tuzhi* (Records and Maps of the World), and all kinds of historical research for the atlas; the thirteenth chapter describes the management, production and delivery of maps, including management of maps before the Sui dynasty, management, production and delivery of maps in the Tang and Song dynasties, and management, production and delivery of maps in the Yuan, Ming and Qing Dynasties.

(6) ***Historical narrative of the Chinese map*** (JIN Yingchun and QIU Fuke. Science Press, Beijing, April 1984)

The book consists of 8 chapters. Chapter one is the legend of ancient maps, including "making paintings of Shi huang", the story of He-bo (River God), *Jiuding Tu* and *Shanhai Jing Tu* (maps of the Classic Mountains and Rivers); chapter two is about the extensive use of maps in the pre-Qin dynasty, including the map for building Luoyang Town by the Duke of Zhou and Zhao, all kinds of pre-Qin maps contained in *Zhou Song* (Ode to Zhou) and *Zhou Li* (Rites of Zhou), the system of enfeoffment and land maps, Jing Ke's presentation of a map to the King of Qin and the origin of the "territory map", the nationwide map—the emergence of "*Jiuzhou Tu*", application of the military map, and the earliest water conservancy projects and mapping; chapter three introduces the rich and colorful geographic maps of the Qin and Han dynasties, including XIAO He's collection and storage of Qin maps and inheritance of the Qin's political system in the Han dynasty, geographic map of the Han dynasty and its application, the amazing discovery of the Han tomb in Mawangdui, the world's earliest military map, the detailed and accurate "*topographic map*" at the beginning of the Han dynasty, the anti-separatism war in the early Han Dynasty and the Garrison Map, the oldest extant town map—"*Chengyi Tu*", and the growth of map surveying technology in the Han dynasty; chapter four is about the creation of the theory of cartography in the Wei and Jin dynasties, including Pei Xiu, a Cartographer of the Western Jin dynasty and his "six principles of cartography", the Yukung map, "*Di Xing Fang Zhang Tu*" (terrain map in one square zhang), and the earliest map models and their evolution; chapter five deals with the cartography achievements in the Tang and Song dynasties, including Jia Dan's contribution to cartography, the meridian measurements by Monk Yixing, the extant stone-engraving maps of the Tang and Song dynasties, and Shen Kuo's contribution to map making; chapter six is about the maps of the Yuan and Ming dynasties and measurements of latitude and longitude, including Zhu siben and his "Yudi Tu", Luo Hongxian and his "*Guangyu Tu*", Guo Shoujing and the concept of elevation, and the introduction of Matteo Ricci's map of the world, as well as Zheng He and Zheng He's Nautical Charts; chapter seven discusses the beginning of the measured map in the Qing dynasty, including "*Huangyu Quantu*" (Complete Imperial Map), the national map presided over by Emperor Kangxi, "*Xiyu Tuzhi*" (maps and records of the Western regions of the Qing dynasty), "*Daqing Yitong Yutu*" (Atlas of the Qing Dynasty) presided over by Emperor Qianlong, the great achievements of surveying and mapping in the early Qing dynasty, introduction to the famous surveyors He Guozong and Ming Antu, "*Haiguo Tuzhi*" (Records and Maps of the World) compiled by Wei Yuan, which is the first world atlas compiled by China, and Yang Shoujing and "*Lidai Yudi Tu*" (geographical maps of through ages); chapter eight is about the surveying and mapping of modern China, including the mapping of basic topographic maps, the development of photogrammetry in China, atlas compilation, and the evolution of China's surveying and mapping institutions.

(7) ***Study of the modern historic chart*** (WANG Jiajun. Surveying and Mapping Press, Beijing, March 1992)

The book consists of 5 chapters. The first chapter is the introduction to the modern historical chart research, including the purpose and nature of modern historical chart study, the contents and methods of modern historical chart study, and the review and prospect of modern historical chart study; the second chapter is the textual research of the modern historical chart, including *Shengjing Qisheng Yangtu* (marine map of seven provinces around Shengjing (now Shenyang)), *Qisheng Yanhai Qauntu* (the whole coastal map of seven provinces around Shengjing), the English version of navigation charts for Chinese seas in the late Qing dynasty, Basheng Yanhai Quantu (a map for the entire coastal area of eight provinces), Yulan Jiangzhemin Yanhai Tu (coastal map of the Jiangsu, Zhejiang and Fujian provinces for imperial reading), navigation charts from the Republic of China period, navigation charts from the early days of the People's Republic of China, and the modern historical charts from various historical stages; the third chapter is the mathematical basis for the modern chart transformation, including plane coordinate transformation, conversion of depth datum, elevation datum plane transformation, map projection transformation, and scale and measurement unit conversion; the fourth chapter is about the compilation of the modern historical marine atlas, including the overall design of the modern historical marine atlas, the technical regulation of the modern historical marine atlas, atlas of the watercourse of the *Yangtze River estuary in past years* (south branch channel), and an overview of the *modern historical marine atlas* (period of the Republic of China). The fifth chapter explores the engineering application of the modern historical chart, including technology preparation for engineering applications of historical charts and engineering applications of modern historical charts.

(8) ***Collection of precious ancient Chinese maps*** (YAN Ping, SUN Guoqing et al. Xi'an Cartographic Publishing House, Xi'an, July 1995)

The book introduces more than 100 precious ancient maps. It presents ancient maps based on three main concepts: First, it reflects the universal law of human thought, that is, hypothesis and speculation, the historical record and gradual appearance of physical examination, formation of the production specification, and the rise of the scientific principle; second, the book reflects the evolution of map works, starting with the original artistic conception map, the measured plane map using rulers, moments and the Pythagorean theorem, and the development of a 3-dimensional map controlled by latitude and longitude; and third, the book reflects the expansion of map application areas. The historical stage of the development of cartographic theory is verified based on physical objects, and then, the map works are evaluated according to theoretical development law.

(9) ***Map surveying and drawing in ancient China*** (GE Jianxiong. Commercial Press, Beijing, December 1998)

There are seven chapters in the book that describe the surveying and mapping of ancient Chinese maps. These are amazing discoveries from the Han tomb in Mawangdui and the Qin tomb in Fangmatan; the original maps and early stage maps are included, as are the maps from the Han dynasty to the Northern and Southern dynasties, the maps from the Sui and Tang dynasties to the Song dynasty, the maps of the Yuan and Ming dynasties, the maps of the Qing dynasty, and what we can learn from the history of cartography in ancient China. This book reflects the mainstream characteristics of Chinese culture from multiple angles and layers. Through this book, readers can understand the basic features of Chinese culture and understand the spirit of the Chinese nation.

(10) ***History of Chinese cartography*** (written by YU Dingguo, translated by JIANG Daozhang. Peking University Press, Beijing, August 2006)

The book is an important part of the *History of World Cartography*, published by the University of Chicago Press. It is divided into 5 chapters. Chapter 1 is the reinterpretation of the traditional geographical map, including whether surveying and mapping in ancient China is considered to be mathematical cartography, and the application and misuse of the history of cartography, including the defects of quantitative methods and the revision of traditional mapping concepts in China. Chapter 2 discusses the Chinese maps of political culture, including maps, rituals, wars, political culture and textual research, maps of the political culture in the Han dynasty, the continuity of map drawing and applications in the Qin and Han dynasties, astrology and astronomy in the political culture, the large increase in the geographic map and book, chorography maps, and the inheritance of maps, academics and cultures. Chapter 3 discusses the measurement of the Earth: Chinese maps between observations and words, including the government's emphasis on measurement, water conservancy and cartography, textual research and cartography, maps, measurements and text descriptions, the digits and texts in Pei Xiu's cartography, the digits and texts in subsequent cartography, and the shape of the Earth, including actual observations and textual research and the map coordinate grid. Chapter 4 is about Chinese cartography in the humanities, including objectivity, subjectivity and exhibition, the relationships among art, literature, maps and the representation of the real world, the dual functions of expression in literature, painting, art, and economy, the common production processing technology, the relationship between cartography and the visual arts in terms of concept and style, the map as a picture and the picture as a map, redefining a map, and combining fact and value. Chapter 5 addresses the problem of traditional Chinese cartography and its westernization, including the introduction of European cartography, European cartography and the mapping of the Qing dynasty, assessment of the extent of western influence and the impact of Europe on the late Qing dynasty. In addition, there are 15 color maps and 131 black and white maps in the book.

(11) ***Modern and contemporary history of Chinese cartography*** (Liao Ke, Yu Cang. Shandong Education Press, Jinan, 2008)

In 5 chapters, the book systematically discusses the development process, main achievements, development background and historical enlightenment of modern and contemporary cartography from the Ming and Qing Dynasties, the Republic of China, and the founding of new China. Chapter one is the historical enlightenment of ancient Chinese cartography, including the definition of the upper and lower bounds of ancient Chinese cartography and the understanding of relevant myths and legends, exploration of the origin of maps, history of the development of ancient Chinese cartography, the far-reaching historical influence of papermaking, printing and the compass on map making and technological progress in China and worldwide, and historical revelations and the main factors in the development of ancient Chinese cartography. Chapter two discusses the main achievements and historical review of cartography during the Ming and Qing dynasties, including the main achievements and historical review of cartography in the Ming dynasty and the main achievements of map making and cartography in the Qing dynasty. Chapter three discusses map making and cartography in the Republic of China period and mainly includes the major achievements of map making and cartography in the Republic of China. Chapter four is about the preliminary development of China's modern cartography during the 1950s and 1960s, including the establishment and development of surveying and mapping institutions, cultivation of surveying and mapping professional talents, surveying and compilation of topographic maps, preparation of general geographical maps, development of thematic maps, compilation of the national atlas and provincial atlas, innovations in mapping techniques, research on cartography theory and the preliminary development of map publishing. Chapter five discusses the further development of modern cartography in China from the 1970s to the end of the twentieth century, including the most recent developments in international cartography over nearly 30 years, the reconstruction and reform of the Organization of the State Bureau of Surveying and Mapping, mapping automation and construction of computer mapping and publishing systems, remote sensing mapping, research and development of multimedia electronic maps and Internet maps, further development and measurement of national basic scale topographic maps, further development of thematic mapping, and further implementation of national atlases and various thematic atlases and regional atlases of China.

(12) ***History of Chinese cartography*** (Yu Cang, Liao Ke. Surveying and Mapping Press, Beijing, October 2009)

The book comprehensively introduces the development history of the map and cartography in China. There are 3 parts in this book: the first part is composed of 7 chapters, which explains the historical trajectory of the development of ancient Chinese maps from the pre-Qin period to the Yuan dynasty and shows that the development and achievements of ancient Chinese maps play an important role in world map history. The second part consists of 4 chapters, which explain the development history of modern Chinese map making and cartography in the Ming dynasty, Qing dynasty and the Republic of China and Liberated Areas. Modern China began to carry out surveying and mapping of national longitude and latitude and use aerial photogrammetry to measure large and medium scales of topographic maps. It was pointed out that modern Chinese mapping technology and cartography lagged behind Western developed countries. The third part consists of 14 chapters and comprehensively discusses the development of new China's map and modern cartography. In particular, this part of the book shows the rapid development of China in terms of remote sensing mapping, computer mapping and publishing systems, multimedia electronic atlases and Internet maps since the 1970s, as well as the realization of digitization and automation of map making and publication, all of which indicates that Chinese cartography has fundamentally changed from traditional cartography to modern cartography, which has reached the advanced levels of developed countries worldwide. Each part of the book concludes with a summary of historical reviews or historical revelations. The book contains illustrations of color maps that reflect the development level of modern maps. Senior expert profiles and cartographic expert list are attached to the final book in memory of their contributions to China's maps and the development of cartography, and a chronology of cartographic events (remote antiquity to 2005 A.D.) is included.

References

Cao, W. (1983). A preliminary study on the theory and method of map making in ancient China. *Studies in the History of Natural Sciences, 2*(3), 246–257.

Cao, W. (1989). Some problems concerning the map unearthed from Qin tomb in Tianshui. *Cultural Relic,* (12), 78–85.

Cao, W. (1990). A research on China's cartography history: Review for the past 40 years. *Studies in the History of Natural Sciences, 9*(3), 283–289.

Cao, W. (1994a). Yan Zhou Tu Jing (Book of Map and Record of Yanzhou) the earliest available Tujing with maps. *Studies in the History of Natural Sciences, 13*(4), 374–382.

Cao, X. (1994b). On ancient maps of Mawangdui and some related problems. *Archaeology,* (4), 69–71.

Chen, S. (1995). The historical honor of ancient Chinese scientists in the history of Surveying and mapping. *Bulletin of Surveying and Mapping,* (2), 33–36.

Ding, C. (2001). Compilation of archaic maps in china. *Bulletin of Surveying and Mapping,* (1), 41–43.

Gao, J. (1962). Survey of compilation of national and provincial atlas in the two dynasties of Ming and Qing Dynasties. *Acta Geodaetica et Cartographica Sinica, 5*(4), 289–306.

Gao, J. (1963). On the evolution characteristics of mathematical elements and representation methods of maps in China. *Acta Geodaetica et Cartographica Sinica, 6*(2), 120–135.

Ge, Z. (2002). Ancient maps and the history of ideas. *China Surveying and Mapping,* (5), 30–33.

Huang, G. (1997). On the origin and development of Chinese maps. *Relics from Southern,* (4), 92–101.

Jiang, D. (1997). Study on the Chinese cartography history in recent ninety years. *Geo-Information,* (3), 56–59.

Jin, Y. (1983). Ancient map of Changsha in Mawangdui and surveying and mapping techniques in Han Dynasty. *Acta Geodaetica et Cartographica Sinica, 12*(1), 3–13.

Li, J. (1995). Three questions about 'the emperor's map. *Map,* (1), 57–59.

Li, X. (1997). The enlightenment of ancient Chinese maps. *Du Shu,* (7), 140–144.

Liu, K., & Yang, S. (1997). Discussion on the graphic records in Zhou Li and its related problems. *Map,* (1), 42–46.

Lu, L. (1982). Evolution and evolution of the base of ancient maps in China. *Acta Geodaetica et Cartographica Sinica, 11*(2), 144–150.

Lu, L. (2002). The probe of the most primitive maps in china. *Journal of Geomatics Science and Technology, 19*(3), 227–229.

Lv, H. (1994). A comparative study of Chinese and Western history of cartography. *Geographical Research, 13*(2), 82–89.

Mawangdui Han silk finishing group. (1975). Map arrangement of Han Tombs in Changsha, Mawangdui. *Bulletin of Surveying and Mapping,* (2), 19–26.

Mao, J. (1984). Pei Xiu: an outstanding map scientist in ancient China. *Zhong Yuan Di Li Yan Jiu,* (2), 38–40.

Niu, R. (1986). A study of ancient place names from the history of Chinese maps. *Bulletin of Surveying and Mapping,* (6), 29–33.
Sun, G. (1997). Try to talk about the emergence and development of China map. *Map,* (1), 34–37.
Wang, H. (1998). A preliminary exploration of map making in prehistoric times. *Journal of Literature and History,* (2), 40–41.
Wang, K. (1990). Inscription of Yi Hou Shi Gui (food utensils of early western Zhou Dynasty)-maps of King Wu's and King Cheng's conquest over Yin Dynasty, East State Map, and comprehensive analysis of the map of the Western Zhou Dynasty. *Map,* (1), 19–52.
Wang, Y. (1989). The earliest extant stone inscription map in China. *Map,* (1), 28.
Wang, Z. (1992). Historical data of ancient map transportation in Mawangdui. *Jianghan Archaeology,* (4), 67–72.
Wu, C. (1990). Mawangdui silk map test. *Map,* (1), 47–50.
Wu, H. (2004). A masterpiece on the study of the history of Chinese cartography: Evaluation of study of the maps on Wooden blocks unearthed from Fangmatan in Tianshui, a book written by YONG Jichun. *Journal of Tianshui Normal University, 24*(1), 109–110.
Yang, X. (2006). Commentary: archives the earliest extant nautical charts in the world—Zheng He's nautical charts. *Archives Science Bulletin,* (2), 91–93.
Yao, S. (1990). Talking about the maps in the Rites of Zhou. *Map,* (2), 39–40.
Yi, J. (2008). Xuanjiao pupil Zhu Siben's thought on cartography science. *Religious Studies,* (2), 7–10.
Yin, C. (2004). The charm of Chinese traditional map drawing. *Map,* (6), 44–47.
Yu, C. (2003). Directions on an ancient map. *Map,* (3), 34–37.
Yu, L. (1987). Representatively great achievement by Shenkuo in quantitative cartography in Song Dynasty of China. *Journal of Nanjing Normal University(Natural Science Edition),* (3), 91–94.
Yu, L. (1991). As regards the history of ancient Chinese quantative cartography. *Studies in the History of Natural Sciences, 10*(2), 177–181.
Zhang, J. (1992). On buddhist monk Qing jun's map and the use of systematical legend symbols in Shui Dong Diary. *Collections of Essays on Chinese Historical Geography,* (1), 245–248.
Zhang, X. (1986). Study on the measured precision and characteristics of the Garrison Map of Han tomb in Mawangdui. *Scientia Geographica Sinica,* (4), 67–77.
Zhang, X. (1991). On the drawing age of the ancient maps of Fangmatan of Tianshui. *Fudan Journal (Social Sciences Edition),* (1), 46–50.
Zhou, S. (1993). The ancient map of Mawangdui is not the river map of the Qin dynasty. *Map,* (3), 47–50.
Zhu, B. (1999a). On the relationship between maps and Chinese traditional culture. *Map,* (1), 39–43.
Zhu, B. (1999b). The cartographer Jia Dan of the Tang Dynasty and his achievements in cartography. *Map,* (1), 43–44.
Zhu, B. (1999c). Zhu Siben: The founder for reaching the peak of China's traditional cartographic achievement. *Map,* (3), 38–39.
Zhu, J. (1990). On the research of ancient maps from New Zheng He's Nautical Chart. *Map,* (3), 38–40.

Postscript

Throughout the entire process of editing this book (to commemorate the 60th anniversary of the founding of the People's Republic of China), I was always very restless. When we rejoice and cheer for the progress and achievements of this subject over the past 60 years, we have not forgotten the historical contributions made by older cartographers, including Zeng Shiying, Wu Zhongxing, Chen Shupeng, Wan Yuxian, Ning Duyi, Li Xuzhi, Li Haichen, Liu Delong, Min Buqiu, Hua Tang, Lin Yuanjie, Yu Cang, Hu Yuju, Lu Quan, Lu Shufen, and Su Shiyu, whose work laid a solid foundation for the restoration and reconstruction of cartography, as well as further development in new China. In addition, we have not forget the well-known experts and scholars in this domain, including Li Guozao Yang Qihe, Zhang Kequan, Yan Mian, Tai Mingzhang, Liu Jidi, Liao Ke, Zhang Mudong, Yao Xurong, He Jianbang, Jiang Jingtong, Zhang Qingpu, Fan Xiaolin, Wu Hehai, Du Gongshun, Mao Zanyou, Gong Jianwen, Chen Zongxing, Zhu Guorui, Liu Yue, Huang Xingyuan, Hu Youyuan, Shi Zuhui, Liu Jiahao, Tian Desen, Fang Bingyan, Fu Suxing, Cui Weihong, Cheng Jicheng, Du Daosheng, Huang Rentao, Zou Yujun, Yin Gongbai, Qian Zhen, Zhang Tianshi, Chen Qinghai, Hu Dingquan, Zhou Kejing, and others, who made unremitting efforts to realize the transformation of traditional cartography to digital cartography. Many of these people have already left their jobs, but many are still contributing to the development of the discipline. More memorable and admired are Gao Jun (academician of the Chinese Academy of Sciences) and Wang Jiayao (academician of the Chinese Academy of Engineering), and although they are both over seventy years old, they still work diligently with many young professors, researchers and senior engineers at the forefront of scientific research and talent cultivation in the discipline. Over the course of the professional construction of cartography and geographic information engineering disciplines, they kept their feet on the ground, learned from the heart, and conducted highly effective work on the subject, which enabled it to become a national key discipline and enter the international advanced ranks; additionally, they achieved fruitful results and laid the foundation for further development in the discipline.

A human's life is limited, but the great cause of discipline development is eternal. Discipline construction, like other undertakings, requires generations of human

J. Wang and F. Wu (eds.), *Advances in Cartography and Geographic Information Engineering*, https://doi.org/10.1007/978-981-16-0614-4

striving and dedication. Experts and scholars of the older generation left us their writings, papers and achievements, as well as the scientific spirit of being realistic, pragmatic, exploratory and innovative; they also left us with the spirit of self-reliance and hard work, dedication to careful research and not seeking fame and wealth, all of which are the most priceless. Today's young and middle-aged experts and scholars are inheriting the past and opening the future, and while they still have a long way to go, with heavy tasks and long road, the spirit of the older generation must be inherited and handed down. In today's impetuous academic environment, this new generation must stay away from utilitarian methods and continue to pursue and explore science. Albert Einstein once said that "most people say that wit is what makes a great scientist; they are wrong, it is personality". This reminds me of two other sayings: "to get what we've never gotten, we have to do what we have never done" and "learning is similar to climbing a mountain; the higher you climb, the more scenery you enjoy". Let us share in this journey.

Editor in chief
November 2009.